Lecture Notes in Computer Science 16217

Founding Editors

Gerhard Goos
Juris Hartmanis

Editorial Board Members

Elisa Bertino, *Purdue University, West Lafayette, IN, USA*
Wen Gao, *Peking University, Beijing, China*
Bernhard Steffen, *TU Dortmund University, Dortmund, Germany*
Moti Yung, *Columbia University, New York, NY, USA*

The series Lecture Notes in Computer Science (LNCS), including its subseries Lecture Notes in Artificial Intelligence (LNAI) and Lecture Notes in Bioinformatics (LNBI), has established itself as a medium for the publication of new developments in computer science and information technology research, teaching, and education.

LNCS enjoys close cooperation with the computer science R & D community, the series counts many renowned academics among its volume editors and paper authors, and collaborates with prestigious societies. Its mission is to serve this international community by providing an invaluable service, mainly focused on the publication of conference and workshop proceedings and postproceedings. LNCS commenced publication in 1973.

Jinguang Han · Yang Xiang · Guang Cheng ·
Willy Susilo · Liquan Chen
Editors

Information and Communications Security

27th International Conference, ICICS 2025
Nanjing, China, October 29–31, 2025
Proceedings, Part I

Editors
Jinguang Han
Southeast University
Nanjing, China

Guang Cheng
Southeast University
Nanjing, China

Liquan Chen
Southeast University
Nanjing, China

Yang Xiang
Swinburne University of Technology
Hawthorn, VIC, Australia

Willy Susilo
University of Wollongong
Wollongong, NSW, Australia

ISSN 0302-9743　　　　　　　ISSN 1611-3349 (electronic)
Lecture Notes in Computer Science
ISBN 978-981-95-3539-2　　　ISBN 978-981-95-3540-8 (eBook)
https://doi.org/10.1007/978-981-95-3540-8

© The Editor(s) (if applicable) and The Author(s), under exclusive license
to Springer Nature Singapore Pte Ltd. 2026

This work is subject to copyright. All rights are solely and exclusively licensed by the Publisher, whether the whole or part of the material is concerned, specifically the rights of translation, reprinting, reuse of illustrations, recitation, broadcasting, reproduction on microfilms or in any other physical way, and transmission or information storage and retrieval, electronic adaptation, computer software, or by similar or dissimilar methodology now known or hereafter developed.
The use of general descriptive names, registered names, trademarks, service marks, etc. in this publication does not imply, even in the absence of a specific statement, that such names are exempt from the relevant protective laws and regulations and therefore free for general use.
The publisher, the authors and the editors are safe to assume that the advice and information in this book are believed to be true and accurate at the date of publication. Neither the publisher nor the authors or the editors give a warranty, expressed or implied, with respect to the material contained herein or for any errors or omissions that may have been made. The publisher remains neutral with regard to jurisdictional claims in published maps and institutional affiliations.

This Springer imprint is published by the registered company Springer Nature Singapore Pte Ltd.
The registered company address is: 152 Beach Road, #21-01/04 Gateway East, Singapore 189721, Singapore

If disposing of this product, please recycle the paper.

Preface

This volume contains the papers that were selected for presentation and publication at the 27th International Conference on Information and Communications Security (ICICS 2025), which was jointly organized by Southeast University (China), Swinburne University of Technology (Australia) and University of Wollongong (Australia), during October 29–31, 2025.

ICICS is one of the mainstream security conferences with the longest history. It started in 1997 and aims to bring together leading researchers and practitioners from both academia and industry to discuss and exchange their experiences, lessons learned and insights related to information and communications security. This year's Program Committee (PC) consisted of 136 members with diverse backgrounds and broad research interests. A total of 357 valid paper submissions were received. After careful checks, 16 submissions were desk rejected due to non-compliance with the submission requirements or obvious low quality. Of the 341 submissions sent for review, each received at least three, and at most four review comments. The review process was double blind, and the papers were evaluated on the basis of their significance, novelty and technical quality. Practically all the papers were reviewed by three or more PC members and then discussed among the Program Committee. The discussions were held online intensively over more than three weeks. Finally, 91 papers were selected for presentation at the conference, giving an acceptance rate of 25.5%.

Following the reviews, The paper "Artemis: Decentralized, Secure, and Efficient Safety Monitoring with Dynamic Trajectories", authored by Meng Li, Zhuangwei Li, Yifei Chen, Yan Qiao and Mauro Conti, was selected for the Best Paper Award, and the paper "FCAL: An Asynchronous Federated Contrastive Semi-Supervised Learning Approach for Network Traffic Classification", authored by Yu Yan, Qingjun Yuan, Weina Niu, Xiangyu Wang, Yanbei Zhu and Yongjuan Wang, was selected for the Best Student Paper Award, respectively. Both awards were generously sponsored by Springer. Additionally, ICICS 2025 was honored to offer three outstanding keynote talks by Liqun Chen, University of Surrey (UK), Sokratis Katsikas, Norwegian University of Science and Technology (Norway) and Gene Tsudik, University of California, Irvine (USA). Our deepest and sincere thanks to them for sharing their knowledge and experience during the conference.

For the success of ICICS 2025, we would like to first thank the authors of all submissions and the PC members for their great effort in selecting the papers. We also thank all the external reviewers for assisting in the reviewing process. For the conference organization, we would like to thank the ICICS Steering Committee, the Publicity Chairs, Weizhi Meng and Yong Yu, the Registration Chair, Chao Sun, the Publication Co-chairs, Ibrahim Khalil and Viet Vo, and the Web Chair, Ge Wu. Finally, we thank everyone else, speakers, session chairs and volunteer helpers, for their contributions to the program of ICICS 2025.

October 2025

Jinguang Han
Yang Xiang
Guang Cheng
Willy Susilo
Liquan Chen

Organization

Steering Committee

Jianying Zhou	Singapore University of Technology and Design, Singapore
Robert Deng	Singapore Management University, Singapore
Dieter Gollmann	Hamburg University of Technology, Germany
Javier Lopez	University of Málaga, Spain
Qingni Shen	Peking University, China
Zhen Xu	Institute of Information Engineering, Chinese Academy of Sciences, China

General Chairs

Guang Cheng	Southeast University, China
Willy Susilo	University of Wollongong, Australia

Program Chairs

Jinguang Han	Southeast University, China
Yang Xiang	Swinburne University of Technology, Australia

Organization Chair

Liquan Chen	Southeast University, China

Publicity Co-chairs

Weizhi Meng	Lancaster University, UK
Yong Yu	Shaanxi Normal University, China

Publication Co-chairs

Ibrahim Khalil Royal Melbourne Institute of Technology, Australia
Viet Vo Swinburne University of Technology, Australia

Registration Chair

Chao Sun Southeast University, China

Web Chair

Ge Wu Southeast University, China

Program Committee

Chuadhry Mujeeb Ahmed Newcastle University, UK
Massimiliano Albanese George Mason University, USA
Cristina Alcaraz University of Málaga, Spain
Saed Alrabaee United Arab Emirates University, UAE
Man Ho Au Hong Kong Polytechnic University, China
Joonsang Baek University of Wollongong, Australia
Guangdong Bai University of Queensland, Australia
Jin Wook Byun Pyeongtaek University, South Korea
Di Cao Swinburne University of Technology, Australia
Rongmao Chen National University of Defense Technology, China
Ting Chen University of Electronic Science and Technology of China, China
Xiao Chen University of Newcastle, Australia
Xiaofeng Chen Xidian University, China
Yuanmi Chen East China Normal University, China
Nathan Clarke University of Plymouth, UK
Mauro Conti University of Padua, Italy
Bruno Crispo University of Trento, Italy
Shujie Cui Monash University, Australia
Jingjing Deng University of Bristol, UK
Changyu Dong Guangzhou University, China
Carmen Fernández-Gago University of Málaga, Spain

Anmin Fu	Nanjing University of Science and Technology, China
Steven Furnell	University of Nottingham, UK
Fei Gao	Beijing University of Posts and Telecommunications, China
Dieter Gollmann	Hamburg University of Technology, Germany
Yong Guan	Iowa State University, USA
Shuai Hao	Old Dominion University, USA
Hongsheng Hu	University of Newcastle, Australia
Zhi Hu	Central South University, China
Qiong Huang	Guangdong University of Finance, China
Xinyi Huang	Fujian Normal University, China
Aditya Japa	Ulster University, UK
Jiaojiao Jiang	University of New South Wales, Australia
Peng Jiang	Beijing Institute of Technology, China
Christos Kalloniatis	University of the Aegean, Greece
Sokratis Katsikas	Norwegian University of Science and Technology, Norway
Georgios Kavallieratos	University of Oslo, Norway
Hyoungshick Kim	Sungkyunkwan University, South Korea
Romain Laborde	Université de Toulouse, France
Jianchang Lai	Southeast University, China
Fagen Li	University of Electronic Science and Technology of China, China
Guyue Li	Southeast University, China
Meng Li	Hefei University of Technology, China
Shane Li	Cardiff University, UK
Shujun Li	University of Kent, UK
Song Li	Nanjing University of Finance and Economics, China
Wanpeng Li	University of Liverpool, UK
Xiaoguo Li	Chongqing University, China
Xin Liao	Hunan University, China
Jingqiang Lin	University of Science and Technology of China, China
Zhen Ling	Southeast University, China
Antonio Lioy	Politecnico di Torino, Italy
Bo Liu	University of Technology Sydney, Australia
Jianghua Liu	Nanjing University of Science and Technology, China
Yang Lu	Nanjing Normal University, China
Bo Luo	University of Kansas, USA
Xiapu Luo	Hong Kong Polytechnic University, China

Wanlun Ma	Swinburne University of Technology, Australia
Jean-Yves Marion	Université de Lorraine, France
Daisuke Mashima	Singapore University of Technology and Design, Singapore
Weizhi Meng	Lancaster University, UK
Yuantian Miao	University of Newcastle, Australia
Atsuko Miyaji	Osaka University, Japan
Siaw-Lynn Ng	Royal Holloway, University of London, UK
Jianting Ning	Wuhan University, China
Takashi Nishide	University of Tsukuba, Japan
Rolf Oppliger	eSECURITY Technologies, Switzerland
Michalis Pavlidis	University of Brighton, UK
Irdin Pekaric	University of Liechtenstein, Liechtenstein
Tran Viet Xuan Phuong	University of Arkansas at Little Rock, USA
Stjepan Picek	Radboud University, The Netherlands
Yanli Ren	Shanghai University, China
Na Ruan	Shanghai Jiao Tong University, China
Sumanta Sarkar	University of Essex, UK
Nitesh Saxena	Texas A&M University, USA
Savio Sciancalepore	Eindhoven University of Technology, The Netherlands
Sevil Sen	Hacettepe University, Turkey
Jun Shao	Zhejiang Gongshang University, China
Qingni Shen	Peking University, China
Yang Shi	Tongji University, China
Chunhua Su	University of Aizu, Japan
Purui Su	Institute of Software, CAS, China
Willy Susilo	University of Wollongong, Australia
Azadeh Tabiban	Concordia University, Canada
Zhiyuan Tan	Edinburgh Napier University, UK
Qiang Tang	Luxembourg Institute of Science and Technology, Luxembourg
Yangguang Tian	University of Surrey, UK
Viet Vo	Swinburne University of Technology, Australia
Ding Wang	Nankai University, China
Huaqun Wang	Nanjing University of Posts and Telecommunications, China
Jianfeng Wang	Xidian University, China
Nan Wang	CSIRO's Data61, Australia
Wei Wang	Xi'an Jiaotong University, China
Jinpeng Wei	University of North Carolina Charlotte, USA
Di Wu	University of Southern Queensland, Australia

Qianhong Wu	Beihang University, China
Tong Wu	University of Science and Technology Beijing, China
Zhe Xia	Wuhan University of Technology, China
Hu Xiong	University of Science and Technology of China, China
Lizhi Xiong	Nanjing University of Information Science and Technology, China
Chungen Xu	Nanjing University of Science and Technology, China
Dongpeng Xu	University of New Hampshire, USA
Guangquan Xu	Tianjin University, China
Peng Xu	Huazhong University of Science and Technology, China
Zhen Xu	Institute of Information Engineering, CAS, China
Hailun Yan	University of Chinese Academy of Sciences, China
Guomin Yang	Singapore Management University, Singapore
Kang Yang	State Key Laboratory of Cryptology, China
Rupeng Yang	University of Wollongong, Australia
Zheng Yang	Southwest University, China
Wun-She Yap	Universiti Tunku Abdul Rahman, Malaysia
Xun Yi	RMIT University, Australia
Zuobin Ying	City University of Macau, China
Yang Yu	Tsinghua University, China
Yong Yu	Shaanxi Normal University, China
Quan Yuan	Shandong University, China
Yachao Yuan	Soochow University, China
Tsz Hon Yuen	Monash University, Australia
Thomas Zacharias	University of Glasgow, UK
Fangguo Zhang	Sun Yat-sen University, China
Futai Zhang	Fujian Normal University, China
Lei Zhang	East China Normal University, China
Leo Zhang	Griffith University, Australia
Mingwu Zhang	Hubei University of Technology, China
Rui Zhang	Chinese Academy of Sciences, China
Yuan Zhang	Nanjing University, China
Zhi Zhang	University of Western Australia, Australia
Zongyang Zhang	Beihang University, China
Liang Zhao	Sichuan University, China
Yunlei Zhao	Fudan University, China
Huiyu Zhou	University of Leicester, UK

Lu Zhou — Nanjing University of Aeronautics and Astronautics, China
Yongbin Zhou — Nanjing University of Science and Technology, China
Sencun Zhu — Pennsylvania State University, USA
Xiaogang Zhu — University of Adelaide, Australia
Youwen Zhu — Nanjing University of Aeronautics and Astronautics, China
Cong Zuo — Beijing Institute of Technology, China

Additional Reviewers

Zhiyuan An
Zijian Bao
Jit Biswas
Nhat Quang Cao
Marco Casagrande
Alberto Castagnaro
Saverio Cavasin
Stefano Cecconello
Decheng Chen
Hanxiao Chen
Jiaqiang Chen
Jinrong Chen
Weihao Chen
Yanyu Chen
Yu Chen
Yumin Chen
Wei Cheng
Jae Hyun Choi
Fuyang Deng
Jianguo Feng
Yu Fu
Ankit Gangwal
Yansong Gao
Yiwen Gao
Matteo Golinelli
Michele Grisafi
Yue Han
Xiaohan Hao
Jinlong He
Lifeng Huang
Yixuan Huang

Meng Jia
Xiangkun Jia
Qin Jiang
Vyron Kampourakis
Neeraj Karamchandani
Chhagan Lal
Yongkang Lang
Tho Thi Ngoc Le
Chen Li
Hongbo Li
Jiawei Li
Jinhui Li
Jun Li
Meng Li
Minghang Li
Yin Li
Junkai Liang
Chao Lin
Yuliang Lin
Haowei Liu
Jiahao Liu
Mengling Liu
Yuejun Liu
Yundong Liu
Zihan Liu
Zhongkai Lu
Pingbin Luo
Kevin Lybarger
Sha Ma
Pierre Marty
Lin Mei

Jingdian Ming
Piyush Nagasubramaniam
Yiming Qi
Zehua Qiao
S. Muhammad Musthafa Roomi
Rahul Saha
Gabriel Sauger
Gang Shen
Jun Shen
Fang Shi
Junbin Shi
Young Ah Shin
Oliwer Sobolewski
Fuyuan Song
Junjie Song
Angelo Spognardi
Chao Sun
Xiaodan Tai
Jiazhuo Tian
Yee Ching Tok
Hoang Dat Tran
Sridhar Venkatesan
Hao Wang
Haoyang Wang
Lulu Wang
Wei Wang
Wenli Wang
Xinqian Wang
Yuzhu Wang
Ahmad Samer Wazan
Gabriel Wechta
Fudong Wu
Ge Wu
Mingli Wu

Weibin Wu
Wenbo Wu
Zhe Xia
Binwu Xiang
Meiyan Xiao
Wenkuan Xiao
Zhikang Xie
Lin Xu
Shengmin Xu
S. J. Yang
Xu Yang
Xuechao Yang
Yang Yang
Weijing You
Awais Yousaf
Peng Yu
Zhen Yu
Marcos Zampieri
Cai Zhang
Chi Zhang
Chiyu Zhang
Gongliang Zhang
He Zhang
Liu Zhang
Qian Zhang
Ruyuan Zhang
Xiaoqi Zhang
Xin Zhang
Yanqi Zhao
Mingmei Zheng
Nan Zhong
Mengjie Zhou
Fei Zhu
Jiajun Zou

Abstracts of Keynotes

Post-Quantum Group-Oriented Anonymous Signatures from Symmetric Primitives

Liqun Chen

University of Surrey, UK

Abstract. Group-oriented anonymous digital signatures, including group signatures, direct anonymous attestation (DAA) and enhanced privacy ID (EPID), have become important cryptographic primitives in information and communications security. Schemes using RSA and elliptic curve cryptography have been integrated into real-world applications and international standards. However, these standardised schemes are insecure against quantum attackers. Research into post-quantum (PQ) anonymous signatures has led to several schemes across various PQ cryptographic families. In this talk, we will focus on designing anonymous signature schemes based on symmetric techniques. For instance, we utilise a hash-based signature as a group membership credential. An anonymous signature is a non-interactive zero-knowledge proof of such a credential. We will also discuss robust design, strong security properties and efficient performance, particularly in relation to accommodating large group sizes, which is essential for rapidly developing applications.

Cyber Ranges and Cyber-Physical Ranges: Progress, Potential, and Future Directions

Sokratis Katsikas

Norwegian University of Science and Technology, Norway

Abstract. A Cyber Range (CR) serves as a specialized environment designed to provide dedicated testbeds and infrastructures for executing immersive training scenarios. Its primary goal is to enhance cybersecurity knowledge among security practitioners and awareness among non-security professionals and the public, while offering a hands-on learning experience for trainees. Over time, CRs have become an indispensable tool, offering a multifaceted approach to strengthening cybersecurity postures. On the other hand, Cyber-Physical Systems (CPSs) are advanced, intelligent systems that integrate physical processes with computational elements. These encompass diverse applications such as smart grids, autonomous vehicles, medical devices, process control systems, and autopilot avionics. As a fundamental pillar of Industry 4.0, CPSs drive the convergence of formerly distinct operational technology and modern information systems. Within this evolving technological landscape, Cyber-Physical Ranges (C-PRs) have emerged as an innovative and cost-effective solution that enable researchers and practitioners to explore vulnerabilities and devise robust defense mechanisms—without compromising real-world systems. This talk will first introduce a comprehensive taxonomy of CR systems, followed by an analysis of existing literature focusing on architecture, scenario development, capabilities, roles, tools, and evaluation criteria. Subsequently, we will present a fine-grained reference architecture for CRs, built upon a rigorous three-step methodology. Additionally, we will propose an evaluation framework that quantifies the alignment of a CR with state-of-the-art practices, offering a standardized method to identify optimal components for implementing the structural, functional, and informational facets of a CR. Finally, we will explore the latest advancements in C-PRs through real-world case studies, uncovering the challenges associated with designing, deploying, and managing these environments. We will also discuss their seamless integration with emerging technologies, illustrating their pivotal role in the future of cybersecurity research and innovation.

Device Awareness and User Privacy in the IoT Ecosystem

Gene Tsudik

University of California, Irvine, USA

Abstract. As many types of IoT devices worm their way into numerous settings in our daily lives, awareness of their presence and functionality becomes a source of major concern. Hidden IoT devices can snoop (via sensing) on unsuspecting nearby users, and impact the environment where unaware users are present, via actuation. This prompts, respectively, privacy and security/safety issues. The dangers of hidden IoT devices have been recognized and prior research suggested some means of mitigation, mostly based on traffic analysis or using specialized hardware to uncover devices. While such approaches are partially effective, there is currently no comprehensive approach to IoT device transparency.

Prompted in part by recent privacy regulations (GDPR and CCPA), this work constructs a privacy-agile Root-of-Trust architecture for IoT devices called PAISA: Privacy-Agile IoT Sensing and Actuation. It guarantees timely and secure announcements of nearby IoT devices' presence and capabilities. PAISA has two components: one on the IoT device that guarantees periodic announcements of its presence even if all device software is compromised, and the other on the user device, which captures and processes announcements. PAISA requires no hardware modifications; it uses a popular off-the-shelf Trusted Execution Environment (TEE) – ARM TrustZone. A follow-on work, DB-PAISA, complements PAISA by offering request-based discovery of IoT devices via BlueTooth. To demonstrate viability, both PAISA and DB-PAISA are available as open-source prototypes. We also address their security properties and performance factors.

Contents – Part I

Cryptography

Multi-signer Locally Verifiable Aggregate Signature from (Leveled) Multilinear Maps 3
   Yuchen Yang, Jie Chen, Qiaohan Chu, Qiuyan Du, and Luping Wang

Conditional Attribute-Based Encryption with Keyword Search for Pay-Per-Query Commercial Model 22
   Zerui Guo, Sha Ma, and Qiong Huang

Lightweight Transparent Zero-Knowledge Proofs for Cross-Domain Statements 40
   Zhengzhou Tu, Min Xie, Junbin Fang, Yong Yu, and Zoe L. Jiang

Public Verifiable Server-Aided Revocable Attribute-Based Encryption 62
   Luqi Huang, Fuchun Guo, Willy Susilo, and Yumei Li

New First-Order Secure AES Implementation Without Online Fresh Randomness Records 82
   Botao Liu and Ming Tang

SM2-VBKE: Achieving Cryptographic Binding Between Verification Integrity and Key Generation 100
   Runze Zhao, Siqi Lu, Yongjuan Wang, Liujia Cai, Wenyi Chen, and Fenghua Jiang

Certificate-Based Quasi-linearly Homomorphic Signatures: Definition, Construction, and Application to Data Integrity Auditing 119
   Jintao Cai, Futai Zhang, Wenjie Yang, Shaojun Yang, Yichi Huang, Rongmao Chen, and Willy Susilo

Zero-Knowledge Protocols with PVC Security: Striking the Balance Between Security and Efficiency 141
   Yi Liu, Yipeng Song, Anjia Yang, and Junzuo Lai

Attribute-Based Adaptor Signature and Application in Control-Based Atomic Swap 160
   Tianyuan Fan, Gang Shen, Yuzhu Wang, Yuntao Wang, and Mingwu Zhang

A Versatile Decentralized Attribute Based Signature Scheme for IoT 181
  Dazhi Xu, Yuejun Liu, Jiabei Wang, Yiwen Gao, and Yongbin Zhou

Post-quantum Cryptography

Cross-Domain Lattice-Based DAA Scheme with Shared Private-Key
for Internet of Things System 203
  *Minzhi Liang, Liquan Chen, Yinghua Jiang, Xuyan Min, Jin Qian,
  and Jun Luo*

Compact Adaptively Secure Identity-Based Encryption
from Middle-Product Learning with Errors 218
  Jingjing Fan, Xingye Lu, Man Ho Au, and Siu Ming Yiu

MDKG: Module-Lattice-Based Distributed Key Generation 237
  Ye Bai, Debiao He, Zhichao Yang, Min Luo, and Cong Peng

Turtle Wins Rabbit Again: Faster Modulus Reduction for RNS-CKKS 257
  Lianglin Yan, Pengfei Zeng, and Mingsheng Wang

A BGV-Subroutined CKKS Bootstrapping Algorithm Without Sine
Approximation 273
  *Jingjing Fan, Chi Zhang, Zejiu Tan, Zoe Lin Jiang, Man Ho Au,
  and Siu Ming Yiu*

PolarKyber: Polished Kyber with Smaller Ciphertexts, Greater Security
Redundancy, and Lower Decryption Failure Rate 286
  Chen An, Ziyao Liu, Xianhui Lu, and Jingnan He

Lion: A New Ring Signature Construction from Lattice Gadget 306
  Yanting Li, Pingbin Luo, Xinjian Chen, and Qiong Huang

Anonymity and Privacy

MagWatch: Exposing Privacy Risks in Smartwatches Through
Electromagnetic Signals 329
  Haowen Xu, Tianya Zhao, Xuyu Wang, Jun Dai, and Xiaoyan Sun

Privacy-Preserving, Secure and Certificate-Based Integrity Auditing
for Cloud Storage 347
  Wenhao Wang, Yu Li, Yinxia Sun, Yuan Zhang, and Sheng Zhong

Unbalanced Private Computation on Set Intersection with Reduced
Computation and Communication 366
    Zelin Tang, Hua Guo, Yewei Guan, and Kaijie Yang

Artemis: Decentralized, Secure, and Efficient Safety Monitoring
with Dynamic Trajectories 387
    Meng Li, Zhuangwei Li, Yifei Chen, Yan Qiao, and Mauro Conti

Privacy-Preserving Framework for k-Modes Clustering Based
on Personalized Local Differential Privacy 405
    *Yuling Luo, Zhangrui Wang, Xue Ouyang, Siyuan Zu, Qiang Fu,
    Sheng Qin, and Junxiu Liu*

AnoST: An Anonymous Optimistic Verification System Based
on Off-Chain State Transition 424
    Qiyuan Gao, Qianhong Wu, Junxiang Nong, and Qi Liu

Privacy-Preserving K-Hop Shortest Path Query on Encrypted Graphs
Based on Graph Pruning 444
    Ya Gao, Chao Mu, Ming Yang, and Xiaoming Wu

TA-PDC: Provable Data Contribution with Traceable Anonymous
for Group Transactions 463
    Xiaocong Lin, Weijing You, Chenchen Wu, Wenmao Liu, and Qi Gu

Fine-Filter: An Effective Defense Against Poisoning Attacks on Frequency
Estimation Under LDP 484
    Yuxia Zhou, Qiao Xue, and Youwen Zhu

BioVite: Efficient and Compact Privacy-Preserving Biometric Verification
via Fully Homomorphic Encryption 503
    Pengfei Zeng, Han Xia, and Mingsheng Wang

Authentication and Authorization

Circulation Control Model and Administration for Geospatial Data 525
    *Heng Li, Fenghua Li, Yunchuan Guo, Lingcui Zhang, Xiao Wang,
    and Ziyan Zhou*

Identifying Unusual Personal Data in Mobile Apps for Better Privacy
Compliance Check 545
    *Jiatao Cheng, Yuhong Nan, Xueqiang Wang, Zhefan Chen,
    and Yuliang Zhang*

Why Biting the Bait? Understanding Bait and Switch UI Dark Patterns
in Mobile Apps 564
 *Yixi Lin, Yue Xu, Zitong Yao, Yuhong Nan, Queping Kong,
and Xueqiang Wang*

Author Index 585

Contents – Part II

Blockchain and Cryptocurrencies

EquinoxBFT: BFT Consensus for Blockchain Emergency Governance 3
    Jialiang Fan, Qianhong Wu, Minghang Li, Decun Luo, Qin Wang, and Bo Qin

fFuzz: A State-Aware Function-Level Fuzzing Framework for Smart Contract Vulnerabilities Detection 21
    Chang Li, Binqin Lu, Wenyang Zhang, Kaixuan Yang, and Huijuan Zhu

TraceBFT: Backtracking-Based Pipelined Asynchronous BFT Consensus for High-Throughput Distributed Systems 39
    Chaofeng Zhuang, Junqing Gong, Zhili Chen, and Haifeng Qian

RADIAL: Robust Adversarial Discrepancy-Aware Framework for Early Detection of Illicit Cryptocurrency Accounts 60
    Kombou Victor, Qi Xia, Jianbin Gao, Hu Xia, Kuiche Sop Brinda Leaticia, and Anto Leoba Jonathan

Enhancing Private Signing Key Protection in Digital Currency Transactions Using Obfuscation 79
    Yang Shi, Jintao Xie, Minyu Teng, Guanxu Liu, Linhai Guo, and Jiangfeng Li

AnsBridge: Towards Secure Cross-Chain Interoperability via Anonymous and Verifiable Validators 100
    Mingming Huang, Xiaodan Zhang, Wei Mi, Huimei Liao, and Yi Sun

TrustBlink: A zkSNARK-Powered On-Demand Relay for PoW Cross-Chain Verification With Low Costs 119
    Bohang Wei, Yang Yang, Shihong Xiong, Minghang Li, Qianhong Wu, and Bo Qin

R1-MFSol: a Smart Contract Vulnerability Detection Model Based on LLM and Multi-modal Feature Fusion 139
    Huibo Yang, Zhize Hao, and Tao Liu

No Place to Hide: An Efficient and Accurate Backdoor Detection Tool for Ethereum ERC-20 Smart Contracts 159
    Shouchen Zhou, Lu Zhou, and Yu Tao

System and Network Security

Batch-Oriented Element-Wise Approximate Activation for Privacy-Preserving Neural Networks 181
Peng Zhang, Ao Duan, Xianglu Zou, and Dongyan Qiu

Social-Aware and Quality-Driven Incentives for Mobile Crowd-Sensing with Two-Stage Game 198
Jun Tao and Hao Zou

A Distributed Privacy Protection Method for Crowd Sensing Based on Trust Evaluation 216
Hai Liu, Maoze Tian, Yadong Peng, and Hongye Peng

DBG-LB: A Trustworthy and Efficient Framework for Data Sharing in the Internet of Vehicles 236
Chaoyue Li, Yongming Zhang, and Xiaolong Xu

Actions Speak Louder Than Words: Evidence-Based Trust Level Evaluation in Multi-agent Systems 255
Nikolaos Fotos, Koffi Ismael Ouattara, Dimitrios S. Karas, Ioannis Krontiris, Weizhi Meng, and Thanassis Giannetsos

Bridging the Interoperability Gaps Among Trusted Architectures in MCUs 274
Sandro Pinto, Luís Cunha, Daniel Oliveira, Michele Grisafi, Emanuele Beozzo, and Bruno Crispo

Security and Privacy of AI

A Dropout-Resilient and Privacy-Preserving Framework for Federated Learning via Lightweight Masking 293
Yufeng Jiang, Jianghua Liu, Chenhao Xu, Cong Zuo, Lei Xu, and Jian Lei

AFedGAN: Adaptive Federated Learning with Generative Adversarial Networks for Non-IID Data 313
Xuyang Zhang, Hua Jin, and Peiyuan Guo

OTTER: Optimized Training with Trustworthy Enhanced Replication via Diffusion and Federated VMUNet for Privacy-Aware Medical Segmentation 331
Haocheng Kan, Yuesheng Zhu, Guibo Luo, and Hanwen Zhang

EAGLE: Ensemble Adaptive Graph Learning for Enhanced Ethereum
Fraud Detection 347
    *Befoum Stephane Richard, Jianbin Gao, Qi Xia, Kombou Victor,
    Eyezo'o Benjamin Fabien, and Mulenga Mukupa Rossini*

BR-CPPFL: A Blockchain-Based Robust Clustered Privacy-Preserving
Federated Learning System 367
    *Yuantong Li, Xiaofen Wang, Ke Zhang, Bo Zhang, Lei Zheng,
    Xiaosong Ding, and Qing Xu*

Efficient Semi-asynchronous Federated Learning with Guided Selective
Participation and Adaptive Aggregation 387
    Chaoyun Wang, Kedong Yan, and Chanying Huang

Improving Byzantine-Resilience in Federated Learning via Diverse
Aggregation and Adaptive Variance Reduction 403
    *Xiuhua Wang, Shikang Li, Fengrui Fan, Shuai Wang, Yiwei Li,
    and Yu Zheng*

Hierarchical Recovery of Convolutional Neural Networks
via Self-embedding Watermarking 424
    Yawen Huang and Huaicong Zhang

Personalized Federated Learning Algorithm Based on User Grouping
and Group Signatures 442
    Hao Lin, Xiaoming Hu, Shuangjie Bai, and Yan Liu

Machine Learning for Security

SPCD: A Shot-Based Partial Copy Detection Method 463
    Yuhan Tao and Danwei Chen

Bayesian-Adaptive Graph Neural Network for Anomaly Detection
(BAGNN) 482
    Yong Ding, Chi Zhang, Shijie Tang, Changsong Yang, and Hai Liang

UzPhishNet Model for Phishing Detection 501
    Bektemir Saydiev, Xiaohui Cui, and Umer Zukaib

CyberNER-LLM: Cyber Threat Intelligence Named Entity Recognition
With Large Language Model 513
    Xinzheng Liu, Wangqun Lin, and Zhaoyun Ding

Provenance-Based Intrusion Detection via Multi-scale Graph
Representation Learning 531
   Xuebo Qiu, Mingqi Lv, Tieming Chen, Tiantian Zhu, and Qijie Song

SADGA: A Self Attention GAN-Based Adversarial DGA with High
Anti-detection Ability 550
   Jiang Luo, ShaoHua Qin, and Zhe Wang

Author Index 569

Contents – Part-III

Attack and Defense

Domain Adaptation for Cross-Device Profiled ML Side-Channel Attacks 3
    Ian Y. Garrett and Ryan M. Gerdes

Find the Clasp of the Chain: Efficiently Locating Cryptographic
Procedures in SoC Secure Boot by Semi-automated Side-Channel Analysis 22
    *Shipei Qu, Yuxuan Wang, Jintong Yu, Cheng Hong, Chi Zhang,
and Dawu Gu*

Full-Phase Distributed Quantum Impossible Differential Cryptanalysis 41
    Kun Zhang, Tao Shang, Yuanjing Zhang, and Jianwei Liu

ProverNG: Efficient Verification of Compositional Masking
for Cryptosystem's Side-Channel Security 57
    *Yiming Yang, Feng Zhou, Yuanyuan Wang, Hua Chen, Limin Fan,
and An Wang*

POWERPOLY: Analyzing Multilingual Programs with the Aid
of WebAssembly 77
    Zhuochen Jiang and Baojian Hua

Not only Spatial, but Also Spectral: Unnoticeable Backdoor Attack on 3D
Point Clouds 97
    Yongzhen Jiang, Haoran Li, Hongjia Liu, Jiageng Pan, and Jian Xu

Permutation-Based Cryptanalysis of the SCARF Block Cipher and Its
Randomness Evaluation 115
    Qi Li, Wenying Zhang, and Xiaomeng Sun

Secure and Scalable TLB Partitioning Against Timing Side-Channel
Attacks 134
    *Tianyi Huang, Xiaolin Zhang, Kailun Qin, Boshi Yuan, Chenghao Chen,
Yipeng Shi, Chi Zhang, and Dawu Gu*

Security Vulnerabilities in AI-Generated Code: A Large-Scale Analysis
of Public GitHub Repositories 153
    Maximilian Schreiber and Pascal Tippe

Vulnerability Analysis

Towards Efficient C/C++ Vulnerability Impact Assessment in Package
Management Systems 175
   Zibo Wang, Xiangkun Jia, Jia Yan, Yi Yang, Huafeng Huang, and Purui Su

AugGP-VD: A Smart Contract Vulnerability Detection Approach Based
on Augmented Graph Convolutional Networks and Pooling 195
   Nianlu Liu, Linlin Zhang, Wenbo Fang, and Kai Zhao

VULDA: Source Code Vulnerability Detection via Local Dependency
Context Aggregation on Vulnerability-Aware Code Mapping Graph 213
   Tao Peng, Ling Gui, Lijun Cai, Junwei Tang, Aoshuang Ye, and Fei Zhu

KVT-Payload: Knowledge Graph-Enhanced Hierarchical Vulnerability
Traffic Payload Generation 232
   Faqi Zhao, Rong Shi, Guoqiao Zhou, Wen Wang, and Feng Liu

Construction and Application of Vulnerability Intelligence Ontology
Under Vulnerability Management Perspective 253
   Guangxiang Dai, Peng Wang, and Duohe Ma

Anomaly Detection

Speaker Inference Detection Using Only Text 277
   Ruoxi Cheng, Yizhong Ding, Shaowei Yuan, and Zhiqiang Wang

DTGAN: Diverse-Task Generative Adversarial Networks for Intrusion
Detection Systems Against Adversarial Examples 295
   Yiyang Wang, Xiabai Wu, and Kun Chen

ConComFND: Leveraging Content and Comment Information
for Enhanced Fake News Detection 312
   Huan Zhang, Chanying Huang, Kedong Yan, and Shan Xiao

Transferable Adversarial Attacks in Object Detection: Leveraging
Ensemble Features and Gradient Variance Minimization 330
   Zhitong Lu, Zhen Xu, Qian Yang, and Kai Chen

VAE-BiLSTM: A Hybrid Model for DeFi Anomaly Detection Combining
VAE and BiLSTM 340
   Shujiang Xu, Xiaomin Luo, Lianhai Wang, Miodrag J. Mihaljević,
   Shuhui Zhang, Wei Shao, and Qizheng Wang

FluxSketch: A Sketch-Based Solution for Long-Term Fluctuating Key
Flow Detection 359
    Jun Xu, Guoju Gao, Yu-E Sun, He Huang, and Yang Du

RUSTGUARD: Detecting Rust Data Leak Issues with Context-Sensitive
Static Taint Analysis 379
    Shanlin Deng, Mingliang Liu, Si Wu, and Baojian Hua

Secure Guard: A Semantic-Based Jailbreak Prompt Detection Framework
for Protecting Large Language Models 398
    Sixin Fang, Ke Cheng, Jixin Zhang, Zheng Qin, and Mingwu Zhang

Traffic Classification

FCAL: An Asynchronous Federated Contrastive Semi-supervised
Learning Approach for Network Traffic Classification 419
    Yu Yan, Qingjun Yuan, Weina Niu, Xiangyu Wang, Yanbei Zhu, and Yongjuan Wang

TetheGAN: A GAN-Based Synthetic Mobile Tethering Traffic Generating
Framework 438
    Xuman Zhang, Guang Cheng, and Li Deng

SPTC: Signature-Based Cross-Protocol Encrypted Proxy Traffic
Classification Approach 457
    Huajie Jia, Yige Chen, and Zhengzhou Tang

Multi-modal Datagram Representation with Spatial-Temporal State
Space Models and Inter-flow Contrastive Learning for Encrypted Traffic
Classification 476
    Xianwen Deng, Ruijie Zhao, Mingwei Zhan, Shaoqian Wu, Yijun Wang, and Zhi Xue

FlowGraphNet: Efficient Malicious Traffic Detection via Graph
Construction 493
    Changsong Yang, Han Wang, Yueling Liu, Yong Ding, Hai Liang, and Zhenyu Li

CascadeGen: A Hybrid GAN-Diffusion Framework for Controllable
and Protocol-Compliant Synthetic Network Traffic Generation 511
    Qingyuan Yu, Chuping Yan, and Xiaoying Liu

Steganography and Watermarking

Towards High-Capacity Provably Secure Steganography via Cascade Sampling 533
Meiyang Lv, Haocheng Fu, Xiaowei Yi, Hongxian Huang, Yun Cao, and Changjun Liu

When There Is No Decoder: Removing Watermarks from Stable Diffusion Models in a No-Box Setting 553
Xiaodong Wu, Tianyi Tang, Xiangman Li, Jianbing Ni, and Yong Yu

Robust Reversible Watermarking for 3D Models Based on Auto Diffusion Function 573
Zixing Lin, Yaolong Song, and Li Rui

Author Index 593

Cryptography

Multi-signer Locally Verifiable Aggregate Signature from (Leveled) Multilinear Maps

Yuchen Yang[1], Jie Chen[1(✉)], Qiaohan Chu[1], Qiuyan Du[1], and Luping Wang[2]

[1] Shanghai Key Laboratory of Trustworthy Computing, School of Software Engineering, East China Normal University, Shanghai, China
s080001@e.ntu.edu.sg
[2] Suzhou University of Science and Technology, Suzhou, China

Abstract. Locally verifiable aggregate signatures (LVAS) enable efficient verification of aggregate signatures. When aggregate signatures reduce the space for storing signatures from linear in N to a fixed constant, local verifiability provides efficient verification. However, its main drawback is that it can only aggregate signatures on messages associated with a single signer.

In this work, we extend LVAS to multi-signer settings, allowing the aggregate signature to verify that all the n users signed their respective n messages. More importantly, the verifier can determine whether a specific message is part of the set without access to the entire message list. The multi-signer locally verifiable aggregate signatures provide more flexibility in collaborative aggregate signatures among multiple users.

Furthermore, we construct a multi-signer locally verifiable aggregate signature scheme from leveled multilinear maps that achieves full security based on the (ℓ, k)-MDHI assumption in the random oracle model.

Keywords: aggregate signature · locally verifiable signature · leveled multilinear maps

1 Introduction

Aggregate Signature. Boneh et al. [1] proposed the aggregate signature that convinces the verifier that the n users sign the n distinct messages. Consider a set of U users, where each user $u \in U$ generates a signature σ_u for their respective message m_u. These individual signatures are then combined into a single aggregate signature $\hat{\sigma}$ by an aggregating party. This party, which can be distinct from and untrusted by the users in U, only has access to the users' verification keys, messages, and corresponding signatures. The aggregate signature $\hat{\sigma}$ has the same length as any individual signature. Given the aggregate signature $\hat{\sigma}$ and the key-message pairs $\{vk_u, m_u\}$, a verifier can check whether each user has signed their respective message.

Since aggregate signatures can significantly reduce storage requirements, they are widely used in various applications. For example, they help reduce storage costs in systems with limited storage space (such as mobile and embedded systems). They are also employed to combine signatures from different transactions prior to their inclusion in blockchains.

Locally Verifiable Aggregate Signature. While aggregate signature readily compress massive partial signatures into a short one independent of the original signature number, the aggregate verifier needs to read the full list of signed messages $\{m_u\}$, resulting in slow verification despite space savings. To address this, Goyal and Vaikuntanathan [2] introduced the locally verifiable aggregate signature (LVAS). In the LVAS scheme, a hint generation algorithm creates an auxiliary hint for the target message(s) to be verified. Now, the verifier can check whether the target message is in the set, without access to the entire message set, but only needs to visit the hint and target message(s) mentioned above, thus improving the efficiency of verification. However, the LVAS scheme only supports the single-signer setting. It is obviously subject to many limitations in practical applications when multiple users participate in the process.

Our Result. We introduce the multi-signer locally verifiable aggregate signature (MLVAS), which simultaneously supports aggregability by multiple signers and local verifiability. Multiple signers combine the signatures of their respective messages into a single signature. By leveraging the locally opening algorithm to generate auxiliary information, verifiers are no longer required to present the entire message list. Instead, they merely need to submit the target messages for verification. Additionally, we define security models for MLVAS based on leveled multilinear maps, whose security relies on the (ℓ, k)-MDHI assumption. Moreover, we prove that our scheme achieves full unforgeability and full aggregated unforgeability against adversarial openings.

We present a concrete comparison in Table 1. The single-signer LVAS scheme [2] only supports single-signer setting. In addition, our partial signature is based on a weakly secure short BB signature [3,4]. Besides, the multi-signer aggregate signature BGLS03 [1] does not support local verifiability.

Table 1. Comparison among prior works.

References	Locally Opening	Aggregatable	Multi-signer	Assumption
BB [3,4]	×	×	×	SDH
BGLS [1]	×	✓	✓	co-CDH
LVAS [2]	✓	✓	×	q-BDHI
Ours	✓	✓	✓	(ℓ, k)-MDHI

1.1 Related Work

Hohenberger et al. [5] introduced the first identity-based aggregate signature that admits unrestricted aggregation. Bagherzandi et al. [6] demonstrates the close relationship between multi-signatures and aggregate signatures. Ağırtaş et al. [7] proposed locally verifiable multi-signatures, which allows multiple signers to collaboratively produce a compact signature for a single message without accessing the entire signers set. Synchronized aggregate signatures [8,9] have been proposed, where the signers do not need to be aware of or interactive with one another, which reduced costly interaction. Qiu et al. [10] proposed predicate aggregate signatures that enable users to aggregate signatures and keys across multiple messages, ensuring that the signers fulfill a specified public predicate while keeping their identities confidential. Brodsky et al. [11] proposed monotone-policy aggregate signatures that allow for the aggregation of signatures based on more expressive signing policies.

Some of the potential attacks that aggregate signatures face, which were previously considered in the context of multi-signer signatures [12,13] and are mentioned in [14,15], can be defended against by compulsorily requiring the signatures on distinct messages. Informally, an adversary is given a challenge public key and attempts to forge an aggregate signature on chosen messages and public keys of its choice [16]. Bellare et al. [17] solved this problem by hashing the message together with the public key, thus achieving unrestricted aggregation. Boneh et al. [18] achieved accountable aggregating multi-signatures, and this work is the most widely used solution in blockchain due to its security and versatility. Omar et al. [19] constructed constant-size multi-signatures and aggregate signatures in the context of multivariate public key cryptography. Mao et al. [20] proposed ID-based Locally Verifiable Efficient Aggregated Signatures and Signcryption, enabling efficient batch authentication and confidentiality in IoMT systems with minimal verification overhead.

We have also conducted a detailed study for multilinear maps [21,22] and regarded it as a valuable construction component. Garg, Gentry and Halevi (GGH) [23] constructed a graded encoding system, which is an approximate version of a multilinear group family. Hohenberger et al. [5] proposed an aggregated signature construction based on leveled multilinear mapping. Meanwhile, they proved the selective security of identity-based construction in the GGH framework. Furthermore, in order to obtain effective evidence in the security proof, we studied about the Diffie-Hellman Inversion assumption by Boneh et al. [24,25] and its variant in multilinear setting by Catalano et al. [26].

2 Technique Overview

In this section, we show how to extend a single-signer LVAS to achieve the multi-signer setting. We are inspired by LVAS of Goyal et al. [2], but our construction is based on different pairing structures.

Recap: Single-Signer LVAS. We start by the scheme of Goyal et al. [2]. They constructed a pairing-based LVAS from a weakly secure BB signature scheme [3,4]. In the single-signer LVAS scheme, the signing key is α, and the verification key is $\{g^{\alpha^i}\}_{i \in [n]}$, where n is the number of messages. The signer outputs a sequence of individual signatures $\{\sigma_i = g^{\frac{1}{\alpha+m_i}}\}_{i \in [n]}$ for messages $m_1, \ldots, m_n$. Given the verification key, the aggregator compresses the above signatures into an aggregate signature $\hat{\sigma} = g^{\prod_{i \in [n]} \frac{1}{(\alpha+m_i)}}$ by Lagrange's inverse polynomial interpolation technique. Concretely, the aggregator can calculate the exponential part $g^{\prod_{i \in [n]} \frac{1}{(\alpha+m_i)}} = g^{\sum_{i \in [n]} \frac{L_i}{(\alpha+m_i)}} = \prod_{i \in [n]} \sigma_i^{L_i}$ without the knowledge of the secret α, where the coefficients $L_1, \ldots, L_n$ can be publicly computed given only $m_1, \ldots, m_n$. As for verification, the verifier can check whether $e(\hat{\sigma}, g^{\prod_{i \in [n]} (\alpha+m_i)}) = e(g,g)$, where the exponential part $g^{\prod_{i \in [n]} (\alpha+m_i)} = g^{\sum_{i \in [n]} \delta_i \alpha^i} = \prod_{i \in [n]} (g^{\alpha^i})^{\delta_i}$, and the coefficients $\delta_1, \ldots, \delta_n$ can also be publicly computed given $m_1, \ldots, m_n$.

For locally opening some message m_j, the aggregator can simply compute auxiliary information $\text{aux}_{j,1} = g^{\prod_{i \in [n] \setminus \{j\}} (\alpha+m_i)}$ and $\text{aux}_{j,2} = g^{\alpha \cdot \prod_{i \in [n] \setminus \{j\}} (\alpha+m_i)}$ by the same method as above. And the verifier can locally verify the message m_j by check whether $e(\hat{\sigma}, \text{aux}_{j,1}^{m_j} \cdot \text{aux}_{j,2}) = e(g,g)$.

Extending to Multi-signer Settings. The key issue is to find an appropriate aggregate structure that supports locally openings and resists adversarial openings. As noted by [2], the verifier needs to check the hint well-formed to avoid being fooled. This demands the verifier to know all underlying messages to recompute the correct hint, conflicting with local verification where only target messages are known. Thus the accumulator-style aggregation (i.e. BGLS signature [1]) cannot resist adversarial openings.

So we still use the above BB-based signature form as our partial signature, mainly considering that the original scheme supports local verifiability. Now we have n individual signatures $\{\sigma_i = g^{\frac{1}{\alpha_i+m_i}}\}$ needed to be aggregated.

Obviously, we cannot use the Lagrange's inverse polynomial interpolation technique to compute the aggregate signature. Because in multi-signer aggregate signatures, each signer holds different signing key $sk_i = \alpha_i$, and the aggregate signatures correspond to distinct signing keys. Thus, the aggregator cannot compute $g^{\prod_{i \in [n]} \frac{1}{\alpha_i+m_i}}$ without the knowledge of $\{\alpha_i\}_i$.

Then the main obstacle is to find the aggregation approach. Our first main observation is that we can obtain a product of all $\{\frac{1}{\alpha_i+m_i}\}_i$ from leveled multilinear maps. In leveled multilinear maps there exists a sequence of groups $(\mathbb{G}_1, \ldots, \mathbb{G}_{2n})$, the canonical generator g_i of $\mathbb{G}_i$, g is denoted as g_1, and a set of bilinear maps $\{e_{i,j} : \mathbb{G}_i \times \mathbb{G}_j \to \mathbb{G}_{i+j}\}_{i,j \geq 1, i+j \leq 2n}$ such that $\forall a, b \in \mathbb{Z}_p$, $e_{i,j}(g_i^a, g_j^b) = g_{i+j}^{ab}$. Since we know $\frac{1}{\alpha_i+m_i}$ belongs to $\mathbb{Z}_p^*$, and σ_i is in $\mathbb{G}_1$. In a nutshell, we aggregate from a multilinear map to have $\hat{\sigma} = e(\sigma_1, \ldots, \sigma_n) = g_n^{\prod_i \frac{1}{(\alpha_i+m_i)}}$. Here all the partial signatures are in the same group $\mathbb{G}_1$, and $\hat{\sigma}$ is an element of $\mathbb{G}_n$, which also ensures compactness. When verifying, we aim to compute an item to match the exponential part of $\hat{\sigma}$. Since each partial signature is $g_1^{\frac{1}{\alpha_i+m_i}}$, we compute a corresponding combination of $g_1^{\alpha_i+m_i}$. Given the verification keys and messages of n signers, we compute an auxiliary item $\text{aux}^{(n)} = e(\text{vk}_1 \cdot g_1^{m_1}, ..., \text{vk}_n \cdot g_1^{m_n}) = g_n^{\prod_{i=1}^{n}(\alpha_i+m_i)}$ and check whether $e_{n,n}(\hat{\sigma}, \text{aux}^{(n)}) = e_{n,n}(g_n, g_n)$ to verify the aggregate signature.

Regarding local openings, our goal is to present only the target message m_j instead of the entire message list during verification. Thus, we need to have the hint generator calculate aux_j in advance. We find it easy to peel off the target message from $\text{aux}^{(n)}$ through the leveled multilinear map. Concretely, we remove the item in $\text{aux}^{(n)}$ that corresponds to target message m_j by removing $g_1^{\alpha_j+m_j}$, ultimately we have $\text{aux}_j = g_{n-1}^{\prod_{i=1,i\neq j}^{n}(\alpha_i+m_i)}$. Since the function of aux_j is analogous to that of auxiliary $\text{aux}^{(n)}$, we give them similar names. In that case we continue verifying $\hat{\sigma}$ with $e_{n,n}(\hat{\sigma}, e_{n-1,1}(\text{aux}_j, \text{vk}_j \cdot g_1^{m_j})) = e_{n,n}(g_n, g_n)$. Under the multi-signer aggregate setting, we use the chosen-key security model defined by Boneh et al. [1]. We prove the aggregate unforgeability with adversarial openings based on $(\ell, 2n)$-MDHI assumption. Briefly, in $(\ell, 2n)$-MDHI assumption, given $\{h^{a^i}\}_{i \in [\ell]}$, we can simulate scheme parameters and components by the polynomial expansion technique. For more details, see Sect. 5.

3 Preliminaries

Notation. We use λ to denote the security parameter. For any integer n, we use $[n]$ to denote the ordered set $\{1, 2, ..., n\}$. For any set S, we use $s \xleftarrow{\$} S$ to indicate that an element s is sampled uniformly randomly from S. A machine is probabilistic polynomial time (PPT) if it is a probabilistic algorithm that runs in poly(λ) time. We also use negl(λ) to denote functions negligible in λ.

Leveled Multilinear Maps. In generic multilinear groups we assume the existence of an algorithm $\mathcal{G}(1^\lambda, k)$ that, on input of security parameter and integer k specifying the number of levels (i.e., the number of allowed pairing operations), generates the description pp of leveled multilinear groups $(\mathbb{G}_1, ..., \mathbb{G}_k)$, each of large prime order $p > 2^\lambda$. We let g_i be a canonical generator of $\mathbb{G}_i$ and we assume pp includes $g_1 \in \mathbb{G}_1$. The groups are such that there exists a set of bilinear maps $e_{i,j} : \mathbb{G}_i \times \mathbb{G}_j \to \mathbb{G}_{i,j \geq 1, i+j \leq k}$ such that $\forall a, b \in \mathbb{Z}_p : e_{i,j}(g_i^a, g_j^b) = g_{i+j}^{ab}$. When it is from the context the indices i, j are dropped from $e_{i,j}$.

Diffie-Hellman Inversion Assumption. We now give a formal definition of the q-DHI computational complexity assumption. The q-DHI assumption states that no PPT adversary, given a sequence$(g, g^a, g^{a^2}, ..., g^{a^q})$ can compute $g^{\frac{1}{a}}$, where a is chosen uniformly at random from $\mathbb{Z}_p^*$. It is called a q-type assumption because the experiment is parameterized by the length of the powers-in-exponent sequence given to the adversary.

Assumption 3.1 ((ℓ, k)-Multilinear Diffie-Hellman Inversion Assumption [26]). Let pp be the description of a set of multilinear groups and $g_1 \in \mathbb{G}_1$ be a random generator. Let $a \xleftarrow{\$} \mathbb{Z}_p$ be chosen at random. We define the advantage of an adversary $\mathcal{A}$ in solving the (ℓ, k)-MDHI problem as

$$\text{Adv}_{\mathcal{A}}^{(\ell,k)\text{-MDHI}(\lambda)} = Pr[\mathcal{A}(g_1, g_1^a, \ldots, g_1^{a^\ell}) = g_k^{\frac{1}{a}}],$$

and we say that the (ℓ, k)-MDHI assumption holds for $\mathcal{G}$ if for every PPT $\mathcal{A}$ and for $\ell = \text{poly}(\lambda)$, $\text{Adv}_{\mathcal{A}}^{(\ell,k)\text{-MDHI}}(\lambda)$ is negligible in λ.

In addition, we also say the (ℓ, k)-MDHI assumption is equivalent to the (ℓ, k)-MDHI* assumption, which is identical to the standard one except that we say the advantage of solving the (ℓ, k)-MDHI* problem as

$$\text{Adv}_{\mathcal{A}}^{(\ell,k)\text{-MDHI}^*}(\lambda) = \Pr[\mathcal{A}(g_1, g_1^a, \ldots, g_1^{a^\ell}) = g_k^{a^{k\ell+1}}]$$

is negligible in λ if for every PPT $\mathcal{A}$ and for $\ell = \text{poly}(\lambda)$. We can simply prove the equivalence of the (ℓ, k)-MDHI and (ℓ, k)-MDHI* problems.

Let $\mathcal{A}$ and $\mathcal{B}$ be the adversary aimed to (ℓ, k)-MDHI and (ℓ, k)-MDHI* problem, respectively. Given $(g_1, g_1^a, \ldots, g_1^{a^\ell})$, $\mathcal{A}$ implicitly sets $b = \frac{1}{a}$, and inputs $(h_1 = g_1^{a^\ell}, \ldots, h_1^{b^\ell} = g_1)$ to $\mathcal{B}$ as an instance of (ℓ, k)-MDHI*. If $\mathcal{B}$ wins, then it outputs $T = h_k^{b^{k\ell+1}}$. It holds that

$$T = h_k^{b^{k\ell+1}} = (g_k^{a^{k\ell}})^{b^{k\ell+1}} = (g_k^{a^{k\ell}})^{(\frac{1}{a})^{k\ell+1}} = g_k^{\frac{1}{a}}$$

Similarly, We can also reduce (ℓ, k)-MDHI* problem to (ℓ, k)-MDHI problem.

3.1 Aggregate Signature

Before defining the multi-signer locally verifiable aggregate signature, we first recall the definition of an aggregate signature.

Syntax

- **Setup**($1^\lambda, n$) → pp. The setup algorithm samples global parameters pp. All the remaining algorithms take pp as input, and for ease of notation, we do not write it explicitly.
- **KeyGen**(1^λ) → (sk, vk). The key generation algorithm, on input of the security parameter λ, outputs a pair of signing and verification keys (vk, sk).
- **Sign**(sk, m) → σ. The signing algorithm takes as input a signing key sk and a message $m \in \mathcal{M}$, and computes a signature σ.
- **Verify**(vk, m, σ) → 0/1. The verification algorithm takes as input a verification key vk, a message $m \in \mathcal{M}$, and a signature σ. It outputs a bit to signal indicating whether the signature is valid or not.
- **Aggregate**($\{\text{vk}_i, m_i, \sigma_i\}_{i \in [n]}$) → $\hat{\sigma}/\bot$. The aggregate algorithm takes as input a sequence of tuples, each containing a verification key vk_i, a message m_i, and a signature σ_i. It outputs either an aggregated signature $\hat{\sigma}$ or a special abort symbol $\bot$.

- **AggVerify**($\{(\text{vk}_i, m_i)\}_{i\in[n]}, \hat{\sigma}) \to 0/1$. The aggregate verification algorithm takes as input a sequence of tuples, each containing a verification key vk_i, a message m_i, and outputs a bit indicating whether the aggregated signature $\hat{\sigma}$ is valid or not.

Correctness and Compactness. An aggregate signature scheme is said to be correct and compact if for all $\lambda, n \in \mathbb{N}$, every verification-signing key pair $(\text{vk}, \text{sk}) \leftarrow \textbf{Setup}(1^\lambda)$, messages m_i for $i \in [n]$ and every signature $\sigma_i \leftarrow \textbf{Sign}(\text{sk}, m_i)$ for $i \in [n]$, the following holds:

Correctness of Sign. For all $i \in [n]$, **Verify**$(\text{vk}_i, m_i, \sigma_i) = 1$.

Correctness of Aggregation. If $\hat{\sigma} = \textbf{Aggregate}(\{(\text{vk}_i, m_i, \sigma_i)\}_i)$, then

$$\textbf{AggVerify}(\{\text{vk}_i, m_i\}_i, \hat{\sigma}) = 1.$$

Compactness of Aggregation. $|\hat{\sigma}| \leq \text{poly}(\lambda)$. That is, the size of an aggregated signature is a fixed polynomial in the security parameter λ, and it remains independent of the number of aggregations n.

Security. We revisit the security notion for both regular signatures and aggregate signatures, adhering to the definition presented in [1], where, more precisely, given only a public key vk_1, no PPT adversary should be able to produce a valid aggregate signature that includes a message m_1^* which was never signed using the corresponding secret key sk_1.

This concept is formalized in what we term the aggregate chosen-key security model. In this model, the adversary $\mathcal{A}$ is provided with a single public key, and his objective is to achieve the existential forgery of an aggregate signature. The model permits the adversary to choose all public keys except for the challenge public key. Additionally, the adversary is granted access to a signing oracle associated with the challenge key. The adversary's advantage is defined to be its probability of success in the following game:

Setup. The aggregate forger $\mathcal{A}$ is provided with a randomly generated public key vk_1.

Queries. $\mathcal{A}$ adaptively requests signatures on messages of his choice using vk_1.

Response. Finally, $\mathcal{A}$ outputs $n-1$ additional public keys $\text{vk}_2, ..., \text{vk}_n$. These keys, together with the initial key vk_1, are used in $\mathcal{A}$'s forged aggregate signature. $\mathcal{A}$ also outputs messages $m_1, ..., m_n$, along with an aggregate signature $\hat{\sigma}^*$ purportedly generated by the n users, each signing his respective message.

Definition 1 (Aggregated Unforgeability). *A multi-signer aggregate signature scheme* (**Setup, Sign, Verify, Aggregate, AggVerify**) *is said to be a secure multi-signer aggregate signature scheme if for every admissible PPT* attacker $\mathcal{A}$, there exists a negligible function $\mathrm{negl}(\cdot)$ such that for all $\lambda \in \mathbb{N}$, the following holds

$$\Pr\left[\mathbf{AggVerify}(\{(\mathrm{vk}_i^*, m_i^*)\}_{i\in[n]}, \hat{\sigma}^*) = 1 : \begin{array}{c}(\mathrm{vk}_1, \mathrm{sk}_1) \leftarrow \mathbf{Setup}(1^\lambda) \\ (\{(\mathrm{vk}_i^*, m_i^*)\}_{i\in[n]}, \hat{\sigma}^*) \\ \leftarrow \mathcal{A}^{\mathbf{Sign}(\mathrm{sk}_1, \cdot)}(1^\lambda, \mathrm{vk}_1)\end{array}\right] \leq \mathrm{negl}(\lambda),$$

where $\mathcal{A}$ is admissible if $\mathrm{vk}_1^* = \mathrm{vk}_1$ and m_1^* was not queried by $\mathcal{A}$ to the $\mathbf{Sign}(\mathrm{sk}_1, \cdot)$ oracle.

Definition 2 (Static Aggregated Unforgeability). *We say the aggregate signature scheme is statically secure if the adversary in the above game is confined to make all of its message queries $\{m_i\}_{i\in[q_s]}$ and declare the challenge message m_1^* at the beginning of the game (defined in Definition 1) before it receives the challenge key vk_1.*

3.2 Multi-signer Locally Verifiable Aggregate Signature

We introduce the concept of local verification for aggregated signatures. In the multi-signer setting, the standard **AggVerify** algorithm checks the entire messages $(m_1, ..., m_n)$ within the aggregated signature $\hat{\sigma}$, with a time complexity linearly to the number of messages n.

To enhance verification efficiency, we propose aggregate signatures with local opening. The local verification algorithm only requires the target message m to verify against the claimed aggregated signature $\hat{\sigma}$, rather than all n messages. Following [2], and without altering the syntax of aggregate signatures, we introduce an auxiliary local opening generator based on leveled multilinear maps. The generator produces auxiliary information specific to the target message m that is being verified locally. We formally define the algorithm below.

- **LocalOpen**$(\hat{\sigma}, \{(\mathrm{vk}_i, m_i)\}_{i\in[n]}, j \in [n]) \to \mathrm{aux}_j$. The local opening algorithm takes as input aggregated signature $\hat{\sigma}$, a sequence of tuples where each contains a verification key vk_i and a message m_i for $i \in [n]$, and an index $j \in [n]$, where n denotes the sequence length. The algorithm outputs auxiliary information aux_j corresponding to the key-message pair (vk_i, m_i).
- **LocalAggVerify**$(\hat{\sigma}, \mathrm{vk}_j, m_j, \mathrm{aux}_j) \to 0/1$. The local aggregate verification algorithm takes as input an aggregated signature $\hat{\sigma}$, a verification key vk, a message m, and auxiliary information aux. It outputs a bit indicating whether the aggregate signature $\hat{\sigma}$ includes a signature for the message m under verification key vk_j.

Correctness and Compactness of Local Opening. An aggregate signature scheme with local openings is considered correct and compact if for all $\lambda, n \in \mathbb{N}$, and for every verification-signing key pair $(\mathrm{vk}, \mathrm{sk}) \leftarrow \mathbf{Setup}(1^\lambda)$, messages m_i for $i \in [n]$, and every signature $\sigma_i \leftarrow \mathbf{Sign}(\mathrm{sk}, m_i)$ for $i \in [n]$, the following conditions are held:

Correctness of Local Opening. For all $j \in [n]$, we have

$$\textbf{LocalAggVerify}(\hat{\sigma}, \text{vk}_j, m_j, \textbf{LocalOpen}(\hat{\sigma}, \{\text{vk}_i, m_i\}_i, j)) = 1.$$

Compactness of Opening. The size of the auxiliary opening is a fixed polynomial in the security parameter λ, independent of the number of aggregations.

Security Against Adversarial Openings. We now define the security notion for aggregate signatures with local openings.

Definition 3 (Aggregated Unforgeability with Adversarial Opening). *A multi-signer locally verifiable aggregate signature scheme* (**Setup, Sign, Verify, Aggregate, AggVerify, LocalOpen, LocalAggVerify**) *is said to be secure against adversarial openings if, for every admissible PPT attacker $\mathcal{A}$, there exists a negligible function* $\text{negl}(\cdot)$ *such that for all* $\lambda \in \mathbb{N}$, *the following holds:*

$$\Pr\left[\textbf{LocalAggVerify}(\text{vk}_1^*, m_1^*, \text{aux}^*, \hat{\sigma}^*) = 1 : \begin{array}{c} (\text{vk}_1, \text{sk}_1) \leftarrow \textbf{Setup}(1^\lambda) \\ (\text{vk}_1^*, m_1^*, \text{aux}^*, \hat{\sigma}^*) \\ \leftarrow \mathcal{A}^{\textbf{Sign}(\text{sk}_1, \cdot)}(1^\lambda, \text{vk}_1) \end{array}\right] \leq \text{negl}(\lambda),$$

where $\mathcal{A}$ is admissible if $\text{vk}_1^* = \text{vk}_1$ *and* m_1^* *was not queried by $\mathcal{A}$ to the* $\textbf{Sign}(\text{sk}_1, \cdot)$ *oracle.*

Definition 4 (Static Aggregated Unforgeability with Adversarial Opening). *We say the locally verifiable aggregate signature scheme is statically secure against adversarial openings if the adversary in the above game (as defined in Definition 3), is restricted to make all of its message queries $\{m_i\}_{i \in [q_s]}$ and declaring the challenge message m_1^* at the outset of the game, before receiving the challenge key* vk_1.

Definition 5 (Fully Public Openings for Aggregate Signatures). *The* **LocalOpen** *algorithm is oblivious to the aggregated signature.* **LocalOpen** *requires only the sequence of key-message pairs, rather than the aggregate signature itself to generate an auxiliary.*

An aggregate signature scheme is said to have fully local public openings if the algorithm **LocalOpen** *has the following syntax*

$$\textbf{LocalOpen}(\{\text{vk}_i, m_i\}_{i \in [n]}, j \in [n]) \to \text{aux}_j.$$

4 Multi-signer Locally Verifiable Aggregate Signature

In this section, we provide a multi-signer locally verifiable aggregate signature scheme with fully public local openings based on the hardness of $(\ell, 2n)$-Multilinear Diffie-Hellman Inversion problem, where n is the number of signing parties.

Injective Message Hashing. An injective mapping from the message space ($\mathcal{M}_\lambda = \{0,1\}^\lambda$) to the prime field $\mathbb{Z}_p$ where $p > 2^\lambda$ is considered. We employ two such simple mappings (HGen, H) to achieve static and full adaptive security respectively in our final construction.

Identity Map. The hash setup $\text{HGen}^\mathcal{I}$ is simply an empty algorithm that outputs $\text{hk} = \epsilon$, and $\text{H}^\mathcal{I}(\epsilon, m) = m$ is interpreted as a field element of $\mathbb{Z}_p$.

RO Map. The RO map is defined by a family of hash functions $\mathcal{H} = \{\mathcal{H}_\lambda\}_\lambda$, where each $h \in \mathcal{H}_\lambda$ takes λ bits as input and produces λ bits as output. The hash setup $\text{HGen}^\mathcal{H}$ samples a hash function $h \in \mathcal{H}_\lambda$ and outputs $\text{hk} = h$. The RO map $\text{H}^\mathcal{H}(\text{hk} = h, m) = h(m)$ then interprets the output $h(m)$ as an element of field $\mathbb{Z}_p$. If h is modeled as a random oracle, the RO map retains this characteristic.

4.1 Construction

- **Setup**($1^\lambda, n$) $\to$ pp. The trusted setup algorithm receives the security parameter and message length as inputs. It begins by running $\mathcal{G}(1^\lambda, 2n)$, producing a sequence of groups $\mathbb{G} = (\mathbb{G}_1, ..., \mathbb{G}_{2n})$ of prime order p, with canonical generators $g_1, ..., g_{2n}$, where g_1 is denoted as g. There exists a set of bilinear maps $e_{i,j} : \mathbb{G}_i \times \mathbb{G}_j \to \mathbb{G}_{i,j \geq 1, i+j \leq k}$ such that $\forall a, b \in \mathbb{Z}_p : e_{i,j}(g_i^a, g_j^b) = g_{i+j}^{ab}$. Subsequent algorithms use the global parameters pp as input, though pp is omitted in the notation for simplicity.
- **KeyGen**(1^λ) $\to$ (vk, sk). The key generation algorithm samples a random $\alpha \leftarrow \mathbb{Z}_p^*$. It also samples the public parameters for message hashing as hk $\leftarrow$ HGen(1^λ). It sets the key pair as vk $= (\text{hk}, g^\alpha)$ and sk $= (\text{hk}, \alpha)$.
- **Sign**(sk, m) $\to \sigma$. The hash of a message m is $h_m = H(\text{hk}, \text{m})$ each signer takes secret key α and h_m to calculate a partial signature $\sigma = g^{\frac{1}{\alpha + h_m}}$.
- **Verify**(vk, m, σ) $\to 0/1$. The partial signature verification algorithm takes as input the verification key vk of a specific signer, a message $m \in \mathcal{M}$, and a signature σ. The verifier computes

$$e(\sigma, \text{vk} \cdot g^{h_m}) = e(g, g).$$

If the verification is successful, the algorithm outputs 1 to indicate that the partial signature is valid, otherwise it outputs 0.
- **Aggregate**($\{(\text{vk}_i, m_i, \sigma_i)\}_i) \to \hat{\sigma} / \perp$. The signature aggregation algorithm takes as input a sequence of signatures σ_i. If any of the partial signature is verified to be invalid, then the aggregate algorithm outputs $\perp$. Otherwise it computes an aggregated signature as

$$\sigma^{(1)} = g_1^{\frac{1}{\alpha_1 + h_{m_1}}},$$

$$from \quad i \in [2, n] \quad \sigma^{(i)} = e_{i-1,1}(\sigma^{(i-1)}, g_1^{\frac{1}{\alpha_i + h_{m_i}}}),$$

ultimately we have the aggregate signature and represent it as

$$\hat{\sigma} = \sigma^{(n)} = g_n^{\prod_{i=1}^{n} \frac{1}{\alpha_i + h_{m_i}}}.$$

- **AggVerify**($\{(\text{vk}_i, m_i)\}_i, \hat{\sigma}) \to 0/1$. As we discussed in the technique overview, we take advantage of multilinear maps to compute an auxiliary, whose index is the reciprocal of the product of the aggregate signature index. Given the aggregate signature $\hat{\sigma}$, a sequence of verification key $\text{vk}_i = g_1^{\alpha_i}$ and messages m_i, we can verify the aggregate signature easily. Below we demonstrate how to conduct the aggregate verification:

$$\text{aux}^{(1)} = \text{vk}_1 \cdot g_1^{h_{m_1}},$$
$$from \quad i \in [2, n], \quad \text{aux}^{(i)} = e_{i-1,1}(\text{aux}^{(i-1)}, \text{vk}_i \cdot g_1^{h_{m_i}}),$$

so we have an auxiliary $\text{aux}^{(n)} = g_n^{\prod_{i=1}^{n}(\alpha_i + h_{m_i})}$, and check whether the following is true or not:

$$e_{n,n}(\hat{\sigma}, \text{aux}^{(n)}) = e_{n,n}(g_n, g_n).$$

- **LocalOpen**($\{(\text{vk}_i, m_i)\}_{i \in [n]}, j \in [n]) \to \text{aux}_j$. We calculate the hint of a target message whose index is j. Note that hint is a set contains all the verification keys and messages except it removes the target message and its vk so that we use leveled multilinear maps to obtain it. For clarity of expression, we divide the algorithm into two situations to deal with the situation where the first message or the j-th message ($j \neq 1$) is the target message:

$j = 1$	$j \neq 1$
$\text{aux}_j = vk_2 \cdot g_1^{h_{m_2}}$ for $i = 3$ to n $\quad \text{aux}_j = e_{i-2,1}(\text{aux}_j, \text{vk}_i \cdot g_1^{h_{m_i}})$ output aux_j	$\text{aux}_j = vk_1 \cdot g_1^{h_{m_1}}$ for $i = 2$ to n $\quad$ if $i < j$ $\quad\quad$ then $\text{aux}_j = e_{i-1,1}(\text{aux}_j, \text{vk}_i \cdot g_1^{h_{m_i}})$ $\quad$ if $i = j$ $\quad\quad$ then $\text{aux}_j = \text{aux}_j$ $\quad$ if $i > j$ $\quad\quad$ then $\text{aux}_j = e_{i-2,1}(\text{aux}_j, \text{vk}_i \cdot g_1^{h_{m_i}})$ output aux_j

the ultimate auxiliary information is

$$\text{aux}_j = g_{n-1}^{\prod_{i=1, i \neq j}^{n}(\alpha_i + h_{m_i})}.$$

– **LocalAggVerify**$(\hat{\sigma}, \text{vk}_j, m_j, \text{aux}_j) \to 0/1$. The local verification algorithm parses the verification key of signers and generates signature on target message m_j. Finally we use the same multilinear pairing as in **AggVerify** to check the following condition:

$$e_{n,n}\left(\hat{\sigma}, e_{n-1,1}(\text{aux}_j, \text{vk}_j \cdot g_1^{h_{m_j}})\right) = e_{n,n}(g_n, g_n).$$

4.2 Correctness

Correctness of Signing. This follows directly from the fact that $e(g^{\frac{1}{\alpha+h_m}}, g^\alpha \cdot g^{h_m}) = e(g,g)$ where $h_m = H(\text{hk}, \text{m})$.

Correctness of Aggregation. If $\hat{\sigma} = \textbf{Aggregate}(\text{pp}, \{\sigma_i\}_{i\in[n]})$, correctness of aggregation indicates that $\textbf{AggVerify}(\{(\text{vk}_i, m_i)\}_{i\in[n]}, \hat{\sigma}) = 1$. Consider any sequence $m_1, ..., m_n$, and corresponding partial signatures $\sigma_i = g_1^{\frac{1}{\alpha_i + h_{m_i}}}$ for $i \in [n]$ where $h_{m_i} = H(\text{hk}, m_i)$. We know that aggregating these signature is done as $\hat{\sigma} = g_n^{\prod_{i=1}^n \frac{1}{\alpha_i + h_{m_i}}}$. Then the verifier computes an $\text{aux}^{(n)}$ by using the verification key and hashes of messages as follows:

$$\text{aux}^{(n)} = e_{n-1,1}(aux^{(n-1)}, \text{vk}_n \cdot g_1^{h_{m_n}})$$
$$= e_{n-1,1}(e_{n-2,1}(\text{aux}^{(n-2)}, \text{vk}_{n-1} \cdot g_1^{h_{m_{n-1}}}), \text{vk}_n \cdot g_1^{h_{m_n}})$$
$$......$$
$$= g_n^{\prod_{i=1}^n (\alpha_i + h_{m_i})}.$$

Finally the aggregated verification checks the following:

$$e_{n,n}(\hat{\sigma}, aux^{(n)}) = e_{n,n}\left(g_n^{\prod_{i=1}^n \frac{1}{\alpha_i + h_{m_i}}}, g_n^{\prod_{i=1}^n (\alpha_i + h_{m_i})}\right)$$
$$= e(g_n, g_n),$$

which is consistent with the content on the right side of the verification equation.

Compactness of Aggregation and Opening. In our scheme, each partial signature is a group element in group $\mathbb{G}_1$. The aggregate signature of n signers for n messages is a group element in group $\mathbb{G}_n$, while the auxiliary corresponding to a particular message is also a group element in group $\mathbb{G}_{n-1}$. So both of their sizes are independent of the number of aggregations, which achieves compactness.

Efficiency. In the multilinear setting, the time of **Aggregate** and **AggVerify** grows with the number of n participants. We need n pairing operations in groups $(\mathbb{G}_1, ..., \mathbb{G}_n)$ to calculate $\hat{\sigma}$ and $\text{aux}^{(n)}$, and an additional $\mathbb{G}_{2n}$ to verify. In **LocalOpen** the hint generator spends $O(n)$ time generating the auxiliary, and the verifier in **LocalAggVerify** only needs $O(1)$ time to complete the verification.

5 Security

Static Aggregated Unforgeability. We demonstrate that if the message hashing function is instantiated as the identity map in our aggregate signature construction, the resulting scheme achieves static unforgeability. We formally prove this as follows.

Theorem 1. *(Static Aggregated Unforgeability). If the $(\ell, 2n)$–MDHI assumption (Assumption 3.2) holds and* (HGen, H) *is an identity hash, then the aggregate signature scheme described above satisfies static unforgeability and static aggregated unforgeability with adversarial openings.*

Proof. The proof for aggregated unforgeability is more general than that for regular unforgeability. Therefore, we first prove static aggregated unforgeability for our signature scheme. Subsequently, we prove static aggregated unforgeability with adversarial openings.

Aggregate Unforgeability. Suppose that there exists a PPT adversary $\mathcal{A}$ that successfully breaks aggregated unforgeability with non-negligible probability ϵ. We then construct a PPT adversary $\mathcal{B}$ that can break the $(\ell, 2n)$-MDHI assumption with probability $\epsilon -$ negl, where negl is a negligible function. The reduction algorithm $\mathcal{B}$ is described below.

In proving static security, the adversary $\mathcal{A}$ must determine all its signature queries $\{m_i\}_{i \in [q_s]}$ and the challenge message m_1^* prior to the beginning of the reduction algorithm $\mathcal{B}$.

$\mathcal{B}$ then breaks the $(\ell, 2n)$-MDHI assumption, where the hardness parameter $(\ell, 2n)$ satisfies $\ell \geq q_s + 1$. The $(\ell, 2n)$-MDHI challenger samples the multilinear group parameters $\Pi = (p, \mathbb{G}_1, ..., \mathbb{G}_{2n}, \{e_{i,j}(\cdot, \cdot)\}_{i+j \leq 2n, i \geq 1, j \geq 1})$, and a sequence of group elements $\{h_1^{(i)} = h^{a^i}\}_{i=0}^{\ell}$ for a randomly chosen exponent $a \in \mathbb{Z}_p^*$ as an instance of $(\ell, 2n)$-MDHI assumption, and sends $(\Pi, h_1^{(0)}, \ldots, h_1^{(\ell)})$ to $\mathcal{B}$.

$\mathcal{B}$ implicitly sets the signing key $\alpha_1 = a - m_1^*$. Then $\mathcal{B}$ sets the base group element g_1 as $h_1^{\prod_{i=1}^{q_s}(\alpha_1 + m_i)} = h_1^{\prod_{i=1}^{q_s}(a - m_1^* + m_i)}$. Based on this, $\mathcal{B}$ similarly compute the remaining group elements as $\{g_s = e_{i, s-i}(g_i, g_{s-i})\}_{s \leq 2n}$.

To compute g_1, $\mathcal{B}$ first computes P_0 to obtain the coefficients $\{\theta^{(i)} \in \mathbb{Z}_p\}_{i=0}^{q_s}$:

$$P_0(X) = \prod_{i \in [q_s]} (X - m_1^* + m_i) = \sum_{i=0}^{q_s} \theta^{(i)} X^i \pmod{p}. \tag{1}$$

Then it samples a random exponent $\delta \in \mathbb{Z}_p^*$ and computes g_1 as

$$g_1 = h_1^{\delta P_0(a)} = h_1^{\delta \sum_{i=0}^{q_s} \theta^{(i)} a^i} = \prod_{i=0}^{q_s} h_1^{\delta \theta^{(i)} a^i} = \prod_{i=0}^{q_s} (h_1^{(i)})^{\delta \theta^{(i)}}. \tag{2}$$

To compute $vk_1 = g_1^{\alpha_1} = h_1^{(a-m_1^*)\prod_{i=1}^{q_s}(a-m_1^*+m_i)}$, $\mathcal{B}$ computes polynomials P_1 to obtain the coefficients $\{\beta^{(i)} \in \mathbb{Z}_p\}_{i=0}^{q_s+1}$:

$$P_1(X) = (X - m_1^*) \prod_{i \in [q_s]} (X - m_1^* + m_i) = \sum_{i=0}^{q_s+1} \beta^{(i)} X^i \pmod{p}. \quad (3)$$

then computes vk_1 as

$$vk_1 = h_1^{\delta P_1(a)} = h_1^{\delta \sum_{i=0}^{q_s+1} \beta^{(i)} a^i} = \prod_{i=0}^{q_s+1} h_1^{\delta \beta^{(i)} a^i} = \prod_{i=0}^{q_s+1} (h_1^{(i)})^{\delta \beta^{(i)}}. \quad (4)$$

$\mathcal{B}$ gives the global parameters and the verification key vk_1 to $\mathcal{A}$ and then replies the signature queries for messages $\{m_i\}_{i \in [q_s]}$.

To compute these signatures, $\mathcal{B}$ first computes some coefficients. Let Q_{-i} ($\forall i \in [q_s]$) be the following polynomial with coefficients $\{\gamma_i^{(j)} \in \mathbb{Z}_p\}_{j=0}^{q_s-1}$:

$$\forall i \in [q_s], \quad Q_{-i}(X) = \prod_{j \in [q_s] \setminus \{i\}} (X - m_1^* + m_j) = \sum_{j=0}^{q_s-1} \gamma_i^{(j)} X^j \pmod{p}, \quad (5)$$

which removes the monomial $(X - m_1^* + m_i)$ from the polynomial P_0. $\mathcal{B}$ then computes the signatures for the queried messages $m_i, ..., m_{q_s}$ as

$$\forall i \in [q_s], \quad \sigma_i = g_1^{\delta(a-m_1^*+m_i)^{-1}} = h_1^{\delta \frac{P_0(a)}{a-m_1^*+m_i}} = h_1^{\delta Q_{-i}(a)} = \prod_{j=0}^{q_s-1} (h_1^{(j)})^{\delta \gamma_i^{(j)}}. \quad (6)$$

$\mathcal{B}$ then sends vk_1 and $\{\sigma_i\}_{i \in [q_s]}$ to the adversary $\mathcal{A}$. Upon receiving these, $\mathcal{A}$ provides a forged signature $\hat{\sigma}^*$, which contains the sequence of key-message pairs $\{(vk_i^*, m_i^*)\}_{i \in [n]}$. $\mathcal{B}$ verifies whether $\hat{\sigma}^*$ is a valid aggregate signature. If $\hat{\sigma}^*$ is invalid, $\mathcal{B}$ aborts. Otherwise $\mathcal{B}$ proceeds to compute the following items:

$$\begin{aligned}\widetilde{aux}^{(1)} &= g_1; \\ \widetilde{aux}^{(i)} &= e_{i-1,1}(\widetilde{aux}^{(i-1)}, vk_i^* \cdot g_1^{m_i^*}) \quad \text{(from } i = 2 \text{ to } n\text{)}.\end{aligned} \quad (7)$$

For the convenience of the reduction, $\mathcal{B}$ compute group elements $h_{2n}^{a^t}$ for $t \in [2n \cdot q_s - 1]$ (which will be used later) by the following equations:

$$h_k^{a^{(i+j)}} = e_{k-s,s}(h_{k-s}^{a^i}, h_s^{a^j}) \quad (8)$$

Then $\mathcal{B}$ computes the polynomials P_{2n} to obtain the coefficients $\{\mu^{(j)} \in \mathbb{Z}_p\}_{j=-1}^{2n \cdot q_s - 1}$:

$$\begin{aligned}P_{2n}(X) &= \frac{P_0^n(X) \cdot P_0^n(X)}{X} = \frac{\left(\prod_{i \in [q_s]}(X - m_1^* + m_i)\right)^{2n}}{X} \\ &= \frac{\mu^{(-1)}}{X} + \sum_{j=0}^{2n \cdot q_s - 1} \mu^{(j)} X^j \pmod{p}.\end{aligned} \quad (9)$$

Finally, it computes

$$Z = \left(e_{n,n}(\hat{\sigma}^*, \widetilde{\text{aux}}^{(n)})^{\frac{1}{\delta^{2n}}} \prod_{j=0}^{2n \cdot q_s - 1} (h_{2n}^{a^j})^{-\mu^{(j)}}\right)^{\frac{1}{\mu^{(-1)}}}$$

and outputs Z as its $(\ell, 2n)$-MDHI solution.

We claim that if $\mathcal{A}$ succeeds with probability ϵ, then $\mathcal{B}$ succeeds with probability ϵ - negl. Note that if $\mathcal{A}$ is an admissible adversary, $\mathcal{B}$ aborts with at most negligible probability. Next we claim that $e_{n,n}(\hat{\sigma}^*, \widetilde{\text{aux}}^{(n)})^{\alpha_1 + m_1^*} = e_{n,n}(\hat{\sigma}^*, \widetilde{\text{aux}}^{(n)})^a = e_{n,n}(g_n, g_n)$ whenever $\hat{\sigma}^*$ is a valid signature for challenge messages $(m_1^*, ..., m_n^*)$ and $\mathcal{A}$ is an admissible adversary. By using Eq. (7), (8), (9), we can compute the following:

$$e_{n,n}(\hat{\sigma}^*, \widetilde{\text{aux}}^{(n)}) = e_{n,n}(g_n, g_n)^{\frac{1}{a}}$$
$$= e_{n,n}\left(h_n^{(\delta P_0(a))^n}, h_n^{(\delta P_0(a))^n}\right)^{\frac{1}{a}}$$
$$= e_{n,n}(h_n, h_n)^{\frac{\delta^{2n} P_0^{2n}(a)}{a}}$$
$$= e_{n,n}(h_n, h_n)^{\delta^{2n}(\frac{\mu^{(-1)}}{a} + \sum_{j=0}^{2n-1} \mu^{(j)} a^j)}$$
$$= e_{n,n}(h_n, h_n)^{\delta^{2n}(\frac{\mu^{(-1)}}{a})} \prod_{j=0}^{2n \cdot q_s - 1} (h_{2n}^{a^j})^{\delta^{2n} \mu^{(j)}}.$$

This can be further simplified as

$$e_{n,n}(\hat{\sigma}^*, \widetilde{\text{aux}}^{(n)})^{\frac{1}{\delta^{2n}}} \prod_{j=0}^{2n \cdot q_s - 1} (h_{2n}^{a^j})^{-\mu^{(j)}} = e_{n,n}(h_n, h_n)^{\frac{\mu^{(-1)}}{a}}$$

The left side of the equation is precisely $Z^{\mu^{(-1)}}$. thus we get that $Z = e_{n,n}(h_n, h_n)^{\frac{1}{a}} = h_{2n}^{\frac{1}{a}}$ whenever $\hat{\sigma}^*$ is a valid signature for m_1^*, and $\mathcal{A}$ is an admissible adversary. Thus aggregated unforgeability follows from hardness of $(\ell, 2n)$-MDHI. □

Aggregated Unforgeability with Adversarial Openings. We demonstrate that our signature scheme is also aggregated unforgeable with adversarial openings. As established in Theorem 1, the unforgeability relies on the hardness of $(\ell, 2n)$-MDHI assumption. Suppose there exists a PPT adversary $\mathcal{A}$ that breaks aggregated unforgeability with adversarial openings with non-negligible probability ϵ. We construct a PPT adversary $\mathcal{B}$ that breaks the $(\ell, 2n)$-MDHI assumption in the multilinear group with probability ϵ - negl, where negl represents the negligible function. The reduction algorithm $\mathcal{B}$ is detailed below.

In static security, $\mathcal{A}$ needs to decide all its signature queries $\{m_i\}_{i \in [q_s]}$ before the beginning of the reduction algorithm $\mathcal{B}$.

$\mathcal{B}$ works as the above proof that breaks the $(\ell, 2n)$-MDHI assumption, where the hardness parameter $(\ell, 2n)$ is such that $\ell \geq q_s + 1$. The $(\ell, 2n)$-MDHI challenger samples the multilinear group parameters $\Pi = (p, \mathbb{G}_1, ..., \mathbb{G}_{2n}, \{e_{i,j}(\cdot, \cdot)\}_{i+j \leq 2n})$, and a sequence of group elements $\{h_1^{(i)} = h^{a^i}\}_{i=0}^{\ell}$ for a randomly chosen exponent $a \in \mathbb{Z}_p^*$ as an $(\ell, 2n)$-MDHI instance, and sends $(\Pi, h_1^{(0)}, \ldots, h_1^{(\ell)})$ to $\mathcal{B}$.

The reduction algorithm $\mathcal{B}$ first computes P_0 (see Eq. (1)) and samples a random exponent $\delta \in \mathbb{Z}_p^*$ and computes the group element g_1 to be included in the verification key as in Eq. (2). Then it computes the verification key vk_1 (see Eq. (4)) that needs to be given to the adversary $\mathcal{A}$. Using the additional verification keys namely $\{vk_2, ..., vk_n\}$ along with the initial key vk_1, adversary $\mathcal{B}$ calculates the signatures $\{\sigma_i\}_{i \in [q_s]}$ on $\mathcal{A}$'s requested messages in the reduction of aggregate unforgeability above (see Eq. (6)). After $\mathcal{A}$ receives vk_1 and $\{\sigma_i\}_i$, he sends forged aggregate signature $\hat{\sigma}^*$ along with a local opening aux^* (which corresponds to an aggregated signature with local opening for the challenge messages m_1^*. $\mathcal{B}$ should check if $\hat{\sigma}^*$ is a valid accepting signature by running the algorithm **LocalAggVerify**. It aborts if it is an invalid signature, otherwise it computes the polynomials P_{2n} to obtain the coefficients $\{\mu^{(j)} \in \mathbb{Z}_p\}_{j=-1}^{2n \cdot q_s - 1}$. Finally, it computes

$$Z = \left(e_{n,n}(\hat{\sigma}^*, \widetilde{aux}^{(n)})^{\frac{1}{\delta^{2n}}} \prod_{j=0}^{2n \cdot q_s - 1} (h_{2n}^{a^j})^{-\mu^{(j)}} \right)^{\frac{1}{\mu^{(-1)}}},$$

and outputs Z as its $(\ell, 2n)$-MDHI solution. From the right side of the equation, $\mathcal{B}$ can calculate $h_{2n}^{\frac{1}{a}}$ in the same way as the above proof, thus aggregated unforgeability with adversarial opening follows from the hardness of $(\ell, 2n)$-MDHI. □

Full Aggregated Unforgeability in ROM. Next, we show that if we instantiate the massage hashing in the ROM, then the above aggregate signature construction satisfies fully unforgeability.

Theorem 2. *(Full Unforgeability). If the $(\ell, 2n)$-MDHI assumption (Assumption 3.2) holds, and $(HGen, H)$ is instantiated in the ROM, then the aggregate signature scheme described above satisfies (full) aggregated unforgeability (defined in Definition 1) and (full) aggregated unforgeability with adversarial openings (defined in Definition 3).*

Proof. The full aggregated unforgeability and aggregated unforgeability with adversarial openings will be proven sequentially below.

Aggregated Unforgeability. Assume there exists a PPT adversary $\mathcal{A}$ breaks full or adaptive aggregated unforgeability with non-negligible probability ϵ. We construct a PPT adversary $\mathcal{B}$ that can break the $(\ell, 2n)$-MDHI assumption with probability ϵ/Q^{RO} - negl, where negl is a negligible function and Q^{RO} is the number of queries made by $\mathcal{A}$ to the random oracle.

The reduction algorithm $\mathcal{B}$ is similar to the one utilized in the proof of (Theorem 1), with the key difference being the added capability of programmability. Specifically, $\mathcal{B}$ guesses the index of the challenge message m_1^* among all the queries made to the random oracle (RO) by $\mathcal{A}$, and programs the hashed value corresponding to m_1^* as a random exponent $h_{m_1^*}$. It then implicitly sets $\alpha = a - h_{m_1^*}$ similar to the approach used in the static security proof. $\mathcal{B}$ samples hash values for all queried messages at the start of the game, and programs them at the time each message is first queried.

The overall reduction algorithm is defined as the static proof, with the addition that $\mathcal{B}$ aborts if it incorrectly guesses the index of the challenge message m_1^*. Since $\mathcal{B}'s$ guess is correct with a probability of at least $1/Q^{RO}$, $\mathcal{B}'s$ final advantage will be ϵ/Q^{RO} - negl. This concludes the proof.

Aggregated Unforgeability with Adversarial openings. With similar modifications as in the above proof, we can reduce full aggregated unforgeability with adversarial openings to $(\ell, 2n)$-MDHI in the ROM. As in the earlier case, the reduction incurs a polynomial loss. □

Acknowledgments. We thank all anonymous reviewers for their helpful comments. This work was supported in part by National Natural Science Foundation of China (62372180).

References

1. Boneh, D., Gentry, C., Lynn, B., Shacham, H.: Aggregate and verifiably encrypted signatures from bilinear maps. In: Biham, E. (ed.) EUROCRYPT 2003. LNCS, vol. 2656, pp. 416–432. Springer, Heidelberg (2003). https://doi.org/10.1007/3-540-39200-9_26
2. Goyal, R., Vaikuntanathan, V.: Locally verifiable signature and key aggregation. In: Dodis, Y., Shrimpton, T. (eds.) CRYPTO 2022, Part II, vol. 13508, pp. 761–791. Springer, Cham (2022)
3. Boneh, D., Boyen, X.: Short signatures without random oracles. In: Cachin, C., Camenisch, J.L. (eds.) EUROCRYPT 2004. LNCS, vol. 3027, pp. 56–73. Springer, Heidelberg (2004). https://doi.org/10.1007/978-3-540-24676-3_4
4. Boneh, D., Boyen, X.: Short signatures without random oracles and the SDH assumption in bilinear groups. J. Cryptol. 21(2), 149–177 (2008)
5. Hohenberger, S., Sahai, A., Waters, B.: Full domain hash from (leveled) multilinear maps and identity-based aggregate signatures. In: Canetti, R., Garay, J.A. (eds.) CRYPTO 2013, Part I. LNCS, vol. 8042, pp. 494–512. Springer, Heidelberg (2013). https://doi.org/10.1007/978-3-642-40041-4_27
6. Bagherzandi, A., Jarecki, S.: Identity-based aggregate and multi-signature schemes based on RSA. In: Nguyen, P.Q., Pointcheval, D. (eds.) PKC 2010. LNCS, vol. 6056, pp. 480–498. Springer, Heidelberg (2010). https://doi.org/10.1007/978-3-642-13013-7_28

7. Ağırtaş, A.R., Gökce, N.Y., Yayla, O.: Enhancing local verification: aggregate and multi-signature schemes. Cryptology ePrint Archive, Paper 2024/1055 (2024)
8. Hohenberger, S., Waters, B.: Synchronized aggregate signatures from the RSA assumption. In: Nielsen, J.B., Rijmen, V. (eds.) EUROCRYPT 2018, Part II. LNCS, vol. 10821, pp. 197–229. Springer, Cham (2018). https://doi.org/10.1007/978-3-319-78375-8_7
9. Tezuka, M., Tanaka, K.: Pointcheval-sanders signature-based synchronized aggregate signature. In: Seo, S.-H., Seo, H. (eds.) ICISC 2022. LNCS, vol. 13849, pp. 317–336. Springer, Cham (2022)
10. Qiu, T., Tang, Q.: Predicate aggregate signatures and applications. In: Guo, J., Steinfeld, R. (eds.) ASIACRYPT 2023, Part II. LNCS, vol. 14439, pp. 279–312. Springer, Singapore (2023)
11. Brodsky, M.F., Choudhuri, A.R., Jain, A., Paneth, O.: Monotone-policy aggregate signatures. In: Joye, M., Leander, G. (eds.) EUROCRYPT 2024, Part IV. LNCS, vol. 14654, pp. 168–195. Springer, Cham (2024)
12. Boldyreva, A.: Efficient threshold signature, multisignature and blind signature schemes based on the gap-Diffie-Hellman-group signature scheme. Cryptology ePrint Archive, Report 2002/118 (2002)
13. Micali, S., Ohta, K., Reyzin, L.: Accountable-subgroup multisignatures: extended abstract. In: Reiter, M.K., Samarati, P. (eds.) ACM CCS 2001: 8th Conference on Computer and Communications Security, Philadelphia, PA, USA, 5–8 November 2001, pp. 245–254. ACM Press (2001)
14. Lysyanskaya, A., Micali, S., Reyzin, L., Shacham, H.: Sequential aggregate signatures from trapdoor permutations. In: Cachin, C., Camenisch, J.L. (eds.) EUROCRYPT 2004. LNCS, vol. 3027, pp. 74–90. Springer, Heidelberg (2004). https://doi.org/10.1007/978-3-540-24676-3_5
15. Xiong, H., Hou, Y., Huang, X., Kumari, S.: Certificate-based parallel key-insulated aggregate signature against fully chosen-key attacks for industrial internet of things. Cryptology ePrint Archive, Report 2020/1027 (2020)
16. Rückert, M., Schroeder, D.: Aggregate and verifiably encrypted signatures from multilinear maps without random oracles. Cryptology ePrint Archive, Report 2013/020 (2013)
17. Bellare, M., Namprempre, C., Neven, G.: Unrestricted aggregate signatures. In: Arge, L., Cachin, C., Jurdziński, T., Tarlecki, A. (eds.) ICALP 2007. LNCS, vol. 4596, pp. 411–422. Springer, Heidelberg (2007). https://doi.org/10.1007/978-3-540-73420-8_37
18. Boneh, D., Drijvers, M., Neven, G.: Compact multi-signatures for smaller blockchains. In: Peyrin, T., Galbraith, S. (eds.) ASIACRYPT 2018, Part II. LNCS, vol. 11273, pp. 435–464. Springer, Cham (2018). https://doi.org/10.1007/978-3-030-03329-3_15
19. Omar, S., Padhye, S., Dey, D.: Multivariate aggregate and multi-signature scheme. In: Roy, B.K., Chaturvedi, A., Tsaban, B., Ul Hasan, S. (eds.) Cryptology and Network Security with Machine Learning, pp. 71–76. Springer, Singapore (2024)
20. Mao, W., Jiang, P., Zhu, L.: Locally verifiable batch authentication in IoMT. IEEE Trans. Inf. Forensics Secur. **19**, 1001–1014 (2024)
21. Boneh, D., Silverberg, A.: Applications of multilinear forms to cryptography. Contemp. Math. **324**(1), 71–90 (2003)
22. Patranabis, S., Mukhopadhyay, D.: Identity-based key aggregate cryptosystem from multilinear maps. Cryptology ePrint Archive, Report 2016/693 (2016)

23. Garg, S., Gentry, C., Halevi, S.: Candidate multilinear maps from ideal lattices. In: Johansson, T., Nguyen, P.Q. (eds.) EUROCRYPT 2013. LNCS, vol. 7881, pp. 1–17. Springer, Heidelberg (2013). https://doi.org/10.1007/978-3-642-38348-9_1
24. Boneh, D., Boyen, X.: Efficient selective-ID secure identity-based encryption without random oracles. In: Cachin, C., Camenisch, J.L. (eds.) EUROCRYPT 2004. LNCS, vol. 3027, pp. 223–238. Springer, Heidelberg (2004). https://doi.org/10.1007/978-3-540-24676-3_14
25. Boneh, D., Boyen, X., Goh, E.-J.: Hierarchical identity based encryption with constant size ciphertext. Cryptology ePrint Archive, Report 2005/015 (2005)
26. Catalano, D., Fiore, D., Gennaro, R., Nizzardo, L.: Generalizing homomorphic MACs for arithmetic circuits. In: Krawczyk, H. (ed.) PKC 2014. LNCS, vol. 8383, pp. 538–555. Springer, Heidelberg (2014). https://doi.org/10.1007/978-3-642-54631-0_31

Conditional Attribute-Based Encryption with Keyword Search for Pay-Per-Query Commercial Model

Zerui Guo[1], Sha Ma[1(✉)], and Qiong Huang[1,2]

[1] South China Agricultural University, Guangzhou 516042, Guangdong, China
guozerui@stu.scau.edu.cn, {martin_deng,qhuang}@scau.edu.cn
[2] Guangzhou Key Laboratory of Intelligent Agriculture, Guangzhou 510642, China

Abstract. These days, cloud service providers have widely implemented the Pay-Per-Query Commercial Model, enabling public access to search services. Data owners utilize leased cloud infrastructure to provide data search services, with billing calculated based on the actual number of searches. Although this business model offers flexibility, convenience, and cost-effectiveness, it also faces the problem that they can obtain query results without consuming query times by colluding paid users with non-paying ones. Attribute-Based Keyword Search (ABKS), due to its access control features, can verify users who haven't paid the fees and prevent their query behaviors. However, it also causes the cloud server to interact with the charging server before each query. To address this issue, We propose the Conditional Attribute-Based Keyword Search (CABKS) paradigm. Based on this paradigm, we have put forward a practical CABKS scheme applicable to Pay-Per-Query Commercial Model. This scheme guarantees that users can search the ciphertext only when they hold valid credentials generated by a specific server. By decoupling the access control structure from the keyword ciphertext, it effectively prevents unauthorized users from consuming cloud server resources. Our scheme satisfies the indistinguishability of ciphertexts under the Generic Group Model. In terms of efficiency, compared with other similar ABKS schemes, as the number of attributes increases, our scheme has the smallest linear growth in both time consumption. Therefore, our scheme demonstrates excellent adaptability and applicability for the Pay-Per-Query Commercial Model, making it a highly suitable choice for practical implementation.

Keywords: Pay-Per-Query Commercial Model · attribute-based encryption · searchable encryption

1 Introduction

1.1 Background

The cloud service [16] has become a cost-effective data storage solution commonly used by individuals and corporate users due to its convenience and reliability. It supports the storage of large-scale data, and users can easily upload, search and update data without worrying about data loss, thereby greatly improving the efficiency and convenience of data management. The Pay-Per-Query Commercial Model, in which data owners charge data users according to the amount of data accessed, has gradually become the mainstream service model of most cloud service providers, such as IBM Cloud [24], Amazon Athena [27], and Google Cloud [23].

Under Pay-Per-Query Commercial Model, the data owner needs to create a paid query project on the cloud and bind their own payment account to this project at the same time. The data owner uploads their ciphertext to the project. When users are interested in the data on this project, they can obtain authorization by paying fees to the data owner. Authorized users can perform query operations on the ciphertext of the project. Conversely, unauthorized users are unable to perform query operations on the ciphertext of the project. However, the authorization of the Pay-Per-Query Commercial Model is mainly based on user-specific indexes, which is prone to data users [4] "free riding". In other words, the potentially malicious users, who are users that have paid the fees normally and obtained authorization, help unauthorized users skip the payment process to obtain authorization. Figure 1 illustrates a typical example of the traditional Pay-Per-Query Commercial Model, which includes three entities: data sender, data storage, and data recipient. Sender 1 and 2 upload their own data to the data storage respectively. For ease of management, data owners can only provide retrieval functions to specific data users. For example, data senders distribute specific keywords to their data recipients so that data recipients can search for files based on these keywords. The data sender limits each data recipient to 100 queries per day. Malicious users may pass their specific keywords to unauthorized users. Unauthorized users retrieve files based on the keywords of malicious users, and the query quota of malicious users is not deducted. The behavior of malicious users undoubtedly harms the interests of data senders.

Hahn et al. [4] pointed out that neither query quotas nor user-specific indexes can effectively prevent potential malicious users from colluding with unauthorized users to harm the interests of data owners and oppose "free riding". Therefore, they proposed a secure multi-key similar data search scheme to solve the "free riding" problem. However, their scheme lacks the verifiability of results and the non-repudiation of costs. Chen et al. [1] proposed a scheme suitable for the Pay-Per-Query Commercial Model and supporting similarity search function. It has the characteristics of non-repudiation of charges and verifiable results, and is also suitable for mobile devices with low processing performance. However, their scheme does not support access control.

Inspired by Attribute-Based Encryption (ABE), Sun et al. [26] and Zheng et al. [25] proposed Attribute-Based Keyword Search (ABKS) schemes respectively. The ABKS scheme can be regarded as an integration of ABE and Public Key Encryption With Keyword Search (PEKS). In this concept, keyword search can be performed only when the user's attributes match the policy in the ciphertext. Due to this feature, ABKS can adapt well to the Pay-Per-Query Commercial Model.

However, it is not feasible to use the ABKS scheme directly in the Pay-Per-Query Commercial Model. As analyzed in [5], when a user sends a search request, the check system will interact with the charging server to verify whether the user has obtained the service permission. Once the number of users increases dramatically, the frequent interaction between the two servers may occupy a lot of communication overhead, resulting in a system bottleneck. Therefore, designing an effective authentication mechanism is the key to access control in the Pay-Per-Query Commercial Model.

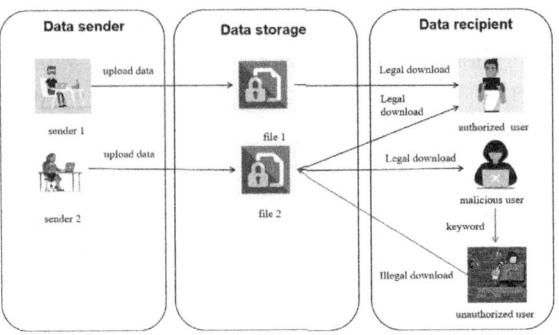

Fig. 1. A typical instance of the conventional Pay-Per-Query Commercial Model.

1.2 Contribution

To better solve the above problems, Our contributions are as follows:

1. we first propose a new paradigm called Conditional Attribute-Based Keyword Search (CABKS). Compared with the traditional ABKS scheme, this paradigm not only inherits the basic properties, but also adds new functional features, allowing the inspector to test the ciphertext only when holding a valid credential generated by another designated server. It reducing the potential bottleneck risk of providing services to a large number of users.
2. We propose the first practical CABKS scheme, addressing key challenges in secure and efficient keyword search. By decoupling access control structures from keyword ciphertext, our scheme effectively prevents unauthorized users from consuming cloud server resources. Moreover, we formally prove that our scheme satisfies Ciphertext Indistinguishability under the Generic Group Model.
3. Compared to related ABKS schemes, the experimental data demonstrates that our CABKS scheme achieves significant efficiency improvements, especially in scenarios with a large number of attributes. This efficiency stems from its support for the large attribute universe, making it highly suitable for Pay-Per-Query Commercial Model.

1.3 Related Work

Waters and Sahai [21] proposed the first Attribute-Based Encryption (ABE) scheme, described as an extension of Identity-Based Encryption (IBE) [22], which enables fine-grained access control over encrypted cloud data files. Ciphertext-Policy ABE (CP-ABE) and Key-Policy ABE (KP-ABE) were introduced by Goyal et al. [18]. Additionally, CP-ABE is suitable for associating ciphertexts with access policies and keys with attributes. Users cannot decrypt the returned ciphertexts unless their attribute lists match the access structure. Conversely, in KP-ABE [19], the secret key incorporates the access policy, while the ciphertext is formulated using user attributes. To date, most CP-ABE solutions provide secure data access policies, but these solutions still have shortcomings in practical applications. Hur [10] proposed a CP-ABE method with a tree-structured access policy, but this method has efficiency issues. Li et al. [12] established an efficient, traceable, and revocable access control scheme, but its drawback is that it does not implement keyword search functionality.

Yin et al. [15] built a keyword search scheme based on CP-ABE. However, most schemes [7, 9, 14] cannot achieve multi-keyword search. Due to the inadequacy of single-keyword search functionality, a large number of irrelevant search results are returned, resulting in low search efficiency and poor user experience. Chen et al. [2] proposed a practical attribute-based conjunctive keyword search scheme, which ensures both the flexibility of the retrieval method and the security and efficiency of the system. Li et al. [11] proposed a fine-grained search solution for cloud data that supports multi-keyword retrieval, improving user practicality and search accuracy.

In most ABKS schemes, users only need to register their attributes to obtain attribute secret keys. Subsequently, they can generate search trapdoors without restrictions using their attribute keys. Once the cloud server receives these search trapdoors, it is required to execute the search algorithm. When malicious users generate trapdoors without limits or the number of malicious users increases, it may lead to the situation where the resources of the cloud server are continuously occupied by some malicious users. Hong et al. [5] proposed a new Conditional Public-Key Encryption with Equality Test (CPKEET) paradigm to address user verification issues. Users need to hold valid credentials generated by specified servers to perform ciphertext testing, solving the problem of unauthorized equality testing and optimizing the number of communication rounds and computational burden, but it does not support keyword search.

2 Preliminaries

2.1 Bilinear Map

Let $\mathbb{G}_1, \mathbb{G}_2$ and $\mathbb{G}_T$ be three multiplicative cyclic groups of prime order p, the generator of $\mathbb{G}_1$ and $\mathbb{G}_2$ are g_1 and g_2 respectively. We define $e : \mathbb{G}_1 \times \mathbb{G}_2 \to \mathbb{G}_T$ as an asymmetric bilinear map if it satisfies the following three conditions:

- Non-degeneracy: $e(g_1, g_2) \neq 1$;

- Bilinearity: $\forall u \in \mathbb{G}_1, v \in \mathbb{G}_2, a, b \in \mathbb{Z}_p^*, e(u^a, v^b) = e(u,v)^{ab}$;
- Computability: $\forall u \in \mathbb{G}_1, v \in \mathbb{G}_2$, there is an efficient algorithm to compute $e(u,v)$.

2.2 Access Structure

A set of participants (or attributes) is presented by $P = \{P_1, P_2, \ldots, P_n\}$. For $\forall B, C$, if $B \in A$ and $B \subseteq C$, then $C \in A$, where $A \subseteq 2^{\{P_1, P_2, \ldots, P_n\}}$ is monotous. If $A \subseteq 2^P \setminus \{\emptyset\}$, the monotonous collection A is access structure. The collection $S \in A$ is called authorized collection.

2.3 LSSS

A secret-sharing scheme Π [20] over a set of parties is called linear over $\mathbb{Z}_p$ if

- The shares of the parties form a vector over $\mathbb{Z}_p$.
- There exists a matrix M with n rows and l columns called the generated matrix for Π. There exists a function ρ which maps each row of the matrix to an associated party. For $i = 1, \ldots, n$, the value $\rho(i)$ is the party associated with row i. Given a column vector $\boldsymbol{v} = (s, v_2, \ldots, v_l)$, where $s \in \mathbb{Z}_p$ is the secret to be shared, and $v_2, \ldots, v_l \in \mathbb{Z}_p$ are randomly chosen, then $M \cdot \boldsymbol{v}$ is the vector of n shares of the secret s according to Π. The share $(M \cdot \boldsymbol{v})_i$ belongs to the party $\rho(i)$. Suppose that Π is an LSSS for the access policy $\mathbb{P} = (M, \rho)$. Let $S \in \mathbb{P}$ be an authorized set, and let $I = \{i : \rho(i) \in S\}$. Then, there exist constants $\{\omega_i\}_{i \in I}$ in $\mathbb{Z}_p$ such that if $\{\mu_i\}$ are valid shares of any secret s according to Π, then $\sum_{i \in I} \omega_i \mu_i = s$.

3 Problem Formulation

3.1 System Model

The system model of the CABKS scheme is illustrated in Fig. 2. It consists of five entities: Trusted Authority, Data Sender, Data Recipient, Cloud Server, and Certification Server. Detailed descriptions of these entities are provided below.

1. **Trusted Authority (TA)**: TA is a fully trusted entity in the system model. It is responsible for generating the public parameters of the system and distributing them to all entities. Secondly, it is responsible for generating attribute secret keys for data recipients.
2. **Data Sender (DS)**: DS is responsible for generating its own key pair and encrypting keywords and access policies using its own private key, and then transmitting the keyword ciphertext and access policy ciphertext to the cloud server and certification server respectively.

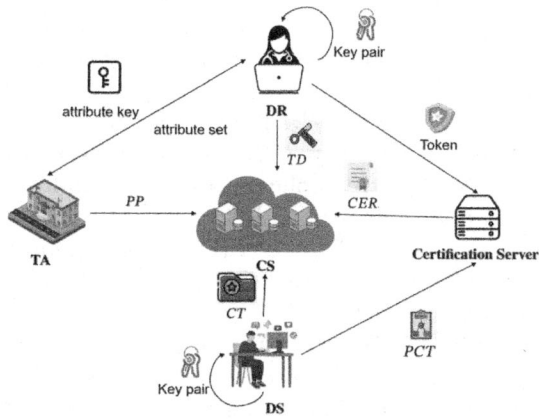

Fig. 2. System model of CABKS in Pay-Per-Query Commercial Model.

3. **Data Recipient (DR)**: DR needs to generate its own key pair. In order to obtain the search keyword permission, DR uses its own attribute key to generate a token and sends the token to the certification server. Then it uses its own private key to generate a trapdoor containing the search keyword. Then it submits the trapdoor to the cloud server.
4. **Certification Server**: The certification server saves the policy ciphertext uploaded by DS. When DR submits the token, the certification server uses the stored policy ciphertext and token as input to match the policy and attributes. If the attributes of DR meet the policy of DS, the certification server will generate a certificate, which will bind the public keys of DR and DS. Then send the certificate to the cloud server.
5. **Cloud Server (CS)**: CS will store the keyword ciphertexts uploaded by the DS and the certificates sent by the certification server. When the DR sends a trapdoor to the CS, the CS will first find the corresponding certificate. If the certificate does not exist, it indicates that the DR is an unauthorized user (who has not passed the access control), and the CS will return $\bot$. If the certificate exists, keyword matching is then performed. Further, if the matching fails, $\bot$ is returned; if it succeeds, Y is returned.

3.2 Algorithm

The scheme proposed in this article includes the following algorithms:

1. $(PP, MSK) \leftarrow \textbf{Setup}(1^\lambda)$: The algorithm is executed by TA, taking the security parameters λ as input, generating system public parameters PP and master secret key MSK.
2. $(PK_r, SK_r) \leftarrow \textbf{KeyGen}_r(PP)$: The algorithm is executed by DR, taking the public parameters PP as input, generating recipient's key pairs (PK_r, SK_r).
3. $(PK_s, SK_s) \leftarrow \textbf{KeyGen}_s(PP)$: The algorithm is executed by DS, taking the public parameters PP as input, generating sender's key pairs (PK_s, SK_s).

4. $SK_a \leftarrow \textbf{KeyGen}_a(MSK, \mathbb{S})$: The algorithm is executed by TA, taking the master secret key MSK and DR's attribute set $\mathbb{S}$ as input, generating attribute key SK_a.
5. $(CT, PCT) \leftarrow \textbf{Enc}(PK_r, SK_s, \mathbb{W}, \mathbb{P})$: The algorithm is executed by the DS, taking the recipient's public key PK_r, the sender's private key SK_s, the keyword set $\mathbb{W}$, access control policy $\mathbb{P}$ as input, and generating keyword ciphertext CT and policy ciphertext PCT.
6. $TK \leftarrow \textbf{TKGen}(SK_a, SK_r)$: The algorithm is executed by the DR, taking the attribute key SK_a and the recipient's private key SK_r as input, and generating attribute token TK.
7. $(CER/\bot) \leftarrow \textbf{CerGen}(PCT, TK)$: The algorithm is executed by the certification server, taking the recipient's attribute token TK as input. If the recipient's attribute set $\mathbb{S}$ matches the sender's access control policy $\mathbb{P}$, the certification server generates the certificate CER, otherwise it will return $\bot$.
8. $TD \leftarrow \textbf{TrapGen}(PK_s, SK_r, \mathbb{Q})$: The algorithm is executed by DU, taking the sender's public key PK_s, the recipient's private key SK_r, the query keyword structure $\mathbb{Q}$ as input to generate a search trapdoor TD.
9. $(Y/\bot) \leftarrow \textbf{Search}(TD, CT, CER)$: The algorithm is executed by the cloud server. If the certificate CER does not exist, the cloud server will return $\bot$. Otherwise, the cloud server will take the search trapdoor TD, keyword ciphertext CT, and certificate CER as input. If the recipient's query keyword structure $\mathbb{Q}$ does not match the sender's keyword set $\mathbb{W}$, the cloud server will return $\bot$, otherwise it will return Y.

3.3 Security Model

Game 1: Ciphertext Indistinguishability

1. **Setup.** Given a security parameter λ, the challenger $\mathcal{C}$ generates the global system parameter PP. Then $\mathcal{C}$ generates a pair of sender's key (PK_s, SK_s) and a pair of recipient's key (PK_r, SK_r). It provides (PP, PK_s, PK_r) to the adversary $\mathcal{A}$. $\mathcal{C}$ generates an access control policy $\mathbb{P}^*$.
2. **Phase 1.** $\mathcal{A}$ is allowed to adaptively issue queries to the following oracles for polynomial many times:
    - **Trapdoor Oracle** $O_T(\mathbb{Q}, PK)$: Given a keyword policy structure $\mathbb{Q}$ and a public key PK (not necessarily the sender's PK_s), the oracle computes a trapdoor $TD \leftarrow \text{TrapGen}(PK, SK_r, \mathbb{Q})$ and returns TD to $\mathcal{A}$.
    - **Secret key Oracle** $O_{SK}(\mathbb{S})$: Given a set of attribute set $\mathbb{S}$, the oracle computes attribute secret key $SK_a \leftarrow \text{KeyGen}_a(MSK, \mathbb{S})$ and returns SK_a to $\mathcal{A}$.
3. **Challenge.** After Phase 1, $\mathcal{A}$ outputs two equal-size keyword sets $\mathbb{W}_0^* = \{type_j, value_j\}_{j \in [1, |\mathbb{W}_0^*|]}$, $\mathbb{W}_1^* = \{type_j, value_j\}_{j \in [1, |\mathbb{W}_1^*|]}$ with the restriction that $\mathbb{W}_0^*$ and $\mathbb{W}_1^*$ have the same keyword type set $\{type_j\}$ and neither of them matches any trapdoor that has been queried for $O_T(., PK_s)$ in Phase 1, and submits them to $\mathcal{C}$. $\mathcal{C}$ randomly chooses a bit $q \in \{0, 1\}$, computes $CT_q^* \leftarrow \textbf{Enc}(PK_r, SK_s, \mathbb{W}_q^*, \mathbb{P}^*)$ and returns CT_q^* to $\mathcal{A}$.

4. **Phase 2.** $\mathcal{A}$ continues to issue queries to O_T and O_{SK} as above, with the restriction that any trapdoor that is queried for $O_T(., PK_s)$ should not be satisfied by $\mathbb{W}_0^*$ and $\mathbb{W}_1^*$.
5. **Guess.** $\mathcal{A}$ outputs $q' \in \{0,1\}$ and wins the game if $q' = q$. We define $\mathcal{A}$'s advantage of successfully distinguishing the ciphertext as

$$\text{Adv}_{CABKS,\mathcal{A}}^{CI}(\lambda) = \left|\Pr[q' = q] - \frac{1}{2}\right|.$$

Definition 1. *A CABKS scheme is fully CI secure if for any PPT adversary $\mathcal{A}$, $Adv_{CABKS,\mathcal{A}}^{CI}(\lambda)$ is negligible for security parameter λ.*

4 The Proposed Scheme

This section covers the detailed construction of the proposed scheme, along with the explanation of each algorithm.

1. **Setup(1^λ):** The security parameter λ is used as input. Let $e: \mathbb{G}_1 \times \mathbb{G}_2 \to \mathbb{G}_T$ be the bilinear map, where $\mathbb{G}_1, \mathbb{G}_2$, and $\mathbb{G}_T$ are cyclic groups of prime order p. Additionally, g_1 is the generator of $\mathbb{G}_1$, and g_2 is the generator of $\mathbb{G}_2$. Two hash function $\mathcal{H}_1$ and $\mathcal{H}_2: \{0,1\}^* \to \mathbb{G}_1$ is defined. Randomly select $d \in \mathbb{Z}_p^*$ The public parameters PP and master secret key MSK are given by

$$PP = \{p, e, \mathbb{G}_1, \mathbb{G}_2, \mathbb{G}_T, g_1, g_2, \mathcal{H}, e(g_1, g_2), g_2^d\}, MSK = \{d\}.$$

2. **KeyGen$_r$(PP):** Randomly select elements $\alpha, b_1, b_2, \delta \in \mathbb{Z}_p$, and generate the data recipient's private key $SK_r = (\alpha, b_1, b_2, \delta)$ and public key $PK_r = (g_2^{b_1}, g_2^{b_2}, g_1^\delta, e(g_1, g_2)^\alpha)$.
3. **KeyGen$_s$(PP):** Randomly select elements $c \in \mathbb{Z}_p$, and generate the data sender's private key $SK_s = c$ and public key $PK_s = (g_1^c, g_2^c)$.
4. **KeyGen$_a$($MSK, \mathbb{S}$):** The recipient submits attribute set $\mathbb{S} = \{x_1, x_2, \cdots, x_n\}$ to TA, TA selects a random value $u \in \mathbb{Z}_p$, and then selects a random value $t \in \mathbb{Z}_p$ for each attribute according to the attribute set $\mathbb{S}$, compute the recipient attribute key $SK_a = (\{SK_{1,k}, SK_{2,k}\}_{x_k \in \mathbb{S}}, SK_3)$, where

$$SK_{1,k} = g_2^{dt}, SK_{2,k} = g_1^u \mathcal{H}_1(x_k)^{-t}, SK_3 = g_2^{du}.$$

5. **Enc($PK_r, SK_s, \mathbb{W}, \mathbb{P}$):** The data sender extracts a keyword set $\mathbb{W} = \{kw_j\} = \{\text{type}_j, \text{value}_j\}_{j \in [1, |\mathbb{W}|]}$, from each file, where each keyword kw_j in the set $\mathbb{W}$ is divided into a keyword type type_j and a keyword value value_j. Randomly select $s_1, s_2, \beta \in \mathbb{Z}_p$ and compute $s = s_1 + s_2$. Calculate the following ciphertext components:

$$CT_1 = g_2^{b_1 s_1}, \quad CT_2 = g_2^{b_2 s_2},$$
$$CT_{3,j} = \mathcal{H}_2(kw_j)^{s/c}, \quad CT_4 = e(g_1, g_2)^{\alpha s} \cdot e(g_1, g_2)^{\delta \beta}.$$

The ciphertext is: $CT = (CT_1, CT_2, \{\text{type}_j, CT_{3,j}\}_{j \in [1, |\mathbb{W}|]}, CT_4)$.

The access policy is represented as an access control structure: $\mathbb{P} = (M_{n \times l}, \rho)$, $M_{n \times l}$ is a matrix with n rows and l columns, derived from the data sender's access policy. ρ is a mapping function that associates the rows of the matrix with specific attributes. Randomly select a vector $\boldsymbol{v} \in \mathbb{Z}_p^{l-1}$. Compute: $\lambda_i = M_i(\beta \| \boldsymbol{v})^\top$, Calculate the following policy ciphertext components:

$$PCT_{1,i} = \mathcal{H}_1(\rho(i))^{\lambda_i}, \quad PCT_{2,i} = g_2^{d \cdot \lambda_i}, \quad PCT_3 = g_1^{\delta \beta}, \quad PCT_4 = g_1^{\delta c}.$$

The policy ciphertext is: $PCT = (M, \rho, \{PCT_{1,i}, PCT_{2,i}\}_{i \in [1,n]}, PCT_3, PCT_4)$.

6. **TKGen**(SK_a, SK_r): The recipient selects a random value $\tau \in \mathbb{Z}_p$ and computes the recipient attribute token $TK = (\{TK_{1,k}, TK_{2,k}\}_{x_k \in \mathbb{S}}, TK_3, \mathbb{S})$, where

$$TK_{1,k} = g_2^{dt\tau}, TK_{2,k} = (g_1^u \mathcal{H}_1(x_k)^{-t})^\tau, TK_3 = g_2^{\frac{du\tau + \delta}{\delta}}.$$

7. **CerGen**(PCT, TK): The certification server uses the mapping function ρ to map the data recipient's attribute set $\mathbb{S}$ to a matrix row set I. When the attribute set $\mathbb{S}$ of the TK satisfies the access policy $\mathbb{P}$ of the PCT, there exists ω such that $\sum_{i \in I} \omega_i \cdot \lambda_i = \beta$. Certification server computes

$$F = \prod_{i \in I} (e(TK_{1,i}, PCT_{1,i}) e(TK_{2,i}, PCT_{2,i}))^{\omega_i} = e(g_1, g_2)^{u\tau d\beta},$$

and computes the certificate $CER = (CER_1, CER2)$, where

$$CER_1 = PCT_4 = g_1^{\delta c}, CER_2 = \frac{e(PCT_3, TK_3)}{F} = e(g_1, g_2)^{\beta \delta}.$$

8. **TrapGen**$(PK_s, SK_r, \mathbb{Q})$: The query statement of the data recipient is converted into the query structure $\mathbb{Q} = (N_{n*l}, \kappa, \{\eta(\kappa(i))\}_{i \in [1,n]})$, N_{n*l} represents a matrix of n rows and l columns, derived from the DR's query statement. κ is a mapping function that maps the row indices of the matrix to keyword categories. η is another mapping function that maps keyword categories to specific keywords. A random vector $\boldsymbol{d} \in \mathbb{Z}_p^{n-1}$ is selected, and $\zeta_i = N_i(\alpha \| \boldsymbol{d})^\top$ is computed, where N_i represents the i-th row of N of the matrix. A random value $r \in \mathbb{Z}_p$ is chosen and the following calculations are performed:

$$TD_{1,i} = (g_1^{\zeta_i} \cdot \mathcal{H}_2(\eta(\kappa(i)))^r)^{\frac{1}{b_1}}, TD_{2,i} = (g_1^{\zeta_i} \cdot \mathcal{H}_2(\eta(\kappa(i)))^r)^{\frac{1}{b_2}},$$
$$TD_3 = g_2^{cr}, TD_4 = g_1^{c\delta}.$$

The trapdoor is: $TD = (N, \kappa, \{TD_{1,i}, TD_{2,i}\}_{i \in [1,n]}, TD_3, TD_4)$.

9. **Search**(TD, CT, CER): The cloud server first checks whether the equation $TD_4 = CER_1$ holds. If the equation does not hold, it means that the user's certificate does not exist, and the cloud server returns $\bot$. Otherwise, it means that the user's certificate exists, and the cloud server continues with the following steps:

From the keyword type set in the CT of the data sender, the mapping function κ is used to obtain the matrix row number set I_1. When the user's search structure $\mathbb{Q}$ satisfies the keyword set $\mathbb{W}$ in the index, there exists $\omega_{1,i}$, so that $\sum_{i \in I_1} \omega_{1,i} \cdot \zeta_i = \alpha$, where I_1 is the matrix row number set. Next, the cloud server checks whether the equation

$$CT_4 = \frac{\prod_{i \in I_1}(e(TD_{1,i}, CT_1) \cdot e(TD_{2,i}, CT_2))^{\omega_{1,i}} \cdot CER_2}{\prod_{i \in I_1}(e(CT_{3,i}, TD_3))^{\omega_{1,i}}}$$

holds. If the equation holds, the search is successful and Y is returned. If the equation does not hold, the search fails and $\perp$ is returned. If the user's attribute set does not meet the access policy or the query structure does not match the keyword set, the algorithm will terminate and $\perp$ will be returned.

Correctness

Right side of the equation:

$$\frac{\prod_{i \in I_1}(e(TD_{1,i}, CT_1) \cdot e(TD_{2,i}, CT_2))^{\omega_{1,i}}}{\prod_{i \in I_1} e(CT_{3,i}, TD_3)^{\omega_{1,i}}} \cdot CER_2$$

$$= \frac{\prod_{i \in I_1} e(g_1^{\zeta_i} \cdot \mathcal{H}_2(\eta(\kappa(i)))^r, g_2^s)^{\omega_{1,i}}}{\prod_{i \in I_1}(e(\mathcal{H}_2(kw_i)^s, g_2^r))^{\omega_{1,i}}} \cdot CER_2$$

$$= \frac{\prod_{i \in I_1} e(g_1^{\zeta_i}, g_2^s)^{\omega_{1,i}} \cdot e(\mathcal{H}_2(\eta(\kappa(i)))^r, g_2^s)^{\omega_{1,i}}}{\prod_{i \in I_1}(e(\mathcal{H}_2(kw_i)^s, g_2^r))^{\omega_{1,i}}} \cdot CER_2$$

$$= \prod_{i \in I_1} e(g_1^{\zeta_i}, g_2^s)^{\omega_{1,i}} \cdot CER_2$$

$$= e(g_1, g_2)^{\alpha s} \cdot e(g_1, g_2)^{\delta \beta}$$

$$= CT_4.$$

5 Security

In this section, we prove the security of our schemes under the Generic Group Model (GGM) and random oracle model. we define the GGM and random oracle separately as follows:

1. **Random oracle:** The challenger $\mathcal{C}$ maintains a list L with entries of the form $\langle x_i, h_i, t_i \rangle$, which is initially empty. When the adversary $\mathcal{A}$ or the simulation inputs an attribute (or keyword) string x_i, $\mathcal{C}$ checks if x_i already appears on the list L in a tuple $\langle x_i, h_i, t_i \rangle$. If yes, then $\mathcal{C}$ responds with $\mathcal{H}(x_i) = h_i \in \mathbb{G}_1$. Otherwise, $\mathcal{C}$ picks $t_i \in \mathbb{Z}_p$ and computes $h_i \leftarrow g^{t_i} \in \mathbb{G}_1$. Then $\mathcal{C}$ adds the tuple $\langle x_i, h_i, t_i \rangle$ to list L and responds to $\mathcal{A}$ by setting $\mathcal{H}(x_i) = h_i$.
2. **Generic group model:** We consider random encodings ψ_1, ψ_2, ψ_T of the additive group $\mathbb{Z}_p$, satisfying injective maps $\psi_1, \psi_2, \psi_T : \mathbb{Z}_p \rightarrow \{0,1\}^m$, where $m > 3\log(p)$. The probability of $\mathcal{A}$ guessing an element in the image of ψ_1, ψ_2, ψ_T is negligible. For $i = 1, 2, T$ we write $\mathbb{G}_i = \{\psi_i(x) : x \in \mathbb{Z}_p\}$.

Theorem 1. *The proposed CABKS scheme is fully CI secure under the generic group model by modeling the hash functions $\mathcal{H}_1$ and $\mathcal{H}_2$ as random oracles.*

Proof. In the fully CI game, the only ciphertext component that is related to the two challenged keyword sets is $CT_{3,j} = \mathcal{H}_2(kw_j)^{\frac{s}{c}}$. Based on the simulation of the random oracle as above, we can simulate $CT_{3,j} = \mathcal{H}_2(kw_j)^{\frac{s}{c}} = g_1^{t_j \cdot \frac{s}{c}}$. Therefore, $\mathcal{A}$ attempts to win the game by distinguishing $\{g_1^{t_{j,0} \cdot \frac{s}{c}}\}_{j \in [1, |\mathbb{W}_0^*|]}$ from $\{g_1^{t_{j,1} \cdot \frac{s}{c}}\}_{j \in [1, |\mathbb{W}_1^*|]}$.

For $\theta \in \mathbb{Z}_p$ and $q \in \{0, 1\}$, the probability of distinguishing $\{g_1^{t_{j,0} \cdot \frac{s}{c}}\}_{j \in [1, |\mathbb{W}_0^*|]}$ from $g_1^{t_{j,q} \cdot \theta}$ is equal to that of distinguishing $g_1^{t_{j,q} \cdot \theta}$ from $\{g_1^{t_{j,1} \cdot \frac{s}{c}}\}_{j \in [1, |\mathbb{W}_1^*|]}$. Therefore, if $\mathcal{A}$ has an advantage ϵ in winning the fully CI game, then it has an advantage $\frac{\epsilon}{2}$ in distinguishing $g_1^{t_{j,q} \cdot \frac{s}{c}}$ from $g_1^{t_{j,q} \cdot \theta}$. Thus, we consider a modified game where $\mathcal{A}$ can distinguish $g_1^{t_{j,q} \cdot \frac{s}{c}}$ from $g_1^{t_{j,q} \cdot \theta}$. The modified game is simulated as follows:

Setup. The challenger $\mathcal{C}$ chooses $\alpha, b_1, b_2, \delta, c, \beta, d \in \mathbb{Z}_p$ and sends the public parameters $PP = (g_1, g_2, g_2^d)$, the challenge data recipient's public key $PK_r = (g_2^{b_1}, g_2^{b_2}, g_1^\delta, e(g_1, g_2)^\alpha)$ and the challenge data sender's public key $PK_s = (g_1^c, g_2^c)$ to $\mathcal{A}$.

Phase 1. In phase 1, $\mathcal{A}$ can query the random oracle, a trapdoor oracle $O_T(\cdot, \cdot)$, and a ciphertext oracle $O_{SK}(\cdot)$ as follows:

- Trapdoor oracle $O_T(\cdot, \cdot)$: When $\mathcal{A}$ makes a trapdoor query for a keyword policy $\mathbb{Q} = (N_{n*l}, \kappa, \{\eta(\kappa(i))\}_{i \in [1,n]})$ and a data sender's public key $\{PK_1, PK_2\}$. $\mathcal{C}$ picks $r \in \mathbb{Z}_p$ and a vector $\boldsymbol{d} \in \mathbb{Z}_p^{n-1}$. Let ζ_i be $N_i(\alpha \| \boldsymbol{d})^\top$. Then $\mathcal{C}$ generates the trapdoor as the following:

$$TD_{1,i} = g_1^{(\zeta_i + t_i r) \cdot \frac{1}{b_1}}, TD_{2,i} = g_1^{(\zeta_i + t_i r) \cdot \frac{1}{b_2}}, TD_3 = PK_1^r, TD_4 = PK_2^\delta.$$

Then $\mathcal{C}$ gives $TD = (\{TD_{1,i}, TD_{2,i}\}_{i \in [1,n]}, TD_3, TD_4)$ to $\mathcal{A}$.
- Secret key oracle $O_{SK}(\cdot)$: When $\mathcal{A}$ issues a secret key query for a set of attribute $\mathbb{S} = \{x_k\}_{k \in [1, |\mathbb{S}|]}$, $\mathcal{C}$ picks $u \in \mathbb{Z}_p$, and then selects a random value $t \in \mathbb{Z}_p$ for each attribute according to the attribute set $\mathbb{S}$. Compute the attribute secret key

$$SK_{1,k} = g_2^{dt}, SK_{2,k} = g_1^u \cdot g_1^{-tt_k}, SK_3 = g_2^{du}.$$

Then $\mathcal{C}$ gives $SK_a = (\{SK_{1,k}, SK_{2,k}\}_{x_k \in \mathbb{S}}, SK_3)$ to $\mathcal{A}$.

Challenge. $\mathcal{A}$ outputs two keyword sets $\mathbb{W}_0^* = \{kw_j\}_{j \in [1, \mathbb{W}_0^*]}$, $\mathbb{W}_1^* = \{kw_j\}_{j \in [1, \mathbb{W}_1^*]}$, which it intends to attack. Note that $\mathbb{W}_0^*, \mathbb{W}_1^*$ must have the same keyword type $\{type_j\}$. $\mathcal{C}$ checks if $\mathbb{W}_0^*$ or $\mathbb{W}_1^*$ satisfies any of the keyword policy $\mathbb{Q}$ queried in Phase 1. If yes, $\mathcal{C}$ rejects $\mathbb{W}_0^*, \mathbb{W}_1^*$, Otherwise, $\mathcal{C}$ chooses $s_1, s_2, \theta \in \mathbb{Z}_p$ and let $s = s_1 + s_2$. Then $\mathcal{C}$ selects $q \in \{0, 1\}$ for encrypting one set of keywords, and chooses $\mu = \{0, 1\}$. If $\mu = 0$, it generates the challenge ciphertext as follows:

$$CT_1 = g_2^{b_1 s_1}, CT_2 = g_2^{b_2 s_2}, CT_{3,j} = g_1^{t_{j,q}\theta}, CT_4 = e(g_1, g_2)^{\alpha s} \cdot e(g_1, g_2)^{\delta \beta}.$$

If $\mu = 1$, it generates the challenge ciphertext as follows:

$$CT_1 = g_2^{b_1 s_1}, CT_2 = g_2^{b_2 s_2}, CT_{3,j} = g_1^{t_{j,q} \cdot \frac{s}{c}}, CT_4 = e(g_1, g_2)^{\alpha s} \cdot e(g_1, g_2)^{\delta \beta}.$$

Then $\mathcal{C}$ gives $CT_q^* = (CT_1, CT_2, \{CT_{3,j,q}\}_{j \in [1, \mathbb{W}_q^*]}, CT_4)$ to $\mathcal{A}$.

Phase 2. It is the same as in Phase 1 with the restriction that any input keyword policy $\mathbb{Q}$ is not allowed to be satisfied by the challenge keyword sets $\mathbb{W}_0^*$ and $\mathbb{W}_1^*$.

For simplicity, we denote $t_{j,q}$ as t_j. We can see that it is impossible for $\mathcal{A}$ to construct $t_j \cdot \frac{s}{c}$ on $\mathbb{G}_1, \mathbb{G}_2$, and $\mathbb{G}_T$. But if $\mathcal{A}$ can construct $e(g_1, g_2)^{\eta t_j s}$ for some $\eta \in \mathbb{Z}_p$ that can be combined from the oracle outputs $\mathcal{A}$ has already queried, then $\mathcal{A}$ can use it to distinguish $g_1^{t_j \theta}$ from $g_1^{t_j \cdot \frac{s}{c}}$ because c and cr occur on $\mathbb{G}_2$.

Therefore, we need to show that $\mathcal{A}$ can construct $e(g_1, g_2)^{\eta t_j s}$ for some η with a negligible probability, which means that $\mathcal{A}$ cannot gain a non-negligible advantage in the fully CI game. Then we consider the probability of $\mathcal{A}$ constructing $e(g_1, g_2)^{\eta t_j s}$ for some $\eta \in \mathbb{Z}_p$ from the oracle outputs $\mathcal{A}$ has queried. Similarly, we first summarize the elements on exponents that could be used in the groups $\mathbb{G}_1, \mathbb{G}_2$ and $\mathbb{G}_T$.

- $\mathbb{G}_1$ elements: $1, u - tt_k, \delta, c, \frac{1}{b_1} \cdot (\zeta_i + t_i r), \frac{1}{b_2} \cdot (\zeta_i + t_i r)$.
- $\mathbb{G}_2$ elements: $1, c, cr, b_1 s_1, b_2 s_2, b_1, b_2, d, c\delta, dt\ du$.
- $\mathbb{G}_T$ elements: $\alpha, \alpha s + \delta \beta$.

Let us consider how to construct $e(g_1, g_2)^{\eta t_j s}$ for some η. The only way that $\mathcal{A}$ can create a term containing s is by pairing $\frac{1}{b_1} \cdot (\zeta_i + t_i r)$ with $b_1 s_1$ and pairing $\frac{1}{b_2} \cdot (\zeta_i + t_i r)$ with $b_2 s_2$ to get the term $(\zeta_i + t_i r) \cdot s_1$ and $(\zeta_i + t_i r) \cdot s_2$ separately, and multiply them together to obtain $(\zeta_i + t_i r) \cdot (s_1 + s_2) = \zeta_i s + t_i r s$ on $\mathbb{G}_T$. Then $\mathcal{A}$ only needs to cancel the term $\zeta_i s$. the only way to cancel $\zeta_i s$ by using existing queries is to reconstruct ζ_i to α since αs is known on $\mathbb{G}_T$. However, it is impossible to reconstruct α since any input keyword policy $\mathbb{Q}$ cannot be satisfied by the keyword sets $\mathbb{W}_q^*$. In other words, it is impossible for $\mathcal{A}$ to construct $\eta t_j s$ on $\mathbb{G}_T$. Finally, we can conclude that $\mathcal{A}$ gains a negligible advantage in the modified game, which means that $\mathcal{A}$ gains a negligible advantage in the fully CI game. Then the proof of theorem 1 is completed.

6 Comparison

In this section, we comprehensively compare the CABKS scheme with related multi-keyword ABKS schemes (Table 1).

Table 1. Functional comparison.

Schemes	Access structure	Multi-keyword expressiveness	KGA resistance	large attribute universe	Fuction1	Fuction2
FAKS [6]	LSSS	AND	×	×	×	×
EMK-ABSE [8]	LSSS	AND	×	×	×	×
TLABKS [13]	LSSS	AND	×	×	×	×
ABKRS-KGA [3]	Tree	AND	✓	✓	×	×
Our scheme	LSSS	AND, OR	✓	✓	✓	✓

Function1: authorized ciphertext retrieval.
Function2: Users do not need to repeat the access control check.

6.1 Theoretical Analysis

Functional Comparison. In Table 3, we compare various multi-keyword ABKS schemes. Evaluated across six criteria, our scheme demonstrates superior functionality, particularly in authorized ciphertext retrieval, making it highly advantageous for practical applications. In terms of keyword expressiveness, our scheme supports both AND and OR expressions, whereas others are limited to AND only. Additionally, only our scheme filters unauthorized users and lets legitimate users bypass repetitive access control checks. Regarding the access structure, all schemes except [3] adopt the LSSS matrix, which is more efficient than the access tree structure. When facing keyword-guessing attacks, only [3] and our scheme are resistant. Only [3] and our scheme support large attribute universes, providing a broader scope of application compared to others.

Table 2. Notations for efficiency comparisons.

Notation	Definition
E_1	Exponential operation in group 1
E_2	Exponential operation in group 2
E_T	Exponential operation in group T
P	Bilinear pairing operation
$|U|$	The number of attribute universe U
$|S|$	Number of DR's attributes
$|A|$	Number of attributes in DS's policy
L_1	Bit-Length of element in group 1
L_2	Bit-Length of element in group 2
L_T	Bit-Length of element in group T

Efficiency Comparison. In Tables 3 and 4, we compare [3,6,8,13], and our scheme in terms of storage and computational costs. These schemes are selected because they are all multi-keyword ABKS schemes and are recently published.

Table 3. Theoretical calculation costs comparison.

Schemes	Setup	KeyGen	Enc	TrapGen	Search
FAKS [6]	$4E_1 + 3P$	$(|S|+3)E_1$	$(3|A|+5)E_1 + P$	$6E_1 + P$	$|S|E_1 + (2|S|+4)P$
EMK-ABSE [8]	$(|U|+2)E_1 + E_T + P$	$(|S|+6)E_1$	$(2|U|+|A|+E_T$	$2E_1$	$4P$
TLABKS [13]	$(|U|+1)E_1 + P$	$(2|S|+3)E_1$	$(|A|+6)E_1$	$3E_1$	$4P$
ABKRS-KGA [3]	$4E_1 + E_2$	$(|S|+1)E_1 + E_2$	$2|A|E_1 + 2E_2$	$|S|E_1 + 2E_2$	$2|S|P + 3E_2 + (|S|+1)E_T$
Our scheme	$P + E_1$	$(|S|+1)E_1 + (|S|+1)E_2$	$(2+|A|)E_1 + (2+|A|)E_2$	$2E_1$	$3P$

Table 4. Theoretical storage costs comparison.

Schemes	PP	SK	CT	TD
FAKS [6]	$3L_T + (4+|U|)L_1$	$(4+|S|)L_1$	$(2+2|P|)L_1 + 2L_T$	$(7+|S|)L_1 + L_T$
EMK-ABSE [8]	$(|U|+3)L_1 + L_T$	$(|S|+4)L_1$	$L_T + (2|P|+4)L_1$	$4L_1$
TLABKS [13]	$(|U|+2)L_1 + L_T$	$(|S|+3)L_1$	$(3|P|+5)L_1 + L_T$	$3L_1$
ABKRS-KGA [3]	$3L_1 + 2L_2$	$L_2 + |S|L_1$	$|P|L_1 + 2L_2$	$2L_2 + |S|L_1$
Our scheme	$L_1 + 2L_2 + L_T$	$(|S|+1)L_2 + |S|L_1$	$(2+|A|)L_1 + (|A|+2)L_2 + L_T$	$2L_2 + 2L_1$

We consider four types of time-consuming operations and three types of storage metrics, with the notations used shown in Table 2.

In Table 3, we compare the time costs of the Setup, KeyGen, Enc, TrapGen, and Search algorithms among the five schemes. Due to varying definitions of encryption algorithms in these schemes, it is difficult to unify the comparison. Therefore, in our analysis, we include both index generation and attribute encryption in the encryption process, and the key generation algorithm refers to the attribute key generation algorithm. The results show that our scheme's time consumption across all algorithms is significantly lower than that of scheme [3]. Compared with [6], except for the KeyGen algorithm, our scheme has lower time consumption in all other aspects. For recipients, the keyGen algorithm only needs to be called once, and the efficiency of the Enc algorithm and the Search algorithm is more important than that of the KeyGen algorithm. Compared to [8,13], our scheme achieves noticeably shorter times in the Setup algorithms. For the KeyGen algorithm, our scheme, similar to [13], is slower than [8]. However, for the Enc algorithm, our scheme is more efficient than [8] and similar to [13]. In the TrapGen and Search algorithms, our scheme is slower because [8,13] do not support multi-keyword OR expressions, whereas our scheme does.

Table 4 compares the PP size, SK size, CT size, and TD size of each scheme, highlighting their respective storage costs. For PP, both [3] and our scheme achieve a constant size, while others scale linearly with $|U|$. For SK, all schemes grow linearly with $|S|$. For CT, the sizes of all schemes increase linearly with $|A|$ and n, but our scheme's increase is smaller. For TD, [3,6] have larger sizes compared to our scheme; [13] and our scheme are of the same level, both scaling linearly with the number of keywords. [8] compresses multiple keyword hashes into a single value, achieving a smaller constant size. However, since [8,13] do not support multi-keyword OR expressions, their sizes are relatively smaller (Fig. 3).

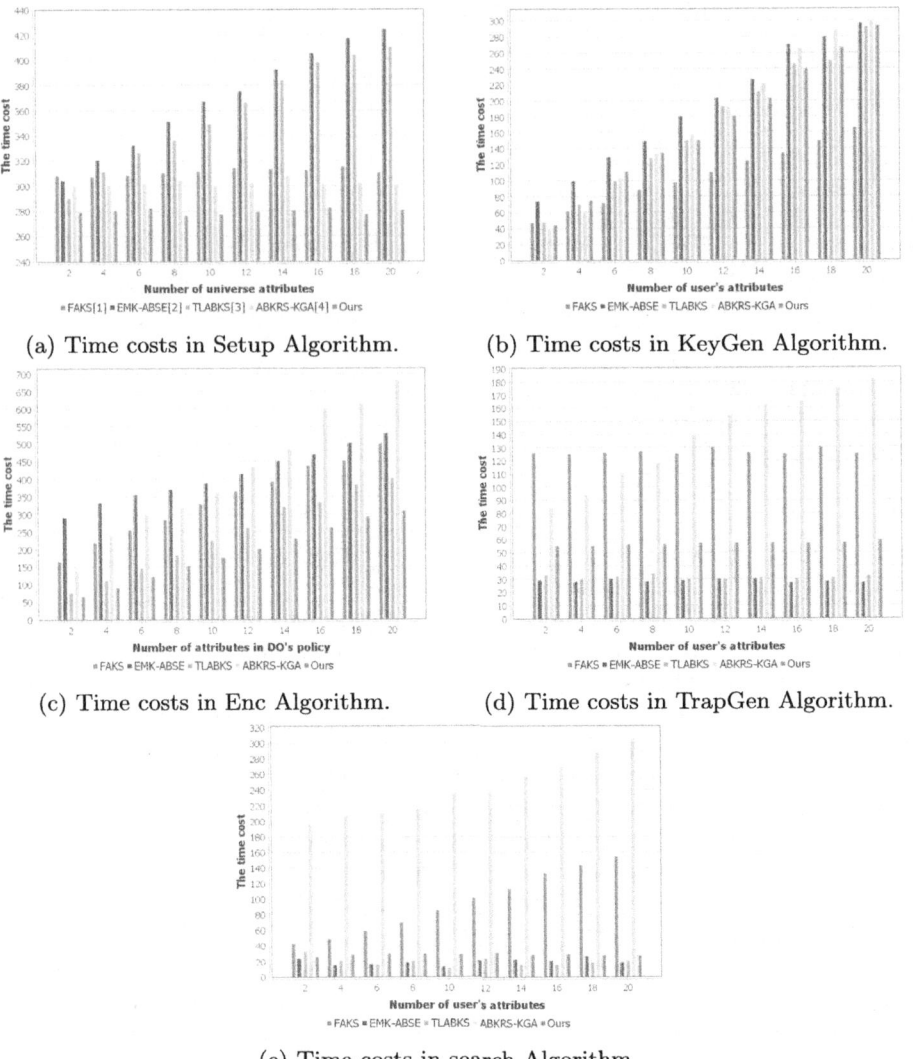

(a) Time costs in Setup Algorithm.
(b) Time costs in KeyGen Algorithm.
(c) Time costs in Enc Algorithm.
(d) Time costs in TrapGen Algorithm.
(e) Time costs in search Algorithm.

Fig. 3. The relationship between attributes and time.

6.2 Experimental Analysis

In Figure (a), we show the computational cost of the Setup algorithm. In this comparison, $|U|$ is set from 2 to 20. Since [8,13] do not support large attribute sets, their time consumption increases linearly with $|U|$, whereas other schemes, including ours, remain constant. Regardless of the value of $|U|$, [8,13] are slower than the other schemes. Among the constant-time schemes, ours takes the least time because the number of common parameters is smaller than in [3,6]. Therefore, our scheme is the most efficient overall, although the total time is still around 280 ms due to the relatively time-consuming construction of bilinear pairings in the Setup phase.

Figure (b) shows the computational cost of the KeyGen algorithm, where $|S|$ ranges from 2 to 20. The time consumption of all schemes increases linearly with $|S|$. Because scheme [6] includes the random numbers of all attributes in the attribute domain in the public parameters, its attribute key only needs to randomize the user's attributes without carrying the random numbers used for randomization through other parameters. But, scheme [6] consumes significantly less time than other schemes, though at the cost of high time consumption in the Setup algorithm. Apart from scheme [6], the time consumption of other schemes is similar.

Figure (c) shows the relationship between the attributes and time of the Enc algorithm, where $|A|$ ranges from 2 to 20. It can be observed that the time consumption of all schemes increases linearly with the number of attributes. In terms of the rate of increase, [3,6] execute six additional E1 operations for every two additional attributes. Although [8,13] exhibit similar growth rates to our solution, their other operations take longer than ours. Therefore, regardless of the value of $|A|$, our solution always has the lowest time consumption.

Figure (d) shows the relationship between attributes and time in the Trap-Gen algorithm, with $|S|$ ranging from 2 to 20. In [3], the number of TD is consistent with the number of attribute, so the time increases linearly with the number of attributes, while other schemes remain constant. Among the constant-time schemes, [6] requires pairing operations, making our scheme faster than [6]. However, [8,13] are faster than our scheme. The reason our scheme is slower than [8,13] is that our hash function maps to the $\mathbb{G}_1$ group, while [8,13] map to $\mathbb{Z}_p$. This difference results in our scheme being about 25 ms slower than [8,13].

Figure (e) shows the relationship between attributes and time in the Search algorithm, where $|S|$ ranges from 2 to 20. Since the number of TD in [3,6] is linearly related to $|S|$, only the time of [3,6] increases linearly with the number of attributes, while the other schemes remain constant. Among them, our scheme performs comparably to [8,13].

7 Conclusion

In this paper, we proposed for the first time a CABKS scheme to utilize the access control mechanism to solve the "free riding" problem existing in the Pay-Per-Query Commercial Model of searchable encryption. Then, we proved the indistinguishability of ciphertexts of this scheme under the Generic Group Model. Finally, it is especially worth mentioning that the experiments conducted on real datasets showed that our scheme has the lowest time consumption. Meanwhile, this scheme not only inherits the characteristics of the traditional ABKS schemes, including resisting keyword guessing attacks and supporting multi-keyword search, but also prevents unauthorized users from accessing the search cloud server and avoids repeated access control.

Acknowledgments. This work is supported by the Guangdong Basic and Applied Basic Research Foundation (2024A1515012666) and the National Natural Science Foundation of China (61872409, 62272174), the Major Program of Guangdong Basic and Applied Research (2019B030302008), and Science and Technology Program of Guangzhou (201902010081).

References

1. Chen, L., Li, J., Li, J., Weng, J.: PAESS: public-key authentication encryption with similar data search for pay-per-query. IEEE Trans. Inf. Forensics Secur. **19**, 9910–9923 (2024)
2. Chen, Y., Li, W., Gao, F., Liang, K., Zhang, H., Wen, Q.: Practical attribute-based conjunctive keyword search scheme. Comput. J. **63**, 1203–1215 (2020)
3. Chen, Y., Li, W., Gao, F., Wen, Q., Zhang, H., Wang, H.: Practical attribute-based multi-keyword ranked search scheme in cloud computing. IEEE Trans. Serv. Comput. **15**, 724–735 (2019)
4. Hahn, C., Yoon, H., Hur, J.: Multi-key similar data search on encrypted storage with secure pay-per-query. IEEE Trans. Inf. Forensics Secur. **18**, 1169–1181 (2023)
5. Hong, H., Sun, Z.: Constructing conditional PKEET with verification mechanism for data privacy protection in intelligent systems. J. Supercomput. **79**, 15004–15022 (2023)
6. Huang, Q., Wei, Q., Yan, G., Zou, L., Yang, Y.: Fast and privacy-preserving attribute-based keyword search in cloud document services. IEEE Trans. Serv. Comput. **16**, 3348–3360 (2023)
7. Li, C., Feng, X., Shen, Q., Wu, Z.: On the security of secure keyword search and data sharing mechanism for cloud computing. IEEE Trans. Dependable Secure Comput. **21**, 4306–4308 (2023)
8. Liu, J., et al.: EMK-ABSE: efficient multikeyword attribute-based searchable encryption scheme through cloud-edge coordination. IEEE Internet Things J. **9**, 18650–18662 (2022)
9. Yang, X., et al.: Multi-client verifiable encrypted keyword search scheme with authorization over outsourced encrypted data. IEEE Trans. Netw. Sci. Eng. **11**, 6356–6371 (2024)
10. Hur, J.: Improving security and efficiency in attribute-based data sharing. IEEE Trans. Knowl. Data Eng. **25**, 2271–2282 (2011)
11. Li, H., Yang, Y., Luan, T.H., Liang, X., Zhou, L., Shen, X.S.: Enabling fine-grained multi-keyword search supporting classified sub-dictionaries over encrypted cloud data. IEEE Trans. Dependable Secure Comput. **13**, 312–325 (2015)
12. Li, Q., Xia, B., Huang, H., Zhang, Y., Zhang, T.: TRAC: Traceable and revocable access control scheme for mHealth in 5G-enabled IIoT. IEEE Trans. Industr. Inf. **18**, 3437–3448 (2021)
13. Varri, U.S., Mallick, D., Das, A.K., Hossain, M.S., Park, Y., Rodrigues, J.J.: TL-ABKS: traceable and lightweight attribute-based keyword search in edge–cloud assisted IoT environment. Alexandria Eng. J. **107**, 757–769 (2024)
14. Yin, H., Li, Y., Deng, H., Zhang, W., Qin, Z., Li, K.: An attribute-based keyword search scheme for multiple data owners in cloud-assisted industrial Internet of Things. IEEE Trans. Industr. Inf. **19**, 5763–5773 (2022)
15. Yin, H., et al.: CP-ABSE: a ciphertext-policy attribute-based searchable encryption scheme. IEEE Access **7**, 5682–5694 (2019)

16. Zhang, X., Huang, C., Su, Y., Qin, J.: Secure, dynamic, and efficient keyword search with flexible merging for cloud storage. IEEE Trans. Serv. Comput. **17**, 2822–2835 (2024)
17. Cheng, L., Meng, F.: Public key authenticated encryption with keyword search from LWE. In: Atluri, V., Di Pietro, R., Jensen, C.D., Meng, W. (eds.) ESORICS 2022. LNCS, vol. 13554, pp. 303–324. Springer, Cham (2022). https://doi.org/10.1007/978-3-031-17140-6_15
18. Goyal, V., Jain, A., Pandey, O., Sahai, A.: Bounded ciphertext policy attribute based encryption. In: Aceto, L., Damgård, I., Goldberg, L.A., Halldórsson, M.M., Ingólfsdóttir, A., Walukiewicz, I. (eds.) ICALP 2008. LNCS, vol. 5126, pp. 579–591. Springer, Heidelberg (2008). https://doi.org/10.1007/978-3-540-70583-3_47
19. Meng, R., Zhou, Y., Ning, J., Liang, K., Han, J., Susilo, W.: An efficient key-policy attribute-based searchable encryption in prime-order groups. In: Okamoto, T., Yu, Y., Au, M.H., Li, Y. (eds.) ProvSec 2017. LNCS, vol. 10592, pp. 39–56. Springer, Cham (2017). https://doi.org/10.1007/978-3-319-68637-0_3
20. Waters, B.: Ciphertext-policy attribute-based encryption: an expressive, efficient, and provably secure realization. In: Catalano, D., Fazio, N., Gennaro, R., Nicolosi, A. (eds.) PKC 2011. LNCS, vol. 6571, pp. 53–70. Springer, Heidelberg (2011). https://doi.org/10.1007/978-3-642-19379-8_4
21. Sahai, A., Waters, B.: Fuzzy identity-based encryption. In: Cramer, R. (ed.) EURO-CRYPT 2005. LNCS, vol. 3494, pp. 457–473. Springer, Heidelberg (2005). https://doi.org/10.1007/11426639_27
22. Shamir, A.: Identity-based cryptosystems and signature schemes. In: Blakley, G.R., Chaum, D. (eds.) CRYPTO 1984. LNCS, vol. 196, pp. 47–53. Springer, Heidelberg (1985). https://doi.org/10.1007/3-540-39568-7_5
23. Challita, S., Zalila, F., Gourdin, C., Merle, P.: 2018 IEEE International Conference on Cloud Engineering (IC2E). IEEE, Piscataway (2018)
24. Zhu, J., et al.: Cloud Computing. Springer, Berlin (2009)
25. Zheng, Q., Xu, S., Ateniese, G.: VABKS: verifiable attribute-based keyword search over outsourced encrypted data. In: IEEE INFOCOM 2014 - IEEE Conference on Computer Communications, pp. 522–530. IEEE, Piscataway (2014)
26. Sun, W., Yu, S., Lou, W., Hou, Y.T., Li, H.: Protecting your right: Attribute-based keyword search with fine-grained owner-enforced search authorization in the cloud. In: IEEE INFOCOM 2014-IEEE conference on computer communications, pp. 226–234. IEEE, Piscataway (2014)
27. Overview of amazon web services. https://docs.aws.amazon.com. Accessed 25 Oct 2014

Lightweight Transparent Zero-Knowledge Proofs for Cross-Domain Statements

Zhengzhou Tu[1], Min Xie[1], Junbin Fang[3], Yong Yu[4], and Zoe L. Jiang[1,2](✉)

[1] Harbin Institute of Technology, Shenzhen, Shenzhen 518055, China
zoeljiang@hit.edu.cn
[2] Key Laboratory of Cyberspace and Data Security, Ministry of Emergency Management, Beijing 100013, China
[3] Jinan University, Guangzhou 510632, China
[4] Shaanxi Normal University, Xi'an 710062, China

Abstract. Commit-prove zero-knowledge proofs (CP-ZKP) efficiently validate cross domain statements spanning both algebraic and non-algebraic components, enabling applications like privacy-preserving credentials and confidential cryptocurrency audits based on standard signatures such as RSA or (EC)DSA. While existing CP-ZKPs using SNARKs offer advantages such as low communication overhead and fast verification, they require provers to perform *group* operations, such as exponentiation, that scales linearly with the statement size. This computational requirement can be challenging for resource-constrained environments like IoT.

To address this, we present CP_{ILC}, a lightweight zero-knowledge proof tailored for cross-domain settings, achieving more efficient prover-side computations dominated with *field* multiplication, rather than *group* exponentiation, that scales linearly with the statement size. The verification is dominated with linear *field* additions, and the communication cost is sublinear. Specifically, we develop CP_{link}, a sub-proof of commitment equivalence, showing that a matrix of Pedersen commitment opens to a matrix of values committed in the ideal linear commitment model. Using the Fiat-Shamir transformation, we can compile CP_{link} and CP_{ILC} into non-interactive. Benchmark results demonstrate that CP_{ILC} reduces proving (verification) time by 59% (55%) for the cross-domain statement "$\exists (w, r) : c = g^w h^r \wedge y = \text{SHA256}(w)$", with even greater efficiency gains as the algebraic component in the statement increases.

Keywords: Zero-knowledge proof · Cross-domain · Commit-and-prove · Ideal linear commitment · Lightweight

1 Introduction

Zero-Knowledge Proofs (ZKPs) were first introduced in 1985 by Goldwasser et al. [19] as an *interactive* protocol enabling a prover to convince a verifier of the validity of a statement without revealing any additional information. Later,

Blum et al. [6] presented a *non-interactive* variant (NIZKP), in which a prover produces a single message—called a *proof*—that suffices to convince any verifier. An important theoretical milestone was established when Goldreich et al. [18] demonstrated that all languages in NP possess zero-knowledge proofs, paving the way for numerous privacy-preserving applications.

ZKPs for Algebraic and Non-algebraic Statements. Over the years, two main categories of statements have emerged in zero-knowledge contexts. *Algebraic statements* (e.g., those involving discrete logarithms) are efficiently handled by *Sigma protocols* [8,22,23,31], which can often be made non-interactive using the Fiat-Shamir heuristic [15]. However, such methods typically rely on group operations and do not extend efficiently to more general computations. In contrast, *non-algebraic statements*, such as those describing arbitrary arithmetic or boolean circuits (e.g., SHA256, AES), are handled by so-called *general-purpose* ZKPs [5,7,11]. These methods can in principle support any computable function but often have higher communication overheads or complex proof generation compared to specialized algebraic-proof systems.

A widely used family of proof systems for non-algebraic statements is Zero-Knowledge Succinct Non-Interactive Arguments of Knowledge (zk-SNARKs) [16,17,20,29]. zk-SNARKs typically offer excellent verification efficiency, with proof size and verification cost often independent of the underlying circuit size. However, their proving cost is dominated by *group exponentiations* (or other group-based operations) that scale linearly with the circuit size, which becomes a bottleneck in resource-constrained settings such as IoT devices. To address these performance concerns, alternative frameworks based on garbled circuits [26], MPC-in-the-head [5,11], or ideal linear commitment model [7] (referred as Π_{ILC} later) have been proposed, enabling provers to focus on more lightweight *field* operations. Some of these systems are also *transparent*, i.e., they require no trusted setup and only rely on public-coin properties.

Cross-Domain Statements. Beyond purely algebraic or purely non-algebraic cases, many real-world applications demand *cross-domain* statements, which involve both an algebraic component (e.g., a discrete-log-based commitment or signature) and a non-algebraic component (e.g., a hash computation). Such statements arise in privacy-preserving credential systems, confidential cryptocurrency audits, and proofs of solvency for Bitcoin exchanges [1,4,12]. One straightforward method is to transform the algebraic part into non-algebraic statements by defining each group operation as a function, but this significantly enlarges the statement size and thus increases both the proof size and the computational effort. Depending on the group size, the circuit for computing a single exponentiation might require thousands or even millions of gates [1,4]. Alternatively, one may implement arithmetic circuits directly as algebraic statements by treating each gate as an algebraic function and proving the relationships among the gates. Although this yields a prover/verifier workload and proof size that grow linearly with the number of gates, it could still entail tens of thousands of public-key operations and group elements for hash functions or block ciphers [4,12].

Existing Approaches and Their Limitations. A number of works have tried to tackle cross-domain statements by combining specialized ZK protocols [1,3,9,10, 28,30,33]. One major line of research relies on zk-SNARKs, leveraging succinct proofs and low verification costs. However, the group exponentiation required for SNARK proving can be impractical for devices with limited computation [27]. Other approaches use garbled-circuit or MPC-in-the-head paradigms to achieve linear *field*-based proving time, at the cost of high communication or proof size [4,11,12]. For instance, Jawurek et al. [26] proposed a garbled-circuit-based ZK protocol that is highly efficient but privately interactive, hindering its direct conversion to a non-interactive scheme via Fiat-Shamir. Meanwhile, ZKB++ [11] and its follow-up works [5] manage to be non-interactive but exhibit communication costs linear in the circuit size. Zhang et al. [33] combine Sigma protocols for discrete-log components with an MPC-in-the-head technique for general statements, resulting in $\mathcal{O}(|C|\log|C|)$ proving time and $\mathcal{O}(\log|C|)$ communication, where $|C|$ is the circuit size. While this is compact in *proof size*, the log-factor in the prover complexity remains non-trivial.

We consider cross-domain statements that there are *shared data* between the algebraic and non-algebraic components[1], which enables the commit-and-prove strategy [4,10,12]. Specifically, algebraic commitments are first generated for the shared data. Then, the cross-domain statement is split into two parts: an algebraic statement (consisting of the algebraic component and the commitments, which remains algebraic) that can be effectively proven by Sigma protocols, and a canonical cross-domain statement where a non-algebraic component is linked to the commitments. Our work focuses on the latter, aiming to reduce both the proverG s computation and communication overhead in a transparent setting. A detailed comparison of our approach with existing zero-knowledge proofs for cross-domain statements is provided in Table 1.

Consequently, no single scheme in the literature simultaneously offers *sublinear communication* and *linear proving* (in field operations) for cross-domain scenarios. For many IoT or low-powered devices that must handle real-time interaction or mutual authentication [27], existing zk-SNARK-based techniques can be computationally prohibitive, whereas solutions with linear or near-linear proving time often suffer from large proofs.

1.1 Our Contribution

We present $\mathsf{CP_{ILC}}$, a non-interactive zero-knowledge proof system specifically designed for cross-domain statements. The key feature is that the proverG s computation relies primarily on *field multiplications* and scales linearly with the size of the circuit, all while achieving *sublinear* communication. Furthermore, $\mathsf{CP_{ILC}}$ is *transparent*, meaning it requires no trusted setup. Concretely, for some secret values committed in algebraic commitment (e.g., Pedersen), the system enables a prover to prove that these secrets satisfy a complex non-algebraic relation (e.g., a hash or block cipher) with high efficiency.

---
[1] Some types do not necessarily involve shared data, such as OR proofs that combine algebraic and non-algebraic components.

Table 1. Comparison of CP-ZKPs for a circuit C with algebraically committed input w, where $|C|$ is the circuit size and $|w|$ is the length of w. Let λ donate the security parameter, pub a public-key operation, sym a symmetric-key operation, and add a symmetric-key addition.

Protocols	Public-coin	Transparent	Prover' work	Verifier' work	Proof size
CGM16 [12] Constr.1	✗	✓	$\mathcal{O}(|w|)$ pub $\mathcal{O}(|C|)$ sym	$\mathcal{O}(|w|)$ pub $\mathcal{O}(|C|)$ sym	$\mathcal{O}(|C|+|w|)$
CGM16 [12] Cons.2	✗	✓	$\mathcal{O}(\lambda)$ pub $\mathcal{O}(|C|+|w|\lambda)$ sym	$\mathcal{O}(\lambda)$ pub $\mathcal{O}(|C|+|w|\lambda)$ sym	$\mathcal{O}(|C|+|w|\lambda)$
AGM18 [1]	✓	✗	$\mathcal{O}(|C|+\lambda)$ pub $\mathcal{O}(|C|+\lambda)$ pub	$\mathcal{O}(|w|+\lambda)$ pub $\mathcal{O}(|w|+\lambda)$ pub	$\mathcal{O}(1)$ $\mathcal{O}(1)$
BHH$^+$19 [4]	✓	✓	$\mathcal{O}(|w|+\lambda)$ pub $\mathcal{O}(|C|\cdot\lambda)$ sym	$\mathcal{O}(|w|+\lambda)$ pub $\mathcal{O}(|C|\cdot\lambda)$ sym	$\mathcal{O}(|C|\lambda+|w|)$
CFQ19 [10] LegoAC1	✓	✗	$\mathcal{O}(|C|)$ pub $\mathcal{O}(|C|\log|C|)$ sym	$\mathcal{O}(|w|)$ pub	$\mathcal{O}(1)$
ABC$^+$22 [3]	✓	✗	$\mathcal{O}(|C|+|w|)$ pub	$\mathcal{O}(|w|)$ pub	$\mathcal{O}(\log|w|)$
ZCYW23 [33]	✓	✓	$\mathcal{O}(\lambda)$ pub $\mathcal{O}(|C|\log|C|)$ sym	$\mathcal{O}(\frac{(|w|+\lambda)^2}{\log(|w|+\lambda)})$ pub $\mathcal{O}(|C|)$ sym	$\mathcal{O}(\text{polylog}(|C|)+\lambda)$
This work	✓	✓	$\mathcal{O}(1)$ pub $\mathcal{O}(|C|+|w|)$ sym	$\mathcal{O}(|w|)$ pub $\mathcal{O}(|C|+|w|)$ add	$\text{poly}(\lambda)\sqrt{|C|+|w|}$

- **Linear-Time Field Operations.** Unlike zk-SNARKs that demand extensive group exponentiations, $\mathsf{CP_{ILC}}$ confines the prover's heavy workload to field multiplications, significantly reducing the cost on resource-constrained devices.
- **Sublinear Communication.** Drawing on ideas from ideal linear commitment model and carefully binding the committed witness, $\mathsf{CP_{ILC}}$ ensures proof sizes that are sublinear in $|C|$. This surpasses protocols whose communication grows linearly with the circuit size.
- **Transparent and Non-Interactive.** Our protocol is public coin and can be rendered non-interactive via the Fiat-Shamir heuristic, obviating any trusted setup. It offers perfect completeness, statistical special honest-verifier zero-knowledge (SHVZK), and computational soundness.
- **Modular Cross-Domain Composition.** For statements like $c = g^w h^r \wedge y = \text{SHA256}(w)$, the prover can efficiently prove consistency between algebraic and non-algebraic parts. Our benchmarks indicate that, compared with Π_{ILC} [7], $\mathsf{CP_{ILC}}$ reduces the proving time by 59% and verification time by 55%. Performance gains grow more pronounced as the ratio of algebraic-to-non-algebraic components increases.

In summary, $\mathsf{CP_{ILC}}$ bridges an important performance gap in CP-ZKPs by providing both sublinear communication and linear-in-$|C|$ *field* operations proving complexity for cross-domain statements, making it well-suited to environments such as IoT or mobile devices where low computational overhead is paramount.

Roadmap. Section 2 covers notation and preliminary definitions. Section 3 introduces our construction, followed by security and efficiency analysis. Section 4 details experimental setup and results. Finally, Sect. 5 concludes the paper.

2 Preliminary

In this paper, λ denotes the security parameter. A function is considered negligible if it becomes smaller than the inverse of any polynomial for sufficiently large inputs, denoted as negl. We define $\mathbb{G}$ as a cyclic group of prime order p and $\mathbb{Z}_p$ as a finite field with the same order. The notation $\mathbb{Z}_p^n$ and $\mathbb{G}^n$ represent the n-dimensional vector spaces over $\mathbb{Z}_p$ and $\mathbb{G}$, respectively. The generators of $\mathbb{G}$ are denoted as g and h.

Uniform sampling from $\mathbb{Z}_p$ is represented by $x \leftarrow_\$ \mathbb{Z}_p$. Vectors are written in bold, e.g., $\boldsymbol{a} = [a_1, \ldots, a_n] \in \mathbb{Z}_p^n$, and matrices are capitalized, e.g., $A \in \mathbb{Z}_p^{m \times n}$, with $A_{i,j}$ denoting the element in the i-th row and j-th column of matrix A. The inner product of two vectors $\boldsymbol{a}, \boldsymbol{b} \in \mathbb{Z}_p^n$ is defined as $\langle \boldsymbol{a}, \boldsymbol{b} \rangle = \sum_{i=1}^n a_i b_i$, and the Hadamard product (element-wise multiplication) of vectors $\boldsymbol{a}$ and $\boldsymbol{b}$ is given by $\boldsymbol{a} \circ \boldsymbol{b} = (a_1 b_1, \ldots, a_n b_n) \in \mathbb{Z}_p^n$.

2.1 Commitment Scheme

A commitment scheme allows one (the committer) to commit a message in a commitment so that it remains hidden to others, while also ensuring that the committed message cannot be changed once it is committed.

Definition 1. *A commitment scheme consists of a tuple of algorithms* (Setup, Com, VerCom) *defined as follows:*

- Setup(1^λ) → ck: *Given a security parameter λ, outputs a commitment key ck that defines the spaces for messages, randomness, and commitments.*
- Com(ck, $m; r$) → c: *Given a message m and randomness r, outputs a commitment c.*
- VerCom(ck, c, m, r) → 0/1: *Verifies whether the commitment c correctly opens to the message m using r.*

A commitment scheme must satisfy two security properties. *(Perfect) Hiding* guarantees that the commitment reveals no information about the message even against unbounded adversaries. *(Computational) Binding* ensures that it is computationally infeasible to open the same commitment with two different messages. A well-known example of a commitment scheme that is perfectly hiding and computationally binding is the *Pedersen commitment*. Here, the Setup outputs a cyclic group $\mathbb{G}$ of prime order p, generators $g, h \in \mathbb{G}$, and a finite field $\mathbb{Z}_p$ for the message and randomness. For a message m and randomness r, the commitment is computed as: Com(ck, $x; r$) $= g^x h^r$.

2.2 Zero-Knowledge Proof

Informally, a *zero-knowledge proof (ZKP)* is a proof system where the verifier learns nothing other than the truth of the statement. A ZKP is defined by three probabilistic polynomial-time (PPT) algorithms $(\mathcal{K}, \mathcal{P}, \mathcal{V})$: the setup algorithm $\mathcal{K}$ generates public parameters pp (including security parameter λ and optional auxiliary data), and both the prover $\mathcal{P}$ and verifier $\mathcal{V}$ use pp. During the protocol, $\mathcal{P}$ and $\mathcal{V}$ exchange messages over a channel $\xleftrightarrow{\text{chan}}$. Given inputs s and t, the verifier's full view is denoted as $\text{view}_\mathcal{V} \leftarrow \langle \mathcal{P}(s) \xleftrightarrow{\text{chan}} \mathcal{V}(t) \rangle$. At the end, the verifier outputs $\langle \mathcal{P}(s) \xleftrightarrow{\text{chan}} \mathcal{V}(t) \rangle \to b \in \{0, 1\}$, where $b = 1$ indicates acceptance and $b = 0$ for rejection. A ZKP is called *public coin* if $\mathcal{V}$ relies solely on public randomness to issue challenges.

We consider a relation $\mathcal{R}$ over tuples (x, w) and define the corresponding language $\mathcal{L}_\mathcal{R} = \{x \mid \exists w : (x, w) \in \mathcal{R}\}$, where x is a statement and w is a witness proving $x \in \mathcal{L}_\mathcal{R}$. The formal definition of a ZKP for $\mathcal{R}$ is given below.

Definition 2. Zero Knowledge Proof. *A zero-knowledge proof $(\mathcal{K}, \mathcal{P}, \mathcal{V})$ is a proof system over a communication channel $\xleftrightarrow{\text{chan}}$ for a relation $\mathcal{R}$, satisfying the properties of* completeness, knowledge soundness, *and* zero-knowledge.
Completeness. *Valid witness always leads to a successful verification:*

$$\Pr\left[pp \leftarrow \mathcal{K}(1^\lambda); (x, w) \in \mathcal{R} : \langle \mathcal{P}(pp, x, w) \xleftrightarrow{\text{chan}} \mathcal{V}(pp, x) \rangle = 1\right] = 1$$

Knowledge Soundness. *A ZKP is (computationally) knowledge-sound if for any PPT prover $\mathcal{P}^*$, there exists an PPT extractor $\mathcal{E}$ such that for all PPT adversaries $\mathcal{A}$, the following holds*

$$\Pr\left[\begin{array}{l} pp \leftarrow \mathcal{K}(1^\lambda); (x, s) \leftarrow \mathcal{A}(pp); w \leftarrow \mathcal{E}^{\text{view}_\mathcal{V} \leftarrow \langle \mathcal{P}^*(s) \xleftrightarrow{\text{chan}} \mathcal{V}(pp,x)\rangle}(pp, x) : \\ \langle \mathcal{P}^*(s) \xleftrightarrow{\text{chan}} \mathcal{V}(pp, x) \rangle = 1 \land (x, w) \notin \mathcal{R} \end{array}\right] = \text{negl}$$

Special Honest-Verifier Zero-Knowledge (SHVZK). *A ZKP is computational SHVZK if there exists a PPT simulator $\mathcal{S}$ such that for all bounded PPT adversaries $\mathcal{A}$ that produce (x, w) with $(x, w) \in \mathcal{R}$ and randomness ρ for $\mathcal{V}$, the distributions of the verifier's view in real and simulated executions are computationally indistinguishable:*

$$\Pr\left[pp \leftarrow \mathcal{K}(1^\lambda); (x, w, \rho) \leftarrow \mathcal{A}(pp); \text{view}_\mathcal{V} \leftarrow \langle \mathcal{P}(pp, x, w) \xleftrightarrow{\text{chan}} \mathcal{V}(pp, x; \rho)\rangle : \mathcal{A}(\text{view}_\mathcal{V}) = 1\right]$$
$$\approx \Pr\left[pp \leftarrow \mathcal{K}(1^\lambda); (x, w, \rho) \leftarrow \mathcal{A}(pp); \text{view}_\mathcal{V} \leftarrow \mathcal{S}(pp, x; \rho) : \mathcal{A}(\text{view}_\mathcal{V}) = 1\right]$$

Intuitively, SHVZK means that if the verifier's challenges are known or even adversarially chosen (yet the verifier still follows the protocol), then the simulator $\mathcal{S}$ can simulate a view indistinguishable from a real $\text{view}_\mathcal{V}$ *without* using the witness. If this property holds against unbounded adversaries and the two probabilities are exactly equal, it is called *perfect SHVZK*. We say a ZKP is *knowledge soundness with straight-line extraction* if an extractor can enforce knowledge soundness by extracting the witness from a *single* transcript of the communication between $\mathcal{P}$ and $\mathcal{V}$ without relying on rewinding.

Using the Fiat-Shamir transform [15], where the verifierG s challenges are computed using a cryptographic hash function applied to the transcript up to the challenge, one can convert a public-coin proof into a non-interactive one.

2.3 Commit and Prove Zero-Knowledge Proof

Infomally, a *Commit and Prove Zero-Knowledge Proof (CP-ZKP)* is a ZKP that can prove knowledge of (x, w) such that $\mathcal{R}(x, w)$ holds w.r.t. a witness $w = (u, \omega)$ and u opens a commitment c_u. We recall the definition of CP-ZKP in [10].

Definition 3. *CP-ZKP*. *Let $\mathcal{R}$ be a relation over $\mathsf{M}_x \times \mathsf{M}_u \times \mathsf{M}_\omega$ such that M_u splits over l arbitrary domains $(\mathsf{M}_1 \times \ldots \times \mathsf{M}_l)$ for some $l \geq 1$. Let* $\mathsf{Com} = (\mathsf{Setup}, \mathsf{Com}, \mathsf{VerCom})$ *be a commitment scheme whose message space is M such that $\mathsf{M}_i \in \mathsf{M}$ for all $i \in [l]$. A CP-ZKP for Com and $\mathcal{R}$ is a ZKP such that:*

- *Every $\mathrm{R} \in \mathcal{R}^{\mathsf{Com}}$ is represented by a pair $(\mathsf{ck}, \mathcal{R})$ where $\mathsf{ck} \in \mathsf{Setup}(1^\lambda)$.*
- *R is over pairs (x, w) where the statement is $\mathrm{x} = (x, (c_j)_{j \in [l]}) \in \mathsf{M}_x \times C^l$, the witness is $\mathrm{w} = ((u_j)_{j \in [l]}, (r_j)_{j \in [l]}, \omega) \in \mathsf{M}_1 \times \ldots \times \mathsf{M}_l \times \mathsf{R}^l \times \mathsf{M}_\omega$, and the relation R hold iff:*

$$\wedge_{j \in [l]} \mathsf{VerCom}(\mathsf{ck}, c_j, u_j, r_j) = 1 \wedge \mathcal{R}(x, (u_j)_{j \in [l]}, \omega) = 1$$

2.4 Ideal Linear Commitment Model

In this paper, we consider the ideal linear commitment (ILC) channel, $\xleftrightarrow{\text{ILC}}$, depicted in Fig. 1. In the standard channel, all messages are directly exchanged between the prover and the verifier, whereas the ILC model introduces structured interactions through three operations: commit, send, and open. Using commit, the prover submits vectors of field elements (of fixed length k), which are securely stored in the channel and hidden from the verifier, who only learns the number of committed vectors. The verifier can use send to issue challenge field elements and

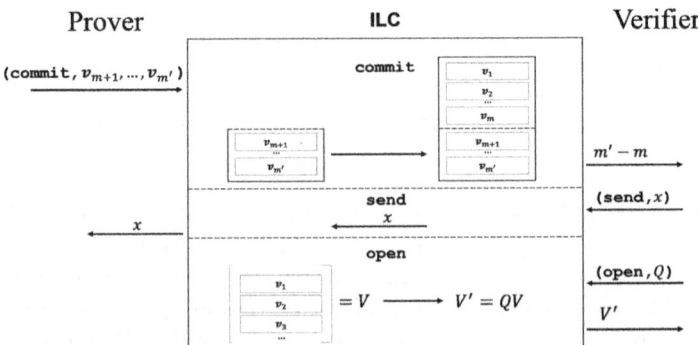

Fig. 1. An illustration of the ILC interaction environment. Each v_i is a committed vector, x is a challenge scalar, V is the matrix formed by all v_i, Q is the querying matrix, and V' is the querying result. The parameters m and m' specify the numbers of committed vectors.

open to request linear combinations of the committed vectors. Multiple linear combinations can be queried in a single open request, as illustrated in Fig. 1.

When translating protocols from the ILC model into real world protocols, the ILC functionalities can be enforced by combining linear error-correcting codes [14,24,32] and commitment schemes [2,25]. In these constructions, no trapdoors or private coins are required, ensuring that the schemes instantiated from the ILC model are transparent. The reader can refer to [7] for more details on the ILC instantiation.

2.5 ZKP for Arithmetic Circuit Satisfiability in ILC

We briefly review an important building block for our construction, i.e., Π_{ILC}, a ZKP for arithmetic circuit satisfiability (AC-SAC) constructed in the ILC model [7]. Consider a circuit with a total of N fan-in 2 gates, which can either be addition gates or multiplication gates over a field $\mathbb{Z}_p$. Each gate has two input (left and right) wires and one output wire. A circuit is said to be satisfiable if there exists an assignment that satisfies all the gates, the wiring, and the known values specified in the instance.

At a high level, the idea of the proof for the prover is to commit to the inputs and outputs of all the circuit gates in the ILC channel, and then prove that these assignments are consistent with the circuit description. This requires performing the following tasks, each achieved by a corresponding sub-proof:

- Π_{eq}: Prove for each public value specified in the circuit instance that this is indeed the value the prover has committed to.
- Π_{sum}: Prove for each addition gate that the committed output is the sum of the committed inputs.
- Π_{prod}: Prove for each multiplication gate that the committed output is the product of the committed inputs.
- Π_{perm}: Prove for each wire that all committed values corresponding to this wire are the same.

The above four sub-proofs constitute the complete Π_{ILC}. The reader can refer to [7] for details on the sub-proofs.

3 NIZK for Non-algebraic Statements on Algebraic Commitments

In this section, we present $\mathsf{CP}_{\mathsf{ILC}}$. For a matrix A committed using an algebraic commitment $\mathsf{Com}(A)$ (e.g. Pedersen commitment) and arbitrary arithmetic circuit instance f (which is non-algebraic), $\mathsf{CP}_{\mathsf{ILC}}$ could perform an efficient proof showing that f is satisfiable and specified wires in f are indeed the values committed to in $\mathsf{Com}(A)$.

3.1 Proof of Commitment Equivalence Between Pedersen and ILC

We first introduce the core building block $\mathsf{CP}_\mathsf{link}$, showing that a Pedersen commitment matrix C opens to a matrix A committed in the ILC channel. Let ck be the commitment key of the Pedersen commitment scheme. Consider $A, B \in \mathbb{Z}_p^{\mathsf{mn} \times k}$, $C \in \mathbb{G}^{\mathsf{mn} \times k}$ where k is determined by the public parameter pp_ILC, m, n are some integers. Without loss of generality, we assume $\mathsf{m} = 2^\mu$ for some integer μ. Let values in square brackets be those have been committed to the ILC channel. Given $([A], [B], C)$, we give the formal relation $\mathcal{R}_\mathsf{link}$ as

$$\mathcal{R}_\mathsf{link} = \left\{ \begin{array}{c} (pp_\mathsf{ILC}, \mathsf{ck}, x) = ((\mathbb{Z}_p, k), (\mathbb{G}, g, h), ([A], [B], C)) : \\ A, B \in \mathbb{Z}_p^{\mathsf{mn} \times k} \wedge g^{A_{r,c}} \cdot h^{B_{r,c}} = C_{r,c}, \forall r \in [1, \mathsf{mn}], \forall c \in [1, k] \end{array} \right\}$$

Relation $\mathcal{R}_\mathsf{link}$ requires that $C_{r,c} = g^{A_{r,c}} h^{B_{r,c}}$ for all specified indices within the given ranges. Intuitively, the high level ideal is to use a random challenge y to embed all constraints into one, with each individual constraint embedded with a different power of y. Recall that in the ILC model, the basic data committed by the prover are vectors in $\mathbb{Z}_p^k$. We parse matrices A and B respectively as collections of row vectors $\boldsymbol{a}_{i,j}$ and $\boldsymbol{b}_{i,j}$ (the $((j-1)\mathsf{m} + (i+1))$-th row), where $0 \leq i \leq \mathsf{m} - 1$ and $1 \leq j \leq \mathsf{n}$. To achieve sub-linear communication, we also utilize μ challenges $X_0, \ldots, X_{\mu-1}$ to compress $2\mathsf{mn}$ vectors $\boldsymbol{a}_{i,j}, \boldsymbol{b}_{i,j}$ of length k, into $2\mathsf{n}$ vectors $\hat{\boldsymbol{a}}_j, \hat{\boldsymbol{b}}_j$ of the same length as follows

$$\hat{\boldsymbol{a}}_j(X_0, \ldots, X_{\mu-1}) = \sum_{i=0}^{\mathsf{m}-1} \boldsymbol{a}_{i,j} X_0^{i_0} X_1^{i_1} \ldots X_{\mu-1}^{i_{\mu-1}}, \tag{1}$$

where $i_{\mu-1}, i_{\mu-2}, \ldots, i_0$ are the digits of the binary expansion of i. The $2\mathsf{n}$ vectors of length k are then obtained by evaluating the multivariate polynomials into challenges $(x_0, x_1, \ldots, x_{\mu-1})$. With similar expressions when a is replaced by b.

We then introduce one more challenge X and embed the compressed vectors into polynomials in X. Again with similar expressions for a replaced by b.

$$\hat{\boldsymbol{a}}(X, x_0, \ldots, x_{\mu-1}) = \sum_{j=1}^{\mathsf{n}} \hat{\boldsymbol{a}}_j(x_0, \ldots, x_{\mu-1}) X^j$$

To isolate the terms we intend to check, we define the vector $\hat{\boldsymbol{w}}$ as follows:

$$\hat{\boldsymbol{w}}(y, X, x_0, \ldots, x_{\mu-1}) = \boldsymbol{y} \sum_{j=1}^{\mathsf{n}} X^{-j} \sum_{i=0}^{\mathsf{m}-1} y^{k((j-1)\mathsf{m}+i)} x_0^{-i_0} x_1^{-i_1} \ldots x_{\mu-1}^{-i_{\mu-1}}$$

where $\boldsymbol{y} = (y^1, y^2, \ldots, y^k)$. Consider the inner product

$$\langle \hat{\boldsymbol{a}}(X, x_0, \ldots, x_{\mu-1}), \hat{\boldsymbol{w}}(y, X, x_0, \ldots, x_{\mu-1}) \rangle.$$

Observe that the *constant term* with respect to $X, x_0, \ldots, x_{\mu-1}$ in this inner product arises solely from the product of the monomial $x_0^{i_0} x_1^{i_1} \cdots x_{\mu-1}^{i_{\mu-1}} X^j$ in $\hat{\boldsymbol{a}}$

with the corresponding monomial $x_0^{-i_0} x_1^{-i_1} \cdots x_{\mu-1}^{-i_{\mu-1}} X^{-j}$ in $\hat{w}$. Moreover, each monomial in $\hat{w}$ is multiplied by a distinct power of y. Since $\hat{w}$ also includes the factor y, the constant term of the inner product is precisely the sum of all entries in the matrix A, with each entry multiplied by a unique power of y.

Careful analysis shows that the constant term with respect to $X, x_0, \ldots, x_{\mu-1}$ of $\hat{a} \cdot \hat{w}$ is $\sum_{r=1}^{mn} \sum_{c=1}^{k} y^{(r-1)k+c} A_{r,c}$. Similarly, the constant term of $\hat{b} \cdot \hat{w}$ is $\sum_{r=1}^{mn} \sum_{c=1}^{k} y^{(r-1)k+c} B_{r,c}$. Denote these constant terms by c_a and c_b, respectively. We observe that

$$g^{c_a} h^{c_b} = \prod_{r=1}^{mn} \prod_{c=1}^{k} C_{r,c}^{y^{(r-1)k+c}},$$

which implies that the matrices A, B, C satisfy the relation $\mathcal{R}_{\mathsf{link}}$ with high probability, by virtue of the binding property of commitments and the Schwarz–Zippel Lemma.

Formal Description. We now detail $\mathsf{CP}_{\mathsf{link}}$. To ensure zero-knowledge in the final protocol, random blinders such as $\hat{a}_0, \hat{b}_0, e_a$, and e_b are introduced into the construction. Communication over a standard channel is denoted by $\leftarrow$ or $\rightarrow$, while "marked arrows" indicate interactions via the ILC channel. We assume that the prover has already committed to A and B using $\mathtt{commit}$.

Proving:

- $\mathcal{P}_{\mathsf{link}} \xrightarrow{\mathtt{commit}} \mathsf{ILC}$: Prover picks $\hat{a}_0, \hat{b}_0 \leftarrow_\$ \mathbb{Z}_p^k, e_a, e_b \leftarrow_\$ \mathbb{Z}_p$, and $\mathtt{commit}$ them.
- $\mathcal{P}_{\mathsf{link}} \longrightarrow \mathcal{V}_{\mathsf{link}}$: Prover sends $E = g^{e_a} h^{e_b}$ to the verifier via standard channel.
- $\mathsf{ILC} \xleftarrow{\mathtt{send}} \mathcal{V}_{\mathsf{link}}$: Verifier picks $y \leftarrow_\$ \mathbb{Z}_p$ and $\mathtt{send}$ y.
- $\mathcal{P}_{\mathsf{link}} \xrightarrow{\mathtt{commit}} \mathsf{ILC}$:
  For $1 \leq j \leq n$, prover constructs vector polynomials $\hat{a}(X, X_0, \ldots, X_{\mu-1})$, $\hat{b}(X, X_0, \ldots, X_{\mu-1})$, and $\hat{w}(y, X, X_0, \ldots, X_{\mu-1})$ as follows:

$$\hat{a}_j(X_0, \ldots, X_{\mu-1}) = \sum_{i=0}^{m-1} a_{i,j} \prod_{t=0}^{\mu-1} X_t^{i_t}, \quad \hat{a}(X, \mathbf{X}) = \hat{a}_0 + \sum_{j=1}^{n} \hat{a}_j(\mathbf{X}) X^j,$$

with $\mathbf{X} = (X_0, \ldots, X_{\mu-1})$ and similar expressions for $\hat{b}$. Then

$$\hat{w}(y, X, \mathbf{X}) = y \sum_{j=1}^{n} X^{-j} \sum_{i=0}^{m-1} y^{k((j-1)m+i)} \prod_{t=0}^{\mu-1} X_t^{-i_t}.$$

Prover computes the following inner product:

$$\hat{a}(X,\mathbf{X})\cdot\hat{w}(y,X,\mathbf{X}) = \sum_{r=1}^{mn}\sum_{c=1}^{k}A_{r,c}\cdot y^{(r-1)k+c} + f_0^+(y)X_0 + f_0^-(y)X_0^{-1}$$
$$+ f_1^+(y,X_0)X_1 + f_1^-(y,X_0)X_1^{-1} + \ldots$$
$$+ f_{\mu-1}^+(y,X_0,\ldots,X_{\mu-2})X_{\mu-1}$$
$$+ f_{\mu-1}^-(y,X_0,\ldots,X_{\mu-2})X_{\mu-1}^{-1}$$
$$+ \sum_{s=-\mathfrak{n},s\neq 0}^{\mathfrak{n}-1} g_s(y,X_0,\ldots,X_{\mu-1})X^s$$

An analogous process is applied to $\hat{b}(X,\mathbf{X})$. Variables bearing prime marks donate the corresponding coefficients in $\hat{b}(X,\mathbf{X})\cdot\hat{w}(y,X,\mathbf{X})$. For example, $f_0^{+'}(y)$ and $f_0^{-'}(y)$ denote the counterparts of $f_0^+(y)$ and $f_0^-(y)$. Then, the prover commit $\{f_0^+(y), f_0^-(y), f_0^{+'}(y), f_0^{-'}(y)\}$ to store them in the ILC channel.

- ILC $\xleftarrow{\text{send}}$ $\mathcal{V}_{\text{link}}$: Verifier chooses $x_0 \leftarrow_\$ \mathbb{Z}_p$ and send x_0 via the ILC.
- $\mathcal{P}_{\text{link}} \xrightarrow{\text{commit}}$ ILC: Prover calculates and commit

$$f_1^\pm = f_1^\pm(y,x_0), f_1^{\pm'} = f_1^{\pm'}(y,x_0)$$

- For $t=1$ to $\mu-2$:
  - ILC $\xleftarrow{\text{send}}$ $\mathcal{V}_{\text{link}}$: Verifier chooses $x_t \leftarrow_\$ \mathbb{Z}_p$ and send x_t via the ILC.
  - $\mathcal{P}_{\text{link}} \xrightarrow{\text{commit}}$ ILC: Prover calculates and commit

$$f_{t+1}^\pm = f_{t+1}^\pm(y,x_0,\ldots,x_t), f_{t+1}^{\pm'} = f_{t+1}^{\pm'}(y,x_0,\ldots,x_t)$$

- ILC $\xleftarrow{\text{send}}$ $\mathcal{V}_{\text{link}}$: Verifier picks $x_{\mu-1} \leftarrow_\$ \mathbb{Z}_p$ and send $x_{\mu-1}$.
- $\mathcal{P}_{\text{link}} \xrightarrow{\text{commit}}$ ILC: Prover calculates

$$g_s = g_s(y,x_0,\ldots,x_{\mu-1}) \text{ for } s \text{ in } [-\mathfrak{n},\mathfrak{n}-1] \text{ and } s\neq 0,$$

and similar for g_s', then commit g_s, g_s' for all s within given range.

Verification

- ILC $\xleftarrow{\text{open}}$ $\mathcal{V}_{\text{link}}$: Verifier selects $x \leftarrow_\$ \mathbb{Z}_p$, and query the ILC channel to get:

$$\hat{a} = \hat{a}_0 + \sum_{i=0,j=1}^{m-1,n}(a_{i,j}x_0^{i_0}x_1^{i_1}\ldots x_{\mu-1}^{i_{\mu-1}})x^j, \quad \hat{b} = \hat{b}_0 + \sum_{i=0,j=1}^{m-1,n}(b_{i,j}x_0^{i_0}x_1^{i_1}\ldots x_{\mu-1}^{i_{\mu-1}})x^j$$

$$\hat{e}_a = e_a + \sum_{t=0}^{\mu-1}f_t^\pm x_t^{\pm 1} + \sum_{s=-\mathfrak{n},s\neq 0}^{\mathfrak{n}-1}g_s x^s, \quad \hat{e}_b = e_b + \sum_{t=0}^{\mu-1}f_t^{\pm'} x_t^{\pm 1} + \sum_{s=-\mathfrak{n},s\neq 0}^{\mathfrak{n}-1}g_s' x^s$$

The verifier evaluate $\hat{w}$ as predefined and check the flowing equation:

$$g^{\hat{a}\hat{w}^T - \hat{e}_a} h^{\hat{b}\hat{w}^T - \hat{e}_b} E = \prod_{r=1}^{mn} \prod_{c=1}^{k} C_{r,c}^{y^{(r-1)k+c}} \qquad (2)$$

The verifier outputs **accept** if (2) holds and **reject** otherwise.

Remarks on ILC Security. In the original ILC model, the prover employs `commit` to submit vectors of field elements. Here, we additionally require the prover to send $E = g^{e_a} h^{e_b}$ over a standard channel. This does not weaken the security guarantees of the ILC model: E is a single group element (essentially a commitment to a pair (e_a, e_b)), so it does not disclose any information about the vectors committed to the ILC channel. Moreover, as the binding properties of the Pedersen commitment scheme hold regardless of how the commitment is transmitted, the protocolG s security (especially zero-knowledge and soundness) in the ILC setting remains intact. Finally, although the protocol requires committing single field elements in some steps, this can be handled by padding (to form length-k vectors, if necessary) without affecting asymptotic efficiency.

Theorem 1. *The zero-knowledge proof* $\mathsf{CP}_{\mathsf{link}} = (\mathcal{K}_{\mathsf{CP}_{\mathsf{ILC}}}, \mathcal{P}_{\mathsf{link}}, \mathcal{V}_{\mathsf{link}})$ *for relation* $\mathcal{R}_{\mathsf{link}}$ *possesses perfect completeness, knowledge soundness, and perfect SHVZK properties.*

We give the proof of Theorem 1 in Appendix A due to space limitation.

Efficiency. Evidently, $\mathsf{CP}_{\mathsf{link}}$ is transparent, which means that it requires no trusted setup and is public-coin. The latter property allows us to apply the Fiat-Shamir transform to convert it into non-interactive. We summarize the efficiency in Table 2.

In $\mathsf{CP}_{\mathsf{link}}$, the prover computes on multivariate polynomials with vector coefficients. Therefore, the prover can save considerable computational effort by computing mostly on vectors, and using challenges $y, x_0, \ldots, x_{\mu-1}$ as they become available to partially evaluate expressions and 'collapse' multiple vectors into fewer vectors. We detail this computation starting from the final round back to the first.

Upon receiving $x_{\mu-1}$, and computing $\hat{a}, \hat{w}$, the prover calculates values g_s. This is done by expressing $\hat{a}, \hat{w}$ as polynomials in X of degree n. Then, g_s are the coefficients of the inner product polynomial, computable using FFT techniques at a cost of $\mathcal{O}(k\mathsf{n}\log \mathsf{n})$. The computation of f_t^+, f_t^- is recursive, following the methods in [7], the dominant cost involves $5k\mathsf{n}(2^\mu - 1)$ multiplications, which is $\mathcal{O}(k\mathsf{mn})$ multiplications in $\mathbb{Z}_p$. The calculations for $f_t^{'+}, f_t^{'}$, and $g_s^{'}$ follow a similar structure, allowing the reuse of the computed $\hat{w}$. Additionally, there are $\mathcal{O}(1)$ exponentiations in $\mathbb{G}$ to compute group element E. Consequently, these steps have a negligible impact on the overall asymptotic efficiency.

The verifier transmits $\mu + 1$ field elements $(y, x_0, \ldots, x_{\mu-1})$ to the prover via the ILC channel. The verifier's total computational cost includes:

- $\mathcal{O}(\mathsf{mn} + k)$ field multiplications to compute $\hat{\boldsymbol{w}}$. Indeed, generating the vector $\boldsymbol{y}$ requires $\mathcal{O}(k)$ multiplications, and the remaining $\mathcal{O}(\mathsf{mn})$ term accounts for combining all constraints.
- $\mathcal{O}(\mathsf{mn})$ field multiplications to produce queries for the ILC channel (e.g., for partial evaluations).
- $\mathcal{O}(\tilde{N})$ group exponentiations to combine $\tilde{N} = \mathsf{mn} \times k$ Pedersen commitments into a single equation. We emphasize that $\tilde{N}$ is the number of committed inputs.

The total number of ILC queries (i.e., the number of open) is $qc = 4$. On the prover's side, the communication cost comprises commitments of $(4\mu + 4\mathsf{n} + 3)$ vectors in $\mathbb{Z}_p^k$ to the ILC, together with one group element E transmitted over the standard channel.

Table 2. Efficiency analysis of $\mathsf{CP}_\mathsf{link}$. The *mult*, *add*, and *exp* correspond to field multiplication, field addition, and group exponentiation respectively.

$T_{\mathcal{P}_\mathsf{link}}$	$T_{\mathcal{V}_\mathsf{link}}$	qc	#rounds	#communication
$\mathcal{O}(k\mathsf{mn} + k\mathsf{n}\log\mathsf{n})$ *mult* $\mathcal{O}(1)$ *exp*	$\mathcal{O}(\mathsf{mn} + k)$ *mult* $+ \mathcal{O}(\tilde{N})$ *exp*	4	$\mu + 2$	$4\mu + 4\mathsf{n} + 3\,\mathbb{Z}_p^k + \mathcal{O}(1)\,\mathbb{G}$

3.2 NIZK for AC-SAT on Algebraic Commitments in ILC

We begin by reviewing Π_{ILC} from Bootle et al. [7], a protocol for proving the satisfiability of an arithmetic circuit. Consider a circuit with N fan-in-2 gates, where each gate has one input wire and two output wire. Following [7], we arrange the $3N$ wires of the circuit as row vectors in $\mathbb{Z}_p^k$, where k is specified by the public parameters. Assume that the numbers of addition and multiplication gates are each divisible by k. Partition these gates into: $A_L, A_R, A_O \in \mathbb{Z}_p^{m_A \times k}$ for addition gates (representing left inputs, right inputs, and outputs), and $M_L, M_R, M_O \in \mathbb{Z}_p^{m_M \times k}$ for multiplication gates. The dimensions $m_A \times k$ and $m_M \times k$ match the respective numbers of addition and multiplication gates, so $N = m_A \times k + m_M \times k$.

We define the matrix V by horizontally concatenating the transposed matrices:

$$V^T = \begin{bmatrix} A_L^T & A_R^T & A_O^T & M_L^T & M_R^T & M_O^T \end{bmatrix} \in \mathbb{Z}_p^{k \times (3m_A + 3m_M)}.$$

Next, we describe how committed values are incorporated. We consider the case where the committed values are not part of the circuit outputs; thus, they serve only as inputs to certain gates. We require that these committed values are always placed on the left (or always on the right) inputs of the gates, so that all committed values can be confined to the row vectors of A_L and M_L (or, equivalently, A_R and M_R).[2] Let m_C be the number of row vectors used for the

---

[2] If an addition (multiplication) gate receives both inputs from commitments, we restructure the circuit by adding two extra dummy gates that each add 0 (multiply by 1). This ensures that every addition (multiplication) gate is fed by at most one committed input.

committed values. We define: $A \in \mathbb{Z}_p^{m_C \times k}, B \in \mathbb{Z}_p^{m_C \times k}, C \in \mathbb{G}^{m_C \times k}$, where A holds the committed values, B stores corresponding randomness, and C stores the Pedersen commitments. Note that B is not included in V.

We then form the matrix T by appending B to V:

$$T^T = \begin{bmatrix} V^T & B^T \end{bmatrix} \in \mathbb{Z}_p^{k \times (3m_A + 3m_M + m_C)}.$$

Denote each row of T by t_i. Let $m = 3m_A + 3m_M$, and define $m' = m + m_C$. Let $\{t_i\}_{i \in S_A}$ be the set of rows in T that represent the committed values, and let $\{t_i\}_{i \in S}$ and $\{t_i\}_{i \in \bar{S}}$ be the sets of rows corresponding to public values and private values, respectively. $S_A, S, \bar{S}$ are the row indices. Finally, let π denote the permutation among the circuit wires. Specifically, π is a matrix of the same dimensions as V, where each entry represents another coordinate, indicating that the values at these two coordinates should be equal (i.e., they correspond to a set of connected wires in the circuit). With this notation, we formalize the relations for AC-SAT on algebraic commitments as follows.

$$\mathcal{R}_{\mathsf{CP}_{\mathsf{ILC}}} = \left\{ \begin{array}{l} (pp_{\mathrm{ILC}}, x, w) = \big((\mathbb{Z}_p, k), (m_A, m_M, m_C, \pi, \{t_i\}_{i \in S}, C), (\{t_i\}_{i \in \bar{S}}) \big) : \\ \text{construct } (A_L, A_R, A_O, M_L, M_R, M_O, V, A, B) \text{ from } \{t_i\}_{i \in [m']} \\ \pi \in \sum_{[m] \times > [k]} \wedge S \subseteq [m] \wedge \bar{S} = [m'] n S \wedge S_A \subseteq [m] \\ \wedge A_L + A_R = A_O \wedge M_L \circ M_R = M_O \\ \wedge V_{i,j} = V_{\pi(i,j)} \quad \forall (i,j) \in [m] \times [k] \\ \wedge g^{A_{i,j}} h^{R_{i,j}} = C_{i,j} \quad \forall (i,j) \in [m_C] \times [k] \end{array} \right\}$$

(3)

In addition to the four requirements for AC-SAT described in Sect. 2.5, $\mathcal{R}_{\mathsf{CP}_{\mathsf{ILC}}}$ further requires a proof of commitment equivalence between Pedersen and ILC. Figure 2 illustrates the sub-protocols employed within $\mathsf{CP}_{\mathsf{ILC}}$. Apart from $\mathsf{CP}_{\mathsf{link}}$, high-level descriptions of the other sub-protocols can be found in Sect. 2.5.

$\mathcal{P}_{\mathsf{CP}_{\mathsf{ILC}}}(pp_{\mathsf{ILC}}, x, w)$	$\mathcal{V}_{\mathsf{CP}_{\mathsf{ILC}}}(pp_{\mathsf{ILC}}, x)$
Parse $x = (m_A, m_M, m_C, \pi, \{t_i\}_{i \in S}, C)$	Parse $x = (m_A, m_M, m_C, \pi, \{t_i\}_{i \in S}, C)$
Parse $w = \{t_i\}_{i \in \bar{S}}$	
Submit (commit, $\{t_i\}_{i \in m'}$) to ILC	
Define following matrices as (3), and let $U = (t_i)_{i \in S}$	
Run $\mathcal{P}_{eq}(pp, \{t_i\}_{i \in S}, [U])$	Run $\mathcal{V}_{eq}(pp, \{t_i\}_{i \in S}, [U])$
Run $\mathcal{P}_{sum}(pp, [A_L], [A_R], [A_O])$	Run $\mathcal{V}_{sum}(pp, [A_L], [A_R], [A_O])$
Run $\mathcal{P}_{prod}(pp, [M_L], [M_R], [M_O])$	Run $\mathcal{V}_{prod}(pp, [M_L], [M_R], [M_O])$
Run $\mathcal{P}_{perm}(pp, \pi, [V], [V])$	Run $\mathcal{V}_{perm}(pp, \pi, [V], [V])$
Run $\mathcal{P}_{link}(pp, [A], [B], C)$	Run $\mathcal{V}_{link}(pp, [A], [B], C)$
	Return 1 if all sub-proofs are accepted, and 0 otherwise

Fig. 2. Proof for AC-SAT on algebraic commitments in the ILC model.

Theorem 2. *The zero-knowledge proof* $\mathsf{CP}_{\mathsf{ILC}} = (\mathcal{K}_{\mathsf{CP}_{\mathsf{ILC}}}, \mathcal{P}_{\mathsf{CP}_{\mathsf{ILC}}}, \mathcal{V}_{\mathsf{CP}_{\mathsf{ILC}}})$ *for relation* $\mathcal{R}_{\mathsf{CP}_{\mathsf{ILC}}}$ *possesses perfect completeness, knowledge soundness, and perfect SHVZK properties.*

Proof. Perfect completeness follows immediately from the perfect completeness of each sub-proof. The perfect special honest-verifier zero-knowledge (SHVZK) property is inherited from the corresponding feature of the sub-proofs, since a simulated transcript can be assembled from the simulators of each individual component. Finally, the overall knowledge soundness is ensured by the knowledge soundness of all sub-proofs. In particular, CP$_{\text{link}}$ guarantees that the Pedersen commitments and the ILC commitments open to the same value, and the straight-line extraction property of the sub-proofs collectively establishes the knowledge soundness of the entire protocol.

Efficiency. The main computational overhead arises from the permutation proof ($\mathcal{P}_{perm}$ and $\mathcal{V}_{perm}$), particularly due to the large dimensions of the relevant matrices. We define $\mathfrak{m}, \mathfrak{n}$ such that $m = 3m_A + 3m_M = \mathfrak{mn}$. The total number of gates is $N = \frac{k\,\mathfrak{mn}}{3}$, and the total number of committed values is $\tilde{N} = k\,m_C$. Without loss of generality, we assume that $\tilde{N} \ll N$. The dominant cost in $\mathcal{P}_{perm}$ comprises $\mathcal{O}(k\,\mathfrak{n}\log\mathfrak{n} + k\,\mathfrak{m}\,\mathfrak{n})$ (which is $\mathcal{O}(N)$) multiplications, while $\mathcal{V}_{perm}$ requires $\mathcal{O}(k\,\mathfrak{m}\,\mathfrak{n})$ additions. Moreover, both the prover and verifier perform group exponentiations in CP$_{\text{link}}$. Table 3 summarizes the asymptotic complexities.

Table 3. Efficiency of CP$_{\text{ILC}}$ in the ILC model. The *mult*, *add*, and *exp* correspond to field multiplication, field addition, and group exponentiation respectively.

$T_{\mathcal{P}_{\text{link}}}$	$T_{\mathcal{V}_{\text{link}}}$	qc	#rounds	#communication
$\mathcal{O}(N)\ mult + \mathcal{O}(1)\ exp$	$\mathcal{O}(N)\ add + \mathcal{O}(\tilde{N})\ exp$	24	$\mathcal{O}(\log\log(N))$	$\mathcal{O}(\sqrt{N})\,\mathbb{Z}_p^k + \mathcal{O}(1)\,\mathbb{G}$

We adopt the instantiation of Π_{ILC} from [7], which relies on the commitment scheme of Applebaum et al. [2] and the linear code of Druk and Ishai [14]. To instantiate CP$_{\text{ILC}}$ over the standard channel, we proceed as follows: 1) Apply the same technique to instantiate CP$_{\text{link}}$; 2) Combine the instantiations of Π_{ILC} and CP$_{\text{link}}$. This construction enables both the prover and verifier to execute computations in linear time, while maintaining compact, square-root-sized commitments that yield sublinear communication costs. Finally, applying the Fiat–Shamir transform to CP$_{\text{ILC}}$ produces a non-interactive zero-knowledge (NIZK) proof. The primary efficiency factors of the resulting protocol are summarized in Table 4.

Table 4. Efficiency of instantiation of CP$_{\text{ILC}}$ using AHI$^+$17 [2]

Prover Computation	$\mathcal{O}(N)\ sym + \mathcal{O}(1)\ exp$
Verifier Computation	$\mathcal{O}(N)\ add + \mathcal{O}(\tilde{N})\ exp$
Communication	$\text{poly}(\lambda)\sqrt{N}\,\mathbb{Z}_p + \mathcal{O}(1)\,\mathbb{G}$
Round Complexity	$\mathcal{O}(\log\log N)$

4 Implementation

We implement $\mathsf{CP}_{\mathsf{ILC}}$ in the ILC model in C++. for field/group operations we use the libraries underlying `libsnark`[3]. We executed our experiments on Ubuntu 20.04. All tests are run single-threaded. To simulate resource-limited IoT devices, tests were run with a maximum memory usage of 512 MB[4].

We note that the efficiency of CP-ZKP largely depends on the underlying ZKP protocol on which they are built. Many existing CP-ZKP schemes leverage zk-SNARK constructions. For instance, LegoGro16 [10] is built on the classic Groth16 scheme [20], in which the prover performs a linear number of group exponentiations. Since our $\mathsf{CP}_{\mathsf{ILC}}$ protocol is based on Π_{ILC}, we initially compared the overheads of Groth16 and Π_{ILC}[5] and present the results in Table 5, where Π_{ILC} achieves a proving time that is roughly 10 times faster than Groth16.

Table 5. Performance comparison of Groth16 and Π_{ILC}. Memory cap is 512 MB. Numbers for verification of the two schemes are in different units. Those for the verification of Π_{ILC} are three orders of magnitude larger.

#multiplication gates	Prove		Verify	
	Groth16 (s)	Π_{ILC} (s)	Groth16 (ms)	Π_{ILC} (ms)
3×10^4	2.9	0.30	2.3	150
6×10^4	6.1	0.59	2.3	290
12×10^4	11.5	1.1	2.3	600

We consider a generic application of proving cross-domain statement where commitments are created using the Pedersen scheme, i.e., for public circuit C, public commitment c, public commitment generator g and h, proving "$\exists (x, r) : c = g^x h^r \wedge y = C(x)$". We conduct experiments under two settings: (1) fixing the circuit size while increasing the number of committed inputs, and (2) keeping the number of committed inputs constant while scaling the circuit size. Additionally, we evaluate Π_{ILC} as a baseline for case (1) under the same conditions. Since Π_{ILC} supports circuits but not algebraic commitments, we represent the cross-domain statement using only circuit gates. As a result, the equivalent circuit C' for Π_{ILC} requires additional gates to simulate commitment operations compared to C. Specifically, representing a single commitment operation required approximately 1.5×10^4 multiplication gates and 1.8×10^4 addition gates using the BN128 curve

---

[3] https://github.com/scipr-lab/libsnark.
[4] Some industrial IoT gateway devices may have several hundred megabytes. For example, Siemens' SIMATIC IOT2050 is available in versions with 1024 MB of memory (https://mall.siemens.com.cn/pcweb/detailIndex/2100012713.html).
[5] Recent progress on zk-SNARKs [13,21] focuses on improvements such as updatable common reference strings, more universal assumptions, or publicly generated parameters during setup. Nonetheless, the underlying proof computation still involves a linear number of group exponentiations, similar to Groth16.

in libsnark. The results are summarized in the following tables and depicted in charts for clearer visualization.

In the first case, the number of gates is fixed at 2.5×10^4. $\mathsf{CP}_{\mathsf{ILC}}$ maintained relatively stable proving and verification times despite the increase in committed inputs. In stark contrast, for Π_{ILC}, when the number of committed inputs exceeds 4, the proving and verification times exhibit a super-linear increase. This trend might stem from memory usage exceeding the limit of 512 MB. Specifically, for a circuit with 25,000 gates approximating the SHA256 scenario "$\exists (m,r) : c = g^m h^r \wedge y = \mathrm{SHA256}(m)$" (SHA256 is about 2.5×10^4 gates and has one committed input m), $\mathsf{CP}_{\mathsf{ILC}}$ demonstrated a 59% reduction in proving time (179.8 vs. 439.6 ms) and a 55% reduction in verification time (102.1 vs. 226.5 ms) compared to Π_{ILC} (Table 6).

Table 6. Case (1) Performance.

#committed inputs	$\mathsf{CP}_{\mathsf{ILC}}$			Π_{ILC}		
	$|C|(\times 10^4)$	$\mathcal{P}$ (ms)	$\mathcal{V}$ (ms)	$|C'|(\times 10^4)$	$\mathcal{P}$ (s)	$\mathcal{V}$ (s)
1	2.5	179.8	102.1	5.8	0.4396	0.2265
2	2.5	182.1	103.6	9.1	0.6552	0.3536
4	2.5	183.0	102.5	15.7	1.033	0.6133
8	2.5	185.5	102.4	28.9	9.334	5.266
16	2.5	183.5	103.4	55.3	30.95	11.41
32	2.5	185.1	102.9	108.1	96.20	34.02

In the second case, we tested the proving and verification efficiency of the $\mathsf{CP}_{\mathsf{ILC}}$ scheme as the circuit size increased, with memory limits set to 128 MB and 512 MB. For circuits up to 5×10^4 gates (or 15×10^4 gates with 512 MB), the proving and verification times scaled linearly with circuit size. Beyond this point, performance exhibited non-linear growth, indicating memory constraints and defining the maximum circuit size $\mathsf{CP}_{\mathsf{ILC}}$ can handle under specific memory limits. This behavior is expected. Zero-knowledge proving/verification inherently involves reading the statement (in this case, the circuit for $\mathsf{CP}_{\mathsf{ILC}}$), and as its size approaches or exceeds available memory, proving/verification naturally slows down (Table 7).

5 Conclusion

This paper introduces $\mathsf{CP}_{\mathsf{ILC}}$, a novel transparent CP-ZKP constructed from Π_{ILC} and $\mathsf{CP}_{\mathsf{link}}$. $\mathsf{CP}_{\mathsf{ILC}}$ enables efficient proofs for statements of the form $\exists (x,r) : c = g^x h^r \wedge y = C(x)$, making it applicable to various cross-domain scenarios, such as confidential cryptographic auditing using standard signatures (e.g., RSA or ECDSA). In $\mathsf{CP}_{\mathsf{ILC}}$, proving costs are dominated by field multiplications proportional to the circuit size, while verification costs are driven by field

Table 7. Case (2) Performance: The number of committed inputs is fixed at 32.

#gates ($\times 10^4$)	128 MB		512 MB	
	$\mathcal{P}$ (s)	$\mathcal{V}$ (s)	$\mathcal{P}$ (s)	$\mathcal{V}$(s)
0.5	0.0465	0.0244	0.0457	0.0244
1.5	0.1086	0.0631	0.1080	0.0659
2.5	0.1875	0.1024	0.1820	0.1019
5.0	0.7052	0.6409	0.3656	0.1959
7.5	2.562	1.546	0.5457	0.2890
10	4.921	2.480	0.6605	0.3875
12.5	8.591	3.147	0.9178	0.4773
15	12.68	4.723	1.007	0.5767
30	34.70	11.33	8.173	5.229
60	94.51	29.14	32.88	13.79
90	170.9	51.76	63.85	25.71

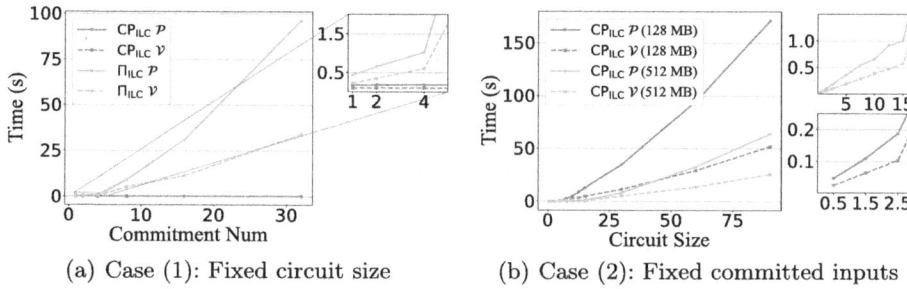

(a) Case (1): Fixed circuit size   (b) Case (2): Fixed committed inputs

Fig. 3. Performance of $\mathsf{CP_{ILC}}$ and Π_{ILC} in different scenarios. We enlarged the details of the images to facilitate observation of the changing trends in time costs.

additions, also linear in circuit size. Communication remains sublinear. These efficiency trade-offs make $\mathsf{CP_{ILC}}$ more suitable for IoT environments with limited computational resources compared to CP-ZKPs based on zk-SNARKs, which is supported by the comparison between Π_{ILC} and Groth16. Additionally, the comparison between $\mathsf{CP_{ILC}}$ and Π_{ILC} confirms our contribution to cross-domain statements in resource-constrained environments (Fig. 3).

Experimental results also reveal that under memory constraints, $\mathsf{CP_{ILC}}$ faces limitations in the size of provable statements. This is expected, as zero-knowledge proving inherently requires reading the entire statement. To advance the applicability of zero-knowledge proofs in IoT, future work could focus on optimizing memory usage for both proof generation and verification.

Acknowledgments. This work is supported by National Natural Science Foundation of China (72442029, U24B20149), Shenzhen Science and Technology Major Project

(KJZD20230923114908017), Guangdong Provincial Key Laboratory of Novel Security Intelligence Technologies (2022B1212010005), Shenzhen Science and Technology Program (KJZD20240903104301003), the Key Research and Development Program of Shaanxi (2024GX-ZDCYL-01-09), and Major Program of Shandong Provincial Natural Science Foundation for the Fundamental Research (ZR2022ZD03).

A Security Proof of Theorem 1

We first introduce a variation of the Schwarz-Zippel Lemma.

Lemma 1. *Let $\mathbb{F}$ be a finite field. Suppose P is a function in the form of (4), where p_{0,i_0} is a constant and $p_{1,i_1}(Z_0), \ldots, p_{u,i_u}(Z_0, \ldots, Z_{u-1})$ are arbitrary functions and are not necessarily polynomial functions.*

$$P(Z_0, \ldots, Z_u) = \sum_{i_0=-d_0}^{d_0} p_{0,i_0} Z_0^{i_0} + \sum_{i_1=-d_1, i_1 \neq 0}^{d_1} p_{1,i_1}(Z_0) Z_1^{i_1}$$
$$+ \ldots + \sum_{i_u=-d_u, i_u \neq 0}^{d_u} p_{u,i_u}(Z_0, \ldots, Z_{u-1}) Z_u^{i_u} \quad (4)$$

Let S be a finite subset of $\mathbb{Z}_p^\times$. Let $z_0, \ldots, z_u$ be selected at random independently and uniformly from S. Let $\mathbf{F}$ be the event that "at least one value among p_{0,i_0} or $p_{s,i_s}(z_0, \ldots, z_{s-1})$ for $s \in [1, u]$, is not zero". Then

$$\Pr\left[P(z_0, \ldots, z_u) = 0 \wedge \mathbf{F}\right] \leq \frac{\sum_{t=0}^u (2d_t + 1)}{|S|} \quad (5)$$

The detailed proof of Lemma 1 can be found in [7].

Proof. **Perfect completeness** follows by careful inspection of the polynomial expressions computed by the prover in the above protocol and the additive homomorphic property of Pedersen commitment scheme.

The straight-line extraction proves the *computational knowledge soundness* of $A, B, \hat{\mathbf{a}}_0, \hat{\mathbf{b}}_0, e_a, e_b, \{f_t^+, f_t^-\}_{t \in [0, \mu-1]}, \{f_t^{+'}, f_t^{-'}\}_{t \in [0, \mu-1]}, \{g_s, g_s'\}_{s \in [-n, n-1], s \neq 0}$. This is demonstrated by the knowledge extractor's access to the matrices submitted and all messages exchanged between $\mathcal{P}_{\mathsf{link}}$ and the ILC. It remains to be shown that for any deterministic malicious prover $\mathcal{P}_{\mathsf{link}}^*$, if the committed vectors are invalid for $\mathcal{R}_{\mathsf{link}}$, the acceptance probability is negligible.

By assumption $\mathcal{P}_{\mathsf{link}}^*$ is deterministic, and we know when it made it's commitments. Hence, $\hat{\mathbf{a}}, \hat{\mathbf{b}}, \hat{e}_a$, and $\hat{e}_b$ are constants; $f_0^+, f_0^-, f_0^{+'}, f_0^{-'}$ depend on y; $f_1^+, f_1^-, f_1^{+'}, f_1^{-'}$ on y and x_0; and so forth. Likewise, g_s and g_s' are functions of $y, x_0, \ldots, x_{\mu-1}$. If an inconsistency exists between an input $A_{r,c}$ and its corresponding commitment $C_{r,c}$, the coefficients of $y^{(r-1)k+c}$ in the exponents of g in Eq. (2) differ, implying the event $\mathbf{F}$ in Lemma 1 occurs. By Lemma 1, the probability of the exponents of g being equal is negligible. Moreover, the binding property of the Pedersen commitment ensures the negligible probability of one commitment opening to two different messages, thus the probability of (2) holding is negligible.

SHVZK refers to the property that given a set of challenges from $\mathcal{V}_{\mathsf{link}}$, it is possible to simulate a proof to answer the challenges without knowing the witness. Given the random values y, x_0, ..., $x_{\mu-1}$, and x from $\mathbb{Z}_p^\times$ used in the protocol, $\hat{\mathbf{w}}$ can be computed. $\mathcal{P}_{\mathsf{link}}$ could just randomly choose $\hat{\mathbf{a}}_0$ and $\hat{\mathbf{b}}_0$, adds them to other random values to obtain $\hat{\mathbf{a}}$ and $\hat{\mathbf{b}}$, and randomly selects e_a, e_b, $\hat{e}_a$, and $\hat{e}_b$. Finally, there is only E left to be simulated by the simulator $\mathcal{S}$, which implies following computation.

$$E = \left(\prod_{r=1}^{mn} \prod_{c=1}^{k} C_{r,c}^{y^{(r-1)k+c}} \right) \left(g^{\hat{\mathbf{a}}\hat{\mathbf{w}}^T - \hat{e}_a} h^{\hat{\mathbf{b}}\hat{\mathbf{w}}^T - \hat{e}_b} \right)^{-1} \tag{6}$$

For an accepting transcript this is easy to compute given all the values on the right-hand side. Therefore, we can simulate the transcript and the proof system $\mathsf{CP}_{\mathsf{link}}$ has special honest verifier zero knowledge.

References

1. Agrawal, S., Ganesh, C., Mohassel, P.: Non-interactive zero-knowledge proofs for composite statements. In: Shacham, H., Boldyreva, A. (eds.) CRYPTO 2018. LNCS, vol. 10993, pp. 643–673. Springer, Cham (2018). https://doi.org/10.1007/978-3-319-96878-0_22
2. Applebaum, B., Haramaty, N., Ishai, Y., Kushilevitz, E., Vaikuntanathan, V.: Low-complexity cryptographic hash functions. Cryptology ePrint Archive (2017)
3. Aranha, D.F., Bennedsen, E.M., Campanelli, M., Ganesh, C., Orlandi, C., Takahashi, A.: Eclipse: enhanced compiling method for Pedersen-committed Zksnark engines. In: IACR International Conference on Public-Key Cryptography, pp. 584–614 (2022)
4. Backes, M., Hanzlik, L., Herzberg, A., Kate, A., Pryvalov, I.: Efficient non-interactive zero-knowledge proofs in cross-domains without trusted setup. In: IACR International Workshop on Public Key Cryptography, pp. 286–313 (2019)
5. Bhadauria, R., Fang, Z., Hazay, C., Venkitasubramaniam, M., Xie, T., Zhang, Y.: Ligero++: a new optimized sublinear IOP. In: Proceedings of the 2020 ACM SIGSAC Conference on Computer and Communications Security, pp. 2025–2038 (2020)
6. Blum, M., Feldman, P., Micali, S.: Non-interactive zero-knowledge and its applications. In: Proceedings of the Twentieth Annual ACM Symposium on Theory of Computing, pp. 103–112 (1988)
7. Bootle, J., Cerulli, A., Ghadafi, E., Groth, J., Hajiabadi, M., Jakobsen, S.K.: Linear-time zero-knowledge proofs for arithmetic circuit satisfiability. Cryptology ePrint Archive (2017)
8. Camenisch, J., Stadler, M.: Efficient group signature schemes for large groups. In: Kaliski, B.S. (ed.) CRYPTO 1997. LNCS, vol. 1294, pp. 410–424. Springer, Heidelberg (1997). https://doi.org/10.1007/BFb0052252

9. Campanelli, M., Faonio, A., Fiore, D., Querol, A., Rodríguez, H.: Lunar: a toolbox for more efficient universal and updatable zkSNARKs and commit-and-prove extensions. In: Tibouchi, M., Wang, H. (eds.) ASIACRYPT 2021. LNCS, vol. 13092, pp. 3–33. Springer, Cham (2021). https://doi.org/10.1007/978-3-030-92078-4_1
10. Campanelli, M., Fiore, D., Querol, A.: LegoSNARK: modular design and composition of succinct zero-knowledge proofs. In: Proceedings of the 2019 ACM SIGSAC Conference on Computer and Communications Security, pp. 2075–2092 (2019)
11. Chase, M., et al.: Post-quantum zero-knowledge and signatures from symmetric-key primitives. In: Proceedings of the 2017 ACM SIGSAC Conference on Computer and Communications Security, pp. 1825–1842 (2017)
12. Chase, M., Ganesh, C., Mohassel, P.: Efficient zero-knowledge proof of algebraic and non-algebraic statements with applications to privacy preserving credentials. In: Robshaw, M., Katz, J. (eds.) CRYPTO 2016. LNCS, vol. 9816, pp. 499–530. Springer, Heidelberg (2016). https://doi.org/10.1007/978-3-662-53015-3_18
13. Daza, V., Ràfols, C., Zacharakis, A.: Updateable inner product argument with logarithmic verifier and applications. In: Kiayias, A., Kohlweiss, M., Wallden, P., Zikas, V. (eds.) PKC 2020, Part I. LNCS, vol. 12110, pp. 527–557. Springer, Cham (2020). https://doi.org/10.1007/978-3-030-45374-9_18
14. Druk, E., Ishai, Y.: Linear-time encodable codes meeting the Gilbert-Varshamov bound and their cryptographic applications. In: Proceedings of the 5th Conference on Innovations in Theoretical Computer Science, pp. 169–182 (2014)
15. Fiat, A., Shamir, A.: How to prove yourself: practical solutions to identification and signature problems. In: Conference on the Theory and Application of Cryptographic Techniques, pp. 186–194 (1986)
16. Gabizon, A., Williamson, Z.J., Ciobotaru, O.: PLONK: permutations over Lagrange-bases for Oecumenical noninteractive arguments of knowledge. Cryptology ePrint Archive (2019)
17. Gennaro, R., Gentry, C., Parno, B., Raykova, M.: Quadratic span programs and succinct NIZKs without PCPs. In: Johansson, T., Nguyen, P.Q. (eds.) EUROCRYPT 2013. LNCS, vol. 7881, pp. 626–645. Springer, Heidelberg (2013). https://doi.org/10.1007/978-3-642-38348-9_37
18. Goldreich, O., Micali, S., Wigderson, A.: How to prove all NP statements in zero-knowledge and a methodology of cryptographic protocol design (extended abstract). In: Odlyzko, A.M. (ed.) CRYPTO 1986. LNCS, vol. 263, pp. 171–185. Springer, Heidelberg (1987). https://doi.org/10.1007/3-540-47721-7_11
19. Goldwasser, S., Micali, S., Rackoff, C.: The knowledge complexity of interactive proof-systems. In: Proceedings of the Seventeenth Annual ACM Symposium on Theory of Computing, pp. 291–304 (1985)
20. Groth, J.: On the size of pairing-based non-interactive arguments. In: Fischlin, M., Coron, J.-S. (eds.) EUROCRYPT 2016. LNCS, vol. 9666, pp. 305–326. Springer, Heidelberg (2016). https://doi.org/10.1007/978-3-662-49896-5_11
21. Groth, J., Kohlweiss, M., Maller, M., Meiklejohn, S., Miers, I.: Updatable and universal common reference strings with applications to ZK-snarks. IACR Cryptol. ePrint Arch. (2018)
22. Groth, J., Sahai, A.: Efficient non-interactive proof systems for bilinear groups. In: Smart, N. (ed.) EUROCRYPT 2008. LNCS, vol. 4965, pp. 415–432. Springer, Heidelberg (2008). https://doi.org/10.1007/978-3-540-78967-3_24
23. Guillou, L.C., Quisquater, J.-J.: A practical zero-knowledge protocol fitted to security microprocessor minimizing both transmission and memory. In: Barstow, D.,

et al. (eds.) EUROCRYPT 1988. LNCS, vol. 330, pp. 123–128. Springer, Heidelberg (1988). https://doi.org/10.1007/3-540-45961-8_11
24. Guruswami, V., Indyk, P.: Linear-time encodable/decodable codes with near-optimal rate. IEEE Trans. Inf. Theory **51**(10), 3393–3400 (2005)
25. Ishai, Y., Kushilevitz, E., Ostrovsky, R., Sahai, A.: Cryptography with constant computational overhead. In: Proceedings of the Fortieth Annual ACM Symposium on Theory of Computing, pp. 433–442 (2008)
26. Jawurek, M., Kerschbaum, F., Orlandi, C.: Zero-knowledge using garbled circuits: how to prove non-algebraic statements efficiently. In: Proceedings of the 2013 ACM SIGSAC Conference on Computer & Communications Security, pp. 955–966 (2013)
27. Jinguo, L., Yaping, L., Rui, L.: Secure anonymous authentication scheme based on elliptic curve and zero-knowledge proof in vanet. J. Commun. **34**(5), 52–61 (2013)
28. Lipmaa, H.: On black-box knowledge-sound commit-and-prove snarks. In: International Conference on the Theory and Application of Cryptology and Information Security, pp. 41–76 (2023)
29. Parno, B., Gentry, C., Howell, J., Raykova, M.: Pinocchio: nearly practical verifiable computation. Cryptology ePrint Archive (2013)
30. Raymond, M., Evers, G., Ponti, J., Krishnan, D., Fu, X.: Efficient zero knowledge for regular language. Cryptology ePrint Archive (2023)
31. Schnorr, C.P.: Efficient identification and signatures for smart cards. In: Brassard, G. (ed.) CRYPTO 1989. LNCS, vol. 435, pp. 239–252. Springer, New York (1990). https://doi.org/10.1007/0-387-34805-0_22
32. Spielman, D.A.: Linear-time encodable and decodable error-correcting codes. In: Proceedings of the Twenty-seventh Annual ACM Symposium on Theory of Computing, pp. 388–397 (1995)
33. Zhang, M., Chen, Y., Yao, C., Wang, Z.: Sigma protocols from verifiable secret sharing and their applications. Cryptology ePrint Archive (2023)

Public Verifiable Server-Aided Revocable Attribute-Based Encryption

Luqi Huang[✉], Fuchun Guo, Willy Susilo, and Yumei Li

Institute of Cybersecurity and Cryptology, School of Computing and Information Technology, University of Wollongong, Wollongong, Australia
`lh852@uowmail.edu.au, {fuchun,wsusilo}@uow.edu.au, leamergo@gmail.com`

Abstract. Revocable Attribute-Based Encryption (RABE) provides a user revocation mechanism to keep the system dynamic and protect data privacy when a user's credentials can be expired or revealed. In verifiable server-aided RABE, where most computations are delegated to an untrusted server, users with private keys and outsourced results can decrypt the ciphertext and then verify message correctness, provided they meet the decryption conditions. This verification process, commonly known as private verification, follows a "decrypt-then-verify" paradigm and is inherently restricted to legitimate users. However, such a model lacks transparency and accountability, as it depends solely on user-side validation. This limited verifiability may undermine trust in the server's behavior, exposing the system to disputes and potential misuse, such as malicious users falsely claiming incorrect computations, thereby leading to operational and financial risks. To address these challenges, we propose an improved server-aided RABE system that enables public verification, allowing any party to verify the untrusted server's computations rather than entirely relying on users. Furthermore, we introduce a "verify-then-decrypt" mechanism, ensuring the correctness of outsourced results before decryption. This approach prevents users from performing redundant or incorrect decryption, thereby improving efficiency and system reliability.

Keywords: Attribute-based encryption · Public verification · Outsourcing computation · User revocation

1 Introduction

ABE plays a key role in enabling fine-grained and scalable access control systems, offering flexibility of managing data access in cloud computing. To maintain privacy and system integrity, RABE enforces periodic key updates to revoke users' private keys when they lose their keys or leave the cloud system. Furthermore, server-aided RABE extends the server's role beyond data storage to include key update processing and outsourced decryption. The common assumption that the fully trustworthy server is often unrealistic in practical scenarios, making verification of the server's actions essential to maintain system security and stability.

Traditional ABE systems with verifiable outsourcing in [9,11,18,19] employ a "decrypt-then-verify" mechanism, where authorized users with private keys first decrypt ciphertexts using their private keys and outsourced results to retrieve the original message and then verify its correctness to ensure the accuracy of the server's computation. However, this private verification approach inherently restricts validation to authorized users with private keys, excluding third parties from confirming whether the server behaved correctly.

This single, vulnerable verification system creates opportunities for malicious user behavior. For example, a user with a valid private key could falsely claim that the server produced incorrect results to seek additional compensation, even when the outsourced computations were accurate. Since no other entities, including the server itself, can prove the correctness of outsourced results, such malicious defamation undermines the credibility of the server—particularly in cloud computing, where trust and reliability are critical.

To address the above concerns, [7,12,20] delegate the verification process of outsourced results to more trusted third-party auditors, such as private clouds or certification authorities. While this reduces dependence on end-users, the verification process still relies on private keys held by these entities. As a result, fully transparent and credibility verification cannot be achieved, and the risk of false claims against server computations persists. Therefore, enabling fully public verification of outsourced results on untrusted servers remains a significant challenge.

Generic Publicly Verifiable Computation (PVC) [13] provide a promising alternative, enabling lightweight clients to outsource computations to untrusted servers and verify the results independently. In 2015, James et al. [1] proposed a hybrid PVC scheme to verify outsourced computations in the ABE system. Their approach compares $H(m)$, the hash value of the outsourced result with the public verification key, where H is a hash function and m is the message, to confirm the correctness of the server's computations.

However, the hybrid PVC scheme presented in [1] has a critical limitation: the outsourced result $H(m)$ computed from the untrusted server, remains constant across verifications aiming at different authorized users. This means that once the server computes $H(m)$, it can repeatedly reuse this static hash value for all subsequent verifications. A robust verification mechanism should ensure that each verification attempt is unique and independently validates the server's current computations. Therefore, a more dynamic approach is needed-one that ensures outsourced results vary with each verification and across different authorized users.

In summary, while private verification in ABE systems lacks transparency and may result in redundant decryption on incorrect outputs, third-party auditing does not fully resolve credibility concerns. Furthermore, existing PVC integrations yield static, non-user-specific outsourced results. Achieving fully public, dynamic, and efficient verification of outsourced computations in server-aided RABE remains a challenging and open problem.

1.1 Our Contribution

In this paper, we present a Public Verifiable Server-Aided Revocable Attribute-Based Encryption (PV-SR-ABE) scheme that enables any party to verify the correctness of outsourced computations from an untrusted server. This public verification capability significantly enhances system transparency and mitigates the risks of fraud and server misconduct. Importantly, the verification process incurs only minimal computational overhead, making it suitable for practical deployment.

The PV-SR-ABE adopts a "verify-then-decrypt" paradigm, which ensures that the correctness of outsourced results is validated before decryption. This approach prevents users from redundantly decrypting incorrect or manipulated outputs, thereby improving efficiency and reliability. Additionally, our scheme guarantees that each verification generates user-specific results rather than static outsourced outputs, enhancing the system's robustness and security.

We formally prove the security of our scheme against chosen selective plaintext attacks (IND-sCPA) of user revocation under the one-user setting security model described in [15]. Moreover, we demonstrate selective public verifiability of the outsourced computations based on the Verifiable Delegable Computation (VDC) security model in [1], ensuring that our scheme meets robust security standards for verification processes.

1.2 Related Work

In 2004, Sahai and Waters proposed ABE in [17]. Since then, numerous variants have been developed to enhance flexibility in fine-grained access control. Typically, ABE schemes are categorized into ciphertext-policy ABE (CP-ABE) and key-policy ABE (KP-ABE). In CP-ABE, a user's attribute private key is linked to an attribute set, while a ciphertext specifies an access policy defined over the attribute universe of the system. Conversely, in KP-ABE, an access policy is embedded in a user's attribute private key, and a ciphertext is generated in relation to an attribute set. The first KP-ABE and CP-ABE schemes were introduced by Goyal et al. and Bethencourt et al. in [8] and [2], respectively.

However, a fundamental drawback of traditional ABE schemes is their high computational cost, which typically increases with the complexity of the access policy. To address this issue, Green et al. [9] introduced ABE with outsourced decryption, introducing a modular architecture that decomposes the decryption process into two stages: a computationally intensive preprocessing phase delegated to an untrusted proxy and a lightweight final decryption performed by the end user. After the proxy performs the heavy computation, the user obtains a simplified ciphertext that can be easily decrypted.

To preserve data privacy and ensure system integrity, revocable ABE schemes [6,10] have been developed, which enable the revocation of a user's private key when it is compromised or the user leaves the system. In 2016, Cui et al. [6] proposed a revocable CP-ABE scheme in which an untrusted server assists non-revoked users by transforming ciphertexts. Later, Qin et al. [15] identified a vulnerability related to decryption key exposure in [6], highlighting the risk of unauthorized access when revocation status is not properly enforced. Then Cheng et al. [5] proposed the first server-aided revocable ABE scheme proven secure in the multi-user setting. This advancement strengthened the practical applicability of revocable ABE in large-scale systems.

In the above server-aided revocable ABE schemes, whether the transformation process is correctly proceeded by the untrusted server that cannot be verified. To address this, Yang et al. [19] proposed a generic construction of verifiable SR-ABE that enables users with private keys to check whether the server behaves as expected. Additionally, Xu et al. [18] presented an adaptively secure scheme with verifiable outsourced decryption but not for user revocation.

2 Preliminaries

2.1 Bilinear Maps

Our design is based on some facts about groups with efficiently computable bilinear maps.

Let $\mathbb{G}$ and $\mathbb{G}_T$ be two multiplicative cyclic groups of prime order p. Let g be a generator of $\mathbb{G}$. A bilinear map is an injective function $e : \mathbb{G} \times \mathbb{G} \to \mathbb{G}_T$ with the following properties:

1. **Bilinearity:** For all $u, v \in \mathbb{G}$ and $a, b \in \mathbb{Z}_p$, we have $e(u^a, v^b) = e(u, v)^{ab}$.
2. **Non-degeneracy:** $e(g, g) \neq 1$.
3. **Computability:** There is an efficient algorithm to compute $e(u, v)$ for $\forall u, v \in \mathbb{G}$.

2.2 Algorithm Definition

Based on the algorithm definition in the [15], we add the Verify algorithm for public verifiability of outsourced results. Our PV-SR-ABE scheme involves five entities: a private key generator (PKG), data owners, data users, public verifiers and an untrusted server. The PV-SR-ABE scheme that is associated with the identity space $\mathcal{I}$, the time period space $\mathcal{T}$ and the message space $\mathcal{M}$, consists of nine algorithms are defined as follows:

- **Setup**$(\lambda, \mathcal{U}, \mathcal{T}) \to (mpk, msk)$. The setup algorithm is executed by the PKG and takes as input the security parameter λ, the universal attribute set $\mathcal{U}$ and the time period space $\mathcal{T}$. It outputs the master public key, master secret key (mpk, msk).
- **UserKG**$(mpk, msk, ID, S) \to (sk_{ID}, vk_{ID}, psk_{ID,S}, st)$. The user key generation algorithm is executed by the PKG and takes as input the master pair keys (mpk, msk), identity ID and the attribute set S. It returns the private key sk_{ID}, public verification key vk_{ID}, long-term transformation key $psk_{ID,S}$, and the updated state st.

- **TKeyUp**$(mpk, msk, R, t) \rightarrow (tuk_t, st)$. The transformation key update algorithm is executed by the PKG and takes as input the master pair keys (mpk, msk), the revocation list R and time period t. It returns the update message key tuk_t and the updated state st.
- **TranKG**$(mpk, psk_{ID,S}, tuk_t) \rightarrow tk_{ID,t}^S / \perp$. The transformation key generation algorithm is executed by the server and takes as input the master public key mpk, the long-term transformation key $psk_{ID,S}$ and key update message tuk_t. It returns the short-term transformation key $tk_{ID,t}^S$ if $ID \notin R$.
- **Encrypt**$(mpk, (\mathbb{M}, \rho), t', M) \rightarrow (Hdr, c)$. The encryption algorithm is executed by a data owner and takes as input the master public key mpk, the access policy $(\mathbb{M}, \rho)$, the time period t' and a message M. It returns the header Hdr and the ciphertext c.
- **Transform**$(mpk, c, tk_{ID,t}^S, ID) \rightarrow \pi$. The transformation algorithm is executed by the server and takes as input master public key mpk, the ciphertext c, short-term transformation key $tk_{ID,t}^S$ and identity ID. It outputs the transformed ciphertext π if S satisfies the access policy $(\mathbb{M}, \rho)$, i.e., $S \in \mathbb{A}$ and $t' = t$.
- **Verify**$(mpk, \pi, c, vk_{ID}, ID) \rightarrow 0/1$. The verify algorithm is executed by a verifier and takes as input the master public key mpk, the transformed ciphertext π, the ciphertext c, the public verification key vk_{ID}, and identity ID. It outputs the verification result 1 if the transformation was indeed done correctly.
- **Decrypt**$(mpk, sk_{ID}, \pi, Hdr, ID) \rightarrow M/\perp$. The decryption algorithm is executed by a data user and takes as input master public key mpk, the private key sk_{ID}, the transformed ciphertext π, the header Hdr and identity ID. It outputs the message M if π is well-formed.
- **Revoke**$(ID, t, R, st) \rightarrow R$. The revoke algorithm is executed by the PKG and takes as inputs the identity ID, time period t, revocation list R and the state st. If the user's identity ID is revoked at time period t, this algorithm adds (ID, t) to R and outputs the updated revocation list R.

Correctness: We require the correctness of PV-SA-ABE are as follows:

- **TranKG Correctness:** For any $(mpk, msk) \leftarrow \text{Setup}(\lambda, \mathcal{U}, \mathcal{T})$, and $(sk_{ID}, vk_{ID}, psk_{ID,S}, st) \leftarrow \text{UserKG}(mpk, msk, ID, S)$, $(tuk_t, st) \leftarrow \text{TKeyUp}(mpk, msk, R, t)$, if $ID \notin R$, the TranKG algorithm always outputs the correct short-term transformation key $tk_{ID,t}^S$, otherwise, the $tk_{ID,t}^S$ can not be derived from $psk_{ID,S}$ and tuk_t.
- **Verify Correctness:** For any $(mpk, msk) \leftarrow \text{Setup}(\lambda, \mathcal{U}, \mathcal{T})$, $(sk_{ID}, vk_{ID}, psk_{ID,S}, st) \leftarrow \text{UserKG}(mpk, msk, ID, S)$, $(Hdr, c) \leftarrow \text{Encrypt}(mpk, (\mathbb{M}, \rho), t', M)$ and $\pi \leftarrow \text{Transform}(mpk, c, tk_{ID,t}^S, ID)$, if the transformation by the server was indeed done correctly, the Verify algorithm always outputs the verification result 1, otherwise outputs 0.
- **Decrypt Correctness:** For any $(mpk, msk) \leftarrow \text{Setup}(\lambda, \mathcal{U}, \mathcal{T})$, $(sk_{ID}, vk_{ID}, psk_{ID,S}, st) \leftarrow \text{UserKG}(mpk, msk, ID, S)$ and $(Hdr, c) \leftarrow \text{Encrypt}(mpk, (\mathbb{M}, \rho), t', M)$, $\pi \leftarrow \text{Transform}(mpk, c, tk_{ID,t}^S, ID)$, if π is well-formed, the Decrypt algorithm always outputs the correctness message M, otherwise, the message cannot be decrypted using sk_{ID}.

2.3 Security Model

We adopt the security notion of *indistinguishability against chosen selective plaintext attacks* (IND-sCPA) for PV-SR-ABE from the security model in [15] for user revocation. The following game is played between an adversary $\mathcal{A}_1$ and the challenger $\mathcal{C}$.

- **Initialization.** $\mathcal{C}$ chooses an attribute universal set $\mathcal{U}$ and the time set $\mathcal{T}$. $\mathcal{A}_1$ declares access policy $\mathbb{A}^*$ and a time period t^*, which it will try to attack, and sends them to $\mathcal{C}$.
- **Setup.** $\mathcal{C}$ runs setup algorithm and gives the public parameter to $\mathcal{A}_1$. $\mathcal{C}$ keeps the master secret key msk, an initially empty revocation R and a state st.
- **Phase 1.** $\mathcal{A}_1$ adaptively issues the following queries to $\mathcal{C}$.
  - Create(ID, S): $\mathcal{C}$ runs user key generation algorithm on (ID, S) to obtain the pair $(sk_{ID}, vk_{ID}, psk_{ID,S})$ and stores in table T the entry $(ID, S, sk_{ID}, vk_{ID}, psk_{ID,S})$. It returns to $\mathcal{A}_1$ the long-term transformation key $psk_{ID,S}$ and the verification key vk_{ID}.
  - Corrupt(ID): If there exists an empty indexed by ID in table T, then $\mathcal{C}$ obtains the entry $(ID, S, sk_{ID}, vk_{ID}, psk_{ID,S})$ and sets $D = D \cup \{(ID, S)\}$. It returns to $\mathcal{A}_1$ the private key sk_{ID}. If no such entry exists, then it returns $\bot$.
  - TKeyUp(t): If $\mathcal{A}_1$ issues a key update query on a time period t, $\mathcal{C}$ runs the transformation key update algorithm TKeyUp(mpk, msk, t, R) and gives the update message key tuk_t to $\mathcal{A}_1$.
  - Revocation(ID, t): When $\mathcal{A}_1$ issues a revocation query on an identity ID and a time period t, $\mathcal{C}$ runs Revoke(ID, t, R, st) and outputs an updated revocation list R.
- **Challenge.** $\mathcal{A}_1$ submits two equal length messages M_0, M_1, an access policy $\mathbb{A}^*$ and a time period t^* satisfying the following constraints (Suppose that the attribute set associated with ID^* is S^*).
  - If there exists a tuple $(ID^*, S^*) \in D$ and $S^* \in \mathbb{A}^*$, then $\mathcal{A}_1$ must query the revocation oracle on (ID^*, t) at or before time period t^*.
  - If there exists a tuple $(ID^*, S^*, sk_{ID^*}, vk_{ID^*}, psk_{ID^*, S^*}) \in T$, $S^* \in \mathbb{A}^*$ and ID^* is not revoked at or before time period t^*, then $\mathcal{A}_1$ cannot query the private key on ID^*.

  $\mathcal{C}$ picks a random bit $b \in \{0, 1\}$ and sends the challenge header and challenge ciphertext (Hdr^*, c^*) to $\mathcal{A}_1$.
- **Phase 2.** $\mathcal{A}_1$ continues issuing queries to $\mathcal{C}$ as in Phase 1, with the same restrictions defined in the challenge phase.
- **Guess.** $\mathcal{A}_1$ makes a guess b' for b and wins the game if $b' = b$. The advantage of $\mathcal{A}_1$ in this game is defined as $|\Pr[b' = b] - 1/2|$.

Definition 1. The PV-SR-ABE scheme is IND-sCPA secure if all polynomial-time adversaries have at most a negligible advantage in the game defined above.

Here we discuss the security notion *selective public verifiability* based on the security definition when the mode setting as Verifiable Delegable Computation (VDC) in [1]. The following game is played between an adversary $\mathcal{A}_2$ and the challenger $\mathcal{C}$.

- **Initialization.** $\mathcal{C}$ chooses an attribute universal set $\mathcal{U}$ and the time set $\mathcal{T}$. $\mathcal{A}_2$ declares access policy $\mathbb{A}^*$ and a time period t^*, which it will try to attack, and sends them to $\mathcal{C}$.
- **Setup.** $\mathcal{C}$ runs the setup algorithm and obtains public parameters to $\mathcal{A}_2$. $\mathcal{C}$ keeps the master secret key to responds to queries from $\mathcal{A}_2$.
- **Challenge.** $\mathcal{C}$ sends the challenge header and ciphertext (Hdr^*, c^*) to $\mathcal{A}_2$.
- **Query.** $\mathcal{A}_2$ adaptively issues the following queries to $\mathcal{C}$.
    - Create(ID, S): $\mathcal{C}$ runs user key generation algorithm on (ID, S) to obtain pair $(sk_{ID}, vk_{ID}, psk_{ID,S})$. It stores in table T the entry $(ID, S, sk_{ID}, vk_{ID}, psk_{ID,S})$ and then returns to $\mathcal{A}_2$.
    - TKeyUp(t): If $\mathcal{A}_2$ issues a key update query on a time period t, $\mathcal{C}$ runs the transformation key update algorithm TKeyUp(mpk, msk, t, R) and gives the update message key tuk_t to $\mathcal{A}_2$.
    - Transform(ID, S, t^*, c^*): $\mathcal{A}_2$ issues a transform query on the challenge cipheretxt c^*, identity ID with attribute set S, (ID, S) and challenge time period t^*. $\mathcal{C}$ runs the transform algorithm Transform$(mpk, c^*, tk_{ID,t^*}^S, ID)$ and gives the transformed ciphertext π to $\mathcal{A}_2$.
    - Revocation(ID, t): When $\mathcal{A}_2$ issues a revocation query on an identity ID and a time period t, $\mathcal{C}$ runs Revoke(ID, t, R, st) and outputs an updated revocation list R.
- **Forgery.** $\mathcal{A}_2$ returns a forged transformed ciphertext π^* on c^* and (ID^*, S^*, t^*) (Suppose that the attribute set associated with ID^* is S^*) and wins game if
    - π^* is valid and has not been queried in the transform query.
    - psk_{ID^*, S^*} has not been queried in the user key query.

Definition 2. The PV-SR-ABE scheme is selective public verifiability if all polynomial-time adversaries have at most a negligible advantage in the selective public verifiability game defined above.

2.4 Binary Tree

We recall the definition of binary-tree data structure in [3]. In the $KUNodes$ algorithm, we use the following notations: BT denotes a binary-tree. $root$ denotes the root node of BT. x denotes a node in the binary tree and θ emphasizes that the node x is a leaf node. The set $Path(BT, \theta)$ stands for the collection of nodes on the path from the leaf θ to the root. If x is a non-leaf node, then x_l, x_r denote the left and right child of x, respectively. The $KUNodes$ algorithm in [3] takes as input a binary tree BT, a revocation list R and a time t, and outputs the minimal set Y of nodes, such that the corresponding key update information can only be used by the non-revoked users to generate a valid short-term transformation key. Specifically, the $KUNodes$ first marks all ancestors of users that were revoked by t as revoked nodes, then outputs all the non-revoked children of revoked nodes. The description of the $KUNodes$ algorithm is as follows:

$KUNodes(BT, R, t)$:
    $X, Y \leftarrow \emptyset$;
    $\forall (\theta_i, t_i) \in R, if\ t_i \leq t,$ then add $Path(\theta_i)$ to X;
    $\forall x \in X, if\ x_l \notin X,$ then add x_l to Y, if $x_r \notin X$ then add x_r to Y;
    If $Y = \emptyset$, then add root to Y; Return Y.

3 Constructions

3.1 Discussion and Overview

Discussion. To ensure message security, private verification using "decrypt-then-verify" restricts access to only authorized users. This is because the original plaintext-associated with the verification tag-can only be recovered by users capable of decrypting the ciphertext. Furthermore, existing data integrity verification protocols in cloud computing, such as Proof of Retrievability (PoR) and Proof of Data Possession (PDP), cannot be directly applied to server-aided RABE schemes, as the proofs generated by servers support only public verification and are not designed to facilitate the decryption process.

In contrast to the above, hybrid PVC selects a random message (instead of the original message) as the outsourced result, allowing anyone to verify its correctness. However, a constant outsourced result renders different verification meaningless. Thus, achieving public verification also requires outsourced results to be dynamically changeable for different users during decryption. To address this challenge, we propose a verify-then-decrypt mechanism that produces user-specific, publicly verifiable results that also enable authorized decryption.

Overview. Our approach separates message decryption from verification, treating them as distinct processes, which ensures that the correctness of outsourced results is validated before decryption. Therefore, the "verify-then-decrypt" process prevents users from redundantly decrypting incorrect outsourced results and introduces minimal computational overhead during public verification.

Specifically, our scheme embeds two secrets, α, β, which are used for message decryption and outsourced results verification, respectively. The encryption is structured on two parallel messages: the original message M, embedded in the header Hdr and a random message M_R, embedded in the ciphertext c. Correspondingly, each user with identity ID is issued two keys: a private key sk_{ID} for decryption and a public verification key vk_{ID} for verification. These two processes correspond to different secrets and are independent of each other.

When the user's attribute set S satisfies the access policy $(\mathbb{M}, \rho)$, such that $\rho(S) = 1$, and the update time period for updates t matches the time period t' in c, the server provides a transformed ciphertext, i.e., outsourced result π. This result π is publicly available and associated with a random message M_R, without revealing any information about the original message M. Additionally, it embeds a user-specific random value r ensuring dynamic changes. The transformed ciphertext π, supports both the following process: (1) Public Verification involving β: using the public verification key vk_{ID} provided by the PKG to obtain M_R and comparing the hashed value of M_R with the checknum in ciphertext c, any entity can verify the server's computations with π. Since the randome M_R can only be derived under the correct π and M_R is not a secret value. (2) Message Decryption relating α: after successful public verification, the authorized user possessing the private key sk_{ID} can decrypt the original message M using the validated outsourced result π and the header Hdr.

3.2 Scheme

Building upon the CP-ABE construction presented in [16] and the binary tree in [14] that is used for user revocations, we present our PV-SR-ABE scheme as follows:

- **Setup**$(\lambda, \mathcal{U}, \mathcal{T})$. The setup algorithm takes as input the security parameter λ, the universal attribute set $\mathcal{U}$ and the time set $\mathcal{T}$. It then generates a bilinear map $e : \mathbb{G} \times \mathbb{G} \to \mathbb{G}_T$ where $\mathbb{G}, \mathbb{G}_T$ are two cyclic groups of prime order $p = p(\lambda)$. Choose a random generator g of group $\mathbb{G}$. Set $\mathcal{G} = (p, \mathbb{G}, \mathbb{G}_T, e, g)$. It selects randoms $\alpha, \beta \in \mathbb{Z}_p$, a hash function $H : \mathbb{G}_T \to \mathbb{G}$ and random elements $g_0, k, k_0, u, h, u_0, h_0, u_1, h_1, w, v \in \mathbb{G}$ and sets $e(g_0, k_0) = e(u_0, g)$. This algorithm initializes the empty revocation list R, the secret state $st = BT$ and keeps the master secret key $msk = (\alpha, \beta, st)$. It also outputs the master public key as

$$mpk = (\mathcal{G}, g_0, k, k_0, u, h, u_0, h_0, u_1, h_1, w, v, e(g,g)^\alpha, e(u_0,g)^\beta, H)$$

- **UserKG**(mpk, msk, ID, S). The user key generation algorithm takes as input the master public key and master secret key (mpk, msk). It also inputs the user's identity ID and attribute set S. The algorithm randomly chooses $r \in \mathbb{Z}_p$ and generates the user's private key sk_{ID} and the public verification key vk_{ID} as follows:

$$sk_{ID} = (g^\alpha (u_0^{ID} h_0)^r, g^r), vk_{ID} = g_0^{\beta + rID}$$

For the user's identity ID, it chooses an undefined leaf node θ from the binary tree BT. Next for each node $x \in Path(\theta)$, the algorithm fetches g_x from the node x. If x has not been defined, it randomly chooses $g_x \in \mathbb{G}$, and stores g_x in the node x. It also picks random exponents $r_x, r_{x,1}, \ldots, r_{x,|S|} \in \mathbb{Z}_p$. The algorithm generates long-term transformation key $psk_{ID,S}$ as follows and updates the state st.

$$psk_{ID,S} = (\{((w^{ID}k)^{r_x}/g_x)u_0^r, g^{r_x}, \{(u^{S_\tau}h)^{r_{x,\tau}} v^{-r_x}, g^{r_{x,\tau}}\}_{\tau \in S}\}_{x \in Path(\theta)})$$

- **TKeyUp**(mpk, msk, R, t). The transformation key update algorithm inputs the master public key and master secret key (mpk, msk). It also inputs the revocation list R, the update time period t. For each $x \in KUNodes(BT, R, t)$, where $KUNodes$ is the algorithm in Sect. 2.4 that outputs the corresponding key update information only from non-revoked users, the algorithm fetches g_x from x (g_x should always be predefined in the above UserKG algorithm), randomly chooses $s_x \in \mathbb{Z}_p$. The algorithm updates the state st and generates the update message key tuk_t as follows:

$$tuk_t = (\{g_x(u_1^t h_1)^{s_x}, g^{s_x}\}_{x \in KUNodes(BT,R,t)})$$

- **TranKG**$(mpk, psk_{ID,S}, tuk_t)$. The transformation key generation algorithm inputs the master public key mpk, long-term transformation key $psk_{ID,S}$ and update message key tuk_t. Suppose that θ is the leaf node storing the identity ID. Let $I = \{x : x \in Path(\theta)\}$ and $J = \{x : x \in KUNodes(BT, R, t)\}$. If $I \cap J = \emptyset$, it returns $\bot$. Otherwise, there must be exactly one node $x \in I \cap J$. The short-term transformation key $tk_{ID,t}^S$ can be computed as follows:

$$tk_{ID,t}^S = ((w^{ID}k)^{r_x} u_0^r (u_1^t h_1)^{s_x}, g^{r_x}, g^{s_x}, \{(u^{S_\tau} h)^{r_{x,\tau}} v^{-r_x}, g^{r_{x,\tau}}\}_{\tau \in S})$$

- **Encrypt**$(mpk, (\mathbb{M}, \rho), t', M)$. The encryption algorithm takes as input a message M, the master public key mpk, the access policy $(\mathbb{M}, \rho)$ with $\mathbb{M} \in \mathbb{Z}_p^{l \times n}$, $\rho : [l] \to \mathbb{Z}_p$ and the time period t'. It sets $\mathbb{A} = (\mathbb{M}, \rho)$ and selects randoms $s, s_\tau \in \mathbb{Z}_p$, a random message $M_R \in \mathcal{M}$ with same length of M. This algorithm also selects $\vec{y} = (s, y_2, \ldots, y_n) \in \mathbb{Z}_p^{n+1}$ and the shares is $\vec{\lambda} = \mathbb{M} \cdot \vec{y}$. It generates the header as $Hdr = (e(g,g)^{\alpha s} M, g^s, h_0^s)$, and ciphertext c as follows:

$$c = (e(u_0, g)^{\beta s} M_R, H(M_R)^s, g^s, k^s, k_0^s, (u_1^{t'} h_1)^s, \{w^{\lambda_\tau} v^{s_\tau}, (u^{\rho(\tau)} h)^{-s_\tau}, g^{s_\tau}\}_{\tau \in l})$$

- **Transform**$(mpk, c, tk_{ID,t}^S, ID)$. The transformation algorithm takes as input the ciphertext c, master public key mpk, short-term transformation key $tk_{ID,t}^S$ and the identity ID. This algorithm computes the constants $\{\omega_i \in \mathbb{Z}_p\}$ such that $\sum_{\rho(i) \in S} \omega_i \mathbb{M}_i = (1, 0, \ldots, 0)$. If S satisfies the access policy $(\mathbb{M}, \rho)$, i.e. $\mathbb{A}(S) = 1$, the server computes

$$\prod_{\rho(i) \in S} (e((u^{A_i} h)^{r_{x,i}} v^{-r_x}, g^{s_i}) \cdot e((u^{\rho(i)} h)^{-s_i}, g^{r_{x,i}}) \cdot e(w^{\lambda_i} v^{s_i}, g^{r_x}))^{\omega_i} = e(w^s, g^{r_x})$$

If $t = t'$, $\pi = \dfrac{e((w^{ID}k)^{r_x} u_0^r (u_1^t h_1)^{s_x}, g^s)}{e(w^s, g^{r_x})^{ID} e(k^s, g^{r_x}) \cdot e((u_1^{t'} h_1)^s, g^{s_x})} = e(u_0^r, g^s)$

- **Verify**$(mpk, c, \pi, vk_{ID}, ID)$. The verify algorithm takes as input the master public key mpk, the transformed ciphertext π, the public verification key vk_{ID} and the identity ID.

If $e(g^s, H(\dfrac{e(u_0,g)^{\beta s} M_R \cdot e(u_0^r, g^s)^{ID}}{e(g_0^{\beta + rID}, k_0^s)})) = e(H(M_R)^s, g)$, return 1, else return 0.

- **Decrypt**$(mpk, sk_{ID}, \pi, Hdr, ID)$. The decryption algorithm takes as input the master public key mpk, the private key sk_{ID}, and the transformed ciphertext π, the header Hdr and the identity ID.

$$\frac{e(g^\alpha(u_0^{ID}h_0)^r, g^s)}{e(u_0^s, g^r)^{ID} \cdot e(h_0^s, g^r)} = e(g,g)^{\alpha s}$$

It outputs the message $M = e(g,g)^{\alpha s} \cdot M/e(g,g)^{\alpha s}$.
- **Revoke**(ID, t, R, st). The revoke algorithm takes as inputs the identity ID, time period t, revocation list R and the state st. If the user's identity ID is revoked at time period t, this algorithm adds (ID, t) to R and outputs the updated revocation list R.

TranKG Correctness: If $ID \notin R$, then $I \cap J = \emptyset$ where $I = \{x : x \in Path(\theta)\}$ and $J = \{x : x \in KUNodes(BT, R, t)\}$. Therefore, there must be exactly one node $x \in I \cap J$ from the definition of the $KUNodes$ algorithm. The first item in the short-term transformation key $tk_{ID,t}^S$ can be derived from $pk_{ID,S}$ and tuk_t as

$$((w^{ID}k)^{r_x}/g_x)u_0^r \cdot g_x(u_1^t h_1)^{s_x} = (w^{ID}k)^{r_x}u_0^r(u_1^t h_1)^{s_x}$$

Verify Correctness: During Transform algorithm, if S satisfies the access policy $\mathbb{A}$ in the ciphertext, i.e. $\mathbb{A}(S) = 1$, we have that $\sum_{\rho(i) \in S} \omega_i \lambda_i = s$, and compute

$$\frac{e((w^{ID}k)^{r_x}u_0^r(u_1^t h_1)^{s_x}, g^s)}{\prod_{\rho(i) \in S}(e((u^{A_i}h)^{r_{1,i}}v^{-r_x}, g^{s_i}) \cdot e((u^{\rho(i)}h)^{-s_i}, g^{r_{1,i}}) \cdot e(w^{\lambda_i}v^{s_i}, g^{r_x}))^{ID\omega_i} \cdot e(k^s, g^{r_x})}$$

$$= \frac{e((w^{ID}k)^{r_x}u_0^r(u_1^t h_1)^{s_x}, g^s)}{\prod_{\rho(i) \in S}(e(v^{-r_x}, g^{s_i}) \cdot e(w^{\lambda_i}v^{s_i}, g^{r_x}))^{ID\omega_i} \cdot e(k^s, g^{r_x})}$$

$$= \frac{e((w^{ID}k)^{r_x}u_0^r(u_1^t h_1)^{s_x}, g^s)}{e(w^s, g^{r_x})^{ID} \cdot e(k^s, g^{r_x})} = e(u_0^r(u_1^t h_1)^{s_x}, g^s)$$

If $t = t'$, we can compute $\pi = \frac{e(u_0^r(u_1^t h_1)^{s_x}, g^s)}{e((u_1^{t'} h_1)^s, g^{s_x})} = e(u_0^r, g^s)$.

Then the verify algorithm computes $M_R = \frac{e(u_0, g)^{\beta s} M_R \cdot e(u_0^r, g^s)^{ID}}{e(g_0^{\beta + rID}, k_0^s)}$ based on $e(g_0, k_0) = e(u_0, g)$ and then test $e(g^s, H(M_R)) \stackrel{?}{=} e(H(M_R)^s, g)$.

Decrypt Correctness: If S satisfies the access policy $\mathbb{A}$ in the ciphertext and $t = t'$, and the transformation by the unstructured server is public verified as correct, we can obtain the transformed ciphertext π. Then we compute

$$\frac{e(g^\alpha(u_0^{ID}h_0)^r, g^s)}{e(u_0^s, g^r)^{ID} \cdot e(h_0^s, g^r)} = e(g,g)^{\alpha s}$$

It outputs the message $M = e(g,g)^{\alpha s} \cdot M/e(g,g)^{\alpha s}$.

4 Security Analysis

4.1 Complexity Assumption

Decisional $(q-1)$ Assumption [16]. For our construction, we will use a $q-1$ type assumption in [16]. This assumption can be proved secure in the following game between $\mathcal{C}$ and $\mathcal{A}$:

Initially $\mathcal{C}$ calls the group generation algorithm with input the security parameter, picks a random group element $g \in \mathbb{G}$, and $q+2$ random exponents $a, s, \vec{b} = \{b_1, \ldots, b_q\} \in \mathbb{Z}_p$. Then it sends to $\mathcal{A}$ the group description $(p, \mathbb{G}, \mathbb{G}_p, e)$ and all of the following terms:

$$g, g^s$$

$$g^{a^i}, g^{b_j}, g^{sb_j}, g^{a^i b_j}, g^{\frac{a^i}{(b_j)^2}}, \forall (i,j) \in [q,q]$$

$$g^{\frac{a^i b_j}{(b_{j'})^2}}, \forall (i,j,j') \in [2q, q, q], j \neq j'$$

$$g^{\frac{a^i}{b_j}}, \forall (i,j) \in [2q, q], i \neq q+1$$

$$g^{\frac{sa^i b_j}{b_{j'}}}, g^{\frac{sa^i b_j}{(b_{j'})^2}} \forall (i,j,j') \in [q,q,q], j \neq j'.$$

$\mathcal{C}$ flips a random coin $b \in \{0,1\}$ and if $b = 0$, it gives to $\mathcal{A}$ the term $T = e(g,g)^{sa^{q+1}}$. Otherwise, it gives a random term $T = Z \in \mathbb{G}_T$. Finally the $\mathcal{A}$ outputs a guess $b' \in \{0,1\}$.

q-Bilinear Diffie-Hellman Exponent Assumption [4]. We define the q-Bilinear Diffie-Hellman Exponent Assumption(q-BDHE) as follows. Choose a group $\mathbb{G}$ of prime order p according to the security parameter. Let $a \in \mathbb{Z}_p, f \in \mathbb{G}$ be chosen and g be a generator of $\mathbb{G}$. If an adversary is given

$$g, g^a, g^{a^2}, \ldots, g^{a^q}, g^{a^{q+2}}, g^{a^{q+3}}, \ldots, g^{a^{2q}}, f$$

it must compute $e(g, f)^{a^{q+1}}$.

4.2 Security Proof

In this section, we provide proof of security for the notions of IND-sCPA in a one-user setting based on [15] and selective verifiability in [1].

Theorem 1 (Selectively IND-CPA Security). *If the $q-1$ assumption in [16] holds then no polynomial adversary can selectively break the PV-SR-ABE scheme with a challenge access policy $(\mathbb{M}^*, \rho^*)$ and a challenge time period t^*.*

We first present the proof of the traditional security notion of IND-sCPA. In the one-user setting, only one user (called "target user") has the capacity to access the challenge ciphertext and the adversary can corrupt either his long-term transformation keys or his private keys. So, the adversary falls into the following two distinct classes.

In Type 1, the target user is revoked at or before the challenge time period t^* and thus the adversary is allowed to corrupt the target user's private key.

In Type 2, the target user is not revoked, so the adversary is not allowed to corrupt the target user's private key.

Proof of Type 1 Adversary. Suppose there exists an adversary $\mathcal{A}_1$ who breaks the proposed scheme with ϵ. We built a simulator $\mathcal{B}$ that has advantage $\frac{\epsilon}{2}$ in solving the $q-1$ assumption in [16]. Given as input an assumption instance, $\mathcal{B}$ plays the role of the challenger $\mathcal{C}$ in the game and interacts with the adversary $\mathcal{A}_1$ as follows.

- **Initialization.** $\mathcal{B}$ first chooses the attribute universal set $\mathcal{U}$, the time set $\mathcal{T}$ and the maximum number of users n_{max}. $\mathcal{A}_1$ outputs a challenge access policy $(\mathbb{M}^*, \rho^*)$ and a challenge time period t^* that it intends to attack. We have that $\mathbb{M}^*$ is an $l \times n$ matrix, where $l \ll q$.
- **Setup.** $\mathcal{B}$ picks randoms $\omega_R, \tilde{\alpha}, \tilde{\beta} \in \mathbb{Z}_p$ and sets $\alpha = a^{q+1} + \tilde{\alpha}, \beta = a + \tilde{\beta}$. Then $\mathcal{B}$ picks the random numbers $\tilde{k}, \tilde{h}_0, \tilde{u}_1, \tilde{h}_1, \tilde{u}, \tilde{v}, \tilde{h} \in \mathbb{Z}_p$ and sets $e(u_0, g) = e(g_0, k_0)$, H be the randome oracle controlled by $\mathcal{B}$. Using the assumption, $\mathcal{B}$ gives to $\mathcal{A}_1$ the following public parameters:

$$g = g, g_0 = g^{\frac{a^q}{b_q}}, k_0 = g^{b_q}, w = g^a, v = g^{\tilde{v}} \cdot \prod_{(j,k) \in [l,n]} (g^{\frac{a^{k+2}}{b_j}})^{M^*_{j,k}},$$

$$k = g^{\frac{b_q a^q}{b_1^2} + \tilde{k}}, u = g^{\tilde{u}} \cdot \prod_{(j,k) \in [l,n]} (g^{\frac{a^{k+1}}{b_j^2}})^{M^*_{j,k}}, u_0 = g^{a^q}, u_1 = g^{a+\tilde{u}_1},$$

$$h = g^{\tilde{h}} \cdot \prod_{(j,k) \in [l,n]} (g^{\frac{a^{k+1}}{b_j^2}})^{-\rho(j)^* M^*_{j,k}}, h_0 = g^{\frac{b_1 a}{b_q^2} + \tilde{h}_0}, h_1 = g^{-t^* a + \tilde{h}_1},$$

$$e(g,g)^\alpha = e(g^a, g^{a^q}) \cdot e(g,g)^{\tilde{\alpha}}, e(u_0, g)^\beta = e(g^a, g^{a^q}) \cdot e(g^{a^q}, g)^{\tilde{\beta}}.$$

Finally, $\mathcal{B}$ gives $\mathcal{A}_1$ mpk and initializes the empty revocation list R and the secret state $st = BT$.

- **Phase 1 and Phase 2 Queries:** $\mathcal{B}$ randomly chooses an unassigned leaf node θ^* for storing the target user ID^* and answers the $\mathcal{A}_1$'s queries as follows.
  - **Create**(ID, S): When $\mathcal{A}_1$ issues a query on (ID, S), $\mathcal{B}$ creates the table T's element $(ID, S, sk_{ID}, vk_{ID}, pk_{ID,S})$ and returns $pk_{ID,S}, vk_{ID}$ to $\mathcal{A}_1$ through the following way.
    * If $S \in \mathbb{A}^*$, $\mathcal{B}$ sets $ID = ID^*$ and stores ID^* in the pre-assigned leaf node θ^*. Then $\mathcal{B}$ randomly picks $\tilde{r} \in \mathbb{Z}_p$ and implicitly sets $r = \tilde{r} + \omega_1 a$, $\omega_1 = \frac{-1}{ID}$. $\mathcal{B}$ generates private key sk_{ID} as follows:

$$g^\alpha (u_0^{ID} h_0)^r = g^{a^{q+1}+\tilde{\alpha}} \cdot ((g^{a^q})^{ID} \cdot g^{\frac{b_1 a}{b_q^2} + \tilde{h}_0})^{\tilde{r}+\omega_1 a}$$
$$= g^{\tilde{\alpha}} \cdot (u_0^{ID} h_0)^{\tilde{r}} \cdot h_0^{\omega_1 a},$$

and $g^r = g^{\widetilde{r}+\omega_1 a}$, $vk_{ID} = g_0^{\beta+rID} = (g^{\frac{a^q}{b_q}})^{\widetilde{\beta}+\widetilde{r}ID}$. For each node $x \in Path(\theta^*)$, $\mathcal{B}$ randomly chooses an exponent $\widetilde{g}_x \in \mathbb{Z}_p$ and implicitly sets $g_x = g^{-\frac{a^{q+1}}{ID}+\widetilde{g}_x}$. It also randomly selects $\widetilde{r}_x, \omega_1' \in \mathbb{Z}_p$ and generates $r_x = \widetilde{r}_x + \omega_1' a^{q-1}$. It computes as follows

$$((w^{ID}k)^{r_x}/g_x)u_0^r = (g^{aID}g^{\frac{b_q a^q}{b_1^2}+\widetilde{k}_0})^{\widetilde{r}_x + \omega_1' a^{q-1}} \cdot g^{\frac{a^{q+1}}{ID}-\widetilde{g}_x} \cdot (g^{a^q})^{\widetilde{r}+\omega_1 a}$$
$$= (w^{ID}k)^{r_x} \cdot g^{-\widetilde{g}_x} \cdot (g^{a^q})^{\widetilde{r}}$$

For each attribute $S_\tau \in S$, $\mathcal{B}$ randomly selects $\widetilde{r}_{x,\tau}, \omega_i' \in \mathbb{Z}_p$ and sets $r_{x,\tau} = \widetilde{r}_{x,\tau} + \sum_{(i,i') \in [n,l], \rho^*(i') \in S} \omega_i' b_{i'} a^{q+1-i}$. Thus $\mathcal{B}$ generates the $(u^{S_\tau}h)^{r_{x,\tau}}v^{-r_x} = (g^{\widetilde{u}S_\tau+\widetilde{h}})^{r_{x,\tau}} \cdot \prod_{(j,k) \in [l,n]} (g^{\frac{(S_\tau - \rho^*(j))M_{j,k}^* a^{k+2}}{b_j^2}})^{r_{x,\tau}} \cdot v^{-r_x}$. Note that $S \in \mathbb{A}^*$, the specific authorized set defined as $\mathcal{S}_1$, for all $S_\tau \in \mathcal{S}_1$, the $S_\tau - \rho^*(i')$ are zero and thus the attributes sets part $(u^{S_\tau}h)^{r_{x,\tau}}v^{-r_x}$ of long-term transformation key can be simulated by $\mathcal{B}$.

* If $S \notin \mathbb{A}^*$, $ID \neq ID^*$, $\mathcal{B}$ randomly chooses an unassigned leaf node θ and stores ID in it. sk_{ID}, vk_{ID} are similar with that when $S \in \mathbb{A}^*$.

1. If $x \in Path(\theta) \cap Path(\theta^*)$, $\mathcal{B}$ selects randoms $\widetilde{r}_x', \theta_0, \ldots, \theta_n \in \mathbb{Z}_p$ and sets $r_x = \widetilde{r}_x' + \theta_0 a^{q-1} + \sum_{i \in [n]} \theta_i a^{q-1-i}$. Then $\mathcal{B}$ builds the long term transformation key $psk_{ID,S}$ similarly with it when $S \in \mathbb{A}^*$. For each attribute $S_\tau \in S$, $\mathcal{B}$ randomly selects $\widetilde{r}_{x,\tau}' \in \mathbb{Z}_p$ and sets $r_{x,\tau} = \widetilde{r}_{x,\tau}' + \frac{\sum_{(i,i') \in [n,l], \rho^*(i') \notin S} \theta_i b_{i'} a^{q-1-i}}{S_\tau - \rho^*(i')}$. Note that $S \notin \mathbb{A}^*$, the specific unauthorized set defined as $\mathcal{S}_2$, for all $S_\tau \in \mathcal{S}_2$, the $S_\tau - \rho^*(i')$ are not zero. Thus $\mathcal{B}$ can only simulate $(u^{S_\tau}h)^{r_{x,\tau}}v^{-r_x}$ for the unauthorized attribute set. Specifically, the first part v^{-r_x} for these terms is the following:

$$v^{-\widetilde{r}_x' - \theta_0 a^{q-1}} \prod_{(i,j,k) \in [n,l,n]} g^{\frac{-\theta_i M_{j,k}^* a^{q-i+k+1}}{b_j}} = \Phi \prod_{\substack{(i,j) \in [n,l] \\ \rho^*(j) \notin S}} g^{\frac{-\langle\theta, M_j^*\rangle a^{q+1}}{b_j}}$$

Then $\mathcal{B}$ generates the $(u^{S_\tau}h)^{r_{x,\tau}}$ as following:

$$(g^{\widetilde{u}S_\tau+\widetilde{h}})^{r_{x,\tau}} \prod_{(j,k) \in [l,n]} (g^{\frac{(S_\tau - \rho^*(j))b_{i'} M_{j,k}^* \theta_i a^{k+2}}{(S_\tau - \rho^*(j))b_j^2}})^{r_{x,\tau}} = \Lambda \prod_{\substack{(i,j) \in [n,l] \\ \rho^*(j) \notin S}} g^{\frac{\langle\theta, M_j^*\rangle a^{q+1}}{b_j}}$$

2. If $x \notin Path(\theta) \cap Path(\theta^*)$, $\mathcal{B}$ knows the value g_x and can compute the following. $\mathcal{B}$ selects randoms $\widetilde{r}_x'', \mu_1, \ldots, \mu_n \in \mathbb{Z}_p$ and sets $r_x = \widetilde{r}_x'' + \mu_0 a^q + \sum_{i \in [n]} \mu_i a^{q-1-i}$, $\mu_0 = \frac{1}{ID^2}$. Then $\mathcal{B}$ builds the long term transformation key $psk_{ID,S}$ as follows:

$$((w^{ID}k)^{r_x}/g_x)u_0^r = (g^{aID})^{\widetilde{r}_x'' + \sum_{i \in [n]} \mu_i a^{q-1-i}} k^{r_x} \cdot g_x^{-1} \cdot (g^{a^q})^{\widetilde{r}}$$

$g^{r_x} = g^{\tilde{r}_x + \mu_0 a^q + \sum_{i \in [n]} \mu_i a^{q-1-i}}$. For each attribute $S_\tau \in S$, $\mathcal{B}$ randomly selects $\tilde{r}''_{x,\tau} \in \mathbb{Z}_p$ and sets $r_{x,\tau} = \tilde{r}''_{x,\tau} + \frac{\sum_{(i,i') \in [n,l], \rho^*(i') \notin S} \mu_i b_{i'} a^{q-1-i}}{S_\tau - \rho^*(i')}$. Note that $S \notin \mathbb{A}^*$, for all $S_\tau \in S$, the $S_\tau - \rho^*(i')$ are not zero. Thus the common part v^{-r_x} for these terms is the following:

$$v^{-\tilde{r}''_x}(g^{\tilde{v}})^{-r_x} \prod_{(j,k) \in [l,n]} ((g^{\frac{a^{k+2}}{b_j}})^{M^*_{j,k}})^{-\mu_0 a^q} \prod_{(i,j,k) \in [n,l,n]} g^{\frac{-\mu_i M^*_{j,k} a^{q+1-i+k}}{b_j}}$$

Thus $\mathcal{B}$ generates the $(u^{S_\tau} h)^{r_{x,\tau}}$ as following:

$$(g^{\tilde{u} S_\tau + \tilde{h}})^{r_{x,\tau}} \prod_{(j,k) \in [l,n]} g^{\frac{(S_\tau - \rho^*(j)) b_{i'} \mu_i M^*_{j,k} a^{k+2}}{(S_\tau - \rho^*(j)) b_j^2}} = \Theta \prod_{\substack{(i,j) \in [n,l] \\ \rho^*(j) \notin S}} g^{\frac{-<\mu, M^*_j> a^{q+1}}{b_j}}$$

- **Corrupt(ID)**: If there exists an entry indexed by ID in table T, then $\mathcal{B}$ obtains the entry $(ID, S, sk_{ID}, vk_{ID}, pk_{ID,S})$ and sets $D = D \cup \{(ID, S)\}$. It returns to $\mathcal{A}_1$ the private key sk_{ID}. If no such entry exists, then it returns $\perp$. Note that, for Type 1 attack, $\mathcal{A}_1$ is allowed to corrupt the target user ID^* and obtain the private key sk_{ID^*}.
- **TKeyUp(t)**: If $\mathcal{A}_1$ issues a key update query on a time period t, $\mathcal{B}$ computes the key update message tuk_t as follows. For all $x \in KUNodes(BT, R, t)$, $\mathcal{B}$ fetches g_x from node x. If x is not defined, it randomly chooses g_x and stores it in the node x.
  * If $x \notin Path(\theta^*)$, $\mathcal{B}$ knows g_x. Thus it chooses a random exponent $s_x \in \mathbb{Z}_p$ and computes $tuk_t = (g_x(u_1^t h_1)^{s_x}, g^{s_x})$.
  * If $x \in Path(\theta^*)$, $\mathcal{B}$ does not know g_x. $\mathcal{B}$ selects a random number $\tilde{s}_x \in \mathbb{Z}_p$ and sets $s_x = \tilde{s}_x + \frac{a^q}{ID(t-t^*)}$. Then $\mathcal{B}$ computes

$$g_x(u_1^t h_1)^{s_x} = g^{-\frac{a^{q+1}}{ID} + \tilde{g}_x} ((g^{a+\tilde{u}_1})^t \cdot g^{-t^* a + \tilde{h}_1})^{\tilde{s}_x + \frac{a^q}{ID(t-t^*)}}$$

$$= g^{\tilde{g}_x} (g^{t\tilde{u}_1} \cdot g^{\tilde{h}_1})^{\frac{a^q}{ID(t-t^*)}} (u_1^t h_1)^{\tilde{s}_x}$$

- **Revoke(ID,t)**: When $\mathcal{A}_1$ issues a revocation query on an identity ID at time period t, $\mathcal{B}$ adds (ID, t) to the revocation list R.
- **Challenge.** When $\mathcal{A}_1$ submits two equal length messages M_0, M_1, $\mathcal{B}$ chooses a random bit $b \in \{0, 1\}$. $\mathcal{B}$ picks a random message M_R with same length of M_0 and M_1. $\mathcal{B}$ randomly selects $\omega_R \in \mathbb{Z}_p$ and sets g^{ω_R} as the response of queries M_R to the random oracle H. Then $\mathcal{B}$ constructs

$$Hdr^* = (T \cdot e(g, g^s)^{\tilde{\alpha}} \cdot M_b, g^s, h_0^s = g^{\frac{sab_1}{b_q^2}} \cdot g^{s\tilde{h}_0}),$$

$$c^* = (T \cdot e(g^{a^q}, g^s)^{\tilde{\beta}} \cdot M_R, H(M_R)^s = g^{s\omega_R}, g^s, k^s = g^{\frac{sa^q b_q}{b_1^2}} \cdot g^{s\tilde{k}},$$

$$k_0^s = g^{sb_q}, (u_1^t h)^s = (g^s)^{\tilde{u}_1 t^* + \tilde{h}_1}, \{w^{\lambda_\tau} v^{s_\tau}, (u^{\rho^*(\tau)} h)^{-s_\tau}, g^{s_\tau}\}_{\tau \in l})$$

$\mathcal{B}$ sets implicitly $\vec{y} = (s, sa + \tilde{y}_2, sa^2 + \tilde{y}_3, \ldots, sa^{n-1} + \tilde{y}_n)$, where $\tilde{y}_2, \ldots, \tilde{y}_n \in \mathbb{Z}_p$. For each row $\mathcal{B}$ sets implicitly $s_\tau = -sb_\tau$ and since $\vec{\lambda} = \mathbb{M}^* \vec{y}$, we have

that $\lambda_\tau = \sum_{i\in[n]} \mathbb{M}^*_{\tau,i} sa^{i-1} + \widetilde{\lambda}_\tau$. Therefore, $\mathcal{B}$ returns the challenge header and challenge ciphertext (Hdr^*, c^*) to $\mathcal{A}_1$.
- **Guess.** Finally. $\mathcal{A}_1$ outputs guess b' for the challenge bit. If $b' = b$, $\mathcal{B}$ outputs 1, it claims that the challenge term is $T = e(g,g)^{sa^{q+1}}$. Otherwise, it outputs 0 meaning $T = Z$ is random in $\mathbb{G}_T$.

When $T = Z$, the ciphertext will give no information about the simulator's choice of b. In this case, we have $\Pr[b' = b|T = Z] = \Pr[b' \neq b|T = Z] = \frac{1}{2}$, because the challenge header will contain only random numbers. Since the simulator outputs a guess $T' = Z$ when $b' = b$, we then have $\Pr[T' = T|T = Z] = \frac{1}{2}$. When $T = e(g,g)^{sa^{q+1}}$, $\mathcal{A}_1$ can sees a valid encryption of M_b. In this situation, the $\mathcal{A}_1$'s advantage is ϵ by definition. The probability is $\Pr[b' = b|T = e(g,g)^{sa^{q+1}}] = \frac{1}{2}+\epsilon$. Since the simulator outputs a guess $T = e(g,g)^{sa^{q+1}}$ of T when $b' = b$, we have $\Pr[T' = T|T = e(g,g)^{sa^{q+1}}] = \frac{1}{2} + \epsilon$. Hence, by putting them all together, $\mathcal{B}$'s advantage in the above game is

$$\Pr[T' = T] - \frac{1}{2}$$
$$= \Pr[T' = T|T = e(g,g)^{sa^{q+1}}]\Pr[T = e(g,g)^{sa^{q+1}}] + \Pr[T' = T|T = Z]\Pr[T = Z] - \frac{1}{2}$$
$$= \frac{1}{2}\Pr[T' = T|T = e(g,g)^{sa^{q+1}}] + \frac{1}{2}\Pr[T' = T|T = Z] - \frac{1}{2}$$
$$= \frac{1}{2}(\frac{1}{2} + \epsilon + \frac{1}{2}) - \frac{1}{2} = \frac{\epsilon}{2}$$

Similarly, we can prove Theorem 1 for *Type 2 adversary*. In the proof of Type 2 adversary, the target user ID^* is not revoked and the adversary is not allowed to corrupt the target user's private key sk_{ID^*}. The **Initialization** and **Setup** are the same as that in the proof of type 1 adversary. The differences lie in the Phase 1 and Phase 2 queries of **Create**(ID,S) as the following: When $ID = ID^*$ for $S \in \mathbb{A}$, $\mathcal{B}$ randomly implicitly sets $r, g_x \in \mathbb{Z}_p$ and embed the unknown item $g^{a^{q+1}}$ to sk_{ID^*} due to the type 2 adversary cannot query the private key sk_{ID^*}. When $S \notin \mathbb{A}$, $\mathcal{B}$ constructs the proof similarly to the proof of type 1 adversary without distinguishing each node $x \in Path(\theta) \cap Path(\theta^*)$ or not. Since the target user ID^* is not revoked in the type 2 adversary, $\mathcal{B}$ can compute the key update message tuk_t as in the scheme for any non-revoked users.

Therefore $\mathcal{B}$ has a non negligible advantage in breaking the $q-1$ assumption in [16] and we obtain Theorem 1 and prove the security of our proposal scheme.

Theorem 2 (Selective Public Verifiability). *If the q-BDHE assumption in [4] hold then the PV-SR-ABE scheme is secure in the sense of selective public verifiability under with a challenge access policy $(\mathbb{M}^*, \rho^*)$ and a challenge time period t^*.*

Proof. Suppose $\mathcal{A}_2$ is an adversary with non-negligible advantage ϵ to break scheme in the selective public verifiability. We construct a simulator $\mathcal{B}$ to solve the q-BDHE assumption. Given as input the assumption instance, $\mathcal{B}$ runs $\mathcal{A}_2$ and works as follows.

- **Initialization.** $\mathcal{B}$ first choose the attribute universal set $\mathcal{U}$ and the time set $\mathcal{T}$. The selective game with $\mathcal{A}_2$ outputs a challenge access policy $(\mathbb{M}^*, \rho^*)$ and a challenge time period t^* that it intends to attack. We have that $\mathbb{M}^*$ is an $l \times n$ matrix, where $l \ll q$.
- **Setup.** To generate the system parameter, $\mathcal{B}$ picks randoms $\widetilde{\alpha}, \widetilde{\beta} \in \mathbb{Z}_p$ and sets $\alpha = \widetilde{\alpha}$, $\beta = \widetilde{\beta}$. $\mathcal{B}$ picks the random numbers $b_k, \widetilde{k}, \widetilde{w}, \widetilde{h}_0, \widetilde{u}_1, \widetilde{h}_1, \widetilde{u}, \widetilde{v}, \widetilde{h} \in \mathbb{Z}_p$, $e(u_0, g) = e(g_0, k_0)$ and sets H be the randome oracle controled by the $\mathcal{B}$. $\mathcal{B}$ defines the output of randome oracle H with M_R query is g^{b_R}, where $b_R \in \mathbb{Z}_p$. $\mathcal{B}$ randomly chooses $z_0, z_1, \ldots, z_q \in \mathbb{Z}_p$ and sets $G(\rho^*(j)) = z_0 + z_1\rho^*(j) + z_2\rho^*(j)^2 + \ldots + z_q\rho^*(j)^q$ according to the challenge access policy $(\mathbb{M}^*, \rho^*)$. Using the assumption, $\mathcal{B}$ generates the following public parameters:

$$g = g, g_0 = g^{\frac{a^q}{\widetilde{h}_k}}, k_0 = g^{b_k}, w = g^{\widetilde{w}}, v = g^{a^q} \prod_{j=1}^{l} G(\rho^*(j)),$$

$$k = g^{\widetilde{k}}, u = g^{\widetilde{u}} \cdot \prod_{(j,k) \in [l,n]} g^{M_{j,k}^*}, u_0 = g^{a^q}, u_1 = g^{a + \widetilde{u}_1},$$

$$h = g^{\widetilde{h}} \cdot \prod_{(j,k) \in [l,n]} g^{-\rho(j)^* M_{j,k}^*}, h_0 = g^{\widetilde{h}_0}, h_1 = g^{-t^* a + \widetilde{h}_1},$$

$\mathcal{B}$ then randomly chooses integers $x_1, \ldots, x_{l+\gamma} \in [0, q-1]$, where γ is the number of digits of identity ID. Aiming at the challenge access policy $(\mathbb{M}^*, \rho^*)$, for the attribute set $S = \{a_1, \ldots, a_{|S|}\}$, if $a_i = \rho^*(j), i = \{1, \ldots, |S|\}, j = \{1, \ldots, l\}$ then $\theta_i = 1$, otherwise, $\theta_i = 0$. When $|S| < l$, then $\theta_i = 0, i = \{|S|, \ldots, l\}$. At last, $\theta_i = ID[i-l+1]$ for $i \in \{l, \ldots, l+\gamma\}$. $\mathcal{B}$ simply set $n = l + \gamma$ and $F(ID, S) = 1$ if $(\sum_{i=1}^{n} \theta_i \cdot x_i) = 0 \bmod q$, otherwise, $F(ID, S) = 0$. Finally, $\mathcal{B}$ gives $\mathcal{A}_1$ the master public key.
- **Challenge.** $\mathcal{B}$ generates challenge header and challenge ciphertext (Hdr^*, c^*) under $(\mathbb{M}^*, \rho^*)$ and t^*. $\mathcal{B}$ randomly chooses random messages $M, M_R \in \mathcal{M}$ and queries the random oracle H. $\mathcal{B}$ computes as the follows:

$$Hdr^* = (e(g, f)^{\widetilde{\alpha}} \cdot M, f, h_0^s = f^{\widetilde{h}_0})$$

$$c^* = (e(g^{a^q}, f)^{\widetilde{\beta}} M_R, H(M_R)^s = f^{b_R}, f, k^s = f^{\widetilde{k}}, k_0^s = f^{b_k},$$
$$(u_1^{t^*} h)^s = f^{\widetilde{u}_1 + \widetilde{h}_1}, \{w^{\lambda_\tau} v^{s_\tau}, (u^{\rho^*(\tau)} h)^{-s_\tau}, g^{s_\tau}\}_{\tau \in l})$$

$\mathcal{B}$ sets implicitly $\overrightarrow{y} = (s, \widetilde{y}_2, \widetilde{y}_3, \ldots, \widetilde{y}_n)$, where $\widetilde{y}_2, \ldots, \widetilde{y}_n \in \mathbb{Z}_p$. For each row $\mathcal{B}$ sets implicitly $s_\tau \in \mathbb{Z}_p$ and since $\overrightarrow{\lambda} = \mathbb{M}^* \cdot \overrightarrow{y}$, we have that $\lambda_\tau = \sum_{i \in [n]} s\mathbb{M}^*_{\tau, i} \widetilde{\lambda}_\tau$. Therefore, $\mathcal{B}$ returns the challenge header and challenge ciphertext (Hdr^*, c^*) to $\mathcal{A}_2$.
- **Query.** $\mathcal{B}$ answers the $\mathcal{A}_2$'s queries as follows:
  - **Create** (ID, S): When $\mathcal{A}_2$ issues a user key query on (ID, S), $\mathcal{B}$ stores ID in the pre-assigned leaf node θ. We set the user key query to be q_1 times.

$\mathcal{B}$ randomly chooses $\tilde{r} \in \mathbb{Z}_p$ and sets $r = \tilde{r} + F(ID,S)a$. $\mathcal{B}$ sets the private key $sk_{ID} = g^\alpha(u_0^{ID}h_0)^r = g^{\tilde{\alpha}}(g^{IDa^q}g^{\widetilde{h_0}})^{\tilde{r}+F(ID,S)a}$ and computes the public verification key vk_{ID} as $g_0^{\beta+rID} = (g^{\frac{a^q}{b_k}})^{\tilde{\beta}+ID(\tilde{r}+F(ID,S)a)}$. For each node $x \in Path(\theta)$, $\mathcal{B}$ randomly chooses random $g_x \in \mathbb{G}$, $r_x \in \mathbb{Z}_p$ and generates the long-term transformation key $psk_{ID,S} = ((w^{ID}k)^{r_x}/g_x)u_0^r = (g^{ID\widetilde{w}}g^{\widetilde{k}})^{r_x} \cdot g_x^{-1} \cdot (g^{a^q})^{\tilde{r}+F(ID,S)a}$. Aiming at each attributes $S_\tau \in S$, $\mathcal{B}$ randomly chooses $r_{x,\tau} \in \mathbb{Z}_p$ and generates the $(u^{S_\tau}h)^{r_{x,\tau}}v^{-r_x}$. When a user key query on (ID,S), if $F(ID,S) \neq 0$, $\mathcal{B}$ aborts. Otherwise, $\mathcal{B}$ returns the private key sk_{ID}, long-term transformation key $psk_{ID,S}$ and public verification key vk_{ID} to $\mathcal{A}_2$.

- **TKeyUp** (t): If $\mathcal{A}_2$ issues a key update query on time period t, $\mathcal{B}$ computes the key update message tuk_t as the scheme.
- **Transform** (ID,S,t^*,c^*): $\mathcal{A}_2$ issues a transformed query on (ID,S,c^*,t^*) and $\mathcal{B}$ computes the transformed ciphertext π as the follows. We set the above query to be q_2 times. For the query $(ID,S = (a_1,\ldots,a_{|S|}),t^*,c^*)$, $\mathcal{B}$ first sets $J(ID,S) = a_1ID + a_2ID^2 + \ldots + a_{|S|}a^{|S|}$ and then defines the random number $r = \frac{\prod_{j=1}^l G(\rho^*(j))}{J(ID,S)} + F(ID,S)a$. When $S \in \mathbb{A}^*$, for a transform query on c^* and tk_{ID,t^*}^S, if $F(ID,S) \neq 0$, $\mathcal{B}$ aborts. Otherwise, $\mathcal{B}$ computes the transformed ciphertext $\pi = e(u_0^r, g^s)$ as

$$= e((g^{a^q})^{\frac{\prod_{j=1}^l G(\rho^*(j))}{J(ID,S)}}, f)e((g^{a^q})^{F(ID,S)a}, f) = e(v^{\frac{1}{J(ID,S)}}, f)e(u_0^{F(ID,S)a}, f)$$

- **Revoke** (ID,t): When $\mathcal{A}_1$ issues a revocation query on an identity ID at time period t, $\mathcal{B}$ adds (ID,t) to the revocation list R.
- **Forgery.** $\mathcal{A}_1$ finishes its query and outputs a guess $\pi^* = e(u_0^{r^*}, f)$

$$= e((g^{a^q})^{\frac{\prod_{j=1}^l G(\rho^*(j))}{J(ID^*,S^*)}}, f)e((g^{a^q})^{F(ID^*,S^*)a}, f) = e(v^{\frac{1}{J(ID^*,S^*)}}, f)e(g^{a^{q+1}}, f)^{F(ID,S)}$$

which it believes to be a valid forgery. $\mathcal{B}$ computes $\dfrac{\pi^*}{e(v^{\frac{1}{J(ID^*,S^*)}}, f)} = e(g^{a^{q+1}}, f)$ as the solution to the q-BDHE problem instance.

According to the proof definition and simulation, there are the following ways to compute the solution to the q-BDHE assumption under $F(ID,S) = 0$ for $(q_s = q_1 + q_2)$ queries and $F(ID^*,S^*) = 1$. We set the number of queries $q = 2q_s$. From the Setup phase, we have $\sum_{i=1}^n \theta_i x_i \in [0, n(q-1)]$, where the range contains integers $0q, 1q, 2q, \ldots, (n-1)q$ $(n < q)$. Noted that the pairs (ID_i, S_i) and (ID^*, S^*) for any i different on at least one bit, $F(ID_i, S_i)$ and $F(ID^*, S^*)$ differ on the coefficient of at least one x_j. Thus the forged transformed ciphertext π^* is reducible with success probability $\Pr[Success] =$

$$\Pr[(\bigwedge_{i=1}^{q_s} F(ID_i, S_i) = 0) \cap F(ID^*, S^*) = 1]$$

$$= \Pr[\bigwedge_{i=1}^{q_s} F(ID_i, S_i) = 0 \, F(ID^*, S^*) = 1] \Pr[F(ID^*, S^*) = 1]$$

$$= (1 - \Pr[\bigvee_{i=1}^{q_s} F(ID_i, S_i) = 1 \, F(ID^*, S^*) = 1]) \Pr[F(ID^*, S^*) = 1]$$

$$\geq (1 - \sum_{i=1}^{q_s} \Pr[F(ID_i, S_i) = 1 \, F(ID^*, S^*) = 1]) \Pr[F(ID^*, S^*) = 1]$$

$$= (1 - \frac{q_s}{q}) \cdot \frac{n}{n(q-1)+1}$$

$$\geq (1 - \frac{q_s}{q}) \cdot \frac{1}{q} = \frac{1}{4q_s}$$

Therefore the adversary $\mathcal{A}_2$ will solve the q-BDHE assumption with the advantage $\frac{\epsilon}{4q_s}$. This completes the proof of Theorem 2 and conclude that the scheme is secure in the sense of selective public verifiability.

5 Conclusion

In this paper, we propose a publicly verifiable server-aided revocable attribute-based encryption scheme that enables any party to verify whether the server behaves as expected. The scheme adopts a "verify-then-decrypt" mechanism, where outsourced results are first verified before being used for decryption, thereby preventing redundant or incorrect decryption attempts. This design enhances the security, transparency, and trustworthiness of server-aided ABE systems while reducing the risk of fraud and misconduct. Nonetheless, several challenges remain, including optimizing the ciphertext size in publicly verifiable server-aided RABE, designing more efficient constructions with shorter transformation keys, and exploring its applicability in practical scenarios.

References

1. Alderman, J., Janson, C., Cid, C., Crampton, J.: Hybrid publicly verifiable computation. In: Sako, K. (ed.) CT-RSA 2016. LNCS, vol. 9610, pp. 147–163. Springer, Cham (2016). https://doi.org/10.1007/978-3-319-29485-8_9
2. Bethencourt, J., Sahai, A., Waters, B.: Ciphertext-policy attribute-based encryption. In: 2007 IEEE Symposium on Security and Privacy (SP 2007) (2007)
3. Boldyreva, A., Goyal, V., Kumar, V.: Identity-based encryption with efficient revocation. In: Proceedings of the 15th ACM Conference on Computer and Communications Security (2008)
4. Boneh, D., Boyen, X., Goh, E.J.: Hierarchical identity based encryption with constant size ciphertext. In: Annual International Conference on the Theory and Applications of Cryptographic Techniques (2005)

5. Cheng, L., Meng, F.: Server-aided revocable attribute-based encryption revised: multi-user setting and fully secure. In: Bertino, E., Shulman, H., Waidner, M. (eds.) ESORICS 2021. LNCS, vol. 12973, pp. 192–212. Springer, Cham (2021). https://doi.org/10.1007/978-3-030-88428-4_10
6. Cui, H., Deng, R.H., Li, Y., Qin, B.: Server-aided revocable attribute-based encryption. In: Askoxylakis, I., Ioannidis, S., Katsikas, S., Meadows, C. (eds.) ESORICS 2016. LNCS, vol. 9879, pp. 570–587. Springer, Cham (2016). https://doi.org/10.1007/978-3-319-45741-3_29
7. Cui, H., Wan, Z., Wei, X., Nepal, S., Yi, X.: Pay as you decrypt: decryption outsourcing for functional encryption using blockchain. IEEE Trans. Inf. Forensics Secur. (2020)
8. Goyal, V., Pandey, O., Sahai, A., Waters, B.: Attribute-based encryption for fine-grained access control of encrypted data. In: Proceedings of the 13th ACM Conference on Computer and Communications Security, pp. 89–98 (2006)
9. Green, M., Hohenberger, S., Waters, B.: Outsourcing the decryption of ABE ciphertexts. In: 20th USENIX Security Symposium (USENIX Security 11) (2011)
10. Hur, J., Noh, D.K.: Attribute-based access control with efficient revocation in data outsourcing systems. IEEE Trans. Parallel Distrib. Syst. (2011)
11. Lai, J., Deng, R.H., Guan, C., Weng, J.: Attribute-based encryption with verifiable outsourced decryption. IEEE Trans. Inf. Forensics Secur. (2013)
12. Miao, Y., et al.: Verifiable outsourced attribute-based encryption scheme for cloud-assisted mobile e-health system. IEEE Trans. Dependable Secure Comput. (2023)
13. Parno, B., Raykova, M., Vaikuntanathan, V.: How to delegate and verify in public: verifiable computation from attribute-based encryption. In: Cramer, R. (ed.) TCC 2012. LNCS, vol. 7194, pp. 422–439. Springer, Heidelberg (2012). https://doi.org/10.1007/978-3-642-28914-9_24
14. Qin, B., Zhao, Q., Zheng, D., Cui, H.: Server-aided revocable attribute-based encryption resilient to decryption key exposure. In: Capkun, S., Chow, S.S.M. (eds.) CANS 2017. LNCS, vol. 11261, pp. 504–514. Springer, Cham (2018). https://doi.org/10.1007/978-3-030-02641-7_25
15. Qin, B., Zhao, Q., Zheng, D., Cui, H.: (Dual) server-aided revocable attribute-based encryption with decryption key exposure resistance. Inf. Sci. (2019)
16. Rouselakis, Y., Waters, B.: Practical constructions and new proof methods for large universe attribute-based encryption (2013)
17. Sahai, A., Waters, B.: Fuzzy identity based encryption. IACR Cryptology ePrint Archive (2004)
18. Xu, S., Han, X., Xu, G., Ning, J., Huang, X., Deng, R.H.: An adaptive secure and practical data sharing system with verifiable outsourced decryption. IEEE Trans. Serv. Comput. (2023)
19. Yang, F., Cui, H., Jing, J.: Generic constructions of server-aided revocable ABE with verifiable transformation. In: Zhou, J., et al. (eds.) ACNS 2023. LNCS, vol. 13907. Springer, Cham (2023). https://doi.org/10.1007/978-3-031-41181-6_25
20. Yu, P., et al.: Decentralized, revocable and verifiable attribute-based encryption in hybrid cloud system. Wireless Pers. Commun. (2019)

New First-Order Secure AES Implementation Without Online Fresh Randomness Records

Botao Liu and Ming Tang(✉)

Key Laboratory of Aerospace Information Security and Trusted Computing, Ministry of Education, School of Cyber Science and Engineering, Wuhan University, Wuhan 430072, China
m.tang@whu.edu.cn

Abstract. The changing of the guards technique achieve uniformity masking by replacing fresh randomness with unrelated parts of the cipher state. In addressing the decomposition and masking of the AES S-box, Askeland et al. proposed the changing of the guards technique with reused randomness. However, this technique does not extend the guard bit deep into the internal structure of the S-box for masking. Building upon this approach, we further enhance the technique by incorporating the guard bit into the internal decomposition of the S-box, thereby reducing the need for reused randomness. Based on this optimization and in combination with the Cagdas Calik AES S-box, we propose a low-randomness, low-area first-order masked AES hardware implementation. This approach eliminates the need for online fresh randomness while achieving minimal shares and providing provable security in the glitch+register-transition-robust probing model. Additionally, we perform leakage detection experiments using PROLEAD and Test Vector Leakage Assessment (TVLA). The masked AES requires only 7671 GE (NanGate 45 nm open cell library) with 236 clock cycles on Application-Specific Integrated Circuit (ASIC). On Xilinx Spartan-6 Field Programmable Gate Arrays (FPGA), it requires only 212 slices with 125 Mbps, which achieves a throughput-per-slice of 0.319 Mbps/slice. To the best of our knowledge, this is the lowest randomness, area, and latency among existing masked AES without online fresh randomness.

Keywords: AES · Side-Channel Attacks · Threshold Implementation · Cagdas Calik AES S-box · Masking

1 Introduction

The Advanced Encryption Standard (AES) [1] remains mathematically secure, with no known method to break it at the theoretical level. However, cryptographic algorithms implemented in hardware are vulnerable to side-channel attacks. For example, in 1999, Kocher et al. introduced a particularly powerful attack known as Differential Power Analysis (DPA) [2], which can recover the secret key by monitoring physical information such as power consumption and electromagnetic radiation during device operation, coupled with statistical methods. As a result, this attack presents a significant threat to cryptographic implementations in hardware. In 1999, Chari et al. introduced the concept of masking, making power consumption independent of the secret values [3]. However, this assumption was later proven to be overly optimistic, as glitches and early signal propagation can introduce security vulnerabilities [4]. In 2006, Nikova et al. proposed the Threshold Implementations (TI) masking scheme, which is inherently resistant to security issues caused by glitches [5]. The TI scheme has since become a well-known method in hardware masking techniques. AES is the de facto security standard for many applications, necessitating efficient and industrially viable masking solutions. Over the years, significant progress has been made in AES masking solutions. However, researchers have identified a hidden cost: each masking operation requires fresh random bits to ensure secure computation, which is practically expensive [6]. These costs primarily arise from the need to generate and use randomness, typically requiring the operation of True Random Number Generators (TRNG) or Pseudo-Random Number Generators (PRNG) in masked solutions. As a result, the objective of masked AES hardware implementations is to minimize the demand for fresh randomness, reduce area overhead, and enhance performance.

1.1 Related Works

The development of Threshold Implementations for AES has evolved with a focus on optimizing the area, randomness requirements, and performance. In 2011, Moradi et al. implemented TI for AES with 3 shares, requiring 48 bits of fresh randomness, 11,114 GE, and 266 clock cycles [7]. In 2014, Bilgin et al. proposed a more efficient 3-shares TI, reducing the randomness requirement to 44 bits, with 9,102 GE and 246 clock cycles [8]. In 2016, Gross et al. introduced the first 2-shares AES TI (AES-simple), which required only 18 bits of fresh randomness, 7,100 GE, and 246 clock cycles [9]. These TI schemes require fresh randomness. However, in 2019, Sugawara utilized the changing of the guards technique to achieve 3 shares AES TI without fresh randomness [10]. The area was 17100 GE and required 266 clock cycles, indicating substantial resource overhead. In 2021, to further advance in TI, Shahmirzadi and Moradi proposed a 2 shares TI for AES without randomness, with 8183 GE and 246 clock cycles. However, the secure S-box is non-uniform [11,12]. In 2022, Askeland et al. introduced a first-order AES TI using generalized changing of the guards technique with 2 shares [13], the Design I (a low-area variant of TI) required 8130 GE and 242 clock cycles.

The Boyar-Matthews-Peralta AES S-box, based on a novel logic minimization technique, is the smallest [14]. However, there has been limited research on hardware masking protection designs for the Boyar-Matthews-Peralta S-box. This paper aims to demonstrate one of the smallest implementations of masked AES without online fresh randomness.

1.2 Our Contributions

Based on the changing of the guards technique with reused randomness proposed by Askeland et al., we further enhance the technique by incorporating the guard bit into the internal decomposition of the S-box, thereby reducing the need for reused randomness. Askeland et al. constructed a masked AES S-box requiring 54 reused random bits, whereas our approach only requires 23 reused random bits. Therefore, compared to the work of Askeland et al., our scheme reduces the reuse of random bits by 57%. Based on this optimization and in combination with the Cagdas Calik AES S-box (an enhanced version of the Boyar-Matthews-Peralta AES S-box), we propose a low-randomness, low-area first-order masked AES hardware implementation. This approach eliminates the need for online fresh randomness while achieving minimal shares and providing provable security in the glitch+register-transition-robust probing model. Additionally, we perform leakage detection experiments using PROLEAD and TVLA. The masked AES requires only 7671 GE with 236 clock cycles on ASIC. On Xilinx Spartan-6 FPGA, it requires only 212 slices with 125 Mbps, which achieves a throughput-per-slice of 0.319 Mbps/slice. To the best of our knowledge, this is the lowest randomness, area, and latency among existing masked AES without online fresh randomness.

2 Preliminaries

2.1 Notations

We denote binary random variables $\in GF(2)$ with lower-case italic x, x_i represents the i-th input variable of a Boolean function, and x^j represents the j-th share of a variable, thus $x = x^0 \oplus x^1 \oplus \cdots \oplus x^j$. In a multi-cycle pipelined implementation, the input variable x_0 is written as $x_{0,0}$ in the first cycle and as $x_{0,1}$ in the second cycle. We denote a Boolean function with lower-case italic sans-serif $f(.)$, $f^l(.)$ represents the l-th component function of $f(.)$, thus $f(.) = f^0(.) \oplus f^1(.) \oplus \cdots \oplus f^l(.)$. The XOR operation $x_0 \oplus x_1$ represents a linear term that is also a linear function, and the AND operation $x_0 x_1$ represents a non-linear term that is also a quadratic function, and the AND operation $x_0 x_1 x_2$ is a cubic function. reg represents the register layer, r denotes random bits, and g stands for guard bits.

2.2 AES

AES is based on an SPN (Substitution-Permutation Networks) structure and 128 bits blocks with 128, 192, and 256 bits key blocks and corresponding 10, 12, and 14 rounds, respectively. According to the length of the three keys, algorithms are denoted as AES-128, AES-192, and AES-256. In this work, we focus on the encryption process of AES-128 as shown in Fig. 1. The round function includes SubBytes, ShiftRows, Mixcolumns and AddRoundKey. The first round requires whitening keys, and the last round removes the Mixcolumns. And the

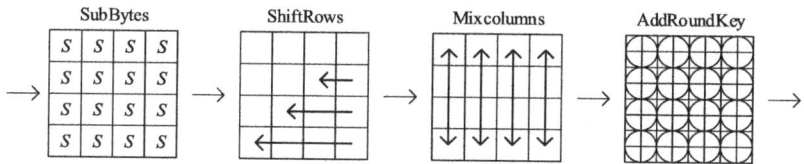

Fig. 1. The encryption process of AES-128

KeyExpansion generates round keys. Each of these operations is described in detail below:

SubBytes: The SubBytes, also called S-boxes, is the only nonlinear transformation in the AES algorithm whose transformation is a multiplicative inverse in the finite field $GF(2^8)$ and an affine transformation over $GF(2)$.

ShiftRows: The ShiftRows is a linear transformation of the AES algorithm, where the transformation is a cyclic left shift of the i-th byte in the i-th ($0 \leq i \leq 3$) row.

MixColumns: The MixColumns is also a linear transformation of the AES algorithm, and the transformation is a matrix multiplication in the finite field $GF(2^8)$.

AddRoundKey: The data block is XORed with the round key of the current round.

The KeyExpansion of AES-128 involves S-boxes and XOR operations.

2.3 First-Order Probing Security

The probing security model, commonly used to assess masking schemes, evaluates the number of probes an adversary can place on circuit signals, reflecting the order of the attack. While effective in software, it fails for hardware due to glitches—undesired transitions in CMOS circuits caused by imbalanced path delays. Faust et al. introduced a glitch-extended probing model to address this, allowing adversaries to observe all ϵ-inputs via extended probes [15]. This article focuses on first-order secure hardware implementations, evaluating them under the first-order glitch-extended probing model, which targets univariate attacks.

2.4 Threshold Implementations

Threshold implementations (TI) have become one of the most widely adopted countermeasures for masking cryptographic primitives. Based on the principles of secure multi-party computation and secret sharing, TI effectively defends against differential power analysis (DPA), even in the presence of glitches. In [5], the authors propose that a nonlinear function with algebraic degree t can be secured against a d-th order side-channel attack by using at least $td + 1$ input shares. However, later research [9] has shown that adhering strictly to the $td + 1$ rule is not a strict requirement for constructing secure masked hardware implementations. Instead, employing $d + 1$ input shares is sufficient to guarantee d-th order probing security for the nonlinear function. TI implementations must satisfy three key properties: Correctness, Non-completeness, and Uniformity.

3 Masked S-Box with the Changing of the Guards

3.1 Analysis of the Changing of the Guards

In the context of threshold implementations, if the masked S-box is non-uniform, remasking operations are typically required. However, remasking often necessitates online fresh randomness. The changing of the guards method proposed by Daemen [16] is a technique that builds a threshold implementation with $d+1$ shares of any invertible S-box layer, ensuring it is correctness, non-completeness, and uniformity. The purpose of the changing of the guards technique is to reduce the reliance on online fresh randomness in threshold implementations. In block ciphers, encryption data is processed in block groups. For example, when the S-box processes one block group of data, it can use unrelated shares from the previous S-box output as online fresh randomness. This approach reduces both the randomness requirements and area resource overhead.

As the algebraic degree of a nonlinear S-box increases, the number of required shares and multiplicative terms for its protection design also grows, ultimately resulting in greater area resource overhead. To reduce the area overhead in protection implementations, a high-degree nonlinear S-box can be decomposed into multiple lower-degree nonlinear functions. For instance, threshold implementations for AES typically employ the Canright AES S-box. In this case, where TI of the S-box requires decomposition to reduce area overhead, the Changing of the Guards technique needs improvement. This improvement would allow it to better accommodate the requirements of masked S-box. Askeland et al. have extended this method by introducing a generalized approach that utilizes randomness—recyclable across different S-boxes—and can be computed in multiple stages, including the whole tower-field decomposition of the AES S-box, while maintaining first-order probing security [13]. The application of this technique is illustrated in Fig. 2 and formally defined in Definition 1.

Definition 1. *The changing of the guards method applied to a masked S-box $\bar{S}$ given inputs $(x^0,\ldots,x^d)$, $(g^0,\ldots,g^{d-1})$, and randomness $\bar{r}$ is calculated as follows*

$$\begin{aligned}
&\bar{r}' = \bar{r}, \\
&x'^0 = \bar{S}(x^0,\ldots,x^d,\bar{r}) \oplus g^0, \ldots, x'^{d-1} = \bar{S}(x^0,\ldots,x^d,\bar{r}) \oplus g^{d-1}, \\
&x'^d = \bar{S}(x^0,\ldots,x^d,\bar{r}) \oplus g^0 \oplus \cdots \oplus g^{d-1}, \\
&g'^0 = x^0, \ldots, g'^{d-1} = x^{d-1}.
\end{aligned} \quad (1)$$

where the $(g^0,\ldots,g^{d-1})$ are the guards of the masked S-box $\bar{S}$.

As shown in Fig. 2, the guard bit is introduced at the output of the masked S-box to ensure uniformity. However, these guard bits are not utilized within the internal operations of the masked S-box, which limits the reduction in reused

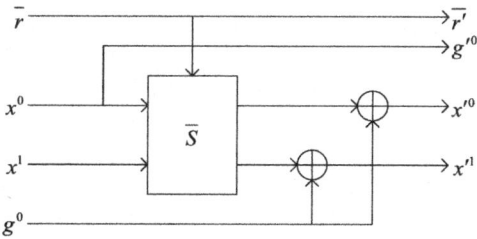

Fig. 2. The changing of the guards method with randomness

randomness. For example, consider an AND gate logic function $t = f(x_0, x_1) = x_0 x_1$. The construction of the masking scheme proposed by Askeland et al., known as the changing of the guards technique with reused randomness, is implemented as follows:

$$\begin{aligned} f^0(x_0^0, x_1^0, r_0) &= x_0^0 x_1^0 \oplus r_0; \\ f^1(x_0^0, x_1^1, r_1) &= x_0^0 x_1^1 \oplus r_1; \\ f^2(x_0^1, x_1^0, r_2) &= x_0^1 x_1^0 \oplus r_2; \\ f^3(x_0^1, x_1^1, r_0, r_1, r_2) &= x_0^1 x_1^1 \oplus r_0 \oplus r_1 \oplus r_2; \end{aligned} \quad (2)$$

After the computation phase is completed, the guard bit g is introduced during the compression phase.

$$\begin{aligned} t^0 &= f^0(.) \oplus f^1(.) \oplus g; \\ t^1 &= f^2(.) \oplus f^3(.) \oplus g. \end{aligned} \quad (3)$$

3.2 Our Design

Building upon the changing of the guards technique with randomness, we propose a low-randomness and low-resource masked design. In our approach, the guard bit is integrated into the internal decomposition of the S-box to perform masked operations, thereby reducing the need for reused randomness. However, incorporating the guard bit into internal S-box calculations must meet certain conditions; otherwise, it may result in transition leakage. Although the guard bits are derived from unrelated shares of the other block group, serving as a substitute for online fresh randomness, issues can arise in the algorithm's pipelined protection implementation. Specifically, shared registers in the pipeline may cause transition leakage. This type of leakage, induced by the pipelined implementation, is often subtle and difficult to detect.

We propose a protected construction for the AND gate logic function $t = f(x_0, x_1) = x_0 x_1$ by the guard bit g and randomness r into the computation.

The detailed protection scheme is as follows:

$$f^0(x_0^0, x_1^0, g) = x_0^0 x_1^0 \oplus g;$$
$$f^1(x_0^0, x_1^1, r) = x_0^0 x_1^1 \oplus r;$$
$$f^2(x_0^1, x_1^0, g) = x_0^1 x_1^0 \oplus g;$$
$$f^3(x_0^1, x_1^1, r) = x_0^1 x_1^1 \oplus r;$$

(4)

Since the guard bit g has already been incorporated during the computation phase, there is no need to introduce it again during the compression phase.

$$t^0 = f^0(.) \oplus f^1(.);$$
$$t^1 = f^2(.) \oplus f^3(.).$$

(5)

Our approach incorporates the guard bit as a random number in the protection operation of the AND logic function, thereby further reducing the need for reused randomness. While the low-resource implementation proposed by Askeland et al. requires three random bits, our solution only requires one random bit. Additionally, our approach involves fewer XOR operations, leading to a more efficient hardware implementation with reduced resource usage and smaller area overhead.

However, when incorporating the guard bit for protection, it is crucial to pay close attention to the implementation details of the algorithm, as neglecting these details could lead to transition leakage. The following examples illustrate such a situation.

In the serial and pipelined implementation of a block cipher, the same S-box operation is executed n times. To simplify the S-box computation, we design the S-box as a pipelined implementation based on a quadratic AND gate logic function. When applying the TI masking scheme for the protection of the AND gate function, random bits are incorporated into the multiplication cross terms to achieve the required uniformity. For instance, the first and second design implementations are presented in reference [9,11].

Let $x_{0,0}^0, x_{0,0}^1, x_{1,0}^0, x_{1,0}^1, x_{0,1}^0, x_{0,1}^1, x_{1,1}^0, x_{1,1}^1$ represent the two shared input data for the two iterative pipelined implementation of the AND logic function $t = f(x_0, x_1) = x_0 x_1$. According to the changing of the guards technique with randomness, in the first clock cycle, the guard bit g is assigned the fresh randomness r; in the second clock cycle, the guard bit g is assigned the value $x_{1,0}^0$ from the input data of the first clock cycle.

The first protected design implementation:
First cycle:

$$reg_0 \leftarrow x_{0,0}^0 x_{1,0}^0;$$
$$reg_1 \leftarrow x_{0,0}^0 x_{1,0}^1 \oplus r;$$
$$reg_2 \leftarrow x_{0,0}^1 x_{1,0}^0 \oplus r;$$
$$reg_3 \leftarrow x_{0,0}^1 x_{1,0}^1;$$

(6)

Second cycle:
$$reg_0 \leftarrow x_{0,1}^0 x_{1,1}^0;$$
$$reg_1 \leftarrow x_{0,1}^0 x_{1,1}^1 \oplus x_{1,0}^0;$$
$$reg_2 \leftarrow x_{0,1}^1 x_{1,1}^0 \oplus x_{1,0}^0;$$
$$reg_3 \leftarrow x_{0,1}^1 x_{1,1}^1;$$
(7)

In the pipelined implementation across two consecutive clock cycles, transition leakage between $x_{1,0}^0$ and $x_{1,0}^1$ occurs at the register reg_1.

The second protected design implementation:
First cycle:
$$reg_0 \leftarrow x_{0,0}^0 x_{1,0}^0 \oplus r;$$
$$reg_1 \leftarrow x_{0,0}^0 x_{1,0}^1;$$
$$reg_2 \leftarrow x_{0,0}^1 x_{1,0}^0;$$
$$reg_3 \leftarrow x_{0,0}^1 x_{1,0}^1 \oplus r;$$
(8)

Second cycle:
$$reg_0 \leftarrow x_{0,1}^0 x_{1,1}^0 \oplus x_{1,0}^0;$$
$$reg_1 \leftarrow x_{0,1}^0 x_{1,1}^1;$$
$$reg_2 \leftarrow x_{0,1}^1 x_{1,1}^0;$$
$$reg_3 \leftarrow x_{0,1}^1 x_{1,1}^1 \oplus x_{1,0}^0;$$
(9)

Transition leakage between $x_{1,0}^0$ and $x_{1,0}^1$ occurs at the register reg_3.
Our protected design implementation:
First cycle:
$$reg_0 \leftarrow x_{0,0}^0 x_{1,0}^0 \oplus r;$$
$$reg_1 \leftarrow x_{0,0}^0 x_{1,0}^1;$$
$$reg_2 \leftarrow x_{0,0}^1 x_{1,0}^0 \oplus r;$$
$$reg_3 \leftarrow x_{0,0}^1 x_{1,0}^1;$$
(10)

Second cycle:
$$reg_0 \leftarrow x_{0,1}^0 x_{1,1}^0 \oplus x_{1,0}^0;$$
$$reg_1 \leftarrow x_{0,1}^0 x_{1,1}^1;$$
$$reg_2 \leftarrow x_{0,1}^1 x_{1,1}^0 \oplus x_{1,0}^0;$$
$$reg_3 \leftarrow x_{0,1}^1 x_{1,1}^1;$$
(11)

In our protected design implementation, no transition leakage occurs at any of the registers.

To reduce the resource overhead in TI of AES, we employed the Cagdas Calik AES S-box (an enhanced version of the Boyar-Matthews-Peralta AES S-box) for masked protection. We have designed and implemented this Cagdas Calik AES S-box. The structure of this S-box is shown in Fig. 3. The first nonlinear computation layer has inputs x_0, x_1, x_2, x_3, x_4, x_5, x_6, and x_7 (the 8-bit input of the S-box), with outputs t_{10}, t_{11}, t_{12}, t_{13}, t_{14}, t_{15}, and t_{16}. The linear XOR operation layer has inputs t_{10}, t_{11}, t_{12}, t_{13}, t_{14}, t_{15}, and t_{16}, with outputs t_{21}, t_{22}, t_{23}, and t_{24}. The second nonlinear computation layer has inputs t_{21}, t_{22}, t_{23}, and t_{24}, with outputs t_{29}, t_{33}, t_{37}, and t_{40}. The third nonlinear computation layer has inputs x_0, x_1, x_2, x_3, x_4, x_5, x_6, x_7, t_{29}, t_{33}, t_{37}, and t_{40}, with outputs s_0, s_1, s_2, s_3, s_4, s_5, s_6, and s_7 (the 8-bit output of the S-box).

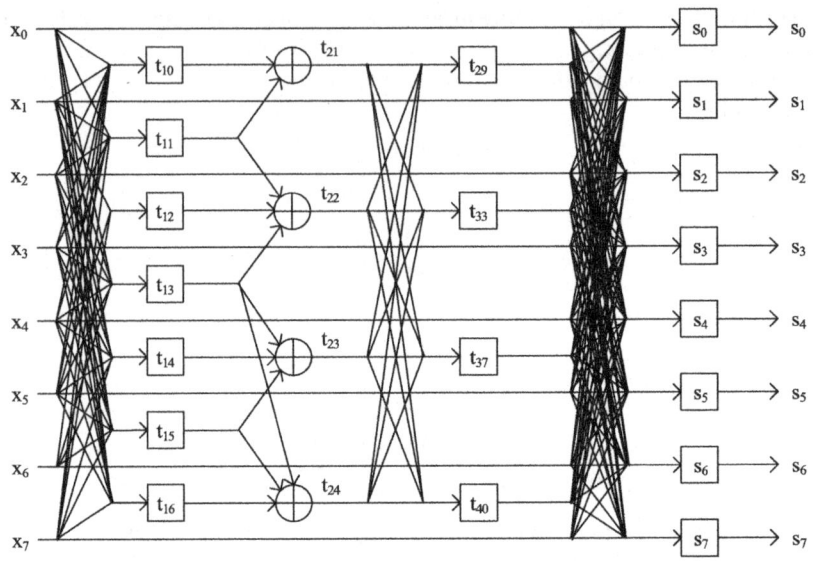

Fig. 3. The unprotected AES S-box

In Fig. 3, the nonlinear computation layers of the Cagdas Calik AES S-box are represented by rectangles. As shown, the S-box comprises three nonlinear computation layers and one linear XOR operation layer.

Protecting AES S-Box Functions. The structure of our protection is illustrated in Fig. 4, where square boxes represent the non-linear calculation layer, circular boxes represent the linear compression layer.

The masked first nonlinear functions are decomposed into four component functions. For instance, consider the quadratic function $t_{12} = f(x_1, x_5, x_7) = x_1 x_5 \oplus (x_5 \oplus x_7)$:

$$\begin{aligned}
f^0(x_1^0, x_5^0, x_7^0) &= x_1^0 x_5^0 \oplus (x_5^0 \oplus x_7^0); \\
f^1(x_1^0, x_5^1, r) &= x_1^0 x_5^1 \oplus r; \\
f^2(x_1^1, x_5^0, r) &= x_1^1 x_5^0 \oplus r; \\
f^3(x_1^1, x_5^1, x_7^1) &= x_1^1 x_5^1 \oplus (x_5^1 \oplus x_7^1); \\
t_{12}^0 &= f^0(.) \oplus f^1(.); \\
t_{12}^1 &= f^2(.) \oplus f^3(.).
\end{aligned} \quad (12)$$

The calculation layer computes component functions $f^0(.)$ to $f^3(.)$, which are stored in registers. This requires one random bit r. The compression layer calculates results t_{12}^0 and t_{12}^1, which also need to be stored in registers.

Similarly, the masked second nonliear functions require decomposition into eight component functions. For instance, $t_{37} = f(t_{21}, t_{22}, t_{23}, t_{24}) = t_{23}t_{22}t_{24} \oplus t_{23}(t_{21} \oplus t_{22}) \oplus (t_{23} \oplus t_{24})$:

$$f^0(t_{23}^0, t_{22}^0, t_{24}^0, t_{21}^0, r_0) = t_{23}^0 t_{22}^0 t_{24}^0 \oplus t_{23}^0(t_{21}^0 \oplus t_{22}^0) \oplus (t_{23}^0 \oplus t_{24}^0) \oplus r_0;$$
$$f^1(t_{23}^0, t_{22}^0, t_{24}^1, t_{21}^1) = t_{23}^0 t_{22}^0 t_{24}^1 \oplus (t_{21}^0 \oplus t_{22}^0);$$
$$f^2(t_{23}^0, t_{22}^1, t_{24}^0, t_{21}^0, r_1) = t_{23}^0 t_{22}^1 t_{24}^0 \oplus t_{23}^0(t_{21}^0 \oplus t_{22}^1) \oplus r_1;$$
$$f^3(t_{23}^0, t_{22}^1, t_{24}^1, t_{21}^1) = t_{23}^0 t_{22}^1 t_{24}^1 \oplus t_{21}^0;$$
$$f^4(t_{23}^1, t_{22}^0, t_{24}^0, t_{21}^0) = t_{23}^1 t_{22}^0 t_{24}^0 \oplus (t_{21}^0 \oplus t_{22}^0);$$
$$f^5(t_{23}^1, t_{22}^0, t_{24}^1, t_{21}^1, r_1) = t_{23}^1 t_{22}^0 t_{24}^1 \oplus t_{23}^1(t_{21}^1 \oplus t_{22}^0) \oplus r_1;$$
$$f^6(t_{23}^1, t_{22}^1, t_{24}^0, t_{21}^0) = t_{23}^1 t_{22}^1 t_{24}^0 \oplus t_{21}^0;$$
$$f^7(t_{23}^1, t_{22}^1, t_{24}^1, t_{21}^1, r_0) = t_{23}^1 t_{22}^1 t_{24}^1 \oplus t_{23}^1(t_{21}^1 \oplus t_{22}^1) \oplus (t_{23}^1 \oplus t_{24}^1) \oplus r_0;$$
$$t_{37}^0 = f^0(.) \oplus f^1(.) \oplus f^2(.) \oplus f^3(.);$$
$$t_{37}^1 = f^4(.) \oplus f^5(.) \oplus f^6(.) \oplus f^7(.).$$

(13)

It requires two random bits r_0 and r_1, and t_{21}^0 and t_{22}^0 are input variables acting as Correction Terms (CTs).

The masked third nonlinear functions are decomposed into four component functions. For instance, $s_3 = f(x_0, x_1, x_2, x_3, x_4, x_5, x_6, x_7, t_{29}, t_{33}, t_{37}, t_{40}) = t_{29}(x_0 \oplus x_2 \oplus x_4 \oplus x_6) \oplus t_{33}(x_0 \oplus x_4) \oplus t_{37}(x_1 \oplus x_2 \oplus x_5) \oplus t_{40}(x_1 \oplus x_3 \oplus x_4 \oplus x_6 \oplus x_7)$:

$$f^0(x_0^0, x_1^0, x_2^0, x_3^0, x_4^0, x_5^0, x_6^0, x_7^0, t_{29}^0, t_{33}^0, t_{37}^0, t_{40}^0, g) = t_{29}^0(x_0^0 \oplus x_2^0 \oplus x_4^0 \oplus x_6^0)$$
$$\oplus t_{33}^0(x_0^0 \oplus x_4^0) \oplus t_{37}^0(x_1^0 \oplus x_2^0 \oplus x_5^0) \oplus t_{40}^0(x_1^0 \oplus x_3^0 \oplus x_4^0 \oplus x_6^0 \oplus x_7^0) \oplus g;$$
$$f^1(x_0^0, x_1^1, x_2^1, x_3^1, x_4^1, x_5^1, x_6^1, x_7^1, t_{29}^0, t_{33}^0, t_{37}^0, t_{40}^0, r) = t_{29}^0(x_0^0 \oplus x_2^1 \oplus x_4^1 \oplus x_6^1)$$
$$\oplus t_{33}^0(x_0^1 \oplus x_4^1) \oplus t_{37}^0(x_1^1 \oplus x_2^1 \oplus x_5^1) \oplus t_{40}^0(x_1^1 \oplus x_3^1 \oplus x_4^1 \oplus x_6^1 \oplus x_7^1) \oplus r;$$
$$f^2(x_0^0, x_1^0, x_2^0, x_3^0, x_4^0, x_5^0, x_6^0, x_7^0, t_{29}^1, t_{33}^1, t_{37}^1, t_{40}^1, g) = t_{29}^1(x_0^0 \oplus x_2^0 \oplus x_4^0 \oplus x_6^0)$$
$$\oplus t_{33}^1(x_0^0 \oplus x_4^0) \oplus t_{37}^1(x_1^0 \oplus x_2^0 \oplus x_5^0) \oplus t_{40}^1(x_1^0 \oplus x_3^0 \oplus x_4^0 \oplus x_6^0 \oplus x_7^0) \oplus g;$$
$$f^3(x_0^1, x_1^1, x_2^1, x_3^1, x_4^1, x_5^1, x_6^1, x_7^1, t_{29}^1, t_{33}^1, t_{37}^1, t_{40}^1, r) = t_{29}^1(x_0^1 \oplus x_2^1 \oplus x_4^1 \oplus x_6^1)$$
$$\oplus t_{33}^1(x_0^1 \oplus x_4^1) \oplus t_{37}^1(x_1^1 \oplus x_2^1 \oplus x_5^1) \oplus t_{40}^1(x_1^1 \oplus x_3^1 \oplus x_4^1 \oplus x_6^1 \oplus x_7^1) \oplus r;$$
$$s_3^0 = f^0(.) \oplus f^1(.);$$
$$s_3^1 = f^2(.) \oplus f^3(.).$$

(14)

It requires one random bit r and one guard bit g.

To construct a secure masked AES S-box, 23 random bits are needed. t_{21}, t_{22}, t_{23}, and t_{24}, being linear functions, do not require any random bit. The seven quadratic functions (t_{10} to t_{16}) require one random bit each, totaling 7 random bits. The four cubic functions (t_{29}, t_{33}, t_{37}, and t_{40}) require two random bits each, totaling 8 random bits. Similarly, the eight quadratic functions (s_0 to s_7) demand one random bits each, totaling 8 random bits. The 23 random bits can be re-used over all the S-boxes. Askeland et al. constructed a masked AES S-box requiring 54 reused random bits, whereas our approach only requires 23 reused random bits. Therefore, compared to the work of Askeland et al., our scheme reduces the reuse of random bits by 57%.

4 First-Order Secure AES Encryption

4.1 Masked AES

Due to the introduction of side-channel protections in the AES algorithm, there is an increase in hardware resource requirements. To reduce the hardware implementation area, a serial and pipelined approach is employed such that only one protected S-box is required for the entire AES algorithm. This serial and pipelined implementation has been widely adopted in numerous works. Our work is also based on this approach. In this implementation, during the first 16 clock cycles, plaintext and key data are loaded into the state registers and key registers. Initially, the plaintext and key data undergo the AddRoundKey. The results of this operation are then input to the masked S-box, which requires 6 clock cycles for computation. The output of the masked S-box is then loaded into registers. Subsequently, the ShiftRows and MixColumns are applied (these are linear transformations and are implemented with combinational logic, so they do not consume additional clock cycles). The data from the MixColumns is then output for the next round of encryption. This round of encryption requires

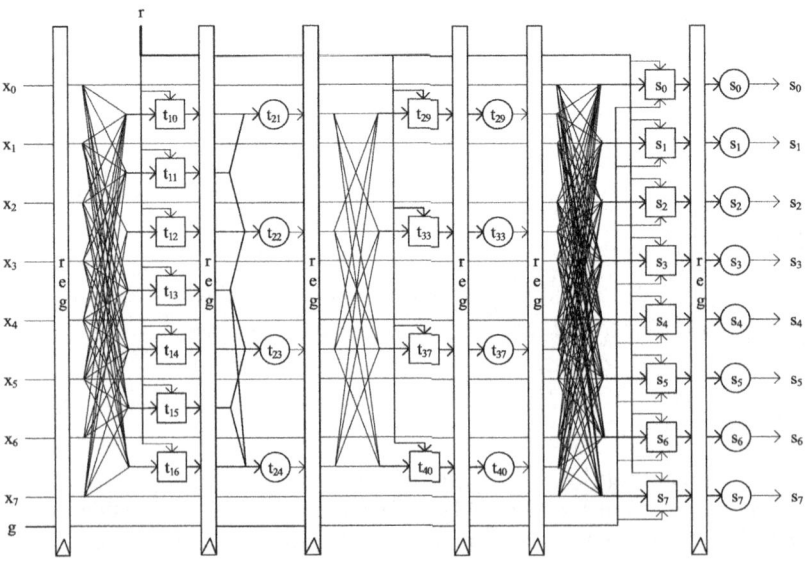

Fig. 4. Our construction for the 2-share masked AES S-box

22 clock cycles, as the 16th cycle of plaintext and key data must wait for the completion of the 6 clock cycles needed for the masked S-box computation to obtain the S-box output. For the KeyExpansion implementation, the four masked S-box operations are carried out immediately after the encryption masked S-box operations, thus avoiding the need for additional clock cycles. The specific implementation process is shown in Fig. 5. Ten rounds of encryption require 22×10 clock cycles, and an additional 16 clock cycles are needed to output the ciphertext, making the total masked AES encryption process require 236 clock cycles.

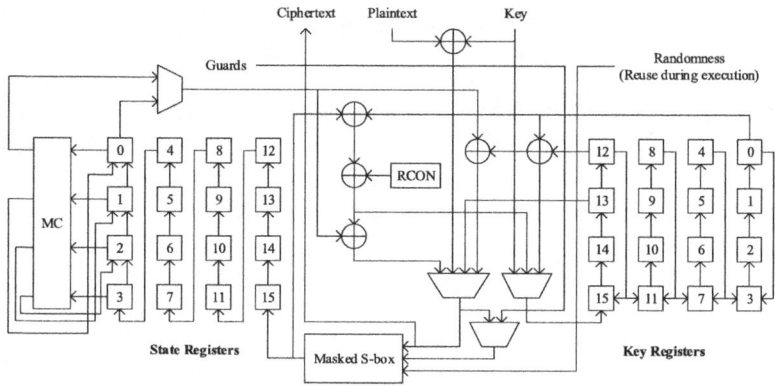

Fig. 5. Our design of the 2-share masked AES-128 encryption.

The generalized changing of the guards method is employed for the masked S-box. This method requires guards and random bits. The specific implementation process can be seen in Fig. 4. The guard bits are initialized with fresh randomness, and subsequent guard bits are taken from the input of the masked S-box. In our scheme, 8 random bits are required to initialize the guards, and these remain static throughout the entire AES protected encryption. The 23 random bits only need to be input once at the beginning, and after that, these random bits can be reused in different masked S-boxes. However, it is essential to note that the random bits in the masked S-boxes are rotated for each S-box application. This rotation is implemented to prevent transitional leakages from occurring in the pipeline registers [13]. Taking the two-shared multiplication of Eq. (15) as an example, other multiplication operations follow a similar process. In a serial and pipelined implementation, when the multiplication of Eq. (15) is performed with two consecutive inputs, the first input data is x_{10}^0, x_{10}^1, x_{50}^0, x_{50}^1, x_{70}^0, x_{70}^1, $r_{0,0}$ and the second input data is x_{11}^0, x_{11}^1, x_{51}^0, x_{51}^1, x_{71}^0, x_{71}^1, $r_{0,1}$. By rotating $r_{0,0}$ and $r_{0,1}$, two consecutive executions calculating register t_{12}^0 produce a Hamming Distance leakage HD $(x_{1,0}^0 x_{5,0} \oplus x_{5,0}^0 \oplus x_{7,0}^0 \oplus r_{0,0}, x_{1,1}^0 x_{5,1} \oplus x_{5,1}^0 \oplus x_{7,1}^0 \oplus r_{0,1})$, which is masked by $r_{0,0}$ and $r_{0,1}$. Initially, the guards and the random bits are input, with no need for fresh random bits during subsequent computational processes, thus eliminating the use of online fresh randomness in our scheme.

4.2 Performance

Metrics in Application-Specific Integrated Circuit (ASIC). For the ASIC synthesis, we used the Synopsis Design Compiler v2016.03 with the NanGate 45nm Open Cell Library. For performance evaluation, measuring latency and maximum clock frequency (FMAX) is essential. By incorporating the block size (BSIZE), maximum throughput (THR) can be calculated, as indicated by Eq. (15) [17].

$$THR = (FMAX \times BSIZE) \div LAT. \tag{15}$$

In our masked S-box implementation, the area is 2686 GE, and the latency (LAT) is 6 clock cycles. In our masked AES implementation, the area is 7921 GE, and the latency is 236 clock cycles. Then, we calculated the ratio of throughput to area. Their detailed performance is shown in Table 1. In contrast to these implementation schemes [7–9,18,19], our scheme does not require online fresh randomness and has advantages in terms of latency. Without fresh randomness, both Sugawara's scheme and Askeland et al.'s schemes employ the changing of the guards [10,13]. Compared to Sugawara's scheme, our scheme demonstrates superior area efficiency and lower latency. In contrast to the low-area scheme proposed by Askeland et al., our scheme reduces area consumption by 459 GE and decreases latency by 6 clock cycles, and reduces delay by 0.27 ns. Moreover, compared to the free fresh randomness TI of AES proposed by Shahmirzadi and Moradi [11], our scheme achieves a 512 GE reduction in area and a 10 clock cycles reduction in latency. Our scheme is inferior in terms of delay and throughput/area. However, we emphasize that their design is based on a non-uniform masked S-box, whereas our masked S-box is uniform allowing for a stronger security argument. Consequently, on the ASIC implementation platform, our work has the lowest area and latency when compared to various state-of-the-art works that do not use online fresh randomness.

Metrics in Field Programmable Gate Array (FPGA). FPGAs are composed of numerous programmable slice (SLC) logic resource units. The resource usage for the masked AES hardware design is obtained from the implementation reports generated by Design Suite 14.7 with Xilinx Spartan-6 (XC6SLX75) FPGAs.

In Spartan-6 FPGAs, the architecture of each slice consists of 4 Look-Up Tables (LUTs) and 8 Flip-Flops (FFs). By utilizing the slice and the throughput, we can calculate the design implementation efficiency (Throughput-per-Slice, THR/SLC) as depicted in Eq. (16). This equation describes the relationship between the implementation size and performance [17].

$$THR/SLC = THR \div SLC. \tag{16}$$

In our masked AES implementation, the area size is 212 slices, the latency is 236 clock cycles, and the maximum clock frequency is 125 MHz. Additionally, we achieved a maximum throughput of 67.8 Mbps, resulting in a Throughput-per-Slice of 0.319 Mbps/slice. The implementation details are provided in Table 2.

Table 1. Performance figures of different (byte-serial) implemetations (with key masking, using Synopsis Design Compiler v2016.03, and NanGate 45 nm Open cell library)

Design	No. of Shares	Online Fresh Masks [bit]	Area (S-box) [GE]	Latency (S-box) [cycles]	Area (AES) [GE]	Delay (AES) [ns]	Latency (AES) [cycles]	THR/Area (AES) [Mbps/GE]
[7][a]	3	48	–	–	11114	–	266	–
[8][a]	3	44	–	–	9102	–	246	–
[18][a]	3	44	–	–	11221	–	246	–
[18][a]	3	32	2840	4	8119	–	246	–
[9][a]	2	18	–	–	7100	–	246	–
[10][b]	3	0	3500	4	17100	–	266	–
[11][b,c]	2	1	–	6	7619	0.25	246	0.27
[11][b,c]	2	0	–	6	8183	0.29	246	0.21
[13][b]	2	0	2710	6	8130	1.21	242	0.05
[19][b]	2	4	–	6	7843	0.94	236	0.07
this work[b]	2	0	2686	6	7671	1.10	236	0.06

[a] Using UMC 180 nm Standard Cell Library.
[b] Using NanGate 45 nm Standard Cell Library.
[c] with a non-uniform masked S-box.

Table 2. Comparison of area-optimized (byte-serial) implementations including PRNG (with key masking, mapped on Xilinx Spartan-6)

Design	SLC	FF	LUT	BRAM	LAT [cycles]	Online Fresh Masks[bit]	FMAX [MHz]	THR [Mbps]	THR/SLC [Mbps/slice]
[9]	433	726	894	0	216	28	110	65.1	0.150
[9]	351	754	792	0	246	18	134	69.7	0.198
[20]	418	690	905	0	276	54	133	61.6	0.147
[21]	254	124	347	0	6852	18	103	1.9	0.007
[14]	140	92	248	0	6852	6	120	2.2	0.015
[11]	631	866	1021	0	246	0	136	70.7	0.112
[22]	203	529	476	108 kb	216	8	138	81.7	0.402
[19]	213	587	700	0	236	4	108	58.6	0.275
this work	212	562	561	0	236	0	125	67.8	0.319

Consequently, on the FPGA implementation platform, when compared to various state-of-the-art works that do not use online fresh randomness, our work also has the best implementation efficiency (THR/SLC).

5 First-Order Robust Probing Security

In this section, we provide a thorough analysis of our masked AES S-box implementation within the glitch+register-transition-robust probing model, emphasizing first-order security. According to the three principles of TI, we have developed a first-order, 2-share TI design for AES.

We construct our masked S-box using the generalized changing of the guards method, which is detailed in Sect. 3 and defined in Definition 1. This method is proven to be correct, first-order probing secure, and uniform. This method enables for a transformation of any first-order glitch-extended robust probing secure masking into a uniform approach, which allows the re-use of the randomness employed in the S-box. For the proof of Definition 1, please refer to [13]. Consequently, our design of the masked AES is first-order glitch-extended robust probing secure.

In addition to the theoretical analysis of our constructions, we also evaluated them using PROLEAD, confirming their first-order security under the glitch+register-transition-robust probing model. Additionally, our masked AES has passed the transitional leakage detection using PROLEAD. PROLEAD is an automated simulation-based tool designed to analyze the statistical independence of simulated intermediate probes in a given masked implementation under the robust probing model [12].

We used Synopsis Design Compiler and the NanGate 45 nm Open Cell library to generate the netlist and performed the evaluation on a machine equipped a 64 core AMD EPTC Bergamo CPU and 288 GB of memory. The tool conducted fixed versus random G-test, with the fixed input being zero vector. In the univariate evaluation, which included all probing sets, we tested the entire masked AES, and found no leakage in the results. The detailed evaluation results reported in this paper can be made publicly available on GitHub (https://github.com/botaoliu/SecAES).

6 Physical Evaluation

We implement the masked AES design on a Xilinx Kintex-7 FPGA, mounted on a Sakura-X evaluation board. The design is programmed, synthesized, and loaded using Xilinx Vivado. Power measurements are amplified using a 30 dB SMA preamplifier, and the traces are captured by an oscilloscope with a sampling rate of 250 MS/s, while the FPGA operates at a clock frequency of 3 MHz.

The TVLA methodology is a qualitative approach used to evaluate the resistance of cryptographic implementations against side-channel attacks [23]. If the calculated t-statistic is greater than 4.5 or less than -4.5, the null hypothesis is rejected, indicating a significant correlation between the power consumption traces and the processed data. If the t-statistic lies within the acceptance region, it suggests that the masking scheme effectively mitigates side-channel leakage.

To evaluate the TVLA trace capture setup and the Signal-to-Noise Ratio (SNR), we conducted an analysis of the unprotected AES-128 implementation [24]. We collected 100,000 power consumption traces for the serialized AES-128 encryption without masking. Figure 6 presents the results of a first-order univariate t-test, where the horizontal axis represents the number of trace points, and the vertical axis shows the t-statistics. The red lines indicate the threshold of ±4.5. From Fig. 6, it is clear that significant side-channel leakage is present.

Figure 7 illustrates the linearly increasing t-statistics as the number of traces increases from 10,000 to 90,000, with the t-statistics values plotted in absolute

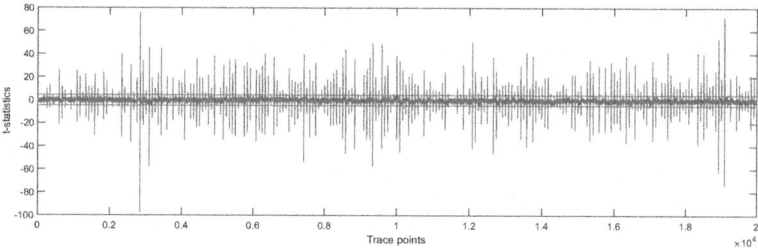

Fig. 6. First-order TVLA of our serialized AES-128 encryption without masking.

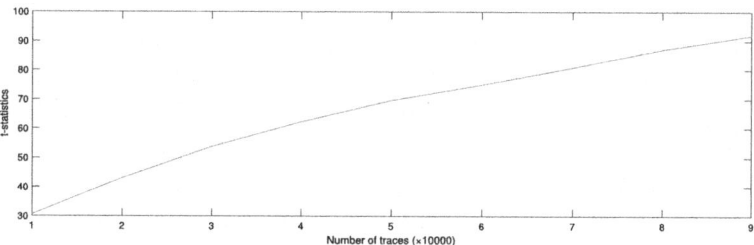

Fig. 7. Incremental TVLA of our serialized AES-128 encryption without masking.

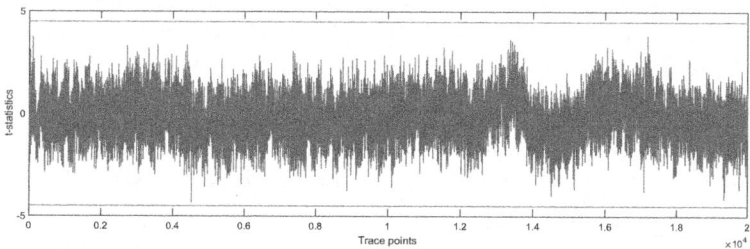

Fig. 8. First-order TVLA of our serialized AES-128 encryption with masking.

terms. Figure 7 demonstrates a sufficiently high SNR for the experimental setup, confirming the robustness of the measurement process.

We collected 10,000,000 power consumption traces for our serialized AES-128 encryption with masking. Figure 8 presents the results of a first-order univariate t-test. As shown in Fig. 8, no first-order leakage was observed.

7 Conclusions

This paper presents a low-randomness, low-area first-order masked AES hardware implementation, minimizing the need for fresh randomness while ensuring security in the glitch+register-transition-robust probing model. By incorporating a guard bit into the S-box decomposition, we reduce the reuse of random bits by 57% compared to Askeland et al.'s work. The implementation requires only

7671 GE and 236 clock cycles on ASIC, and 212 slices with 125 Mbps on Xilinx Spartan-6 FPGA, achieving a throughput of 0.319 Mbps/slice. To our knowledge, this is the most efficient masked AES without online fresh randomness in terms of randomness, area, and latency.

Acknowledgments. This work was supported in part by the National Key R&D Program of China (No. 2022YFB3103800).

References

1. Dworkin, M., et al.: Advanced encryption standard (AES), 26 November 2001
2. Kocher, P., Jaffe, J., Jun, B.: Differential power analysis. In: Wiener, M. (ed.) CRYPTO 1999. LNCS, vol. 1666, pp. 388–397. Springer, Heidelberg (1999). https://doi.org/10.1007/3-540-48405-1_25
3. Chari, S., Jutla, C.S., Rao, J.R., Rohatgi, P.: Towards sound approaches to counteract power-analysis attacks. In: Wiener, M. (ed.) CRYPTO 1999. LNCS, vol. 1666, pp. 398–412. Springer, Heidelberg (1999). https://doi.org/10.1007/3-540-48405-1_26
4. Mangard, S., Pramstaller, N., Oswald, E.: Successfully attacking masked AES hardware implementations. In: Rao, J.R., Sunar, B. (eds.) CHES 2005. LNCS, vol. 3659, pp. 157–171. Springer, Heidelberg (2005). https://doi.org/10.1007/11545262_12
5. Nikova, S., Rechberger, C., Rijmen, V.: Threshold implementations against side-channel attacks and glitches. In: Ning, P., Qing, S., Li, N. (eds.) ICICS 2006. LNCS, vol. 4307, pp. 529–545. Springer, Heidelberg (2006). https://doi.org/10.1007/11935308_38
6. Dhooghe, S., Shahmirzadi, A.R., Moradi, A.: Second-order low-randomness d+ 1 hardware sharing of the AES. In: Proceedings of the 2022 ACM SIGSAC Conference on Computer and Communications Security, pp. 815–828 (2022)
7. Moradi, A., Poschmann, A., Ling, S., Paar, C., Wang, H.: Pushing the limits: a very compact and a threshold implementation of AES. In: Paterson, K.G. (ed.) EUROCRYPT 2011. LNCS, vol. 6632, pp. 69–88. Springer, Heidelberg (2011). https://doi.org/10.1007/978-3-642-20465-4_6
8. Bilgin, B., Gierlichs, B., Nikova, S., Nikov, V., Rijmen, V.: A more efficient AES threshold implementation. In: Pointcheval, D., Vergnaud, D. (eds.) AFRICACRYPT 2014. LNCS, vol. 8469, pp. 267–284. Springer, Cham (2014). https://doi.org/10.1007/978-3-319-06734-6_17
9. Groß, H., Mangard, S., Korak, T.: Domain-oriented masking: compact masked hardware implementations with arbitrary protection order. Cryptology ePrint Archive (2016)
10. Sugawara, T.: 3-share threshold implementation of AES S-box without fresh randomness. IACR Trans. Cryptographic Hardware Embed. Syst., 123–145 (2019)
11. Shahmirzadi, A.R., Moradi, A.: Re-consolidating first-order masking schemes: nullifying fresh randomness. IACR Trans. Cryptographic Hardware Embed. Syst., 305–342 (2021)
12. Müller, N., Moradi, A.: PROLEAD: a probing-based hardware leakage detection tool. IACR Trans. Cryptographic Hardware Embed. Syst., 311–348 (2022)

13. Askeland, A., Dhooghe, S., Nikova, S., Rijmen, V., Zhang, Z.: Guarding the first order: the rise of AES Maskings. In: Buhan, I., Schneider, T. (eds.) CARDIS 2022. LNCS, vol. 13820, pp. 103–122. Springer, Cham (2022). https://doi.org/10.1007/978-3-031-25319-5_6
14. Wegener, F., Meyer, L., Moradi, A.: Spin me right round rotational symmetry for FPGA-specific AES: extended version. J. Cryptol. **33**, 1114–1155 (2020)
15. Faust, S., Grosso, V., Merino Del Pozo, S., Paglialonga, C., Standaert, F.-X.: Composable masking schemes in the presence of physical defaults & the robust probing model. IACR Trans. Cryptographic Hardware Embed. Syst. **2018**(3), 89–120 (2018)
16. Daemen, J.: Changing of the guards: a simple and efficient method for achieving uniformity in threshold sharing. In: Fischer, W., Homma, N. (eds.) CHES 2017. LNCS, vol. 10529, pp. 137–153. Springer, Cham (2017). https://doi.org/10.1007/978-3-319-66787-4_7
17. Lara-Nino, C.A., Diaz-Perez, A., Morales-Sandoval, M.: Lightweight hardware architectures for the present cipher in FPGA. IEEE Trans. Circuits Syst. I Regul. Pap. **64**(9), 2544–2555 (2017)
18. Bilgin, B., Gierlichs, B., Nikova, S., Nikov, V., Rijmen, V.: Trade-offs for threshold implementations illustrated on AES. IEEE Trans. Comput. Aided Des. Integr. Circuits Syst. **34**(7), 1188–1200 (2015)
19. Yao, F., Chen, H., Wei, Y., Pasalic, E., Zhou, F., Fan, L.: Optimizing AES threshold implementation under the glitch-extended probing model. IEEE Trans. Comput.-Aided Des. Integr. Circuits Syst. (2024)
20. De Cnudde, T., Reparaz, O., Bilgin, B., Nikova, S., Nikov, V., Rijmen, V.: Masking AES with $d+1$ shares in hardware. In: Gierlichs, B., Poschmann, A.Y. (eds.) CHES 2016. LNCS, vol. 9813, pp. 194–212. Springer, Heidelberg (2016). https://doi.org/10.1007/978-3-662-53140-2_10
21. Meyer, L., Moradi, A., Wegener, F.: Spin me right round: rotational symmetry for FPGA-specific AES. IACR Trans. Cryptographic Hardware Embed. Syst. **2018**(3), 596–626 (2018)
22. Shahmirzadi, A.R., Božilov, D., Moradi, A.: New first-order secure AES performance records. IACR Trans. Cryptographic Hardware Embed. Syst., 304–327 (2021)
23. Becker, G., et al.: Test vector leakage assessment (TVLA) methodology in practice. In: International Cryptographic Module Conference, vol. 1001.sn, p. 13 (2013)
24. Jati, A., Gupta, N., Chattopadhyay, A., Sanadhya, S.K., Chang, D.: Threshold implementations of GIFT: a trade-off analysis. IEEE Trans. Inf. Forensics Secur. **15**, 2110–2120 (2019)

SM2-VBKE: Achieving Cryptographic Binding Between Verification Integrity and Key Generation

Runze Zhao[1,2], Siqi Lu[1,2(✉)], Yongjuan Wang[1,2(✉)], Liujia Cai[1,2], Wenyi Chen[1], and Fenghua Jiang[1]

[1] Henan Key Laboratory of Network Cryptography Technology, Information Engineering University, Zhengzhou 450000, China
080lusiqi@sina.com, pingkwyj@163.com
[2] Key Laboratory of Cyberspace Security, Ministry of Education, Zhengzhou 450000, China

Abstract. We present SM2-VBKE, the first elliptic curve key exchange protocol that cryptographically enforces verification integrity through intrinsic binding with key material generation. Addressing systemic vulnerabilities in cryptographic implementations—where verification flaws persistently undermine theoretical security guarantees—our protocol introduces three fundamental innovations: (1) mandatory elliptic curve point validation through confirmation codes derived from intermediate verification values, (2) cryptographic binding between verification integrity and session keys via the key derivation function, and (3) reusing of verification intermediates avoid introducing additional overhead. Security analysis under the eCK model demonstrates that SM2-VBKE reduces protocol security to the ECCDH assumption while providing automatic error detection through key divergence—any deviation in verification steps alters confirmation codes. Experimental evaluation shows the design adds only 1.29 ms to baseline SM2 execution while eliminating weakness to invalid curve attacks and implementation errors that plagued prior schemes. By transforming verification from an external process to an intrinsic cryptographic property, the SM2-VBKE protocol binds the verification step with key generation and can be proven to be secure against adaptive adversaries.

Keywords: SM2-VBKE · Cryptographic binding · eCK security · Verification Integrity

1 Introduction

The integrity of verification procedures constitutes the cornerstone of cryptographic protocol security. Verification primitives emerge as the primal safeguard in cryptographic protocol design, where their correct instantiation represents a necessary convergence point for security guarantees; these active security predicates—encompassing parametric compliance checks to hierarchical certificate validation—establish the inductive basis for adversarial manipulation resistance. However, implementation flaws in verification mechanisms remain prevalent across cryptographic ecosystems, as evidenced by seminal vulnerabilities including OpenSSL's certificate validation bypass in 2012 [1], Apple's catastrophic "goto fail" bug in 2014 [2], and Amazon's TLS session ticket encryption failure in 2023 [3]. Recent studies on cryptographic techniques for big data security [4] have emphasized the importance of holistic life cycle verifiable protection, including secure storage and ciphertext-based computations. Such incidents collectively demonstrate a systemic weakness: verification steps, though theoretically sound, often succumb to implementation errors that evade conventional testing methodologies.

This vulnerability becomes particularly acute in elliptic curve cryptography (ECC), where point validation forms the bedrock of security guarantees. Historical attacks—spanning Biehl's differential fault attacks in 2000 [5], Jager's invalid-curve assaults on TLS-ECDH in 2015 [6], to the invalid-curve point attacks targeting Bluetooth protocols in 2020 [7]—have all surgically exploited this cryptographic Achilles' heel. In 2024, Lu et al. [8] demonstrated a user identifier stealing attack in Private-Join-and-Compute protocols, where improper binding between verification steps and key derivation allowed adversaries to infer private data via summation results. This vulnerability becomes particularly acute in elliptic curve cryptography (ECC), where point validation forms the bedrock of security guarantees. The SM2 key exchange protocol is based on elliptic curves [9], released by the China State Cryptography Administration in 2010. Previous analyses have identified several attack vectors against the SM2 key exchange, all of which rely on flawless execution of the verification steps—an assumption frequently compromised in real-world deployments due to developer errors or lapses in maintenance.

Our work bridges this foundational gap through the SM2-VBKE, a novel key exchange protocol featuring cryptographically enforced self-verification. Unlike traditional protection mechanisms that rely on validation steps, the SM2-VBKE protocol offers the following advantages:

- **Mandatory Elliptic Curve Point Verification:** Innovatively incorporates the correctness of the validation step as a prerequisite for shared key generation. By using a "confirmation code" cryptographically bound to the process, the protocol ensures that key unavailability results from validation failure through a self-destructive security design.
- **Reuse of Intermediate Results in Elliptic Curve Point Verification:** Breaks away from the traditional paradigm where verification results are discarded. Instead, it reuses the intermediate values generated during the elliptic curve equation validation to produce the confirmation code. This technique transforms the mathematical relations, initially only used for ephemeral validation, into persistent cryptographic credentials. Alterations to the validation

process trigger avalanche-like changes in subsequent keys, ensuring cryptographic traceability of the validation process.
- **Lightweight Verification Paradigm:** The SM2-VBKE protocol overcomes the resource consumption bottleneck of traditional redundant verification (such as incremental verification) by directly reusing the algebraic results of point verification as the confirmation code. This enables the verification step to be validated with an additional overhead of only 0.0012 s per execution.

1.1 Related Work

Existing Security Analyses of SM2: In 2011, Xu et al. [10] found that there are two unknown key sharing attacks in the SM2 key exchange protocol in a common computing environment. In a similar vein, Zhao et al. [11] discovered that flaws in the design of the application programming interface within the TPM2.0 environment could potentially result in key sharing and key leakage simulation attacks. In 2019, Wei et al. [12] proposed an attack that exploits partial information leakage from intermediate variables to undermine the SM2 key exchange protocol. In 2022, Cao et al. [13] found that the algorithmic structure and code quality of the SM2 algorithm implementation are problematic, and proposed an extended key recovery attack in a targeted manner. It is evident that SM2 may encounter potential issues in real-world implementations due to the insufficient cryptographic expertise of developers and the lack of regular maintenance and thorough professional testing of the code.

Mitigating Implementation Errors: Conventional approaches to address cryptographic implementation flaws including Necula's Proof-Carrying Code [14], NIST's CAVP [15], and Mouha-Celi's LDT [16] rely on external validation frameworks that inherently introduce verification overhead while failing to prevent runtime errors. Emerging blockchain approaches [17] employ network-aware agreement mechanisms to dynamically adjust verification processes, demonstrating adaptive efficiency improvements. Our SM2-VBKE protocol fundamentally transcends these limitations through cryptographically enforced self-verification, where validation integrity becomes an inseparable component of key material generation. Unlike PCC's proof burdens or CAVP's post-hoc testing, SM2-VBKE achieves automatic error detection by entangling verification correctness with session key derivation—any deviation in validation steps propagates through the confirmation code to disrupt key consistency. This intrinsic security mechanism eliminates redundant verification paths, addressing the critical gap between theoretical security and practical implementation vulnerabilities that persists in existing methodologies.

Traditional Verification Paradigms: In 2003, Marc Fischlin introduced incremental verification [18], linking verification reliability to the time spent on the process. In 2019, Duc V. Le et al. expanded this with flexible signatures [19], quantifying signature validity based on the number of verification executions. Building on this, Abdul Rahman Taleb and Damien Vergnaud advanced asymptotic verification in 2021 [20], proposing an efficient probabilistic procedure for signature schemes like RSA and ECDSA, where the probability of verifier error decreases exponentially with increased runtime. While asymptotic verification boosts reliability by associating time with accuracy, it adds computational overhead, increasing execution costs. In 2023, Marc Fischlin et al. expanded this with verifiability verification [21], enhancing the RSA-PSS and HMAC schemes and suggesting improvements for key exchange protocols, though lacking specific examples for practical implementation. Inspired by the work of Fischlin et al. [21], we implemented the transformation of verification from an external process to an intrinsic cryptographic property in the SM2 key exchange protocol and provided a comprehensive security proof.

1.2 The Article Structure

Section 2 outlines the difficult problem assumptions, collision resistance of functions, and offers a comprehensive overview of the eCK model. Section 3 presents the SM2-VBKE protocol proposed. Section 4 provides a security proof in the eCK model and a proof of self checking for the verification steps. Section 5 analyzes the performance of the SM2-VBKE protocol and includes a simulation of its implementation. Section 6 concludes this work.

For convenience of representation, the notations and definitions used in this paper are presented in Table 1.

Table 1. Notation and Definition

Notation	Definition
A, B	Clients
d_A, d_B	Long-term private key of A and B
P_A, P_B	Long-term public key of A and B
ecc	Elliptic curve parameters such as $\mathbb{F}_q, E(\mathbb{F}_q), G, n$
$\mathbb{F}_q$	The finite field containing q elements
$E(\mathbb{F}_q)$	The set of all points on an elliptic curve E defined over $\mathbb{F}_q$, including the point at infinity $\mathcal{O}$
$\#E(\mathbb{F}_q)$	The number of points on $E(\mathbb{F}_q)$
G	A distinguished point on an elliptic curve
n	The prime order of G
$KDF(Z, klen)$	A key derivation function whose output length is $klen$
$x\|\|y$	Concatenation of two strings x and y
$\&$	Bitwise AND operator
$y \xleftarrow{\$} \mathcal{F}(x)$	An algorithm $\mathcal{F}$ outputting a random result y
$y \leftarrow \mathcal{F}(x)$	An algorithm $\mathcal{F}$ outputs the determined result y
h	Co-factor, $h = \#E(\mathbb{F}_q)/n$
$[d]P$	Scalar multiplication of a point P, where $[d]P = \underbrace{P + P + \cdots + P}_{d \text{ times}}$

2 Preparatory Knowledge

In this section, we describe the foundational concepts, including the difficult problem assumption, collision resistance of functions, and the definition of the eCK model, as referenced in this paper.

2.1 Difficult Problem Assumptions

To facilitate the security proof of the scheme in subsequent sections, this subsection introduces the computational elliptic curve Diffie-Hellman (ECCDH) problem.

Definition 1 (Elliptic Curve Computational Diffie-Hellman (ECCDH) Problem). *The computational elliptic curve Diffie-Hellman (ECCDH) problem on a cyclic group G of prime order q is the problem of computing $[xy]P$ for any $P \in G$, $[x]P, [y]P \in G$ where $x, y \in Z_q^*$.*

2.2 Collision Resistance for Functions

To support the functional analysis of the scheme in subsequent sections, this subsection presents the definitions of collision resistance for functions.

Definition 2 (Collision resistance). *Let the function $\mathcal{F} : X \to Y$ be given. The advantage of an adversary $\mathcal{A}$ against the collision resistance of the function $\mathcal{F}$ is defined as:*

$$Adv_{\mathcal{F},\mathcal{A}}^{CR} = Pr\left[\mathcal{F}(x) = \mathcal{F}(x') \wedge x \neq x' | (x, x') \xleftarrow{\$} \mathcal{A}\right]$$

2.3 eCK Model

We focus on protocol security analysis using the eCK model [22], which allows adversaries to more attacks closely resemble real-world environments.

Let $P = \{P_1, P_2, \ldots, P_n\}$ denote the set of users in the eCK model, where each user P_i is represented as a probabilistic polynomial-time Turing machine. An instance $\prod_{i,j}^{sid}$ of user P_i and user P_j is a session, where sid is the session identifier, P_i is the protocol initiator, and P_j is the protocol responder. We define $\prod_{j,i}^{sid^*}$ as the matching session of $\prod_{i,j}^{sid}$, where $sid^* = sid$, and both sessions have completed execution.

Under the eCK model, the adversary $\mathcal{A}$ is endowed with the following capabilities:

- Ephemeral-Key-Reveal(P_i,sid) Query: An adversary $\mathcal{A}$ may obtain the ephe-meral private key generated by user P_i by issuing this query.
- Static-Key-Reveal(P_i) Query: An adversary $\mathcal{A}$ may acquire the long-term private key of user P_i by issuing this query.

- Send(*sid*) Query: This query simulates a scenario where the adversary controls network communication and can send arbitrary messages. The adversary $\mathcal{A}$ may send a message to session *sid* and receive the corresponding response according to the protocol rules.
- Test(*sid*) Query: Session must remain fresh for this query. It can only be issued once, where a random bit $b \in \{0,1\}$ is chosen. If $b = 0$, the adversary $\mathcal{A}$ receives the session key. Otherwise, if $b = 1$, the adversary receives a random value of the same length as the session key.

Definition 3 (Session freshness). *Let $\prod_{i,j}^{sid}$ denote the session established between two honest users P_i and P_j during session sid. We say that $\prod_{i,j}^{sid}$ is fresh if none of the following conditions hold:*

- If a session $\prod_{i,j}^{sid}$ exists, the adversary $\mathcal{A}$ may perform a Session-Key-Reveal (*sid*) Query on $\prod_{i,j}^{sid}$ or on $\prod_{j,i}^{sid}$.
- If no matching session $\prod_{j,i}^{sid}$ exists, the adversary $\mathcal{A}$ may perform a Static-Key-Reveal(P_i) Query and an Ephemeral-Key-Reveal(P_i,*sid*) Query, or a Static-Key-Reveal(P_j) Query on user P_j.
- If a matching session $\prod_{j,i}^{sid}$ exists, the adversary $\mathcal{A}$ may perform a Static-Key-Reveal(P_i) Query and an Ephemeral-Key-Reveal(P_i,*sid*) Query on P_i, or a Static-Key-Reveal(P_i) Query and an Ephemeral-Key-Reveal(P_j,*sid*) Query on P_j.

In the scenario of a Test(*sid*) Query, if the adversary $\mathcal{A}$ is considered to win the game if $b' = b$. Let λ be the system security parameter, which typically determines the bit length of the keys involved in the cryptographic protocol. The advantage of the adversary $\mathcal{A}$ in winning the game is defined as the probability difference between the adversary's success rate and a random guesser, formalized as:

$$Adv_{\mathcal{A}}(\lambda) = \left| \Pr[\mathcal{A} \ wins] - \frac{1}{2} \right|$$

Definition 4 (eCK Secure Key Exchange Protocol). *It is considered that a key exchange protocol is secure within the eCK framework when it meets the following criteria:*

(1) The sessions $\prod_{i,j}^{sid}$ and $\prod_{j,i}^{sid}$ derive an identical session key.
(2) The probability that any probabilistic polynomial-time adversary $\mathcal{A}$ wins the game is negligible.

If all the conditions stipulated in Definition 4 are satisfied, it is our conviction that the protocol achieves eCK security.

3 SM2-VBKE Protocol

3.1 Confirmation Code Definition and Role

The SM2 key exchange protocol involves two elliptic curve point validity checks:

1. The responder verifies the validity of the initiator's temporary public key on the curve.
2. The initiator verifies the validity of the responder's temporary public key on the curve.

These checks ensure that the elliptic curve point (x, y) satisfies the equation $y^2 \equiv x^3 + ax + b \pmod{n}$, where a, b, and n are curve parameters.

In our protocol, we adopt Fischlin et al.'s method [21], defining the sender's confirmation code as (y^2, y^2) and the verifier's confirmation code as $(x^3 + ax + b, y^2)$. If the verification steps are correct, both parties generate the same confirmation code. Any discrepancies, due to invalid verification or malicious attacks, prevent the protocol from proceeding, ensuring the integrity of the exchanged information.

3.2 SM2-VBKE Protocol Description

The SM2-VBKE protocol involves two entities: the protocol initiator, denoted as A, and the protocol responder, denoted as B. The protocol flow, illustrated in Fig. 1, begins with both parties performing the Setup algorithm for initialization, followed by the Temporary key generation algorithm to produce a temporary key, and the Exchange algorithm to generate the shared keys, K_A and K_B. This protocol is an improvement and extension of the original SM2 key exchange protocol, utilizing the same system parameters as those in SM2. As defined by the confirmation code, both parties are required to calculate this code, which is then converted into a string and incorporated into the key derivation function, as specified in the official SM2 documentation. The steps of the protocol are as follows:

- **Setup:** Both parties generate their key pairs (d_A, P_A) and (d_B, P_B), compute user identifiers Z_A and Z_B, and negotiate the elliptic curve parameter ecc to be used in the protocol. Additionally, the length of the generated shared key $klen$ is agreed upon.
- **Temporary key generation:** Both parties compute their respective temporary key pairs $(r_A, R_A), (r_B, R_B)$ and corresponding confirmation codes $\overrightarrow{c_{A1}}, \overrightarrow{c_{B2}}$. According to the protocol implementation requirements, both parties need to calculate $w = \left\lceil \frac{\lceil \log_2 n \rceil}{2} \right\rceil - 1$.
- **Exchange:** Both parties interact, verify that the other party's temporary public key satisfies the conditions, generate a confirmation code using the other party's temporary public key $\overrightarrow{c_{A2}}, \overrightarrow{c_{B1}}$. Further calculate the confirmation code $\overrightarrow{c_A}, \overrightarrow{c_B}$. Finally, compute the shared secret key K_A, K_B.

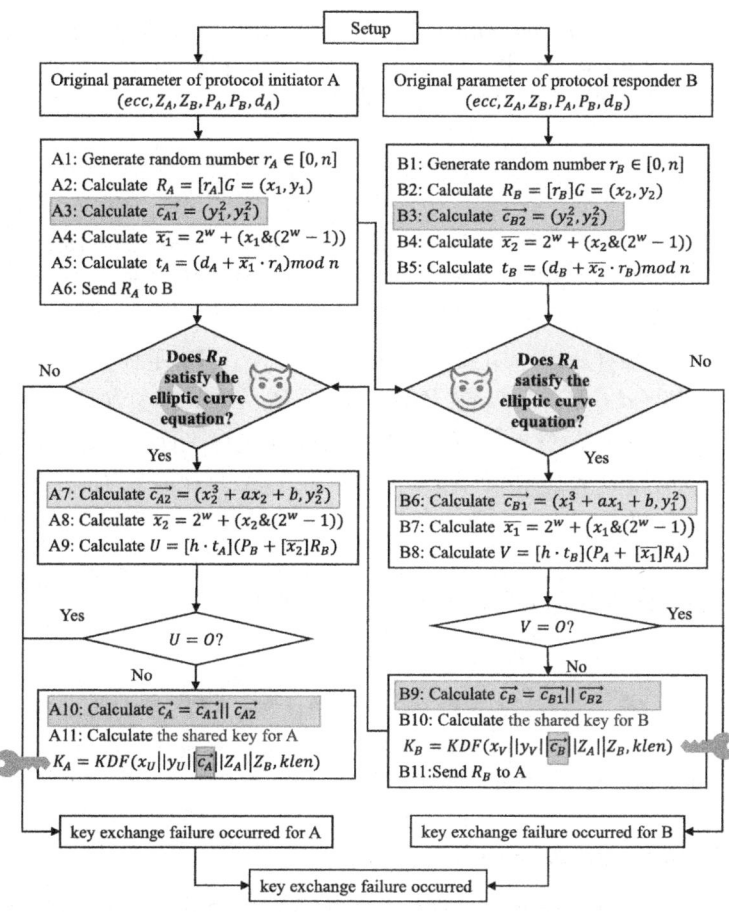

Fig. 1. SM2-VBKE protocol

It is important to note that under normal circumstances, the protocol correctly performs the verification of the temporary public key. However, in the subsequent security analysis, we model the implementation errors of the verification step and malicious attacks that bypass the verification step as the adversary's ability to circumvent the verification.

4 Security Proof

4.1 Correctness Analysis

In this subsection, we conduct a correctness analysis of the protocol, specifically examining whether both parties are capable of generating an identical shared key in the event of a correct implementation.

Given a correct implementation of the protocol, both parties are able to generate temporary public keys that satisfy the elliptic curve equations. Consequently, the following relations must hold:

$$\vec{c_A} = \overrightarrow{c_{A1}} \| \overrightarrow{c_{A2}} = \left(y_1^2, y_1^2, x_2^3 + ax_2 + b, y_2^2\right)$$
$$= \left(x_1^3 + ax_1 + b, y_1^2, y_2^2, y_2^2\right) = \overrightarrow{c_{B1}} \| \overrightarrow{c_{B2}} = \vec{c_B}$$

That is, both parties are able to compute the same confirmation code. Then if we want to get the same shared key, we need to have the same KDF input, and because the

$$U = [h \cdot t_A](P_B + [\overline{x_2}] R_B) = [h][d_A + \overline{x_1} r_A](P_B + [\overline{x_2}] R_B)$$
$$= [h]([d_A] P_B + [d_A \overline{x_2}] R_B + [\overline{x_1} r_A] P_B + [\overline{x_1} r_A \overline{x_2}] R_B)$$
$$= [h]([d_A d_B] G + [d_A \overline{x_2} r_B] G + [\overline{x_1} r_A d_B] G + [\overline{x_1} r_A \overline{x_2} r_B] G)$$
$$= [h]([d_B] P_A + [r_B \overline{x_2}] P_A + [\overline{x_1} d_B] R_A + [\overline{x_1} r_B \overline{x_2}] R_A)$$
$$= [h][d_B + \overline{x_2} r_B](P_A + [\overline{x_1}] R_A) = [h \cdot t_B](P_A + [\overline{x_1}] R_A) = V$$

that $(x_U, y_U) = (x_V, y_V)$. So that both parties to the protocol get the same input to the KDF function, then we have the following:

$$K_A = KDF(x_U \| y_U \| \vec{c_A} \| Z_A \| Z_B, klen)$$
$$= KDF(x_V \| y_V \| \vec{c_B} \| Z_A \| Z_B, klen) = K_B$$

To sum up, as long as the protocol is carried out properly, the two parties will be able to acquire an identical shared key.

4.2 eCK Security Analysis

In this section, we demonstrate the security of the SM2-VBKE protocol under the eCK model, specifically relying on the ECCDH security assumption within the context of the elliptic curve cyclic additive group. We assume that the key derivation function (KDF) is modeled as an independent random oracle.

Theorem 1 (eCK Security for SM2-VBKE protocols). *The SM2-VBKE protocol is secure under the eCK model if the KDF is treated as an independent random oracle and the ECDDH assumption holds.*

Proof. First, let the security parameters of the system be denoted by λ, which includes n honest users and s sessions. Since the KDF is modeled as a random oracle KDF, there are only three ways for an adversary $\mathcal{A}$ to obtain the session key after engaging in the test game:

- **Guessing Attack:** The adversary $\mathcal{A}$ acquires the correct session key by guessing.
- **Session Construction Attack:** The adversary $\mathcal{A}$ constructs a session that diverges from the test session, but generates the same shared key as the test session, and uses the `Session-Key-Reveal` (*sid*) `Query` capability to acquire key.

– **Forgery Attack:** Adversary $\mathcal{A}$ performs the same query as the test session on the random oracle KDF.

Analysis of Each Attack

1. Guessing Attack: If the adversary aims to successfully get key, they must correctly guess the output of the random oracle KDF. The probability of a successful guess is $O\left(\frac{1}{2^\lambda}\right)$. Thus, guessing attacks do not pose a significant threat to the protocol's security.

2. Session Construction Attack: Since the input to the random oracle KDF encompasses the information of the session identifier, and two mismatched sessions cannot share the same temporary public key, compromising this scenario is equivalent to breaking the collision resistance of the key derivation function. As KDF is modeled as a random oracle, the probability of a collision in its output is $O\left(\frac{s^2}{2^\lambda}\right)$. Consequently, session construction attacks do not present a significant threat to the protocol security.

3. Forgery Attack: In the context of a forgery attack, we categorize the following two scenarios based on the definition of session freshness:

– **Scenario 1:** A corresponding session exists for the test session.
– **Scenario 2:** No corresponding session is found for the test session.

Referring to the definition of session freshness and considering the query patterns of the adversary, we can further subdivide Scenario 1 into four distinct scenarios:

– **Scenario 1.1:** The adversary $\mathcal{A}$ simultaneously queries temporary keys for the test session and its corresponding session.
– **Scenario 1.2:** The adversary $\mathcal{A}$ simultaneously queries the long-term private keys for both the test session and its corresponding session.
– **Scenario 1.3:** The adversary $\mathcal{A}$ queries the long-term keys for the test session while querying the temporary keys for the corresponding session.
– **Scenario 1.4:** The adversary $\mathcal{A}$ queries the temporary key for the test session while querying the long-term key for the corresponding session.

Scenario 1.1

In this scenario, the adversary $\mathcal{A}$ queries the temporary private keys for both the test session and its corresponding session simultaneously. Given an ECCDH instance aG, bG, the following simulation procedure $\mathcal{S}$ can solve the ECCDH problem with a probability that is not negligible. $\mathcal{S}$ can accurately predict with at least $\frac{1}{n^2}$ which user the adversary $\mathcal{A}$ selects as A and B. Then, $\mathcal{S}$ simulates the execution environment of the protocol, ensuring that the adversary $\mathcal{A}$ cannot distinguish the simulated environment from the actual environment with a non-negligible probability. For queries involving users other than A and B, $\mathcal{S}$ responds in accordance with the protocol specification. When the query targets A or B, the response is as follows:

- Static-Key-Reveal(C) Query: If the user is A or B, $\mathcal{S}$ aborts the current simulation run. Otherwise, $\mathcal{S}$ provides the long-term private key of user C as the response, where the long-term private key is generated by $\mathcal{S}$ during the initial phase of the simulation.
- Ephemeral-Key-Reveal(C, sid) Query: $\mathcal{S}$ provides the ephemeral private key of user C for session sid as the response.
- Session-Key-Reveal(C, sid) Query: $\mathcal{S}$ provides the session key for the current query response as follows. Since the adversary can access any message transmitted over the network, we assume that the adversary can successfully obtain the public keys exchanged during the session, and thereby compute the corresponding confirmation codes. Let the session identifier be $(A, C, \vec{c_A}, \vec{c_C}, Z_A, Z_C)$, then the corresponding session key is $SK = KDF(\sigma)$, where the input to the random oracle is $\sigma = (x \parallel y \parallel \vec{c} \parallel Z_A \parallel Z_B)$ where

$$(x, y) = (h\,(\text{ECCDH}(aG, bG) + \bar{x}_2 \cdot \text{ECCDH}(r_B G, aG) + \bar{x}_1 \cdot \text{ECCDH}(r_A G, bG) + \bar{x}_1 \bar{x}_2 \cdot \text{ECCDH}(r_A G, r_B G))), \quad \mathbf{c} \in \{\vec{c_A}, \vec{c_C}\}.$$

$\mathcal{S}$ queries the random oracle KDF to check if the input has been queried before. If the input is new, it produces a random value that follows the distribution pattern of the session key. Otherwise, it outputs the session key previously stored by the random oracle.

Thus, the simulator $\mathcal{S}$ successfully simulates the execution environment of $\mathcal{A}$. Since the adversary $\mathcal{A}$ performs ephemeral private key queries for both the test session and the matching session, the simulator $\mathcal{S}$ obtains the corresponding values of r_A and r_B, implying that $\mathcal{A}$ can successfully compute

$$\text{ECCDH}(r_B G, aG), \quad \text{ECCDH}(r_A G, bG), \quad \text{ECCDH}(r_A G, r_B G).$$

If the adversary $\mathcal{A}$ wishes to win the forgery attack, then $\mathcal{A}$ must query the KDF for contents involving $\text{ECCDH}(aG, bG)$, which indicates that $\mathcal{S}$ has solved the ECCDH problem. In addition to this, it is necessary to consider that $\mathcal{A}$ must solve the ECDLP problem within time t.

In summary, the advantage of successfully solving the ECCDH problem in scenario 1.1 is given by the following expression:

$$Adv_G^{ECCDH}(\lambda, t, \mathcal{S}) \geq \frac{1}{n^2} \cdot P_1(\lambda, t) - Adv_G^{ECDLP}(\lambda, t, \mathcal{S}),$$

where $P_1(\lambda, t)$ denotes the probability that scenario 1.1 occurs and $\mathcal{A}$ succeeds.

Scenario 1.2

The adversary $\mathcal{A}$ performs long-term key queries on the test session and the matching session. Given an ECCDH instance $aG, bG \in G$, we construct a simulation algorithm $\mathcal{S}$ that solves the ECCDH problem with a non-negligible advantage. The algorithm $\mathcal{S}$ can guess the session identifier sid of the test session with at least $\frac{1}{n^2}$ probability, where the test session corresponds to user A and the matching session corresponds to user B.

The simulation algorithm $\mathcal{S}$ mimics the protocol environment such that $\mathcal{A}$ cannot distinguish it from the real environment with non-negligible probability. The simulation of various queries in Scenario 1.2 follows the same structure as Scenario 1.1. Therefore, the advantage of solving the ECCDH problem in Scenario 1.2 is given by:

$$Adv_G^{ECCDH}(\lambda, t, \mathcal{S}) \geq \frac{1}{n^2} \cdot P_2(\lambda, t) - Adv_G^{ECDLP}(\lambda, t, \mathcal{S}),$$

where $P_2(\lambda, t)$ represents the probability of Scenario 1.2 occurring and the success of $\mathcal{A}$.

Scenario 1.3

The adversary $\mathcal{A}$ performs long-term key queries on the test session and temporary key queries on the matching session. Given an ECCDH instance $aG, bG \in G$, we construct a simulation algorithm $\mathcal{S}$ that solves the ECCDH problem with a non-negligible advantage. The algorithm $\mathcal{S}$ can guess the session identifier sid of the test session with at least $\frac{1}{ns}$ probability, where the test session corresponds to user A and the matching session corresponds to user B.

The simulation algorithm $\mathcal{S}$ mimics the protocol environment such that $\mathcal{A}$ cannot distinguish it from the real environment with non-negligible probability. The simulation of various queries in Scenario 1.3 follows the same structure as Scenario 1.1. Therefore, the advantage of solving the ECCDH problem in Scenario 1.3 is given by:

$$Adv_G^{ECCDH}(\lambda, t, \mathcal{S}) \geq \frac{1}{ns} \cdot P_3(\lambda, t) - Adv_G^{ECDLP}(\lambda, t, \mathcal{S}),$$

where $P_3(\lambda, t)$ represents the probability of Scenario 1.3 occurring and the success of $\mathcal{A}$.

Scenario 1.4

The adversary $\mathcal{A}$ performs temporary key queries on the test session and long-term key queries on the matching session. This Scenario is similar to Scenario 1.3, so the advantage in scenario 1.4 is given by the following expression:

$$Adv_G^{ECCDH}(\lambda, t, \mathcal{S}) \geq \frac{1}{ns} \cdot P_4(\lambda, t) - Adv_G^{ECDLP}(\lambda, t, \mathcal{S}),$$

where $P_4(\lambda, t)$ represents the probability of scenario 1.4 occurring and the success of $\mathcal{A}$.

Based on the definition of the matching session and session freshness, we can divide scenario 2 into two sub-scenarios:

- **Scenario 2.1**: $\mathcal{A}$ performs temporary private key queries on user A.
- **Scenario 2.2**: $\mathcal{A}$ performs long-term private key queries on user A.

Scenario 2.1

$\mathcal{A}$ performs temporary key queries on the test session. Since the test session fails to have an associated congruous session, we assume that when $\mathcal{A}$ successfully obtains the temporary key of the test session owner, it can also successfully obtain the temporary key of the communication partner. This effectively allowing $\mathcal{A}$ to perform a temporary key query on the matching session owner. Therefore, this situation is analogous to Scenario 1.1. Hence, the advantage in Scenario 2.1 is given by the following expression:

$$Adv_G^{ECCDH}(\lambda, t, \mathcal{S}) \geq \frac{1}{ns} \cdot P_5(\lambda, t) - Adv_G^{ECDLP}(\lambda, t, \mathcal{S}),$$

where $P_5(\lambda, t)$ represents the probability of Scenario 2.1 occurring and the success of $\mathcal{A}$.

Scenario 2.2

The adversary $\mathcal{A}$ performs long-term key queries on the test session. Since the test session does not have an associated matching session, we assume that when $\mathcal{A}$ successfully obtains the long-term key of the test session owner, it can also successfully obtain the temporary key of the communication partner. This effectively allows $\mathcal{A}$ to perform a temporary key query on the matching session owner. Therefore, this situation is analogous to Scenario 1.3. Hence, the advantage in Scenario 2.2 is given by the following expression:

$$Adv_G^{ECCDH}(\lambda, t, \mathcal{S}) \geq \frac{1}{ns} \cdot P_6(\lambda, t) - Adv_G^{ECDLP}(\lambda, t, \mathcal{S}),$$

where $P_6(\lambda, t)$ represents the probability of scenario C.2 occurring and the success of $\mathcal{A}$.

Derivation of the Overall Advantage

Finally, the advantage of algorithm $\mathcal{S}$ in solving the ECCDH problem is derived as follows:

$$Adv_G^{ECCDH}(\lambda, t, \mathcal{S}) \geq \max \begin{Bmatrix} \frac{1}{n^2} \cdot p_1(\lambda, t), \frac{1}{ns} \cdot p_2(\lambda, t), \\ \frac{1}{n^2} \cdot p_3(\lambda, t), \frac{1}{s^2} \cdot p_4(\lambda, t), \\ \frac{1}{ns} \cdot p_5(\lambda, t), \frac{1}{ns} \cdot p_6(\lambda, t) \end{Bmatrix} - Adv_G^{ECDLP}(\lambda, t, \mathcal{S})$$

In conclusion, if $\mathcal{A}$ can successfully win the Test game, it implies that the simulator $\mathcal{S}$ can solve the ECCDH problem with a non-negligible advantage. This goes against the premise that the ECCDH problem is difficult. Therefore, we can assert that the SM2-VBKE protocol exhibits robust security characteristics in the eCK model. □

4.3 The Implementation Error Detection of SM2-VBKE

In an ideal execution of the SM2 protocol, if the temporary public key sent by the other party does not reside on the elliptic curve, the protocol will fail to proceed, and a shared key will not be generated. However, in the scenario of an erroneous implementation, there may be situations where the temporary public key is not on the elliptic curve, but a shared key is still successfully computed, thereby introducing potential security vulnerabilities. In contrast, in our protocol, if the receiver of the temporary public key fails to properly execute the verification step, and the temporary public key does not satisfy the elliptic curve equation, the computed confirmation values of both parties will differ. This discrepancy will result in the generation of distinct shared keys during key derivation, which effectively reveals the implementation error. Such a difference in shared keys can be readily uncovered during the non-interactive evaluation of the protocol.

We now formalize the detection procedure and rigorously prove the self-verifiability of the protocol's verification steps. We first introduce a non-adaptive adversary $\mathcal{A}$, who is allowed to non-adaptively introduce errors during certain verification steps performed by the protocol participants. This means that the actions of $\mathcal{A}$ are independent of the execution of the protocol. Based on this setup, we define the adversary $\mathcal{A}$ advantage as follows:

Definition 5 (the Attack Advantage of a Non-Adaptive Adversary $\mathcal{A}$).
The adversary $\mathcal{A}$ wins the game if, despite the erroneous verification step, both parties are still able to compute the same shared key. In this scenario, the advantage of the adversary is defined as

$$Adv_{\mathcal{A}} = Pr[K_A = K_B \neq \perp \mid faulty\ verification]$$

We assert that the verification error in the SM2-VBKE protocol can only be detected if, and only if, a non-adaptive adversary $\mathcal{A}$ is able to win the game with non-negligible probability. We now formally prove that the SM2-VBKE protocol implements error feedback.

Theorem 2 (the Non-Adaptive Adversary Advantage in the SM2-VBKE protocol). *Let $\mathcal{A}$ be a non-adaptive adversary against the SM2-VBKE protocol, and let $\mathcal{B}$ be the adversary against the collision resistance of the key derivation function. Then, the advantage of the adversary $\mathcal{A}$ is bounded by*

$$Adv_{\mathcal{A}}^{SM2-VBKE} \leq Adv_{\mathcal{B}}^{KDF-CR} + \frac{12}{n}$$

Proof. We assume that both parties in the SM2-VBKE protocol can always compute the same elliptic curve point, denoted as $U = V \neq \mathcal{O}$, meaning the shared key will not result in $\perp$.

Game 0 → Game 1: Game 0 represents the initial game, where the adversary's advantage is denoted as $Adv_{\mathcal{A}}^{SM2-VEBK}$. In Game 1, the adversary skips the verification step and non-adaptively generates elliptic curve points to replace the temporary public keys of both parties. According to Theorem 4.3 in Fischlin et al.'s 2023 paper [21],

$$Pr[\overrightarrow{c_{\mathcal{A}}} = \overrightarrow{c_{A1}}] \leq \frac{12}{n}$$

Thus, the probability difference between Game 1 and Game 0 is the probability of generating $\overrightarrow{c_{A1}} = \overrightarrow{c_{B1}}$ or $\overrightarrow{c_{A2}} = \overrightarrow{c_{B2}}$, given by

$$|Pr[G_0] - Pr[G_1]| \leq \frac{12}{n}$$

Game 1 → Game 2: The shared key is computed as $K_A = KDF(x_U||y_U||\vec{C_A}||Z_A||Z_B)$. In Game 2, the adversary again replaces the Hash function with the oracle $\mathcal{O}^{KDF(\cdot)}$. Therefore, the probability difference between Game 1 and Game 2 is

$$|Pr[G_1] - Pr[G_2]| \leq Adv^{CR}_{KDF,\mathcal{B}}$$

At this point, the adversary's probability of winning in **Game 2** is 0. In summary, the adversary's advantage is

$$Adv^{SM2-VBKE}_{\mathcal{A}} \leq Adv^{CR}_{KDF,\mathcal{B}} + \frac{12}{n}$$

Therefore, we have reduced the advantage of the non-adaptive adversary to the collision resistance of KDF and the unpredictability of the confirmation code. This successfully proves that the non-adaptive adversary cannot win the game, demonstrating that the SM2-VBKE protocol incorporates error feedback into the shared key and achieves the design objective. □

5 Performance Analysis

We provide a comprehensive analysis of the SM2-VBKE protocol, focusing on its security features and empirical performance. By comparing it to the original SM2 protocol, we highlight the unique aspects of our solution. The choice of SM2 as a benchmark is driven by the goal of enhancing an existing protocol. Specifically, our improvements optimize the verification step in shared key generation, only adding minimal computational overhead reducing computational overhead. As a result, our solution preserves the strengths of the original SM2 protocol while addressing implementation vulnerabilities and protecting against malicious attacks on the verification step.

5.1 Security Properties Analysis

We conduct an informal yet comprehensive analysis of the SM2-VBKE protocol's security characteristics:

- **Shared Key Confidentiality:** As the proof of the Theorem 1, the protocol's security under the eCK model requires adversaries to solve the ECCDH problem to compromise session keys. This computational hardness guarantee ensures the confidentiality of derived shared keys.
- **Bidirectional Authentication:** The key generation process cryptographically binds both ephemeral and long-term private keys, creating mutual implicit authentication. This dual-key dependency ensures both parties authenticate each other's identities during the exchange.
- **Forward Secrecy:** Session-specific ephemeral keys guarantee that compromise of long-term keys does not reveal past session keys. The protocol's design ensures temporary keys are never reused, providing strong forward secrecy guarantees.

- **Binding mechanism between verification steps and key generation:** The protocol establishes cryptographic binding between validation outcomes and key material through confirmation code generation. Specifically, the verification residuals $(x^3 + ax + b, y^2)$ from elliptic curve equation checks are hashed into the confirmation code $\vec{c}$, which directly feeds into the key derivation function. Theorem 2 proves that any validation failure alters $\vec{c}$, causing detectable key mismatch through KDF output divergence.
- **Comparison with the SM2 Key Exchange Protocol:** By establishing a binding mechanism between verification steps and key generation, SM2-VBKE protocol enables detection of whether verification steps have been correctly executed, thereby resisting various existing attacks targeting the verification steps.

5.2 Experimental Evaluation

The design concept of the SM2-VBKE protocol binds the verification step with key generation, while only slightly increasing the computational cost. To more clearly demonstrate the advantages of this approach, we conduct an experimental evaluation of the SM2-VBKE protocol, comparing its performance with that of the original SM2 key exchange protocol. In Table 2, we present our experimental results, where both protocols were executed 1,000 times.

From a theoretical standpoint, in terms of communication overhead, the computation of confirmation codes is performed locally and no comparison of these codes between the two parties is required. Furthermore, the transmission data in the SM2-VBKE protocol is identical to that in the standard SM2 key exchange protocol, meaning no additional communication burden is introduced. In terms of computational cost, the generation of confirmation codes in the SM2-VBKE protocol does not introduce extra computational overhead, as the correct SM2 protocol execution also requires the computation of $(x^3 + ax + b, y^2)$. However, since the KDF input is expanded during the key derivation process, there is a slight increase in the computational cost for key derivation.

The experiments were conducted on a Legion R9000P notebook, equipped with an AMD Ryzen 7 5800H processor with Radeon Graphics, clocked at 3.20 GHz, and running Windows 11. We measured the execution times of the Setup, Temporary Key Generation, and Exchange phases for both the SM2-VBKE and SM2 protocols. Since both protocols transmit identical messages, we excluded transmission time due to network latency in order to avoid discrepancies caused by network fluctuations.

Experimental results show that the overhead during the initialization phase remains essentially unchanged, confirming the conclusion in the theoretical analysis that there is no additional computation in the parameter pre-calculation. The 0.103 s increase in the temporary key generation phase is due to the use of the y-coordinate of the elliptic curve point to generate the user's local confirmation code. The 1.118 s increase in the exchange process is due to hash block padding caused by the KDF input expansion.

Although the protocol introduces an increase of approximately 1.290 s, SM2-VBKE achieves a breakthrough in verifying integrity through cryptographic binding, while also being able to detect errors in the verification step or malicious attacks by adversaries. Therefore, the increase is acceptable.

Table 2. Performance Testing

	Setup(s)	Temporary key generation(s)	Exchange(s)	Sum(s)
SM2-VBKE	1.370	2.144	10.771	14.285
SM2-KE	1.367	1.998	9.630	12.995
time+	0.003	0.103	1.118	1.290

6 Conclusion

The SM2-VBKE protocol introduces a groundbreaking approach to elliptic curve key exchange by cryptographically binding the integrity of verification with session key generation, bridging the gap between theoretical security assurances and practical implementation vulnerabilities. Our design utilizes intermediate values from elliptic curve point verification to derive confirmation codes, enforcing point verification as an indispensable prerequisite for key derivation, while reusing verification intermediate results to maintain computational efficiency. Security proof under the eCK model demonstrates that the protocol's security is reduced to the Elliptic Curve Discrete Logarithm (ECCDH) assumption, while proving that any deviation in the verification step triggers detectable key divergence through mismatched KDF outputs. Compared to the standard SM2, the overhead increases by only 1.29 ms, while mitigating a whole class of attacks exploiting verification flaws, such as invalid curve attacks and fault injection attacks. This work provides a novel perspective by transforming the verification step from an external check into an intrinsic cryptographic property, fundamentally reshaping the relationship between protocol design and practical implementation. Future extensions may generalize this binding mechanism to other types of protocols.

References

1. Georgiev, M., Iyengar, S., Jana, S., et al.: The most dangerous code in the world: validating SSL certificates in non-browser software. In: Proceedings of the 2012 ACM Conference on Computer and Communications Security, pp. 38–49 (2012)
2. Poulsen, K.: Behind iPhone's critical security bug, a single bad 'Goto', February 2014. https://www.wired.com/2014/02/gotofail/
3. Hebrok, S., Nachtigall, S., Maehren, M., et al.: We really need to talk about session tickets: a large-scale analysis of cryptographic dangers with TLS session tickets. In: 32nd USENIX Security Symposium (USENIX Security 23), pp. 4877–4894 (2023)

4. Siqi, L., Zheng, J., Cao, Z., Wang, Y., Chunxiang, G.: A survey on cryptographic techniques for protecting big data security: present and forthcoming. Sci. China Inf. Sci. **65**(10), 201301 (2022)
5. Biehl, I., Meyer, B., Müller, V.: Differential fault attacks on elliptic curve cryptosystems. In: Bellare, M. (ed.) CRYPTO 2000. LNCS, vol. 1880, pp. 131–146. Springer, Heidelberg (2000). https://doi.org/10.1007/3-540-44598-6_8
6. Jager, T., Schwenk, J., Somorovsky, J.: Practical invalid curve attacks on TLS-ECDH. In: Pernul, G., Ryan, P.Y.A., Weippl, E. (eds.) ESORICS 2015. LNCS, vol. 9326, pp. 407–425. Springer, Cham (2015). https://doi.org/10.1007/978-3-319-24174-6_21
7. Biham, E., Neumann, L.: Breaking the Bluetooth pairing – the fixed coordinate invalid curve attack. In: Paterson, K.G., Stebila, D. (eds.) SAC 2019. LNCS, vol. 11959, pp. 250–273. Springer, Cham (2020). https://doi.org/10.1007/978-3-030-38471-5_11
8. Lu, S., Dong, H., Li, Z., Yang, L.T.: Not just summing: the identifier leakage of private-join-and-compute and its improvement. IEEE Trans. Dependable Secure Comput. (2024)
9. Beijing Huada Xinan Technology Co., Ltd., Chinese People's Liberation Army Information Engineering University, and Chinese Academy of Sciences Data and Communication Protection Research and Education Center. Information security technology SM2 elliptic curve public key cryptographic algorithm part 1: General provisions. National Quality Supervision of the People's Republic of China General Administration of Inspection and Quarantine, Standardization Administration of China (2016)
10. Xu, J., Feng, D.: Comments on the SM2 key exchange protocol. In: Lin, D., Tsudik, G., Wang, X. (eds.) CANS 2011. LNCS, vol. 7092, pp. 160–171. Springer, Heidelberg (2011). https://doi.org/10.1007/978-3-642-25513-7_12
11. Zhao, S., Xi, L., Zhang, Q., et al.: Security analysis of SM2 key exchange protocol in TPM2.0. Secur. Commun. Netw. **8**(3), 383–395 (2015)
12. Wei, W., Chen, J., Li, D., Wang, B.: Partially known information attack on SM2 key exchange protocol. Sci. China Inf. Sci. **62**(3), 1–14 (2019). https://doi.org/10.1007/s11432-018-9515-9
13. Cao, J., Cheng, Q., Weng, J.: EHNP strikes back: analyzing SM2 implementations. In: Batina, L., Daemen, J. (eds.) AFRICACRYPT 2022. LNCS, vol. 13503, pp. 576–600. Springer, Cham (2022). https://doi.org/10.1007/978-3-031-17433-9_25
14. Necula, G.C., Lee, P.: Research on proof-carrying code for untrusted-code security. In: IEEE Symposium on Security and Privacy, p. 204. IEEE Computer Society (1997)
15. National Institute of Standards and Technology: Cryptographic algorithm validation program (2023). https://csrc.nist.gov/projects/cryptographic-algorithm-validation-program. Accessed 4 May 2023
16. Mouha, N., Celi, C.: Extending NIST's CAVP testing of cryptographic hash function implementations. In: Jarecki, S. (ed.) CT-RSA 2020. LNCS, vol. 12006, pp. 129–145. Springer, Cham (2020). https://doi.org/10.1007/978-3-030-40186-3_7
17. Yang, H., Yu, Y., Zhu, Y., Tao, X., Yu, J.: Towards trustworthy 6g networks: a trust-based consensus scheme. IEEE Netw. (2024)
18. Fischlin, M.: Progressive verification: the case of message authentication. In: Johansson, T., Maitra, S. (eds.) INDOCRYPT 2003. LNCS, vol. 2904, pp. 416–429. Springer, Heidelberg (2003). https://doi.org/10.1007/978-3-540-24582-7_31

19. Le, D.V., Kelkar, M., Kate, A.: Flexible Signatures: making authentication suitable for real-time environments. In: Sako, K., Schneider, S., Ryan, P.Y.A. (eds.) ESORICS 2019. LNCS, vol. 11735, pp. 173–193. Springer, Cham (2019). https://doi.org/10.1007/978-3-030-29959-0_9
20. Taleb, A.R., Vergnaud, D.: Speeding-up verification of digital signatures. J. Comput. Syst. Sci. **116**, 22–39 (2021)
21. Fischlin, M., Günther, F.: Verifiable verification in cryptographic protocols. In: Proceedings of the 2023 ACM SIGSAC Conference on Computer and Communications Security, pp. 3239–3253 (2023)
22. LaMacchia, B., Lauter, K., Mityagin, A.: Stronger security of authenticated key exchange. In: Susilo, W., Liu, J.K., Mu, Y. (eds.) ProvSec 2007. LNCS, vol. 4784, pp. 1–16. Springer, Heidelberg (2007). https://doi.org/10.1007/978-3-540-75670-5_1

Certificate-Based Quasi-linearly Homomorphic Signatures: Definition, Construction, and Application to Data Integrity Auditing

Jintao Cai[1], Futai Zhang[1(✉)], Wenjie Yang[1(✉)], Shaojun Yang[2], Yichi Huang[1], Rongmao Chen[3], and Willy Susilo[4]

[1] College of Computer and Cyber Security, Fujian Normal University, Fuzhou 350117, China
`futai@fjnu.edu.cn, njnuywj@163.com`
[2] School of Mathematics and Statistics, Fujian Normal University, Fuzhou 350117, China
[3] College of Computer Science and Technology, National University of Defense Technology, Changsha 410073, China
[4] School of Computing and Information Technology, University of Wollongong, Wollongong, NSW 2522, Australia

Abstract. With the rise of digital transformation, cloud storage has become a mainstream solution for managing large-scale data due to its scalability and cost-efficiency. However, outsourcing data to the cloud deprives users of physical control, introducing significant risks of data tampering or loss, which in turn motivates the development of cloud storage auditing. Due to the advantages of certificate-based cryptography, many certificate-based cloud storage data integrity auditing (CB-CSDIA) protocols have been proposed in recent years. Nevertheless, existing CB-CSDIA protocols suffer from limitations in terms of security or efficiency. In this paper, we analyze the core techniques and underlying issues of current CB-CSDIA constructions. To provide a rigorous theoretical foundation for CB-CSDIA, we formally define certificate-based quasi-linearly homomorphic signatures (CB-QLHS) and establish a comprehensive security model. A generic construction from CB-QLHS to CB-CSDIA is then proposed. Furthermore, we present an efficient instantiation of CB-QLHS and prove its security in the random oracle model. Performance evaluations demonstrate that auditing protocol built upon our CB-QLHS construction outperform existing solutions.

Keywords: Cloud storage · Data integrity auditing · Certificate-based cryptography · Bilinear pairing

1 Introduction

With the accelerating global digital transformation, the volume of data is growing exponentially, rendering traditional local storage insufficient for efficient management and flexible access. Cloud storage, which offers high scalability, cost efficiency, and remote accessibility, has emerged as a cornerstone infrastructure for data-intensive applications [5]. However, the loss of physical control over outsourced data introduces significant security risks, including potential data tampering and loss. Cloud service providers (CSPs) may, driven by self-interest, conceal data loss incidents or even tamper with user data [10]. To address these concerns, cloud storage data integrity auditing (CSDIA) has been developed, enabling users to verify the integrity of their data without direct possession.

In CSDIA, the user partitions the outsourced file into blocks and generates homomorphic verifiable tags (HVTs), uploading both the file and corresponding tags to the cloud. To verify data integrity, the user or a delegated third party auditor (TPA) can randomly samples file blocks and their tags, prompting the CSP to perform a designated linear homomorphic operation, which yields a constant-size proof of data possession equal in size to a single block-tag pair. This constant-size proof enables efficient integrity verification with minimal communication overhead, eliminating the need to retrieve the entire file. The HVT essentially functions as a quasi-linearly homomorphic signature (QLHS) [20]. As the theoretical foundation of CSDIA protocols, QLHS provides formal security guarantees and underpins the reliability of such protocols.

Certificate-based cryptography (CBC) [6] simplifies certificate management in traditional public key infrastructures (PKI), avoids the key escrow problem in identity-based cryptography (IBC) [14], and eliminates the need for secure channels in certificateless cryptography (CLC) [1]. Owing to these advantages, a number of certificate-based cloud storage data integrity auditing (CB-CSDIA) protocols have been proposed. However, existing CB-CSDIA protocols typically adopt a certificate-based short signature as the file tag, encapsulating a commitment to a temporary private key. The HVTs are then computed using this temporary key, without involving the user's private key or certificate. While this design circumvents public key replacement attacks on the HVTs by relying on the file tag's validity, it introduces several critical limitations. First, the user must manage a large number of temporary private keys, unless dynamic data updates are not supported. Second, the overall security of the auditing protocol relies on the certificate-based short signature used for the file tag, most of which are instantiated using the scheme proposed by Liu et al. [11], later shown to have security weaknesses by Cheng et al. [4]. Third, generating the HVT for each file block requires map-to-point (MTP) operations, which not only increases the user's computational overhead but also causes the TPA's verification cost to grow with the number of challenged blocks.

At the core of these issues lies the lack of a systematic study on certificate-based quasi-linearly homomorphic signature (CB-QLHS), which serves as the foundational building block for such protocols. Motivated by the above observations, this paper makes the following contributions:

– We introduce a formal definition of CB-QLHS that combines the essential properties of CBC and QLHS. In addition, we establish a rigorous security model that captures four types of adversaries.

- We propose a generic construction of CB-CSDIA protocols from CB-QLHS schemes, and prove that the security of the resulting CB-CSDIA protocols can be reduced to the unforgeability of the underlying CB-QLHS schemes.
- We present an efficient instantiation of CB-QLHS and prove its security. The construction avoids costly MTP operations during signature generation and reduces the verification overhead for derived signatures. Theoretical analysis and experimental evaluation demonstrate that the auditing protocol instantiated with this construction outperforms existing solutions.

Related Work. Ateniese et al. [2] introduced the provable data possession, enabling probabilistic auditing of data integrity through random sampling. Juels and Kaliski [9] proposed the Proof of Retrievability (PoR), incorporating error-correcting codes to ensure data retrievability. To improve efficiency, Shacham and Waters [13] presented a compact PoR scheme based on BLS signatures. Subsequent research has explored various cryptographic paradigms, including traditional PKI [3,16], IBC [7,15], CLC [8,19], and CBC [10,17,18,21,22].

CBC has received increasing attention due to its advantages. As a result, numerous CB-CSDIA protocols have been proposed in recent years. Li et al. [10] proposed a certificate-based auditing protocol that employs two-dimensional data partitioning and a linearly homomorphic hash function (LHHF), significantly improving the efficiency. Building on Li et al.'s approach, Wang et al. [18] extended the protocol to support batch auditing. In their protocols, the LHHF is $\mathcal{H}(\boldsymbol{f}_i = \{f_{ij}\}_{j=1}^l) = g^{\sum_{j=1}^l H(ID,j)f_{ij}}$, where g is a generator of a cyclic group of prime order p, and H is a hash function mapping to $\mathbb{Z}_p^*$. This function satisfies the homomorphic property $\mathcal{H}(\sum_{i=1}^n \omega_i \boldsymbol{f}_i) = \prod_{i=1}^n \mathcal{H}(\boldsymbol{f}_i)^{\omega_i}$. However, as Tian et al. [17] pointed out, this construction has a critical weakness that allows collisions to be easily found. An adversary can generate two distinct blocks $\boldsymbol{f} \neq \boldsymbol{f}'$ satisfying $\sum_{j=1}^l H(ID,j)f_j = \sum_{j=1}^l H(ID,j)f_j'$, leading to $\mathcal{H}(\boldsymbol{f}) = \mathcal{H}(\boldsymbol{f}')$. To address this issue, Tian et al. [17] proposed replacing $H(ID,j)$ with random values α_j, and publishing corresponding commitments $g_j = g^{\alpha_j}$ for $j = 1$ to l. This allows only users possessing the trapdoor α_j can efficiently compute $\mathcal{H}(\boldsymbol{f}_i) = g^{\sum_{j=1}^l \alpha_j f_{ij}}$, while others must compute the hash via $\prod_{j=1}^l g_j^{f_{ij}}$. Zhang et al. [21] addressed this problem using a similar approach, and realized privacy preservation. Zhou et al. [22] proposed a certificate-based multi-replica cloud storage auditing protocol supporting dynamic data updates, but their one-dimensional data partitioning results in substantially higher computational and storage overhead.

Organization. Section 2 covers preliminaries. Section 3 formalizes the CB-QLHS definition and security model. Section 4 presents a generic CB-CSDIA construction from CB-QLHS. Section 5 provides an instantiation of CB-QLHS, its security proof and performance comparison. Section 6 concludes the paper.

2 Preliminaries

2.1 Bilinear Pairing

Suppose $\mathbb{G}_1$ and $\mathbb{G}_T$ are two multiplicative cyclic groups of the same prime order p, with $\mathbb{G}_1$ having a generator g. A bilinear map $e\colon \mathbb{G}_1 \times \mathbb{G}_1 \to \mathbb{G}_T$ satisfies the following properties:

- *Bilinearity*: $e(x^\alpha, y^\beta) = e(x,y)^{\alpha\beta}$ holds for any $x, y \in \mathbb{G}_1$ and $\alpha, \beta \in \mathbb{Z}_p$.
- *Non-degeneracy*: There exists an element $g \in \mathbb{G}_1$ such that $e(g,g) \neq 1_{\mathbb{G}_T}$.
- *Computability*: For any $x, y \in \mathbb{G}_1$, $e(x, y)$ can be efficiently computed.

2.2 Hardness Assumptions

Let g be a generator of a multiplicative cyclic group $\mathbb{G}$ of prime order p.

Definition 1 (Discrete Logarithm (DL) Assumption). *The DL problem in $\mathbb{G}$ is formulated as follows: Given g, $g^x \in \mathbb{G}$, the objective is to compute $x \in \mathbb{Z}_p$. The DL assumption in $\mathbb{G}$ states that the advantage of any probabilistic polynomial-time (PPT) algorithm in solving the DL problem in $\mathbb{G}$ is negligible.*

Definition 2 (Computational Diffie-Hellman (CDH) Assumption). *The CDH problem in $\mathbb{G}$ is formulated as follows: Given $g, g^a, g^b \in \mathbb{G}$, the objective is to compute $g^{ab} \in \mathbb{G}$. The CDH assumption in $\mathbb{G}$ states that the advantage of any PPT algorithm in solving the CDH problem in $\mathbb{G}$ is negligible.*

Definition 3 (q-Collusion Attack Algorithm (q-CAA) Assumption [12]). *The q-CAA problem in $\mathbb{G}$ is formulated as follows: Given g, $g^x \in \mathbb{G}$, and q pairs $\{(z_i, g^{1/(x+z_i)})\}_{i=1}^q$ where $z_i \in \mathbb{Z}_p^*$ for $i = 1, \ldots, q$, the objective is to compute a new pair $(z^*, g^{1/(x+z^*)})$ for some $z^* \notin \{z_1, z_2, \ldots, z_q\}$. The q-CAA assumption in $\mathbb{G}$ states that the advantage of any PPT algorithm in solving the q-CAA problem in $\mathbb{G}$ is negligible.*

Definition 4 (Weak Modified q-CAA Assumption [11]). *The weak modified q-CAA problem in $\mathbb{G}$ is formulated as follows: Given g, g^x, g^a, $g^b \in \mathbb{G}$, and q pairs $\{(z_i, (g^{ab})^{1/(x+z_i)})\}_{i=1}^q$ where $z_i \in \mathbb{Z}_p^*$ for $i = 1, \ldots, q$, the objective is to output g^{ab} or a new pair $(z^*, (g^{ab})^{1/(x+z^*)})$ for some $z^* \notin \{z_1, z_2, \ldots, z_q\}$. The Weak Modified q-CAA assumption in $\mathbb{G}$ states that the advantage of any PPT algorithm in solving the weak modified q-CAA problem in $\mathbb{G}$ is negligible.*

2.3 Formal Definition of CB-CSDIA

Before utilizing cloud services, the data owner (DO) submits its identity and public key to the certifier, who then issues a certificate to the DO. In some CB-CSDIA protocols, the CSP acts as the certifier. Given a file $\boldsymbol{F}$, the DO first divides it into n blocks and further partitions each block into l sectors, representing the file as a matrix $\boldsymbol{F} = (\boldsymbol{f}_1, \ldots, \boldsymbol{f}_n) \in \mathbb{Z}_p^{l \times n}$, where the i-th column

$\boldsymbol{f}_i = (f_{i1}, \ldots, f_{il})^\top \in \mathbb{Z}_p^l$ corresponds to a file block. Subsequently, the DO uses its private key and certificate to generate a HVT σ_i for each $\boldsymbol{f}_i$, and uploads the file along with $\{\sigma_i\}_{i=1}^n$ to the CSP. It delegates the TPA to perform periodic integrity audits. During the auditing phase, the TPA issues a challenge to the CSP by randomly sampling file blocks and their corresponding tags. The CSP computes a linear combination of the specified block-tag pairs and returns a proof, which the TPA verifies using the DO's public key to confirm data integrity.

Definition 5. *A CB-CSDIA protocol consists of the following four phases:*

- *Initialization*$(1^\lambda) \to (msk, pp)$: The CSP initializes the system with security parameter λ, generating the public parameters pp and a master secret key msk. The parameters pp are made public, while msk remains private.
- *Registration*: This phase involves two algorithms:
  - *UserCreate*$(pp) \to (sk_{ID}, PK_{ID})$: The user with identity ID generates a secret-public key pair (sk_{ID}, PK_{ID}).
  - *Certify*$(msk, ID, PK_{ID}) \to Cert_{ID}$: The CSP issues a certificate $Cert_{ID}$ binding the user's identity ID and public key PK_{ID}.
- *Outsourcing*: This phase includes two algorithms:
  - *TagGen*$(sk_{ID}, Cert_{ID}, \boldsymbol{F}, Fid) \to (\tau, \{\sigma_i\}_{i=1}^n)$: Given a file $\boldsymbol{F} = \{\boldsymbol{f}_i\}_{i=1}^n$ with identifier Fid, the user produces a file tag τ and block-tag pairs using its private key sk_{ID} and certificate $Cert_{ID}$.
  - *TagVerify*$(PK_{ID}, \tau, i, \boldsymbol{f}_i, \sigma_i) \to \{0, 1\}$: The verifier checks whether σ_i is a valid block tag on $\boldsymbol{f}_i$ under PK_{ID}, where $\boldsymbol{f}_i$ is the i-th block of a file with tag τ. It outputs 1 if the verification succeeds, and 0 otherwise.
- *Auditing*:
  The TPA interacts with the CSP through:
  - *Challenge*$(ID, \tau) \to chal$:
    The TPA generates a challenge $chal$ for the file associated with tag τ outsourced by the user with identity ID.
  - *ProofGen*$(\tau, \{(\boldsymbol{f}_i, \sigma_i)\}_{i=1}^n, chal) \to PF$: Given an auditing challenge $chal$, the CSP computes a proof PF in response, based on the block-tag pairs $\{(\boldsymbol{f}_i, \sigma_i)\}_{i=1}^n$, where $\boldsymbol{f}_i$ is the i-th block of a file with tag τ.
  - *ProofVerify*$(PK_{ID}, \tau, PF, chal) \to \{0, 1\}$:
    The TPA verifies whether PF is a valid proof of possession, in response to challenge $chal$, for the file associated with tag τ under PK_{ID}. It outputs 1 if valid, and 0 otherwise.

Correctness. We require that for any $(msk, pp) \leftarrow$ *Initialization*(1^λ), any $(sk_{ID}, PK_{ID}) \leftarrow$ *KeyGen*(pp) and $Cert_{ID} \leftarrow$ *Certify*(msk, ID, PK_{ID}), the following conditions hold:

- For any file $\boldsymbol{F} = \{\boldsymbol{f}_i\}_{i=1}^n$ with identifier Fid, if $(\tau, \{\sigma_i\}_{i=1}^n) \leftarrow$ *TagGen*$(sk_{ID}, Cert_{ID}, \boldsymbol{F}, Fid)$, then *TagVerify*$(PK_{ID}, \tau, i, \boldsymbol{f}_i, \sigma_i) = 1$ for all i.
- For any file $\boldsymbol{F} = \{\boldsymbol{f}_i\}_{i=1}^n$ with identifier Fid, and any challenge $chal \leftarrow$ *Challenge*(ID, τ), if $(\tau, \{\sigma_i\}_{i=1}^n) \leftarrow$ *TagGen*$(sk_{ID}, Cert_{ID}, \boldsymbol{F}, Fid)$ and $PF \leftarrow$ *ProofGen*$(\tau, \{(\boldsymbol{f}_i, \sigma_i)\}_{i=1}^n, chal)$, then *ProofVerify*$(PK_{ID}, \tau, PF, chal) = 1$.

3 Certificate-Based Quasi-linearly Homomorphic Signature

3.1 Formal Definition

Definition 6. *A CB-QLHS scheme is a tuple of probabilistic polynomial-time algorithms with the following functionalities:*

- $Setup(1^\lambda) \to (msk, pp)$: Given a security parameter 1^λ, the certifier initializes the system and generates a master secret key msk that is kept secret, along with system parameters pp that are made publicly available.
- $KeyGen(pp) \to (sk_{ID}, PK_{ID})$: Taking as input the system parameters pp, the algorithm outputs a key pair (sk_{ID}, PK_{ID}).
- $CertGen(msk, ID, PK_{ID}) \to Cert_{ID}$: Taking the master secret key msk, a user's identity ID, and its public key PK_{ID} as inputs, the certifier generates a certificate $Cert_{ID}$.
- $Sign(sk_{ID}, Cert_{ID}, \boldsymbol{D}, Did) \to (\tau, \{\sigma_i\}_{i=1}^n)$: Given the signer's secret key sk_{ID}, its certificate $Cert_{ID}$, and a raw data sequence $\boldsymbol{D} = (\boldsymbol{m}_1, \ldots, \boldsymbol{m}_n) \in \mathbb{Z}_p^{l \times n}$ with an identifier Did, where each datum $\boldsymbol{m}_i = (m_{i1}, \ldots, m_{il})^\top$ is an l-dimensional vector, the signer generates a unique tag τ for the metadata of sequence $\boldsymbol{D}$, and a set of signatures $\{\sigma_i\}_{i=1}^n$ for the raw data $\{\boldsymbol{m}_i\}_{i=1}^n$.
- $Verify(PK_{ID}, \tau, i, \boldsymbol{m}_i, \sigma_i) \to \{0, 1\}$: With the public key PK_{ID}, a tag τ of a raw data sequence $\boldsymbol{D}$, the i-th datum $\boldsymbol{m}_i$ in $\boldsymbol{D}$, and its signature σ_i as input, the verifier checks whether σ_i is a valid signature on $\boldsymbol{m}_i$ under PK_{ID}, outputting 1 if the verification succeeds and 0 otherwise.
- $Combine(\tau, (\boldsymbol{D}, \{\sigma_i\}_{i=1}^n), \boldsymbol{\omega}) \to (\tilde{\boldsymbol{m}}, \tilde{\sigma})$: Given a raw data sequence $\boldsymbol{D} = \{\boldsymbol{m}_i\}_{i=1}^n$ associated with tag τ, its corresponding signatures $\{\sigma_i\}_{i=1}^n$, and a combination coefficient vector $\boldsymbol{\omega} = (\omega_1, \omega_2, \ldots, \omega_n)^\top \in \mathbb{Z}_p^l$, the combiner generates a derived datum-signature pair $(\tilde{\boldsymbol{m}} = \sum_{i=1}^n \omega_i \boldsymbol{m}_i, \tilde{\sigma})$.
- $DSVerify(PK_{ID}, \tau, (\tilde{\boldsymbol{m}}, \tilde{\sigma}), \boldsymbol{\omega}) \to \{0, 1\}$: Given the public key PK_{ID}, a tag τ of a raw data sequence $\boldsymbol{D}$, a derived datum-signature pair $(\tilde{\boldsymbol{m}}, \tilde{\sigma})$, and a coefficient vector $\boldsymbol{\omega}$, the verifier checks whether $\tilde{\sigma}$ is a valid derived signature on $\tilde{\boldsymbol{m}}$ under PK_{ID}, outputting 1 if the verification succeeds and 0 otherwise.

Correctness. We require that for any $(msk, pp) \leftarrow Setup(1^\lambda)$, any $(sk_{ID}, PK_{ID}) \leftarrow KeyGen(pp)$ and $Cert_{ID} \leftarrow CertGen(msk, ID, PK_{ID})$, the following hold:

- For any raw data sequence $\boldsymbol{D} = \{\boldsymbol{m}_i\}_{i=1}^n$ with identifier Did, if $(\tau, \{\sigma_i\}_{i=1}^n) \leftarrow Sign(sk_{ID}, Cert_{ID}, \boldsymbol{D}, Did)$, then $Verify(PK_{ID}, \tau, i, \boldsymbol{m}_i, \sigma_i) = 1$ for all i.
- For any raw data sequence $\boldsymbol{D} = \{\boldsymbol{m}_i\}_{i=1}^n$ with identifier Did, and any coefficient vector $\boldsymbol{\omega}$, if $(\tau, \{\sigma_i\}_{i=1}^n) \leftarrow Sign(sk_{ID}, Cert_{ID}, \boldsymbol{D}, Did)$ and $(\tilde{\boldsymbol{m}}, \tilde{\sigma}) \leftarrow Combine(\tau, (\boldsymbol{D}, \{\sigma_i\}_{i=1}^n), \boldsymbol{\omega})$, then $DSVerify(PK_{ID}, \tau, (\tilde{\boldsymbol{m}}, \tilde{\sigma}), \boldsymbol{\omega}) = 1$.

3.2 Security Model

In CB-QLHS, generating a valid raw datum signature necessitates the simultaneous use of both the private key and the corresponding certificate. Furthermore, a valid derived datum-signature pair can be produced through linear combination only from valid raw datum-signature pairs. Accordingly, a CB-QLHS scheme must satisfy two requirements: raw datum signature unforgeability and derived datum unforgeability. Therefore, we classify adversaries along two orthogonal dimensions: (i) *Attack capability*: whether the adversary can replace public keys or acts as a malicious certifier with access to the master secret key; and (ii) *Attack goal*: whether the adversary aims to forge raw datum signatures or derived datum-signature pairs after tampering with the raw data. This leads to four adversary types, summarized in Table 1, with detailed descriptions as follows:

1. $\mathcal{A}_1$: Replaces the user's public key PK_{ID} with a self-chosen key PK'_U, but cannot obtain the certificate for PK'_U.
2. $\mathcal{A}_2$: Possesses the master secret key and can issue certificates for any identity and public key, but is not allowed to replace public keys.
3. $\mathcal{A}_3$: Combines the key replacement ability of $\mathcal{A}_1$ with the goal of forging a valid derived datum-signature pair after tampering with the raw data.
4. $\mathcal{A}_4$: Combines capability of $\mathcal{A}_2$ with the forgery goal of $\mathcal{A}_3$.

Table 1. Adversary Classification in CB-QLHS

Capability \ Goal	Raw Datum Signature	Derived Datum
Public Key Replacement	$\mathcal{A}_1$	$\mathcal{A}_3$
Master Secret Key Possession	$\mathcal{A}_2$	$\mathcal{A}_4$

The security of CB-QLHS is formalized through the following games between the challenger $\mathcal{C}$ and different types of adversaries.

Definition 7. *A CB-QLHS scheme satisfies raw datum signature unforgeability against public key replacement attacks if, for any PPT adversary $\mathcal{A}_1$, the advantage of winning the following game is negligible.*

Game 1. $\mathcal{A}_1$ interacts with the challenger $\mathcal{C}$ as follows.

Initialization: The challenger $\mathcal{C}$ generates the system public parameters pp and the master secret key msk, sends pp to $\mathcal{A}_1$ and keeps msk secret.

Query: $\mathcal{A}_1$ can adaptively query the following oracles in polynomial time.

- $PK_{ID} \leftarrow UserKeyGen\langle ID\rangle$: $\mathcal{C}$ generates a secret-public key pair (sk_{ID}, PK_{ID}) and returns PK_{ID}.

- $(sk'_{ID}, PK'_{ID}) \leftarrow PKReplace\langle ID, PK'_{ID}, sk'_{ID}\rangle$: $\mathcal{C}$ replaces the user's original secret-public key pair (sk_{ID}, PK_{ID}) with (sk'_{ID}, PK'_{ID}).
- $sk_{ID} \leftarrow Corruption\langle ID\rangle$: $\mathcal{C}$ returns the secret key sk_{ID}.
- $Cert_{ID} \leftarrow Certificate\langle ID, PK_{ID}\rangle$: $\mathcal{C}$ generates a certificate $Cert_{ID}$ for (ID, PK_{ID}) and returns $Cert_{ID}$.
- $(\tau, \{\sigma_i\}_{i=1}^n) \leftarrow Sign\langle ID, \boldsymbol{D} = \{\boldsymbol{m}_i\}_{i=1}^n, Did\rangle$: $\mathcal{C}$ generates a tag τ and signatures $\{\sigma_i\}_{i=1}^n$ for the raw data sequence $\boldsymbol{D}$ with the identifier Did, and returns $(\tau, \{\sigma_i\}_{i=1}^n)$. It records $(\boldsymbol{D}, \tau)$ in a set $\mathcal{L}$.

Forgery: $\mathcal{A}_1$ chooses a target identity ID^* and a raw data sequence $\boldsymbol{D}^* = \{\boldsymbol{m}_i^*\}_{i=1}^n$, then outputs $(PK^*, \tau^*, i^*, \boldsymbol{m}^*, \sigma^*)$. $\mathcal{A}_1$ wins Game 1 if:

- $Verify(PK^*, \tau^*, i^*, \boldsymbol{m}^*, \sigma^*) = 1$.
- (ID^*, PK^*) has never been submitted to the $Certificate$ oracle.
- Either $(\boldsymbol{D}^*, \tau^*) \in \mathcal{L}$ and $\boldsymbol{m}^* \neq \boldsymbol{m}_{i^*}^*$; or $(\boldsymbol{D}^*, \tau^*) \notin \mathcal{L}$.

Definition 8. *A CB-QLHS scheme satisfies raw datum signature unforgeability against the malicious certifier if, for any PPT adversary $\mathcal{A}_2$, the advantage of winning the following game is negligible.*

Game 2. $\mathcal{A}_2$ interacts with the challenger $\mathcal{C}$ as follows.

Initialization: The challenger $\mathcal{C}$ generates the system public parameters pp and the master secret key msk, then sends (pp, msk) to $\mathcal{A}_2$.

Query: $\mathcal{A}_2$ can adaptively query the oracles $UserKeyGen$, $Corruption$, and $Sign$ within polynomial time, as in Game 1.

Forgery: $\mathcal{A}_2$ chooses a target identity ID^* and a raw data sequence $\boldsymbol{D}^* = \{\boldsymbol{m}_i^*\}_{i=1}^n$, then outputs $(\tau^*, i^*, \boldsymbol{m}^*, \sigma^*)$.
$\mathcal{A}_2$ wins Game 2 if:

- $Verify(PK_{ID^*}, \tau^*, i^*, \boldsymbol{m}^*, \sigma^*) = 1$.
- ID^* has never been submitted to the $Corruption$ oracle.
- Either $(\boldsymbol{D}^*, \tau^*) \in \mathcal{L}$ and $\boldsymbol{m}^* \neq \boldsymbol{m}_{i^*}^*$; or $(\boldsymbol{D}^*, \tau^*) \notin \mathcal{L}$.

Definition 9. *A CB-QLHS scheme satisfies derived datum unforgeability against public key replacement attacks if, for any PPT adversary $\mathcal{A}_3$, the advantage of winning the following game is negligible.*

Game 3. $\mathcal{A}_3$ interacts with the challenger $\mathcal{C}$ as follows.

Initialization: The challenger $\mathcal{C}$ generates the system public parameters pp and the master secret key msk, sends pp to $\mathcal{A}_3$, and keeps msk secret.

Query: $\mathcal{A}_3$ can adaptively query the oracles $UserKeyGen$, $PKReplace$, $Corruption$, $Certificate$ and $Sign$ within polynomial time, as in Game 1. It attempts to tamper with any queried raw data, generates derived datum-signature pairs using the *Combine* algorithm, and checks their validity via *DSVerify* algorithm.

Forgery: $\mathcal{A}_3$ chooses a target identity ID^*, a raw data sequence $\boldsymbol{D}^* = \{\boldsymbol{m}_i^*\}_{i=1}^n$ with tag τ^* from the **Query** phase, and specifies an index set $I^* \subseteq \{1,\ldots,n\}$, indicating the indices of data in $\boldsymbol{D}^*$ that it may tamper with. $\mathcal{C}$ then provides a coefficient vector $\boldsymbol{\omega}^* = (\omega_1^*, \omega_2^*, \ldots, \omega_n^*)^\top$, where $\exists i \in I^*$ such that $\omega_i^* \neq 0$. Finally, $\mathcal{A}_3$ outputs $(PK^*, \tilde{\boldsymbol{m}}^*, \tilde{\sigma}^*)$, it wins Game 3 if:

- $DSVerify(PK^*, \tau^*, (\tilde{\boldsymbol{m}}^*, \tilde{\sigma}^*), \boldsymbol{\omega}^*) = 1$ and $\tilde{\boldsymbol{m}}^* \neq \sum_{i=1}^n \omega_i \boldsymbol{m}_i^*$.
- (ID^*, PK^*) has never been submitted to the *Certificate* oracle.

Definition 10. *A CB-QLHS scheme satisfies derived datum unforgeability against the malicious certifier if, for any PPT adversary $\mathcal{A}_4$, the advantage of winning the following game is negligible.*

Game 4. $\mathcal{A}_4$ interacts with the challenger $\mathcal{C}$ as follows.

Initialization: The challenger $\mathcal{C}$ generates the system public parameters pp and the master secret key msk, then sends (pp, msk) to $\mathcal{A}_4$.

Query: $\mathcal{A}_4$ can adaptively query the oracles *UserKeyGen, Corruption* and *Sign* within polynomial time, as in Game 2. It also generates and verifies derived datum-signature pairs using the *Combine* and *DSVerify* algorithms as $\mathcal{A}_3$ does.

Forgery: $\mathcal{A}_4$ chooses a target identity ID^*, a raw data sequence $\boldsymbol{D}^* = \{\boldsymbol{m}_i^*\}_{i=1}^n$ with tag τ^* from the **Query** phase, and specifies an index set $I^* \subseteq \{1,\ldots,n\}$, indicating the indices of data in $\boldsymbol{D}^*$ that it may tamper with. $\mathcal{C}$ then provides a coefficient vector $\boldsymbol{\omega}^* = (\omega_1^*, \omega_2^*, \ldots, \omega_n^*)^\top$, where $\exists i \in I^*$ such that $\omega_i^* \neq 0$. Finally, $\mathcal{A}_4$ outputs a derived data-signature pair $(\tilde{\boldsymbol{m}}^*, \tilde{\sigma}^*)$, it wins Game 4 if:

- $DSVerify(PK_{ID^*}, \tau^*, (\tilde{\boldsymbol{m}}^*, \tilde{\sigma}^*), \boldsymbol{\omega}^*) = 1$ and $\tilde{\boldsymbol{m}}^* \neq \sum_{i=1}^n \omega_i^* \boldsymbol{m}_i^*$.
- ID^* has never been submitted to the *Corruption* oracle.

4 Generic Construction of CB-CSDIA from CB-QLHS

In this section, we demonstrate how a secure CB-CSDIA protocol can be constructed from a secure CB-QLHS scheme. In CB-QLHS, the signer generates signatures on raw data, the combiner performs linear homomorphic operations to produce derived datum-signature pairs, and the verifier checks the validity of both the raw and derived signatures. In CB-CSDIA, the DO generates HVTs for outsourced file blocks, the CSP responds to the TPA's challenge with a proof, and the TPA verifies it. This correspondence maps the DO to the signer, the CSP to the combiner, and the TPA to the verifier of derived datum-signature pairs. The raw data signatures (i.e., HVTs) can be publicly verified during transmission to ensure data integrity against tampering.

Let $\mathcal{CBA}$ and $\mathcal{CBQ}$ denote the certificate-based auditing protocol described in Definition 5 and the CB-QLHS scheme described in Definition 6, respectively. We now present a generic construction of CB-CSDIA from CB-QLHS:

- $\mathcal{CBA}.Initialization(1^\lambda) \to (\mathcal{CBA}.msk, \mathcal{CBA}.pp)$:
  1. $(\mathcal{CBQ}.msk, \mathcal{CBQ}.pp) \leftarrow \mathcal{CBQ}.Setup(1^\lambda)$;
  2. $(\mathcal{CBA}.msk, \mathcal{CBA}.pp) \leftarrow (\mathcal{CBQ}.msk, \mathcal{CBQ}.pp)$.
- $\mathcal{CBA}.Registration$: This phase involves two algorithms:
  - $\mathcal{CBA}.UserCreate(\mathcal{CBA}.pp) \to (\mathcal{CBA}.sk_{ID}, \mathcal{CBA}.PK_{ID})$:
    1. $(\mathcal{CBQ}.sk_{ID}, \mathcal{CBQ}.PK_{ID}) \leftarrow \mathcal{CBQ}.KeyGen(\mathcal{CBA}.pp)$;
    2. $(\mathcal{CBA}.sk_{ID}, \mathcal{CBA}.PK_{ID}) \leftarrow (\mathcal{CBQ}.sk_{ID}, \mathcal{CBQ}.PK_{ID})$.
  - $\mathcal{CBA}.Certify(\mathcal{CBA}.msk, ID, \mathcal{CBA}.PK_{ID}) \to \mathcal{CBA}.Cert_{ID}$:
    1. $\mathcal{CBQ}.Cert_{ID} \leftarrow \mathcal{CBQ}.CertGen(\mathcal{CBA}.msk, ID, \mathcal{CBA}.PK_{ID})$;
    2. $\mathcal{CBA}.Cert_{ID} \leftarrow \mathcal{CBQ}.Cert_{ID}$.
- $\mathcal{CBA}.Outsourcing$: This phase includes two algorithms:
  - $\mathcal{CBA}.TagGen(\mathcal{CBA}.sk_{ID}, \mathcal{CBA}.Cert_{ID}, \boldsymbol{F} = \{\boldsymbol{f}_i\}_{i=1}^n, Fid) \to (\tau, \{\sigma_i\}_{i=1}^n)$:
    $(\tau, \{\sigma_i\}_{i=1}^n) \leftarrow \mathcal{CBQ}.Sign(\mathcal{CBA}.sk_{ID}, \mathcal{CBA}.Cert_{ID}, \boldsymbol{F} = \{\boldsymbol{f}_i\}_{i=1}^n, , Fid)$.
  - $\mathcal{CBA}.TagVerify(\mathcal{CBA}.PK_{ID}, \tau, i, \boldsymbol{f}_i, \sigma_i) \to \{0, 1\}$:
    Output $\mathcal{CBQ}.Verify(\mathcal{CBQ}.PK_{ID}, \tau, i, \boldsymbol{f}_i, \sigma_i)$.
- $\mathcal{CBA}.Auditing$: This phase consists of three algorithms:
  - $\mathcal{CBA}.Challenge(ID, \tau) \to \mathcal{CBA}.chal$:
    1. Randomly select $c_i \in \mathbb{Z}_p^*$ for each $i \in \mathcal{N}$, where $\mathcal{N} \subseteq \{1, \ldots, n\}$;
    2. $\mathcal{CBA}.chal \leftarrow \{(i, c_i)\}_{i \in \mathcal{N}}$.
  - $\mathcal{CBA}.ProofGen(\tau, \{(\boldsymbol{f}_i, \sigma_i)\}_{i=1}^n, \mathcal{CBA}.chal = \{(i, c_i)\}_{i \in \mathcal{N}}) \to \mathcal{CBA}.PF$:
    1. $\mathcal{CBQ}.\boldsymbol{\omega} \leftarrow \{\omega_i\}_{i=1}^n$, where $\omega_i = c_i$ if $i \in \mathcal{N}$, and $\omega_i = 0$ otherwise;
    2. $\mathcal{CBA}.PF \leftarrow \mathcal{CBQ}.Combine(\tau, (\boldsymbol{F} = \{\boldsymbol{f}_i\}_{i=1}^n, \{\sigma_i\}_{i=1}^n), \mathcal{CBQ}.\boldsymbol{\omega})$.
  - $\mathcal{CBA}.ProofVerify(\mathcal{CBA}.PK_{ID}, \tau, \mathcal{CBA}.PF, \mathcal{CBA}.chal) \to \{0, 1\}$:
    1. $\mathcal{CBQ}.\boldsymbol{\omega} \leftarrow \{\omega_i\}_{i=1}^n$, where $\omega_i = c_i$ if $i \in \mathcal{N}$, and $\omega_i = 0$ otherwise;
    2. Output $\mathcal{CBQ}.DSVerify(\mathcal{CBA}.PK_{ID}, \tau, \mathcal{CBA}.PF, \mathcal{CBQ}.\boldsymbol{\omega})$.

The correctness of the resulting CB-CSDIA protocol is guaranteed by the correctness of the underlying CB-QLHS scheme.

Theorem 1. *Let $\mathcal{CBQ}$ = (Setup, KeyGen, CertGen, Sign, Verify, Combine, DSVerify) be a secure CB-QLHS scheme. Then the resulting $\mathcal{CBA}$ = (Initialization, Registration = (UserCreate, Certify), Outsourcing = (TagGen, TagVerify), Auditing = (Challenge, ProofGen, ProofVerify)) is a secure CB-CSDIA protocol.*
Proof. Based on the above observations, the security of CB-CSDIA is guaranteed by two aspects: the unforgeability of block tags and the unforgeability of proofs. The former derives from the unforgeability of raw data signatures, while the latter relies on the unforgeability of derived datum-signature pairs in the underlying CB-QLHS scheme. These relationships give rise to two key conclusions:

1. If an adversary can forge a block tag in the CB-CSDIA protocol with non-negligible advantage, then the adversary can forge a raw datum signature in the underlying CB-QLHS scheme with the same advantage.
2. If an adversary can forge a proof of data possession in the CB-CSDIA protocol with non-negligible advantage, then the adversary can forge the derived datum in the underlying CB-QLHS scheme with the same advantage.

Remark 1. In CB-CSDIA protocols where the CSP serves as the certifier, the adversary $\mathcal{A}_3$ can be disregarded. This is because the certifier cannot replace user public keys, and public key replacement attacks launched by external adversaries to forge valid auditing proofs are meaningless.

5 An Efficient Instantiation of CB-QLHS

5.1 Scheme Description

- $Setup(1^\lambda)$: Take the security parameter λ as input, the certifier executes the following steps:
  (1) Let p be a prime number such that $p > 2^\lambda$, and consider two multiplicative cyclic groups $\mathbb{G}_1$ and $\mathbb{G}_T$ of order p.
  (2) Choose a bilinear pairing $e: \mathbb{G}_1 \times \mathbb{G}_1 \to \mathbb{G}_T$, and a generator g of $\mathbb{G}_1$.
  (3) Pick four hash functions $H_0: \{0,1\}^* \times \mathbb{G}_1 \to \mathbb{G}_1$, $H_1: \{0,1\}^* \to \mathbb{Z}_p^*$, $H_2: \{0,1\}^* \to \mathbb{Z}_p^*$, $H_3: \{0,1\}^* \to \mathbb{Z}_p^*$.
  (4) Randomly select $s_1, s_2 \in \mathbb{Z}_p^*$, set the master secret key $msk = (s_1, s_2)$, and compute the master public key $MPK = (MPK_1, MPK_2) = (g^{s_1}, g^{s_2})$.
  (5) Securely store (s_1, s_2) and publish the system public parameters $pp = \{\mathbb{G}_1, \mathbb{G}_T, p, e, g, H_0, H_1, H_2, H_3, MPK\}$.
- $KeyGen(pp)$: The user with identity ID selects a random value $x \in \mathbb{Z}_p^*$ as the secret key sk_{ID}, computes the corresponding public key as $PK_{ID} = g^x$, and transmits the tuple (ID, PK_{ID}) to the certifier.
- $CertGen(msk, ID, PK_{ID})$: Given (ID, PK_{ID}), the certifier generates a certificate for the user through the following steps.
  (1) Compute $Q_{ID} = H_0(ID, PK_{ID})$, $Cert_1 = Q_{ID}^{s_1}$, $Cert_2 = Q_{ID}^{s_2}$.
  (2) Set the certificate $Cert_{ID} = (Cert_1, Cert_2)$ and send $Cert_{ID}$ to the user.
- $Sign(sk_{ID}, Cert_{ID}, \boldsymbol{D}, Did)$: For a raw data sequence $\boldsymbol{D} = (\boldsymbol{m}_1, \ldots, \boldsymbol{m}_n) \in \mathbb{Z}_p^{l \times n}$ with an identifier Did, where each $\boldsymbol{m}_i = (m_{i1}, \ldots, m_{il})^\top \in \mathbb{Z}_p^l$, the signer generates a tag τ that encapsulates metadata such as the identifier, the sequence length, and auxiliary information for verification, along with a signature σ_i for each datum $\boldsymbol{m}_i$. The signing process proceeds as follows:
  (1) Select l random elements $\gamma_j \in \mathbb{Z}_p^*$ for $1 \leq j \leq l$, and compute the commitments $\{g_j = (MPK_2)^{\gamma_j}\}_{j=1}^l$.
  (2) Construct the metadata string $\hat{\tau} = Did \parallel n \parallel \{g_j\}_{j=1}^l$.
  (3) Compute $\sigma_{\hat{\tau}} = (Cert_1 \cdot Cert_2^{\gamma_1})^{\frac{1}{x+H_1(ID \parallel \hat{\tau})}}$ and set the sequence tag as $\tau = (\hat{\tau}, \sigma_{\hat{\tau}})$.
  (4) Calculate signature $\sigma_i = (Cert_1^{H_2(Did,i)} \cdot Cert_2^{\sum_{j=1}^l \gamma_j m_{ij} + \gamma_1 H_3(i)})^{\frac{1}{x+H_1(ID \parallel \hat{\tau})}}$ for each raw datum $\boldsymbol{m}_i = (m_{i1}, m_{i2}, \ldots, m_{il})^\top$.
  (5) Output the tag and signatures $(\tau, \{\sigma_i\}_{i=1}^n)$.
- $Verify(PK_{ID}, \tau, i, \boldsymbol{m}_i, \sigma_i)$: To validate the signature σ_i on the i-th datum $\boldsymbol{m}_i$ in the sequence $\boldsymbol{D}$ with tag τ, the verifier proceeds as follows.
  (1) Parse τ to extract Did, n, and $\{g_j\}_{j=1}^l$.
  (2) If this is the first time verifying a signature on any datum in $\boldsymbol{D}$, validate the tag τ by checking: $e(\sigma_{\hat{\tau}}, PK_{ID} \cdot g^{H_1(ID \parallel \hat{\tau})}) = e(Q_{ID}, MPK_1 \cdot g_1)$.
  (3) Validate the signature σ_i via the equation:

$$e\left(\sigma_i, PK_{ID} \cdot g^{H_1(ID \parallel \hat{\tau})}\right) = e\left(Q_{ID}, MPK_1^{H_2(Did,i)} \cdot \prod_{j=1}^l g_j^{m_{ij}} \cdot g_1^{H_3(i)}\right).$$

(4) Output 1 (accept) if the equations hold; otherwise, output 0 (reject).

- $Combine(\tau, (\boldsymbol{D}, \{\sigma_i\}_{i=1}^n), \boldsymbol{\omega})$: Given a raw data sequence $\boldsymbol{D} = \{\boldsymbol{m}_i\}_{i=1}^n \in \mathbb{Z}_p^{l \times n}$ associated with tag τ, its corresponding signatures $\{\sigma_i\}_{i=1}^n$, and a combination coefficient vector $\boldsymbol{\omega} = (\omega_1, \omega_2, \ldots, \omega_n)^\top$, the combiner performs the following operations to generate a derived datum-signature pair.
  (1) Compute $\tilde{\boldsymbol{m}} = \sum_{i=1}^n \omega_i \boldsymbol{m}_i = (\tilde{m}_1, \tilde{m}_2, \ldots, \tilde{m}_l)^\top$ and $\tilde{\sigma} = \prod_{i=1}^n \sigma_i^{\omega_i}$.
  (2) Output a derived datum-signature pair $(\tilde{\boldsymbol{m}}, \tilde{\sigma})$.

- $DSVerify(PK_{ID}, \tau, (\tilde{\boldsymbol{m}}, \tilde{\sigma}), \boldsymbol{\omega})$: The verifier performs the following operations to validate the derived datum-signature pair.
  (1) Check the validity of the tag τ if it has not been verified previously.
  (2) Validate the derived datum-signature pair as follows:

$$e\left(\tilde{\sigma}, PK_{ID} \cdot g^{H_1(ID\|\hat{\tau})}\right) = e\left(Q_{ID}, MPK_1^{\sum_{i=1}^n \omega_i H_2(Did,i)} \prod_{j=1}^l g_j^{\tilde{m}_j} \cdot g_1^{\sum_{i=1}^n \omega_i H_3(i)}\right).$$

(3) Output 1 (accept) if the equation holds; otherwise, output 0 (reject).

Correctness. We prove the correctness of our CB-QLHS instantiation by showing that the following verification equations hold.

The verification of a raw datum signature:

$$e\left(\sigma_i, PK_{ID} \cdot g^{H_1(ID\|\hat{\tau})}\right)$$
$$= e\left((Cert_1^{H_2(Did,i)} \cdot Cert_2^{\sum_{j=1}^l \gamma_j m_{ij} + \gamma_1 H_3(i)})^{\frac{1}{x+H_1(ID\|\hat{\tau})}}, g^x \cdot g^{H_1(ID\|\hat{\tau})}\right)$$
$$= e\left(Q_{ID}^{s_1 H_2(Did,i) + s_2(\sum_{j=1}^l \gamma_j m_{ij} + \gamma_1 H_3(i))}, g\right)$$
$$= e\left(Q_{ID}, MPK_1^{H_2(Did,i)} \cdot \prod_{j=1}^l g_j^{m_{ij}} \cdot g_1^{H_3(i)}\right).$$

The verification of a derived datum-signature pair:

$$e\left(\tilde{\sigma}, PK_{ID} \cdot g^{H_1(ID\|\hat{\tau})}\right) = e\left(\prod_{i=1}^n \sigma_i^{\omega_i}, PK_{ID} \cdot g^{H_1(ID\|\hat{\tau})}\right)$$
$$= e\left(\prod_{i=1}^n (Cert_1^{H_2(Did,i)} \cdot Cert_2^{\sum_{j=1}^l \gamma_j m_{ij} + \gamma_1 H_3(i)})^{\frac{\omega_i}{x+H_1(ID\|\hat{\tau})}}, g^x \cdot g^{H_1(ID\|\hat{\tau})}\right)$$
$$= e\left(Q_{ID}^{s_1 \sum_{i=1}^n \omega_i H_2(Did,i) + s_2(\sum_{j=1}^l \gamma_j \tilde{m}_j + \gamma_1 \sum_{i=1}^n \omega_i H_3(i))}, g\right)$$
$$= e\left(Q_{ID}, MPK_1^{\sum_{i=1}^n \omega_i H_2(Did,i)} \cdot \prod_{j=1}^l g_j^{\tilde{m}_j} \cdot g_1^{\sum_{i=1}^n \omega_i H_3(i)}\right).$$

5.2 Security Analysis

Theorem 2. *Assume that the weak modified q-CAA assumption holds in $\mathbb{G}_1$, then our CB-QLHS scheme is secure against the adversary $\mathcal{A}_1$. Specifically, if $\mathcal{A}_1$ is an adversary capable of forging a raw datum signature with a non-negligible advantage ϵ, then there exists a PPT algorithm $\mathcal{B}$ that can solve the weak modified q-CAA problem with a non-negligible advantage $\epsilon' \geq (1 - \frac{1}{q_u})^{q_c+q_k} \cdot (1 - \frac{1}{q_s+1})^{q_s} \cdot \frac{\epsilon}{q_u \cdot (q_s+1)}$, where q_u, q_k, q_c, and q_s denote the number of queries made to the oracles UserKeyGen, Corruption, Certificate, and Sign, respectively.*

Proof. Given a random instance of the weak modified q-CAA problem $(g, g^x, g^a, g^b, \{(z_i, (g^{ab})^{1/(x+z_i)})\}_{i=1}^{q})$, where g is a generator of $\mathbb{G}_1$, the goal of $\mathcal{B}$ is to utilize the capability of $\mathcal{A}_1$ to compute g^{ab} or a new pair $(z^*, (g^{ab})^{1/(x+z^*)})$ for some $z^* \notin \{z_1, \ldots, z_q\}$. We regard the hash functions H_0, H_1 as random oracles, and $H_2, H_3 : \{0,1\}^* \to \mathbb{Z}_p^*$. $\mathcal{B}$ simulates the random oracle as follows:

Initialization: $\mathcal{B}$ randomly selects $s_1, s_2 \in \mathbb{Z}_p^*$ and sets $MPK = (MPK_1, MPK_2) = ((g^a)^{s_1}, (g^a)^{s_2})$. $\mathcal{B}$ then sends $pp = \{\mathbb{G}_1, \mathbb{G}_T, p, e, g, MPK, H_2, H_3\}$ to $\mathcal{A}_1$, and randomly chooses an integer k^* from $\{1, \ldots, q_u\}$.

Query:

- *UserKeyGen*: $\mathcal{B}$ maintains a list L_U consisting of tuples $(ID_k, sk_{ID_k}, PK_{ID_k})$.
  If ID_k already exists in L_U, $\mathcal{B}$ directly returns PK_{ID_k}. Otherwise:
  - If $k \neq k^*$, $\mathcal{B}$ selects $x_k \in \mathbb{Z}_p^*$, sets $sk_{ID_k} = x_k$ and $PK_{ID_k} = g^{x_k}$.
  - If $k = k^*$, $\mathcal{B}$ sets $sk_{ID_k} = \bot$ and $PK_{ID_k} = g^x$.

  $\mathcal{B}$ adds $(ID_k, sk_{ID_k}, PK_{ID_k})$ to the list L_U and returns PK_{ID_k}.

- H_0: $\mathcal{B}$ maintains a list L_{H_0} consisting of tuples $(ID_k, PK_{ID_k}, Q_k, \mu_k)$. On input (ID_k, PK_{ID_k}), if (ID_k, PK_{ID_k}) already exists in L_{H_0}, $\mathcal{B}$ directly returns Q_k. Otherwise:
  - If $ID_k \neq ID_{k^*}$, $\mathcal{B}$ chooses $\mu_k \in \mathbb{Z}_p^*$ and sets $Q_k = g^{\mu_k}$.
  - If $ID_k = ID_{k^*}$, $\mathcal{B}$ sets $\mu_k = \bot$ and $Q_k = g^b$.

  $\mathcal{B}$ adds $(ID_k, PK_{ID_k}, Q_k, \mu_k)$ to the list L_{H_0} and returns Q_k.

- H_1: $\mathcal{B}$ maintains a list L_{H_1} consisting of tuples $(ID_k, \hat{\tau}_k, \beta_k, c_k)$. On input $(ID_k, \hat{\tau}_k)$, if $(ID_k, \hat{\tau}_k)$ already exists in L_{H_1}, $\mathcal{B}$ directly returns β_k. Otherwise:
  - If $ID_k = ID_{k^*}$,
    $\mathcal{B}$ throws a biased coin $c_k \in \{0, 1\}$, where $c_k = 1$ with probability η and $c_k = 0$ with probability $1 - \eta$. If $c_k = 1$, $\mathcal{B}$ selects $\beta_k \in \mathbb{Z}_p^*$ such that $\beta_k \notin \{z_1, \ldots, z_q\}$. If $c_k = 0$, $\mathcal{B}$ selects $z_k \in \{z_1, \ldots, z_q\}$ that has not been chosen before, and sets $\beta_k = z_k$.
  - Otherwise, $\mathcal{B}$ chooses $\beta_k \in \mathbb{Z}_p^*$ and sets $c_k = \bot$.

  $\mathcal{B}$ then adds $(ID_k, \hat{\tau}_k, \beta_k, c_k)$ to the list L_{H_1} and returns β_k.

- *PKReplace*: On input (ID, PK'_{ID}, sk'_{ID}), if PK'_{ID} is a valid public key for sk'_{ID}, $\mathcal{B}$ replaces the original key pair (sk_{ID}, PK_{ID}) with (sk'_{ID}, PK'_{ID}).

- *Corruption*: Upon inputting an identity ID_k, $\mathcal{B}$ terminates if $ID_k = ID_{k^*}$; otherwise, $\mathcal{B}$ returns sk_{ID_k}.

- *Certificate*: Given an identity ID_k, $\mathcal{B}$ terminates if $ID_k = ID_{k^*}$. Otherwise, it initiates an H_0 Query to obtain $(ID_k, PK_{ID_k}, Q_k, w_k)$, and returns $Cert_{ID_k} = ((g^a)^{\mu_k s_1}, (g^a)^{\mu_k s_2})$.
- *Sign*: Given $(ID_k, \boldsymbol{D} = \{\boldsymbol{m}_i\}_{i=1}^n, Did)$, $\mathcal{B}$ retrieves $(ID_k, sk_{ID_k}, PK_{ID_k})$ from L_U, $(ID_k, PK_{ID_k}, Q_k, w_k)$ from L_{H_0}, and $(ID_k, \tau_k, \beta_k, c_k)$ from L_{H_1}. Then, $\mathcal{B}$ randomly selects $\gamma_1, \ldots, \gamma_l \in \mathbb{Z}_p^*$ and computes $\{g_j = MPK_2^{\gamma_j}\}_{j=1}^l$. $\mathcal{B}$ constructs $\hat{\tau} = Did \parallel n \parallel \{g_j\}_{j=1}^l$ and proceeds as follows:
  - If $ID_k = ID_{k^*}$ and $c_k = 1$, $\mathcal{B}$ terminates.
  - Else if $ID_k = ID_{k^*}$, $c_k = 0$, and the corresponding public key PK_{ID_k} in L_U has not been replaced, $\mathcal{B}$ calculates $\tau = \left[(g^{ab})^{1/(x+z_k)}\right]^{(s_1+\gamma_1 s_2)}$, $\sigma_i = \left[(g^{ab})^{1/(x+z_k)}\right]^{(s_1 H_2(Did,i)+s_2(\sum_{j=1}^l \gamma_j m_{ij}+\gamma_1 H_3(i)))}$ for $i = 1$ to n, where $z_k = \beta_k$. It then returns $(\tau, \{\sigma_i\}_{i=1}^n)$.
  - Otherwise, $\sigma_i = (g^a)^{\mu_k(s_1 H_2(Did,i)+s_2(\sum_{j=1}^l \gamma_j m_{ij}+\gamma_1 H_3(i)))(sk_{ID_k}+\beta_k)^{-1}}$, $\tau = (g^a)^{\mu_k(s_1+\gamma_1 s_2)(sk_{ID_k}+\beta_k)^{-1}}$.

Forgery: $\mathcal{A}_1$ outputs a forgery $(ID^*, PK^*, \tau^*, i^*, \boldsymbol{m}^*, \sigma^*)$. If $ID^* \neq ID_{k^*}$, $\mathcal{B}$ terminates. Otherwise, $\mathcal{B}$ obtains (ID^*, sk^*, PK^*) from L_U and $(ID^*, \hat{\tau}_{k^*}, \beta_{k^*}, c_{k^*})$ from L_{H_1}. Then, $\mathcal{B}$ performs the following operations to solve the given weak modified q-CAA problem:

- If $c_{k^*} = 0$, $\mathcal{B}$ terminates.
- Else if $c_{k^*} = 1$, then $\beta_{k^*} \notin \{z_1, \ldots, z_q\}$.
  If PK^* is the original public key output from *UserKeyGen*, $\sigma^* = \left[(g^{ab})^{1/(x+\beta_{k^*})}\right]^{s_1 H_2(Did,i)+s_2(\sum_{j=1}^l \gamma_j m_j^*+\gamma_1 H_3(i))}$, $\mathcal{B}$ calculates $(g^{ab})^{1/(x+\beta_{k^*})} = (\sigma^*)^{1/(s_1 H_2(Did,i)+s_2(\sum_{j=1}^l \gamma_j m_j^*+\gamma_1 H_3(i)))}$. If it has been replaced, $\sigma^* = \left[(g^{ab})^{1/(sk^*+\beta_{k^*})}\right]^{s_1 H_2(Did,i)+s_2(\sum_{j=1}^l \gamma_j m_j^*+\gamma_1 H_3(i))}$, $\mathcal{B}$ calculates $g^{ab} = (\sigma^*)^{(sk^*+\beta_{k^*})/(s_1 H_2(Did,i)+s_2(\sum_{j=1}^l \gamma_j m_j^*+\gamma_1 H_3(i)))}$.
  Thus, $\mathcal{B}$ obtains either g^{ab} or a new pair $(g^{ab})^{1/(x+\beta_{k^*})}$ as the solution to the given weak modified q-CAA problem, where $\beta_{k^*} \notin \{z_1, \ldots, z_q\}$.

Probability Analysis. $\mathcal{B}$ solves the hard problem if the following events occur: E_1: $\mathcal{B}$ does not terminate during the simulation; E_2: The output of $\mathcal{A}_1$ satisfies the winning condition; E_3: $ID^* = ID_{k^*}$ and $c_{k^*} = 1$. Thus, $\epsilon' = Pr[E_1 \wedge E_2 \wedge E_3] = Pr[E_1] \cdot Pr[E_2 \mid E_1] \cdot Pr[E_3 \mid E_1 \wedge E_2]$. To ensure that $\mathcal{B}$ does not terminate in the simulation, it must respond successfully to the *Corruption*, *Certificate*, and *Sign* queries, with the probabilities given by $(1 - 1/q_u)^{q_k}$, $(1 - 1/q_u)^{q_c}$, $(1 - \eta/q_u)^{q_s} \geq (1 - \eta)^{q_s}$, respectively. Thus, we have $Pr[E_1] \geq (1 - 1/q_u)^{q_k+q_c} \cdot (1 - \eta)^{q_s}$. Furthermore, $Pr[E_2 \mid E_1] = \epsilon$, $Pr[E_3 \mid E_1 \wedge E_2] = \eta/q_u$. We derive $\epsilon' \geq (1 - 1/q_u)^{q_k+q_c} \cdot \eta(1-\eta)^{q_s} \cdot (\epsilon/q_u)$, where $\eta(1-\eta)^{q_s}$ reaches its maximum at $\eta = 1/(q_s + 1)$. Finally, we obtain $\epsilon' \geq (1 - \frac{1}{q_u})^{q_k+q_c} \cdot (1 - \frac{1}{q_s+1})^{q_s} \cdot \frac{\epsilon}{q_u \cdot (q_s+1)}$. This completes the proof of Theorem 2.

Theorem 3. *Assume that the q-CAA assumption holds in $\mathbb{G}_1$, then our CB-QLHS scheme is secure against the adversary $\mathcal{A}_2$. Specifically, if $\mathcal{A}_2$ is an* adversary capable of forging a raw datum signature with a non-negligible advantage ϵ, then there exists a PPT algorithm $\mathcal{B}$ that can solve the q-CAA problem with a non-negligible advantage $\epsilon' \geq (1 - \frac{1}{q_u})^{q_k} \cdot (1 - \frac{1}{q_s+1})^{q_s} \cdot \frac{\epsilon}{q_u \cdot (q_s+1)}$.

Proof. Given a random instance of the q-CAA problem $(g, g^x, \{(z_i, g^{1/(x+z_i)})\}_{i=1}^{q})$, where g is a generator of $\mathbb{G}_1$, the goal of $\mathcal{B}$ is to utilize the capability of $\mathcal{A}_2$ to compute a new pair $(z^*, g^{1/(x+z^*)})$ for some $z^* \notin \{z_1, z_2, \ldots, z_q\}$. We regard the hash functions H_0, H_1 as random oracles, and $H_2, H_3 : \{0,1\}^* \to \mathbb{Z}_p^*$. $\mathcal{B}$ simulates the random oracle and interacts with $\mathcal{A}_2$ as follows:

Initialization: $\mathcal{B}$ randomly selects $s_1, s_2 \in \mathbb{Z}_p^*$ and sets $msk = (s_1, s_2)$, $MPK = (MPK_1, MPK_2) = (g^{s_1}, g^{s_2})$. $\mathcal{B}$ then sends $pp = \{\mathbb{G}_1, \mathbb{G}_T, p, e, g, MPK, H_2, H_3\}$ and msk to $\mathcal{A}_2$, and randomly chooses an integer k^* from $\{1, \ldots, q_u\}$.

Query: The *UserKeyGen*, H_1 and *Corruption* queries follows the proof of Theorem 2.

- H_0: $\mathcal{B}$ maintains a list L_{H_0} consisting of tuples $(ID_k, PK_{ID_k}, Q_k, \mu_k)$. On input (ID_k, PK_{ID_k}), if (ID_k, PK_{ID_k}) already exists in L_{H_0}, $\mathcal{B}$ directly returns Q_k. Otherwise, $\mathcal{B}$ chooses $\mu_k \in \mathbb{Z}_p^*$ and sets $Q_k = g^{\mu_k}$. $\mathcal{B}$ adds $(ID_k, PK_{ID_k}, Q_k, \mu_k)$ to the list L_{H_0} and returns Q_k.
- *Sign*: Given $(ID_k, \boldsymbol{D} = \{\boldsymbol{m}_i\}_{i=1}^{n}, Did)$, $\mathcal{B}$ retrieves $(ID_k, sk_{ID_k}, PK_{ID_k})$ from L_U, $(ID_k, PK_{ID_k}, Q_k, \mu_k)$ from L_{H_0}, and $(ID_k, \tau_k, \beta_k, c_k)$ from L_{H_1}. Then, $\mathcal{B}$ randomly selects $\gamma_1, \ldots, \gamma_s \in \mathbb{Z}_p^*$ and computes $\{g_j = MPK_2^{\gamma_j}\}_{j=1}^{l}$. Subsequently, it constructs $\hat{\tau} = Did \parallel n \parallel \{g_j\}_{j=1}^{l}$ and proceeds as follows:
  - If $ID_k = ID_{k^*}$ and $c_k = 1$, $\mathcal{B}$ terminates.
  - Else if $ID_k = ID_{k^*}$ and $c_k = 0$, $\mathcal{B}$ calculates $\tau = \left[g^{1/(x+z_k)}\right]^{(s_1+\gamma_1 s_2)}$, $\sigma_i = \left[g^{1/(x+z_k)}\right]^{(s_1 H_2(Did,i)+s_2(\sum_{j=1}^{l}\gamma_j m_{ij}+\gamma_1 H_3(i)))}$ for $i = 1$ to n, where $z_k = \beta_k$. It then returns $(\tau, \{\sigma_i\}_{i=1}^{n})$.
  - Otherwise, $\sigma_i = g^{\mu_k(s_1 H_2(Did,i)+s_2(\sum_{j=1}^{l}\gamma_j m_{ij}+\gamma_1 H_3(i)))(sk_{ID_k}+\beta_k)^{-1}}$, $\tau = g^{\mu_k(s_1+\gamma_1 s_2)(sk_{ID_k}+\beta_k)^{-1}}$.

Forgery: $\mathcal{A}_2$ outputs a forgery $(ID^*, \tau^*, i^*, \boldsymbol{m}^*, \sigma^*)$. If $ID^* \neq ID_{k^*}$, $\mathcal{B}$ terminates. Otherwise, $\mathcal{B}$ obtains (ID^*, sk^*, PK^*) from L_U and $(ID^*, \hat{\tau}_{k^*}, \beta_{k^*}, c_{k^*})$ from L_{H_1}. It proceeds as follows to solve the given q-CAA problem:

- If $c_{k^*} = 0$, $\mathcal{B}$ terminates.
- Else if $c_{k^*} = 1$, $\sigma^* = \left[g^{1/(x+\beta_{k^*})}\right]^{s_1 H_2(Did,i)+s_2(\sum_{j=1}^{l}\gamma_j m_j^* + \gamma_1 H_3(i))}$, where $\beta_{k^*} \notin \{z_1, \ldots, z_q\}$, $g^{1/(x+\beta_{k^*})} = (\sigma^*)^{1/(s_1 H_2(Did,i)+s_2(\sum_{j=1}^{l}\gamma_j m_j^* + \gamma_1 H_3(i)))}$. $\mathcal{B}$ obtains a new pair $g^{1/(x+\beta_{k^*})}$ as the solution to the given q-CAA problem.

Probability Analysis. $\mathcal{B}$ solves the hard problem if the following events occur: E_1: $\mathcal{B}$ does not terminate during the simulation; E_2: The output of $\mathcal{A}_2$ satisfies the winning condition; E_3: $ID^* = ID_{k^*}$ and $c_{k^*} = 1$. Thus, $\epsilon' = Pr[E_1 \wedge E_2 \wedge E_3] = Pr[E_1] \cdot Pr[E_2 \mid E_1] \cdot Pr[E_3 \mid E_1 \wedge E_2]$. To ensure that $\mathcal{B}$ does not terminate in the simulation, it must not terminate when responding to the *Corruption*, and *Sign* queries, with the probabilities given by $(1 - 1/q_u)^{q_k}$, and $(1 - \eta/q_u)^{q_s} \geq (1 - \eta)^{q_s}$, respectively. Thus, we have $\Pr[E_1] \geq (1 - 1/q_u)^{q_k} \cdot (1 - \eta)^{q_s}$. Furthermore, $\Pr[E_2 \mid E_1] = \epsilon$, $\Pr[E_3 \mid E_1 \wedge E_2] = \eta/q_u$. We derive $\epsilon' \geq (1 - 1/q_u)^{q_k} \cdot \eta(1 - \eta)^{q_s} \cdot (\epsilon/q_u)$, where $\eta(1 - \eta)^{q_s}$ reaches its maximum at $\eta = 1/(q_s + 1)$. Finally, we obtain $\epsilon' \geq (1 - \frac{1}{q_u})^{q_k} \cdot (1 - \frac{1}{q_s+1})^{q_s} \cdot \frac{\epsilon}{q_u \cdot (q_s+1)}$. This completes the proof of Theorem 3.

Theorem 4. *Assuming that our CB-QLHS scheme satisfies raw datum signature unforgeability and the CDH assumption holds in $\mathbb{G}_1$, then our CB-QLHS scheme is secure against the adversary $\mathcal{A}_3$.*

Specifically, if $\mathcal{A}_3$ is an adversary capable of forging a derived datum-signature pair with a non-negligible advantage ϵ, then there exists a PPT algorithm $\mathcal{B}$ that can solve the CDH problem with a non-negligible advantage $\epsilon' = (1 - \frac{1}{q_u})^{q_k+q_c+q_k} \cdot \frac{\epsilon}{q_u}$.

Proof. Given a CDH instance (g, g^a, g^b), where g is a generator of $\mathbb{G}_1$, the goal of $\mathcal{B}$ is to utilize the capability of $\mathcal{A}_3$ to compute $g^{ab} \in \mathbb{G}_1$. We regard the hash function H_0 as random oracle, and $H_1, H_2, H_3: \{0,1\}^* \to \mathbb{Z}_p^*$. $\mathcal{B}$ simulates the random oracle and interacts with $\mathcal{A}_3$ as follows:

Initialization: $\mathcal{B}$ randomly selects $s_1, s_2 \in \mathbb{Z}_p^*$ and sets $MPK = (MPK_1, MPK_2) = ((g^a)^{s_1}, (g^a)^{s_2})$. It then sends $pp = \{\mathbb{G}_1, \mathbb{G}_T, p, e, g, MPK, H_1, H_2, H_3\}$ to $\mathcal{A}_3$, and randomly chooses an integer k^* from $\{1, \ldots, q_u\}$.

Query: As in the proof of Theorem 2, $\mathcal{B}$ responds to *PKReplace*, H_0, *Corruption*, *Certificate* queries in the same way. The remaining queries are handled as follows.

- *UserKeyGen*: $\mathcal{B}$ maintains a list L_U consisting of tuples $(ID_k, sk_{ID_k}, PK_{ID_k})$. Given a user identity ID_k, if ID_k already exists in L_U, $\mathcal{B}$ directly returns PK_{ID_k}. Otherwise, $\mathcal{B}$ selects $x_k \in \mathbb{Z}_p^*$, sets $sk_{ID_k} = x_k$ and $PK_{ID_k} = g^{x_k}$. $\mathcal{B}$ adds $(ID_k, sk_{ID_k}, PK_{ID_k})$ to the list L_U and returns PK_{ID_k}.
- *Sign*: Given $(ID_k, \boldsymbol{D} = \{\boldsymbol{m}_i\}_{i=1}^n, Did)$, $\mathcal{B}$ retrieves $(ID_k, sk_{ID_k}, PK_{ID_k})$ from L_U, $(ID_k, PK_{ID_k}, Q_k, w_k)$ from L_{H_0}, and $(ID_k, \tau_k, \beta_k, c_k)$ from L_{H_1}. Then, $\mathcal{B}$ randomly selects $\gamma_1, \ldots, \gamma_l \in \mathbb{Z}_p^*$ and computes $\{g_j = MPK_2^{\gamma_j}\}_{j=1}^l$. $\mathcal{B}$ constructs $\hat{\tau} = Did \parallel n \parallel \{g_j\}_{j=1}^l$ and proceeds as follows:
  - If $ID_k = ID_{k^*}$, $\mathcal{B}$ terminates.
  - Otherwise, $\sigma_i = (g^a)^{\mu_k(s_1 H_2(Did,i) + s_2(\sum_{j=1}^l \gamma_j m_{ij} + \gamma_1 H_3(i)))(sk_{ID_k}+\beta_k)^{-1}}$, $\tau = (g^a)^{\mu_k(s_1+\gamma_1 s_2)(sk_{ID_k}+\beta_k)^{-1}}$.

Forgery: $\mathcal{A}_3$ chooses a target identity ID^*, a raw data sequence $\boldsymbol{D}^* = \{\boldsymbol{m}_i^*\}_{i=1}^n$ with tag τ^* from the **Query** phase, and specifies an index set $I^* \subseteq \{1, \ldots, n\}$, indicating the indices of data in $\boldsymbol{D}^*$ that it may tamper with. $\mathcal{B}$ then provides a coefficient vector $\boldsymbol{\omega}^* = (\omega_1^*, \ldots, \omega_n^*)^\top$, where $\exists i \in I^*$ such that $\omega_i^* \neq 0$. Finally, $\mathcal{A}_3$ outputs $(PK^*, \tilde{\boldsymbol{m}}^*, \tilde{\sigma}^*)$ based on $\boldsymbol{\omega}^*$ for $\boldsymbol{D}^*$. If PK^* has not been replaced, $\mathcal{B}$ retrieves sk^* from L_U; otherwise, $\mathcal{A}_3$ needs to provide the corresponding secret key sk^*. If $ID^* \neq ID_{k^*}$, $\mathcal{B}$ terminates. Otherwise, we have

$\tilde{\sigma}^* = (g^{ab})^{(s_1 \sum_{i=1}^n \omega_i^* H_2(Did,i) + s_2(\sum_{j=1}^l \gamma_j \tilde{m}_j^* + \gamma_1 H_3(i)))/(sk^* + \beta_{k^*})}$. $\mathcal{B}$ can calculate $g^{ab} = (\tilde{\sigma}^*)^{(sk^* + \beta_{k^*})/(s_1 \sum_{i=1}^n \omega_i^* H_2(Did,i) + s_2(\sum_{j=1}^l \gamma_j \tilde{m}_j^* + \gamma_1 H_3(i)))}$ as the solution.

Probability Analysis. $\mathcal{B}$ solves the hard problem if the following events occur: E_1: $\mathcal{B}$ does not terminate during the simulation; E_2: The output of $\mathcal{A}_3$ satisfies the winning condition; E_3: $ID^* = ID_{k^*}$. Thus, $\epsilon' = Pr[E_1 \wedge E_2 \wedge E_3] = Pr[E_1] \cdot Pr[E_2 \mid E_1] \cdot Pr[E_3 \mid E_1 \wedge E_2]$. To ensure that $\mathcal{B}$ does not terminate in the simulation, it must respond successfully to the *Corruption*, *Certificate*, and *Sign* queries, with the probabilities given by $(1 - 1/q_u)^{q_k}$, $(1 - 1/q_u)^{q_c}$, $(1 - 1/q_u)^{q_s}$, respectively. Thus, we have $Pr[E_1] = (1 - 1/q_u)^{q_k + q_c + q_s}$. Furthermore, $Pr[E_2 \mid E_1] = \epsilon$, $Pr[E_3 \mid E_1 \wedge E_2] = 1/q_u$. We derive $\epsilon' = (1 - 1/q_u)^{q_k + q_c + q_k} \cdot \epsilon/q_u$. This completes the proof of Theorem 4.

Theorem 5. *Assuming that our CB-QLHS scheme satisfies raw datum signature unforgeability and that the DL assumption holds in $\mathbb{G}_1$, then our CB-QLHS scheme is secure against the adversary $\mathcal{A}_4$. Specifically, if $\mathcal{A}_4$ is an adversary capable of forging a derived datum-signature pair with a non-negligible advantage ϵ, then there exists a PPT algorithm $\mathcal{B}$ that can solve the DL problem with a non-negligible advantage $\epsilon' = \epsilon \cdot (1 - \frac{1}{p})$, where p is the order of the group $\mathbb{G}_1$.*

Proof. Given a DL instance (g, g^x), where g is a generator of $\mathbb{G}_1$, the goal of $\mathcal{B}$ is to utilize the capability of $\mathcal{A}_4$ to compute x. We regard the hash function H_0 as random oracle, and $H_1, H_2, H_3 : \{0,1\}^* \to \mathbb{Z}_p^*$. $\mathcal{B}$ simulates the random oracle and interacts with $\mathcal{A}_4$ as follows:

Initialization: $\mathcal{B}$ selects $s_1, s_2 \in \mathbb{Z}_p^*$, sets $msk = (s_1, s_2)$, $MPK = (MPK_1, MPK_2) = (g^{s_1}, g^{s_2})$, then sends $(pp = \{\mathbb{G}_1, \mathbb{G}_T, p, e, g, MPK, H_1, H_2, H_3\}, msk)$ to $\mathcal{A}_4$.

Query: The *UserKeyGen* query follows the proof of Theorem 4, while the H_0 and *Corruption* queries follow the proof of Theorem 3.

- *Sign*: Given $(ID_k, \boldsymbol{D} = \{\boldsymbol{m}_i\}_{i=1}^n, Did)$, $\mathcal{B}$ retrieves $(ID_k, sk_{ID_k}, PK_{ID_k})$ from L_U, and $(ID_k, PK_{ID_k}, Q_k, w_k)$ from L_{H_0}.
  It then selects $\gamma_j, \alpha_j \in \mathbb{Z}_p^*$ and computes $g_j = (g^{\gamma_j}(g^x)^{\alpha_j})^{s_2}$ for $j = 1$ to l. Subsequently, it constructs $\hat{\tau} = Did \parallel n \parallel \{g_j\}_{j=1}^l$ and calculates $\tau = g^{\mu_k(s_1 + \gamma_1 s_2)(sk_{ID_k} + \beta_k)^{-1}}$, $\sigma_i = (g^{s_1 H_2(Did,i)} \cdot \prod_{j=1}^l g_j^{m_{ij}} \cdot g_1^{H_3(i)})^{\mu_k(sk_{ID_k} + \beta_k)^{-1}}$ for $i = 1$ to n, where $\beta_k = H_1(ID_k \parallel \hat{\tau}_k)$. Finally, $\mathcal{B}$ returns $(\tau, \{\sigma_i\}_{i=1}^n)$.

Forgery: $\mathcal{A}_4$ chooses a target identity ID^*, a raw data sequence $\boldsymbol{D}^* = \{\boldsymbol{m}_i^*\}_{i=1}^n$ with tag τ^* from the **Query** phase, and specifies an index set $I^* \subseteq \{1, \ldots, n\}$, indicating the indices of data in $\boldsymbol{D}^*$ that it may tamper with. $\mathcal{B}$ then provides a coefficient vector $\boldsymbol{\omega}^* = (\omega_1^*, \ldots, \omega_n^*)^\top$, where $\exists i \in I^*$ such that $\omega_i^* \neq 0$. Finally, $\mathcal{A}_4$ outputs a forged derived data-signature pair $(\tilde{\boldsymbol{m}}^*, \tilde{\sigma}^*)$ based on $\boldsymbol{\omega}^*$ for $\boldsymbol{D}^*$.

We employ the $Combine(\tau^*, (\boldsymbol{D}^*, \{\sigma_i^*\}_{i=1}^n), \boldsymbol{\omega}^*)$ algorithm to compute the correct derived pair $(\tilde{\boldsymbol{m}}, \tilde{\sigma})$, where $\tilde{\boldsymbol{m}} = \sum_{i=1}^n \omega_i^* \boldsymbol{m}_i = (\tilde{m}_1, \ldots, \tilde{m}_l)$. Based on the process described above, we know that the $\mathcal{A}_4$'s goal is to tamper with the raw data, while the raw data signature is unforgeable, the adversary can only obtain the raw data signature through *Sign* query. Therefore, $\tilde{\sigma} = \tilde{\sigma}^*$. According to the verification equation, we have:

$$e\Big(H_0(ID^*, PK_{ID^*}), MPK_1^{\sum_{i=1}^n \omega_i^* H_2(Did,i)} \cdot \prod_{j=1}^{l} g_j^{\tilde{m}_j^*} \cdot g_1^{\sum_{i=1}^n \omega_i^* H_3(i)}\Big)$$

$$= e\Big(\tilde{\sigma}^*, PK_{ID^*} \cdot g^{H_1(ID\|\hat{\tau})}\Big) = e\Big(\tilde{\sigma}, PK_{ID^*} \cdot g^{H_1(ID\|\hat{\tau})}\Big)$$

$$= e\Big(H_0(ID^*, PK_{ID^*}), MPK_1^{\sum_{i=1}^n \omega_i^* H_2(Did,i)} \cdot \prod_{j=1}^{l} g_j^{\tilde{m}_j} \cdot g_1^{\sum_{i=1}^n \omega_i^* H_3(i)}\Big).$$

From this, we can conclude that $\prod_{j=1}^{l} g_j^{\tilde{m}_j^*} = \prod_{j=1}^{l} g_j^{\tilde{m}_j}$. Let $\Delta \tilde{m}_j = \tilde{m}_j^* - \tilde{m}_j$ for $j = 1$ to l. Then, we obtain: $1 = \prod_{j=1}^{l} g_j^{\Delta \tilde{m}_j} = \prod_{j=1}^{l} (g^{\gamma_j} g^{\alpha_j x})^{s_2 \Delta \tilde{m}_j} = g^{s_2 \sum_{j=1}^{l} \gamma_j \Delta \tilde{m}_j} \cdot (g^x)^{s_2 \sum_{j=1}^{l} \alpha_j \Delta \tilde{m}_j}$. Thus, $\mathcal{B}$ can obtain the solution to the DL instance: $x = -\frac{\sum_{j=1}^{l} \gamma_j \Delta \tilde{m}_j}{\sum_{j=1}^{l} \alpha_j \Delta \tilde{m}_j}$ (mod p), provided that $\sum_{j=1}^{l} \alpha_j \Delta \tilde{m}_j \neq 0$ (mod p).

Probability Analysis. From the winning condition of $\mathcal{A}_4$, we know that $\tilde{m}^* \neq \tilde{m}$, implying that $\{\Delta \tilde{m}_j\}_{j=1}^{l}$ are not all zero. Since $\{\alpha_j\}_{j=1}^{l}$ are randomly selected from $\mathbb{Z}_p^*$, the probability that $\sum_{j=1}^{l} \alpha_j \Delta \tilde{m}_j \neq 0$ (mod p) is $1 - 1/p$. Therefore, the advantage of $\mathcal{B}$ in solving the DL problem is $\epsilon' = \epsilon \cdot (1 - 1/p)$. This completes the proof of Theorem 5.

5.3 Performance Comparison

We compare the auditing protocol instantiated from our CB-QLHS construction with several state-of-the-art counterparts [17,21,22], through both theoretical and experimental analyses of computational, communication, and storage cost.

Theoretical Analysis. Compared to operations in $\mathbb{G}_1$ and $\mathbb{G}_T$, the computational cost of operations in $\mathbb{Z}_p^*$ and ordinary hash functions (i.e., those mapping to $\mathbb{Z}_p^*$) is negligible. Therefore, we omit these operations in the theoretical analysis of computational cost. As shown in Table 2, we conduct a theoretical analysis in terms of computational cost, storage complexity, and the communication overhead of audit proofs (from the CSP to the TPA). Given that the CSP typically possesses enterprise-level equipment and abundant computational resources, our analysis focuses on the computational cost of tag generation by DO and proof verification by the TPA. In our protocol, tag generation and proof verification do not require MTP hash functions, and the verification cost remains constant regardless of the number of challenged blocks. In contrast, the verification cost in other protocols increases with the number of challenged blocks. Unlike protocol [22], which adopts one-dimensional partitioning without dividing data blocks into sectors, our protocol, as well as protocols [17,21], employs two-dimensional data partitioning. Although this leads to a proof size that scales with l, our protocol maintains low computational and storage overhead. In comparison, protocol [22] achieves storage complexity of $O(1)$, but at the expense of significantly higher computational and communication overhead, approximately l times greater than those of our protocol.

Table 2. Comparison on Computation, Storage, and Communication Complexity

Protocols	Computation Cost		Storage Complexity	Proof Communication Complexity (CSP→TPA)
	DO	TPA	CSP	
[17]	$n(H_{G_1} + 2E_{G_1} + M_{G_1}) + lE_{G_1}$	$2P + cH_{G_1} + (l+c)E_{G_1} + (l+c-1)M_{G_1}$	$\mathcal{O}(n)$	$\mathcal{O}(l)$
[21]	$n(H_{G_1} + 2E_{G_1} + M_{G_1}) + lE_{G_1}$	$2P + cH_{G_1} + (l+c)E_{G_1} + (l+c)M_{G_1}$	$\mathcal{O}(n)$	$\mathcal{O}(l)$
[22]	$nl(H_{G_1} + 2E_{G_1} + M_{G_1})$	$2P + cH_{G_1} + (c+1)E_{G_1} + cM_{G_1}$	$\mathcal{O}(nl)$	$\mathcal{O}(1)$
Ours	$n(3E_{G_1} + M_{G_1}) + lE_{G_1}$	$2P + (l+3)E_{G_1} + (l+2)M_{G_1}$	$\mathcal{O}(n)$	$\mathcal{O}(l)$

* n denotes the number of blocks into which the file is divided, l represents the number of sectors per block, and c signifies the number of data blocks being challenged. E_{G_1} and E_{G_T} represent exponentiation operations on groups G_1 and G_T, respectively. M_{G_1} denotes multiplication operations on groups G_1. H_{G_1} denotes a MTP hash function from $\{0,1\}^*$ to G_1. P represents a pairing operation from $G_1 \times G_1$ to G_T.

Experimental Evaluation. All experiments are conducted on an Ubuntu 22.04 system with an AMD Ryzen 7 8845HS CPU @ 3.80GHz and 24GB RAM. The implementation is based on the java pairing-based cryptography (JPBC) 2.0.0 library. We use the Type A pairing from the JPBC library, which is constructed on the elliptic curve $E(F_q) : y^2 = x^3 + x$, where q is a prime number of length 512 bits. Type A pairing is a symmetric bilinear pairing $e: G_1 \times G_1 \to G_T$ that has been optimized for acceleration. The multiplicative groups G_1 and G_T are cyclic subgroups with the same prime order p on the curve $E(F_q)$, where the length of p is 160 bits. The sizes of an element in G_1, $\mathbb{Z}_p$, and G_T are 128 bytes, 20 bytes, and 128 bytes, respectively. Following the parameter setting adopted in [20], we also set the number of sectors per block to $l = 8$ in our experiments.

As shown in Fig. 1a, our protocol exhibits lower computational cost for tag generation compared to other approaches, with the file size ranging from 200 KB to 1000 KB. Fig. 1b shows the proof verification cost under varying numbers of challenged blocks (200 to 1000). Our protocol achieves a constant and substantially lower verification time compared to the others. Figure 1c and Fig. 1d compares the storage and proof communication overhead when outsourcing a 200 KB file. As shown in Fig. 1c, the additional storage cost for tags in our protocol is identical to that of [17] and [21], and one-eighth of that in [22]. As illustrated in Fig. 1d, the communication cost of our protocol matches that of [17], is significantly lower than that of [21], and slightly higher than that of [22].

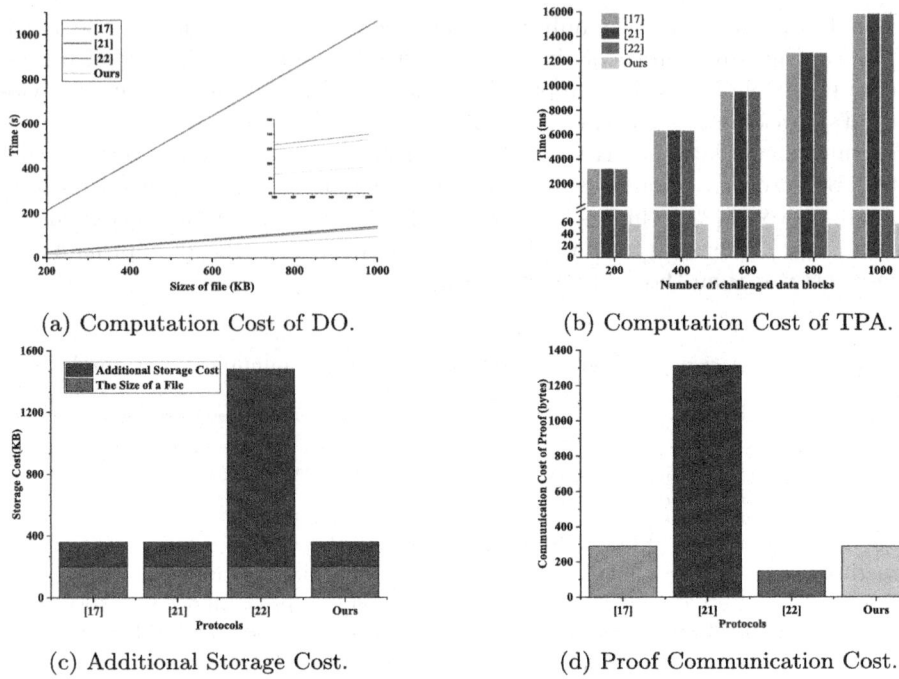

Fig. 1. Comparison of Computation, Storage, and Communication Costs.

6 Conclusion

In this paper, We investigates the core techniques and challenges in existing CB-CSDIA protocols. To provide a theoretical foundation for CB-CSDIA, we formally define CB-QLHS and present a comprehensive security model. We further explore the connection between CB-QLHS schemes and CB-CSDIA protocols, and demonstrate that a secure CB-QLHS scheme can be used to construct a secure CB-CSDIA protocol. In addition, we propose an efficient instantiation of CB-QLHS using bilinear map. Performance comparisons between the CB-CSDIA protocol derived from our CB-QLHS scheme and several representative existing protocols show that our construction offers comprehensive advantages in terms of computation, communication, and storage. This work reduces the design of secure CB-CSDIA protocols to the construction of secure CB-QLHS schemes, thereby facilitating the study of data integrity auditing in cloud storage.

Acknowledgments. We thank the anonymous reviewers for their valuable comments and suggestions which helped us to improve the content and presentation of this paper. This work was supported in part by the National Natural Science Foundation of China (Grant Nos. 62172096, 62202101, and 62272104).

References

1. Al-Riyami, S.S., Paterson, K.G.: Certificateless public key cryptography. In: International Conference on the Theory and Application of Cryptology and Information Security, pp. 452–473. Springer (2003)
2. Ateniese, G., et al.: Provable data possession at untrusted stores. In: Proceedings of the 14th ACM Conference on Computer and Communications Security, pp. 598–609 (2007)
3. Chen, J., Zhou, T., Ji, S., Tan, H., Zheng, W.: Efficient public auditing scheme for non-administrator group with secure user revocation. J. Inf. Secur. Appl. **80**, 103676 (2024)
4. Cheng, L., Xiao, Y., Wang, G.: Cryptanalysis of a certificate-based on signature scheme. Procedia Eng. **29**, 2821–2825 (2012)
5. Dewan, H., Hansdah, R.: A survey of cloud storage facilities. In: 2011 IEEE World Congress on Services, pp. 224–231. IEEE (2011)
6. Gentry, C.: Certificate-based encryption and the certificate revocation problem. In: International Conference on the Theory and Applications of Cryptographic Techniques, pp. 272–293. Springer (2003)
7. Hu, X., Chang, J., Ahmad, T., Zhang, F., Zhang, Y.: Identity-based integrity auditing scheme with sensitive information hiding for proxy-server-assisted cloud storage applications. IEEE Internet Things J. **12**(6), 6673–6684 (2025)
8. Huang, Y., Shen, W., Qin, J., Hou, H.: Privacy-preserving certificateless public auditing supporting different auditing frequencies. Comput. Secur. **128**, 103181 (2023)
9. Juels, A., Kaliski Jr, B.S.: PORs: proofs of retrievability for large files. In: Proceedings of the 14th ACM Conference on Computer and Communications Security, pp. 584–597 (2007)
10. Li, Y., Zhang, F.: An efficient certificate-based data integrity auditing protocol for cloud-assisted WBANs. IEEE Internet Things J. **9**(13), 11513–11523 (2021)
11. Liu, J.K., Bao, F., Zhou, J.: Short and efficient certificate-based signature. In: International Conference on Research in Networking, pp. 167–178. Springer (2011)
12. Mitsunari, S., Sakai, R., Kasahara, M.: A new traitor tracing. IEICE Trans. Fundam. Electron. Commun. Comput. Sci. **85**(2), 481–484 (2002)
13. Shacham, H., Waters, B.: Compact proofs of retrievability. J. Cryptol. **26**(3), 442–483 (2013)
14. Shamir, A.: Identity-based cryptosystems and signature schemes. In: Advances in Cryptology: Proceedings of CRYPTO 84 4, pp. 47–53. Springer (1985)
15. Shen, W., Yu, J., Yang, M., Hu, J.: Efficient identity-based data integrity auditing with key-exposure resistance for cloud storage. IEEE Trans. Dependable Secur. Comput. **20**(6), 4593–4606 (2023)
16. Tian, H., Gan, N., Peng, F., Quan, H., Chang, C., Vasilakos, A.V.: Smart contract-based public integrity auditing for cloud storage against malicious auditors. Future Gener. Comput. Syst. **166**, 107709 (2025)
17. Tian, Y., Zhou, X., Zhou, T., Zhong, W., Li, R., Yang, X.: A secure certificate-based data integrity auditing protocol with cloud service providers. Mathematics **12**(13), 1964 (2024)
18. Wang, W., Sun, Y., Li, Y.: Security-enhanced certificate-based remote data integrity batch auditing for cloud-IoT. Secur. Commun. Netw. **2022**(1), 7882662 (2022)

19. Yang, X., Wei, L., Li, M., Du, X., Wang, C.: Backdoor-resistant certificateless-based message-locked integrity auditing for computing power network. J. Syst. Archit. **154**, 103244 (2024)
20. Zhang, F., Huang, Y., Yang, W., Tian, J.: Quasi-linearly homomorphic signature for data integrity auditing in cloud storage. In: Information Security and Cryptology, pp. 44–65. Springer, Singapore (2025)
21. Zhang, Y., Chang, J., Chen, Y., Xu, R.: Certificate-based remote auditing protocol with privacy protection and deduplication functions for cloud-assisted applications. Comput. Netw. 111083 (2025)
22. Zhou, H., Shen, W., Liu, J.: Certificate-based multi-copy cloud storage auditing supporting data dynamics. Comput. Secur. **148**, 104096 (2025)

Zero-Knowledge Protocols with PVC Security: Striking the Balance Between Security and Efficiency

Yi Liu, Yipeng Song, Anjia Yang, and Junzuo Lai(✉)

College of Cyber Security, Jinan University, Guangzhou 510632, China
`liuyi@jnu.edu.cn, yipeng93857@gmail.com, anjiayang@gmail.com, laijunzuo@gmail.com`

Abstract. Zero-knowledge protocols allow a prover to prove possession of a witness for an NP-statement without revealing any information about the witness itself. This kind of protocol has found extensive applications in various fields, including secure computation and blockchain. However, in certain scenarios (*e.g.*, when the statements are complicated), existing zero-knowledge protocols may not be well-suited due to their limited applicability or high computational overhead.

We address these limitations by incorporating the notion of *publicly verifiable covert (PVC) security* into zero-knowledge protocols. PVC security, recently emerging from secure computation, effectively balances security and efficiency in practical scenarios. With PVC security, while a malicious party may attempt to cheat, such cheating will be detected and become publicly verifiable with a significant probability (called deterrence factor, *e.g.*, >90%). This notion is well-suited for practical scenarios involving reputation-conscious parties (*e.g.*, companies) and offers substantial efficiency improvements.

In this paper, we present the *first* definition of *zero-knowledge protocols with PVC security*. We then propose a generic transformation to convert Sigma protocols with 1-bit challenge, a kind of protocol widely used for zero-knowledge, into efficient zero-knowledge protocols with PVC security. By applying our transformation, we can substantially improve the efficiency of existing protocols for reputation-sensitive parties. For instance, applying the transformation to achieve a deterrence factor of 93.75% incurs a cost of only around 20% compared to the original protocol. Therefore, our results contribute to significant advancements in practical zero-knowledge protocols.

Keywords: Publicly verifiable covert security · Zero-knowledge · Generic transformation

1 Introduction

The notion of zero-knowledge proofs was introduced by Goldwasser, Micali, and Rackoff in 1985 [16]. This cryptographic paradigm enables a prover to convincingly demonstrate the validity of a statement to a verifier without revealing any information beyond the truth of the statement itself.

Zero-knowledge *proof of knowledge* is a widely applicable notion closely related to zero-knowledge proofs. It enables the prover to demonstrate *possession of specific knowledge* (referred to as a witness) regarding a statement while avoiding the disclosure of any information about the knowledge itself. Zero-knowledge proof of knowledge protocols have found extensive applications in various fields, including secure computation [14], identification [28], and blockchain [4]. For simplicity and conciseness, we will use the term "zero-knowledge protocols" hereinafter to refer to zero-knowledge proof of knowledge protocols[1].

Despite their broad applicability, zero-knowledge protocols can be ineffective in certain contexts. Specific zero-knowledge protocols, such as those based on the structure of Schnorr protocols [28], may achieve high efficiency but are constrained to proving a limited range of statements. In contrast, generic zero-knowledge protocols, such as zk-SNARKs [12,17,27] and Bulletproofs [5], can become computationally intensive when dealing with complex representations of statements. Notably, many interesting and useful zero-knowledge protocols that are not efficient enough follow a similar repetitive structure, where a sub-protocol is invoked multiple times, and in each iteration, the prover responds to a 1-bit challenge $e \in \{0, 1\}$ chosen by the verifier, *e.g.*, quadratic residues [16], graph isomorphism [15], twin-ciphertext proofs [7,25], etc. On the one hand, unlike Schnorr-like protocols with large challenge spaces, the challenge spaces for this kind of protocol are only 1 bit, and thus, repeating the sub-protocol many times can be computationally expensive. On the other hand, the statements for this kind of protocol are complicated and often ineffective in proving the statements directly via generic zero-knowledge protocols.

In cases where efficient specific zero-knowledge proofs are inapplicable and existing generic zero-knowledge protocols are inefficient, we, fortunately, find that modest relaxation of security requirements could lead to significant efficiency improvements in some certain real-life scenarios, such as the protocols with the repetitive structure mentioned above.

Our approach builds on the notion of publicly verifiable covert (PVC) security that has recently emerged in secure computation. Introduced by Asharov and Orlandi in 2012 [2], PVC security strikes a balance between semi-honest and malicious security while providing a publicly verifiable cheating detection mechanism. A protocol with PVC security guarantees that any deviation of cheaters will be detected by an honest party with a fixed probability ϵ, called the *deterrence factor*. Upon detection, PVC protocols enable the honest party to generate a *publicly verifiable certificate* as *proof of cheating* with respect to the cheater. This certificate can convince all entities and holds utility in legal proceedings or smart contracts that financially penalize the cheater. This PVC security property is particularly compelling in practice due to its efficiency[2].

---

[1] In this paper, we do not explicitly differentiate between the notions of *proof* and *argument*, and we confine our considerations to scenarios where both the prover and the verifier run within probabilistic polynomial-time.

[2] Recent studies have demonstrated that a two-party PVC protocol with a deterrence factor of 50% incurs only 20–40% overhead compared to state-of-the-art semi-honest protocols based on garbled circuits [21].

Moreover, it permanently exposes the cheater's misbehavior publicly, imposing reputational risk and accountability. Due to this practical property, the notion of PVC security has gained increasing attention in recent years, leading to advances in protocol design [2,3,9,10,21,22,24,29] and many applications, including financially backed protocols [11,30], secure computation involving commitment on the blockchain [1], and private function evaluation [26]. In this work, we find that the notion of PVC security is also applicable to certain zero-knowledge protocol scenarios in addition to secure computation. We outline relevant scenario characteristics as follows.

Reputation-conscious parties. In many secure computation protocols, zero-knowledge protocols may serve as sub-protocols to prove the correctness of data or operations. If parties in the protocols are reputation-conscious (*e.g.*, companies like Google and Microsoft), though they may have the incentive to cheat in order to gain advantages, they are also concerned about the risk of being caught and publicly exposed. In these scenarios, having a noticeable probability (*e.g.*, 93.75%) of being detected and publicly exposed is sufficient to deter malicious behaviors.

Repeated sub-protocols. Many interesting zero-knowledge protocols are constructed by repeating a sub-protocol sequentially or in parallel, including protocols for quadratic residues [16], graph isomorphism [15], and twin-ciphertext proofs [7,25]. Typically, the sub-protocol involved has high knowledge error (*i.e.*, the challenge space is small), necessitating many repetitions to prevent a malicious party from guessing the challenge. However, for reputation-conscious parties, if cheating would lead to detection and public exposure, the number of sub-protocol repetitions could be significantly reduced (*e.g.*, from 40 to 4). PVC security is thus applicable.

The primary objective of this paper is to explore the approach of integrating the notion of PVC security into zero-knowledge protocols with respect to reputation-conscious parties. By leveraging the deterrent effect of publicly verifiable cheating detection, this incorporation aims to improve efficiency.

1.1 Our Contributions

In this paper, we provide a formal definition of zero-knowledge protocols with PVC security and propose a generic transformation method to convert protocols with specific structures into zero-knowledge protocols with PVC security. Our main contributions can be summarized as follows.

New notion. To address scenarios involving reputation-conscious parties where efficient protocols are lacking for certain statements, we introduce the notion of *zero-knowledge protocols with PVC security*. This novel notion strikes a balance between security and efficiency, and we provide a formal definition for it.

New protocols. We propose a generic transformation approach to convert protocols with specific structures, namely Sigma protocols with 1-bit challenge, into zero-knowledge protocols with PVC security. This class of protocols encompasses a variety of interesting statements, including quadratic residues [16], graph isomorphism [15], and twin-ciphertext proofs [7,25]. By applying our approach, we *significantly* improve the efficiency of the original protocols in scenarios involving reputation-conscious parties. For instance, achieving a deterrence factor of 93.75% incurs a cost of only around 20% compared to the original protocol (which typically requires numerous repetitions to achieve low soundness error).

Overall, our contributions make significant advancements in the field of zero-knowledge protocols, providing a practical approach to improving the efficiency of existing protocols with specific structures when reputation-conscious parties are involved.

1.2 Technical Overview

In this subsection, we provide a concise introduction to zero-knowledge protocols with PVC security and outline the fundamental idea of designing such protocols from Sigma protocols with a 1-bit challenge.

Informally, given a statement x and a NP language L (with relation R), zero-knowledge protocols with PVC security allow a prover to prove to a verifier that she possesses a witness w, such that $(x, w) \in R$, while the verifier learns no additional information beyond the fact that the prover knows a witness w satisfying $(x, w) \in R$. Leveraging the notion of PVC security, a malicious prover, lacking the witness w, can convince the verifier that she possesses the witness for x *only with a small probability* (denoted as $1 - \epsilon$, *e.g.*, 6.25%). However, if the malicious prover fails to persuade the verifier with a significant probability ϵ (*e.g.*, 93.75%, called *deterrence factor*), the verifier should be capable of generating a publicly verifiable certificate to attribute blame to the malicious prover. This certificate serves as evidence of cheating by the malicious prover and can be verified by any party external to the protocol.

It is crucial to emphasize that once the malicious prover becomes aware that her dishonest actions have been detected by the verifier, she is unable to prevent the verifier from generating and disseminating the certificate. Thus, the malicious prover's decision to proceed with the protocol and cheat without possessing the witness is made independently of the event of being caught. Consequently, the design of zero-knowledge protocols with PVC security presents a significant challenge. Achieving a mechanism that enables the verifier to gather sufficient evidence for blaming cheating before the malicious prover becomes aware of being caught is a highly nontrivial task.

We observe that a class of zero-knowledge protocols constructed by repeating a sub-protocol is well-suited for extension to PVC security in scenarios involving reputation-conscious parties. To illustrate our idea, we consider as an example the well-known zero-knowledge protocol for quadratic residues [16]. In this protocol, a prover aims to prove to the verifier that she knows $w \in \mathbb{Z}_N^*$, such that $x = w^2 \mod N$, in zero-knowledge. The protocol proceeds as follows.

1. The prover selects a random value r in $\mathbb{Z}_N^*$, computes $y = r^2 \mod N$, and sends y to the verifier.
2. The verifier returns a random bit $b \in \{0, 1\}$ to the prover.
3. If $b = 0$, the prover sends $z = r$ to the verifier. If $b = 1$, the prover sends $z = wr \mod N$ to the verifier.
4. The verifier accepts the proof if and only if $z^2 \equiv x^b y \mod N$.

The insight that the prover possesses w (*i.e.*, the protocol is proof-of-knowledge) is that if we obtain two accepting transcripts for the same y and different b, we can easily extract w. It is also easy to see that if the prover does not possess w, she can still guess a bit $\hat{b}$ in Step 1 and provide $y = z^2 x^{-\hat{b}} \mod N$ for a random z in $\mathbb{Z}_n^*$. If $b = \hat{b}$, the verifier will accept the proof for the given z. Hence, to ensure the prover's possession of w, the protocol is typically executed multiple times, sequentially or in parallel, with slight modifications [13]. In the setting of reputation-conscious provers, we can introduce PVC security to largely reduce the number of sub-protocol executions. For example, a deterrence factor of 93.75% can be achieved with only 4 instances to be verified by the verifier, and a deterrence factor of approximately 99% requires only 7 instances. This is sufficient for most real-world scenarios, especially when the publicly verifiable cheating detection mechanism is also available.

To achieve PVC security, the challenge bit b must be unknown to the prover in Step 4. If a malicious prover without possessing w is aware of b, she can simply abort the protocol, claiming a network error, and the verifier will be unable to generate a publicly verifiable certificate for the abortion. However, if the prover answers both challenges $b = 0$ and $b = 1$ (without knowing b now), the verifier can extract the witness w, thereby compromising the zero-knowledge property. To address this issue, we employ an oblivious transfer (OT) protocol. Using the OT protocol, the prover's input becomes (z_0, z_1), representing her answers for $b = 0$ and $b = 1$, respectively, while the verifier's input is either $b = 0$ or $b = 1$. The verifier's output in the OT protocol is z_b, allowing for only checking the correctness of z_b, while the prover is not aware of b.

To generate a publicly verifiable certificate for blaming a malicious prover providing incorrect (z_0, z_1), the verifier requires the prover's signature on a cheating message. However, directly having the prover sign z_0 and z_1 and include the signature in the OT protocol is ineffective as the prover can generate an invalid signature for the incorrect z_i. Our basic idea is to use a derandomization approach. Initially, the verifier selects a random seed, commits to it, and sends the commitment to the prover. The verifier then employs the seed to derive the randomness used during the protocol execution. After the execution of the (perfectly correct) OT protocol, the prover, unaware of the challenge bit, signs the complete transcript of the OT protocol, the messages she has sent, and the commitment of the verifier's seed. In case of detecting cheating, the verifier can include the OT protocol transcript, the prover's signature and sent messages, and the opening of the commitment in the certificate. With the seed, the prover's message, and the OT protocol transcript, any party can now verify the signature, simulate the protocol execution of the verifier, and verify whether the prover was cheating.

We note that the definition and protocol design also involves several technical details. For instance, while zero-knowledge protocols can be seen as a specific class of secure computation, we cannot directly combine the definition of PVC security for secure computation with that of zero-knowledge protocols. Instead, we need to carefully describe the security requirements when the verifier has no input and only outputs a single bit. Another technical detail is that, as our definition of zero-knowledge protocols does not impose restrictions on the deterrence factor, we choose to use a commitment scheme that supports equivocation (*e.g.*, implemented using a random oracle) in the protocol. Consequently, the protocol is modified to let the prover commit to the first messages of the sub-protocols and subsequently open them after the OT protocol. This modification enables us to prove that the protocol achieves the zero-knowledge property when the challenge space of the protocol becomes exponentially large. For a detailed definition, description of the protocol, and its security proof, we refer the reader to Sect. 3 and 4.

2 Preliminaries

For a set S, we denote the size of S as $|S|$ and use $x \leftarrow_\$ S$ to represent the uniform sampling of an element x from S. Moreover, we define $[n] = \{1, \ldots, n\}$ for a positive integer n.

Let κ denote the computational security parameter, which is provided as input in unary format to all algorithms. A function f in κ, mapping natural numbers to the interval $[0, 1]$, is *negligible* if $f(\kappa) = \mathcal{O}(\kappa^{-c})$ for every constant $c > 0$.

Given a seed $\in \{0,1\}^\kappa$, we can employ a pseudorandom function with seed as the key in Counter (CTR) mode to generate a sufficient number of pseudorandom numbers, which can be utilized as local randomness in protocol executions.

In our protocol, we utilize a (non-interactive) commitment scheme denoted as Com. This scheme employs random coins decom for both commitment generation and opening. It is required to satisfy (computational) binding and hiding properties while also supporting equivocation. This scheme can be implemented by a random oracle $H\colon \{0,1\}^* \to \{0,1\}^\kappa$ via defining $\mathsf{Com}(m) = H(m, \mathsf{decom})$, where $\mathsf{decom} \leftarrow_\$ \{0,1\}^\kappa$. Additionally, our protocol incorporates the signature scheme (KGen, Sig, Verify), which is existentially unforgeable under chosen-message attacks (EUF-CMA).

An execution transcript of a two-party protocol is of the form trans = $(m_1, m_2, m_3, \ldots)$, where the parties send their messages alternately.

Let Π_{OT} represent the protocol that securely realizes a parallel version of the OT functionality $\mathcal{F}_{\mathsf{OT}}$ below with perfect correctness [9], ensuring that the receiver cannot "equivocate" its view by finding a random tape that produces a different output from the output in a real execution.

Functionality $\mathcal{F}_{\mathsf{OT}}$

Private inputs: The sender has input $\{(z_j^0, z_j^1)\}_{j \in [\lambda]}$ and the receiver has input $e \in [\lambda]$.

Upon receiving $\{(z_j^0, z_j^1)\}_{j \in [\lambda]}$ from the sender, send $z_j^{e_j}$ to the receiver, where e_j is the jth bit of e.
---

3 Zero-Knowledge Protocols with PVC Security

Before providing the definition of zero-knowledge protocols with PVC security, we first define the task of zero-knowledge with the ideal functionality $\mathcal{F}_{\mathsf{zk}}^R$ below, as in [6] (for further details, refer to [19,20]).

Functionality $\mathcal{F}_{\mathsf{zk}}^R$

Public inputs: Both parties has input an NP statement x.
Private inputs: P has input w.
Upon receiving input (prove, x, w) from P and (verify, x) from V, if $x = x'$ and $(x, w) \in R$, send (accepted) to V. Otherwise, send (rejected) to V.

This definition approach offers the convenience of defining all the desired properties of a zero-knowledge protocol, including the *proof-of-knowledge* property, in a concise manner. It is worth mentioning that this simulation-based definition aligns with our definition of a zero-knowledge protocol with PVC security (Definition 1), which is better suited for simulation-based definitions.

Now, we modify $\mathcal{F}_{\mathsf{zk}}^R$ to accommodate PVC security and propose the functionality $\mathcal{F}_{\mathsf{pvc-zk}}^R$ as introduced in Sect. 1.2.

We would like to highlight that our definition, as described above, differs slightly from the definition of PVC security in the setting of secure two-party computation [2,21,22]. In zero-knowledge protocols, verifiers do not possess private input, and their output is constrained to a single bit of information, specifically either accepted or rejected when no abortion occurs. When cheating goes undetected, the prover can only cheat the verifier to output accepted, without gaining knowledge of the honest party's input or modifying the output to an arbitrary value, as is the case in secure computation.

Functionality $\mathcal{F}^R_{\text{pvc-zk}}$ **with deterrence** ϵ

Public inputs: Both parties have input an NP statement x.
Private inputs: P has input w.
Upon receiving the message (prove, x, y) from P, where $y \in \{0,1\}^n$, $y = \bot$, $y = (\text{cheat}, \hat{\epsilon})$ for $\hat{\epsilon} \geq \epsilon$, or $y = (\text{stupidCheat}, w, \tilde{\epsilon})$ (where the last three are sent by P* corrupted by the adversary) and (verify, x') from V:

- If $y = \bot$ or $x = x'$, then send $\bot$ to both parties.
- If $y \in \{0,1\}^n$, then if $(x, y) \in R$, send (accepted) to V. Otherwise, if $(x, y) \notin R$, send $\bot$ to both parties.
- If $y = (\text{cheat}, \hat{\epsilon})$, then:
  - With probability $\hat{\epsilon}$, send (failed) to the adversary and (rejected) to V.
  - With probability $1 - \hat{\epsilon}$, send (successful) to the adversary and (accepted) to V.
- If $y = (\text{stupidCheat}, w, \tilde{\epsilon})$, such that $(x, w) \in R$, then:
  - With probability $\tilde{\epsilon}$, send (failed) to the adversary and (rejected) to V.
  - With probability $1 - \tilde{\epsilon}$, send (successful) to the adversary and (accepted) to V.

The formal definition of zero-knowledge protocols with PVC security is presented below.

Definition 1. *A protocol $\Pi^R_{\text{pvc-zk}}$ along with algorithms* Blame *and* Judge *is a zero-knowledge protocol for the* NP*-relation R that achieves PVC security with deterrence ϵ if the following hold:*

Simulatability *The protocol $\Pi^R_{\text{pvc-zk}}$, where the honest verifier* V *might generate a certificate* cert *if cheating is detected, securely realizes the ideal functionality $\mathcal{F}^R_{\text{pvc-zk}}$ with deterrence ϵ.*
Public Verifiability *If the honest* V *outputs* rejected *(together with a certificate* cert*), then* Judge(cert) $= 1$*, except for a negligible probability.*
Defamation Freeness *For every* PPT *verifier* V* *corrupted by the adversary, if the prover is honest, the probability that* V* *generates a certificate* cert* *that blames the honest prover and leads to* Judge(cert*) $= 1$ *is negligible.*

Moreover, in the description of secure computation protocols with PVC security, if malicious behaviors are detected, the honest party should generate a certificate and send it to the adversary. Sending the certificate to the adversary indicates that the certificate does not reveal any information about the honest party's private input in the security proof. Conversely, in zero-knowledge protocols, since the verifier has no private input, there is no need to send the generated certificate to the adversary in the protocol description.

Furthermore, we introduce stupidCheat message in $\mathcal{F}^R_{\text{pvc-zk}}$ in order to capture the scenario where the prover knows the witness but still intend to provide invalid responses. It is important to note that this additional instruction does not compromise the security of the zero-knowledge protocol.

Additionally, we observe that setting the deterrence ϵ to be negligible in $\mathcal{F}^R_{\mathsf{pvc\text{-}zk}}$ achieves malicious security. Based on our definition, once the prover P causes the verifier V to output rejected in the maliciously secure protocols (i.e., ϵ is negligible), the honest verifier can also generate a certificate to attribute blame to the prover's misconduct.

4 Our Approach

In this section, we introduce our transformation method to convert Sigma protocols with 1-bit challenge into zero-knowledge protocols with PVC security. We first present the definition of Sigma protocols with 1-bit challenge based on [8].

Definition 2. *A two-party protocol Π between a prover P and a verifier V is said to be a Sigma protocol with 1-bit challenge for the NP-relation R if*

- *The protocol Π is of three-move form for a statement x as follows.*

> *1. P sends a message a.*
> *2. V sends a challenge $e \leftarrow_\$ \{0,1\}$ to P.*
> *3. P sends a reply message z, and V decides to accept or reject P based on the message (x, a, e, z).*

If $(x, w) \in R$ and P follows the protocol honestly with the correct witness w, then V always accepts.

- *For any statement x and any pair of accepting transcripts for x, $(a, 0, z)$ and $(a, 1, z')$, we can efficiently compute w, such that $(x, w) \in R$. This property is sometimes called* special soundness.
- *There exists a PPT simulator $\mathcal{S}$ that, on input, a statement x and a challenge $e \in \{0, 1\}$, outputs an accepting transcript (a, e, z) that is (computationally) indistinguishable from the transcript in a real execution of Π on x. This property is sometimes called* special honest-verifier zero-knowledge.

Now, given a two-party Sigma protocol Π with a 1-bit challenge, we introduce a protocol $\Pi^R_{\mathsf{pvc\text{-}zk}}$ based on Π, following the idea presented in Sect. 1.2. Note that we let an honest prover check the correctness of her witness at the beginning of the protocol. This is to facilitate the security proof as in [20]. Additionally, we propose two algorithms, Blame and Judge, for $\Pi^R_{\mathsf{pvc\text{-}zk}}$. We then prove that the protocol $\Pi^R_{\mathsf{pvc\text{-}zk}}$ along with the two algorithms Blame and Judge achieves PVC security.

Protocol $\Pi_{\text{pvc-zk}}^R$ with deterrence ϵ

Public inputs: Both parties know the two-party Sigma protocol Π with 1-bit challenge for the NP-relation R. Both parties have input an NP statement x. Both parties agree on the parameters κ and λ.

Private inputs: P has input the witness w and a private key sk for the EUF-CMA-secure signature scheme (KGen, Sig, Verify). Both parties know the corresponding public key pk of the signature scheme.

1. If $(x, w) \notin R$, P aborts. V chooses seed $\leftarrow_\$ \{0,1\}^\kappa$ and uses seed to derive a string $e \leftarrow_\$ \{0,1\}^\lambda$ and $\text{seed}_j \leftarrow_\$ \{0,1\}^\kappa$ for all $j \in [\lambda]$. We denote the ith bit of e by e_i. Then V computes the commitment $c_{\text{seed}} \leftarrow, (\text{seed})$ and sends c_{seed} to P.
2. For each $j \in [\lambda]$, P initiates an instance of Π and generates the first message a_j. Then P sends $c_a \leftarrow, (\{a_j\}_{j \in [\lambda]})$ to V.
3. For each $j \in [\lambda]$, P compute z_j^0 and z_j^1 for different challenge 0 and 1, respectively, according to Π. P and V perform λ executions of Π_{OT} in parallel. In the jth execution of Π_{OT}, the input of P as the sender is (z_j^0, z_j^1) while the input of V as the receiver is e_j. The randomness used by V in the jth execution of Π_{OT} are derived from seed_j. After the executions of Π_{OT}, V obtains $\{z_j^{e_j}\}$. We denote the transcript of the jth execution of Π_{OT} by trans_j.
4. P generates the signature $\sigma \leftarrow \text{Sig}_{\text{sk}}(x, \lambda, c_{\text{seed}}, \{a_j, \text{trans}_j\}_{j \in [\lambda]})$. Then P sends $\sigma, \{a_j\}_{j \in [\lambda]}$, and the decommitment of c_a (denoted by decom_a) to V.
5. V verifies the signature σ (with respect to $x' = x$) using pk, and the opening of c_a, and aborts with output $\bot$ if σ or the opening is invalid. V checks the correctness of $\{z_j^{e_j}\}$ for all $j \in [\lambda]$ based on (x, a_j, e_j) according to Π. If all $\{z_j^{e_j}\}$ are correct, V outputs `accepted`. Otherwise, V outputs `rejected` and can execute the algorithm Blame to generate the certificate cert as follows.
    (a) Randomly select an index $\hat{j}$ such that $z_{\hat{j}}^{e_{\hat{j}}}$ is incorrect.
    (b) Generate $\text{cert} = (x, \lambda, \sigma, c_{\text{seed}}, \text{seed}, \text{decom}_{\text{seed}}, \hat{j}, \{a_j, \text{trans}_j\}_{j \in [\lambda]})$.

Theorem 1. *If Π is a Sigma protocol with 1-bit challenge for the NP-relation R, then the protocol $\Pi_{\text{pvc-zk}}^R$ along with the algorithms Blame and Judge is a zero-knowledge protocol for the NP-relation R that achieves PVC security with deterrence $\epsilon = 1 - 1/2^\lambda$.*

Proof. We separately prove that the protocol $\Pi_{\text{pvc-zk}}^R$ achieves simulatability, public verifiability, and defamation freeness below.

Simulatability. We first focus on the simulatability. Let $\mathcal{A}$ be an adversary corrupting the prover. Given the auxiliary input z, we construct the simulator $\mathcal{S}$ that holds pk and runs $\mathcal{A}(z)$ as a subroutine while playing the role of the verifier V in the ideal world interacting with the ideal functionality $\mathcal{F}_{\text{pvc-zk}}^R$ as follows.

Algorithm Judge

Inputs: A public key pk and certificate cert.
1. Parse cert as $(x, \lambda, \sigma, c_{\text{seed}}, \text{seed}, \text{decom}_{\text{seed}}, \hat{\jmath}, \{a_j, \text{trans}_j\}_{j \in [\lambda]})$.
2. If $\text{Verify}_{\text{pk}}((x, \lambda, c_{\text{seed}}, \{a_j, \text{trans}_j\}_{j \in [\lambda]}), \sigma) = 0$, output 0.
3. If $c_{\text{seed}} \neq, (\text{seed}; \text{decom}_{\text{seed}})$, output 0.
4. Use seed to derive $e \leftarrow_\$ \{0,1\}^\lambda$ and $\text{seed}_j \leftarrow_\$ \{0,1\}^\kappa$ for all $j \in [\lambda]$ as V done in $\Pi^R_{\text{pvc-zk}}$.
5. Use $\text{seed}_{\hat{\jmath}}$ and $e_{\hat{\jmath}}$ to simulate the execution of V in Π_{OT} based on $\text{trans}_{\hat{\jmath}}$ and derive the output $z_{\hat{\jmath}}^{e_{\hat{\jmath}}}$. If the message that should be sent by V does not match the one in $\text{trans}_{\hat{\jmath}}$, output 0.
6. Check the correctness of $z_{\hat{\jmath}}^{e_{\hat{\jmath}}}$ based on $(x, a_{\hat{\jmath}}, e_{\hat{\jmath}})$ according to Π. If it is correct, output 0. Otherwise, output 1.
---

1. $\mathcal{S}$ computes the commitment $c_{\text{seed}} \leftarrow \text{Com}(0)$ and sends c_{seed} to $\mathcal{A}$.
2. $\mathcal{S}$ receives c_a.
3. $\mathcal{S}$ uses the simulator $\mathcal{S}_{\text{OT}}$ for Π_{OT} and extracts P^*'s input $\{(z_j^0, z_j^1)\}_{j \in [\lambda]}$ from $\mathcal{A}$.
4. $\mathcal{S}$ receives the signature σ, $\{a_j\}_{j \in [\lambda]}$, and decom_a.
5. $\mathcal{S}$ verifies the signature σ using pk and verify the opening of c_a. If one is invalid, $\mathcal{S}$ sends $\bot$ to $\mathcal{F}^R_{\text{pvc-zk}}$ and simulate the abortion of V.
   $\mathcal{S}$ initiates an empty list CheatIndex. For each $j \in [\lambda]$, $\mathcal{S}$ verifies the correctness of both z_j^0 and z_j^1 based on x, e_j, and a_j.
   – If both z_j^0 and z_j^1 are accepted, compute w_j, such that $(x, w_j) \in R$.
   – If both z_j^0 and z_j^1 are rejected, $\mathcal{S}$ directly sends $(\texttt{prove}, x, (\texttt{cheat}, 1))$ to $\mathcal{F}^R_{\text{pvc-zk}}$ and halts.
   – If one is accepted while the other is rejected, let $w_j = \bot$ and store j in the list CheatIndex.
   After the verification, if CheatIndex is empty, $\mathcal{S}$ sends $(\texttt{prove}, x, w_j)$ for a witness w_j with $j \leftarrow_\$ [\lambda]$ to $\mathcal{F}^R_{\text{pvc-zk}}$.
   If CheatIndex is not empty, let $m = |\text{CheatIndex}|$ be the number of elements in CheatIndex. If $m = \lambda$, then $\mathcal{S}$ sends $(\texttt{prove}, x, (\texttt{cheat}, 1 - 1/2^\lambda))$ to $\mathcal{F}^R_{\text{pvc-zk}}$ and receives (failed) or (successful) to conclude the simulation. If $m < \lambda$, $\mathcal{S}$ randomly picks $w_j \neq \bot$. Then $\mathcal{S}$ sends $(\texttt{prove}, x, (\texttt{stupidCheat}, w_j, 1 - 1/2^m))$ to $\mathcal{F}^R_{\text{pvc-zk}}$ and receives (failed) or (successful) to conclude the simulation.

Now, it remains to show that the joint distribution of the view of the adversary $\mathcal{A}$ and the output of V in the ideal world is computationally indistinguishable from the view of $\mathcal{A}$ and the output of V in the real world. In the following, we consider a sequence of hybrids, where the output of each hybrid is the view of $\mathcal{A}$ and the output of V.

Hybrid$_0$ This is the execution in the ideal world above. In this hybrid, $\mathcal{S}$ and the verifier V interact with the functionality $\mathcal{F}^R_{\text{pvc-zk}}$. We inline the actions of $\mathcal{S}$, $\mathcal{F}^R_{\text{pvc-zk}}$, and V, and rewrite the hybrid as follows.
1. Compute the commitment $c_{\text{seed}} \leftarrow \text{Com}(0)$ and send c_{seed} to $\mathcal{A}$.

2. Receive c_a.
3. Use the simulator $\mathcal{S}_{\mathsf{OT}}$ for λ instances of Π_{OT} and extract P*'s input $\{(z_j^0, z_j^1)\}_{j \in [\lambda]}$ from $\mathcal{A}$.
4. Receive the signature σ, $\{a_j\}_{j \in [\lambda]}$, and decom_a.
5. Verify the signature σ using pk and verify the opening of c_a. If it is invalid, output $\perp$ with abortion.
    Initiate an empty list CheatIndex. For each $j \in [\lambda]$, verify the correctness of both z_j^0 and z_j^1 based on x, e_j, and a_j.
    – If both z_j^0 and z_j^1 are accepted, compute w_j, such that $(x, w_j) \in R$.
    – If both z_j^0 and z_j^1 are rejected, output rejected.
    – If one is accepted while the other is rejected, let $w_j = \perp$ and store j in the list CheatIndex.
    After the verification, if CheatIndex is empty, output accepted.
    If CheatIndex is not empty, let $m = |\mathsf{CheatIndex}|$ be the number of elements in CheatIndex. If $m = \lambda$, then with probability $1 - 1/2^\lambda$ output rejected and with the remaining probability output accepted. If $m < \lambda$, then with probability $1 - 1/2^m$ output rejected and with the remaining probability output accepted.

Hybrid$_1$ We modify the operations after the verification of Step 5: let $m = |\mathsf{CheatIndex}|$ be the number of elements in CheatIndex. If $m = 0$, output accepted. Otherwise, with probability $1 - 1/2^m$ output rejected and with the remaining probability output accepted.

It is easy to see that the distribution for the output of **Hybrid$_1$** is identical to the distribution for the output of **Hybrid$_0$**.

Hybrid$_2$ We modify Step 5 as follows.
5. Verify the signature σ using pk. If it is invalid, output $\perp$ with abortion. Initiate an empty list CheatIndex. Pick $e \leftarrow_\$ \{0,1\}^\lambda$. For each $j \in [\lambda]$, verify the correctness of $z_j^{e_j}$ based on x, e_j, and a_j.
    – If $z_j^{e_j}$ is rejected, store j in the list CheatIndex.
    After the verification, if CheatIndex is empty, output accepted.
    If CheatIndex is not empty, let $m = |\mathsf{CheatIndex}|$ be the number of elements in CheatIndex. If $m \neq 0$, then output rejected. Otherwise, output accepted.

The distribution for the output of **Hybrid$_2$** is identical to the distribution for the output of **Hybrid$_1$**. To check this fact, we consider the corresponding cases in **Hybrid$_1$** and show that they are identical to the cases in **Hybrid$_2$** in the following.

– In **Hybrid$_1$**, if all z_j^0 and z_j^1 are accepted, output accepted. In **Hybrid$_2$**, the output is also accepted in this case.
– In **Hybrid$_1$**, if there exists an index j, such that both z_j^0 and z_j^1 are rejected, output rejected. In **Hybrid$_2$**, no matter what e_j is chosen, the output is also rejected.
– In **Hybrid$_1$**, if there are m instances of z_j^0 and z_j^1, such that one is accepted while the other is rejected, with probability $1 - 1/2^m$, the output is rejected. In **Hybrid$_2$**, since e is uniformly chosen, the probability that the output is rejected is also $1 - 1/2^m$.

Therefore, the distribution for the output of **Hybrid**$_2$ is identical to the distribution for the output of **Hybrid**$_1$.

Hybrid$_3$ In this hybrid, we modify Step 1: Choose seed $\leftarrow_\$ \{0,1\}^\kappa$ and use seed to derive a string $e \leftarrow_\$ \{0,1\}^\lambda$ and $\mathsf{seed}_j \leftarrow_\$ \{0,1\}^\kappa$ for all $j \in [\lambda]$ as in the protocol. Then compute the commitment $c_{\mathsf{seed}} \leftarrow \mathsf{Com}(0)$ and send c_{seed} to $\mathcal{A}$. Since the committed value is unchanged, c_{seed} is identical to that in **Hybrid**$_2$. The challenge e is now derived from seed. Since e now is pseudorandom, the output of **Hybrid**$_3$ is computationally indistinguishable from the output of **Hybrid**$_2$.

Hybrid$_4$ We modify Step 1 by computing c_{seed} as $c_{\mathsf{seed}} \leftarrow \mathsf{Com}(\mathsf{seed})$. Since the commitment scheme is (computationally) hiding, the output of **Hybrid**$_4$ is (computationally) indistinguishable from the output of **Hybrid**$_3$.

Hybrid$_5$ In Step 3 of this hybrid, execute λ instances of Π_{OT} honestly with true randomness instead of using $\mathcal{S}_{\mathsf{OT}}$. Based on the security of Π_{OT}, the output of **Hybrid**$_5$ is computationally indistinguishable from the output of **Hybrid**$_4$.

Hybrid$_6$ In this hybrid, the randomness used in the execution Π_{OT} are derived from seed_j. Since the randomness used in Π_{OT} is now pseudorandom instead of true randomness, the output of **Hybrid**$_6$ is computationally indistinguishable from the output of **Hybrid**$_5$. We state **Hybrid**$_6$ as follows.

1. Choose seed $\leftarrow_\$ \{0,1\}^\kappa$ and use seed to derive a string $e \leftarrow_\$ \{0,1\}^\lambda$ and $\mathsf{seed}_j \leftarrow_\$ \{0,1\}^\kappa$ for all $j \in [\lambda]$. Compute $c_{\mathsf{seed}} \leftarrow \mathsf{Com}(\mathsf{seed})$ and send c_{seed} to $\mathcal{A}$.
2. Receive c_a.
3. Execute λ instances of Π_{OT} with $\mathcal{A}$ using input $\{e_j\}_{j\in[\lambda]}$ with randomness derived from seed_j and receive $\{z_j^{e_j}\}_{j\in[\lambda]}$.
4. Receive the signature σ, $\{a_j\}_{j\in[\lambda]}$, and decom_a.
5. Verify the signature σ using pk and verify the opening of c_a. If it is invalid, output $\bot$ with abortion. For each $j \in [\lambda]$, verify the correctness of $z_j^{e_j}$ based on x, e_j, and a_j. If all $\{z_j^{e_j}\}$ are correct, output accepted. Otherwise, output rejected.

It is easy to see that **Hybrid**$_6$ is identical to the execution of the protocol for an honest verifier V. Since **Hybrid**$_6$ is computationally indistinguishable from **Hybrid**$_0$, the joint distribution of the view of the adversary $\mathcal{A}$ and the output of V in the ideal world is computationally indistinguishable from the view of $\mathcal{A}$ and the output of V in the real world.

We now focus on the case where the verifier V is corrupted by the adversary $\mathcal{S}$. Given the auxiliary input z, we construct the simulator $\mathcal{S}$ that runs $\mathcal{A}(z)$ as a subroutine while playing the role of the prover P in the ideal world interacting with the ideal functionality $\mathcal{F}^R_{\mathsf{pvc\text{-}zk}}$ as follows.

1. $\mathcal{S}$ generates a pair of keys $(\mathsf{pk}, \mathsf{sk})$ for the signature scheme using KGen and sends pk to $\mathcal{A}$.
2. If $\mathcal{S}$ receives $\bot$ from $\mathcal{F}^R_{\mathsf{pvc\text{-}zk}}$, $\mathcal{S}$ simulates the abortion of P in the protocol. Otherwise, $\mathcal{S}$ receives c_{seed} from $\mathcal{A}$.
3. $\mathcal{S}$ generates $c_a \leftarrow \mathsf{Com}(0)$ and sends it to $\mathcal{A}$.

4. $\mathcal{S}$ uses the simulator $\mathcal{S}_{\mathsf{OT}}$ for Π_{OT} and extracts the challenge $\{e_j\}_{j\in[\lambda]}$. Given e_j, $\mathcal{S}$ uses the simulator $\mathcal{S}_\Sigma$ for the protocol Π to generate the accepting transcript (a_j, e_j, z_j). Then $\mathcal{S}$ returns $\{z_j\}_{j\in[\lambda]}$ as output of Π_{OT} to $\mathcal{A}$.
5. $\mathcal{S}$ generates the signature σ as in the protocol. $\mathcal{S}$ equivocates the commitment c_a to be the commitment to $\{a_j\}_{j\in[\lambda]}$. Then $\mathcal{S}$ sends σ, $\{a_j\}_{j\in[\lambda]}$, and the decommitment of c_a to $\mathcal{A}$ to conclude the simulation.

Now it remains to show that the joint distribution of the view of the adversary $\mathcal{A}$ and the output of P (P may output $\bot$) in the ideal world is computationally indistinguishable from the view of $\mathcal{A}$ and the output of P in the real world. In the following, we consider a sequence of hybrids, where the output of each hybrid is the view of $\mathcal{A}$ and the output of P. We first denote the simulation in the ideal world above by **Hybrid**$_0$.

Hybrid$_1$ Assume that $\mathcal{S}$ is given the witness w. Then, if $(x, w) \notin R$, $\mathcal{S}$ simulates the abortion of the prover in Step 1. It is easy to see that the distribution of the output for **Hybrid**$_1$ is identical to the distribution for the output of $hybrid_0$.

Hybrid$_2$ In Step 4, $\mathcal{S}$ generates the accepting transcripts for the cases $e_j = 0$ and $e_j = 1$, i.e., $(a_j^0, 0, z_j^0)$ and $(a_j^1, 1, z_j^1)$, and returns the corresponding $\{z_j^{e_j}\}_{j\in[\lambda]}$ to $\mathcal{A}$. Then in Step 4, $\mathcal{S}$ equivocates c_a to be the commitment to $\{a_j^{e_j}\}_{j\in[\lambda]}$. Since the transcripts for $1 - e_j$'s are not used, it is easy to see that the distribution of the output for **Hybrid**$_2$ is identical to the distribution for the output of $hybrid_1$.

Hybrid$_3$ For this hybrid, $\mathcal{S}$ generates a_j for all $j \in [\lambda]$. Then in Step 3, $\mathcal{S}$ computes z_j^0 and z_j^1 as in the protocol based on a_j. Since for the protocol Π, each accepting transcript (a_j, e_j, z_j) simulated by $\mathcal{S}_\Sigma$ is (computationally) indistinguishable from that in a real execution and $\mathcal{A}$ only receive $z_j^{e_j}$ for each $j \in [\lambda]$, the output of **Hybrid**$_3$ is (computationally) indistinguishable from the output of **Hybrid**$_2$.

Hybrid$_4$ $\mathcal{S}$ generates c_a honestly as in the protocol based on $\{a_j\}_{j\in[\lambda]}$. Now, $\mathcal{S}$ does not need to equivocate c_a in Step 4. Based on the security of the commitment scheme, the output of **Hybrid**$_4$ is computationally indistinguishable from the output of **Hybrid**$_3$.

Hybrid$_5$ In this hybrid, $\mathcal{S}$ honestly executes the protocol Π_{OT}. We restate this hybrid as follows.

0. Generate a pair of keys $(\mathsf{pk}, \mathsf{sk})$ for the signature scheme using KGen and send pk to $\mathcal{A}$.
1. If $(x, w) \notin R$, abort. Otherwise, receive c_{seed} from $\mathcal{A}$.
2. For each $j \in [\lambda]$, initiate an instance of Π and generate a_j. Compute $c_a \leftarrow \mathsf{Com}(\{a_j\}_{j\in[\lambda]})$ and sends it to $\mathcal{A}$.
3. For each $j \in [\lambda]$, compute z_j^0 and z_j^1 for different challenge 0 and 1, respectively, according to Π. Execute λ Π_{OT} with $\mathcal{A}$ using input (z_j^0, z_j^1).
4. Generate the signature σ as in the protocol. Then send σ, $\{a_j\}_{j\in[\lambda]}$, and the decommitment of c_a to $\mathcal{S}$ to conclude the simulation.

It is easy to see that **Hybrid**$_5$ is identical to the execution of the protocol for an honest prover P. Since **Hybrid**$_5$ is computationally indistinguishable from **Hybrid**$_0$, the joint distribution of the view of the adversary $\mathcal{A}$ and the output of P in the ideal world is computationally indistinguishable from the view of $\mathcal{A}$ and the output of P in the real world. Therefore, the simulatability of the protocol is proved.

Public Verifiability. We now argue that whenever an honest verifier V outputs `rejected`, V is capable of generating a valid certificate for the malicious prover P. In Step 5 of the protocol, without receiving a valid signature and a valid opening of c_a, V will not continue the execution of the protocol. At this point, given the security of $\varPi_{\mathsf{OT}}$, a malicious P cannot make her decision to continue the protocol (*i.e.*, provide a valid signature and opening) or abort based on the challenge e. When the malicious P chooses not to abort and proceeds to provide the signature and opening, V now possesses sufficient information to generate the certificate. Subsequently, it can be readily verified that the algorithm Judge is applicable to a certificate cert generated by an honest V in the protocol using Blame. This verification is feasible since the algorithm Judge simulates the operations carried out by V and employs the same verification procedure as V in the protocol.

Defamation Freeness. We show that the protocol achieves defamation freeness by contradiction. We assume that an honest prover P is accused by a verifier V corrupted by the adversary via a valid certificate. This implies that the adversary can provide a valid signature of P for the message

$$(x, \lambda, c_{\mathsf{seed}}, \{a_j, \mathsf{trans}_j\}_{j \in [\lambda]})\,.$$

Since P is honest, she only generates a signature for the message she has seen and generated. Considering the EUF-CMA-secure signature scheme, the adversary cannot forge the signature except with negligible probability. Furthermore, as the commitment scheme is binding, the adversary cannot provide a different opening for c_{seed} from the one used in Step 1 to generate the signature. Therefore, the simulation performed in the algorithm Judge will employ the exact seed. Given that $\varPi_{\mathsf{OT}}$ is perfectly correct, with the signed transcript of $\varPi_{\mathsf{OT}}$, there exists exactly one valid output for V consistent with $\mathsf{trans}_{\hat{j}}$, regardless of V's randomness and input. Thus, the simulated transcript in the algorithm Judge will match $\mathsf{trans}_{\hat{j}}$ only if seed_j is the one used by V to derive randomness during the execution of $\varPi_{\mathsf{OT}}$. In this scenario, the simulation will yield the same $z_{\hat{j}}^{e_j}$ generated by the honest P, and consequently, the algorithm Judge will not output 1. Hence, the protocol achieves defamation freeness, and a malicious verifier cannot generate a valid certificate cert to blame an honest prover except with negligible probability.

In summary, if $\varPi$ is a Sigma protocol with 1-bit challenge for the NP-realation R, then the protocol $\varPi^R_{\mathsf{pvc\text{-}zk}}$ along with the algorithms Blame and Judge is a zero-knowledge protocol for the NP-realation R that achieves PVC security with deterrence $\epsilon = 1 - 1/2^\lambda$.

Remark 1. It is worth mentioning that extending the challenge space of a Sigma protocol Π in Definition 2 to accommodate other small challenge spaces can be easily accomplished, along with slight modifications to $\Pi_{\text{pvc-zk}}^R$ to support such Sigma protocols.

5 Performance

In this section, we conduct a performance analysis of our protocol $\Pi_{\text{pvc-zk}}^R$ in comparison to an original protocol, denoted as Π_o, which repetitively executes the Sigma protocol Π with 1-bit challenge to minimize the knowledge error.

It is important to note that traditionally, Π_o requires the execution of at least 40 instances of Π. However, in scenarios involving reputation-conscious parties, our $\Pi_{\text{pvc-zk}}^R$ could significantly reduce the number of instances required. Table 1 presents the required number of instances for Π to achieve specific deterrence factors. Although our protocol requires the prover to perform the OT protocol, sending two answers for the challenge bit (while the verifier only derives one answer), the communication cost of each instance in comparison to Π is at most doubled, considering that only the number of answers is doubled. Moreover, it is important to mention that the commitment and signature have constant sizes, as we employ a random oracle to instantiate the commitment scheme and compress the signed messages. Additionally, the OT protocol is quite efficient nowadays [18]. For the number of rounds, assuming that the random OTs have already been generated[3], our protocol consists of 5 rounds. In comparison, the constant-round Goldreich-Kahan protocol [13,23] also consists of 5 rounds. Thus, the cost of our protocol is primarily determined by the cost of Π and the number of instances required for repetition.

Table 1. Number of instances for Π required to achieve the specified deterrence factors.

Deterrence Factors	50%	75%	87.5%	93.75%	96.88%	98.44%	99.22%
Number of Instances	1	2	3	4	5	6	7

Therefore, we observe that even achieving a deterrence factor of 99.22%, our protocol only requires 7 instances, resulting in a cost increase of at most 14× compared to Π. In contrast, the traditional protocol incurs a cost of at least 40×. Consequently, our protocol $\Pi_{\text{pvc-zk}}^R$ costs less than 35% of Π_o in the scenarios involving reputation-conscious parties. For a deterrence factor of 93.75%, the cost of $\Pi_{\text{pvc-zk}}^R$ is approximately 20% of the cost of Π_o. This substantial reduction in cost proves particularly advantageous in real-world scenarios.

---

[3] It is common practice to pre-generate random OTs before executing the protocol. Since zero-knowledge protocols are often used as sub-protocols, it is reasonable to assume that random OTs have been generated.

6 Conclusion

In this paper, we explore the approach of integrating zero-knowledge protocols into the PVC security model to address the significant overhead challenges present in many useful and interesting zero-knowledge protocols. Specifically, we introduce a new notion called zero-knowledge protocols with PVC security, which allows honest parties to generate publicly verifiable certificates against detected misbehaviors. This approach offers the potential to simultaneously achieve high efficiency and strong security guarantees, effectively balancing efficiency and security. Building on the notion, we propose a generic transformation that converts any Sigma protocol with 1-bit challenge into a zero-knowledge protocol with PVC security. Our transformation achieves a high cheating detection accuracy (e.g. 99.22%) while significantly reducing overhead (e.g. by approximately 65%) compared to the original protocol, thereby contributing to significant advancements in the practical application of zero-knowledge protocols.

Acknowledgements. This work was supported in part by the National Key Research and Development Program of China under Grant No. 2021ZD0112802, in part by the National Natural Science Foundation of China under Grant Nos. 62302194, 62072215 and 62472198, in part by Guangdong Basic and Applied Basic Research Foundation under Grant Nos. 2023B1515040020 and 2019B030302008, in part by Guangzhou Basic and Applied Basic Research Foundation under Grant Nos. 2025A04J2146 and 2024A04J3458, in part by the Guangzhou Basic Research Plan City-School Joint Funding Project under Grant No. 2024A03J0405, and in part by National Joint Engineering Research Center of Network Security Detection and Protection Technology, Guangdong Key Laboratory of Data Security and Privacy Preserving, Guangdong Hong Kong Joint Laboratory for Data Security and Privacy Protection, and Engineering Research Center of Trustworthy AI, Ministry of Education.

References

1. Agrawal, N., Bell, J., Gascón, A., Kusner, M.J.: MPC-friendly commitments for publicly verifiable covert security. In: Kim, Y., Kim, J., Vigna, G., Shi, E. (eds.) CCS 2021: 2021 ACM SIGSAC Conference on Computer and Communications Security, Virtual Event, Republic of Korea, 15–19 November 2021, pp. 2685–2704. ACM, New York (2021)
2. Asharov, G., Orlandi, C.: Calling out cheaters: covert security with public verifiability. In: Wang, X., Sako, K. (eds.) ASIACRYPT 2012. LNCS, vol. 7658, pp. 681–698. Springer, Heidelberg (2012). https://doi.org/10.1007/978-3-642-34961-4_41
3. Attema, T., Dunning, V., Everts, M.H., Langenkamp, P.: Efficient compiler to covert security with public verifiability for honest majority MPC. In: Ateniese, G., Venturi, D. (eds.) Applied Cryptography and Network Security - ACNS 2022. Lecture Notes in Computer Science, vol. 13269, pp. 663–683. Springer, Cham (2022)
4. Ben-Sasson, E., et al.: Zerocash: decentralized anonymous payments from bitcoin. In: 2014 IEEE Symposium on Security and Privacy, SP, Berkeley, CA, USA, pp. 459–474. IEEE Computer Society (2014)

5. Bünz, B., Bootle, J., Boneh, D., Poelstra, A., Wuille, P., Maxwell, G.: Bulletproofs: short proofs for confidential transactions and more. In: 2018 IEEE Symposium on Security and Privacy, SP 2018, Proceedings, San Francisco, California, USA, pp. 315–334. IEEE Computer Society (2018)
6. Canetti, R.: Universally composable security: a new paradigm for cryptographic protocols. In: 42nd Annual Symposium on Foundations of Computer Science, FOCS 2001, 14–17 October 2001, Las Vegas, Nevada, pp. 136–145. IEEE Computer Society (2001)
7. Couteau, G., Peters, T., Pointcheval, D.: Encryption switching protocols. In: Robshaw, M., Katz, J. (eds.) CRYPTO 2016. LNCS, vol. 9814, pp. 308–338. Springer, Heidelberg (2016). https://doi.org/10.1007/978-3-662-53018-4_12
8. Damgård, I.: On σ-protocols. Lecture Notes, University of Aarhus, Department for Computer Science (2002)
9. Damgård, I., Orlandi, C., Simkin, M.: Black-box transformations from passive to covert security with public verifiability. In: Micciancio, D., Ristenpart, T. (eds.) CRYPTO 2020. LNCS, vol. 12171, pp. 647–676. Springer, Cham (2020). https://doi.org/10.1007/978-3-030-56880-1_23
10. Faust, S., Hazay, C., Kretzler, D., Schlosser, B.: Generic compiler for publicly verifiable covert multi-party computation. In: Canteaut, A., Standaert, F.-X. (eds.) EUROCRYPT 2021. LNCS, vol. 12697, pp. 782–811. Springer, Cham (2021). https://doi.org/10.1007/978-3-030-77886-6_27
11. Faust, S., Hazay, C., Kretzler, D., Schlosser, B.: Financially backed covert security. In: Hanaoka, G., Shikata, J., Watanabe, Y. (eds.) PKC 2022, Part II. LNCS, vol. 13178, pp. 99–129. Springer, Cham (2022)
12. Gennaro, R., Gentry, C., Parno, B., Raykova, M.: Quadratic Span Programs and Succinct NIZKs without PCPs. In: Johansson, T., Nguyen, P.Q. (eds.) EUROCRYPT 2013. LNCS, vol. 7881, pp. 626–645. Springer, Heidelberg (2013). https://doi.org/10.1007/978-3-642-38348-9_37
13. Goldreich, O., Kahan, A.: How to construct constant-round zero-knowledge proof systems for NP. J. Cryptol. **9**(3), 167–189 (1996). https://doi.org/10.1007/BF00208001
14. Goldreich, O., Micali, S., Wigderson, A.: How to play any mental game or A completeness theorem for protocols with honest majority. In: Aho, A.V. (ed.) Proceedings of the 19th Annual ACM Symposium on Theory of Computing, pp. 218–229. ACM, New York (1987)
15. Goldreich, O., Micali, S., Wigderson, A.: Proofs that yield nothing but their validity for all languages in NP have zero-knowledge proof systems. J. ACM **38**(3), 691–729 (1991)
16. Goldwasser, S., Micali, S., Rackoff, C.: The knowledge complexity of interactive proof-systems (extended abstract). In: Sedgewick, R. (ed.) Proceedings of the 17th Annual ACM Symposium on Theory of Computing, Providence, Rhode Island, USA, pp. 291–304. ACM (1985)
17. Groth, J.: On the size of pairing-based non-interactive arguments. In: Fischlin, M., Coron, J.-S. (eds.) EUROCRYPT 2016. LNCS, vol. 9666, pp. 305–326. Springer, Heidelberg (2016). https://doi.org/10.1007/978-3-662-49896-5_11
18. Guo, C., Katz, J., Wang, X., Weng, C., Yu, Yu.: Better concrete security for half-gates garbling (in the multi-instance setting). In: Micciancio, D., Ristenpart, T. (eds.) CRYPTO 2020. LNCS, vol. 12171, pp. 793–822. Springer, Cham (2020). https://doi.org/10.1007/978-3-030-56880-1_28
19. Hazay, C., Lindell, Y.: Efficient Secure Two-Party Protocols - Techniques and Constructions. Information Security and Cryptography. Springer, Berlin (2010)

20. Hazay, C., Lindell, Y.: A note on zero-knowledge proofs of knowledge and the ZKPOK ideal functionality. IACR Cryptol. ePrint Arch. (2010). http://eprint.iacr.org/2010/552
21. Hong, C., Katz, J., Kolesnikov, V., Lu, W., Wang, X.: Covert security with public verifiability: faster, leaner, and simpler. In: Ishai, Y., Rijmen, V. (eds.) EUROCRYPT 2019. LNCS, vol. 11478, pp. 97–121. Springer, Cham (2019). https://doi.org/10.1007/978-3-030-17659-4_4
22. Kolesnikov, V., Malozemoff, A.J.: Public verifiability in the covert model (almost) for free. In: Iwata, T., Cheon, J.H. (eds.) Advances in Cryptology - ASIACRYPT 2015, Part II. LNCS, vol. 9453, pp. 210–235. Springer (2015)
23. Lindell, Y.: How to simulate it – a tutorial on the simulation proof technique. In: Tutorials on the Foundations of Cryptography. ISC, pp. 277–346. Springer, Cham (2017). https://doi.org/10.1007/978-3-319-57048-8_6
24. Liu, Y., Lai, J., Wang, Q., Qin, X., Yang, A., Weng, J.: Robust publicly verifiable covert security: Limited information leakage and guaranteed correctness with low overhead. In: Guo, J., Steinfeld, R. (eds.) Advances in Cryptology - ASIACRYPT 2023, Part I. LNCS, vol. 14438, pp. 272–301. Springer, Cham (2023)
25. Liu, Y., Wang, Q., Yiu, S.-M.: Blind polynomial evaluation and data trading. In: Sako, K., Tippenhauer, N.O. (eds.) ACNS 2021. LNCS, vol. 12726, pp. 100–129. Springer, Cham (2021). https://doi.org/10.1007/978-3-030-78372-3_5
26. Liu, Y., Wang, Q., Yiu, S.: Making private function evaluation safer, faster, and simpler. In: Hanaoka, G., Shikata, J., Watanabe, Y. (eds.) Public-Key Cryptography - PKC 2022, Part I. LNCS, vol. 13177, pp. 349–378. Springer, Cham (2022)
27. Parno, B., Howell, J., Gentry, C., Raykova, M.: Pinocchio: nearly practical verifiable computation. In: 2013 IEEE Symposium on Security and Privacy, SP 2013, Berkeley, CA, USA, pp. 238–252. IEEE Computer Society (2013)
28. Schnorr, C.P.: Efficient identification and signatures for smart cards. In: Brassard, G. (ed.) CRYPTO 1989. LNCS, vol. 435, pp. 239–252. Springer, New York (1990). https://doi.org/10.1007/0-387-34805-0_22
29. Scholl, P., Simkin, M., Siniscalchi, L.: Multiparty computation with covert security and public verifiability. Cryptology ePrint Archive, Report 2021/366 (2021). https://ia.cr/2021/366
30. Zhu, R., Ding, C., Huang, Y.: Efficient publicly verifiable 2pc over a blockchain with applications to financially-secure computations. In: Cavallaro, L., Kinder, J., Wang, X., Katz, J. (eds.) Proceedings of the 2019 ACM SIGSAC Conference on Computer and Communications Security, CCS 2019, London, UK, pp. 633–650. ACM (2019)

Attribute-Based Adaptor Signature and Application in Control-Based Atomic Swap

Tianyuan Fan[1,2], Gang Shen[1,2], Yuzhu Wang[1,2], Yuntao Wang[3], and Mingwu Zhang[1,2(✉)]

[1] School of Computer Science, Hubei University of Technology, Wuhan 430068, China
[2] Hubei Provincial Key Laboratory of Green Intelligent Computing Power Network, Wuhan 430068, China
mzhang@hbut.edu.cn
[3] Graduate School of Informatics and Engineering, The University of Electro-Communications, Tokyo 182-8585, Japan

Abstract. Adaptor signatures (AS) provide a crucial cryptographic technique that can be used to solve the problems of poor scalability and low transaction throughput cross-chain such as cryptocurrency and atomic swap. To address the limitations of existing attribute-based signature (ABS) schemes lacking conditional payment capabilities for the policy constraints of traditional adaptor signature schemes, this paper proposes the first Linear Secret Sharing Scheme (LSSS)-compatible Attribute-Based Adaptor Signature (ABAS) scheme. The ABAS scheme embeds a hard relation witness y and innovatively integrates access control matrix $\mathbf{M}$ with a pre-signature mechanism to unify attribute privacy preservation, which attains the correctness of both pre-signature and completed signature. The ABAS scheme provides the security of Pre-signature Adaptability, Adaptor Existential Unforgeability in the selective policy model (sP-aEUF-CMA security), Witness Extractability under the q-BDHE assumption, and Attributes Anonymity. We provide a concrete control-based atomic swap protocol that employs our proposed attribute-based adaptor signature scheme as a security primitive to implement multiple-party atomic swap for satisfying the control policy in cross-chain systems.

Keywords: Adaptor signature · Witness extractability · Adaptability · Linear secret sharing · Atomic swap

1 Introduction

Adaptor signatures (AS) [1] is a class of cryptographic primitive extended from traditional digital signature, which allows users to create a pre-signature that

---

This work is supported by the National Natural Science Foundation of China under grant 62472150, the Major Research Plan of Hubei Province under grant 2023BAA027, and partially supported by JSPS KAKENHI Grant Number JP21K11751, Japan.

© The Author(s), under exclusive license to Springer Nature Singapore Pte Ltd. 2026
J. Han et al. (Eds.): ICICS 2025, LNCS 16217, pp. 160–180, 2026.
https://doi.org/10.1007/978-981-95-3540-8_9

implicitly encodes a statement of a hard relation. This pre-signature can only be converted into a valid full/complete signature by utilizing the witness/solution of the hard relation. Crucially, the witness can be extracted by comparing the pre-signature and its corresponding full signature. AS effectively enable blockchain applications such as cross-chain fair exchange [2,3], atomic swaps [4,5] and payment channel network [6–8], etc. By reducing on-chain computation and enhancing transaction fungibility, they significantly improve interoperability and efficiency in decentralized systems. Aumayr et al. [9] formally defined adaptor signature as a standalone primitive, constructing their scheme using Schnorr and ECDSA. Malavolta et al. [10] proposed a two-party threshold AS protocol based on Schnorr and ECDSA, and leveraged it to build an anonymous multi-hop lock mechanism. Subsequently, Erwig et al. [11] formalized two-party adaptor signature and demonstrated a generic framework to construct AS from authentication schemes. Qin et al. [12] proposed a universal approach to build AS from identification schemes, though their construction primarily suits lattice-based instantiations. Concurrently, they introduced blind adaptor signature and linkable ring adaptor signature. Dai et al. [13] formalized the unlinkability property of AS and designed a generic approach for constructing unlinkable variants. With the rise of quantum computing, conventional AS faces existential threats. Esgin et al. [6] addressed this by designing lattice-based AS, implementing the post-quantum secure AS relying on standard lattice assumptions. AS have versatile applications, spanning various domains such as Payment Channel Networks(PCNs) [14], private coin mixing [15], and oracle-based payments [16], etc. Notably, most existing adaptor signature remains algorithm-specific adaptations rather than generalized framework.

Attribute-based signature(ABS) [17] is categorized into two classes: signature-policy ABS[18] and key-policy ABS[19]. In signature-policy ABS, a user's private key is generated based on their attribute set, while the signature corresponds to access control policy. In key-policy ABS, a user's private key is associated with an access control policy, where the signature is linked to an attribute set. By decoupling authorization logic from cryptographic operations, ABS schemes enable fine-grained data access control and provide enhanced privacy protection-signers remain anonymous unless their attribute violate the verification policy. This duality makes ABS indispensable for privacy-preserving systems like decentralized credential sharing and policy-driven blockchain transactions. Maji et al. [17] pioneered an ABS scheme enabling users to generate signatures under fine-grained access control policies, achieving policy expressiveness equivalent to linear secret-sharing schemes. Okamoto et al. [20] introduced a non-monotonic predicate-supporting ABS that significantly optimized verification complexity to $O(\log n)$ for $n-$attribute system, while demonstrating full security in the standard model against chosen-message attacks. Gu et al. [21] constructed a monotonic-predicate ABS build upon the Waters signature framework [17,22]. Their scheme guarantees: perfect privacy and existential unforgeability.

While ABS enables fine-grained access control over signing privileges through linear secret-sharing schemes(LSSS) [23], is inherently lacks flexible mechanisms to enforce conditional payment execution– a critical requirement for cross-chain atomic swaps and decentralized escrow systems. Conversely, adaptor signature, as advanced derivatives of Schnorr signatures [24], elegantly link signature validly to predefined conditions through witness extraction, yet suffer from rigid policy expression and attribute exposure risks. Existing attempts to reconcile these requirements remain inadequate. Traditional ABS frameworks (e.g., Waters-ABS[22]) enforce complex policies **M** while preserving signer anonymity but can not bind transaction validity to external conditions (e.g., time-locked payments). Meanwhile, state-of-the-art AS[25] will conditionally disclose witness y through divergence proofs but operate under fixed policy structure, exposing sensitive attributes during verification. This fundamental mismatch impedes privacy-sensitive applications like medical data monetization and anonymous voting, where both policy-driven authorization and conditional fund release must coexist cryptographically.

Our contribution. The contribution of this work provides a novel attribute-based adaptor signature(ABAS) scheme, which for the first time integrates attribute-based cryptography with adaptor signature technology to achieve policy-controlled signature flexibility and enhanced security. Concretely, the contribution is: (1) We construct an ABAS scheme supporting LSSS, organically combining attribute authorization, policy verification, and adaptor functionality through pre-signature mechanism; (2) We formalize the security model for ABAS by defining three crucial security requirements: pre-signature adaptability, selective-policy existential unforgeability against adaptive chosen-message attacks(sP-aEUF-CMA), and witness extractability, while introducing additional strong anonymity to protect signers' attribute privacy; (3) Based on the q-BDHE hardness assumption in random oracle model, we give the security proof, while ensuring attribute indistinguishability through randomized bilinear pairing operations. Finally, this work provides theoretical foundations for blockchain applications such as cross-chain transactions and privacy-preserving smart contracts, offering both fine-grained access control and atomic swap security.

The remaining of this paper is structured as follows. In Sect. 2, we review the related work and primitives of AS and ABS schemes. In Sect. 3, we introduce algorithms definitions and security model. In Sect. 4, we give the concrete construction of the ABAS scheme. In Sect. 5, we provide the formal proof and analysis of the security, and in Sect. 6 give the performance evaluation. In Sect. 7, we introduce its application in cross-chain atomic swaps, and finally, Sect. 8 concludes this paper.

2 Preliminaries

2.1 Hard Relation

Let R be a binary relation, and $\mathcal{L}_R$ denote the language describing this relation, i.e., $\mathcal{L}_R = \{Y | \exists\ y : (Y,y) \in R\}$. A relation R is termed a hard relation if it satisfies the following requirements:

1. There exists a probabilistic polynomial-time(PPT) algorithm GenR that generates an instance (Y, y) for the relation, denoted as $(Y, y) \leftarrow \mathsf{GenR}(1^\lambda)$, where λ is the security parameter.

2. There exists a deterministic polynomial-time(DPT) algorithm VerifyR that verifies whether (Y, y) satisfies R. Formally, $0/1 \leftarrow$ VerifyR(Y, y), where it outputs 1 if $(Y, y) \in R$, and 0 otherwise.
3. For all PPT adversaries $\mathcal{A}$, the probability of computing y given Y is negligible in λ. Specifically,

$$\Pr[(Y, y) \leftarrow \mathsf{GenR}(1^\lambda), \mathcal{A}(Y) = y : (Y, y) \in R] \leq \epsilon(\lambda). \quad (1)$$

The paper gives a hard relation $R = \{(Y, y) | Y = g^y \in \mathbb{G}, y \leftarrow \mathbb{Z}_p\}$, based on the discrete logarithm hard relationship, which satisfies the three essential properties outlined.

Definition 1. *(Discrete logarithm hard problem) In a finite group $\mathbb{G}$ of order p, the discrete logarithm problem is stated as: Given $\mathbb{G}$ of order p with a generator g, for any element $Y \in \mathbb{G}$, to find an integer x such that $g^x = Y$ is computationally hard.*

2.2 Bilinear Maps

Let $(\mathbb{G}_1, \mathbb{G}_2, \mathbb{G}_T)$ be a set of cyclic groups with the same prime order p. A bilinear pairing map $e : \mathbb{G}_1 \times \mathbb{G}_2 \to \mathbb{G}_T$ satisfies the following properties [26]:

1. **Bilinearity**: For any $g \in \mathbb{G}_1, h \in \mathbb{G}_2$, and $a, b \in \mathbb{Z}_p$, it has $e(g^a, h^b) = e(g, h)^{ab}$.
2. **Non-degeneracy**: There exists $g \in \mathbb{G}_1, h \in \mathbb{G}_2$ such that $e(g, h) \neq 1_{\mathbb{G}_T}$.
3. **Computability**: There exists a PPT algorithm to compute $e(g, h)$ for all valid inputs.

Bilinear pairings are primarily categorized into two types: **symmetric** and **asymmetric**. The symmetric type satisfies the condition $\mathbb{G}_1 = \mathbb{G}_2$ with a mapping denoted as $e : \mathbb{G}_1 \times \mathbb{G}_1 \to \mathbb{G}_T$, while the asymmetric type operates between distinct groups $\mathbb{G}_1$ and $\mathbb{G}_2$, characterized by the mapping $e : \mathbb{G}_1 \times \mathbb{G}_2 \to \mathbb{G}_T$. In this work, we use the symmetric bilinear groups to provide the construction of the scheme.

2.3 q-Bilinear Diffie-Hellman Exponent (q-BDHE) Assumption

Let $\mathbb{G}$ be a cyclic group of prime order p with generator g, and $\alpha \in \mathbb{Z}_p$ be a uniformly sampled integer. The (t, ϵ) q-BDHE assumption [24] holds in $\mathbb{G}$ if for all PPT adversaries $\mathcal{A}$ given the elements: $(g, h, g^\alpha, g^{\alpha^2}, \ldots, g^{\alpha^q}, g^{\alpha^{q+2}}, \ldots, g^{\alpha^{2q}}) \in \mathbb{G}^{2q+1}$, the advantage in computing the target value $g^{\alpha^{q+1}}$ has:

$$Adv^{\mathcal{A}}_{\mathsf{q\text{-}BDHE}} = \Pr[\mathcal{A}(g, h, g^\alpha, g^{\alpha^2}, \ldots, g^{\alpha^q}, g^{\alpha^{q+2}}, \ldots, g^{\alpha^{2q}}) = g^{\alpha^{q+1}}] \leq \epsilon(\lambda) \quad (2)$$

where λ denotes the security parameter and the adversary's running time is bounded by t.

2.4 Linear Secret-Sharing Scheme (LSSS)

An LSSS-based access structure can be formally represented as $\{\mathbf{M}_{l\times n}, \rho\}$ [23].

1. Matrix and Mapping Function: $\mathbf{M}_{l \times n}$ denotes a share-generating matrix with l rows and n columns. Each row $i(1 \leq i \leq l)$ in matrix $\mathbf{M}_{l \times n}$ is corresponding to an attribute. The mapping function $\rho : i \to W_i$ assigns a row index i to a specific attribute, such that $\rho(i) = W_i$.
2. Share Distribution Algorithm: Input secret $\alpha \in \mathbb{Z}_p^*$, generates attribute-bound shares λ_i.
    - Randomly select $\{a_2, a_3, \ldots, a_n \xleftarrow{R} \mathbb{Z}_p^*\}$ and generates a $n-$dimensional column vector as $\mathbf{v} = (\alpha, a_2, a_3, \ldots, a_n)^\top$.
    - For the $i-$th row $\mathbf{M}_i$ of matrix $\mathbf{M}$, compute: $\lambda_i = \langle \mathbf{M}_i, \mathbf{v} \rangle$. Share λ_i for attribute $\rho(i)$.
3. Secret Reconstruction: Input attribute set W. If S satisfies the access structure $\mathbb{A}$ (denoted $W \subseteq \mathbb{A}$), outputs the recovered secret α.
    - There exists constants $\{\omega_i\}_{i \in S}$ satisfying: $\sum_{i \in W} \omega_i \mathbf{M}_i = (1, 0, \ldots, 0)$.
    - Recover the secret via linear combination: $\alpha = \sum_{i \in S} \omega_i \lambda_i = \sum_{i \in S} \omega_i \cdot \mathbf{M}_i \cdot \mathbf{v} = (1, 0, \ldots, 0) \cdot (\alpha, a_2, \ldots, a_n)$.

3 Algorithm Definition and Security Model

3.1 Definition of Attributed-Based Adaptor Signature

Definition 2 (Attribute-Based Adaptor Signature Scheme). *Let ABS= (Setup, KGen, Sign, Verify) be signature-policy attribute-based signature scheme and R be a hard relation on language $\mathcal{L}_R = \{Y | \exists y : (Y, y) \in R\}$. We define the attribute-based adaptor signature scheme ABAS=(Setup, KGen, GenR, pSign, pVerify, Adapt, Extract) as follows:*

- $(pp, msk) \leftarrow$ Setup(1^λ): *The system setup algorithm takes as input the security parameter λ, and outputs the system's public parameters pp and the master secret key msk.*
- $sk_W \leftarrow$ KGen(pp, msk, W): *The key generation algorithm takes as input the system's public parameters pp, the master secret key msk, and an attribute set W, outputs the secret key sk_W.*
- $(Y, y) \leftarrow$ GenR(1^λ): *The hard relation algorithm takes as input the security parameter λ, and outputs the statement-witness pair (Y, y).*
- $\tilde{\sigma} \leftarrow$ pSign$(sk_W, m, (\mathbf{M}, \rho), Y)$: *The pre-signature algorithm takes as input the secret key sk_W, message $m \in \{0,1\}^*$, the access policy $(\mathbf{M}, \rho)$ and a statement $Y \in \mathcal{L}_R$, outputs a pre-signature $\tilde{\sigma}$.*
- $1/0 \leftarrow$ pVerify$(pp, m, (\mathbf{M}, \rho), \tilde{\sigma}, Y)$: *The pre-verification algorithm takes as input a system's public parameters pp, message $m \in \{0,1\}^*$, the access policy $(\mathbf{M}, \rho)$, the pre-signature $\tilde{\sigma}$, and the statement $Y \in \mathcal{L}_R$, it outputs 1 if the pre-signature $\tilde{\sigma}$ passes the verification. Otherwise, it outputs 0.*

- $\sigma \leftarrow \mathsf{Adapt}(\tilde{\sigma}, y)$: The adaptor algorithm takes as input the pre-signature $\tilde{\sigma}$ and witness y for the statement Y, outputs an adapted signature σ.
- $y'/\perp \leftarrow \mathsf{Extract}(\tilde{\sigma}, \sigma, Y)$: The extract algorithm takes as input the pre-signature $\tilde{\sigma}$, signature σ and statement $Y \in \mathcal{L}_R$, outputs a witness y' such that $(Y, y') \in R$ or $\perp$.

Besides the signature correctness of ABS= (Setup, KGen, Sign, Verify) scheme, we give the additional correctness for attribute-based adaptor signature:

Definition 3 (Pre-signature correctness). *For any message* $m \in \{0,1\}^*$, *any* (pp, msk) *generated by* $\mathsf{Setup}(1^\lambda)$, *any attribute sets* W, sk_W *generated by* $\mathsf{KGen}(pp, msk, W)$, *any access structure* $(\boldsymbol{M}, \rho)$ *and* $(Y, y) \in R$, *the ABAS scheme satisfies pre-signature correctness if the following equation holds:*

$$\Pr\left[\begin{array}{ll} & (pp, msk) \leftarrow \mathsf{Setup}(1^\lambda); \\ \mathsf{pVerify}(pp, m, (\boldsymbol{M}, \rho), \tilde{\sigma}, Y) = 1; & sk_W \leftarrow \mathsf{KGen}(pp, msk, W); \\ \wedge\ \mathsf{Verify}(pp, m, (\boldsymbol{M}, \rho), \sigma) = 1;\ | & \tilde{\sigma} \leftarrow \mathsf{pSign}(sk_W, m, (\boldsymbol{M}, \rho), Y); \\ \wedge\ \mathsf{VerifyR}(Y, y) = 1. & \sigma \leftarrow \mathsf{Adapt}(\tilde{\sigma}, y); \\ & y \leftarrow \mathsf{Extract}(\tilde{\sigma}, \sigma, Y); \end{array}\right] = 1.$$

3.2 Security Model of Attribute-Based Adaptor Signature

In this section, we present three security properties of attribute-based adaptor signature. The first property, i.e., *pre-signature adaptability*, which guarantees that any valid pre-signature w.r.t hard instance Y can deterministically convert this pre-signature into a full signature that verifies under the original signature verification algorithm using a witness y with $(Y, y) \in R$. The formal definition of pre-signature adaptability is described as follows:

Definition 4 (Pre-signature adaptability). *An ABAS scheme* Π_{ABAS} *satisfies pre-signature adaptability if for any message* $m \in \{0,1\}^*$, *any* (pp, msk), *any attribute sets* W, *any access structure* $(\boldsymbol{M}, \rho)$, *any statement/witness pair* $(Y, y) \in R$, *and any pre-signature* $\tilde{\sigma} \in \{0,1\}^*$ *with* $\mathsf{pVerify}(pp, m, (\boldsymbol{M}, \rho), \tilde{\sigma}, Y) = 1$, *we have* $\mathsf{Verify}(pp, m, (\boldsymbol{M}, \rho), \mathsf{Adapt}(\tilde{\sigma}, y)) = 1$.

The second security requirement, i.e., *adaptor existential unforgeability* in the selective policy model under the adaptive chosen-message attack (sP-aEUF-CMA), which will provide the guarantee of the security for the signer. It guarantees that any party who does not possess the witness y cannot feasibly forge a full signature for a message m, even when given access to valid pre-signatures corresponding to m. In the existential unforgeability security model, a PPT adversary $\mathcal{A}$, which simulates a realworld attack, aims to forge a valid signature without possessing the secret. The challenger $\mathcal{B}$, emulating the system's operation, leverages the adversary's computation power to solve a mathematically hard problem. The security model is formally defined through the following interactive game between $\mathcal{A}$ and $\mathcal{B}$.

Definition 5. *(sP-aEUF-CMA unforgeability)* An ABAS scheme Π_{ABAS} is sP-aEUF-CMA secure if for every PPT adversary $\mathcal{A}$ there exists a negligible function ϵ such that: $\Pr[aSigForge(\lambda) = 1] \leq \epsilon(\lambda)$, where is defined as follows:

1. **Initialization Phase**: The adversary $\mathcal{A}$ declares a challenge access policy (M^*, ρ^*) to challenger $\mathcal{B}$.
2. **System Setup Phase**: Challenger $\mathcal{B}$ runs Setup(1^λ) to generate public parameters pp and a master secret key msk, executes GenR(1^λ) to produce $(Y, y) \in R$, sends pp and Y to adversary $\mathcal{A}$, and retains msk to answer $\mathcal{A}$'s private key, pre-signature, and signature queries.
3. **Private Key Query Phase**: When adversary $\mathcal{A}$ select an attribute set W and requests the corresponding private key, the set W must not satisfy the target access policy (M^*, ρ^*). Challenger $\mathcal{B}$ executes KGen to generate sk_W and returns it to $\mathcal{A}$.
4. **Pre-signature Query Phase**: The adversary $\mathcal{A}$ adaptively selects message m_i to request pre-signatures. For each query, challenger $\mathcal{B}$ performs the following steps: First adds m_i to the pre-signature query list $\mathcal{Q}$, then executes the pSign algorithm to generate the corresponding pre-signature $\tilde{\sigma}_i$, and finally sends both m_i and $\tilde{\sigma}_i$ to $\mathcal{A}$.
5. **Signature Query Phase**: The adversary $\mathcal{A}$ adaptively selects message m_i and access policy (M, ρ) to request signature. Challenger $\mathcal{B}$ adds each queries message m_i to signature query list $\mathcal{Q}$, runs the Sign algorithm to generate the corresponding signature σ_i and returns the tuple (m_i, σ_i) to $\mathcal{A}$.
6. **Signature Forgery Phase**: The adversary $\mathcal{A}$ selects a target message m^* to forge a full signature and submits it to the challenger $\mathcal{B}$. The challenger first executes KGen(pp, msk, W) to generate the private key sk_W, then runs pSign($sk_W, m^*, (M^*, \rho^*), Y$) to produce a pre-signature $\tilde{\sigma}^*$, which is sent back to $\mathcal{A}$. Subsequently, $\mathcal{A}$ outputs a forged signature σ^* for m^*. The game concludes as follows:
   - If Verify($pp, m^*, (M^*, \rho^*), \sigma^*) = 1$ and $m^* \notin \mathcal{Q}, (M^*, \rho^*)$ has never been submitted as a query, $\mathcal{A}$ wins the game(output 1);
   - Otherwise, the challenger prevails(output 0);

The third security requirement is called *witness extractability*, which guarantees that a valid pre-signature/signature $(\tilde{\sigma}, \sigma)$ can effectively extract the secret witness y. The security of witness extractability is formally defined through an interactive game $WitExt(1^\lambda)$ between a challenger $\mathcal{B}$ and an adversary $\mathcal{A}$, which is described as follows:

Definition 6 (Witness extractability). An ABAS scheme Π_{ABAS} is witness extractable if for every PPT adversary $\mathcal{A}$ there exists a negligible function ϵ such that: $\Pr[WitExt(\lambda) = 1] \leq \epsilon(\lambda)$, where is defined as follows:

1. **Initialization**: Challenger $\mathcal{B}$ runs Setup(1^λ) to generate public parameters pp and a master secret key msk, sends pp to adversary $\mathcal{A}$, and retains msk to answer $\mathcal{A}$'s private key, pre-signature, and signature queries. The adversary $\mathcal{A}$ executes GenR(1^λ) to generate $(Y, y) \in R$, publishes the statement Y, and retains the witness y as secret.

2. **Private Key Query**: When adversary $\mathcal{A}$ select an attribute set W and requests the corresponding private key, the set W must not satisfy the target access policy $(\boldsymbol{M}^*, \rho^*)$. Challenger $\mathcal{B}$ executes KGen to generate sk_W and returns it to $\mathcal{A}$.
3. **Pre-signature Query**: The adversary $\mathcal{A}$ adaptively selects message m_i to request pre-signatures. For each query, challenger $\mathcal{B}$ performs the following steps: First adds m_i to the pre-signature query list $\mathcal{Q}$, then executes the pSign algorithm to generate the corresponding pre-signature $\tilde{\sigma}'_i$, and finally sends both m_i and $\tilde{\sigma}'_i$ to $\mathcal{A}$.
4. **Signature Query**: The adversary $\mathcal{A}$ adaptively selects message m_i to request signature. Challenger $\mathcal{B}$ adds each queries message m_i to signature query list $\mathcal{Q}$, runs the Sign algorithm to generate the corresponding signature σ_i and returns the tuple (m_i, σ_i) to $\mathcal{A}$.
5. **Witness Extract**: The adversary $\mathcal{A}$ selects a target message m^* and statement Y^*, sends them to the challenger $\mathcal{B}$. The challenger $\mathcal{B}$ runs $\mathsf{KGen}(pp, msk, W)$ to generate the private key sk_W, then runs $\mathsf{pSign}(sk_W, m^*, Y)$ to produce a pre-signature $\tilde{\sigma}^*$, which is sent back to $\mathcal{A}$. Subsequently, $\mathcal{A}$ generates a full signature σ^* for m^* and submits it to $\mathcal{B}$. The challenger $\mathcal{B}$ then executes $\mathsf{Extract}(\tilde{\sigma}^*, \sigma^*, Y^*)$ to extract a witness y'. The game concludes as follows:
    - If $\mathsf{Verify}(pp, m^*, (\boldsymbol{M}^*, \rho^*), \sigma^*) = 1$, while $m^* \notin \mathcal{Q}$, $(Y^*, y') \notin R$ and $(\boldsymbol{M}^*, \rho^*) \notin$, $\mathcal{A}$ wins the game(output 1);
    - Otherwise, the challenger prevails(output 0);

We also give the anonymity of the security of ABAS: given two attribute sets W_1, W_2 satisfying policy $(\mathbf{M}, \rho)$, no PPT adversary can distinguish(with non-negligible advantage) whether a signature was generated under W_1 or W_2.

Definition 7 (Anonymity). *An ABAS scheme Π_{ABAS} is anonymous if for any PPT adversary $\mathcal{A}$, the advantage $Adv_{\mathcal{A}}^{anon}(\lambda)$ in winning the anonymity game is negligible in the security parameter λ, where is defined as follows:*

1. **System Setup**: Challenger $\mathcal{B}$ executes the Setup algorithm to initialize the cryptographic framework. The algorithm accepts a security parameter λ as input, generates the public key pp and master secret key msk, and explicitly discloses both pp and msk to the adversary $\mathcal{A}$. Since adversary $\mathcal{A}$ possesses the master secret key msk, it can generate corresponding secret keys and signatures autonomously.
2. **Challenge**: Adversary $\mathcal{A}$ selects a target message m^* along with two distinct attribute sets W_1^* and W_2^*, both rigorously satisfying the same access control policy $(\boldsymbol{M}^*, \rho^*)$. $\mathcal{A}$ submits tuple (m^*, W_1^*, W_2^*) to challenger $\mathcal{B}$. Challenger $\mathcal{B}$ randomly selects a bit $b \in \{0, 1\}$, executes the Sign algorithm to generate a signature σ^* on message m^* with respect to the attribute set W_b^*, and sends σ^* back to $\mathcal{A}$.
3. **Guess Phase**: adversary $\mathcal{A}$ attempts to determine which attribute set(W_1^* or W_2^*) was used to generate the signature σ^*. Adversary $\mathcal{A}$ outputs a guess

$b' \in \{0,1\}$. If $b' = b$, $\mathcal{A}$ wins the security game. The advantage of adversary $\mathcal{A}$ in winning the security game is formally defined as: $Adv_{\mathcal{A}}^{anony}(\lambda) = |\Pr[b' = b] - \frac{1}{2}|$.

Definition 8. *An ABAS scheme Π_{ABAS} is secure, if it is sP-aEUF-CMA secure, pre-signature adaptable, witness extractable and anonymous.*

4 Attribute-Based Adaptor Signature Scheme

This section first constructs an attribute-based adaptor signature ABAS based on the Waters signature, then formally proves that it achieves all security objectives of the definition of adaptor signature in Sect. 3.2.

4.1 Construction of Attribute-Based Adaptor Signature Scheme

The concrete construction of our ABAS scheme(as shown in Fig. 1.) is as follows:

- $(pp, msk) \leftarrow \text{Setup}(1^\lambda)$: The Attribute Authority(AA) does: (1) take a security parameter λ as intake and generate two multiplicative cyclic groups $\mathbb{G}, \mathbb{G}_T$ with prime order p to define a bilinear pairing $e : \mathbb{G} \times \mathbb{G} \to \mathbb{G}_T$. (2) randomly select $g, h_1, h_2, \ldots, h_n$ from $\mathbb{G}$ (g is a generator of $\mathbb{G}$) and select a collision-free hash function $H : \{0,1\}^* \to \mathbb{G}$. (3) set $U = \{1, 2, \ldots, n\}$ as the attribute universe. (4) at random, pick $a, \alpha \in \mathbb{Z}_p^*$ and compute $g_1 = g^a, msk = g^\alpha$ and $z = e(g,g)^\alpha$. Thus, the public parameters are $pp = (g, g_1, e, p, \mathbb{G}, z, h_1, \ldots, h_n, H)$, and the master key as $msk = g^\alpha$.
- $(Y, y) \leftarrow \text{GenR}(1^\lambda)$: The verifier randomly selects $y \in \mathbb{Z}_p^*$, computes $Y = g^y$, and outputs the statement-witness pair $(Y, y) \in R$.
- $sk_W \leftarrow \text{KGen}(pp, msk, W)$: The AA combines the system public key pp and the master private key msk, randomly selects a element $t \in \mathbb{Z}_p$, and computes the private key $sk_W = (K_1, K_2, \{K_x\}_{x \in W})$: $K_1 = g^\alpha g^{at}, K_2 = g^t, K_x = h_x^t, \forall x \in W$.
- $\tilde{\sigma} \leftarrow \text{pSign}(sk_W, m, (\mathbf{M}, \rho), Y)$: The signer does: (1) select an access control structure $\mathbb{A} = (\mathbf{M}, \rho)$, where $\mathbf{M}$ is an $l \times n$ matrix, and each row of the matrix corresponds to an attribute $\rho(i)$ in the access expression. (2) compute a vector $\mathbf{a} = (a_1, a_2, \ldots, a_l)$ such that $\sum_{i \in [l]} a_i \mathbf{M}_i = (1, 0, \ldots, 0)$ ($\mathbf{M}_i$ denotes the $i-th$ row of matrix $\mathbf{M}$), with the constraint that $a_i = 0$ if $\rho(i) \notin W$. Additionally, it selects another vector $\mathbf{b} = (b_1, b_2, \ldots, b_l)$ such that $\sum_{i \in [l]} b_i \mathbf{M}_i = (0, 0, \ldots, 0)$. (3) randomly select $\theta, r \in \mathbb{Z}_p$, compute the pre-signature $\tilde{\sigma} = (\tilde{\sigma}_1, \tilde{\sigma}_2, \tilde{\sigma}_3, \{\tilde{\sigma}_i'\}_{i \in [l]})$: $\tilde{\sigma}_1 = g^\theta, R = g^r, \tilde{\sigma}_2 = K_1 \cdot \prod_{i \in [l]} K_x^{a_i} h_x^{b_i} \cdot H(m\|R \cdot Y)^\theta, \tilde{\sigma}_3 = r, \tilde{\sigma}_i' = g^{b_i} \cdot K_2^{a_i}, \forall i \in [l]$.
- $1/0 \leftarrow \text{pVerify}(pp, m, (\mathbf{M}, \rho), \tilde{\sigma}, Y)$: The verifier does: (1) Randomly select $c_2, \ldots, c_n \in \mathbb{Z}_p^*$, set $c_1 = 1$, and compute $\omega_i = \sum_{j \in [n]} c_j \mathbf{M}_{i,j}, \forall i \in [l]$. (2) if the following equation (3) holds, then the verification is successful; otherwise, it will be rejected, that is,

$$\frac{e(\tilde{\sigma}_2, g)}{(\prod_{i \in [l]} e(g_1^{\omega_i} h_{\rho(i)}, \tilde{\sigma}_i')) e(H(m\|g^{\tilde{\sigma}_3} \cdot Y), \tilde{\sigma}_1)} = z. \quad (3)$$

- $\sigma \leftarrow \mathsf{Adapt}(\tilde{\sigma}, y)$: Let $\tilde{\sigma} = (\tilde{\sigma}_1, \tilde{\sigma}_2, \tilde{\sigma}_3, \{\tilde{\sigma}'_i\}_{i \in [l]})$. Given a pre-signature $\tilde{\sigma}$ and a witness y, the verifier outputs a signature as follows: $\sigma_3 = \tilde{\sigma}_3 + y, \sigma_1 = \tilde{\sigma}_1, \sigma_2 = \tilde{\sigma}_2, \sigma'_i = \tilde{\sigma}'_i$. Return completed signature $\sigma = (\sigma_1, \sigma_2, \sigma_3, \{\sigma'_i\}_{i \in [l]})$.
- $y'/\perp \leftarrow \mathsf{Extract}(\tilde{\sigma}, \sigma, Y)$: Given a pre-signature $\tilde{\sigma}$, a completed signature σ and a statement Y, the signer computes the witness y': $y' = \sigma_3 - \tilde{\sigma}_3$. If $\mathsf{VerifyR}(Y, y) = 1$, return y'; otherwise, return $\perp$.

Lemma 1 (Pre-signature correctness). *The ABAS scheme satisfies pre-signature correctness.*

Proof. Let us fix arbitrary $\alpha, y \in \mathbb{Z}_p^*$, $m \in \{0,1\}^*$, and define $msk = g^\alpha$ and $Y = g^y$. For any attribute sets W, private keys $sk_W \leftarrow \mathsf{KGen}(pp, msk, W)$, access structure $(\mathbf{M}, \rho)$, and pre-signature $\tilde{\sigma} \leftarrow \mathsf{pSign}(sk_W, m, (\mathbf{M}, \rho), Y)$ it holds that equation (3), for some $\theta, r \in \mathbb{Z}_p$. Since

$$e(\tilde{\sigma}_2, g) = e(K_1 \cdot \prod_{i \in [l]} K_x^{a_i} h_x^{b_i} \cdot H(m\|R \cdot Y)^\theta, g)$$

$$= e(g^\alpha g^{at}, g) \cdot e(\prod_{i \in [l]} h_x^{ta_i + b_i}, g) \cdot e(H(m\|R \cdot Y)^\theta, g) \quad (4)$$

$$= e(g,g)^\alpha \cdot e(g^{at}, g) \cdot e(\prod_{i \in [l]} h_x^{ta_i + b_i}, g) \cdot e(H(m\|R \cdot Y)^\theta, g).$$

$$\prod_{i \in [l]} e(g_1^{\omega_i} h_{\rho(i)}, \tilde{\sigma}'_i) = \prod_{i \in [l]} e(g_1^{\omega_i} h_{\rho(i)}, g^{b_i} \cdot K_2^{a_i})$$

$$= \prod_{i \in [l]} e(g^{a\omega_i}, g^{b_i + ta_i}) e(h_{\rho(i)}, g^{b_i + ta_i})$$

$$= \prod_{i \in [l]} e(h_{\rho(i)}, g^{b_i + ta_i}) \cdot e(g^a, g)^{\sum_{i \in [l]} (b_i + ta_i)\omega_i} \quad (5)$$

$$= \prod_{i \in [l]} e(h_{\rho(i)}, g^{b_i + ta_i}) \cdot e(g^a, g)^{\sum_{i \in [l]} c_i (b_i + ta_i) \mathbf{M}_{i,j}}$$

$$= \prod_{i \in [l]} e(h_\rho(i), g^{b_i + ta_i}) \cdot e(g^a, g)^t.$$

then, the pre-signature verification equation has:

$$\frac{e(\tilde{\sigma}_2, g)}{\prod_{i \in [l]} e(g_1^\omega h_{\rho(i)}, \tilde{\sigma}'_i) \cdot e(H(m\|g^{\tilde{\sigma}_3} \cdot Y), \tilde{\sigma}_1)}$$

$$= \frac{e(g,g)^\alpha \cdot e(g^{at}, g) \cdot e(\prod_{i \in [l]} h_x^{ta_i + b_i}, g) \cdot e(H(m\|R \cdot Y)^\theta, g)}{\prod_{i \in [l]} e(h_\rho(i), g^{b_i + ta_i}) \cdot e(g^a, g)^t \cdot e(H(m\|R \cdot Y), g^\theta)} \quad (6)$$

$$= e(g,g)^\alpha = z.$$

Noticely, we have $\mathsf{pVerify}(pp, m, (\mathbf{M}, \rho), \tilde{\sigma}, Y) = 1$, from equation (6). By the requirement of pre-signature adaptability, this implies that $\mathsf{Verify}(pp, m, (\mathbf{M}, \rho), \sigma) = 1$ for the adaptor: $\mathsf{Adapt}(\tilde{\sigma}, y)$ outputs σ where $(\sigma_1, \sigma_2, \sigma_3, \sigma'_i) = (\sigma_1, \sigma_2, r + y, \sigma'_i)$. Finally, $\mathsf{Extract}(\tilde{\sigma}, \sigma, Y) = (r + y) - r = y$.

$(pp, msk) \leftarrow \mathsf{Setup}(\lambda)$	$sk_W \leftarrow \mathsf{KGen}(pp, msk, W)$
$1: a, \alpha \leftarrow \mathbb{Z}_p^*$	$1: t \leftarrow \mathbb{Z}_p$
$2: g \leftarrow \mathbb{G} \;//\; \text{a generator of } \mathbb{G}$	$2: K_1 = g^\alpha g^{at}$
$3: g_1 = g^a$	$3: K_2 = g^t$
$4: h_1, h_2, \ldots, h_n \leftarrow \mathbb{G}$	$4: K_x = h_x^t, \forall x \in W$
$5: H: \{0,1\}^* \leftarrow \mathbb{G}$	$5: sk_W = (K_1, K_2, \{K_x\}_{x \in W})$
$6: z = e(g,g)^\alpha$	
$7: pp = (g, g_1, e, p, \mathbb{G}, \mathbb{G}_T, z, h_1, \ldots, h_n, H)$	$(Y, y) \leftarrow \mathsf{GenR}(1^\lambda)$
$8: msk = g^\alpha$	$1: y \leftarrow \mathbb{Z}_p^*$
	$2: Y = g^y$

$\tilde{\sigma} \leftarrow \mathsf{pSign}(sk_W, m, (\mathbf{M}, \rho), Y)$

1 : compute a vector $\mathbf{a} = (a_1, a_2, \ldots, a_l)$,
   s.t. $\sum_{i \in [l]} a_i \mathbf{M}_i = (1, 0, \ldots, 0)$

$1/0 \leftarrow \mathsf{pVerify}(pp, m, (\mathbf{M}, \rho), \tilde{\sigma}, Y)$

1 : $c_2, \ldots, c_n \leftarrow \mathbb{Z}_p^*, c_1 = 1$

2 : compute a vector $\mathbf{b} = (b_1, b_2, \ldots, b_l)$,
   s.t. $\sum_{i \in [l]} b_i \mathbf{M}_i = (0, 0, \ldots, 0)$

2 : $\omega_i = \sum_{j \in [n]} c_j \mathbf{M}_{i,j}, \forall i \in [l]$

3 : if
$$\frac{e(\tilde{\sigma}_2, g)}{\prod_{i \in [l]} e(g_1^{\omega_i} h_{\rho(i)}, \tilde{\sigma}_i') e(H(m \| g^{\tilde{\sigma}_3} \cdot Y), \tilde{\sigma}_1)} = z$$
   return 1; else return 0

3 : $\theta, r \leftarrow \mathbb{Z}_p$
4 : $\tilde{\sigma}_1 = g^\theta, R = g^r$
5 : $\tilde{\sigma}_2 = K_1 \cdot \prod_{i \in [l]} K_x^{a_i} h_x^{b_i} \cdot H(m \| R \cdot Y)^\theta$
6 : $\tilde{\sigma}_3 = r$
7 : $\tilde{\sigma}_i' = g^{b_i} \cdot K_2^{a_i}, \forall i \in [l]$
8 : $\tilde{\sigma} = (\tilde{\sigma}_1, \tilde{\sigma}_2, \tilde{\sigma}_3, \{\tilde{\sigma}_i'\}_{i \in [l]})$

$\sigma \leftarrow \mathsf{Adapt}(\tilde{\sigma}, y)$

1 : $\sigma_3 = \tilde{\sigma}_3 + y$
2 : $\sigma_1 = \tilde{\sigma}_1, \sigma_2 = \tilde{\sigma}_2, \sigma_i' = \tilde{\sigma}_i'$
3 : $\sigma = (\sigma_1, \sigma_2, \sigma_3, \{\sigma_i'\}_{i \in [l]})$

$y' \leftarrow \mathsf{Extract}(\tilde{\sigma}, \sigma, Y)$

1 : $y' = \sigma_3 - \tilde{\sigma}_3$
2 : if $\mathsf{VerifyR}(Y, y) = 1$, return y'
   else return $\bot$

Fig. 1. Construction of ABAS adaptor signature.

5 Security

In this section, we give the formal proof of the security properties of our scheme.

Lemma 2 (Pre-signature adaptability). *The ABAS scheme satisfies the pre-signature adaptability.*

Proof. Let fix arbitrary $\alpha, y \in \mathbb{Z}_p^*, m \in \{0,1\}^*$, $\mathsf{msk} = g^\alpha$, and $(\tilde{\sigma}_1, \tilde{\sigma}_2, \tilde{\sigma}_3, \tilde{\sigma}_i') \in \mathbb{G} \times \mathbb{G} \times \mathbb{Z}_p \times \mathbb{G}$. We have $Y = g^y$, and $\sigma_3 = \tilde{\sigma}_3 + y$. Assuming that $\mathsf{pVerify}(pp, m, (\mathbf{M}, \rho), \tilde{\sigma}, Y) = 1$, we have

$$\frac{e(\tilde{\sigma}_2, g)}{\prod_{i \in [l]} e(g_1^\omega h_{\rho(i)}, \tilde{\sigma}_i') \cdot e(H(m \| g^{\tilde{\sigma}_3} \cdot Y), \tilde{\sigma}_1)}$$
$$= \frac{e(g,g)^\alpha \cdot e(g^{at}, g) \cdot e(\prod_{i \in [l]} h_x^{ta_i + b_i}, g) \cdot e(H(m \| R \cdot Y)^\theta, g)}{\prod_{i \in [l]} e(h_\rho(i), g^{b_i + ta_i}) \cdot e(g^a, g)^t \cdot e(H(m \| g^{\sigma_3}), g^\theta)} \quad (7)$$
$$= \frac{e(\sigma_2, g)}{\prod_{i \in [l]} e(g_1^\omega h_{\rho(i)}, \sigma_i') \cdot e(H(m \| g^{\sigma_3}), \sigma_1)}$$
$$= e(g,g)^\alpha = z.$$

which implies that $\mathsf{Verify}(pp, m, \sigma, (\mathbf{M}, \rho)) = 1$ from equation (7).

Lemma 3. *(sP-aEUF-CMA security) Assuming that the hash function H is modeled as a random oracle, q-BDHE problem is hard, and R is a hard relation, the ABAS scheme is sP-aEUF-CMA secure.*

Proof. If there exists a PPT adversary $\mathcal{A}$ capable of winning the aforementioned security game under chosen-message attack with non-negligible advantage ϵ, we can construct a PPT challenger $\mathcal{B}$ that leverages $\mathcal{A}$'s capability to solve the q-BDHE problem. Specifically, upon receiving a random q-BDHE instance $(g, h, g^a, g^{a^2}, \ldots, g^{a^q}, g^{a^{q+2}}, \ldots, g^{a^{2q}}) \in \mathbb{G}^{2q+1}$, $\mathcal{B}$ must compute $g^{a^{q+1}}$ within polynomial time. The adversarial interaction between $\mathcal{A}$ and $\mathcal{B}$ proceeds as follows:

1. **Initialization**: The adversary $\mathcal{A}$ gives the challenger $\mathcal{B}$ the challenge access structure $(\mathbf{M}^*, \rho^*)$.
2. **System Setup**: The challenger $\mathcal{B}$ first defines the attribute universe $U = \{1, 2, \ldots, n\}$, computes $g^i = g^{a^i}$ for all $i \in U$, randomly selects $\alpha' \in \mathbb{Z}_p$, and sets $z = e(g_1, g_n) \cdot e(g,g)^{\alpha'}$ by letting $\alpha = a^{n+1} + \alpha'$. Next, let $c = g^a = g_1$, and $\mathcal{B}$ selects a hash function $H : \{0,1\}^* \to \mathbb{G}$. For each attribute $x \in U$, the $\mathcal{B}$ randomly selects $z_x \in \mathbb{Z}_p$. If attribute x corresponds to the i-th row in $\mathbf{M}^*$ (i.e., $\rho^*(i) = x$), then compute: $h_x = g^{z_x} \prod_{j \in [n]} g_j^{-\mathbf{M}_{i,j}^*}$, otherwise, set $h_x = g^{z_x}$. Then, the $\mathcal{B}$ runs $\mathsf{GenR}(1^\lambda)$ to generate a statement-witness pair (Y, y) satisfying $Y = g^y$. Finally, $\mathcal{B}$ provides $pp = (U, n, g, c, z, H, \{h_x\}_{x \in U})$ and Y to the adversary $\mathcal{A}$. The message query set $\mathcal{Q}_2$ is initialized as empty.
3. **Hash Query**: The $\mathcal{A}$ may issue hash queries to the $\mathcal{B}$. The $\mathcal{B}$ maintains an initially empty list $\mathcal{H}_{List}$. When $\mathcal{A}$ queries $H(m_i \| R_i)$, $\mathcal{B}$ first checks if the pair (m_i, R_i) exists in $\mathcal{H}_{List}$. If recorded, $\mathcal{B}$ returns the precomputed value $H(m_i \| R_i) = g^{\zeta_i}$ directly from $\mathcal{H}_{List}$. Otherwise, $\mathcal{B}$ randomly select $\zeta_i \in \mathbb{Z}_p$, computes $H(m_i \| R_i) = g_q g^{\zeta_i} = g^{a^q + \zeta_i}$, sends this value to $\mathcal{A}$, and appends the tuple $(m_i, R_i, H(m_i, R_i), T_1)$ to $\mathcal{H}_{List}$, where the tag T_1 explicitly indicates this entry was created during hash query operations. This simulation ensures consistency across adaptive queries while preserving the random oracle's programmability in security reductions.

4. **Private Key**: The $\mathcal{A}$ submits an attribute set W to $\mathcal{B}$ to request the corresponding private key. The queries attribute set W must not satisfy the access policy $(\mathbf{M}^*, \rho^*)$ that $\mathcal{A}$ intends to attack. To generate the private key sk_W for W, $\mathcal{B}$ proceeds as follows:
Since W does not satisfy $(\mathbf{M}^*, \rho^*)$, there exists a vector $\boldsymbol{\omega} = (\omega_1, \omega_2, \ldots, \omega_n)$ with $\omega_1 = -1$ such that $\boldsymbol{\omega} \cdot \mathbf{M}_i^* = 0$. The $\mathcal{B}$ randomly selects $r \in \mathbb{Z}_p$, compute $t = r + \omega_1 a^q + \omega_2 a^{q-1} + \ldots + \omega_n \cdot a^{q-n+1}$, and defines the private key components as shown in equations (8)(9)

$$K_2 = g^t = g^r \cdot \prod_{i=1,\ldots,n} (g^{a^{q+1-i}})^{\omega_i}, \tag{8}$$

$$K_1 = g^{\alpha'} g^{ar} \cdot \prod_{i=2,\ldots,n} (g^{a^{q+2-i}})^{\omega_i}. \tag{9}$$

From K_x, its computation diverges based on the following two cases:
Case 1: If for every row vector i in $\mathbf{M}^*$, $\rho(i) \neq x$(i.e., attribute x is not associated with $\mathbf{M}^*$), then: $K_x = h_x^t = (g^{z_t})^t = (g^t)^{z_x} = K_2^{z_x}$;
Case 2: If there exists a row vector i in $\mathbf{M}^*$ such that $\rho(i) = x$(i.e., x is embedded in the target policy), then: $K_x = K_2^{z_x} \cdot \prod_{j=1,\ldots,n}(g^{a_j \cdot r} \prod_{k=1,\ldots,n k \neq j}(g^{a^{q+1+j-k}})^{\omega_k})^{-\mathbf{M}_{i,j}^*}$.
Thus, the private key w.r.t attribute set W is: $sk_W = (K_2, K_1, \{K_x\}_{x \in W})$.

5. **Pre-signature Query**: The adversary $\mathcal{A}$ query $\mathcal{B}$ for pre-signatures associated with an access control policy $(\mathbf{M}, \rho)$ and a message m. Then, $\mathcal{B}$ generates an attribute set W satisfying $(\mathbf{M}, \rho)$, and verifies whether W satisfies the Challenge policy $(\mathbf{M}^*, \rho^*)$ selected by $\mathcal{A}$.
Case 1: If W fails to satisfy $(\mathbf{M}^*, \rho^*)$, $\mathcal{B}$ obtains the corresponding private key sk_W via a private key query. Subsequently, $\mathcal{B}$ executes the pSign algorithm by taking sk_W and $(\mathbf{M}, \rho)$ as inputs, running the pre-signature algorithm defined in the scheme to pre-signature the message m and returning the resulting pre-signature to $\mathcal{A}$.
Case 2: If W satisfies $(\mathbf{M}^*, \rho^*)$, $\mathcal{B}$ generates a pre-signature for message m under access control policy $(\mathbf{M}, \rho)$ as follows:
First, $\mathcal{B}$ computes a vector $\mathbf{a} = (a_1, a_2, \ldots, a_l)$ such that $\sum_{i \in [l]} a_i \mathbf{M}_i = 1$, where $a_i = 0$ for all induces i where $\rho(i) \notin W$. Concurrently, $\mathcal{B}$ constructs a complementary vector $\mathbf{b} = (b_1, b_2, \ldots, b_l)$ satisfying $\sum_{i \in [l]} b_i \mathbf{M}_i = 0$, ensuring algebraic consistency with the policy structure.
Next, $\mathcal{B}$ randomly selects $r_i, \zeta_i \in \mathbb{Z}_p$ for each i, computes $R_i = g^{r_i}$, and verifies whether the tuples $(R_i Y, m_i, T_1)$ or $(R_i Y, m_i, T_3)$ already exist in the hash query log $\mathcal{H}_{List}$. If either tuple is found, the simulation is aborted to prevent cryptographic collisions. Otherwise, $\mathcal{B}$ programmatically defines the hash output as $H(m_i || R_i Y) = g_q g^{\zeta_i} = g^{a^q + \zeta_i}$, embedding the trapdoor a into the hash response, and records $(R_i Y, m_i, \zeta_i, H(m_i || R_i Y), T_2)$ in $\mathcal{H}_{List}$ to maintain random oracle consistency, where T_2 marks the corresponding entry to be added to $\mathcal{H}_{List}$ during the pre-signature query phase.
Subsequently, $\mathcal{B}$ selects θ' and t uniformly from $\mathbb{Z}_p$, setting $\theta = \theta' - a$ to blind the master secret α during pre-signature computation. Finally, the

pre-signature $\tilde{\sigma}$ is generated by algorithmically combining these components under the scheme's prescribed algebraic constraints, ensuring binding to $(\mathbf{M}, \rho)$ while concealing critical secrets such as sk_W and α. This process rigorously enforces security against adaptive chosen-policy attacks while preserving simulation indistinguishability in the standard model. The pre-signature is then generated as follows equations (10):

$$\tilde{\sigma}_1 = g^\theta = g^{\theta'} g_1^{-1},$$
$$\tilde{\sigma}_2 = g^\alpha g^{at} \cdot \prod_{i \in [l]} h_x^{ta_i + b_i} \cdot H(m\|R \cdot Y)^\theta$$
$$= g^{\alpha'} \cdot \prod_{i \in [l]} h_x^{ta_i + b_i} \cdot g^{(a^q + \zeta)\theta'} \cdot g_1^{t-\zeta}, \qquad (10)$$
$$\tilde{\sigma}_3 = r,$$
$$\tilde{\sigma}'_i = g^{b_i} K_2^{a_i} = g^{b_i + ta_i}.$$

Finally, $\mathcal{B}$ returns the pre-signature $\tilde{\sigma} = (\tilde{\sigma}_1, \tilde{\sigma}_2, \tilde{\sigma}_3, \{\tilde{\sigma}'_i\}_{i \in W})$ corresponding to message m.

6. **Signature Query**: The adversary $\mathcal{A}$ query $\mathcal{B}$ for signatures associated with an access control policy $(\mathbf{M}, \rho)$ and a message m. Then, $\mathcal{B}$ generates an attribute set W satisfying $(\mathbf{M}, \rho)$, and verifies whether W satisfies the Challenge policy $(\mathbf{M}^*, \rho^*)$ selected by $\mathcal{A}$. **Case 1**: The same as case 1 of the pre-signature. **Case 2**: If W satisfies $(\mathbf{M}^*, \rho^*)$, $\mathcal{B}$ generates a signature for message m under access control policy $(\mathbf{M}, \rho)$ as follows: First, the same as case 2 of the pre-signature.

Next, $\mathcal{B}$ randomly selects $r_i, \zeta_i \in \mathbb{Z}_p$ for each i, computers $R_i = g^{r_i}$, and verifies whether the tuples (R_i, m_i, T_1) or (R_i, m_i, T_2) already exist in the hash query log $\mathcal{H}_{List}$. If either tuple is found, the simulation is aborted to prevent cryptographic collisions. Otherwise, $\mathcal{B}$ programmatically defines the hash output as $H(m_i\|R_i) = g_q g^{\zeta_i} = g^{a^q + \zeta_i}$, embedding the trapdoor a into the hash response, and records $(R_i, m_i, \zeta_i, H(m_i\|R_i), T_3)$ in $\mathcal{H}_{List}$ to maintain random oracle consistency, where T_3 marks the corresponding entry to be added to $\mathcal{H}_{List}$ during the signature query phase.

Subsequently, $\mathcal{B}$ selects θ' and t uniformly from $\mathbb{Z}_p$, setting $\theta = \theta' - a$ to blind the master secret α during signature computation. The signature is then generated as follows equations (11):

$$\sigma_1 = g^\theta = g^{\theta'} g_1^{-1},$$
$$\sigma_2 = g^\alpha g^{at} \cdot \prod_{i \in [l]} h_x^{ta_i + b_i} \cdot H(m\|R \cdot Y)^\theta$$
$$= g^{\alpha'} \cdot \prod_{i \in [l]} h_x^{ta_i + b_i} \cdot g^{(a^q + \zeta)\theta'} \cdot g_1^{t-\zeta}, \qquad (11)$$
$$\sigma_3 = r + y,$$
$$\sigma'_i = g^{b_i} K_2^{a_i} = g^{b_i + ta_i}.$$

Finally, $\mathcal{B}$ returns the signature $\sigma = (\sigma_1, \sigma_2, \sigma_3, \{\sigma'_i\}_{i \in W})$ corresponding to message m.

7. **Forgery Phase**: The adversary $\mathcal{A}$ sends the message m^* and the access control policy $(\mathbf{M}^*, \rho^*)$ to the challenger $\mathcal{B}$. $\mathcal{B}$ generates a valid pre-signature for m and $(\mathbf{M}^*, \rho^*)$ using the method defined in the pre-signature query phase, and sends it to $\mathcal{A}$, then $\mathcal{A}$ outputs a forged signature $\sigma = (\sigma_1, \sigma_2, \sigma_3, \{\sigma'_i\}_{i \in [l]})$ for m^* and $(\mathbf{M}^*, \rho^*)$.

If $\sigma^* = (\sigma_1^*, \sigma_2^*, \sigma_3^*, \{\sigma'^*_i\}_{i \in [l]})$ pass verification and the message m^* and $(\mathbf{M}^*, \rho^*)$ have never been queried, then the $\mathcal{A}$ successfully forges a signature. Since σ^* is a valid signature, it satisfies the verification algorithm as (12):

$$\sigma^* = (\sigma_1^*, \sigma_2^*, \sigma_3^*, \{\sigma'^*_i\}_{i \in [l]}) = (g^{\theta^*}, g^{\alpha} g^{at^*} \cdot \prod_{i \in [l]} h_x^{t^* a_i^* + b_i^*} \cdot g^{\zeta^* \theta^*}, r, g^{t^* a_i^* + b_i^*}). \tag{12}$$

where $H(m^* \| g^{\sigma_3^*}) = g^{\zeta^*}$. The $\mathcal{B}$ retrieves ζ^* frame $\mathcal{H}_{List}(R^*, m^*, \zeta^*, H(m^* \| R^*), T_1)$ and computes $g^{\alpha^{q+1}}$ as follows (13):

$$g^{\alpha^{q+1}} = \frac{\sigma_2^*}{g^{\alpha'}(\sigma_1^*)^{\zeta^*} \prod_{i \in [l]} (\sigma'^*_i)^{z_i}} = \frac{g^{\alpha^{q+1}} g^{\alpha'} \prod_{i \in [l]} (g^{z_i})^{t^* a_i^* + b_i^*} g^{\zeta^* \theta^*}}{g^{\alpha'} (g^{\theta^*})^{\zeta^*} \prod_{i \in [l]} (g^{t^* a_i^* + b_i^*})^{z_i}} = g^{\alpha^{q+1}}. \tag{13}$$

Thus, if there exists a PPT adversary $\mathcal{A}$ that can win the security game withe non-negligible probability ϵ, then the challenger $\mathcal{B}$ can leverage $\mathcal{A}$'s capability to solve the q-BDHE problem with probability ϵ in PPT.

Lemma 4 (Witness extractability). *The ABAS scheme obtains witness extractability.*

Proof. If there exist a PPT adversary $\mathcal{A}$ that wins the game $WitExt(\lambda)$, then the challenger $\mathcal{B}$ can find a collision pair for the hash function H in PPT. The $\mathcal{B}$ simulates the game $WitExt(\lambda)$ for $\mathcal{A}$, and the simulation proceeds as follows:

1. **Initialization**: Challenger $\mathcal{B}$ executes the Setup and KGen algorithm, delivers the pp to adversary $\mathcal{A}$, and retains the sk_W to answer $\mathcal{A}$'s pre-signature and signature queries. The $\mathcal{A}$ runs the GenR algorithm to generate a statement-witness pair (Y, y) satisfying $Y = g^y$, and submits the Y to $\mathcal{B}$. The message query set $\mathcal{Q}$ is initialized as empty.
2. **Pre-signature Query**: If $\mathcal{A}$ requests a pre-signature for message m_i, the $\mathcal{B}$ runs the pSign algorithm to generate a pre-signature $\tilde{\sigma}$ for m_i, sends $\tilde{\sigma}$ to $\mathcal{A}$, and adds m_i to the set $\mathcal{Q}$.
3. **Signature**: If $\mathcal{A}$ submits a signature query for message m_i, $\mathcal{B}$ runs the Sign algorithm to generate a signature σ for m_i, sends σ to $\mathcal{A}$ and adds m_i to the set $\mathcal{Q}$.
4. **Witness Extraction**: The $\mathcal{A}$ selects a message $m^* \in \{0, 1\}^*$, sends it to $\mathcal{B}$. The $\mathcal{B}$ runs the pSign algorithm to generate a pre-signature $\tilde{\sigma}^*$ for m^* and sends $\tilde{\sigma}^*$ to $\mathcal{A}$. Then $\mathcal{A}$ produces a signature σ^* for m^* and submits it to $\mathcal{B}$. Finally, $\mathcal{B}$ executes the Extract algorithm to extract the witness y'.

If $\mathsf{Verify}(pp, m, \sigma, (\mathbf{M}, \rho)) = 1, m^* \notin \mathcal{Q}$, and $Y \neq g^{y'}$, the adversary $\mathcal{A}$ wins the game(output 1); otherwise, output 0. Let $\tilde{\sigma}_3^* = r$, extracting the witness y', $\mathcal{B}$ sends the bitstrings $m^*||g^{r+y'}$ and $m^*||g^r Y$ to $\mathcal{B}$ as a collision pair for the hash function H.

Given that $(g^\theta, K_1 \cdot \prod_{i \in [l]} K_x^{a_i} h_x^{b_i} \cdot H(m^*||g^r Y)^\theta, r, g^{b_i} K_2^{a_i})$ is a valid pre-signature provided by $\mathcal{B}$, if the game outputs 1, $\mathcal{B}$ successfully extracts y' such that $Y \neq g^{y'}$, and $(g^\theta, K_1 \cdot \prod_{i \in [l]} K_x^{a_i} h_x^{b_i} \cdot H(m^*||g^r Y)^\theta, r + y', g^{b_i} K_2^{a_i})$ is a valid full signature. Concurrently, $(g^\theta, K_1 \cdot \prod_{i \in [l]} K_x^{a_i} h_x^{b_i} \cdot H(m^*||g^{r+y'})^\theta, r, g^{b_i} K_2^{a_i})$ must also be a valid full signature. If $H(m^*||g^{r+y'}) \neq H(m^*||g^r Y)$, the non-degeneracy of the bilinear pairing e would fail. Thus, when the game outputs 1, it must hold that: $H(m^*||g^{r+y'}) = H(m^*||g^r Y)$. Since $Y \neq g^{y'}$, the bitstrings $m^*||g^{r+y'}$ and $m^*||g^r Y$ are distinct, forming a collision for H. Therefor, the probability that $\mathcal{B}$ successfully produces a hash collision is:

$$\Pr[(x_1, x_2) \leftarrow B(\lambda): x_1 \neq x_2, H(x_1) = H(x_2)] = \Pr[WitExt(\lambda) = 1].$$

If H is collision-resistant, the ABAS scheme satisfies witness extractability.

Lemma 5 (Anonymity). *The ABAS scheme satisfies the anonymity, meaning it achieves privacy preservation of the signer's attributes.*

Proof. In the ABAS scheme, the verifier can only infer from the signature that the signer possesses an attribute set satisfying the access structure, but cannot discern which specific attributes were used to generate the signer's private key. The security game interaction between adversary $\mathcal{A}$ and challenger $\mathcal{B}$ proceeds as follows:

1. **System Setup**: The $\mathcal{B}$ first executes the Setup algorithm to initialize the system, then discloses the public key pp and master secret key msk to $\mathcal{A}$. The $\mathcal{A}$ randomly selects the challenge access structure $(\mathbf{M}^*, \rho^*)$ and sends it to $\mathcal{B}$.
2. **Challenge**: The $\mathcal{A}$ selects a message m^* and two attribute sets W_1^*, W_2^* satisfying $(\mathbf{M}^*, \rho^*)$, then sends them to $\mathcal{B}$. The $\mathcal{B}$ executes the KGen algorithm on W_1^* and W_2^* to generate the corresponding secret keys $sk_{W_1^*}$ and $sk_{W_2^*}$. The detailed construction proceeds as follows (14):

$$sk_{W_1^*} = \{K_1 = g^\alpha g^{at_1}, K_2 = g^{t_1}, K_x = h_x^{t_1}, \forall x \in W_1^*\},$$
$$sk_{W_2^*} = \{K_1 = g^\alpha g^{at_2}, K_2 = g^{t_2}, K_x = h_x^{t_2}, \forall x \in W_2^*\}. \quad (14)$$

The $\mathcal{B}$ randomly selects $b \in \{0, 1\}$, executes the Sign algorithm to generate a signature σ_b^* for the message m^* under the attribute set W_b^* as (15), and sends σ_b^* to $\mathcal{A}$.

$$\sigma_b^* = \{\sigma_{1_b} = g^\theta, \sigma_{2_b} = K_1 \cdot \prod_{i \in [l]} K_x^{a_i} h_x^{b_i} \cdot H(m||RY)^\theta,$$
$$\sigma_{3_b} = r + y, \sigma'_{i_b} = g^{b_i} \cdot K_2^{a_i}, \forall i \in [l]\}. \quad (15)$$

3. **Gussing**: After receiving the signature, $\mathcal{A}$ attempts to guess the value of b. If $b = 0$, $\mathcal{A}$ randomly selects $t_b, r_b, \{a_{i_b}, b_{i_b}\}_{i \in [l]}$, and θ_b. When $t_0 = t_1$ and $\theta_0 = \theta_1$, the signatures generated using $sk_{W_1^*}$ and $sk_{W_2^*}$ are identical, as demonstrated below (16).

$$\sigma_{1_0} = g^{\theta_0} = g^{\theta_1} = \sigma_{1_1},$$

$$\sigma_{2_0} = K_{10} \cdot \prod_{i \in [l]} K_x^{a_{i_0}} h_x^{b_{i_0}} \cdot H(m\|R_0Y)^{\theta_0}$$

$$= g^{\alpha} g^{at_1} \prod_{i \in [l]} h_x^{t_1 a_{i_1} + b_{i_1}} \cdot H(m\|R_1Y)^{\theta_1} = \sigma_{2_1}, \quad (16)$$

$$\sigma_{3_0} = r_0 + y = r_1 + y = \sigma_{3_1},$$

$$\sigma'_{i_0} = g^{b_i} K_2^{a_i} = g^{t_0 a_{i_0} + b_{i_0}} = g^{t_1 a_{i_1} + b_{i_1}} = \sigma'_{i_1}, \forall i \in [l].$$

By the same reasoning, if $b = 1$, the signatures generated using $sk_{W_1^*}$ and $sk_{W_2^*}$ are also identical. The signature σ_b^* could have been produced by either W_1^* or W_2^*. Consequently, $\mathcal{A}$ cannot determine which attribute set's secret key was used to generate the signature, and its probability of winning the anonymity security game is exactly $\frac{1}{2}$.

6 Performance

In terms of key size and signature length, d denotes the number of attributes in the user's attribute set, $|G|$ represents the order of the group $\mathbb{G}$ in a symmetric bilinear map setting, our ABAS introduces an additional storage overhead of $|G|$ (from $(2+d)|G|$ in Waters-ABS to $(3+d)|G|$ in ours) due to the adaptor private key y and the binding term σ_3. For signature computation, T_e^G and $T_e^{G_T}$ represent the computation time for one exponentiation operation in group $\mathbb{G}$ and group $\mathbb{G}_T$, respectively, the generation of adaptor-related parameter $RY = g^r \cdot Y$ will add the computation cost in T_e^G (from $(d+3)T_e^G$ in Waters-ABS to $(d+4)T_e^G$ in ours).

In verification computation time, T_p denotes the computation time for one symmetric bilinear pairing operation $e: \mathbb{G} \times \mathbb{G} \leftarrow \mathbb{G}_T$, an extra bilinear pairing operation $e(H(m\|RY), \sigma_1)$ increases the pairing count from $3T_p$ to $4T_p$, while the exponentiation overhead in the target group G_T retains unchanged($2T_e^{G_T}$).

Overall, ABAS achieves AS functionality with a constant-level performance overhead(+1 group operation/pairing) while maintaining linear complexity($O(d)$), balancing functional extensibility and performance trade-offs as shown in Table 1.

Table 1. Performance.

Scheme	Key size	Signature length	Signature computation	Verify computation
Waters-ABS[22]	$(2+d) \cdot |G|$	$(2+d) \cdot |G|$	$(d+3) \cdot T_e^G$	$3 \cdot T_p + 2T_e^{G_T}$
ABAS	$(3+d) \cdot |G|$	$(3+d) \cdot |G|$	$(d+4) \cdot T_e^G$	$4 \cdot T_p + 2T_e^{G_T}$

7 Application in Access-Control Based Atomic Swap in Cross-Chain

To demonstrate the use of ABAS scheme, we deploy its application in cross-chain atomic swaps in Fig. 2. In a cross-chain atomic exchange scenario, Alice seeks to swap her asset c_A on chain A with Bob's asset c_B on chain B, requiring both parties to fulfill predefined compliance obligations: Bob must satisfy a KYC Tier of at least 2 and reside in a jurisdiction excluded from sanctioned country lists, while Alice must provide proof of a valid issuance license. The protocol aims to ensure the simultaneous and conditional exchange of c_A and c_B only when all regulatory and contractual requirements are cryptographically verified and met by both participants.

1. **Asset Locking**: Two parties Alice and Bob first set the time-lock for c_A with the timeout t_A and c_B with the timeout t_B on-chain($t_A > t_B$). If a timeout occurs, the assets will automatically revert to their original accounts. The time-lock mechanism primarily provides Bob with sufficient time to complete the asset swap, preventing Alice from withdrawing Bob's funds first and subsequently attempting to reclaim his own assets.
2. **Attribute Registration and Distribution**: The Attribute Authority(AA) assigns attribute sets W_A and W_B to Alice and Bob based on their identities and access privileges, respectively. It then generates corresponding private keys sk_{W_A} and sk_{W_B} tied to these attribute sets, which are securely distributed to Alice and Bob.
3. **Alice sends $\tilde{\sigma}_A, tx_A,$ and Y to Bob**: Alice chooses a hard relation $(Y, y) \in R$ and pre-signing a transaction tx_A for spending the coins c_A to Bob, and then sends the pre-signature $\tilde{\sigma}_A, tx_A, Y$ to Bob.
4. **Bob sends $\tilde{\sigma}_B$ and tx_B to Alice**: Bob can check the validity of $\tilde{\sigma}_A$ and pre-signing a transaction tx_B for spending the coins c_B to Alice and then sends the pre-signature $\tilde{\sigma}_B, tx_B$ to Alice.
5. **Publish σ_A**: Alice can check the validity of $\tilde{\sigma}_B$ and adapts $\tilde{\sigma}_B$ into the full signature σ_B by the witness y, and then publishes σ_A on the blockchain to get the coin c_B within t_B.
6. **Extract y and publish σ_A**: Bob can extract the witness y from σ_B and $\tilde{\sigma}_B$ and adapts $\tilde{\sigma}_A$ into σ_A, then publishes σ_A on the blockchain to get the coin c_A within t_A.

Our ABAS scheme ingeniously integrates ABS with adaptor signature, enabling atomic swaps with fine-grained access control. This approach is particularly suited for scenarios demanding both access control and atomicity guarantees, such as restricted digital asset transactions and privacy-preserving data markets. Compared to conventional AS, it introduces an additional policy verification layer, while surpassing standard ABS by embedding transaction coordination capabilities.

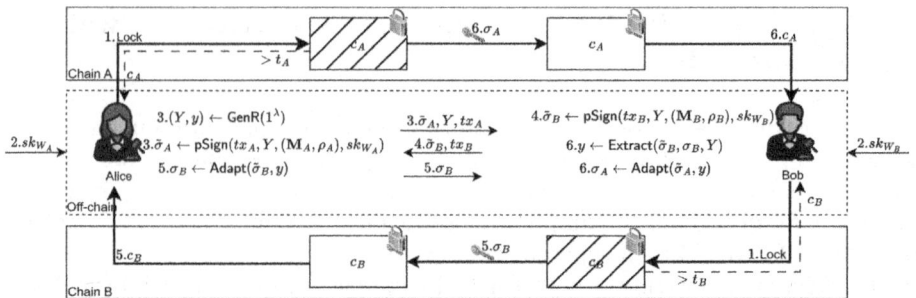

Fig. 2. The control-based atomic swap protocol in cross-chain based on ABAS scheme.

8 Conclusion

This paper proposes an attribute-based adaptor signature scheme, achieving deep integration of access control and atomic exchange through cryptographic innovation. Specifically, an attribute authorization mechanism is introduced during the key generation phase, while adaptor statements are bound to access policies through bilinear pairing operations in the pre-signing phase. We formally prove that our scheme satisfies sP-aEUF-CMA security, pre-signature adaptability, witness extractability and anonymity. It also provides a practical scenario in access-control based cross-chain atomic swaps.

References

1. Gerhart, P., Schröder, D., Soni, P., et al.: Foundations of adaptor signatures. In: Proceedings of EUROCRYPT 2024. LNCS, vol. 14645, pp. 161–189. Springer, Cham (2024)
2. Bursuc, S., Mauw, S.: Contingent payments from two-party signing and verification for abelian groups. In: IEEE CSF, pp. 195–210. IEEE (2022)
3. Vanjani N., Soni P., Thyagarajan S.A.K.: Functional adaptor signatures: beyond all-or-nothing blockchain-based payments. In: Proceedings of CCS 2024, pp. 1493–1507. ACM, New York (2024)
4. Deshpande, A., Herlihy, M.: Privacy-preserving cross-chain atomic swaps. In: Bernhard, M., et al. (eds.) FC 2020. LNCS, vol. 12063, pp. 540–549. Springer, Cham (2020). https://doi.org/10.1007/978-3-030-54455-3_38
5. Thyagarajan, S.A.K., Malavolta, G., Moreno-Sanchez, P.: Universal atomic swaps: secure exchange of coins across all blockchains. In: IEEE S&P, pp. 1299–1316. IEEE (2022)
6. Esgin, M.F., Ersoy, O., Erkin, Z.: Post-quantum adaptor signatures and payment channel networks. In: Chen, L., Li, N., Liang, K., Schneider, S. (eds.) ESORICS 2020. LNCS, vol. 12309, pp. 378–397. Springer, Cham (2020). https://doi.org/10.1007/978-3-030-59013-0_19
7. Moreno-Sanchez, P., Blue, A., Le, D.V., Noether, S., Goodell, B., Kate, A.: DLSAG: non-interactive refund transactions for interoperable payment channels in Monero. In: Bonneau, J., Heninger, N. (eds.) FC 2020. LNCS, vol. 12059, pp. 325–345. Springer, Cham (2020). https://doi.org/10.1007/978-3-030-51280-4_18

8. Yu, X., Wang, Y., Huang, X.: Quantum-resistant ring signature-based authentication scheme against secret key exposure for VANETs. Comput. Netw. **262**, 111213 (2025)
9. Aumayr, L., et al.: Generalized channels from limited Blockchain scripts and adaptor signatures. In: Tibouchi, M., Wang, H. (eds.) ASIACRYPT 2021. LNCS, vol. 13091, pp. 635–664. Springer, Cham (2021). https://doi.org/10.1007/978-3-030-92075-3_22
10. Malavolta, G., Moreno-Sanchez, P., Schneidewind, C., et al.: Anonymous multi-hop locks for blockchain scalability and interoperability. Cryptology ePrint Archive, Report 2018/472 (2018). https://eprint.iacr.org/2018/472
11. Erwig, A., Faust, S., Hostáková, K., Maitra, M., Riahi, S.: Two-party adaptor signatures from identification schemes. In: Garay, J.A. (ed.) PKC 2021. LNCS, vol. 12710, pp. 451–480. Springer, Cham (2021). https://doi.org/10.1007/978-3-030-75245-3_17
12. Qin, X., Cui, H., Yuen, T.H.: Generic adaptor signature. Cryptology ePrint Archive, Report 2021/161 (2021). https://eprint.iacr.org/2021/161
13. Dai, W., Okamoto, T., Yamamoto, G.: Stronger security and generic constructions for adaptor signatures. In: Proceedings of INDOCRYPT 2022. LNCS, vol. 13774, pp. 52–77. Springer, Cham (2022)
14. Miller, A., Bentov, I., Bakshi, S., Kumaresan, R., McCorry, P.: Sprites and state channels: payment networks that go faster than lightning. In: Goldberg, I., Moore, T. (eds.) FC 2019. LNCS, vol. 11598, pp. 508–526. Springer, Cham (2019). https://doi.org/10.1007/978-3-030-32101-7_30
15. Qin, X., Pan, S., Mirzaei A., et al.: BlindHub: Bitcoin-compatible privacy-preserving payment channel hubs supporting variable amounts. In: IEEE SP, pp. 2462–2480. IEEE (2023)
16. Madathil, V., Thyagarajan, S.A.K., Vasilopoulos, D., et al.: Cryptographic oracle-based conditional payments. Cryptology ePrint Archive, Report 2022/499 (2022). https://eprint.iacr.org/2022/499
17. Maji, H.K., Prabhakaran, M., Rosulek, M.: Attribute-based signatures. In: Kiayias, A. (ed.) CT-RSA 2011. LNCS, vol. 6558, pp. 376–392. Springer, Heidelberg (2011). https://doi.org/10.1007/978-3-642-19074-2_24
18. Bellare, M., Fuchsbauer, G.: Policy-based signatures. In: Krawczyk, H. (ed.) PKC 2014. LNCS, vol. 8383, pp. 520–537. Springer, Heidelberg (2014). https://doi.org/10.1007/978-3-642-54631-0_30
19. Li, J., Kim, K.: Hidden attribute-based signatures without anonymity revocation. Inf. Sci. **180**(9), 1681–1689 (2010)
20. Okamoto, T., Takashima, K.: Efficient attribute-based signatures for non-monotone predicates in the standard model. IEEE Trans. Cloud Comput. **2**(4), 409–421 (2014)
21. Gu, K., Jia, W., Wang, G., et al.: Efficient and secure attribute-based signature for monotone predicates. Acta Informatica **54**, 521–541 (2017)
22. Waters, B.: Ciphertext-policy attribute-based encryption: an expressive, efficient, and provably secure realization. In: Proceedings of PKC 2011. LNCS, vol. 6571, pp. 53–70. Springer, Berlin, Heidelberg (2011)
23. Xiong, H., Bao, Y., Nie, X., et al.: Server-aided attribute-based signature supporting expressive access structures for industrial internet of things. IEEE Trans. Industr. Inf. **16**(2), 1013–1023 (2019)
24. Maxwell, G., Poelstra, A., Seurin, Y., et al.: Simple Schnorr multi-signatures with applications to Bitcoin. Des. Codes Crypt. **87**(9), 2139–2164 (2019)

25. Aumayr, L., Maffei, M., Ersoy O., et al.: Bitcoin-compatible virtual channels. In: IEEE S& P, pp. 901–918. IEEE (2021)
26. Boneh, D., Gentry, C., Lynn, B., Shacham, H.: Aggregate and verifiably encrypted signatures from bilinear maps. In: Biham, E. (ed.) EUROCRYPT 2003. LNCS, vol. 2656, pp. 416–432. Springer, Heidelberg (2003). https://doi.org/10.1007/3-540-39200-9_26
27. Oberko, P.S.K., Obeng, V.H.K.S., Xiong, H., et al.: A survey on attribute-based signatures. J. Syst. Architect. **124**, 102396 (2022)

A Versatile Decentralized Attribute Based Signature Scheme for IoT

Dazhi Xu[1], Yuejun Liu[1(✉)], Jiabei Wang[1], Yiwen Gao[1], and Yongbin Zhou[1,2]

[1] School of Cyber Science and Engineering, Nanjing University of Science and Technology, Nanjing, China
{xudazhi,liuyuejun,wangjiabei,gaoyiwen,zhouyongbin}@njust.edu.cn
[2] Institute of Information Engineering, Chinese Academy of Sciences, Beijing, China

Abstract. Attribute-based signature (ABS) allows a signer whose attribute set satisfies a given signing policy to endorse a message. The verifier checks whether the signature was generated by a signer whose attributes match the signing policy, enabling flexible anonymous authentication in Internet of Things environments. However, existing ABS schemes face challenges such as high computational overhead, signature misuse, key exposure, and the presence of a central attribute authority. To address these issues, we propose an efficient, forward-secure, and traceable decentralized attribute-based signature (EFSTD-ABS) scheme. We prove its forward security under the computational Diffie-Hellman assumption in the random oracle model. Our scheme achieves low computational overhead by employing only modular exponentiations for signing, mitigates the impact of key exposure, and enables the trust authority to trace the signer's real identity in cases of abusive behavior. Experimental results demonstrate that the EFSTD-ABS scheme is efficient in terms of computational overhead.

Keywords: Attribute-based signature · Efficient · Forward-secure · Traceable · Decentralized

1 Introduction

With the continuous development of the Internet of Things (IoT), new security vulnerabilities and attack vectors continue to emerge. Thus, It is important to guarantee the authentication and integrity of data. In medical IoT applications, for instance, when healthcare devices collect and transmit patient data to servers, the sensitive nature of such data necessitates identity authentication while preserving anonymity. To address this, ABS [5,9,12,14] have been proposed for data authentication. In ABS, the signer generates a signature based on their attributes and the signing policy, while the verifier checks whether the signer's attributes satisfy the signing policy, thereby preserving the signer's identity and attribute privacy.

However, earlier ABS schemes face several challenges in IoT environments. First, due to limited computational capabilities of IoT devices, they are unable to bear the computational overhead introduced by pairing operations in the ABS. To address this issue, we introduce a random number associated with the access control structure to randomize the signing key and generate signatures, thereby reducing the computational cost during the signing process by only requiring exponentiation operations. Second, malicious signers can illegitimately endorse messages without accountability, particularly, erroneous medical data may lead to severe medical incidents. To address this, Traceable Attribute-Based Signatures (TABS) [3,4,7,8,10] were introduced, enabling the traceability of the signer's real identity in the event of signature misuse. Third, the aforementioned schemes do not address key exposure, where an attacker could forge signatures and invalidate previously generated ones. To mitigate this, forward-secure signatures (FSS) [1] were proposed. In FSS, time is divided into discrete periods. The verifier can validate the signature by checking the time period ensuring that even if the key for a given period is compromised, previous signatures remain valid. Finally, ABS schemes relying on a central attribute authority are vulnerable to attacks if the authority is corrupted. To mitigate this, Okamoto and Takashima [13] proposed the decentralized multi-authority attribute-based signature (DMA-ABS) scheme, where multiple independent attribute authorities generate public and private keys based on global parameters. Even if one authority is corrupted, others remain secure.

In this paper, we propose an Efficient, Forward-Secure, and Traceable Decentralized Attribute-Based Signature scheme that simultaneously addresses these challenges. Concretely, our contributions are as follows.

- We first present an EFSTD-ABS scheme supports decentralized deployment fulfills the security properties for traceability and forward security. The proposed scheme utilizes the trust authority to trace the signer's authentic identity when the signer occurs abusing behavior. Furthermore, the provided scheme alleviates the damage of key exposure. The proposed scheme exploits the Kunode algorithm to update the signing secret key of diverse time periods.
- The designed EFSTD-ABS scheme is proved existentially unforgeable against selective predicate attack under the random oracle model, with its security reduced to the computational Diffie-Hellman (CDH) assumption. Furthermore, we demonstrate that the proposed EFSTD-ABS scheme ensures signer privacy.
- Based on the experimental results, with 20 attributes, our scheme improves signature generation efficiency by 81.92% compared to scheme [17], and enhances traceability efficiency by 37.8% compared to Scheme [8]. Additionally, the complexity of the update algorithm is reduced from $\mathcal{O}(n^2)$ to $\mathcal{O}(n)$.

1.1 Related Work

Attribute-based signature Maji et al. [11] first formalized the definition of ABS. However, the security of their scheme is limited because it relies on the generic group model for proof. Guo and Zeng [16] proposed a novel ABS scheme, but

Maji et al. [12] criticized its security definition for neglecting signer privacy. In response, Maji et al. [12] introduced an ABS scheme that guarantees perfect privacy. Unfortunately, Zhang et al. [21] identified a forgery vulnerability in their third instantiation [16]. To reduce computational overhead, Gu et al. [6] presented an efficient framework for monotone predicate ABS by leveraging Waters [18].

Traceable attribute-based signature Escala et al. [5] introduced a TABS scheme that utilizes structure-preserving signatures to ensure signer traceability. However, El Kaafarani et al. [4] highlighted the inefficiency of the scheme [5], due to its reliance on composite-order groups. Ding et al. [3] proposed a more efficient TABS scheme. Nonetheless, it does not achieve perfect privacy.

Forward-secure attribute-based signature To address the risks associated with key exposure in ABS schemes, Yuen et al. [20] presented a universal framework for forward-secure attribute-based signatures (FS-ABS). However, in their scheme, updating each user's secret key essentially requires the attribute authority to generate a new secret key for the user in each time period. To construct a practical FS-ABS scheme, Wei et al. [19] proposed a construction supporting threshold predicates and employed a binary tree structure to update signing keys across different time periods. Subsequently, Kang et al. [8] proposed an ABS scheme that simultaneously supports traceability and forward security. However, the scheme relies on a single attribute authority, which degrades both security and reliability of the system.

2 Preliminaries

The preliminary knowledge exploited in this article is presented in this section.

2.1 Bilinear Map

Let G_1 and G_T be two multiplicative cyclic groups of prime order p and g be the generator of G_1. $e : G_1 \times G_1 \to G_T$ is a bilinear map if it fulfills the following properties.

- Bilinearity: For $\forall g_1, g_2 \in G_1$ and $\forall a, b \in Z_p^*$, we have $e(g_1{}^a, g_2{}^b) = e(g_1, g_2)^{ab}$.
- Non-degeneracy: There exists $g_1, g_2 \in G_1$ such that $e(g_1, g_2) \neq 1$.
- Computability: For $\forall g_1, g_2 \in G_1$, there exists an efficient algorithm to calculate $e(g_1, g_2)$.

2.2 Complexity Assumption

Definition 1 (CDH problem). *Suppose that G_1 is a multiplicative cyclic group of prime order p and g is a generator of G_1. Given a tuple (g, g^a, g^b) for unknown $a, b \in Z_p$ to calculate g^{ab}.*

Definition 2 (CDH Assumption). *We consider that the CDH problem in G_1 is (t, ϵ)-computationally infeasible if for any PPT adversary $\mathcal{A}$ who executes in time at most t and has the advantage $\Pr[g^{ab} \leftarrow \mathcal{A}(g, g^a, g^b)] \leq \epsilon$.*

2.3 Linear Secret-Sharing Schemes

Definition 3 (LSSS). *Supposing that U is an attribute universe. The secret-sharing scheme Π_Γ over an access policy Γ On U is said to be linear if Π_Γ contains two polynomial time algorithms as below, in which M is $l \times k$ matrix and ρ maps a row for the matrix M to an attribute $\rho(i)$. M is referred to as the share-generating matrix over Π_Γ.*

- *Distribute(M, ρ, s): The algorithm is utilized to share a secret α ground on Γ. Set the vector $v = (s, v_2, v_3, \cdots, v_k) \in Z_p^k$ are chosen at random. The algorithm returns a collection $\{M_i \cdot v : i \in [l]\}$ for l shares, in which $M_i \in Z_p^k$ is the ith row for M. $\lambda_{\rho(i)} = M_i \cdot v$ is a share of secret s which belongs to an attribute $\rho(i)$.*
- *Reconstruct(M, ρ, S): The algorithm is utilized to reconstruct s from secret shares. It inputs M, ρ and an attribute set $S \in \Gamma$. Set $I = \{i \in [l] : \rho(i) \in S\}$. The algorithm outputs a constant set $\{\omega_i \in Z_p : i \in I\}$ about secret reconstruction such that $\sum_{i \in I} M_i \cdot \omega_i = (1, 0, \cdots, 0)$, i.e., $\sum_{i \in I} \lambda_{\rho(i)} \omega_i = s$, provided that $\{\lambda_{\rho_i} : i \in I\}$ is a valid shares set on s according to Π_Γ.*

2.4 KUNode Algorithm

To achieve scalable user revocation, we mainly follow Boldyrevaet et al. [2] strategy that defines a KUNode algorithm by using a binary. For convenience, we first give several notations. Let $\mathcal{BT}$ be a binary tree with N leaf nodes and root be the **root** node. For each non-leaf node θ, denote by θ_l and θ_r its left and right child, respectively. Each user is associated with a leaf node η of $\mathcal{BT}$, and denote by **Path**(η) the set of all nodes over the path from root to η (including **root** and η).

The KUNode algorithm takes as input a binary tree $\mathcal{BT}$, a revocation list **RL** consisting of two tuples recorded in the form of(η_i, t_i) as well as a time period t. It outputs a minimal set **Y** of nodes of $\mathcal{BT}$ such that for any leaf node η listed in **RL**, it holds that **Path**(η) $\cap$ **Y** $= \emptyset$. On the other hand, for each non-revoked leaf node there exactly exists a node $\theta \in$ **Y** such that θ is an ancestor of this leaf node.

3 Syntax and Security Model of EFSTD-ABS Scheme

3.1 Syntax of EFSTD-ABS Scheme

Definition 4. *An EFSTD-ABS scheme consists of the following polynomial-time algorithms:*

GlobalSetup(λ) $\rightarrow GP$. *The algorithm is executed by the third party. It inputs security parameter λ, and returns the global parameter GP of the system, which is available for all AAs and users in the system. Meanwhile, the tracing authority generates a tracing key tk to reveal the real identity of malicious users through signature.*

Algorithm 1. KUNode ($\mathcal{BT}, RL, t$)

1: $\mathbf{X, Y} \leftarrow \emptyset$
2: **for all** $(\eta_i, t_i) \in RL$ **do**
3:    **if** $t_i \leq t$ **then**
4:       Add **Path**(η_i) to $\mathbf{X}$
5:    **end if**
6: **end for**
7: **for all** $\theta \in \mathbf{X}$ **do**
8:    **if** $\theta_l \notin \mathbf{X}$ **then**
9:       Add θ_l to $\mathbf{Y}$
10:    **end if**
11:    **if** $\theta_r \notin \mathbf{X}$ **then**
12:       Add θ_r to $\mathbf{Y}$
13:    **end if**
14: **end for**
15: **if** $\mathbf{Y} = \emptyset$ **then**
16:    Add **root** to $\mathbf{Y}$
17: **end if**
18: **return** $\mathbf{Y}$

AuthSetup$(GP, T, N) \to (apk_\delta, ask_\delta)_{\delta \in AA}$. *The algorithm is executed by each attribute authority $\delta \in AA$. It inputs the global public parameter GP, the total number of system time periods T, and the number of system users. It produces the attribute authority public key apk_δ and the attribute authority secret key ask_δ of this authority. In addition, it creates a revocation list RL_δ initialized as an empty set and the state information st_δ used to manage users' identifiers.*

UskGen$(GP, ask_\delta, GID, S_\delta, st_\delta) \to (usk_{GID,S_\delta}, st_\delta)$. *The algorithm is executed by each attribute authority $\delta \in AA$. It inputs the global public parameter GP, the attribute authority secret key ask_δ, the user's identifier GID, the corresponding attribute set $S_\delta \subseteq S$ where S_δ represents the subset of attributes S within the user's attribute set that are controlled by δ, and the current state information st_δ. The algorithm generates a secret key component usk_{GID,S_δ} for the user, and also outputs an updated state information st_δ.*

UkGen$(GP, apk_\delta, ask_\delta, RL_\delta, st_\delta, t) \to uk_{\delta,t}$. *The algorithm is executed by each attribute authority $\delta \in AA$. It inputs the global public parameter GP, the attribute authority public key apk_δ, the attribute authority secret key ask_δ, the current revocation list RL_δ, and the update time period t. The algorithm produces an update key component $uk_{\delta,t}$, which is only useful for non-revoked users.*

UUskGen$(GP, apk_\delta, usk_{GID,S_\delta}, uk_{\delta,t}) \to uusk^t_{GID,S_\delta}$. *The algorithm is executed by signer. It inputs the the global public parameter GP, the attribute authority public key apk_δ, the secret key component usk_{GID,S_δ}, and the update key component $uk_{\delta,t}$. It derives a update user's key component $uusk^t_{GID,S_\delta}$ for the non-revoked user.*

Sign$(GP, uusk^t_{GID,S}, S, m, \Gamma(\cdot)) \to \sigma$. *The algorithm is implemented via the signer who owns an attribute set S. It inputs the global public parameter GP, the updated user's key $uusk^t_{GID,S}$ with regard to the attribute set S along with the* current time period t, the attribute set S, a message m, a predicate $\varGamma = (\boldsymbol{M}, \rho)$. If $\varGamma(S) = 1$, the algorithm generates signature σ associated with the access structure $\varGamma(\cdot)$ and message m at the time period t.

Verify($GP, \{apk_\delta\}_{\delta \in AA}, m, \sigma, \varGamma(\cdot), t$) $\to$ 0 or 1. The algorithm is executed by the verifier. It inputs the global public parameter GP, the attribute authority public key $\{apk_\delta\}_{\delta \in AA}$, message m, signature σ under the predicate $\varGamma(\cdot)$ and time period t. It outputs value '1' if σ is a valid signature. Otherwise, it outputs value '0'.

Revoke($GID, RL_\delta, st_\delta, t$) $\to RL_\delta$. The algorithm is executed by each attribute authority $\delta \in AA$. It inputs the user's identifier GID to be revoked, the current revocation list RL_δ, and state information st_δ as well as a revocation time t. It outputs an updated revocation list.

Trace($GP, m, \varGamma(\cdot), \sigma, tk$) $\to GID$. The Trace algorithm is implemented by TA. It is exploited to trace the signer's actual identity on the signature σ. The algorithm takes the Global parameters GP, the message m, the signature under signing $\varGamma(\cdot)$ and tracing key tk as input, then it outputs the signer's real identity.

Correctness An EFSTD-ABS scheme is correct if for $GP \leftarrow$ GlobalSetup(λ), all $(apk_\delta, ask_\delta) \leftarrow$ AuthSetup(GP, T, N), attribute S, all $usk_{GID, S_\delta} \leftarrow$ USkGen($GP, ask_\delta, RL_\delta, st_\delta, t$), time period t, all $uk_{\delta, t} \leftarrow$ UkGen($GP, apk_\delta, ask_\delta, RL_\delta, st_\delta, t$), all $uusk^t_{GID, S_\delta} \leftarrow$ UUskGen($GP, apk_\delta, usk_{GID, S_\delta}, uk_{\delta, t}$), message m, signing predicates $\varGamma(\cdot)$ such that $\varGamma(S) = 1$, and a signature $\sigma \leftarrow$ Sign($GP, uusk^t_{GID, S}, S, m, \varGamma$), We have

$$\Pr[\textsf{Verify}(GP, \{apk_\delta\}_{\delta \in AA}, m, \sigma, \varGamma, t) \to 1] \geq 1 - \epsilon.$$

3.2 Security Model

An EFSTD-ABS scheme needs to fulfill the fundamental security requirements for forward security, privacy and traceability. The security model is referred to the security model in [8].

Forward Security: The formal definition for forward security under selective predicate (SP) attack is provided in the game between the adversary $\mathcal{A}$ and the challenger $\mathcal{C}$ as below.

Initial $\mathcal{A}$ claims a challenge predicate $\varGamma^*(\cdot)$ and a time period t^* which are utilized in undermentioned game.

Setup Given security parameter λ, $\mathcal{C}$ executes GlobalSetup algorithm and AuthSetup algorithm to produce global public parameters GP, $\{apk_\delta\}_{\delta \in AA}$, and $\{ask_\delta\}_{\delta \in AA}$. $\mathcal{A}$ set $AA_c \subseteq AA$ of corrupted attribute authorities. then $\mathcal{C}$ outputs GP, $\{apk_\delta\}_{\delta \in AA}$, and $\{ask_\delta\}_{\delta \in AA_c}$ to $\mathcal{A}$ and sets initial time period $t_0 = 0$, while the $\{ask_\delta\}_{\delta \in AA \setminus AA_c}$ is kept confidential for himself.

Queries $\mathcal{M}$ is permitted to issue polynomially bounded-time queries to following Oracle.

- UskGen Oracle: $\mathcal{A}$ requests the user's secret key $\{usk_{GID_i,S_i}\}$, under the constraint $\Gamma(S_i \bigcup\{\bigcup_{\delta \in AA_c} \mathcal{U}_\delta\} = 0$. $\mathcal{C}$ executes UskGen algorithm then outputs the user's secret key usk_{GID_i,S_i} to $\mathcal{A}$.
- UKGen Oracle: $\mathcal{A}$ picks a new time period t_j and requests $\mathcal{C}$ to run the UKGen algorithm such that the current time period t_i is increased to the time period t_j, $\mathcal{C}$ generates a update key components UK_{δ,t_j} and forwards it to $\mathcal{A}$.
- UUskGen Oracle: $\mathcal{A}$ picks GID_i, S_i at any time period t_i under the constraint $\Gamma(S_i \bigcup\{\bigcup_{\delta \in AA_c} \mathcal{U}_\delta\} \neq 1$, or $t_i > t^*$ and $\Gamma(S_i \bigcup\{\bigcup_{\delta \in AA_c} \mathcal{U}_\delta\} = 1$, then requests $\mathcal{C}$ to run the UUskGen algorithm, $\mathcal{C}$ generates a updated user's secret key components $UUsk_{GID_i,S_i,\delta}^{t_i}$ and forwards it to $\mathcal{A}$.
- Revocation Oracle: $\mathcal{A}$ picks GID_i, δ_i and t_i, then requests $\mathcal{C}$ to run the *Revoke* algorithm, $\mathcal{C}$ generates a updated revocation list RL_{δ_i} and forwards it to $\mathcal{A}$.
- Sign Oracle: $\mathcal{A}$ issues Sign Oracle query for any attribute set S upon any message m as well as signing predicate $\Gamma(\cdot)$ to $\mathcal{C}$. In answer to the query, $\mathcal{C}$ executes Sign algorithm and gets a signature σ associated with the signing predicate $\Gamma(\cdot)$ at the current time period t_i, which is sent to $\mathcal{A}$.

Forgery Eventually, $\mathcal{A}$ returns the signature σ^* of message m^* related to $\Gamma^*(\cdot)$ at current time period t^*. If the following requirements are fulfilled, $\mathcal{A}$ succeeds in the aforementioned game.

- The adversary $\mathcal{A}$ does not submit the UskGen Oracle on (S, t_i) such that $\Gamma(S \bigcup\{\bigcup_{\delta \in AA_c} \mathcal{U}_\delta\} = 1$ and $t_i \leq t^*$;
- The adversary $\mathcal{A}$ does not query the Sign Oracle upon $(m^*, \Gamma^*(\cdot))$ at the exact time period t^*;
- σ^* is a valid signature, which means Verify(GP, $\{apk_\delta\}_{\delta \in AA}$, m^*, σ^*, $\Gamma^*(\cdot)$, t^*) $\rightarrow 1$.

The advantage $Adv_{ERTD-ABS,\mathcal{A}}^{SP-FS}(\lambda)$ is described over the probability which $\mathcal{A}$ succeeds in the aforementioned game.

Definition 5 (Forward security). *The adversary $\mathcal{A}(t, q_{usk}, q_{uk}, q_{uusk}, q_r, q_s, \epsilon)$ breaks the forward security of an EFSTD-ABS scheme provided that $\mathcal{A}$ executes within time at most t, and launches at most q_{usk} times UskGen Oracle queries, q_{uk} times UkGen Oracle queries, q_{uusk} times UuskGen Oracle queries, q_r times Revocation Oracle queries, q_s times Sign Oracle queries, while the advantage $Adv_{EFSTD-ABS}^{SP-FS}(\lambda) \geq \epsilon$. An EFSTD-ABS scheme is $(t, q_{usk}, q_{uk}, q_{uusk}, q_r, q_s, \epsilon)$-forward secure if for any PPT adversary $\mathcal{A}$, the advantage $Adv_{EFSTD-ABS,\mathcal{A}}^{SP-FS}(\lambda) < \epsilon$.*

Privacy. To ensure the privacy (Priv) of EFSTD-ABS scheme, the signature discloses nothing about the signer's identity. The formal definition about privacy is provided via the undermentioned game between the adversary $\mathcal{A}$ and the challenger $\mathcal{C}$.

Setup Given security parameter λ, $\mathcal{C}$ executes GlobalSetup algorithm and AuthSetup algorithm to produce global public parameters GP, $\{apk_\delta\}_{\delta \in AA}$, and $\{ask_\delta\}_{\delta \in AA}$. $\mathcal{C}$ outputs them to $\mathcal{A}$.

Queries The challenger $\mathcal{C}$ provides UskGen Oracle, Update Oracle and Sign Oracle queries as the aforementioned game of forward security but the scope of queried authorities is now expanded from $\delta \in AA \backslash AA_c$ to $\delta \in AA$.

Challenge. $\mathcal{A}$ extracts two attribute sets S_0, S_1 and a signing policy $\Gamma^*(\cdot)$ such that $\Gamma^*(S_0) = 1$ and $\Gamma^*(S_1) = 1$, and sends them to $\mathcal{C}$. The challenger $\mathcal{C}$ picks $\zeta \in \{0,1\}$ at random. Then, $\mathcal{C}$ calculates the signature σ^* exploiting the signing secret key with regard to the attribute set S_ζ via running Sign algorithm, and returns σ^* to $\mathcal{A}$.

Guess $\mathcal{A}$ returns $\zeta' \in \{0,1\}$ and succeeds in the aforementioned simulation if $\zeta' = \zeta$. The probability for $\mathcal{A}$ in winning the aforementioned game is determined by the advantage $Adv^{Priv}_{ERTD-ABS,\mathcal{A}}(\lambda) = |Pr[\zeta' = \zeta] - \frac{1}{2}|$.

Definition 6 (Privacy). *An EFSTD-ABS scheme fulfills the requirement of the signer's privacy if there exists no PPT adversary $\mathcal{A}$ can win the aforementioned game at negligible advantage $Adv^{Priv}_{EFSTD-ABS,\mathcal{A}}(\lambda)$.*

Traceability. To ensure the traceability of EFSTD-ABS scheme, for all valid signatures generated correctly, the TA can always break the anonymity of the signatures and identify the corresponding signer.

Definition 7 (Traceability). *Let EFSTD-ABS = (GlobalSetup, AuthSetup, UskGen, UkGen, UUskGem, Sign, Verify, Revoke, Trace) be the expeditious, revocable and traceable decentralized attribute based signature scheme, which satisfies the requirements of privacy and forward security. The ERTD-ABS scheme has traceability if for $(GP) \leftarrow$ GlobalSetup(λ), all $(apk_\delta, ask_\delta) \leftarrow$ AuthSetup(λ), message m, attribute set S, a time period t, all $(Usk_{GID,S_\delta}) \leftarrow$ UskGen$(GP, ask_\delta, GID, S_\delta, st_\delta)$, all $(RL'_\delta) \leftarrow$ Revoke$(GID, RL_\delta, st_\delta, t)$, all $(uk_{\delta,t}) \leftarrow$ UkGen$(GP, apk_\delta, ask_\delta, RL'_\delta, st_\delta, t)$, all $uusk^t_{GID,S_\delta} \leftarrow$ UUskGen$(GP, apk_\delta, usk_{GID,S_\delta}, uk_{\delta,t})$, Signing predicates $\Gamma(\cdot)$ such that $\Gamma(S) = 1$, and signature $\sigma \leftarrow$ Sign$(GP, uusk^t_{GID,S}, S, m, \Gamma(\cdot))$, the TA can find out the signer's identity ID.*

4 Our Construction

We provide an efficient, forward-secure and traceable decentralized attribute-based signatures scheme.

GlobalSetup$(\lambda) \rightarrow GP$. Given a security parameter λ, this algorithm produces the global public parameter as follows.

1) Choose two cyclic groups G_1 and G_2 of prime order p, g is the generator of G_1, and a bilinear mapping $e: G_1 \times G_1 \rightarrow G_2$.

2) Let $\mathcal{I}$ be the identifier space, AA be the set of authorities, and $\mathcal{U}$ be the attribute universe. For each authority $\delta \in AA$, denote by $\mathcal{U}_\delta$ the attribute universe managed by δ. Select exponent $a \in Z_p^*$ and compute $P = g^a$, then send a to TA as trace key. Select three hash functions: $H_1 : \mathcal{I} \to G_1$, $H_2 : \mathcal{I} \to Z_p$ and $F : \mathcal{U} \to G_1$.
3) Publish the global public parameter as

$$GP = \{G_1, G_2, e, p, g, e(g,g), P, H_1, H_2, F\}.$$

AuthSetup$(GP, T, N) \to (apk_\delta, ask_\delta)_{\delta \in AA}$. The authority setup algorithm takes as input the global parameter GP, the number of system users N, and the number of time periods $T = 2^d$. For an authority δ, it generates the public key and secret key by conducting the following steps.

1) Generate a binary tree $\mathcal{BT}_\delta$ with N leaf nodes, initialize $RL_\delta = \emptyset$ and set st_δ to $\mathcal{BT}_\delta$. Denote by $\mathcal{N}_\delta$ the set of all nodes of $\mathcal{BT}_\delta$. For each node $\theta \in \mathcal{N}_\delta$, select a random integer $r_\theta \in Z_p^*$.
2) Choose random exponents $\beta_\delta \in Z_p^*$ and random group elements $f_{\delta,0}, f_{\delta,1}, \cdots, f_{\delta,d} \in G_1$, set the public key and secret key as

$$apk_\delta = \{g^{\beta_\delta}, f_{\delta,0}, f_{\delta,1}, \cdots, f_{\delta,d}\}$$

and

$$ask_\delta = \{\beta_\delta, \{r_\theta\}_{\theta \in \mathcal{N}_\delta}\}.$$

In addition, for each authority $\delta \in AA$, we define a function $W_\delta(t) : \{0, \cdots, T-1\} \to G_1$ as follows:

$$W_\delta(t) = f_{\delta,0} \prod_{j=1}^{d} f_{\delta,j}^{t[j]}$$

where $t[j]$ is the jth bit of t in the form of binary string.

UskGen$(GP, ask_\delta, GID, S_\delta, st_\delta) \to usk_{GID, S_\delta}$. The user's secret key generation algorithm takes as input the global public parameter GP, the authority secret key ask_δ, a global identifier GID, an attribute set $S_\delta \subseteq \mathcal{U}_\delta$, and the state information st_δ. It produces a user's secret key for (GID, S_δ) as follows.

1) Select an unassigned leaf node η of $\mathcal{BT}_\delta$ and store GID in it.
2) For each node $\theta \in \mathbf{Path}(\eta)$, choose a random integer $r_u \in Z_p^*$ for each attribute $u \in S_\delta$, and create a key component $usk_\theta = \{(K_{\theta,\delta,u,GID}, K'_{\theta,\delta,u,GID},\}_{u \in S_\delta}$, where

$$K_{\theta,\delta,u,GID} = P^{H_2(GID)} g^{-r_\theta} H_1(GID)^{\beta_\delta} F(u)^{r_u}$$

and

$$K'_{\theta,\delta,u,GID,} = g^{r_u}.$$

3) Output the user's secret key $usk_{GID,S_\delta} = \{usk_\theta\}_{\theta \in \mathbf{Path}(\eta)}$ and updated state information st_δ.

$\mathsf{UkGen}(GP, apk_\delta, ask_\delta, RL_\delta, st_\delta, t) \rightarrow uk_{\delta,t}$. The update key generation algorithm takes as input the global public parameter GP, the authority public key apk_δ, the authority secret key ask_δ, the current revocation list RL_δ and state information st_δ, and the update time period t. It generates an update key for non-revoked users as follows.

1) Run the algorithm $\mathbf{KUNode}(\mathcal{BT}_\delta, RL_\delta, t)$ to get a node set $Y_\delta \subseteq \mathcal{N}_\delta$.
2) For each node $\theta \in Y_\delta$, choose a random exponent $\gamma_t \in Z_p$ and compute

$$uk_{\theta,\delta} = (U_{\theta,\delta}, U'_{\theta,\delta}) = (g^{r_\theta} \cdot W_\delta(t)^{\gamma_t}, g^{\gamma_t}).$$

3) Output the update key $uk_{\delta,t} = \{uk_{\theta,\delta}\}_{\theta \in Y_\delta}$.

$\mathsf{UUskGen}(GP, apk_\delta, usk_{GID,S_\delta}, uk_{\delta,t}) \rightarrow uusk^t_{GID,S_\delta}$. The updated user's key generation algorithm takes as input the global public parameter GP, the authority public key apk_δ, a user's secret key usk_{GID,S_δ}, and an update key $uk_{\delta,t}$. It generates a updated user's key by conducting the following steps.

1) Find out $\theta \in Y_\delta \cap \mathbf{Path}(\eta)$, and parse the two components as $usk_\theta = \{(K_{\theta,\delta,u,GID}, K'_{\theta,\delta,u,GID}\}_{u \in S_\delta}$ and $uk_{\theta,\delta} = (U_{\theta,\delta}, U'_{\theta,\delta})$.
2) Choose a random exponent $\gamma'_t \in Z_p$, and for each attribute $u \in S_\delta$ compute

$$D_{\delta,u.GID} = K_{\theta,\delta,u,GID} \cdot U_{\theta,\delta} \cdot W_\delta(t)^{\gamma'_t}$$

and

$$D'_{\delta,u.GID} = K'_{\theta,\delta,u,GID}, \quad D_{\delta,t} = U'_{\theta,\delta} \cdot g^{\gamma'_t}.$$

3) Return the updated user's key

$$uusk^t_{GID,S_\delta} = \{D_{\delta,t}, \{D_{\delta,u.GID}, D'_{\delta,u.GID}\}_{u \in S_\delta}\}.$$

$\mathsf{Sign}(GP, uusk^t_{GID,S}, S, m, \Gamma(\cdot)) \rightarrow \sigma$. The signature generation algorithm takes as input the global public parameter GP, the updated user's key $uusk^t_{GID,S_\delta}$ with regard to the attribute set S along with the current time period t, the attribute set S, a message m, a predicate $\Gamma = (\mathbf{M}, \rho, \psi)$, where $\mathbf{M}$ is an $l \times n$ access matrix associated with two mappings $\rho : [l] \rightarrow \mathcal{U}$ and $\psi : [l] \rightarrow \mathcal{A}$ [i.e., the attribute $\rho(i)$ belongs to the authority $\psi(i)$], and $\Gamma(S) = 1$. The signer proceeds as below.

1) choose a random vector $\vec{v} = \{s, v_2, \cdots, v_n\}$ and computes the shares of s as $\vec{\lambda} = (\lambda_1, \cdots, \lambda_l)^\top$.
2) For each $i \in \mathbf{I}$ where $\mathbf{I} = \{i \in [l] : \rho(i) \in S\}$, compute

$$\sigma_0 = g^{s \cdot H_2(GID)}, \sigma_{1,i} = D^{-\lambda_i \cdot m}_{\psi(i),\rho(i),GID}, \sigma_{2,i} = H_1(GID)^{\lambda(i)},$$

$$\sigma_{3,i} = D'^{\lambda_i}_{\psi(i),\rho(i),GID}, \quad \sigma_{4,i} = D^{\lambda_i}_{\psi(i),t}.$$

3) Finally, the signer outputs the signature $\sigma = \{\sigma_0, \{\sigma_{1,i}, \sigma_{2,i}, \sigma_{3,i}, \sigma_{4,i}\}_{i \in \mathbf{I}}\}$ and transmits σ to the verifier.

Verify$(GP, \{apk_\delta\}_{\delta \in AA}, m, \sigma, \Gamma(\cdot), t) \to 0 \ or \ 1$. The verify algorithm takes as input the the global public parameter GP, the attribute authority public key $\{apk_\delta\}_{\delta \in AA}$, message m, signature σ under the predicate $\Gamma(\cdot)$ and time period t. The verify computes $\omega_i : i \in \mathbf{I}$ for secret reconstruction constants satisfying $\sum_{i \in \mathbf{I}} M_i \cdot \omega_i = (1, 0, \cdots, 0)$. The Verifier outputs the result "1" if the following equation holds; otherwise, the verifier outputs the result "0".

$$e(\sigma_0, P)^{-m} = \prod_{i \in \mathbf{I}} [e(\sigma_{1,i}, g) \cdot e(\sigma_{2,i}, g^{\beta_{\psi(i)} m}) \cdot e(\sigma_{3,i}, F(\rho(i))^m) \cdot e(\sigma_{4,i}, W_\delta(t)^m)]^{\omega_i}.$$

Revoke$(GID, RL_\delta, st_\delta, t) \to RL_\delta$. The revocation algorithm takes as input an identifier GID to be revoked at a time period t, the current revocation list RL_δ, and the state information st_δ. It returns the updated revocation list $RL_\delta \leftarrow RL_\delta(\eta, t)$, where η is the leaf node of $\mathcal{BT}_\delta$ used to store GID.

Trace$(GP, m, \Gamma(\cdot), \sigma, tk) \to GID$. TA can track the signer's authentic identity when a signer misuses the signing behavior. The *Trace* algorithm inputs the message Global parameters GP, the message m, the signature $\sigma = \{\sigma_{0,1}, \{\sigma_{1,i}, \sigma_{2,i}, \sigma_{3,i}, \sigma_{4,i}\}_{i \in \mathbf{I}}\}$ upon the message m about the predicate $\Gamma(\cdot)$ at the time period t_i and tracing key $tk = a$. For any possible identity GID, the algorithm calculates

$$e(\prod_{i \in I} \sigma_{2,i}^{\omega_i}, P^{H_2(GID)}) = e(H_1(GID)^a, \sigma_0).$$

If the aforementioned equation holds, the algorithm outputs the actual identity GID for the signer.

Correctness The correctness on the presented EFSTD-ABS scheme is demonstrated in this section.

In accordance with the Verify algorithm, we check the validity for σ on m by utilizing the equation as below. From the scheme, we can conclude that for attribute set S satisfying the access structure $(\mathbf{M}, \rho)$, $\sum_{i \in \mathbf{I}} M_i \cdot \vec{v} \cdot \omega_i = s$, for $\sum_{i \in \mathbf{I}} M_i \cdot \omega_i = (1, 0, \cdots, 0)$. Then

$$\prod_{i \in \mathbf{I}} [e(\sigma_{1,i}, g) \cdot e(\sigma_{2,i}, g^{\beta_{\psi(i)} m}) \cdot e(\sigma_{3,i}, F(\rho(i))^m) \cdot e(\sigma_{4,i}, W_\delta(t)^m)]^{\omega_i}$$

$$= \prod_{i \in \mathbf{I}} [e(D_{\psi(i),\rho(i),GID}^{-\lambda_i \cdot m}, g) \cdot e(H_1(GID)^{\lambda(i)}, g^{\beta_{\psi(i)} \cdot m}) \cdot e(D'^{\lambda_i}_{\psi(i),\rho(i),GID}, F(\rho(i))^m) \cdot$$

$$e(D_{\psi(i),t}^{\lambda_i}, W_{\psi(i)}(t)^m)]^{\omega_i}$$

$$= \prod_{i \in \mathbf{I}} [e(g^{aH_2(GID)-r_\theta} H_1(GID)^{\beta_{\psi(i)}} F(rho(i))^{rho(i)} \cdot g^{r_\theta} \cdot W_{\psi(i)}(t)^{\gamma_t} \cdot$$

$$W_{\psi(i)}(t)^{\gamma'_t}, g)^{-\lambda_i \cdot m} \cdot e(H_1(uid), g)^{\lambda_i \cdot \beta_{\psi(i)} \cdot m} \cdot e(g, F(\rho(i)))^{m \cdot rho(i) \cdot \lambda_i} \cdot$$

$$e(g^{\gamma t} \cdot g^{\gamma'_t}, W_{\psi(i)}(t))^{m \cdot \lambda_i}]^{\omega_i}$$

$$= \prod_{i \in \mathbf{I}} e(g, g)^{-aH_2(GID) \cdot \lambda_i \cdot \omega_i m} = e(g, g)^{-aH_2(GID) \cdot s \cdot m} = e(\sigma_0, P)^{-m}$$

5 Security Analysis

5.1 Forward Security

In the section, we reduce the security of the designed EFSTD-ABS scheme to the CDH assumption.

Theorem 1. *Provided that there is an adversary $\mathcal{A}$ that breaks the forward security of the designed ERTD-ABS scheme with a non-negligible probability ϵ, then there is a challenger $\mathcal{C}$ who is capable of addressing the CDH problem at probability ϵ', where*

$$\epsilon' \geq \frac{\epsilon}{q_{usk} \cdot q_{uusk} \cdot q_s \cdot q_{H_2} \cdot T \cdot p_1 \cdot (p_1 + p_2 \cdot (T - t^*)) \cdot (p_1 + p_2 \cdot (T - 1))}$$

Here, q_{usk} is the maximum quantity of UskGen Oracle, q_{uusk} is the maximum quantity of UuskGen Oracle, q_s is the maximum quantity of Sign Oracle, q_{H_2} is the maximum quantity of H_2-Oracle queries queried by $\mathcal{A}$, $p_1 = l^* \cdot (|U| - |S_i|)$ and $p_2 = \binom{|S_i|}{l^*}$.

Proof. Intuitively, our scheme updates the user's signing key at the end of each time period. During this process, a time-related parameter W(t) is randomized and embedded into the user's updated signing secret key. This update is designed to be one-way, meaning that even if a user obtains the signing key for the current time period, it is computationally infeasible to derive the signing key from any previous period. See Appendix A for details.

5.2 Privacy

Theorem 2. *The presented EFSTD-ABS scheme satisfies the privacy of the signer.*

Proof. Intuitively, in our scheme, signatures are generated by randomizing the user's signing secret key using randomness associated with the access control structure. This process is one-way, making it computationally infeasible for an attacker to reverse-engineer the user's identity from a signature. See Appendix B for details.

5.3 Traceability

Theorem 3. *The presented EFSTD-ABS scheme in Sect. 4 satisfies traceability.*

Proof. In the proposed EFSTD-ABS scheme, for any valid signature, the TA can always use the tk to recover the actual identity of the signer. Thus, it is sufficient to prove that the Trace algorithm can correctly retrieve the signer's identity.

$$\begin{aligned} e(\prod_{i \in I} \sigma_{2,i}^{\omega_i}, P^{H_2(GID)}) &= e(\prod_{i \in I} H_1(GID)^{\lambda_i \cdot \omega_i}, g^{a \cdot H_2(GID)}) \\ &= e(H_1(GID), g)^{s \cdot a \cdot H_2(GID)} \\ &= e(H_1(GID)^a, \sigma_0) \end{aligned}$$

Table 1. Properties comparison

Schemes	Decentralized	Traceability	Forward Security	Privacy
[8]	×	✓	✓	✓
[17]	✓	×	×	✓
ours	✓	✓	✓	✓

Table 2. Comparisons of computational cost

Schemes	Signature Generation	Signature Verification	Update	Tracing
[8]	$(4d-k+8)C_e + C_p$	$4C_p$	$(N(l^2+|U|))C_e$	$5C_P$
[17]	$(6k+1)C_e + (4k+1)C_p$	$(3k+1)C_e + 2kC_p$	\	\
ours	$(4k+1)C_e$	$5C_e + (4k+1)C_p$	$5C_e$	$(k+2)C_e + 2C_p$

6 Performance Analysis

We theoretically analyze the performance of the provided EFSTD-ABS with the schemes [8] [17] in terms of the computational overhead about signature generation, verification, update and tracing in Table 2. Furthermore, we also reveal that the proposed EFSTD-ABS scheme is efficient in terms of computational overhead based on experimental simulations.

6.1 Theoretical Analysis

In Table 1, we compare the relevant properties in the proposed EFSTD-ABS scheme with the schemes [8] [17]. It is obvious that the provided EFSTD-ABS scheme supports extended functionality and improved security.

We compare the computational cost for the provided EFSTD-ABS scheme and schemes [8,17] in Table 2. N represents the number of attributes, k denotes the number of attributes that the signer must hold to fulfill the signing predicate, $|U|$ represents the size of the attribute universe managed by the AA, There exists a parameter d, which permits the value k to change from 1 to d. In addition, we denote C_e, C_p as the time of exponentiation computation and pairing computation, respectively, and let l be the depth of the binary tree. As shown in Table 2, the computational cost of signature generation for the EFSTD-ABS scheme, and the schemes in [8,17], it is evident that our scheme incurs a lower signature generation cost than the schemes in [8,17], since no pairing operations are required during the signature generation in our scheme. Furthermore, the time complexity of the update algorithm is reduced from $\mathcal{O}(n^2)$ to $\mathcal{O}(n)$, as our scheme only updates the secret values associated with specific nodes in the tree, while in the scheme of [8], the update process depends on the attributes to be revoked. Moreover, the computation cost of tracing in our scheme is significantly lower than that of scheme [8] due to the reduced number of pairing operations involved in the traceability process.

6.2 Experimental Analysis

The following simulation experiments were conducted on a system running Ubuntu 24.04, equipped with a 4-core vCPU and 8GB of RAM. The target scheme was implemented using the C++ pairing-based cryptography (PBC) library, version pbc-0.5.14, with the type g curve provided by the PBC library. Furthermore, parallel computing is feasible for modern IoT devices, and we leverage this capability to enhance the efficiency of our scheme. The number of attributes ranges from 1 to 25, and the total number of time periods is 2^5.

Based on the theoretical analysis presented in Table 2 and the experimentally measured time consumption for pairing and exponentiation operations where the average time for one pairing operation is 0.205 ms and for one exponentiation operation is 4.054 ms, it can be concluded that, with 20 attributes, our scheme improves signature generation efficiency by 81.92% and traceability efficiency by 37.8% compared to scheme [8].

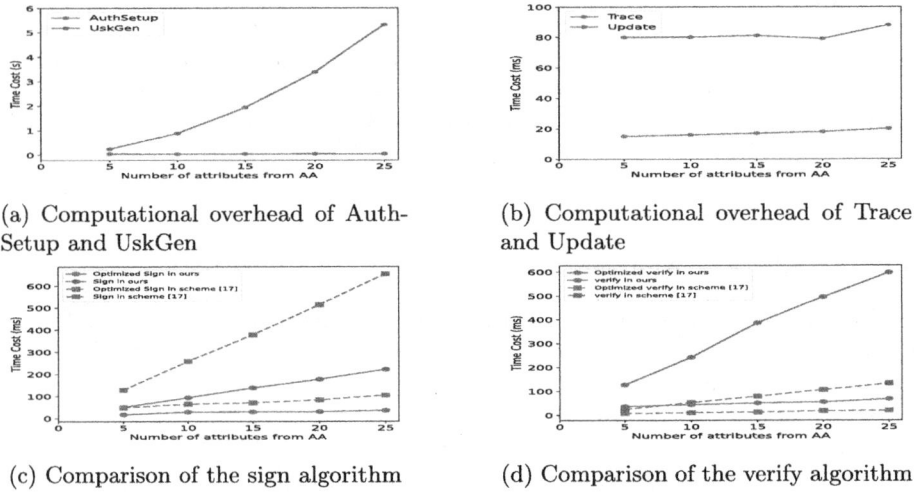

(a) Computational overhead of AuthSetup and UskGen

(b) Computational overhead of Trace and Update

(c) Comparison of the sign algorithm

(d) Comparison of the verify algorithm

Fig. 1. Experimental Analysis.

As shown in Fig. 1(a), the time overhead for the attribute authority to generate its key pair using global parameters is relatively low, while the computational overhead for generating the user's secret key is higher. Due to the decentralized design, this significant computational overhead is distributed among the attribute authorities responsible for the user's attributes, thereby alleviating the burden on individual attribute authorities.

As shown in Fig. 1(b), the computational overhead for both trace and update operations remains relatively constant as the number of attributes increases. In IoT environments, where each attribute authority manages a large number of attributes, our scheme enables efficient traceability of malicious signers, revocation of their identities, and updating of legitimate users' private keys, thereby minimizing potential losses.

As shown in Figures Fig. 1(c) and Fig. 1(d), regardless of whether multi-threading optimization is employed in the signing process, our scheme demonstrates superior signature generation efficiency compared to Scheme [17]. Furthermore, parallel computing enables a reduction in the signature generation time from 221 ms to 36 ms and a decrease in verification time from 597 ms to 68 ms achieving comparable efficiency to Scheme [17] when 25 attributes are involved, thereby significantly enhancing the scheme's efficiency.

7 Conclusion

We first present an EFSTD-ABS scheme that supports LSSS access structures. The proposed scheme simultaneously achieves the security properties of forward security, privacy, and traceability. It has been proven secure against selective predicate attacks under the random oracle model. Furthermore, the security of the EFSTD-ABS scheme is reduced to the CDH assumption. Experimental results demonstrate that the scheme is efficient.

Acknowledgments. This work was supported by Yunnan Province Science and Technology Major Project (NO.202302AD080002), National Natural Science Foundation of China (No.62202230, 62202231, No.62302224, No.62302226) and Yunnan Provincial New R&D Institution Cultivation Project (202404BQ040148).

A Forward Security

Proof. Suppose that $\mathcal{A}$ can break the proposed EFSTD-ABS scheme. We create a PPT challenger $\mathcal{C}$ who solves the CDH problem at the probability ϵ'. $\mathcal{C}$ takes a random CDH tuple (g, g^a, g^b) as input, in which $a, b \in Z_p$ is a random integer, g is a generator of G_1. The objective of $\mathcal{C}$ is to calculate g^{ab} $\mathcal{C}$ interacts with $\mathcal{A}$ in the undermentioned game.

Initial . $\mathcal{A}$ first declares a challenge signing predicate $\Gamma^*(\cdot)$ and forwards it to the challenger $\mathcal{C}$. What's more, $\mathcal{C}$ guesses t^* at random as the time period during which $\mathcal{A}$ will forge a signature.

Setup Let $c = |AA_c|$, $n' = |AA| - c$, and denote $R' = \{i \mid i \in [l^*] \wedge \psi^*(i) \in AA_c\}$. The algorithm $\mathcal{C}$ first constructs an LSSS matrix M' of size $l^* \times n^*$ in the following way: For all $i \in R'$ and all $j \in [n']$, assign $M'_{i,j} = 0$; in other cases, let $M'_{i,j} = M^*_{i,j}$. As shown in [15], the distribution of the shares $\{\lambda_i\}_{i \in [l^*]}$ sharing a secret $s \in Z_p$ according to M^* is exactly the same as the distribution of the shares $\{\lambda'_i\}_{i \in [l^*]}$ sharing the same secret s according to M'.

$\mathcal{C}$ set $P = g^a$ and publishes the global public parameter as $GP = \{G_1, G_2, p, e, g, P\}$. Then $\mathcal{C}$ set $AA_c \subseteq AA$ of corrupted attribute authorities. To generate the public parameter for each attribute authority, $\mathcal{C}$ considers the following two cases:

- $\delta \in AA_c$. $\mathcal{C}$ calls the algorithm AuthSetup, sends apk_δ and ask_δ to $\mathcal{A}$.
- $\delta \in AA \backslash AA_c$. $\mathcal{C}$ Choose random exponents $f'_{\delta,0}, f'_{\delta,1}, \cdots, f'_{\delta,l} \in Z_p^*$, sets $F'_{\delta,i} = g^{f'_{\delta,i}}$, picks random exponents $\beta'_\delta \in Z_p$ and sets the public parameter as $aok_\delta = \{g^{\beta'_\delta}, F'_{\delta,0}, F'_{\delta,1}, \cdots, F'_{\delta,l}\}$. In addition, $\mathcal{C}$ sets

$$W'_\delta(t) = F'_{\delta,0} \prod_{j=1}^{d} F'^{t[j]}_{\delta,j}$$

and generates generates a binary tree $\mathcal{BT}_\delta$ with N leaf nodes. Denote by $\mathcal{N}_\delta$ the set of all nodes of $\mathcal{BT}_\delta$, for each node $\theta \in \mathcal{N}_\delta$, $\mathcal{C}$ chooses a random exponent r'_θ. The secret secret key of the authority δ is kept as $ask_\delta = \{\beta'_\delta, \{r'_\theta\}_{\theta \in \mathcal{N}_\delta}\}$

Queries $\mathcal{A}$ is permitted to issue polynomially bounded-time queries to H_1-Oracle, H_2-Oracle, F-Oracle, UskGen Oracle, UkGen Oracle, UUskGen Oracle, Revocation Oracle, and Sign Oracle. $\mathcal{C}$ responses these queries as follows.

- H_1-Oracle: The simulator $\mathcal{C}$ maintains the list L_{H_1} to store the result $(GID_i, H_{1,i})$. If receiving a query request GID_i from $\mathcal{A}$, $\mathcal{C}$ checks L_{H_1} and returns the result if the request had been received. Otherwise, $\mathcal{C}$ selects a random parameter $h_{i,1} \in Z_p^*$, and returns the result $F_u = g^{h_{i,1}}$ to $\mathcal{A}$, then adds $(GID_i, H_{i,1})$ into L_{H_1}.
- H_2-Oracle: The $\mathcal{C}$ maintains the list L_{H_2} to store the result $(GID_i, H_{2,i})$, chooses a random parameter $k \in [1, q_{H_2}]$. If receiving a query request GID_i from $\mathcal{A}$, $\mathcal{C}$ checks L_{H_2} and returns the result if the request had been received. Otherwise, C processes as follows:
  (a) If $i = k$, it selects a random $h_{2,k} \in Z_p^*$ and publishes $H_{2,k} = h_{i,k}$ to $\mathcal{A}$. Then it adds $(GID_i, H_{2,i})$ into L_{H_2}.
  (b) If $i \neq k$, it selects a random $h_{2,i}, h'_{2,i} \in Z_p^*$ and publishes $H_{2,i} = h_{2,i} \cdot h'_{2,i}$ to $\mathcal{A}$. Then it adds $(GID_i, H_{2,i})$ into L_{H_2}.
- F-Oracle: The $\mathcal{C}$ maintains the list L_F to store the result (u, F_u). If receiving a query request u from $\mathcal{A}$, $\mathcal{C}$ checks L_F and returns the result if the request had been received. Otherwise, $\mathcal{C}$ selects a random parameter $f_u \in Z_p^*$, and returns the result $F_u = g^{f_u}$ to $\mathcal{A}$, then adds (u, F_u) into L_F.
- USKGen Oracle: Below we show how $\mathcal{C}$ generates a secret key for (GID_i, S_i). The $\mathcal{C}$ considers the following cases:
  (a) If $\Gamma^*(S_i \bigcup \{\bigcup_{\delta \in AA_c} \mathcal{U}_\delta\} \neq 1$, for each attribute $u \in S_i$, if $\psi(u) \in AA_c$, $\mathcal{C}$ calls the algorithm UskGen, and sends usk_{Gid_i, S_i} to $\mathcal{A}$, else $\mathcal{C}$ chooses a random integer r'_u and sets $r_u = r'_u - \frac{1}{f_u}(h_{1,i} \cdot \beta'_{\psi(u)})$, then gets $usk_\theta = (K_{\theta,\psi(u),u,GID_i}, K'_{\theta,\psi(u),u,GID_i})$, where

$$K_{\theta,\psi(u),u,GID_i} = P^{h_{2,i}} g^{-r'_\theta} g^{h_{1,i} \cdot \beta_\delta} g^{f_u \cdot (r'_u - \frac{1}{f_u}(h_{1,i} \cdot \beta'_{\psi(u)}))}$$

and

$$K'_{\theta,\delta,u,GID_i} = g^{r'_u - \frac{1}{f_u}(h_{1,i} \cdot \beta'_{\psi(u)})}.$$

(b) If $\Gamma^*(S_i \bigcup\{\bigcup_{\delta \in AA_c} \mathcal{U}_\delta\} = 1$, it stops and defines the process event as E_1.
- UKGen Oracle: Given (RL, t_j), the $\mathcal{C}$ first runs algorithm $\mathsf{KUNode}(\mathcal{BT}_\delta, RL_\delta, t_j)$ to obtain a set of nodes Y_δ and generates the corresponding update key component as follows:
  (a) If $\delta \in AA_c$, the $\mathcal{C}$ calls the algorithm UkGen and generates the update key component UK_{δ, t_j}.
  (b) If $\delta \in AA \setminus AA_c$, the $\mathcal{C}$ chooses a random exponent $\gamma \in Z_p$ for each node $\theta \in Y_\delta$ and computes

$$uk_{\theta,\delta} = (U_{\theta,\delta}, U'_{\theta,\delta}) = (g^{r'_\theta + \gamma_t \cdot (f'_{\delta,0} + \sum_{i=0}^d f'_{\delta,j} t[j])}, g^{\gamma_t}).$$

- UUskGen Oracle: given (GID_i, S_i, t_i), the $\mathcal{C}$ generates the corresponding updated user's key component as follows:
  (a) If $\Gamma^*(S_i \bigcup\{\bigcup_{\delta \in AA_c} \mathcal{U}_\delta\} \neq 1$ or $t_i > t^*$ and $\Gamma^*(S_i \bigcup\{\bigcup_{\delta \in AA_c} \mathcal{U}_\delta\} = 1$, the $\mathcal{C}$ calls the algorithm UUskGen, and generates a updated user's secret key to $\mathcal{A}$.
  (b) In other cases, it stops and defines the process event as E_2.

- Revocation Oracle: Given (GID_i, δ_i, t_i), the algorithm $\mathcal{C}$ updates the revocation list as in the original construction.
- Sign Oracle: Given an attribute set S, considering the query on signature about m upon predicate $\Gamma(S)$ at time t_j from $\mathcal{A}$. $\mathcal{C}$ generates signature as below.
  (a) If $\Gamma^*(S_i \bigcup\{\bigcup_{\delta \in AA_c} \mathcal{U}_\delta\} \neq 1$ or $\Gamma^*(S_i \bigcup\{\bigcup_{\delta \in AA_c} \mathcal{U}_\delta\} = 1$ and $t_i \neq t^*$, the $\mathcal{C}$ obtains the updated user's secret key by utilizing the aforementioned UUskGen Oracle queries. $\mathcal{C}$ gets signature with predicate $\Gamma(S) = 1$ and message m employing these updated user's secret key.
  (b) In other cases, it stops and defines the process event as E_3.

Forgery If the challenger $\mathcal{C}$ does not terminate in the aforementioned queries, $\mathcal{A}$ returns a valid forged signature $\sigma^* = \{\sigma_0^*, \{\sigma_{1,i}^*, \sigma_{2,i}^*, \sigma_{3,i}^*, \sigma_{4,i}^*\}_{i \in I}\}$ upon a message m^* related to the objective signing predicate $\Gamma^*(S)$ and time period t_j^*. If $t_j^* \neq t^*$ terminates. If not, $\mathcal{C}$ defines $ab = (a \cdot h_{2,i} + f_u \cdot r'_u) \cdot s$, and the calculation is as follows.

$$\sigma_{1,i}^* = P^{h_{2,i} \cdot \lambda_i} g^{h_{1,i} \cdot \beta_\delta \cdot \lambda_i} g^{f_u \cdot (r'_u - \frac{1}{f_u}(h_{1,i} \cdot \beta'_{\psi(u)})) \cdot \lambda_i} g^{(\gamma_t + \gamma'_t) \cdot (f'_{\delta,0} + \sum_{i=0}^d f'_{\delta,j} t[j]) \cdot \lambda_i}$$
$$= P^{h_{2,i} \cdot \lambda_i} g^{f_u \cdot r'_u \cdot \lambda_i} \sigma_{3,i}^{* \, f'_{\delta,0} + \sum_{i=0}^d f'_{\delta,j} t[j]}$$

for each $i \in [l^*]$, if $\psi[i] \in AA_c$, we have that $M'_{i,j} = 0$ for $j \in [n']$, which implies $\lambda_i = 0$ where $\overrightarrow{v} = \left(s, \overbrace{0, 0, \ldots, 0}^{n^*-1}\right)$, else $\lambda_i = s \cdot M'_{i,1}$. Then

$$\sigma_{1,i}^* = P^{h_{2,i} \cdot \lambda_i} g^{f_u \cdot r'_u \cdot \lambda_i} \sigma_{3,i}^{* \, f'_{\delta,0} + \sum_{i=0}^d f'_{\delta,j} t[j]}$$
$$= g^{(a \cdot h_{2,i} + f_u \cdot r'_u) \cdot s \cdot M'_{i,1}} \sigma_{3,i}^{* \, f'_{\delta,0} + \sum_{i=0}^d f'_{\delta,j} t[j]}$$

So, $g^{a \cdot b} = \dfrac{\sigma_{1,i}^*}{\sigma_{3,i}^* f'_{\delta,0} + \sum_{i=0}^d f'_{\delta,j} t[j]}^{\frac{1}{M'_{i,1}}}$. The probability for the challenger $\mathcal{C}$ not terminating is

$$\Pr(E) \geq (\dfrac{1}{q_{usk} \cdot l^* \cdot (|U| - |S_i|)} \cdot \dfrac{1}{q_{uusk} \cdot [l^* \cdot (|U| - |S_i|) + \binom{|S_i|}{l^*}) \cdot (T - t^*)]} \cdot \dfrac{1}{q_s \cdot [l^* \cdot (|U| - |S_i|) + \binom{|S_i|}{l^*}(T - 1)]} \cdot \dfrac{1}{q_{H_2}} \cdot \dfrac{1}{T}).$$

Sets $p_1 = l^* \cdot (|U| - |S_i|)$ and $p_2 = \binom{|S_i|}{l^*}$,

$$\Pr(E) \geq \dfrac{1}{q_{usk} \cdot q_{uusk} \cdot q_s \cdot q_{H_2} \cdot T \cdot p_1 \cdot (p_1 + p_2 \cdot (T - t^*)) \cdot (p_1 + p_2 \cdot (T - 1))}$$

Hence, if $\mathcal{A}$ wins at probability ϵ, $\mathcal{C}$ addresses the CDH problem at probability

$$\epsilon' \geq \dfrac{\epsilon}{q_{usk} \cdot q_{uusk} \cdot q_s \cdot q_{H_2} \cdot T \cdot p_1 \cdot (p_1 + p_2 \cdot (T - t^*)) \cdot (p_1 + p_2 \cdot (T - 1))}$$

B Privacy

Proof. In the proposed EFSTD-ABS scheme, the signature does not disclose the attributes of signer's attribute set S which is utilized to endorse the message m. The signer will construct a signature when his/her attribute set S satisfies the signing predicate $\Gamma^*(\cdot)$. We only need to prove that the provided EFSTD-ABS scheme fulfills the privacy of signer when S meets the $\Gamma^*(\cdot)$.

Setup $\mathcal{C}$ picks security parameter λ, T, N then executes GlobalSetup and AuthSetup algorithm to produce $GP = \{G_1, G_2, e, p, g, e(g,g), P, H, F\}$, $\{apk_\delta\}_{\delta \in AA}, \{ask_\delta\}_{\delta \in AA}$. The challenger $\mathcal{C}$ returns $GP, \{apk_\delta\}_{\delta \in AA}, \{ask_\delta\}_{\delta \in AA}$ to $\mathcal{A}$.

Queries $\mathcal{A}$ extracts two attribute sets S_0, S_1 satisfying $\Gamma^*(\cdot)$, then issues queries to USKGen Oracle, UKGen Oracle and UUSKGen Oracle. The challenger $\mathcal{C}$ responds to run USKGen, UKGen, UUSKGen algorithm and obtains updated user's secret key $uusk_{GID,S_\zeta}^t = \{D_{\delta,t}, \{D_{\delta,u.GID}, D'_{\delta,u.GID}\}_{u \in S_\zeta}\}$, where $\zeta \in \{0,1\}$. the challenger $\mathcal{C}$ randomly chooses r_u for each $u \in S_\zeta$ and γ_t, γ'_t, then computes $D_{\delta,u.GID} = g^{aH_2(GID)} H_1(GID)^{\beta_{\psi(i)}} F(rho(i))^{r_{rho(i)}} \cdot W_{\psi(i)}(t)^{\gamma_t} \cdot W_{\psi(i)}(t)^{\gamma'_t}$, $D'_{\delta,u.GID} = g^{r_{rho(i)}}$, $D_{\delta,t} = g^{\gamma_t} \cdot g^{\gamma'_t}$. Then the challenger $\mathcal{C}$ sends $uusk_{GID,S_\zeta}^t = \{D_{\delta,t}, \{D_{\delta,u.GID}, D'_{\delta,u.GID}\}_{u \in S_\zeta}\}$ to $\mathcal{A}$.

Challenge The adversary $\mathcal{A}$ makes Sign Oracle query on message m, which is relevant to an attribute set satisfying $\Gamma(\cdot)$ from either $uusk_{GID,S_0}^t$ or $uusk_{GID,S_1}^t$. $\mathcal{C}$ randomly selects $\zeta \in \{0,1\}$, a random vectors $\overrightarrow{v} = \{s_\zeta, v_{2,\zeta}, \cdots, v_{n,\zeta}\}$ and computes the shares of s_ζ as $\overrightarrow{\lambda} = (\lambda_{1,\zeta}, \cdots, \lambda_{l,\zeta})^\top$. Then $\mathcal{C}$ calculates the signature σ^* for the updated user's secret key

$$UUsk_{GID,S_\zeta}^t = \{D_{\delta,t}, \{D_{\delta,u.GID}, D'_{\delta,u.GID}\}_{u \in S_\zeta}\}.$$

by running Sign algorithm, and outputs a signature

$$\sigma_0 = g^{s_\zeta H_2(GID)}, \sigma_{1,i} = D_{\psi(i),\rho(i),GID}^{-\lambda_{i,\zeta} \cdot m}, \sigma_{2,i} = H_1(GID)^{\lambda_{i,\zeta}},$$

$$\sigma_{3,i} = D'^{\lambda_{i,\zeta}}_{\psi(i),\rho(i),GID}, \sigma_{4,i} = D^{\lambda_{i,\zeta}}_{\psi(i),t}.$$

Guess $\mathcal{A}$ returns guess $\zeta' \in \{0,1\}$ about ζ. Since r_u for each $u \in S_\zeta$ and $\gamma_t, \gamma'_t, \vec{v}$ are randomly chosen, this distribution of the signature produced via $uusk^t_{GID,S_0}$ or $uusk^t_{GID,S_1}$ is accordant. Hence, we have proved that the signature produced from the signing secret key $uusk^t_{GID,S_0}$ with attribute set S_0, it could also be generated by $uusk^t_{GID,S_1}$ with attribute set S_1. Thus, the presented EFSTD-ABS scheme fulfills the signer's privacy.

References

1. Anderson, R.: Two remarks on public key cryptology, Technical report, University of Cambridge, Computer Laboratory (2002). https://doi.org/10.48456/tr-549
2. Boldyreva, A., Goyal, V., Kumar, V.: Identity-based encryption with efficient revocation. In: Proceedings of the 15th ACM Conference on Computer and Communications Security, pp. 417–426 (2008). https://doi.org/10.1145/1455770.1455823
3. Ding, S., Zhao, Y., Liu, Y.: Efficient traceable attribute-based signature. In: 13th IEEE International Conference on Trust, Security and Privacy in Computing and Communications, TrustCom 2014, Beijing, China, September 24–26, 2014, pp. 582–589 (2014). https://doi.org/10.1109/TRUSTCOM.2014.74
4. El Kaafarani, A., Ghadafi, E., Khader, D.: Decentralized traceable attribute-based signatures. In: Topics in Cryptology - CT-RSA 2014 - The Cryptographer's Track at the RSA Conference 2014, San Francisco, CA, USA, February 25–28, 2014. Proceedings, vol. 8366, pp. 327–348 (2014). https://doi.org/10.1007/978-3-319-04852-9_17
5. Escala, A., Herranz, J., Morillo, P.: Revocable attribute-based signatures with adaptive security in the standard model. In: Progress in Cryptology - AFRICACRYPT 2011 - 4th International Conference on Cryptology in Africa, Dakar, Senegal, July 5–7, 2011. Proceedings, vol. 6737, pp. 224–241 (2011). https://doi.org/10.1007/978-3-642-21969-6_14
6. Gu, K., Jia, W., Wang, G., Wen, S.: Efficient and secure attribute-based signature for monotone predicates. Acta Informatica **54**(5), 521–541 (2017). https://doi.org/10.1007/S00236-016-0270-5
7. Gu, K., Wang, K., Yang, L.: Traceable attribute-based signature. J. Inf. Secur. Appl. **49**, 102400 (2019). https://doi.org/10.1016/J.JISA.2019.102400
8. Kang, Z., Li, J., Shen, J., Han, J., Zuo, Y., Zhang, Y.: TFS-ABS: traceable and forward-secure attribute-based signature scheme with constant-size. IEEE Trans. Knowl. Data Eng. **35**(9), 9514–9530 (2023). https://doi.org/10.1109/TKDE.2023.3241198
9. Li, J., Au, M.H., Susilo, W., Xie, D., Ren, K.: Attribute-based signature and its applications. In: Proceedings of the 5th ACM Symposium on Information, Computer and Communications Security, ASIACCS 2010, Beijing, China, April 13–16, 2010, pp. 60–69 (2010). https://doi.org/10.1145/1755688.1755697

10. Lu, Y., Wang, X., Hu, C., Li, H., Huo, Y.: A traceable threshold attribute-based Signcryption for Mhealthcare social network. Int. J. Sens. Netw. **26**(1), 43–53 (2018). https://doi.org/10.1504/IJSNET.2018.10009293
11. Maji, H.K., Prabhakaran, M., Rosulek, M.: Attribute-based signatures: achieving attribute-privacy and collusion-resistance. IACR Cryptology ePrint Archive, p. 328 (2008)
12. Maji, H.K., Prabhakaran, M., Rosulek, M.: Attribute-based signatures. IACR Cryptology ePrint Archive, p. 595 (2010)
13. Okamoto, T., Takashima, K.: Decentralized attribute-based signatures. In: Public-Key Cryptography - PKC 2013 - 16th International Conference on Practice and Theory in Public-Key Cryptography, Nara, Japan, February 26–March 1, 2013. Proceedings, vol. 7778, pp. 125–142 (2013). https://doi.org/10.1007/978-3-642-36362-7_9
14. Okamoto, T., Takashima, K.: Efficient attribute-based signatures for non-monotone predicates in the standard model. IEEE Trans. Cloud Comput. **2**(4), 409–421 (2014). https://doi.org/10.1109/TCC.2014.2353053
15. Seo, J.H., Emura, K.: Revocable identity-based encryption revisited: Security model and construction. In: Public-Key Cryptography–PKC 2013: 16th International Conference on Practice and Theory in Public-Key Cryptography, Nara, Japan, February 26–March 1, 2013. Proceedings 16, pp. 216–234 (2013). https://doi.org/10.1007/978-3-642-36362-7_14
16. Shanqing, G., Yingpei, Z.: Attribute-based signature scheme. In: 2008 International Conference on Information Security and Assurance (ISA 2008), pp. 509–511 (2008). https://doi.org/10.1109/ISA.2008.111
17. Sun, Y., Zhang, R., Wang, X., Gao, K., Liu, L.: A decentralizing attribute-based signature for healthcare blockchain. In: 2018 27th International Conference on Computer Communication and Networks (ICCCN), pp. 1–9 (2018). https://doi.org/10.1109/ICCCN.2018.8487349
18. Waters, B.R.: Efficient identity-based encryption without random oracles. IACR Cryptology ePrint Archive, p. 180 (2004)
19. Wei, J., Liu, W., Hu, X.: Forward-secure threshold attribute-based signature scheme. Comput. J. **58**(10), 2492–2506 (2015). https://doi.org/10.1093/COMJNL/BXU095
20. Yuen, T.H., Liu, J.K., Huang, X., Au, M.H., Susilo, W., Zhou, J.: Forward secure attribute-based signatures. In: Information and Communications Security - 14th International Conference, ICICS 2012, Hong Kong, China, October 29–31, 2012. Proceedings, vol. 7618, pp. 167–177 (2012). https://doi.org/10.1007/978-3-642-34129-8_15
21. Zhang, Y., Feng, D., Zhang, Z., Zhang, L.: On the security of an efficient attribute-based signature. In: Network and System Security - 7th International Conference, NSS 2013, Madrid, Spain, June 3–4, 2013. Proceedings, vol. 7873, pp. 381–392 (2013). https://doi.org/10.1007/978-3-642-38631-2_28

Post-quantum Cryptography

Cross-Domain Lattice-Based DAA Scheme with Shared Private-Key for Internet of Things System

Minzhi Liang[1(✉)], Liquan Chen[1,2], Yinghua Jiang[1], Xuyan Min[1], Jin Qian[3], and Jun Luo[3]

[1] Southeast University, Nanjing, China
{220235401,Lqchen,220235311}@seu.edu.cn, yhjiang@xzmu.edu.cn
[2] Purple Mountain Laboratories for Network Communication and Security, Nanjing, China
[3] Hangzhou Branch State Grid Zhejiang Electric Power, Nanjing, China

Abstract. With the rapid development of Internet of Things (IoT) technology, there has been exponential growth in the number of IoT systems, making domain-based management increasingly critical. As the number of domains continues to proliferate, the demand for inter-domain communication has become more frequent. Consequently, the development of an efficient and secure cross-domain authentication protocol to ensure the security and efficiency of inter-domain communication has emerged as a pressing issue within the IoT domain that requires immediate attention. Concurrently, the rise of quantum computers has rendered traditional cryptographic systems based on complex mathematical problems insecure, as these problems become solvable in the context of quantum computing. Therefore, the resistance of cryptographic systems to quantum attacks has become particularly important, and cryptographic systems built on lattice theory have shown significant advantages in dealing with quantum attacks. To address these issues, this paper proposes a new cross-domain Direct Anonymous Authentication (DAA) scheme with shared private key for Internet of Things System (SPCD-LDAA). The scheme relies on post-quantum key exchange protocols and anonymous authentication protocols, aiming to provide a new cross-domain communication solution for IoT systems in the era of quantum computing. The paper compares the proposed scheme with existing schemes, demonstrating improvements in both efficiency and security.

Keywords: Internet of Things · Quantum Resistance · Key Exchange · DAA · Cross-Domain Authentication

1 Introduction

The Internet of Things encompasses the interconnection of physical devices to the internet, including a spectrum of entities such as sensors, actuators, and embedded systems. The proliferation of IoT has expanded into various sectors and domains, including but not limited to smart homes, telemedicine, and intelligent transportation systems. As the constellation of devices expands, the significance of domain-specific device management becomes increasingly pronounced. Concurrently, the necessity for inter-domain communication has escalated, underscoring the fundamental role of cross-domain authentication in IoT. This authentication mechanism is pivotal for safeguarding the security and efficiency of communications across different domains, thus emerging as a critical area of focus within the IoT landscape.

Direct Anonymous Authentication [2] is a method of verifying user identity without disclosing user identity, which can be applied to devices with trusted computing platform (TPM). With the development of quantum computers, the security of traditional single-domain DAA schemes [7] or cross-domain DAA schemes [9] based on RSA/ECC has been challenged. Post-quantum DAA schemes, exemplified by Lattice-based DAA (LDAA), have emerged [8, 11], and they have been widely studied on IoT system [10]. In IoT, single-domain authentication DAA schemes can no longer meet the needs. Based on DAA, many scholars have extended the functionality of DAA to cross-domain authentication. Chen et al. [5] proposed a scheme for cross-domain authentication by pre-storing trust relationships within verifier entities, this approach risks exposing trust relationships across domains if one domain is compromised and lacks flexibility in establishing such relationships. Subsequently, they proposed a scheme in [6] that achieves cross-domain authentication through the issuance of passports by a third party, as depicted in the system framework shown in Fig. 1. Wang et al. also proposed a scheme in [16] that achieves cross-domain authentication with the help of a trusted third party. This scheme requires the assistance of an additional trusted third party. If the number of IoT devices increases, the load on the third party will become significant. Key exchange protocols enable both parties to securely share a common secret without the involvement of a third party, making it an effective solution for establishing cross-domain trust relationships.

Key exchange protocols refer to protocols where both parties establish a common key over an insecure channel, such as the Diffie-Hellman key exchange protocol [12]. With the development of post-quantum encrypted communication demands, post-quantum key exchange protocols have also been widely studied. Peikert et al. proposed a post-quantum key exchange protocol based on lattices in [14]. Based on their research, Alkim et al. proposed a more efficient post-quantum key exchange protocol [1], which has better performance in practical applications.

The contributions of this paper are as follows:

1. This paper proposes cross-domain lattice-based DAA scheme with shared private-key for Internet of Things systems. The scheme leverages a key exchange protocol to establish a pre-trust relationship, and then utilizes this pre-trust relationship to achieve cross-domain authentication.
2. The paper proposes SPCD-LDAA that employs a key exchange protocol to share private keys among different domains, and then uses this private key to generate identical parameters for issuing certificates, which are valid across these domains. It addresses the need for a central authority in cross-domain authentication and ensures the security, flexibility and efficiency of cross-domain authentication. Additionally, a trust relationship revocation process is introduced to mitigate the risk of trust relationship exposure.

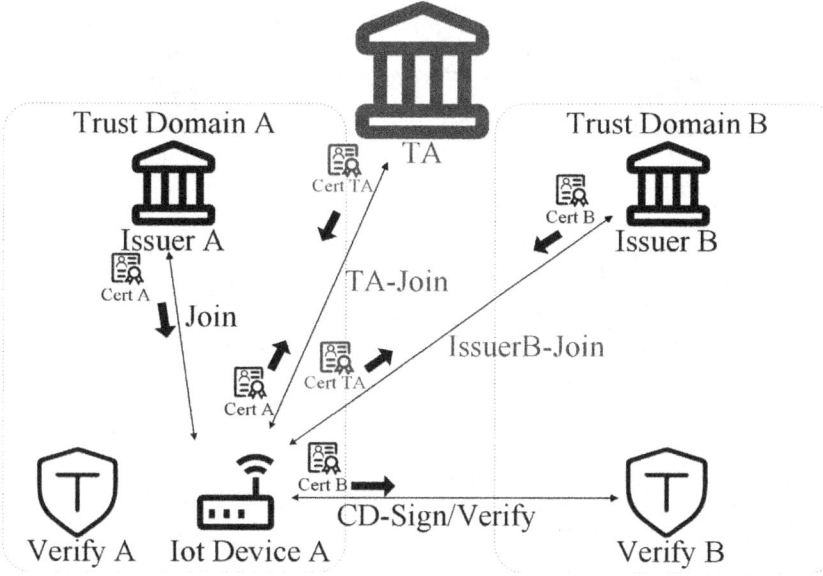

Fig. 1. Cross-domain authentication scheme proposed in [6]

3. This paper conducts a security analysis of the proposed scheme, demonstrating its security features. Comparative experiments are also performed under the same experimental conditions with existing schemes to illustrate the advantages of the proposed design in terms of security and efficiency.

In Sect. 2 of this paper, the design of the system architecture is described. In Sect. 3, the specific design of the SPCD-LDAA scheme is detailed. In Sect. 4, the security analysis of the SPCD-LDAA scheme is discussed. In Sect. 5, experimental results and analysis are presented and the paper is concluded in Sect. 6.

2 Framework of SPCD-LDAA

In this section, the system framework and authentication protocol of SPCD-LDAA will be described. The specific process of SPCD-LDAA discussed in this paper will be presented, followed by an introduction to the cross-domain authentication procedure through an illustrative example.

This paper proposes a SPCD-LDAA protocol system framework, as shown in Fig. 2. The SPCD-LDAA scheme proposed in this paper does not require an additional trusted third party and can evaluate the other domain to decide whether to agree to the other party's cross-domain application. We use IoT devices in domain A and domain B as an example for cross-domain authentication.

1. $Issuer_A$ applies to $Issuer_B$ for a shared cross-domain private key. If $Issuer_B$ agrees, both issuers share the private key and publicly disclose the corresponding key parameters in their respective domains. Consequently, both can issue certificates that are considered valid in both domains.

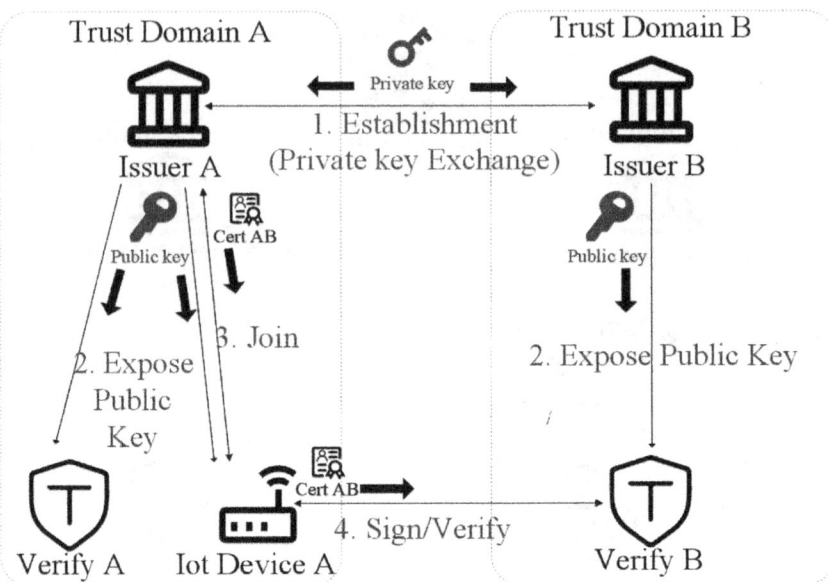

Fig. 2. Cross-domain authentication scheme of SPCD-LDAA

2. IoT Device A initiates a cross-domain communication request to $Issuer_A$ in its domain, applying for a valid identity certificate that is mutually recognized in domains A and B. $Issuer_A$ verifies the identity of the IoT Device and issues a certificate $Cert_{AB}$ that is valid in both domains A and B.
3. IoT Device A proves its valid identity to $Verifier_B$ through the issued certificate $Cert_{AB}$.

IoT devices consist of two parts: TPM and Host. The TPM stores the endorsement key to verify its validity, while the Host stores the identity certificate issued by the issuer. The subsequent trust relationship for the IoT device transfers from the endorsement key to the identity certificate. In the latter part of this paper, TPM&HOST will be used to represent the IoT device.

3 The Proposed SPCD-LDAA Scheme

In this section, the detailed process of SPCD-LDAA will be introduced, including the Setup stage, Establishment stage, Join stage, Sign/Verify stage, and Revoke stage. The symbols used in this section are defined in Table 1.

3.1 Setup Stage

In the Setup stage, the public parameters of the entire system are generated first. These parameters are used for the subsequent cross-domain authentication process. Generate

Table 1. Symbol Definition Table

Symbol	Definition
$\mathbb{Z}[x]$	Integers Ring
$\mathbf{R}_q$	Polynomial ring of modulo q
ξ	Gaussian distribution
q	Large prime number
d	The degree of the cyclotomic polynomial (Power of 2)
u, b, H	Public parameters of Issuer
R	Private key of Issuer
$Cert(\tau, s)$	Certificate of Host
τ	Unique identifier of certificate
bsn	Base name
Hash	Hash function
nym	Pseudo name
a_h, a_t	Public parameters of Host, TPM
e_h, e_t	Private parameters of Host, TPM
D_ξ	Gaussian distribution
ψ	Error distribution

system public parameters $(q, d, \mathbf{R}_q, \xi)$, where q is a large prime number, d is a power of 2 and also the dimension of the polynomial ring, $\mathbf{R}_q = \mathbb{Z}[x]/(X^d + 1)$ is the polynomial ring modulo q, and ξ is the standard deviation of the Gaussian distribution.

3.2 Establishment Stage

In the Establishment stage, a trusted relationship between two domains is established. This relationship is based on shared secret values calculated by both issuers. The sharing is completed through the Post-quantum Key Exchange protocol in [1]. In this paper, the establishment of a trusted relationship between two domains is discussed, using the example of domain A and domain B, with the detailed process shown in Table 2.

In which Parse, Shake, HelpRec, Rec definitions can be found in [1]. After both issuers exchange R, they can generate the same secret value and public parameters. If two issuers issue certificates with these parameters, the certificate is considered valid in both domains. The public key of the issuer is (u, b, H), and the private key is R, where $H = [h\ 1]$, $b = h \cdot R$.

3.3 Join Stage

In the Join stage, even if the Host has a single-domain identity certificate, using this certificate to prove its identity allows the issuer to query the Host's identity by comparing the issued certificate number, thus compromising the Host's anonymity to some

Table 2. Establish Trust Stage

ISSUER$_A$		ISSUER$_B$
$u, h \leftarrow \mathbf{R}_q$	$\xrightarrow{u,h}$	IF accept, store u, h
seed $\leftarrow \{0,1\}^{256}$		
$\alpha \leftarrow$ Parse(Shake(seed))		
$s_R, e_R \leftarrow \psi$		$s'_R, e'_R, e''_R \leftarrow \psi$
$b_R \leftarrow \alpha \cdot s_R + e_R$	$\xrightarrow{b,\text{seed}}$	$\alpha \leftarrow$ Parse(Shake(seed))
		$u_R \leftarrow \alpha s'_h + e'_h$
		$v_R \leftarrow b_h s'_h + e''_h$
$v'_R = u_R s_R$	$\xleftarrow{u_R, r_R}$	$r_R \leftarrow$ HelpRec(v_h)
$R \leftarrow$ Rec(v'_R, t_R)		$R \leftarrow$ Rec(v_R, t_R)

extent. The TPM and Host must cooperate to provide a zero-knowledge proof of their secret values to the issuer in order to prove their valid identity. Although embedded in the Host, the TPM will not disclose its secret values to the Host; thus, a portion of the computation is performed within the TPM. Manufacturers pre-install secret values on the TPM and the public parameters of the endorsement key on the issuer. The issuer will only permit valid TPMs to apply for identity certificates. In subsequent communication processes, the trust relationship is established by this identity certificate. The specific Join process is shown in Table 3.

Table 3. Join stage

TPM	HOST		ISSUER
Public parameters: a_t	Public parameters: a_h		Public parameters: (u, b, h)
Private parameters: e_t	Private parameters: e_h		Private parameters: R
$c_t = c \cdot e_t$	$\xleftarrow{c}$	$\xleftarrow{c}$	$c \leftarrow \mathbf{Z}$
$u_t = a_t \cdot e_t$	$c_h = c \cdot e_h$		
$\xrightarrow{u_t, c_t}$	$u_h = a_h \cdot e_h$		
	$y \leftarrow \mathbf{R}_q^{2 \times 1}$		
	$u_1 = u_t + u_h, \quad c_1 = [c_t \; c_h]$		
	$a = [a_t \; a_h]$		
	$H_a =$ Hash(a)		
	$w = a \cdot y$		
	$z = c_1 + y$	$\xrightarrow{(u_1, H_a, w, z)}$	IF $z \cdot a \neq c \cdot u_1 + w$, reject;
			IF $H_a \neq$ Hash$_a$, reject;
	Store Cert(τ, s)	$\xleftarrow{\text{Cert}(\tau,s)}$	Generate Cert(τ, s)

After sending the join request to the issuer, the issuer generates an integer c to enable TPM&HOST to conduct a zero-knowledge proof. The zero-knowledge proof process follows the Stern-type scheme in [16]. The TPM calculates the values of u_t, c_t, while the Host calculates the values of u_h, c_h, and the rest of the zero-knowledge proof is completed by the Host.

Upon receiving the joint zero-knowledge proof from TPM&HOST, the issuer first checks whether the zero-knowledge proof passes with its public parameters. If it passes, a certificate $Cert(\tau, s)$ is generated for TPM&HOST, and then the identity certificate is stored in the Host. Here, $\tau \leftarrow \mathbf{Z}$ is used to ensure the uniqueness of the identity certificate, and the MP-sample sampling algorithm [13] is employed to generate $s \leftarrow \mathbf{R}_q^{4 \times 1}$, which must satisfy (1).

$$(H\ b + \tau g) \cdot s = u + u_1 \qquad (1)$$

3.4 Sign/Verify Stage

In the Sign stage, the Host must prove to the verifier that it possesses a valid identity certificate while ensuring that the identity certificate and secret values are not disclosed, thus achieving anonymous authentication. Therefore, the TPM uses the pseudonym generated by its secret values to complete the zero-knowledge proof process to the verifier. The verifier subsequently identifies the same TPM&HOST based on the pseudonym. The pseudonym generation process is shown in Algorithm 1.

Algorithm 1. Pseudo-name generation algorithm

1: **if** bsn not exists **then**
2:     bsn $\leftarrow \mathbf{R}_q$
3: **else**
4:     bsn get from storage
5: **end if**
6: sk $\leftarrow \mathbf{Z}_q$
7: $d = \text{Hash}(\text{bsn}), e_n = \text{Hash}([\text{sk bsn}])$
8: nym $= d \cdot e_t + e_n$

After generating the pseudonym, TPM&HOST must still provide a zero-knowledge proof for the identity certificate to demonstrate possession of a valid identity certificate issued by the issuer. Here, the improved Sign scheme in [5] is chosen. The proof satisfies (2).

$$[H\ b\ g\ a_t\ a_h] \cdot [s_1\ s_2\ \tau s_2\ -e_t\ -e_h]^T = u \qquad (2)$$

where s_1, s_2 satisfy $s = [s_1\ s_2]$, multiplying the identity identifier by s_1 can prevent the disclosure of the identity identifier. The specific Sign/Verify process is shown in Table 4.

The verifier first checks whether the public parameters of the TPM&HOST domain are still valid. If they are valid, this indicates that the domain remains a trusted domain. If not, it will be rejected directly without further verification. After that, the zero-knowledge proof is verified. Only a TPM&HOST that passes all verifications indicates successful cross-domain communication and is permitted to communicate across domains.

Table 4. Sign/Verify stage

TPM&HOST		VERIFIER
		IF domain of (u,g,b) is not trusted, reject
$y \leftarrow D_\xi$	$\xleftarrow{c_1,c_2}$	$c_1, c_2 \leftarrow \mathbf{Z}$
$w = [H\ b\ g\ a_t\ a_h] \cdot y$		
$z = c_1 \cdot [s_1\ s_2\ \tau s_2\ -e_t\ -e_h]^T + y$		
$y_1, y_2 \leftarrow D_\xi$		
$w_1 = d \cdot y_1 + y_2$		
$z_1 = c_2 \cdot [e_t\ e_h] + y_1$		
$z_2 = c_2 \cdot e_n + y_2$		
IF $\text{rej}([z|z_1|z_2], [c_1 t | c_2 [e_t\ e_h] | c_2 e_n], \xi) = 1$, abort	$\xrightarrow[\text{nym},d,\text{bsn}]{z,z_1,z_2}$	IF $\|z\|, \|z_1\|, \|z_2\| > \text{Threshold}_1$, reject
		IF $\|d \neq \text{Hash}(bsn)\|$, reject
		IF $[H\ b\ g\ a_t\ a_h] \cdot z - c_1 \cdot u \neq w$, reject
		IF $d \cdot z_1 + z_2 - c_2 \cdot \text{nym} \neq w_1$, reject
		IF $\|\text{nym} - de\| \geq \text{Threshold}_2$, reject

3.5 Revoke Stage

If a domain, after quantifying the trust assessment of another domain, determines that the security of the other domain does not meet the requirements or concludes that there is no need for further cross-domain communication, it can choose to revoke the trust relationship. The issuer will refuse to recognize the mutual public secret values and public parameters and will no longer use these parameters to issue new certificates. The verifier will reject certificates issued based on these parameters and delete TPM&HOST that had been authenticated by this certificate. If cross-domain communication is needed again, it will be necessary to start from the Establishment stage once more.

4 Security Analysis

In this section, the security of SPCD-LDAA will be proven. The proof process uses the same proof method as [5], specifically employing the ideal/real model [3,15] to prove the security of the SPCD-LDAA protocol. Additionally, the unforgeability and anonymity of the proposed scheme will be proven, and the security of the mechanism will be analyzed.

The ideal/real model consists of the environment ε, the ideal system, and the real system. The ideal system includes the ideal function $\mathbf{F}$, the ideal protocol participants $\mathbf{P}_i$, and the simulator $\mathbf{S}$. The real system consists of the real protocol participants $\mathbf{P}_i$ and the adversary $\mathbf{A}$. Here, $\mathbf{S}$ corresponds to $\mathbf{A}$, and $\mathbf{F}$ implements the various modules of the real protocol. If ε cannot distinguish whether the ideal system or the real system is running, then the security of the SPCD-LDAA protocol can be proven.

The basic framework of this paper is based on the lattice-based DAA scheme. Building on this, two new processes, the Establishment stage and the Revoke stage, are introduced to improve the security and efficiency of cross-domain authentication, with security being proven. The security analysis will be extended based on the security analysis of [5], with a focus on analyzing the security of the Establishment and Revoke stages.

4.1 Ideal System Model

The ideal system model established in [5, 11] already incorporates the stages of Join, Sign, and Verify. This section will focus on analyzing the ideal functionality model for the Establish and Revoke stages.

1. **Game1:** The real-world protocol Π is implemented.
2. **Game2:** In this game, **F** receives all inputs and forwards them to **S**. Subsequently, **S** simulate the Π and send the output back to **F**. This game is equivalent to Game1.
3. **Game3:** Here, **F** implements the setup stage. It generates all necessary public parameters and transmits them to **S**. In this current game, **F** does not require any inputs. Consequently, Game3 is equivalent to Game2.
4. **Game4:** This game involves **F** implementing the Establish stage. The process is divided into two phases:
   (a) **Phase A (performed by I_A):** **F** generates and stores a random seed, then executes the process for I_A. It subsequently sends (b, seed) to **S**, which merely forwards these values. This phase is equivalent to Game3.
   (b) **Phase B (performed by I_B):** The behavior of I_B depends on whether it is honest or not. If I_B is honest, **S** simply forward the inputs to **F**. However, if I_B is dishonest, **S** can utilize the seed from Phase A to complete the process independently of I_B. **Phase B** is equivalent to **Phase A**
   Therefore, Game4 is equivalent to Game3.
5. **Game5:** In this game, **F** implements the Revoke stage. Whether it is I_A or I_B requesting to terminate the validity period of the cross-domain authentication, they must send a revocation request through **S**. Regardless of whether the issuer is honest or not, the authentication's validity period should be terminated in this round of communication. Consequently, **F** only needs to delete the corresponding authentication data, such as private keys. Thus, Game5 is equivalent to Game4.
6. **Game6:** Here, **F** implements the Verify stage. In this game, all necessary information for **F** has already been stored in previous stages. Since **F** does not require the secret keys of other entities, it can independently complete its task and forward the outputs to **S**. Therefore, Game6 is equivalent to Game5.

As demonstrated in [11], **Game16** (in [11]) corresponds step by step to **Game5** (in [11], and equivalently **Game6** in this paper), as both implement the Verify process. Consequently, starting from the definitions established in this paper, we can progress through these games sequentially to reach **Game16** in [11]. Since ε is unable to distinguish between the ideal system and the real system throughout this progression, it follows that **F** in this paper possesses all the functionalities of an ideal function. Therefore, the real protocol can be considered secure under the ideal function model.

4.2 Unforgeability and Anonymity

In the real system, to obtain a forged cross-domain identity certificate, the verification process is the same as in single-domain authentication. If **A** wants to forge a single-domain authentication certificate, the difficulty has been proven in [4,11]. I checks whether the identity information of the joiner is credible; **A** cannot forge identity information to obtain a valid identity certificate without solving the Ring-SIS problem. Additionally, in the Establishment stage, it involves the zero-knowledge proof of TPM&HOST's two secret values, not the direct verification of its identity identifier τ, making it impossible to infer the identity information of the Host, thereby ensuring the anonymity of the Host.

In the ideal system, **F** participates in the verification of the joiner's identity information and informs I. **S** attempting to forge a certificate cannot do so without solving the Ring-SIS problem. [5] has proven that **S** cannot obtain the Host's identity information through the signature, and the signature information in the SPCD-LDAA protocol is the same as in [5], thus guaranteeing the anonymity of the SPCD-LDAA protocol.

Combining the above proofs, the signature scheme of this proposal demonstrates unforgeability, and the identity information of the Host maintains anonymity. Unless the attacker can solve the Ring-SIS problem, it is impossible to forge a valid identity certificate or obtain the Host's identity information.

4.3 Security of Mechanism

In the cross-domain mechanism of this paper, the security proofs for other stages have been established in various Lattice-Based DAA articles. Thus, the security issues of the mechanism in this paper are focused on the Establishment stage and Revoke stage. The security proof of the Establishment stage primarily has two aspects: the security of the key exchange agreement and the establishment of the trust relationship. The security of the key exchange agreement has been proven in [1], while the security of the trust relationship establishment has been verified in articles on Lattice-Based DAA, as once the key is exchanged, it aligns with the security proof of the Setup stage.

This paper also considers scenarios where the trusted domain is compromised and becomes less credible, introducing the Revoke stage, which enables the revocation of the trust relationship. In [5], the pre-trusted relationship establishment is accomplished by embedding the public matrix into the verifier entity of the other domain. However, if the other domain is attacked, the public matrix may be leaked. The scheme proposed in this paper generates the trust relationship randomly. If the trust relationship is compromised, the domain can revoke this trust relationship at any time through the Revoke stage and regenerate the trust relationship when communication is needed next.

In summary, the security of the SPCD-LDAA protocol proposed in this paper can be attributed to the security of the single-domain LDAA scheme and the key exchange agreement scheme, both of which have been proven by many scholars, ensuring that the SPCD-LDAA protocol of this paper is secure.

5 Performance Analysis and Comparison

In this section, we conduct a thorough performance evaluation of our proposed scheme against existing methods, analyzing its time and memory consumption under various parameters. Additionally, we compare the mechanisms in terms of post-quantum security, trust flexibility, and the elimination of third-party requirements.

To explore the performance and advantages of this paper's scheme, a simulation experiment was designed to compare it with the schemes in [5,6,16]. Since each method implements the details of DAA in different ways, to better reflect the work efficiency of the cross-domain mechanism in this paper, each protocol's cross-domain mechanism was re-evaluated according to the various stages proposed herein. The experimental equipment is shown in Table 5.

Table 5. Experimental device

Device	Configuration
Central Processing Unit	11th Gen Intel(R) Core(TM) i5-11500 @ 2.70 GHz
Random Access Memory	32 GB (Lenovo DDR4 2666 MHz 16 GB $\times$ 2)
Operating System	Windows Subsystem for Linux (Ubuntu 22.04 TLS)
Programming Language	Python 3.11
Mathematical Library	Sagemath

5.1 Performance of Mechanism

For the establishment of trust relationships, since [5] involves pre-storing entities, the time for this part cannot be estimated, and scheme in [6,16] do not establish trust relationships in advance. Given that the focus of this paper's experiment is on the cross-domain time consumption of a Host, the establishment time of trust relationships is not considered. The theoretical calculation of time consumption for each protocol is shown in Table 6, where t_{join} represents the time for the Join stage, t_{sign} represents the time consumed by signing, and t_{verify} represents the time consumed by verification.

Table 6. Time Consumption of Each Scheme

Protocol	Cross-domain Time
CD-LDAA in [5]	$t_{\text{join}} + t_{\text{sign}} + t_{\text{verify}}$
CD-LDAA in [6]	$3t_{\text{join}} + t_{\text{sign}} + t_{\text{verify}}$
CD-DAA in [16]	$t_{\text{join}} + 3t_{\text{sign}} + 3t_{\text{verify}}$
SPCD-LDAA	$t_{\text{join}} + t_{\text{sign}} + t_{\text{verify}}$

This section focuses on the time consumption for a TPM&HOST to perform a single cross-domain authentication. The schemes mentioned in this paper only necessitate a single Join, sign/verify stage, as does the [5]. In contrast, the literature [6] requires additional TA-Join and IssuerB-Join stages, each of which consumes time equivalent to a single Join stage. The mechanism in [16] is similar to [6], but it employs a different signing method, necessitating multiple signings and authentications. The experiment evaluates the total time and memory consumption of each protocol under various parameters, as shown in Table 7.

Table 7. Experimental parameters

Parameter	Para. 1	Para. 2	Para. 3	Para. 4
d	512	1024	2048	4096
q	2^{43}	2^{51}	2^{62}	2^{70}

Both d and q are parameters defined in the Setup stage. The test was conducted to assess the total time consumption and memory usage of each protocol under different parameters. Figure 3 shows the time consumption of the three protocols under various parameters, and Fig. 4 illustrates the memory usage of the three protocols under these parameters.

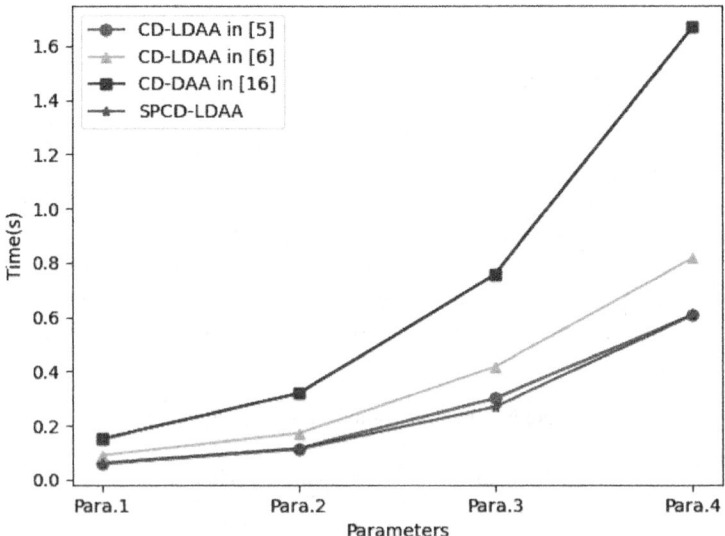

Fig. 3. Time consumption of each scheme

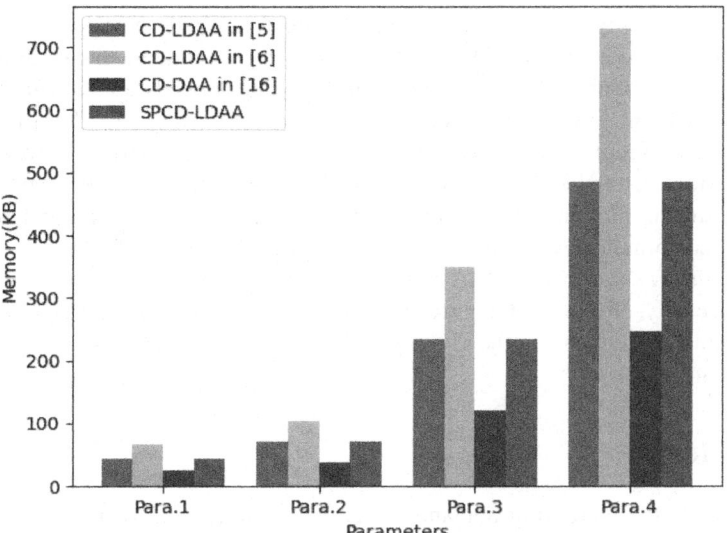

Fig. 4. Memory usage of each scheme

The proposed scheme in this paper exhibits comparable time consumption to that in reference [5], while being less than that in references [6,16]. In terms of memory consumption, the proposed scheme is also comparable to reference [5], but lower than that in reference [6]. Reference [16] employs a message-signing approach in the subsequent signing process rather than issuing certificates, resulting in lower memory consumption; however, this leads to significantly higher time consumption.

5.2 Comparison of Mechanism

Table 8. Results of Mechanism Comparison

	CD-LDAA in [5]	CD-LDAA in [6]	CD-DAA in [16]	SPCD-LDAA
Post-quantum	✓	✓	✗	✓
Pre-established trust	✓	✗	✗	✓
Revocable	✗	✗	✓	✓
Flexible Trust	✗	✗	✓	✓
Single Host Trust	✓	✓	✓	✓
No Third Party Required	✓	✗	✗	✓

SPCD-LDAA is based on lattices and has post-quantum capabilities. For members in the domain needing to communicate across domains, they must obtain cross-domain certificates separately, allowing them to authenticate a single host instead of blindly trusting the entire domain. SPCD-LDAA achieves cross-domain authentication by pre-establishing trust relationships between domains, which speeds up authentication efficiency. The scheme outlined in this paper adds a stage for revoking the trust relationship, improving security. The trust relationship can be established and revoked at any time, providing strong flexibility. The scheme does not require an additional third-party trusted platform, allowing all domains to establish trust relationships independently, enhancing performance when facing a large number of domains (Table 8).

In summary, the SPCD-LDAA proposed in this paper improves efficiency while ensuring security, eliminates the need for a third party, and achieves the revocability of trust relationships, enhancing the security and efficiency of the entire system in complex environments.

6 Conclusion

In response to the increasing demand for cross-domain authentication in IoT due to the excessive number of domains, this paper proposes a new cross-domain authentication solution, SPCD-LDAA. This scheme combines a key exchange agreement with a lattice-based cross-domain DAA scheme, which can resist quantum attacks. The original DAA framework introduces Establishment and Revoke stages to improve the security and efficiency of cross-domain authentication and proves its security. In the experimental process, the cross-domain mechanisms of multiple studies were compared in terms of time and space efficiency across multiple sets of parameters, and the differences among various mechanisms were analyzed.

Acknowledgment. This work was supported by the National Natural Science Foundation of China (U22B2026).

References

1. Alkim, E., Ducas, L., Pöppelmann, T., Schwabe, P.: Post-quantum key {Exchange—A} new hope. In: 25th USENIX Security Symposium (USENIX Security 16), pp. 327–343 (2016)
2. Brickell, E., Camenisch, J., Chen, L.: Direct anonymous attestation. In: Proceedings of the 11th ACM Conference on Computer and Communications Security, pp. 132–145 (2004)
3. Canetti, R.: Studies in secure multiparty computation and applications. Thesis (1995)
4. Chen, L., Tu, T., Yu, K., Zhao, M., Wang, Y.: V-LDAA: a new lattice-based direct anonymous attestation scheme for VANETs system. Secur. Commun. Netw. **2021**(1), 4660875 (2021)
5. Chen, L., Yin, B., Wang, J., Chen, Z., Meng, X.: An efficient cross-domain lattice-based direct anonymous authentication scheme. In: 2022 IEEE 8th International Conference on Computer and Communications (ICCC), pp. 1222–1228. IEEE (2022)
6. Chen, L., Zhu, W., Tu, T., Wang, Z.: Lattice-based DAA cross-domain authentication protocol for internet of things. In: 2023 9th International Conference on Computer and Communications (ICCC), pp. 1315–1321. IEEE (2023)
7. Chen, L., Dong, C., El Kassem, N., Newton, C.J., Wang, Y.: Hash-based direct anonymous attestation. In: Johansson, T., Smith-Tone, D. (eds.) PQCrypto 2023. LNCS, vol. 14154, pp. 565–600. Springer, Cham (2023). https://doi.org/10.1007/978-3-031-40003-2_21

8. Chen, L., El Kassem, N., Lehmann, A., Lyubashevsky, V.: A framework for efficient lattice-based DAA. In: Proceedings of the 1st ACM Workshop on Workshop on Cyber-Security Arms Race, pp. 23–34 (2019)
9. Chen, X., Feng, D.: A direct anonymous attestation scheme in multi-domain environment. Chin. J. Comput. **31**(7), 1122–1130 (2008)
10. Kassem, N., et al.: More efficient, provably-secure direct anonymous attestation from lattices. Futur. Gener. Comput. Syst. **99**, 425–458 (2019)
11. Kassem, N.E., et al.: Lattice-based direct anonymous attestation (LDAA). Cryptology ePrint Archive (2018)
12. Krawczyk, H.: HMQV: a high-performance secure Diffie-Hellman protocol. In: Shoup, V. (ed.) CRYPTO 2005. LNCS, vol. 3621, pp. 546–566. Springer, Heidelberg (2005). https://doi.org/10.1007/11535218_33
13. Micciancio, D., Peikert, C.: Trapdoors for lattices: simpler, tighter, faster, smaller. In: Pointcheval, D., Johansson, T. (eds.) EUROCRYPT 2012. LNCS, vol. 7237, pp. 700–718. Springer, Heidelberg (2012). https://doi.org/10.1007/978-3-642-29011-4_41
14. Peikert, C.: Lattice cryptography for the internet. In: Mosca, M. (ed.) PQCrypto 2014. LNCS, vol. 8772, pp. 197–219. Springer, Cham (2014). https://doi.org/10.1007/978-3-319-11659-4_12
15. Pfitzmann, B., Waidner, M.: Composition and integrity preservation of secure reactive systems (2000)
16. Wang, X., Cheng, H., Zhang, R.: One kind of cross-domain DAA scheme from bilinear mapping. In: 2014 IEEE 13th International Conference on Trust, Security and Privacy in Computing and Communications, pp. 237–243 (2014). https://doi.org/10.1109/TrustCom.2014.62

Compact Adaptively Secure Identity-Based Encryption from Middle-Product Learning with Errors

Jingjing Fan[1], Xingye Lu[2](✉), Man Ho Au[2], and Siu Ming Yiu[1]

[1] University of Hong Kong, Pokfulam, Hong Kong
{jjfan,smyiu}@cs.hku.hk
[2] The Hong Kong Polytechnic University, Kowloon, Hong Kong
{xing-ye.lu,mhaau}@polyu.edu.hk

Abstract. Identity-Based Encryption (IBE) is a cryptographic primitive where any string, such as an email address, can serve as a public key. With the advent of quantum computing, post-quantum secure IBE constructions have become critical for ensuring long-term data security. The state-of-the-art construction based on MPLWE introduced by Fan et al. significantly advanced the field by achieving adaptive security under standard assumptions, however the size of the master public key (MPK) grows linearly with the identity length, posing scalability challenges for real-world applications. In this work, we build on Fan et al.'s construction by employing a fully homomorphic trapdoor function to optimize the number of polynomials required for generating secret keys. This approach significantly reduces the MPK size from $O(\ell)$ polynomial vectors to $O(\ell^{1/d})$, where d is a constant. Despite this compactness, our scheme retains the same secret key and ciphertext sizes as Fan et al.'s construction and introduces no additional security assumptions.

Keywords: Lattice-Based Cryptography · Middle-Product LWE · Compact Identity-Based Encryption

1 Introduction

Identity-Based Encryption (IBE), introduced by Shamir in 1984 [17], is a public-key encryption paradigm where any arbitrary string, such as a mobile phone number or an email address, can function as a public key. This eliminates the need for a Public Key Infrastructure (PKI), which is a major bottleneck in traditional public-key cryptosystems. However, practical IBE schemes were not realized until 2001, when Boneh and Franklin [5], and independently Cocks [7], proposed the first IBE constructions.

The importance of IBE in its practical applications lies in simplifying key management. By leveraging identities directly as public keys, IBE reduces the complexity of certificate management in scenarios such as secure email, access control, and device authentication. As communication systems become increasingly interconnected and complex, IBE provides a scalable and streamlined solution to security challenges.

With the advent of quantum computing, the security assumptions that underpin traditional cryptographic primitives are threatened. Quantum algorithms, such as Shor's algorithm, can efficiently solve the integer factorization and discrete logarithm problems, undermining the foundations of widely used IBE schemes. This highlights the urgent need for post-quantum secure IBE constructions, particularly those based on quantum-resistant problems such as lattice-based cryptography.

Lattice-based IBE schemes began with Gentry et al.'s construction in 2008 [9], which was selectively secure in the random oracle model. This was followed by Agrawal et al. in 2010 [3], who achieved adaptive security in the standard model but at the cost of a large master public key (MPK). Subsequent research has focused on reducing the MPK size while maintaining security and efficiency. In standard lattice setting, Yamada's 2016 scheme [20], Zhang et al. [22], and Yamada [21] achieved logarithmic MPK size with respect to the identity length ℓ, and the MPK size further shortened by Abla [1] to $\omega(\frac{\log(\ell)}{\log\log\log(\ell)})$. For ring-based constructions, Katsumata and Yamada [10] first achieved a MPK size (number of ring vectors in MPK) sublinear to ℓ, and later Abla et al. [2] proposed a scheme with an MPK size that is independent of the identity length.

Despite these advancements, lattice-based IBE schemes still face challenges in balancing efficiency and security, particularly when adapting to real-world constraints. Recent constructions Middle-product LWE (MPLWE) have shown promise, offering robustness and efficiency. However, earlier MPLWE-based IBE schemes either relied on the random oracle model or only achieved selective security in the standard model. The first adaptively secure MPLWE-based IBE in the standard model [8] built on these efforts but inherited the inefficiency of a linearly growing MPK size.

In this paper, we address the challenge of constructing an adaptively secure IBE scheme in the standard model based on MPLWE with a shorter MPK size. By improving the abort-hash function from prior work [20], we encode identities more efficiently, achieving a sublinear growth of MPK size relative to identity length. Our proposed scheme not only reduces the MPK size but also maintains comparable secret key and ciphertext size to the state-of-the-art construction [8]. This result marks a significant step toward practical, quantum-secure IBE schemes, balancing compactness, efficiency, and adaptive security in the standard model.

1.1 Our Contribution

Our Main Construction. This paper introduces a compact, adaptively secure Identity-Based Encryption (IBE) scheme based on the Middle-Product Learning with Errors (MPLWE) problem. Our scheme significantly reduces the master public key size to $O(n\ell^{1/c} \log n)$, while maintaining $O(n \log n)$ size for both

Table 1. Comparison of Lattice-Based Adaptively Secure IBEs in the Standard Model.

LWE type	Scheme	$|\mathsf{mpk}|$	$|C|, |\mathsf{sk}_{\mathsf{id}}|$
Standard LWE	CHKP10 [6]	$\tilde{O}(n^2\ell)$	$\tilde{O}(n\ell)$
	ABB10 [3]	$\tilde{O}(n^2\ell)$	$\tilde{O}(n)$
	Yam16: Type 1 [20]	$\tilde{O}(n^2\ell^{\frac{1}{c}})$	$\tilde{O}(n)$
	Yam16: Type 2 [20]	$\tilde{O}(n^2\ell^{\frac{1}{c}})$	$\tilde{O}(n)$
	ZCZ16 [22]	$\tilde{O}(n^2\log\ell)$	$\tilde{O}(n)$
	Yam17: Type 1 [21]	$\tilde{O}(n^2\log^3\ell)$	$\tilde{O}(n)$
	Yam17: Type 2 [21]	$\tilde{O}(n^2\log^2\ell)$	$\tilde{O}(n)$
	Abla24 [1]	$\tilde{O}(n^2\frac{\log\ell}{\log\log\ell})$	$\tilde{O}(n)$
RLWE	KY16: Type 1 [10]	$\tilde{O}(n\ell^{\frac{1}{c}})$	$\tilde{O}(n)$
	KY16: Type 2 [10]	$\tilde{O}(n\ell^{\frac{1}{c}})$	$\tilde{O}(n)$
	ALWW21: Type 1 [2]	$\tilde{O}(n\log\ell)$	$\tilde{O}(n)$
	ALWW21: Type 2 [2]	$\tilde{O}(n)$	$\tilde{O}(n)$
MPLWE	FLA23 [8]	$\tilde{O}(n\ell)$	$\tilde{O}(n)$
	Our IBE	$\tilde{O}(n\ell^{\frac{1}{c}})$	$\tilde{O}(n)$

c is a flexible constant (recommended to be 2 or 3 in [10]).
$\tilde{O}(f(n)) = O(f(n)\log^k n)$ for some constant k.

ciphertexts and private keys. As shown in Table 1, this is a marked improvement over existing MPLWE-based IBE schemes. This reduction is achieved through a tuple representation of identity, leveraging the specific structure of isomorphic sets to enable a more compact encoding. Furthermore, taking a broader view, our scheme, compared to standard LWE-based constructions, offers reduced public key bandwidth, while avoiding the dependency on the specific algebraic structure of polynomial rings present in RLWE assumptions, thus striking a balance between the two.

Comparison with Existing IBEs. Table 1 demonstrates a comparison between our IBE scheme and existing lattice-based adaptively secure IBEs. For our analysis, we set the identity length as $\ell = O(n)$. Specifically, compare with [8], our IBE scheme is the most compact adaptively secure IBE scheme from the MPLWE. The size of the public parameters, ciphertexts, and secret keys are $\tilde{O}(n\ell^{\frac{1}{c}}), \tilde{O}(n)$, and $\tilde{O}(n)$, respectively. Among the RLWE-based and standard LWE-based IBE schemes, [2] has the smallest public key size. Despite the efficiency, the security of the schemes based on RLWE crucially depend on the choice of the underlying ring. Based on MPLWE, our scheme is at least as secure as the schemes in [10], while enjoying a broader selection of number fields. Compared with the schemes from standard lattice, i.e. schemes in [3,6,20–22] our scheme has shorter master public key size when assuming the length of the identity ℓ to be $O(n)$.

1.2 Overview of Our Construction and Security Proof

Our work builds on the KY16 framework for constructing compact IBE schemes from lattice problems [10]. While KY16 uses RLWE, we instead employ MPLWE - a variant offering stronger security guarantees for broader number fields. Our scheme preserves KY16's identity compression methodology but introduces key adaptations for MPLWE compatibility. Notably, KY16's security reduction requires invertible ring elements Fy(id), which becomes problematic in MPLWE settings as inverses rarely exist. Our critical modification replaces ring elements with scalars for Fy(id), ensuring invertibility while maintaining functionality. This migration to MPLWE constitutes our main contribution, overcoming fundamental algebraic constraints through novel parameterization and structural adjustments. The resulting scheme retains KY16's compactness while enhancing security foundations and efficiency through optimized parameter selection.

In detail, the security proof prepares the master public keys as follows: $\overrightarrow{b_0} = \overrightarrow{S_0}\overrightarrow{a} + y_0 \overrightarrow{g}, \overrightarrow{b_{i,j}} = \overrightarrow{S_{i,j}}\overrightarrow{a} + y_{i,j}\overrightarrow{g}$ for $(i,j) \in [2] \times [\kappa]$. We use $R^{<d}[x]$ to denote polynomials in the ring R with a degree less than d. Let $\mathbf{d} = (d_1, \cdots, d_n)^T \in \mathbb{Z}^n$ be an integer vector with positive coefficients. We use $R^{<\mathbf{d}}[x]$ to denote the polynomial vector, whose i-th entry is in $R^{<d_i}[x]$ for $i \in [n]$. In the above equation, $\overrightarrow{a} \in (\mathbb{Z}_q^{<n}[x])^t \times (\mathbb{Z}_q^{<n+d-1}[x])^{\gamma\tau}$, short $\overrightarrow{S_0}, \overrightarrow{S_{i,j}} = [\overrightarrow{s_1}, \overrightarrow{s_2}, \cdots, \overrightarrow{s_{t'}}]^T$, where $\overrightarrow{s_j} \in (\mathbb{Z}_q^d[x])^t \times \mathbb{Z}_q^{\gamma\tau}$ for $j \in [t']$, and $\overrightarrow{b_0}, \overrightarrow{b_{i,j}} \in (\mathbb{Z}_q^{<n+d-1}[x])^{\gamma\tau}$ are uniformly random polynomial vectors, $y_0, y_{i,j} \in \mathbb{Z}_q$ are uniformly random scalars and $\overrightarrow{g} \in (\mathbb{Z}_q^{<n+d-1}[x])^{\gamma\tau}$ is an analogue of the matrix $\mathbf{G}$ in [13]. We can write $h(\text{id}) = \overrightarrow{S_{\text{id}}}\overrightarrow{a} + F_{\mathbf{y}}(\text{id})\overrightarrow{g}$, where $F_{\mathbf{y}}(\text{id}) = y_0 + \sum_{(i,j) \in \mathfrak{F}(\text{id})} y_{1,i}y_{2,j}$.

We partition identities by following method: non-zero $F_{\mathbf{y}}(\text{id})$ enable trapdoor-based key extraction via $\overrightarrow{g}$, while $F_{\mathbf{y}}(\text{id}^*) = 0$ for the challenge identity id^* allows for MPLWE challenge embedding. The security proof requires non-negligible probability that all queried identities satisfy $F_{\mathbf{y}}(\text{id}^*) = 0$ for the challenge identity id^*. This introduces a parameterization constraint: Since $y_{i,j}$ scales with query count, and $\mathbf{S}_{i,j}$ scales accordingly, both require super-polynomial modulus q. To maintain polynomial q, we must bound Q - a restriction absent in KY16's RLWE-based scheme where $R_q = \mathbb{Z}_q^n$ naturally supports exponential identity space. To deal with this subtlety, we have to restrict the number of queries Q to n^c for some constant c.

2 Preliminaries

2.1 Notations

In this paper, the security parameter is denoted by n, and the growth of functions will be classified using the big-O notation. Also, we write $f(n) = \tilde{O}(g(n))$ if $f(n) = O(g(n) \cdot \log^c n)$ for some constant c. poly(n) is used to denote a random function $f(n) = O(n^c)$ for some constant c. A function $f(n)$ is negligible, denoted by negl(n), if for every polynomial g, $\exists N \in \mathbb{N}$ s.t. $f(n) < \frac{1}{g(n)}, \forall n > N$. We say that a probability is overwhelming if it is $1 - \text{negl}(n)$. In this paper, we will work on the polynomials in the ring $R = \mathbb{Z}[x]/(x^N+1)$ or $R_q = R/qR = \mathbb{Z}_q[x]/(x^N+1)$ for some positive integer N.

2.2 Identity-Based Encryption

Syntax. An identity-based encryption(IBE) scheme consists of four algorithms, KeyGen, Extract, Enc, Dec, which runs as follows: KeyGen(1^n) → (mpk, msk): On input the security parameter 1^n output the master public key mpk and the master secret key msk. Extract(id, mpk, msk) → sk$_{\text{id}}$: On input the identity id, the master public key mpk and the master secret key msk, output the secret key for the identity sk$_{\text{id}}$. Enc(mpk, id, m) → c: On input the master public key mpk, the identity id and the message m, output the ciphertext c. Dec(sk$_{\text{id}}$, c) → m: On input the ciphertext c and the secret key for the identity sk$_{\text{id}}$, output the message m.

An IBE scheme IBE = (KeyGen, Extract, Enc, Dec) possesses correctness if for all the identities id and message m, $\Pr[\text{Dec}(\text{sk}_{\text{id}}, \text{Enc (mpk, id}, m)) = m] = 1 - \text{negl}(n)$, where the probability is taken over the randomness of KeyGen, Extract, Enc and Dec.

IND-CPA Security. The IND-CPA security for an IBE is defined as follows: **Setup**: The challenger $\mathcal{C}$ runs KeyGen(1^n) → (mpk, msk) and sends mpk to the adversary $\mathcal{A}$. **Key Extraction Query**: The adversary $\mathcal{A}$ sends an identity id to challenger $\mathcal{C}$. $\mathcal{C}$ runs the extraction algorithm Extract(mpk, msk, id) → sk$_{\text{id}}$ and sends sk$_{\text{id}}$ back to $\mathcal{A}$. **Challenge**: $\mathcal{A}$ chooses two messages m_0, m_1 and an identity id* to be challenged. $\mathcal{A}$ sends (id*, m_0, m_1) to $\mathcal{C}$. If id* has been queried, $\mathcal{C}$ aborts and sends $\perp$ to $\mathcal{A}$. Otherwise, $\mathcal{C}$ picks coin $\xleftarrow{\$} \{0, 1\}$ and sends $c \leftarrow$ Enc(mpk, id*, m_{coin}) to $\mathcal{A}$. **Key Extraction Query**: $\mathcal{A}$ can make further key extraction queries on identity id $\neq$ id*. **Output**: $\mathcal{A}$ outputs coin* $\in \{0, 1\}$.

$\mathcal{A}$ wins the game if coin* = coin. The advantage of $\mathcal{A}$ winning the IND-CPA game is defined to be $Adv_\mathcal{A} = \Pr[\mathcal{A} \text{ wins IND-CPA game}] - \frac{1}{2}$.

Definition 1 (IND-CPA). *An IBE scheme is IND-CPA secure if, for any polynomial-time adversary $\mathcal{A}$, it has $Adv_\mathcal{A}^{IBE} = \text{negl}(n)$.*

2.3 Lattices and Gaussian Distributions

An n-dimensional lattice in m-dimensional Euclidean space is an additive discrete set defined as follows: $\Lambda(\mathbf{b_1}, \ldots, \mathbf{b_n}) = \{\sum_{i=1}^n x_i \mathbf{b}_i | x_i \in \mathbb{Z}\}$, where $\mathbf{b_1}, \ldots, \mathbf{b_n} \in \mathbb{Z}^m$ are linearly independent column vectors. Λ is full-rank if and only if $m = n$. $\Lambda^* := \{\mathbf{y} \in \text{span}_\mathbb{R}(\Lambda) | \langle \mathbf{y}, \mathbf{x} \rangle \in \mathbb{Z}, \forall \mathbf{x} \in \Lambda\}$ is defined to be the dual lattice of Λ, denoted as $\Lambda^\perp$. For positive integers n, q and any matrix $\mathbf{A} \in \mathbb{Z}^{n \times m}$, let $\Lambda^\perp := \{\mathbf{z} \in \mathbb{Z}^m | \mathbf{A}\mathbf{z} = \mathbf{0} \mod q\}$. For $\mathbf{u} \in \mathbb{Z}_q^n$ s.t. $\exists \mathbf{t} \in \mathbb{Z}_q^m$ satisfying $\mathbf{A}\mathbf{t} = \mathbf{u}$, let $\Lambda_\mathbf{u}^\perp(\mathbf{A}) := \{\mathbf{z} \in \mathbb{Z}^m | \mathbf{A}\mathbf{z} = \mathbf{u} \mod q\}$, $\Lambda_\mathbf{u}^\perp(\mathbf{A}) = \Lambda^\perp(\mathbf{A}) + \mathbf{t}$. Next, we will recap some well-known definitions concerning Gaussian distribution.

Continuous Gaussian Distribution: For a positive semi-definite matrix $\Sigma \in \mathbb{R}^{n \times n}$, the continuous Gaussian distribution D_Σ is the probability distribution over $\mathbb{R}^n$ whose density is proportional to $\rho_\Sigma(\mathbf{x}) = \exp(-\pi \mathbf{x}^T \Sigma^{-1} \mathbf{x})$.

Discrete Gaussian Distribution: Given a countable set $S \subset \mathbb{R}^n$ and $\sigma > 0$, the discrete Gaussian distribution $D_{S,\sigma,\mathbf{c}}$ is the probability distribution over S whose density is proportional to $\rho_{\sigma,\mathbf{c}}(x) := \exp(\frac{-\pi \cdot \|\mathbf{x}-\mathbf{c}\|^2}{\sigma^2})$. That is, for $x \in S$, $D_{S,\sigma,\mathbf{c}} := \frac{\rho_{\sigma,\mathbf{c}}(x)}{\rho_{\sigma,\mathbf{c}}(S)}$. If $\mathbf{c} = 0$, we can omit $\mathbf{c}$ and write $D_{S,\sigma}$ instead.

Next, we will illustrate some inequalities that will be used in our proof.

For any n-dimensional lattice Λ and real $\epsilon > 0$, the smoothing parameter $\eta_\epsilon(\lambda)$ is defined to be the smallest real $s > 0$ such that $\rho_{1/s}(\Lambda^* \setminus \{\mathbf{0}\}) \leq \epsilon$ [14].

Gaussian Tail Inequality: For any $\epsilon > 0$, any $\sigma \geq \eta_\epsilon(\mathbb{Z})$, and any $t > 0$, we have $\Pr_{x \leftarrow D_{\mathbb{Z},\sigma,c}}[|x - c| \geq t \cdot \sigma] \leq 2e^{-\pi t^2} \cdot \frac{1+\epsilon}{1-\epsilon}$. In particular, for $\epsilon \in (0, 1/2)$ and $t \geq \omega(\sqrt{\log n})$, the probability that $|x - c| \geq t \cdot \sigma$ is negligible in n [9].

2.4 Polynomials, Matrices and Middle Product of Polynomials

First, we will recap some notations and operations for common vectors and matrices. In this paper, a vector refers to a column vector. For a matrix $\mathbf{A}$, $A_{i,j}$ denotes its (i,j)-th entry. For a vector $\mathbf{v} \in \mathbb{R}^n$, its Euclidean and sup norm is denoted by $\|\mathbf{v}\|$ and $\|\mathbf{v}\|_\infty$, respectively. We define the largest singular value of a matrix $\mathbf{A} \in \mathbb{R}^{m \times n}$ as $\sigma_1(\mathbf{A}) := max_{\|\mathbf{u}\|=1} \|\mathbf{Au}\|$. Also, we may define $\sigma_1(\mathbf{A})$ and the tensor product and introduce the noise rerandomization algorithm.

Lemma 1. *For any matrix $\mathbf{A} \in \mathbb{R}^{m \times n}$, we have $\sigma_1(\mathbf{A}) \leq \sqrt{mn} \max_{i,j} |A_{i,j}|$*

Definition 2 (Tensor Product). *For $m_1, m_2, n_1, n_2 > 0$, the tensor product $\otimes : \mathbb{R}^{m_1 \times n_1} \times \mathbb{R}^{m_2 \times n_2} \mapsto \mathbb{R}^{m_1 m_2 \times n_1 n_2}$ is defined to be the map: $\mathbf{A} \otimes \mathbf{B} = \begin{bmatrix} A_{1,1}\mathbf{B} & \cdots & A_{1,n}\mathbf{B} \\ \vdots & \ddots & \vdots \\ A_{m,1}\mathbf{B} & \cdots & A_{m,n}\mathbf{B} \end{bmatrix}$, for matrix $\mathbf{A} \in \mathbb{R}^{m_1 \times n_1}$ and matrix $\mathbf{B} \in \mathbb{R}^{m_2 \times n_2}$.*

Next, we will define the polynomials to be used in this paper. For any $d > 0$ and any set $S \subset R$, we will use $S^{<d}[x]$ to denote the set of polynomials in $R[x]$ of degree $< d$ whose coefficients are in S. For any distribution χ defined over R, we set $\chi^{<d}[x]$ to be the set of polynomials in $R^{<d}[x]$, with coefficients sampled independently according to χ.

Let $a = \sum_{i=0}^{d-1} \mathbf{a}_i x^i$ be a polynomial in $R^{<d}[x]$. We define the coefficient vector of a to be $\mathbf{a} = (\mathbf{a}_0, \mathbf{a}_1, \ldots, \mathbf{a}_{d-1})$ and for $0 \leq i \leq d-1$, $\mathbf{a}_i$ denotes the coefficient of x^i in a. Also, we set $\overline{\mathbf{a}} = (\mathbf{a}_{d-1}, \ldots, \mathbf{a}_1, \mathbf{a}_0)$. Naturally, we have $\overline{\overline{\mathbf{a}}} = \mathbf{a}$. Let $\mathbf{d} \in \mathbb{Z}^n = (d_1, \cdots, d_n)^T$ be an integer vector with positive coefficients. The polynomial vector, which is a column vector, is defined to be $\vec{a} := (a_1, \cdots, a_n)^T \in R^{<\mathbf{d}}[x]$, where a_i is a polynomial in $R[x]^{<d_i}$ for $i = 1, \cdots, n$.

Let $\mathbf{d} = (d_1, \cdots, d_n)^T \in \mathbb{Z}^n$ be an integer vector with positive coefficients and $t > 0 \in \mathbb{N}$, we define the polynomial matrix $\vec{A} := [\vec{a_1}, \cdots, \vec{a_t}]^T \in R^{t \times <\mathbf{d}}[x]$, where $\vec{a_i}$ is a polynomial vector in $R[x]^{<\mathbf{d}}[x]$ for $i = 1, \cdots, t$. We recap the inner product between polynomial vectors, multiplication between polynomial matrix and polynomial vector as in [8].

Definition 3 ([8], **Definition 2**). *Let* $\mathbf{d}, \mathbf{k} \in \mathbb{Z}^n$ *be integer vectors with positive coefficients. Let* $\overrightarrow{a} = (a_1, \ldots, a_n) \in R^{\mathbf{d}}[x]$, $\overrightarrow{b} = (b_1, \ldots, b_n) \in R^{\mathbf{k}}[x]$. *The inner product of* $\overrightarrow{a}$ *and* $\overrightarrow{b}$ *is defined to be* $\overrightarrow{a} \cdot \overrightarrow{b} := \sum_{i=1}^{n} a_i b_i$.

Definition 4 ([8], **Definition 2**). *Let* $\mathbf{d}, \mathbf{k} \in \mathbb{Z}^n$ *be integer vectors with positive coefficients. Let* $\overrightarrow{A} = [\overrightarrow{a_1}, \ldots, \overrightarrow{a_t}]^T \in R^{t \times <\mathbf{d}}[x]$ *where* $\overrightarrow{a_1}, \ldots, \overrightarrow{a_t} \in R^{\mathbf{d}}[x]$. *Let* $\overrightarrow{b} \in R^{\mathbf{k}}[x]$. *The multiplication between polynomial matrix and polynomial vector is defined to be* $\overrightarrow{A} \cdot \overrightarrow{b} := [\overrightarrow{a_1} \cdot \overrightarrow{b}, \ldots, \overrightarrow{a_t} \cdot \overrightarrow{b}]^T$.

For simplicity, in calculation and analysis, we tend to transform polynomial vectors and matrices into scalar vectors and matrices. Next, we will introduce the way we transform vectors into Toepliz matrices. Specifically, we make remarks on the special polynomial vectors and matrices that will be used in this paper.

Remark 1. In this paper, without further notation, we set gadget polynomial vector to be $\overrightarrow{g} = (g_1, g_2, \ldots, g_{\gamma\tau}) \in (\mathbb{Z}_q^{<n+d-1}[x])^{\gamma\tau}$ s.t. $g_j = 2^u x^{dv}$, for $j = v\tau + u + 1$, where $u \in \{0, \cdots, \tau - 1\}, v \in \{0, \cdots, \gamma - 1\}$ and its corresponding Topeliz matrix as $\mathbf{G} = \mathbf{I}_{\gamma d} \otimes [1 \cdots 2^{\tau-1}] \in \mathbb{Z}_q^{\gamma d \times \gamma d \tau}$. Here $\mathbf{I}_{\gamma d}$ is the identity matrix with size $\gamma d \times \gamma d$.

Definition 5 ([12], **Definition 6**). *Let R be a ring and $d, k > 0$ be positive integers. For any polynomial $a \in R^{<k}[x]$ of degree less than k, let $\mathsf{T}^{k,d}(a)$ denote the matrix in $R^{(k+d-1) \times d}$ whose ith column, for $i = 1, \cdots, d$ is given by the coefficients of $x^{i-1} \cdot a$, listed from the lowest to the highest degree.*

Lemma 2 ([13]). *For a gadget polynomial vector* $\overrightarrow{g} = (g_1, g_2, \ldots, g_{\gamma\tau}) \in (\mathbb{Z}_q^{<n+d-1}[x])^{\gamma\tau}$ *s.t.* $g_j = 2^t x^{dv}$, *for* $j = v\tau + t + 1$, *where* $t \in \{0, \cdots, \tau - 1\}, v \in \{0, \cdots, \gamma - 1\}$. *There exists a deterministic efficiently computable function* $\overrightarrow{g_b}^{-1}$ *that takes as input a polynomial vector* $\overrightarrow{u} \in (\mathbb{Z}_q^{n+2d-2}[x])^{\gamma\tau}$ *outputs a polynomial matrix* $\overrightarrow{R}$ *s.t.* $\overrightarrow{g_b}^T \overrightarrow{R} = \overrightarrow{u}^T$ *and the coefficients of the polynomials in* $\overrightarrow{R}$ *are in the interval* $[-b, b]$.

Proof. Set $\overrightarrow{u} = (u_1, u_2, \ldots, u_{\gamma\tau})$. The problem can be transformed to find polynomial vectors $\overrightarrow{r_1}, \ldots, \overrightarrow{r_{\gamma\tau}}$ s.t. $\overrightarrow{g} \cdot \overrightarrow{r_i} = u_i$. And the Topeliz matrix corresponding to $\overrightarrow{g}$ is $\mathbf{G} = \mathbf{I}_{\gamma d} \otimes [1 \cdots 2^{\tau-1}] \in \mathbb{Z}_q^{\gamma d \times \gamma d \tau}$ by Remark 1. Without loss of generality, we will take $\overrightarrow{r_i}$ and u_i as an example. Set $\mathbf{u_i} = \mathsf{T}^{n+d-1,d}(u_i)$. By [13] and [11], we are able to find $\mathbf{e}$ s.t. $\mathbf{Ge} = \mathbf{u_i}$. Then we write $\mathbf{e}$ as $(\mathsf{T}^{d,1}(r_{i,1})^T | \mathsf{T}^{d,1}(r_{i,2})^T | \ldots | \mathsf{T}^{d,1}(r_{i,\gamma\tau})^T)^T$ where $\deg r_{i,j} < d$ for $1 \leq j \leq \gamma\tau$. Set $\sigma = \omega(d\tau\gamma\sqrt{\gamma})$, the distribution of $(r_{i,j})_{j=1}^{\gamma\tau}$ is exactly $(D_{\mathbb{Z}^d, \sigma}[x])^{\gamma\tau} | \sum_{j=1}^{\gamma\tau} g_j \cdot r_{i,j} = u_i$. So we can find a deterministic computable function $\overrightarrow{g_b}^{-1}$ that on input a polynomial vector $u \in (\mathbb{Z}_q^{n+2d-2})^{\gamma\tau}$ outputs a polynomial matrix $\overrightarrow{R}$ s.t. $\overrightarrow{g}^T \overrightarrow{R} = \overrightarrow{u}^T$. Moreover, we require the coefficients of each polynomial in $\overrightarrow{R}$ in the interval $[-b, b]$.

After introducing the definitions of polynomial vectors and matrices and operations on them, we will introduce the middle-product operation. For our purpose, we focus on the polynomials with $\mathsf{poly}(n)$-bounded expansion factors. One such class [12] is the family of all $f = x^m + h$ where $\deg(h) \leq m/2$ and $\|h\|_\infty \leq \mathsf{poly}(n)$.

Definition 6 (Middle-Product [16], Definition 3.1). *Let d_a, d_b, d, k be integers such that $d_a + d_b - 1 = d + 2k$. The middle product $\odot_d : R^{<d_a}[x] \times R^{<d_b}[x] \to R^{<d}[x]$ is defined to be the map: $(a,b) \mapsto a \odot b = \lfloor \frac{(a \cdot b) \mod x^{k+d}}{x^k} \rfloor = \sum_{k \leq i+j \leq k+d-1} (a_i b_j) x^{i+j-k}$, where $\mathbf{a} = (a_0, \cdots a_{d_a-1})$ and $\mathbf{b} = (b_0, \cdots b_{d_b-1})$ are the coefficient vectors of a and b, respectively.*

From Definition 6, we can get that the middle product is commutative trivially, i.e., $a \odot_d b = b \odot_d a$ for all polynomials a, b. Besides, the middle product also satisfies a "quasi-associative" property defined below in Lemma 3.

Lemma 3 ([16]). *Let $d, k, n > 0$. For all $r \in R^{<k+1}[x], a \in R^{<n}[x], s \in R^{<n+d+k-1}[x]$, we have $r \odot_d (a \odot_{d+k} s) = (r \cdot a) \odot_d s$.*

Next, we exhibit the transformation of the middle-product between polynomials into the product of scalar matrices.

Lemma 4 ([15], Lemma 3.2). *Let $d, k > 0$. Let $r \in R^{<k+1}[x]$ and $a \in R^{<k+d}[x]$ and $b = r \odot_d a$. Then $\overline{\mathbf{b}} = (\mathsf{T}^{k+1,d}(r))^T \cdot \overline{\mathbf{a}}$. In other words, we have $\mathbf{b} = \overline{(\mathsf{T}^{k+1,d}(r))^T \cdot \overline{\mathbf{a}}}$.*

Finally, we come to the definition of MPLWE.

Definition 7 (MP-LWE [11], Definition 9). *Let $n > 0$, $q > 2$, $m > 0$, $\mathbf{d} \in [\frac{n}{2}]^t$, $\alpha \in (0,1)$ and let $\chi = D_{\alpha \cdot q}$ be a distribution over $\mathbb{R}_q$. For $s \in \mathbb{Z}_q^{<n-1}[x]$, we define the distribution $\mathrm{MP}_{q,n,\mathbf{d},\chi}(s)$ over $\Pi_{i=1}^t U(\mathbb{Z}_q^{n-d_i} \times \mathbb{R}_q^{<d_i}[x])$ as follows.*
1. For each $i \in [t]$, sample $a_i \xleftarrow{\$} \mathbb{Z}_q^{n-d_i}[x]$ and sample $e_i \leftarrow \chi^{<d_i}[x]$ 2. Output $(a_i, b_i := a_i \odot_{d_i} s + e_i)$ The MP-LWE problem consists of distinguishing between arbitrarily many samples from $\mathrm{MP}_{q,n,\mathbf{d},\chi}(s)$ (denoted as $\mathrm{MPLWE}_{q,n,\mathbf{d},\chi}(s)$) and the same number of samples from $\Pi_{i=1}^t U(\mathbb{Z}_q^{n-d_i} \times \mathbb{R}_q^{<d_i}[x])$ with non-negligible probability over the choice of $s \xleftarrow{\$} \mathbb{Z}_q^{<n-1}[x]$.

Theorem 1 (Hardness of MPLWE [11], Theorem 2). *Let $n > 0$, $q \geq 2$, $t > 0$, $\mathbf{d} \in [\frac{n}{2}]^t$, and $\alpha \in (0,1)$. For $S > 0$, let $\mathcal{F}(S, \mathbf{d}, n)$ be the set of polynomials in $\mathbb{Z}[x]$ that are monic, have constant coefficient coprime with q, have degree m in $\cap_{i=1}^t [d_i, n - d_i]$ and satisfy $\mathrm{EF}(f) < S$. Then there exists a ppt reduction from $\mathrm{PLWE}_{D_{\alpha \cdot q}}^{(f)}$ for any $f \in \mathcal{F}(S, \mathbf{d}, n)$ to $\mathrm{MPLWE}_{q,n,\mathbf{d},D_{\alpha' \cdot q}}$ with $\alpha' = \alpha \cdot \sqrt{\frac{n}{2}} \cdot S$. Here $\mathrm{EF}(f) := \max_{g \in \mathbb{Z}^{<2m-1}[x]} \frac{\|g \mod f\|_\infty}{\|g\|_\infty}$.*

It has been proven that solving $\mathrm{PLWE}_{q,\chi}^{(f)}(s)$ is as hard as solving the Shortest Vector Problem (SVP) over the ideal lattice in $\mathbb{Z}_q[x]/f$ [18].

3 Trapdoor Generators and Related Operations

In this section, we will introduce the trapdoor generator and the sampling algorithms used in this paper. We will use the trapdoor generation and preimage sampling algorithm proposed in [8].

Theorem 2 ([13], Definition 5.2, Theorem 4). Let $\mathbf{G} := \mathbf{I}_k \otimes [1\ 2\ \ldots\ 2^{\tau-1}] \in \mathbb{Z}_q^{k \times k\tau}$. Then given a matrix $\mathbf{A} \in \mathbb{Z}^{k \times (m+k\tau)}$, a matrix $\mathbf{R} \in \mathbb{Z}^{m \times k\tau}$ is a *G-trapdoor* for $\mathbf{A}$ if $\mathbf{A}\left[\mathbf{R}^T | \mathbf{I}_{k\tau}\right]^T = \mathbf{G}$. And there exists an efficient algorithm $\mathcal{C}(\mathbf{A}, \mathbf{R}, \mathbf{u})$ that operates as follows: On input matrices $\mathbf{A}, \mathbf{R}$ and a vector $\mathbf{u}$, sample from $D_{\Lambda_u^\perp(\mathbf{A}), \sigma}$, as long as $\sigma \geq \omega(\sqrt{\log k})\sqrt{7(\sigma_1(\mathbf{R}))^2 + 1)}$. Moreover, the running time of $\mathcal{C}$ is the time to compute $\mathbf{Rx}$ for $\mathbf{x} \in \mathbb{Z}^{k\tau}$ plus $\tilde{O}(m + k\tau)$.

Theorem 3 (TrapGen [11], Theorem 5). Let $q = \text{poly}(n)$ be a prime, $n \in \mathbb{Z}$, $d \leq n$, $dt/n = \Omega(\log n)$, $\sigma = c\ell \cdot \omega(n \log^{3/2} n)$ for some constant c, $\gamma = \lceil \log_2 q \rceil$ and $\gamma = \frac{n+2d-2}{d}$ be an integer. Then there exists a PPT algorithm TrapGen that generates a polynomial vector $\overrightarrow{a} = (a_1, a_2, \cdots, a_t, a_{t+1}, \cdots, a_{t+\gamma\tau}) \in (\mathbb{Z}_q^{<n}[x])^t \times (\mathbb{Z}_q^{<n+d-1}[x])^{\gamma\tau}$ together with a trapdoor td able to be stored in $O(n\tau t)$ space, where $\overrightarrow{a} \approx_s U((\mathbb{Z}_q^{<n}[x])^t \times (\mathbb{Z}_q^{<n+d-1}[x])^{\gamma\tau})$

Theorem 4 (SamplePre($\overrightarrow{a}$, td, u, σ) [11], Theorem 5). Let $q = \text{poly}(n)$ be a prime, $n \in \mathbb{Z}$, $d \leq n$, $dt/n = \Omega(\log n)$, $\sigma = \mathfrak{d}n^{c(\mathfrak{d}-1)} \cdot \omega(n^2 \log^2 n)$ for some constant c, $\gamma = \frac{n+2d-2}{d}$ be an integer, $\tau = \lceil \log_2 q \rceil$ and $t' = t + \gamma\tau$. Then there exists a ppt algorithm SamplePre that on input polynomial vector $\overrightarrow{a} = (a_1, a_2, \cdots, a_t, a_{t+1}, \cdots, a_{t+\gamma\tau})$ with a trapdoor td and a syndrome $u \in \mathbb{Z}_q^{n+2d-2}[x]$, outputs $\overrightarrow{r} = (r_i)_{i=1}^{t+\gamma\tau}$ satisfying $\sum_{i=1}^{t+\gamma\tau} a_i \cdot r_i = u$ in $\tilde{O}(nt)$ time. And the output distribution of $(r_i)_{i=1}^{t'}$ is exactly the conditional distribution $(D_{\mathbb{Z}^{2d-1}, \sigma}[x])^t \times (D_{\mathbb{Z}^d, \sigma}[x])^{\gamma\tau} | \overrightarrow{a} \cdot \overrightarrow{r} = u$.

After illustrating the algorithms, we would like to adjust the SampleLeft algorithm and the SampleRight algorithm in [8] to fit in our new setting.

Algorithm SampleLeft($\overrightarrow{a}$, td, u, σ). Inputs: a polynomial vector $\overrightarrow{a}$ in $(\mathbb{Z}_q^{<n}[x])^t \times (\mathbb{Z}_q^{<n+d-1}[x])^{\gamma\tau}$, a trapdoor td, a polynomial u, a parameter $\sigma = \ell n^{c(\mathfrak{d}-1)} \cdot \omega(n^2 \log^2 n)$. Outputs: $(r_i)_{i=1}^{t+2\gamma\tau} \approx_s (D_{\mathbb{Z}^{2d-1}, \sigma}[x])^t \times (D_{\mathbb{Z}^d, \sigma}[x])^{2\gamma\tau} | \sum_{i=1}^{t+\gamma\tau} a_i \cdot r_i + \sum_{i=1}^{\gamma\tau} h_i \cdot r_{i+t'} = u$. The correctness of this algorithm is shown in [8] Lemma 6.

Algorithm SampleRight $(\overrightarrow{g}, y, \overrightarrow{a}, \overrightarrow{S}, u, \sigma)$. Inputs: a polynomial vector $\overrightarrow{a}$ in $(\mathbb{Z}_q^{<n}[x])^t \times (\mathbb{Z}_q^{<n+d-1}[x])^{\gamma\tau}$, a polynomial vector $\overrightarrow{g}$ acting as a trapdoor together with a non-zero scalar y, a polynomial u, a polynomial matrix $\overrightarrow{S}$ a parameter $\sigma = \ell n^{c(\mathfrak{d}-1)} \cdot \omega(n^2 \log^2 n)$. Outputs: $(r_i)_{i=1}^{t+2\gamma\tau} \approx_s (D_{\mathbb{Z}^{2d-1}, \sigma}[x])^t \times (D_{\mathbb{Z}^d, \sigma}[x])^{2\gamma\tau} | \sum_{i=1}^{t+\gamma\tau} a_i \cdot r_i + \sum_{i=1}^{\gamma\tau} h_i \cdot r_{i+t'} = u$. The correctness of this algorithm will be shown in Lemma 5.

Lemma 5 (SampleRight($\overrightarrow{g}, y, \overrightarrow{a}, \overrightarrow{S}, u, \sigma$) [8], Lemma 7). Let $q = \text{poly}(n)$ be a prime, $3d \leq n$, $dt/n = \Omega(\log n)$, $\sigma = \ell n^{c(\mathfrak{d}-1)} \cdot \omega(n^2 \log^2 n)$ for some constant c, $\gamma = \frac{n+2d-2}{d}$ is an integer, $\tau = \lceil \log_2 q \rceil$ and $t' = t + \gamma\tau$. Then there exists a ppt algorithm SampleRight that on input polynomial vector $\overrightarrow{g} \in (\mathbb{Z}_q^{<n+d-1}[x])^{\gamma\tau}$, a non-zero integer $y \in \mathbb{Z}^*$, polynomial vector $\overrightarrow{a} = (a_1, a_2, \cdots, a_{t+\gamma\tau})$, and polynomial matrix $\overrightarrow{S} = (\overrightarrow{s_1}, \overrightarrow{s_2}, \cdots, \overrightarrow{s_{t'}})^T$ where $\overrightarrow{s_j} \in (\mathbb{Z}_q^d)^t \times (\mathbb{Z}_q)^{\gamma\tau}$ for $1 \leq j \leq \gamma\tau$

and a syndrome $u \in \mathbb{Z}_q^{n+2d-2}[x]$, outputs polynomial vector $\vec{r} = (r_i)_{i=1}^{t+2\gamma\tau}$ satisfying $\sum_{i=1}^{t+\gamma\tau} a_i \cdot r_i + \sum_{i=1}^{\gamma\tau} h_i \cdot r_{t+i+\gamma\tau} = u$, where $h_i = \vec{s_i} \cdot \vec{a} + y\vec{g}$ for $i \in [\gamma\tau]$ in $\tilde{O}(nt)$ time. And the output distribution of (r_i) is exactly the conditional distribution $(D_{\mathbb{Z}^{2d-1},\sigma}[x])^t \times (D_{\mathbb{Z}^d,\sigma}[x])^{2\gamma\tau} | \sum_{i=1}^{t+\gamma\tau} a_i \cdot r_i + \sum_{i=1}^{\gamma\tau} h_i \cdot r_{i+t+\gamma\tau} = u'$.

Proof. The proof is exactly the same as that of Lemma 7 in [8] except for the value of σ. Let $\mathbf{S}$ be the matrix corresponding to $\vec{S}$. We can get that $\sigma_1(\tilde{\mathbf{S}}) = \sqrt{\gamma d\tau((2d-1)t+d\gamma\tau)} + \ell\omega(\gamma d\tau((2d-1)t+d\gamma\tau))\delta^{\mathfrak{d}-1}$. So this algorithm is correct by setting $\sigma = \ell n^{c(\mathfrak{d}-1)} \cdot \omega(n^2 \log^2 n)$. □

4 Our Middle-Product IBE

Injective Map. Let $\mathfrak{d}$ and ℓ be some integers, and $\kappa = \lceil \ell^{\frac{1}{\mathfrak{d}}} \rceil$. For any element in $[1, \ell]$, there exists an injective map that maps the element to an element in $[1, \kappa]^{\mathfrak{d}}$. Hence, there exists an injective map that maps any element in $\{0,1\}^{\ell}$ to a subset of $\{[1, \kappa]^{\mathfrak{d}}\}$. The latter injective map can be achieved by naturally extending the previous one. In this way, we can define an effectively computable injective function $\mathfrak{F}$ that maps a bit-string $\text{id} \in \mathcal{ID} = \{0,1\}^{\ell}$ to $\mathfrak{F}(\text{id}) \subset \{[1, \kappa]^{\mathfrak{d}}\}$.

Homomorphic Computation. Let $\mathfrak{d} \in \mathbb{N}$ be a constant. Let $\vec{h_1}, \vec{h_2}, \ldots, \vec{h_{\mathfrak{d}}} \in \mathbb{Z}_q^{<n+d-1}$ be polynomial vectors. We introduce a function $\mathsf{PubEval}_{\mathfrak{d}} : (\mathbb{Z}_q^{<n+d-1})^{\mathfrak{d}} \mapsto \mathbb{Z}_q^{<n+d-1}$ as follows: $\mathsf{PubEval}_{\mathfrak{d}}(\vec{h_1}, \vec{h_2}, \ldots, \vec{h_{\mathfrak{d}}}) = \vec{h_1}$, if $\mathfrak{d} = 1$; $\mathsf{PubEval}_{\mathfrak{d}}(\vec{h_1}, \vec{h_2}, \ldots, \vec{h_{\mathfrak{d}}}) = (\vec{h_1}^T \cdot \vec{g_b}^{-1}(\mathsf{PubEval}(\vec{h_2}, \ldots, \vec{h_{\mathfrak{d}}})^T))^T$ if $\mathfrak{d} \geq 2$.

Remark 2. In this paper, we define a series of matrices (denoted by $\vec{S}, \vec{S'}, \vec{S''}, \vec{S_{i,j}}, \ldots$) which enjoy the following properties: Without loss of generality, here we will denote the polynomial matrix by $\vec{S}$. 1. $\vec{S} = [\vec{s_1}, \vec{s_2}, \cdots, \vec{s_{t'}}]^T$, where $\vec{s_j} \leftarrow (\Psi^d[x])^t \times \Psi^{\gamma\tau}$ for $\Psi := U(\{-1, 0, 1\})$, $j \in [t']$. 2. We use the scalar matrix $\tilde{\mathbf{S}}$ to denote the Toepliz matrix corresponding to $\vec{S}$. Here $\tilde{\mathbf{S}}$ is defined as follows:

$$\tilde{\mathbf{S}} = \begin{bmatrix} \mathsf{T}^{d,d}(s_{1,1}) & \cdots & \mathsf{T}^{d,d}(s_{1,\gamma\tau}) \\ \vdots & \vdots & \vdots \\ \mathsf{T}^{d,d}(s_{t,1}) & \cdots & \mathsf{T}^{d,d}(s_{t,\gamma\tau}) \\ \hline \mathsf{T}^{1,d}(s_{t+1,t+1}) & \cdots & \mathsf{T}^{1,d}(s_{t+1,\gamma\tau}) \\ \vdots & \vdots & \vdots \\ \mathsf{T}^{1,d}(s_{t',t+1}) & \cdots & \mathsf{T}^{1,d}(s_{t',\gamma\tau}) \end{bmatrix}.$$

Lemma 6 ([10], Lemma 2, Lemma 6). *Let polynomial vector* $\vec{a} = (a_1, a_2, \cdots, a_{t+\gamma\tau})$, *and polynomial matrices* $\vec{S_1}, \ldots \vec{S_{\mathfrak{d}}}$, *where for* $0 \leq i \leq \mathfrak{d}$, $\vec{S_i}$ *is generated following Remark 2. Let* $y_1, y_2, \cdots, y_{\mathfrak{d}} \in \mathbb{Z}_q$. *Let* $\vec{h_i} = \vec{S_i} \cdot \vec{a} + y_i \cdot \vec{g}$. *Further, let* $\tilde{\mathbf{S}}_\mathbf{i}$ *be the Toepliz matrix corresponding to* $\vec{S_i}$ *generated following Remark 2, we can get that* $\sigma_1(\tilde{\mathbf{S}}_\mathbf{i}) \leq \sqrt{\gamma d\tau((2d-1)t+d\gamma\tau)}$ *and assume* $|y_i| \leq \delta$

for $i \in \mathfrak{d}$, and $\delta > \sqrt{\gamma d\tau((2d-1)t + d\gamma\tau)}$. Then there exists an efficient algorithm $\mathsf{TrapEval}_\mathfrak{d}$ that on input $\vec{S_1}, \ldots, \vec{S_\mathfrak{d}}, y_1, \ldots, y_\mathfrak{d}$ outputs $\vec{S'}$ s.t.

$$\mathsf{PubEval}_\mathfrak{d}(\vec{h_1}, \vec{h_2}, \ldots, \vec{h_\mathfrak{d}}) = \vec{S'} \cdot \vec{a} + y_1 y_2 \cdots y_\mathfrak{d} \vec{g}. \tag{1}$$

Let $\tilde{\mathbf{S}}'$ be the Toepliz matrix corresponding to $\vec{S'}$.

Moreover, we have $\sigma_1(\tilde{\mathbf{S}}') \leq \sqrt{\gamma d\tau((2d-1)t + d\gamma\tau)}\mathfrak{d}\delta^{\mathfrak{d}-1}$

Proof. We will prove it by mathematical induction. **Case 1** When $\mathfrak{d} = 1$, Eq. 1 holds trivially. **Case 2** When $\mathfrak{d} \geq 2$, we may assume that $\mathsf{PubEval}_\mathfrak{n}(\vec{h_1}, \vec{h_2}, \ldots, \vec{h_\mathfrak{n}}) = \vec{S''} \cdot \vec{a} + y_1 y_2 \cdots y_\mathfrak{n} \vec{g}$, holds for all $\mathfrak{n} \leq \mathfrak{d}$. We have $\mathsf{PubEval}_\mathfrak{d}(\vec{h_1}, \vec{h_2}, \ldots, \vec{h_\mathfrak{d}})^T = (\vec{S_1}^T \vec{g_b}^{-1}(\mathsf{PubEval}(\vec{h_2}, \ldots, \vec{h_\mathfrak{d}})^T) + y_1 \vec{S''^T})^T \vec{a} + y_1 y_2 \cdots y_\mathfrak{d} \vec{g}$. Set $\vec{S'} = \vec{g_b}^{-1}(\mathsf{PubEval}(\vec{h_2}, \ldots, \vec{h_\mathfrak{d}}))\vec{S_1} + y_1 \vec{S''}$, we can get that Eq. 1 holds. We may set $\vec{g_b}^{-1}(\mathsf{PubEval}(\vec{h_2}, \ldots, \vec{h_\mathfrak{d}})^T) = R$, and $\tilde{\mathbf{R}}$ be its corresponding Toepliz matrix. From Lemma 2 and 6 in [10], $\sigma_1(\tilde{\mathbf{S}}') \leq \sigma_1(\tilde{\mathbf{S}}_1)\delta^{\mathfrak{d}-1} + \sigma_1(\tilde{\mathbf{S}}_1)\sigma_1(\tilde{\mathbf{R}})(\delta^{\mathfrak{d}-1} - 1)/(\delta - 1)$. In our setting, we have $\sigma_1(\tilde{\mathbf{R}}) = b\sigma_1(\tilde{\mathbf{S}}_1)$, we have $\sigma_1(\tilde{\mathbf{S}}') = \omega(\gamma d\tau((2d-1)t + d\gamma\tau))\delta^{\mathfrak{d}-1}$. □

4.1 Our IBE Scheme

Overview of Our Construction. In our scheme, let $\mathfrak{d} \in \mathbb{N}$ be a flexible constant. Set the identity space to be $\mathcal{ID} = \{0,1\}^\ell$ for some constant $\ell \in \mathbb{N}$. There exists an efficiently computable injective mapping $\mathfrak{F}$ that maps an identity $\mathsf{id} \in \mathcal{ID}$ to a subset of $\{[1, \kappa]^\mathfrak{d}\}$ where $\kappa = \lceil \ell^{\frac{1}{\mathfrak{d}}} \rceil$. Then we define a hash function $h(\cdot)$ using homomorphic computation defined in Sect. 4. $\vec{h_\mathsf{id}} = h(\mathsf{id}) = \vec{h_0} + \sum_{(j_1, j_2, \ldots, j_\mathfrak{d}) \in \mathfrak{F}(\mathsf{id})} \mathsf{PubEval}(\vec{h_{1,j_1}}, \ldots, \vec{h_{\mathfrak{d}, j_\mathfrak{d}}})$. The master public key contains a polynomial u, a polynomial vector $\vec{a} \in (\mathbb{Z}_q^{<n}[x])^t \times (\mathbb{Z}_q^{<n+d-1}[x])^{\gamma\tau}$ together with its trapdoor td generated by $\mathsf{TrapGen}$ algorithm introduced in Theorem 3 and polynomial vectors $\{\vec{h_0}, \{\vec{h_{i,j}}_{(i,j) \in [\mathfrak{d},\kappa]}\}\}$. Set $t' = t + \gamma\tau$ and $t'' = t' + \gamma\tau$, the secret key for the identity is a polynomial vector $\vec{r}$ s.t. $[\vec{a}^T | \vec{h_\mathsf{id}}^T]^T \cdot \vec{r} = u$. The polynomial vector $\vec{r}$ can be sampled by $\mathsf{SampleLeft}$ algorithm defined in [8] Lemma 6. using trapdoor td. Let $q = q(n)$ be prime, $\tau := \lceil \log_2 q \rceil$, $n, d, k \in \mathbb{N}$ satisfying $\gamma = \frac{n+2d-2}{d} \in \mathbb{N}$ and $2d + k \leq n$. Let $t > 0$, $t' = t + \gamma\tau$, $t'' = t' + \gamma\tau$. Set $\chi := \lfloor D_{\alpha \cdot q} \rceil, \chi_1 := \lfloor D_{\alpha_1 \cdot q} \rceil$ be the distributions over $\mathbb{Z}$ where $\epsilon \leftarrow D_{\alpha \cdot q}$, or $\epsilon_1 \leftarrow D_{\alpha_1 \cdot q}$ is sampled and then rounded to the nearest integer respectively. Let $0 < s \leq d/2$ be an integer. Let $\sigma = c\ell \cdot \omega(n \log^{3/2} n)$ for some constant c. We define an IBE scheme with message space $\mathcal{M} = \{0,1\}^{<k+1}[x]$.

Key Generation. $\mathsf{KeyGen}(1^n)$: Generate $\vec{a} = (a_1, \ldots, a_n) \in (\mathbb{Z}_q^{<n}[x])^t \times (\mathbb{Z}_q^{<n+d-1}[x])^{\gamma\tau}$ and its trapdoor td using $\mathsf{TrapGen}$. Sample $\vec{h_0} \xleftarrow{\$} \mathbb{Z}_q^{<n+d-1}$, $\vec{h_{i,j}} \xleftarrow{\$} \mathbb{Z}_q^{<n+d-1}$ for $(i,j) \in [\mathfrak{d}] \times [\kappa]$. Sample $u \xleftarrow{\$} \mathbb{Z}_q^{n+2d-2}$. Define $h(\mathsf{id}) = \vec{h_0} + \sum_{(j_1, j_2, \ldots, j_\mathfrak{d}) \in \mathfrak{F}(\mathsf{id})} \mathsf{PubEval}(\vec{h_{1,j_1}}, \ldots, \vec{h_{\mathfrak{d}, j_\mathfrak{d}}})$. Output $\mathsf{mpk} = \{\vec{a}, \vec{h_0}, \{\vec{h_{i,j}}_{(i,j) \in [\mathfrak{d},\kappa]}\}, u\}$, $\mathsf{msk} = \mathsf{td}$.

Extraction. Extract(mpk, msk, id) : Let $\vec{h_{\text{id}}} = h(\text{id})$. Use SampleLeft $(\vec{a}, \vec{h_{\text{id}}}, \text{td}, u, \sigma)$ to generate $(r_1, \ldots, r_{t''})$ s.t. $\sum_{i=0}^{t'} a_i\, r_i + \sum_{i=1}^{\gamma\tau} h_{\text{id},i}\, r_{i+t'} = u$. Output $\text{sk}_{\text{id}} = (r_1, \cdots, r_{t''})$. Note that $[\vec{a}^T | \vec{h_{\text{id}}}^T]\text{sk}_{\text{id}} = u$.

Encryption. Enc(mpk, id, m) : Compute $\vec{h_{\text{id}}} = h(\text{id})$. Sample $p \xleftarrow{\$} \mathbb{Z}_q^{<n+2d+k-1}[x]$. For $1 \le i \le t$, sample $e_i \leftarrow \chi^{2d+k}[x]$, and compute $c_i = a_i \odot_{2d+k} p + e_i$. For $t+1 \le i \le t'$, sample $e_i \leftarrow \chi^{d+k+1}[x]$, and compute $c_i = a_i \odot_{d+k+1} p + e_i$. For $1 \le i \le \gamma\tau$, sample $e_{i+t'} \leftarrow \chi_1^{d+k+1}[x]$, compute $c_{i+t'} = h_{\text{id},i} \odot_{d+k+1} p + e_{i+t'}$. Sample $e_0 \leftarrow \chi^{k+1}$, and compute $c_0 = m\lfloor \frac{q}{2} \rfloor + u \odot_{k+1} p + e_0$. Output $c = (c_0, c_1, \cdots, c_{t''})$.

Decryption. Dec(sk_{id}, c) : $w = c_0 - \sum_{i=1}^{t''} c_i \odot_{k+1} r_i \mod q$. Output $\lfloor \frac{w}{q/2} \rfloor$.

Lemma 7. *Our IBE scheme satisfies* $1 - \mathsf{negl}(n)$ *correctness when* $\alpha = (n^{2c(\mathfrak{d}-1)+4}l^2\mathfrak{d}^2 \cdot \omega(\log^{\frac{9}{2}} n) + 1)^{-1}$ *and* $\alpha_1 = (n^{c(\mathfrak{d}-1)+\frac{5}{2}}\ell\mathfrak{d} \cdot \omega(\log^3 n) + 1)^{-1}$.

Proof. We aim to show that $\mathsf{Dec}(\mathsf{sk}_{\text{id}}, \mathsf{Enc}(\text{mpk}, \text{id}, m)) = m$ with probability $1 - \mathsf{negl}(n)$ over the randomness of KeyGen and Enc. Consider a random key pair (mpk, msk) and a random identity $\text{id} = \{0,1\}^\ell$. Let $\mathsf{sk}_{\text{id}} \leftarrow$ Extract(mpk, msk, id) and compute ciphertext $c = (c_0, (c_i)_{1 \le i \le t''}) \leftarrow$ Enc(mpk, id, m). By Lemma 3 (the quasi-associative law for middle products), we have $c_0 = m + (\sum_{i=0}^{t'} a_i r_i + \sum_{i=1}^{\gamma\tau} h_{\text{id},i} r_{i+t'}) \odot_{k+1} p + e' = m + \sum_{i=1}^{t} r_i \odot_{k+1} (a_i \odot_{2d+k} p) + \sum_{i=t+1}^{t'} r_i \odot_{k+1} (a_i \odot_{d+k+1} p) + \sum_{i=1}^{\gamma\tau} r_{i+t'} \odot_{k+1} (h_{\text{id},i} \odot_{d+k+1} p) + e'$, which leads to the following equation $c_0 - \sum_{i=1}^{t''} c_i \odot_{k+1} r_i = m + (e_0 - \sum_{i=1}^{t''} r_i \odot_{k+1} e_i)$. So if $\|m + (e_0 - \sum_{i=1}^{t''} r_i \odot_{k+1} e_i)\|_\infty < q/2$, Dec($\mathsf{sk}_{\text{id}}, c$) will output the message m correctly. Then, we will calculate the bound of the coefficients in $\sum_{i=1}^{t''} r_i \odot_{k+1} e_i$ to make the correctness criterion of our scheme satisfied. The coefficient of x^l in $r_i \odot_{k+1} e_i$ equals $\sum_{w \in [0, deg(r_i)] \cap [l+k-deg(e_i), z+k]} (\mathbf{y}_i)_w (\mathbf{e}_i)_{l+k-w}$. Applying the Gaussian tail inequality and a union bound, we can get the following bounds on $\|\mathbf{y}_i\|_\infty$ and $\|\mathbf{e}_i\|_\infty$: $\Pr[\|\mathbf{y}_i\|_\infty > \omega((\sqrt{\log n})\sigma)] = \mathsf{negl}(n)$ and $\Pr[\|\mathbf{e}_i\|_\infty > \omega((\sqrt{\log n})\alpha_1 q)] = \mathsf{negl}(n)$ as $\alpha_1 > \alpha$. Therefore, again by union bound, except with $\mathsf{negl}(n)$ probability $\|e_0 - \sum_{i=1}^{t''} r_i \odot_{k+1} e_i\|_\infty < K\omega((\sqrt{\log n})\sigma)\omega((\sqrt{\log n})\alpha_1 q) + \omega((\sqrt{\log n})\alpha_1 q)$ for $K := (2d-1)t + 2\gamma\tau d \ge \sum_{i=1}^{t''}(deg(r_i)+1)$. Setting $\alpha_1 < (4\sigma K \cdot \omega(\log n) + 1)^{-1} = (n^{c(\mathfrak{d}-1)+3}\ell \cdot \omega(\log^4 n) + 1)^{-1}$, the above is less than $q/4$ and thus the scheme is $(1 - \mathsf{negl}(n))$-correct. In this case, $\alpha = (n^{2c(\mathfrak{d}-1)+6}\ell^2 \cdot \omega(\log^{\frac{13}{2}} n) + 1)^{-1}$.
□

5 Security Proof

In this section, we are to prove the security of our IBE scheme. The main idea of our proof is the partition technique following [3,4,8,19–21].

Lemma 8 ([20], Lemma 8 Claim 1). *Let q be a prime and $0 \le Q \le \frac{n^{\bar{c}}}{2} - 1$. Define the hash function $F_{\text{Wat}} : \{F_\mathbf{y} : (\mathbb{Z}_q)^\ell \mapsto \mathbb{Z}_q\}$ as : $F_\mathbf{y}(\text{id}) = y_0 +$*

$\sum_{(j_1,j_2,\ldots,j_\mathfrak{d})\in\mathfrak{F}(\mathsf{id})} y_{1,j_1}\cdots y_{\mathfrak{d},j_\mathfrak{d}}$. The hash function F_{Wat} is $(Q,\alpha_{\min},\alpha_{\max})$ abort-resistant, with $\alpha_{\min}=\frac{1}{\kappa+1}(\frac{1}{\mathfrak{d}n^{\tilde{c}}})^{\mathfrak{d}}(1-\frac{Q}{n^{\tilde{c}}})$, $\alpha_{\max}=\frac{1}{\kappa+1}(\frac{1}{\mathfrak{d}n^{\tilde{c}}})^{\mathfrak{d}}$.

For identity queries $\{\mathsf{id}_1,\cdots,\mathsf{id}_Q\}$ and challenge identity id^*, we further define the probability that for all key extraction queries, $F_{\mathbf{y}}(\mathsf{id}_i)\ne 0$ and for the challenge identity id^*, $F_{\mathbf{y}}(\mathsf{id}^*) = 0$ to be $\Upsilon(\mathsf{id}^*,\mathsf{id}_1,\cdots,\mathsf{id}_Q)$. With artificial abort, according to [10], $\Upsilon(\mathsf{id}^*,\mathsf{id}_1,\cdots,\mathsf{id}_Q)$ is non-negligible for all possible combinations of $(\mathsf{id}^*,\mathsf{id}_1,\cdots,\mathsf{id}_Q)$. In our setting $\overrightarrow{g}$ is equipped with a trapdoor, which is publicly known [13]. Then we transform the polynomial vectors $\overrightarrow{a}$, $\overrightarrow{g}$ and $\overrightarrow{S_{\mathsf{id}}}$ to matrices $\mathbf{A}$, $\mathbf{G}$ and $\mathbf{S}$ respectively. The goal of the key extraction query is to find $\mathbf{e}$, the vector representation of $\mathsf{sk}_{\mathsf{id}}$, satisfying $[\mathbf{A}|\mathbf{S}\mathbf{A}+F_{\mathbf{y}}(\mathsf{id})\mathbf{G}]\mathbf{e}=\mathbf{u}$. Here $\mathbf{u}$ is the vector representation of the polynomial u predefined in mpk. As $[\mathbf{A}|\mathbf{S}\mathbf{A}+F_{\mathbf{y}}(\mathsf{id})\mathbf{G}]\left[-\mathbf{S}^T|\mathbf{I}\right]^T = F_{\mathbf{y}}(\mathsf{id})\mathbf{G}$, when $F_{\mathbf{y}}(\mathsf{id})\ne 0$, SampleRight algorithm works and the key extraction queries can be answered. When it comes to the challenge identity id^*, $F_{\mathbf{y}}(\mathsf{id}^*)=0$ and $\overrightarrow{h_{\mathsf{id}}}=\overrightarrow{S_{\mathsf{id}}}\cdot\overrightarrow{a}$. Set $(c_i)_{i=1}^{t'}$ and $c_0 - \lfloor\frac{q}{2}\rfloor m_{\text{coin}}$ to be the MPLWE samples, where m_{coin} coin $\xleftarrow{\$}\{0,1\}$ is the challenge message. Then the rest of ciphertext, i.e. $(c_i)_{i=t'+1}^{t''}$, will be generated using $(c_i)_{i=1}^{t'}$ and the rerandomization algorithm ReRand from [10]. At this final stage, the ciphertext contains no information about coin under the MPLWE assumption.

Our IBE is IND-CPA secure with the parameter setting given below.

Parameter Settings. The parameters used in our IBE are listed below.

$$d=\Theta(n) \qquad\qquad t=\Omega(\log n)$$
$$q = n^{2c(\mathfrak{d}-1)+c'+7}l^2\cdot\omega(\log^{\frac{13}{2}}) \qquad \sigma = \ell n^{c(\mathfrak{d}-1)}\cdot\omega(n^2\log^2 n)$$
$$\alpha = (\ell^2(n^{2c(\mathfrak{d}-1)+6}\cdot\omega(\log^{\frac{13}{2}} n)+1)^{-1} \quad \alpha_1 = (\ell n^{c(\mathfrak{d}-1)+3}\cdot\omega(\log^4 n)+1)^{-1}$$

Theorem 5. *Assume that* $\sigma = \ell\mathfrak{d}n^{c(\mathfrak{d}-1)+\frac{3}{2}}\cdot\omega(\log n)$ *for some constant* c, $dt/n = \Omega(\log n)$, $q = \text{poly}(n)$ *is a prime number*, $q = \Omega(\alpha^{-1}n^{1+c})$ *and* $q = \sigma\cdot\omega(\log^{1/2} n)$, $\alpha_1 < (4\sigma K\cdot\omega(\log n)+1)^{-1}$, *and* $\alpha = \alpha_1/(2\sqrt{t'}\ell\mathfrak{d}n^{c(\mathfrak{d}-1)+3/2}\cdot\omega(\log n))$ *where* $K := (2d-1)t + 2\gamma\tau d$, *the IBE system is IND-CPA adaptively secure assuming* $\text{MPLWE}_{q,n+2d+k,\mathbf{d},D_{\alpha\cdot q}}$ *is hard with degree vector* $\mathbf{d} = (d_i)_{i=1}^{t'+1}$ *where* $d_i = 2d+k$, *for* $1\le i\le t$; $d_i = d+k+1$, *for* $t+1\le i\le t'$; $d_i = k+1$, *for* $i = t''+1$.

Proof. Let $\mathcal{A}$ be a PPT adversary to break our scheme. Let $\epsilon = \epsilon(n)$ $\mathcal{A}$'s advantage. Let $Q = Q(n)$ be the upper bound of the number of key extraction queries $\mathcal{A}$ makes. Here, without the loss of generality, we may assume that $\mathcal{A}$ always makes Q queries. Then we define constant $\tilde{c}$ s.t. $Q \le \frac{n^{\tilde{c}}}{2} - 1$. We can see that such $\tilde{c}$ always exists. Our proof consists of a sequence of games. Among them, the first game is identical to the IND-CPA adaptive game from Sect. 2.2. And in the last game in the sequence, $\mathcal{A}$ has a negligible advantage to win. In Game 1, we employ the artificial abort to ensure that the queried identities $(\mathsf{id}_1,\ldots,\mathsf{id}_Q)$

and the challenge id id* selected by the adversary $\mathcal{A}$ satisfy our requirement. From Game 2 to Game 4, the way mpk is generated is changed gradually. After the alternation of mpk, $\mathcal{C}$ is able to answer all the key extraction queries with the new embedded trapdoor, while for the challenge identity id*, the trapdoor does not exit. From Game 5 to Game 7, the generation method of the ciphertext is gradually changed. In Game 7, the $\mathcal{A}$ has no advantage to win the game. The MPLWE assumption is used to prove that Game 6 and Game 7 are indistinguishable. In each game, two values coin′, $\widetilde{\text{coin}} \in \{0, 1\}$ are defined. While we set coin′ $= \widetilde{\text{coin}}$ in the first game, these values might be different in the later guess when artificial abort occurs. Finally, we define X_i to be the event that coin′ $=$ coin at the end of event Game i.

Game 0. This is the original game, identical to the IND-CPA adaptive game in Sect. 2.2, between a PPT adversary $\mathcal{A}$ and a challenger $\mathcal{C}$. In the challenge phase, the ciphertext is set to be $c = (c_0, (c_i)_{i \leq t''}) \leftarrow \mathsf{Enc}(\mathsf{mpk}, \mathsf{id}^*, m_{\mathsf{coin}})$. Here mpk is generated in the setup phase and id* is the challenge identity. At the end of the game, $\mathcal{A}$ outputs a guess $\widetilde{\text{coin}}$ for coin. Finally, $\mathcal{C}$ sets coin′ $= \widetilde{\text{coin}}$. By definition, we have $|\Pr[X_0] - \frac{1}{2}| = |\Pr[\text{coin}' = \text{coin}] - \frac{1}{2}|$.

Game 1. Game 1 is identical to Game 0 except that we add an artificial abort at the beginning of the game. The challenger $\mathcal{C}$ picks $\mathbf{y} = (y_0, \{y_{i,j}\}_{(i,j) \in [\mathfrak{d}] \times [\kappa]})$ where $y_0 \xleftarrow{\$} [-(\kappa+1)(\mathfrak{d}n^{\tilde{c}})+1, 0]$ and $y_{i,j} \xleftarrow{\$} [1, \mathfrak{d}n^{\tilde{c}}]$ for $(i,j) \in [\mathfrak{d}] \times [\kappa]$. Then define the function $F_\mathbf{y} : \mathcal{ID} \mapsto \mathbb{Z}_q$ as follows: $F_\mathbf{y}(\mathsf{id}) = y_0 + \sum_{(j_1, j_2, \ldots, j_\mathfrak{d}) \in \mathfrak{F}(\mathsf{id})} y_{1,j_1} \cdots y_{\mathfrak{d}, j_\mathfrak{d}}$. Recall that by Lemma 8, F_{Wat} is a $(Q, \alpha_{\min}, \alpha_{\max})$ abort-resistant hash function. Then $\mathcal{C}$ checks whether the following condition holds:

$$F_\mathbf{y}(\mathsf{id}^*) = 0 \wedge F_\mathbf{y}(\mathsf{id}_1) \neq 0 \wedge F_\mathbf{y}(\mathsf{id}_2) \neq 0 \wedge \cdots \wedge F_\mathbf{y}(\mathsf{id}_Q) \neq 0 \quad (2)$$

where id* is the challenge identity and $\mathsf{id}_1, \cdots \mathsf{id}_Q$ are the identities on which $\mathcal{A}$ has made key extraction queries. And set $\Upsilon(\mathsf{id}^*, \mathsf{id}_1, \cdots, \mathsf{id}_Q) = \Pr_\mathbf{y}[F_\mathbf{y}(\mathsf{id}^*) = 0 \wedge F_\mathbf{y}(\mathsf{id}_1) \neq 0 \wedge F_\mathbf{y}(\mathsf{id}_2) \neq 0 \wedge \cdots \wedge F_\mathbf{y}(\mathsf{id}_Q) \neq 0]$. Here we call $F_\mathbf{y}(\mathsf{id})$ an abort-resistant hash function. Then $\mathcal{C}$ does as follows. (I) *Abort Check*: If (2) does not hold, $\mathcal{C}$ sets coin′ $\xleftarrow{\$} \{0, 1\}$. In this case, we say that the challenger aborts. If condition (2) holds, the challenger $\mathcal{C}$ sets coin′ $= \widetilde{\text{coin}}$. (II) *Artificial Abort*: As in [3], with probability $\Upsilon(\mathsf{id}^*, \mathsf{id}_1, \cdots, \mathsf{id}_Q)$, $\mathcal{C}$ ignores $\mathcal{A}$'s output, and sets coin′ $\xleftarrow{\$} \{0, 1\}$, otherwise, $\mathcal{C}$ sets coin′ $= \widetilde{\text{coin}}$. For a $(Q+1)$ tuple of identities $\mathfrak{I} = (\mathsf{id}^*, \mathsf{id}_1, \ldots, \mathsf{id}_Q)$, let $\epsilon(\mathfrak{I})$ be the probability that an abort (either real or artificial) does not happen when extraction queries are made. Then we set $\epsilon_{\min}$ and $\epsilon_{\max}$ to be scalars s.t. $\epsilon(\mathfrak{I}) \in [\epsilon_{\min}, \epsilon_{\max}]$ for all $\mathfrak{I}$. We have the following lemma. The probability that Game 0 and Game 1 give a different output is bounded by [3] Lemma 28. Here, F_{Wat} is a $(Q, \alpha_{\min}, \alpha_{\max})$ abort-resistant hash function. If no abort occurs, $\epsilon_{\min} = \alpha_{\min}$ and $\epsilon_{\max} = \alpha_{\max}$. But then $\epsilon_{\max} - \epsilon_{\min} = \frac{1}{(\kappa+1)}(\frac{1}{\mathfrak{d}n^{\tilde{c}}})^{\mathfrak{d}} \frac{Q}{n^{\tilde{c}}}$, which is non-negligible. To obtain a good lower bound on $|\Pr[X_1 - \frac{1}{2}]|$, we opt for the Waters approach [19]. We introduce the artificial abort and define $\gamma(\cdot)$ as in [19]. With the artificial abort, $(\epsilon_{\max} - \epsilon_{\min}) \leq \alpha_{\min}$. So we can get $|\Pr[X_1] - \frac{1}{2}| \geq \frac{1}{2}\alpha_{\min}|\Pr[X_0 - \frac{1}{2}]| \geq \frac{1}{2(\kappa+1)}(\frac{1}{\mathfrak{d}n^{\tilde{c}}})^{\mathfrak{d}}(1 - \frac{Q}{n^{\tilde{c}}})|\Pr[X_0 - \frac{1}{2}]|$.

Game 2. Game 2 is identical to Game 1 except that we change the way that $\mathcal{C}$ generates the polynomial vectors $\vec{h_0}, \{\vec{h}_{i,j}\}_{(i,j)\in[\mathfrak{d}]\times[\kappa]}$. Instead of selecting them uniformly at random from $(\mathbb{Z}_q^{<n+d-1}[x])^{\gamma\tau}$, $\mathcal{C}$ first generates a polynomial vector $\vec{g} = (g_1, g_2, \cdots, g_{\gamma\tau}) \in (\mathbb{Z}_q^{<n+d-1}[x])^{\gamma\tau}$ s.t. $g_j = 2^u x^{dv}$, for $j = v\tau + u + 1$, where $u \in \{0, \cdots, \tau - 1\}, v \in \{0, \cdots, \gamma - 1\}$. After that the challenger $\mathcal{C}$ picks $\mathbf{y} = (y_0, \{y_{i,j}\}_{(i,j)\in[\mathfrak{d}]\times[\kappa]})$ where $y_0 \xleftarrow{\$} [-(\kappa+1)(\mathfrak{d}n^{\tilde{c}})+1, 0]$ and $y_{i,j} \xleftarrow{\$} [1, \mathfrak{d}n^{\tilde{c}}]$ for $(i,j) \in [\mathfrak{d}] \times [\kappa]$. Then $\mathcal{C}$ generates polynomial matrices $\vec{S_0}, \{\vec{S}_{i,j}\}_{(i,j)\in[\mathfrak{d}]\times[\kappa]}$ according to Remark 2. Next, set $\vec{h_0} = \vec{S_0} \cdot \vec{a} + y_0 \vec{g}$, and $\vec{h}_{i,j} = \vec{S}_{i,j} \cdot \vec{a} + y_{i,j} \vec{g}$. Then by [8] Theorem 5, the distribution of h_0 and $h_{i,j}$ are indistinguishable in Game 2 and Game 1, i.e. $\vec{h_0} \approx_c U((\mathbb{Z}_q^{<n+d-1}[x])^{\gamma\tau})$ and $\vec{h_{i,j}} \approx_c U((\mathbb{Z}_q^{<n+d-1}[x])^{\gamma\tau})$ for $(i,j) \in [\mathfrak{d}] \times [\kappa]$. Therefore, Game 2 is computationally indistinguishable from Game 1, i.e. $|\Pr[X_2] - \Pr[X_1]| = \mathsf{negl}(n)$.

Game 3. In this game, $\mathcal{C}$ changes the time the abortion is made. In game 2, $\mathcal{C}$ aborts at the end of the game if the condition (2) does not hold. Here, the challenger aborts as soon as the abort condition is satisfied. Since this is only a conceptual change, we have $\Pr[X_3] = \Pr[X_2]$.

Game 4. In this game, instead of generating $\vec{a}$ together with a trapdoor td according to KeyGen, $\mathcal{C}$ picks $\vec{a} \xleftarrow{\$} (\mathbb{Z}_q^{<n}[x])^t \times (\mathbb{Z}_q^{<n+d-1}[x])^{\gamma\tau}$. By Theorem 2, this only differs negligibly with game 3. Then $\mathcal{C}$ sets $\vec{S_{\mathsf{id}}} = \vec{S_0} + \sum_{(j_1,j_2,\ldots,j_{\mathfrak{d}})\in \mathfrak{F}(\mathsf{id})}$ TrapEval$(\vec{S_{1,j_1}}, \ldots, \vec{S_{\mathfrak{d},j_{\mathfrak{d}}}}, y_{1,j_1}, \ldots, y_{\mathfrak{d},j_{\mathfrak{d}}})$. It holds that $\vec{h_{\mathsf{id}}} = \vec{S_{\mathsf{id}}} \cdot \vec{a} + F_{\mathbf{y}}(\mathsf{id}) \vec{g}$. When $\mathcal{A}$ makes a key extraction query. If the abortion does not happen, it runs SampleRight$(\vec{g}, \mathsf{td}, F_{\mathbf{y}}(\mathsf{id}), \vec{a}, \vec{S_{\mathsf{id}}}, u, \sigma) \to \vec{r}$ and returns $\vec{r}$ to $\mathcal{A}$. Compared with Game 3, where the key $\vec{r}$ is sampled using SampleLeft$(\vec{a}, \vec{h_{\mathsf{id}}}, \mathsf{td}, u, \sigma) \to \vec{r}$. By [8] Lemma 6., Lemma 5 and the choice of σ, the difference in the output distributions of SampleRight and SampleLeft are $\mathsf{negl}(n)$-close. Therefore, we have $|\Pr[X_4] - \Pr[X_3]| = \mathsf{negl}(n)$.

Game 5. In this game, $\mathcal{C}$ changes the way the ciphertext is generated. If $F_{\mathbf{y}}(\mathsf{id}^*) = 0$ (i.e. if it does not abort), to create the challenge ciphertext, $\mathcal{C}$ chooses $p \xleftarrow{\$} \mathbb{Z}_q^{<n+2d+k-1}[x]$, $\chi \leftarrow \lfloor D_{\alpha \cdot q} \rceil$, $\chi_1 \leftarrow \lfloor D_{\alpha_1 \cdot q} \rceil$, and noise $e_0, e_1, e_2, \cdots, e_{t''}$, where $e_i \leftarrow \chi^{k+1}$ for $i = 0$, $e_i \leftarrow \chi^{2d+k}[x]$ for $1 \leq i \leq t$, $e_i \leftarrow \chi^{d+k+1}[x]$ for $t+1 \leq i \leq t'$, $e_i \leftarrow \chi_1^{d+k+1}[x]$ for $t'+1 \leq i \leq t''$. Next, for $1 \leq i \leq t$ set $c_i = a_i \odot_{2d+k} p + e_i$. For $t+1 \leq i \leq t'$ set $c_i = a_i \odot_{d+k+1} p + e_i$. For $t = 0$, set $c_0 = \lfloor \frac{q}{2} \rfloor m_{\mathsf{coin}} + u \odot_{k+1} p + e_i$. For $t'+1 \leq i \leq t''$, WLOG, c_ι, where $t'+1 \leq \iota \leq t''$, will be taken as an example. Since $c_\iota = h_{\mathsf{id},\iota} \odot_{d+k+1} p + e_\iota$, and $\vec{h_{\mathsf{id}}} = \vec{S_{\mathsf{id}}} \cdot \vec{a}$ as $F_{\mathbf{y}}(\mathsf{id}^*) = 0$, the following equations hold: $c_\iota = h_{\mathsf{id},\iota} \odot_{d+k+1} p + e_\iota = (\sum_{i=1}^{t'} s_i \cdot a_i) \odot_{d+k+1} p + e_\iota = \sum_{i=1}^{t}(s_i \cdot a_i) \odot_{d+k+1} p + \sum_{i=t+1}^{t'}(s_i \cdot a_i) \odot_{d+k+1} p + e_\iota = \sum_{i=1}^{t} s_i \odot_{d+k+1} (a_i \odot_{2d+k} p) + \sum_{i=t+1}^{t'} s_i(a_i \odot_{d+k+1} p) + e_\iota$, where $(s_1, s_2, \cdots, s_{t'})$ are the entries on the ι-th row of the polynomial matrix $\vec{S_{\mathsf{id}}}$. The fourth equality holds by Lemma 3, and $s_i \in \mathbb{Z}_q$ for $t+1 \leq i \leq t'$. Next, $\mathcal{C}$ sets $\phi_i = a_i \odot_{2d+k} p$, for $1 \leq i \leq t$, $\phi_i = a_i \odot_{d+k+1} p$, for $t+1 \leq i \leq t'$, and uses ϕ_i to represent its

coefficient vector. And the coefficient vector of e_ι is denoted by $\mathbf{e}_\iota$. Then $\mathcal{C}$ sets $\overline{\mathbf{c}_\iota} = \sum_{i=1}^{t}(\mathsf{T}^{d,d+k+1}(s_i))^T \overline{\phi_i} + \sum_{i=t+1}^{t'} s_i \overline{\phi_i} + 2\overline{\mathbf{e}_\iota}$. By Lemma 4, we can see that $\mathbf{c}_\iota = \overline{\mathbf{c}_\iota}$ is the coefficient vector of c_ι. So Game 4 and Game 5 are identical and $\Pr[X_5] = \Pr[X_4]$.

Game 6. In this game, we slightly change the way the ciphertext is generated. If the game does not abort, $\mathcal{C}$ first chooses $p \xleftarrow{\$} \mathbb{Z}_q^{<n+2d+k-1}[x]$, $\chi \leftarrow \lfloor D_{\alpha \cdot q} \rceil$, $\chi_1 \leftarrow \lfloor D_{\alpha_1 \cdot q} \rceil$, and noise $e_0, e_1, e_2, \cdots, e_{t''}$, where $e_i \leftarrow \chi^{k+1}$ for $i = 0$, $e_i \leftarrow \chi^{2d+k}[x]$ for $1 \leq i \leq t$, $e_i \leftarrow \chi^{d+k+1}[x]$ for $t+1 \leq i \leq t'$. Next, for $1 \leq i \leq t$ set $c_i = a_i \odot_{2d+k} p + e_i$. For $t+1 \leq i \leq t'$ set $c_i = a_i \odot_{d+k+1} p + e_i$. For $t = 0$, set $c_0 = \lfloor \frac{q}{2} \rfloor m_{\text{coin}} + u \odot_{k+1} p + e_i$. For $t'+1 \leq i \leq t''$, WLOG, we will take c_ι, where $t'+1 \leq \iota \leq t''$, as an example. First, since $c_i = a_i \odot_{2d+k} p + e_i$ for $1 \leq i \leq t$ and $c_i = a_i \odot_{d+k+1} p + e_i$, for $t+1 \leq i \leq t'$. And the coefficients vectors are denoted as $\mathbf{c}_i$. Then, $\mathcal{C}$ sets $\overline{\mathbf{c}_\iota} = \sum_{i=1}^{t} \mathsf{ReRand}((\mathsf{T}^{d,d+k+1}(s_i))^T, \overline{\mathbf{c}_i}, \alpha q, \frac{\alpha_1}{2\sqrt{t'}\alpha})$ $+ \sum_{i=t+1}^{t'} \mathsf{ReRand}(s_i I_{d+k+1}, \overline{\mathbf{c}_i}, \alpha q, \frac{\alpha_1}{2\sqrt{t'}\alpha})$. Set c_ι to be the polynomial whose coefficient vector is $\mathbf{c}_\iota = \overline{\mathbf{c}_\iota}$. Since $\alpha_1 = 2\sqrt{t'} \ell n^{c(\mathfrak{d}-1)} \cdot \omega(n^2 \log^2 n)\alpha$, $\frac{\alpha_1}{2\sqrt{t'}\alpha} > \ell n^{c(\mathfrak{d}-1)} \cdot \omega(n^2 \log^2 n) \geq \sigma_1(\mathsf{T}^{d,d+k+1}(s_i))$, from [10] Lemma 1, we can see that c_ι in this game is distributed statistically close to c_ι in Game 5. So we have $|\Pr[X_6] - \Pr[X_5]| = \mathsf{negl}(n)$.

Game 7. In Game 7, changes are made to the generation of ciphertext. If the game does not abort, $\mathcal{C}$ sets $c_1, c_2, \cdots, c_t \xleftarrow{\$} \mathbb{Z}_q^{2d+k}$, and $c_{t+1}, \cdots, c_{t'} \xleftarrow{\$} \mathbb{Z}_q^{d+k+1}$ and $\mathfrak{c} \xleftarrow{\$} \mathbb{Z}_q^{k+1}$, $c_0 = \mathfrak{c} + \lfloor \frac{q}{2} \rfloor m_{\text{coin}}$. Then $\mathcal{C}$ generates the rest of the ciphertexts as in Game 6. We can see that $|\Pr[X_7] - \frac{1}{2}| = \mathsf{negl}(n)$ as it contains no information about the coin. By Lemma 9, we can see that $|\Pr[X_7] - \Pr[X_6]| = \mathsf{negl}(n)$, assuming $\mathsf{MPLWE}_{q,n+2d+k,\mathbf{d},D_{\alpha \cdot q}}$ is hard with degree vector $\mathbf{d} = (d_i)_{i=1}^{t'+1}$ where $d_i = 2d+k$, for $1 \leq i \leq t$; $d_i = d+k+1$, for $t+1 \leq i \leq t'$; $d_i = k+1$, for $i = t''+1$. Analysis From the previous games, we can get that when $|\Pr[X_7] - \frac{1}{2}| = \mathsf{negl}(n)$, $|\Pr[X_0] - \frac{1}{2}| = \mathsf{negl}(n)$. The previous equation holds trivially as $\mathcal{A}$ has no information about the message m. □

Lemma 9 (Lemma11, [8]). *For any PPT adversary $\mathcal{A}$, there exists another PPT adversary $\mathcal{B}$ such that $|\Pr[X_7] - \Pr[X_6]| \leq \mathsf{Adv}_{\mathcal{B}}^{\mathsf{MPLWE}_{q,n+2d+k,\mathbf{d},D_{\alpha \cdot q}}}$. In particular, under the $\mathsf{MPLWE}_{q,n+2d+k,\mathbf{d},D_{\alpha \cdot q}}$ assumption, we have $|\Pr[X_7] - \Pr[X_6]| = \mathsf{negl}(n)$.*

Proof. Suppose $\mathcal{A}$ has non-negligible advantage in distinguishing Game 6 and 7. We use $\mathcal{A}$ to construct an MPLWE algorithm, denoted $\mathcal{B}$. Recall from Theorem 1 that an MPLWE problem instance is provided as a sampling oracle $\mathcal{O}$ that can be either truly random $\mathcal{O}_\$$ or noisy pseudo-random $\mathcal{O}_s$ for some secret $\mathbb{Z}_q^{<n+2d+k-1}[x]$. The simulator $\mathcal{B}$ uses the adversary $\mathcal{A}$ to distinguish between the two, and proceeds as follows: **Instance.** $\mathcal{B}$ requests from $\mathcal{O}$ and receives fresh pairs $(a_1^*, w_1), \cdots, (a_{t'}^*, w_{t'}), (u^*, v^*)$, where $a_i^* \in \mathbb{Z}_q^{<n}$, $w_i^* \in \mathbb{Z}_q^{<2d+n}$ for $1 \leq i \leq t$, $a_i^* \in \mathbb{Z}_q^{<n+d-1}$, $w_i^* \in \mathbb{Z}_q^{<d+k+1}$ for $t+1 \leq i \leq t'$, and $u^* \in \mathbb{Z}_q^{n+2d-2}$,

$v^* \in \mathbb{Z}_q^{<k+1}$. **Setup** To construct master public key mpk, as follows: 1. Set $a_i = a_i^*$ for $i = 1, ... t'$ 2. $\mathcal{B}$ picks **y** as in Game 1. 3. Generates polynomial vectors $\vec{h_1}, \cdots, \vec{h_\partial}$ by picking polynomial matrices $\vec{S_1}, \vec{S_2}, \cdots, \vec{S_\partial}$ and generating polynomial vector $\vec{b}$ with trapdoor td as in Game 6. 4. Set $u = u^*$. $\mathcal{C}$ Then returns mpk to $\mathcal{A}$. $\mathcal{B}$ also picks a random bit coin $\leftarrow \{0,1\}$ and keeps it secret. **Queries** When $\mathcal{A}$ makes a key extraction for id, $\mathcal{B}$ first calculates $F_\mathbf{y}(\mathsf{id})$. It aborts and sets coin' $\xleftarrow{\$} \{0,1\}$ if $F_\mathbf{y}(\mathsf{id}) = 0$. Otherwise, $\mathcal{B}$ generates the private key as in Game 4. **Challenge** When $\mathcal{A}$ makes a key extraction query for the challenge identity id^*, $\mathcal{B}$ first computes $F_\mathbf{y}(\mathsf{id})$ and aborts and sets coin' $\xleftarrow{\$} \{0,1\}$ if $F_\mathbf{y}(\mathsf{id}) \neq 0$. Otherwise, it proceeds as follows: If coin $= 0$, it computes $\vec{S}_{\mathsf{id}^*} = [\vec{s_1}, \vec{s_2}, \cdots, \vec{s_{t'}}]^T$ where $\vec{s_j} \leftarrow (\chi^d[x])^t \times 0^{\gamma\tau}$ for $j = 1, \cdots, t$, $\vec{s_j} \leftarrow (\chi^d[x])^t \times \chi^{\gamma\tau}$ for $j = t+1, \cdots, t'$. And we can write $\vec{s_j}$ as $(s_{j,1}, \cdots, s_{j,t'})$. Set $\chi := \lfloor D_{\alpha \cdot q} \rceil, \chi_1 := \lfloor D_{\alpha_1 \cdot q} \rceil$. Let $e_i \leftarrow \chi^{k+1}$ for $i = 0$, $e_i \leftarrow \chi^{2d+k}[x]$ for $1 \le i \le t$, $e_i \leftarrow \chi^{d+k+1}[x]$ for $t+1 \le i \le t'$, $e_i \leftarrow \chi_1^{d+k+1}[x]$ for $t'+1 \le i \le t'+t$, $e_i \leftarrow \chi_1^{d+k+1}[x]$ for $t'+t+1 \le i \le 2t'$. $c_0 = m + v^*$, $c_i = w_i^*$ for $1 \le i \le t'$. For $t'+1 \le i \le 2t'$, generate c_i as follows: For $t'+1 \le i \le t'+t$, WLOG, we will take c_ι, where $t'+1 \le \iota \le t'+t$, as an example. First, we have $c_i = a_i \odot_{2d+k} p + e_i$, for $1 \le i \le t$. We will use $\mathbf{c}_i$ to represent the coefficient vectors. Then, we set $\overline{\mathbf{c}_\iota} = \sum_{i=1}^{t} \mathsf{ReRand}((\mathsf{T}^{d,d+k+1}(s_i))^T, \overline{\mathbf{c}_i}, \alpha q, \frac{\alpha_1}{2\sqrt{t\alpha q}})$. And set c_ι to be the polynomial whose coefficient vector is $\mathbf{c}_\iota$. From [10] Lemma 1, we can see that c_ι in this game is distributed statistically close to c_ι in Game 5. For $t'+t+1 \le i \le 2t'$, WLOG, we will take c_ι, where $t'+t+1 \le \iota \le 2t'$, as an example. First, since $c_i = a_i \odot_{2d+k} p + e_i$ for $1 \le i \le t$, $c_i = a_i \odot_{d+k+1} p + e_i$, for $t+1 \le i \le t'$. Use $\mathbf{c}_i$ to represent the coefficient vectors. Then, we set $\overline{\mathbf{c}_\iota} = \sum_{i=1}^{t} \mathsf{ReRand}((\mathsf{T}^{d,d+k+1}(s_i))^T, \overline{\mathbf{c}_i}, \alpha q, \frac{\alpha_1}{2\sqrt{t'\alpha}}) + \sum_{i=t+1}^{t'} \mathsf{ReRand}(s_i I_{d+k+1}, \overline{\mathbf{c}_i}, \alpha q, \frac{\alpha_1}{2\sqrt{t'\alpha}})$. And set c_ι to be the polynomial whose coefficient vector is $\mathbf{c}_\iota$. Finally, it returns $(c_1, (c_i)_{i=1}^{2t'})$ to $\mathcal{A}$. **Guess** At last, $\mathcal{A}$ outputs its guess $\widetilde{\mathsf{coin}}$ if the abort condition has not been satisfied. Then $\mathcal{B}$ sets coin' = coin. Finally, $\mathcal{B}$ outputs 1 if coin = coin' and 0 otherwise. **Analysis** We can see that $\mathcal{B}$ perfectly simulates the view of $\mathcal{A}$ in Game 6 if $\mathcal{O} = \mathcal{O}_s$ and Game 7 if $\mathcal{O} = \mathcal{O}_\$$. Note that both games only differ in the generation of the challenge ciphertext in the case of coin = 0. Furthermore, we can see that the generation of mpk, the queries, and the challenge phase are perfect for the case of both coin = 1. Therefore, $|\Pr[X_6] - \Pr[X_7]| = \mathsf{Adv}_\mathcal{B}^{\mathsf{dMPLWE}_{q,n+2d+k,\mathbf{d},D_{\alpha \cdot q}}} = \mathsf{negl}(n)$ as desired. □

6 Conclusion

In this paper, we propose a new compact, adaptively secure IBE from MPLWE. Our scheme inherits the typical advantage of MPLWE. Namely, it has comparable efficiency to its RLWE counterparts while offering stronger security. Its master public key size is smaller than the state-to-art MPLWE-based construction.

References

1. Abla, P.: Identity-based encryption from LWE with more compact master public key. In: Oswald, E. (ed.) Topics in Cryptology - CT-RSA 2024 - Cryptographers' Track at the RSA Conference 2024, San Francisco, CA, USA, 6–9 May 2024, Proceedings. Lecture Notes in Computer Science, vol. 14643, pp. 319–353. Springer (2024). https://doi.org/10.1007/978-3-031-58868-6_13
2. Abla, P., Liu, F.-H., Wang, H., Wang, Z.: Ring-based identity based encryption – asymptotically shorter MPK and tighter security. In: Nissim, K., Waters, B. (eds.) TCC 2021. LNCS, vol. 13044, pp. 157–187. Springer, Cham (2021). https://doi.org/10.1007/978-3-030-90456-2_6
3. Agrawal, S., Boneh, D., Boyen, X.: Efficient lattice (H)IBE in the standard model. In: Gilbert, H. (ed.) EUROCRYPT 2010. LNCS, vol. 6110, pp. 553–572. Springer, Heidelberg (2010). https://doi.org/10.1007/978-3-642-13190-5_28
4. Bellare, M., Ristenpart, T.: Simulation without the artificial abort: simplified proof and improved concrete security for waters' IBE scheme. In: Joux, A. (ed.) EUROCRYPT 2009. LNCS, vol. 5479, pp. 407–424. Springer, Heidelberg (2009). https://doi.org/10.1007/978-3-642-01001-9_24
5. Boneh, D., Franklin, M.: Identity-based encryption from the weil pairing. In: Kilian, J. (ed.) CRYPTO 2001. LNCS, vol. 2139, pp. 213–229. Springer, Heidelberg (2001). https://doi.org/10.1007/3-540-44647-8_13
6. Cash, D., Hofheinz, D., Kiltz, E., Peikert, C.: Bonsai trees, or how to delegate a lattice basis. In: Gilbert, H. (ed.) EUROCRYPT 2010. LNCS, vol. 6110, pp. 523–552. Springer, Heidelberg (2010). https://doi.org/10.1007/978-3-642-13190-5_27
7. Cocks, C.: An identity based encryption scheme based on quadratic residues. In: Honary, B. (ed.) Cryptography and Coding 2001. LNCS, vol. 2260, pp. 360–363. Springer, Heidelberg (2001). https://doi.org/10.1007/3-540-45325-3_32
8. Fan, J., Lu, X., Au, M.H.: Adaptively secure identity-based encryption from middle-product learning with errors. In: Simpson, L., Baee, M.A.R. (eds.) Information Security and Privacy - 28th Australasian Conference, ACISP 2023, Brisbane, QLD, Australia, 5–7 July 2023, Proceedings. Lecture Notes in Computer Science, vol. 13915, pp. 320–340. Springer (2023). https://doi.org/10.1007/978-3-031-35486-1_15
9. Gentry, C., Peikert, C., Vaikuntanathan, V.: Trapdoors for hard lattices and new cryptographic constructions. In: Dwork, C. (ed.) Proceedings of the 40th Annual ACM Symposium on Theory of Computing, Victoria, British Columbia, Canada, 17–20 May 2008, pp. 197–206. ACM (2008). https://doi.org/10.1145/1374376.1374407
10. Katsumata, S., Yamada, S.: Partitioning via non-linear polynomial functions: more compact IBEs from ideal lattices and bilinear maps. In: Cheon, J.H., Takagi, T. (eds.) ASIACRYPT 2016. LNCS, vol. 10032, pp. 682–712. Springer, Heidelberg (2016). https://doi.org/10.1007/978-3-662-53890-6_23
11. Lombardi, A., Vaikuntanathan, V., Vuong, T.D.: Lattice trapdoors and IBE from middle-product LWE. In: Hofheinz, D., Rosen, A. (eds.) TCC 2019. LNCS, vol. 11891, pp. 24–54. Springer, Cham (2019). https://doi.org/10.1007/978-3-030-36030-6_2
12. Lyubashevsky, V.: Digital signatures based on the hardness of ideal lattice problems in all rings. In: Cheon, J.H., Takagi, T. (eds.) ASIACRYPT 2016. LNCS, vol. 10032, pp. 196–214. Springer, Heidelberg (2016). https://doi.org/10.1007/978-3-662-53890-6_7

13. Micciancio, D., Peikert, C.: Trapdoors for lattices: simpler, tighter, faster, smaller. In: Pointcheval, D., Johansson, T. (eds.) EUROCRYPT 2012. LNCS, vol. 7237, pp. 700–718. Springer, Heidelberg (2012). https://doi.org/10.1007/978-3-642-29011-4_41
14. Micciancio, D., Regev, O.: Worst-case to average-case reductions based on gaussian measures. In: 45th Symposium on Foundations of Computer Science (FOCS 2004), 17–19 October 2004, Rome, Italy, Proceedings, pp. 372–381. IEEE Computer Society (2004). https://doi.org/10.1109/FOCS.2004.72
15. Roşca, M., Sakzad, A., Stehlé, D., Steinfeld, R.: Middle-product learning with errors. In: Katz, J., Shacham, H. (eds.) CRYPTO 2017. LNCS, vol. 10403, pp. 283–297. Springer, Cham (2017). https://doi.org/10.1007/978-3-319-63697-9_10
16. Rosca, M., Stehlé, D., Wallet, A.: On the ring-LWE and polynomial-LWE problems. In: Nielsen, J.B., Rijmen, V. (eds.) EUROCRYPT 2018. LNCS, vol. 10820, pp. 146–173. Springer, Cham (2018). https://doi.org/10.1007/978-3-319-78381-9_6
17. Shamir, A.: Identity-based cryptosystems and signature schemes. In: Blakley, G.R., Chaum, D. (eds.) CRYPTO 1984. LNCS, vol. 196, pp. 47–53. Springer, Heidelberg (1985). https://doi.org/10.1007/3-540-39568-7_5
18. Stehlé, D., Steinfeld, R., Tanaka, K., Xagawa, K.: Efficient public key encryption based on ideal lattices. In: Matsui, M. (ed.) ASIACRYPT 2009. LNCS, vol. 5912, pp. 617–635. Springer, Heidelberg (2009). https://doi.org/10.1007/978-3-642-10366-7_36
19. Waters, B.: Efficient identity-based encryption without random oracles. In: Cramer, R. (ed.) EUROCRYPT 2005. LNCS, vol. 3494, pp. 114–127. Springer, Heidelberg (2005). https://doi.org/10.1007/11426639_7
20. Yamada, S.: Adaptively secure identity-based encryption from lattices with asymptotically shorter public parameters. In: Fischlin, M., Coron, J.-S. (eds.) EUROCRYPT 2016. LNCS, vol. 9666, pp. 32–62. Springer, Heidelberg (2016). https://doi.org/10.1007/978-3-662-49896-5_2
21. Yamada, S.: Asymptotically compact adaptively secure lattice IBEs and verifiable random functions via generalized partitioning techniques. In: Katz, J., Shacham, H. (eds.) CRYPTO 2017. LNCS, vol. 10403, pp. 161–193. Springer, Cham (2017). https://doi.org/10.1007/978-3-319-63697-9_6
22. Zhang, J., Chen, Yu., Zhang, Z.: Programmable hash functions from lattices: short signatures and IBEs with small key sizes. In: Robshaw, M., Katz, J. (eds.) CRYPTO 2016. LNCS, vol. 9816, pp. 303–332. Springer, Heidelberg (2016). https://doi.org/10.1007/978-3-662-53015-3_11

MDKG: Module-Lattice-Based Distributed Key Generation

Ye Bai[1], Debiao He[1,2(✉)], Zhichao Yang[3], Min Luo[1], and Cong Peng[1]

[1] Key Laboratory of Aerospace Information Security and Trusted Computing Ministry of Education, School of Cyber Science and Engineering, Wuhan University, Wuhan 430072, China
{mluo,cpeng}@whu.edu.cn
[2] Key Laboratory of Computing Power Network and Information Security, Ministry of Education, Shandong Computer Science Center, Qilu University of Technology (Shandong Academy of Sciences), Jinan 250014, China
hedebiao@163.com
[3] Department of Information Security, Naval University of Engineering, Wuhan 430000, China

Abstract. The issue of a single point of failure has grown increasingly severe due to the widespread adoption of blockchain systems and voting protocols. Fortunately, threshold cryptography can solve this problem by distributing the key storage. However, a key issue of such a scheme is how to generate a secret shared by all participants without revealing their secrets. Most solutions are based on a trusted third party, while a few solutions rely on all participants together. The first approach imposes significant demands on the third party, making its implementation challenging and resource-intensive. The second method, known as distributed key generation (DKG), operates without relying on a trusted third party, offering a more efficient and decentralized solution. Compared with existing DKG protocols which rely on cryptographic foundations such as discrete logarithm problems and bilinear pairings, lattice-based cryptography offers notable benefits, including superior efficiency and quantum resistance, making it a promising alternative. Recognizing these advantages, we proposed a module-lattice-based DKG (MDKG) scheme based on zero-knowledge proofs. Our MDKG scheme does not require broadcasting shared secrets between users, satisfying the security requirements of robust correctness and secrecy, and can resist quantum attacks. Experiments show that our MDKG scheme reduces the secret sharing size by 74% and reduces the time cost by at least 90% compared to the existing lattice-based DKG scheme.

Keywords: distributed key generation · module-lattice · zero-knowledge proofs · robust correctness · secrecy · efficiency

1 Introduction

As technology advances rapidly, voting protocols [1] and blockchain systems [36] have garnered significant attention from professionals. The practical implementation of these systems is bound to carry the inherent risk of a single point of failure. Currently, the most widely adopted approach involves utilizing tools like threshold cryptography [15,34] and secure multi-party computation [14,23]. In threshold cryptography, e.g. threshold encryption [27] and threshold signature [10], the core problem is how to securely generate a shared secret key by all participants without revealing their secrets. Thus, distributed key generation (DKG) [22,31] was studied to overcome this problem.

A DKG protocol enables every participant to securely acquire a share of the secret key and the associated public key without disclosing any additional information. Initially proposed by Pedersen [31], this approach relies on the discrete logarithm problem and is constructed upon Feldman's verifiable secret share (VSS) protocol [19]. However, Gennaro et al. [22] observed that Pedersen's scheme is biased and proposed a new DKG to make it unbiasedly. Later, some 1-round DKG schemes [21,24] based on bilinear pairing ciphers were proposed. Unfortunately, Katz [26] discovered that all 1-round DKGs can be biased, and proposed an optimal DKG. Recently, Gurkan et al. [25] proposed an aggregatable DKG based on bilinear pairings, which effectively reduces both verification and communication complexities by aggregating all valid (partial) transcripts. Cascudo and David [13] proposed a 1-round DKG along with a two-round DKG based on discrete logarithm which improves on the concrete communication/computational complexity and the round complexity.

Regarding post-quantum DKG, Beullens et al. [9] introduced the first isogeny-based DKG. They achieved robust correctness and secrecy within the framework of very hard homogeneous spaces (VHHS) by introducing piecewise verifiable proofs (PVP). Currently, Atapoor et al. [4] constructed a new PVP for structured public keys and proposed two DKGs that have better performance. Compared to isogeny-based cryptography, lattice-based cryptography has better performance, but there have been very few proposals for lattice-based DKGs [18,35].

Therefore, we propose a module-lattice-based DKG (MDKG) scheme, which offers improved efficiency over current state-of-the-art DKGs.

1.1 Our Contributions

In this paper, we propose MDKG, a DKG scheme that utilizes Shamir secret sharing within the module-lattice framework. Built upon the design principles of Gennaro et al. [22], our MDKG ensures robust correctness. Furthermore, we propose a non-interactive zero-knowledge (NIZK) proof system based on module-lattice cryptography, derived through the Fiat-Shamir transformation [20]. This innovation ensures that participants' secret shares remain confidential even when complaints arise, eliminating the risk of malicious adversaries gaining access to additional secret information. As a result, our MDKG is highly efficient when compared to existing DKG schemes [18,35]. Leveraging recent security advancements [8,17], we prove the security of the module-lattice-based NIZK proof. And we establish the secrecy of our MDKG through a sequence of security games.

1.2 Organization

In Sect. 2, we review the key concepts of lattices, Shamir secret sharing, NIZK, and DKG. The module-lattice-based NIZK proof is introduced in Sect. 3. In Sect. 4, we propose the construction of MDKG. Next, we provide a comprehensive security analysis and evaluate the performance of the scheme in Sect. 5 and Sect. 6, respectively. Finally, Sect. 7 offers a conclusion to our work.

2 Preliminaries

In this paper, $\mathbb{Z}_q$ refers to the integer ring within the range $\left[-\frac{q-1}{2}, \frac{q-1}{2}\right]$. Additionally, $\mathbb{Z}_q[X]$ denotes the polynomial ring where the coefficients are elements of $\mathbb{Z}_q$. On this basis, we set R_q to represent the ring $\mathbb{Z}_q[X]/(X^N+1)$. Elements in R_q (or $\mathbb{Z}_q$) are written in regular font letters. Similarly, vectors are represented by bold lowercase letters, while matrices are denoted by bold uppercase letters.

The security parameter is represented by λ. An unspecified function is expressed as $\mathsf{poly}(\lambda)$, while a negligible function is denoted by $\mathsf{negl}(\lambda)$. A distribution is termed overwhelming when its probability is $1 - \mathsf{negl}(\lambda)$.

The l_2-norm and l_∞-norm of a m-dimensional vector $\mathbf{c} = (c_1, \ldots, c_m)^T$ is defined as $\|\mathbf{c}\| = \sqrt{\sum_{i=1}^m c_i^2}$, $\|\mathbf{c}\|_\infty = \max_i |c_i|$.

Similarly, the l_2-norm and l_∞-norm of a m-dimensional N-degree polynomial vector $\mathbf{c} = (c_1, \ldots, c_m)^T$ where $\forall i \in \{1, \ldots, m\}, c_i = c_{i0} + c_{i1}X + \cdots + c_{i,N-1}X^{N-1}$ is defined as $\|\mathbf{c}\| = \sqrt{\sum_{i=1}^m \sum_{j=0}^{N-1} c_{ij}^2}$, $\|\mathbf{c}\|_\infty = \max_i \max_j |c_{ij}|$.

2.1 Lattice

Consider an n-dimensional Euclidean space $\mathbb{R}^n$ and m linearly independent vectors $\mathbf{b}_1, \ldots, \mathbf{b}_m \in \mathbb{R}^n$. A lattice $\Lambda = \mathcal{L}(\mathbf{B})$ is defined as the set $\{\mathbf{Bc} = \sum_{i=1}^m \mathbf{b}_i c_i \mid \mathbf{c} \in \mathbb{Z}^m\}$, which is generated by the $n \times m$ matrix $\mathbf{B} = (\mathbf{b}_1, \ldots, \mathbf{b}_m)$. The vectors $\mathbf{b}_1, \ldots, \mathbf{b}_m$ are referred to as the basis of the lattice Λ. When these basis vectors are sampled from the ring R_q^n, we call the lattice Λ a module-lattice.

For a distribution χ, we write $c \leftarrow \chi$ to indicate that c is sampled from the distribution χ. The uniform distribution over a set S is denoted by $\mathcal{U}(S)$, and $c \xleftarrow{\$} S$ represents the process of uniformly and randomly selecting c from the set S. Furthermore, the Gaussian function centered at $\mathbf{c} \in \mathbb{R}^k$ with a parameter $s > 0$ is defined as $\rho_{\mathbf{c},s}(\mathbf{x}) = \exp\left(-\pi\|\mathbf{x} - \mathbf{c}\|^2/s^2\right)$. The discrete Gaussian distribution over a lattice Λ is defined as $D_{\Lambda,\mathbf{c},s}(\mathbf{x}) = \frac{\rho_{\mathbf{c},s}(\mathbf{x})}{\rho_{\mathbf{c},s}(\Lambda)}$, for any $\mathbf{x} \in \Lambda$, where the quantity $\rho_{\mathbf{c},s}(\Lambda) = \sum_{\mathbf{x} \in \Lambda} \rho_{\mathbf{c},s}(\mathbf{x})$ acts as a scaling factor to ensure that $D_{\Lambda,\mathbf{c},s}(\mathbf{x})$ is a valid probability distribution. When $\mathbf{c}$ is set to $\mathbf{0}$, we can simplify the notation by omitting $\mathbf{c}$ and writing $D_{\Lambda,\mathbf{c},s}$ as $D_{\Lambda,s}$. The smoothing parameter for a lattice Λ, denoted as $\eta_\epsilon(\Lambda)$, is defined for a positive real $\epsilon > 0$ as $\eta_\epsilon(\Lambda) = \min\left\{s > 0 \mid \rho_{1/s}(\Lambda^* \setminus \{\mathbf{0}\}) \leq \epsilon\right\}$ [30].

Lemma 1. *([30], Lemma 4.4) For any n-dimensional lattice and $s \geq \eta_\epsilon(\Lambda)$, the probability that a sample $\mathbf{x}$ drawn from the distribution $D_{\Lambda,s}$ satisfies $\|\mathbf{x}\| > s\sqrt{n}$ is bounded as $\Pr_{\mathbf{x} \sim D_{\Lambda,s}}[\|\mathbf{x}\| > s\sqrt{n}] \leq 2^{-n}$.*

MLWE [11] and MSIS [28] are currently widely studied lattice-based hard problem assumptions which are generalizations of LWE, Ring-LWE [12,33], SIS and Ring-SIS [2] on module-lattice, and they are defined as follows.

Definition 1. *(Module-LWE distribution, [11]) Given a security parameter λ, two positive integers m, n, and a modulus $q \geq 2$, let χ be an error distribution over R_q and $\mathbf{s} \xleftarrow{\$} R_q^n$ be a fixed secret vector. The Module-LWE distribution is defined as $\text{MLWE}_\mathbf{s}(m, n, q, \chi) := \left\{ (\mathbf{A}, \mathbf{A}^T\mathbf{s} + \mathbf{e}) \mid \mathbf{A} \xleftarrow{\$} R_q^{n \times m}, \mathbf{e} \leftarrow \chi^m \right\}$.*

Definition 2 (Decision-MLWE, [11]). *Given m, n, q, and χ as defined in the Module-LWE distribution, the decision-MLWE problem, denoted d-$\text{MLWE}_\mathbf{s}(m, n, q, \chi)$, is the task of distinguishing between the uniform distribution $\mathcal{U}(R_q^{n \times m} \times R_q^m)$ and the Module-LWE distribution $\text{MLWE}_\mathbf{s}(m, n, q, \chi)$.*

Definition 3 (MSIS distribution, generated from SIS distribution [29]). *Given two positive integer m, n, the modulus $q \geq 2$, a parameter $0 < \beta < q$, and a fixed secret vector $\mathbf{s} \in R_q^m$ with $0 < \|\mathbf{s}\|_\infty < \beta$, the MSIS distribution $\text{MSIS}(m, n, q, \beta)$ is defined as $\text{MSIS}_\mathbf{s}(m, n, q, \beta) := \left\{ (\mathbf{A}, \mathbf{A}\mathbf{s}) \middle| \mathbf{A} \xleftarrow{\$} R_q^{n \times m} \right\}$.*

The MSIS search problem, denoted s-$\text{MSIS}_\mathbf{s}(m, n, q, \beta)$, is defined as the task of finding a vector $\mathbf{s} \in R_q^m$ such that $\mathbf{A}\mathbf{s} = \mathbf{0} \mod q$ where $0 < \|\mathbf{s}\| < \beta$ Similarly, the decision-MSIS problem, denoted d-$\text{MSIS}_\mathbf{s}(m, n, q, \beta)$, is defined as below.

Definition 4 (Decision-MSIS, generated from Decision-SIS [29]). *Given m, n, q, and β as defined in the MSIS distribution, the decision-MSIS problem d-$\text{MSIS}_\mathbf{s}(m, n, q, \beta)$ involves distinguishing the distribution $\text{MSIS}_\mathbf{s}(m, n, q, \beta)$ from the uniform distribution $\mathcal{U}(R_q^{n \times m} \times R_q^n)$.*

Note that the Decision-MSIS instance d-$\text{MSIS}_\mathbf{s}(m, n, q, \beta)$, can be transformed into a Decision-SIS instance d-$\text{SIS}_\mathbf{s}(N \cdot m, N \cdot n, q, \chi)$ [29], by considering a matrix rotation $\text{rot}(\mathbf{A}) \in \mathbb{Z}_q^{N \cdot n \times N \cdot m}$ [16]. Consequently, the Decision-MSIS problem is as computationally difficult as the Decision-SIS problem.

2.2 Shamir Secret Sharing

In a (T, K)-threshold Shamir secret sharing scheme [34], the secret value s is distributed among K participants over the ring $\mathbb{Z}_q$. This is achieved by polynomial interpolation over the ring $\mathbb{Z}_q$. For a given secret s, a polynomial $f(x)$ of degree T is constructed such that $f(0) = s$. Each participant $\mathcal{P}_i$, where $i \in \{1, \ldots, K\}$, is assigned a share $s_i := f(i)$. To reconstruct the secret s, any group S of at least T participants can utilize Lagrange interpolation, computing

$s = f(0) = \sum_{i \in S} s_i L_i^S$, where L_i^S defined in Eq. (1) represents the Lagrange coefficients corresponding to the subset S.

$$L_i^S := L_{0,i}^S = \prod_{j \in S \setminus \{i\}} \frac{j}{j-i} \mod q \qquad (1)$$

For any subset S' of less than T participants, the reconstruction will fail, each participant can get nothing about s.

2.3 Zero-Knowledge Proof and Σ-Protocols

The zero-knowledge proof [5] enables the prover to demonstrate to the verifier that they possess a specific secret value without disclosing any details about the secret itself. The Σ-protocol [7] is a specific form of zero-knowledge proof, for a specified security parameter λ, consider a two-party protocol $(\mathcal{P}, \mathcal{V})$, where $\mathcal{V}$ represents a probabilistic polynomial-time (PPT) algorithm. Let L denote a set of languages and R_L a collection of binary relations associated with L. The protocol $(\mathcal{P}, \mathcal{V})$ is referred to as a Σ-protocol for the relation R_L with a challenge set $\mathcal{C}$, a public input x, and a private witness input w, provided it adheres to the following properties [8,17]:

1. *Three-move form*: The protocol follows a structured interaction consisting of three sequential steps. First, $\mathcal{P}$, utilizing the inputs $(x, w) \in R_L$, computes and transmits an initial commitment a to the verifier $\mathcal{V}$. Next, $\mathcal{V}$ responds with a challenge c selected from the set $\mathcal{C}$. Finally, $\mathcal{P}$ provides a response z based on the challenge c, completing the interaction. $\mathcal{V}$ returns accepts or rejects according to the transcript (a, c, z).
2. *Completeness*: For all $(x, w) \in R_L$, the honest verifier $\mathcal{V}$ accepts the transcript with an overwhelming probability $1 - \mathsf{negl}(\lambda)$.
3. *Soundness*: For any PPT adversary $\mathcal{A}$ that does not know the witness w of the statement x where $(x, w) \in R_L$, the probability that $\mathcal{A}$ outputs an acceptable transcript (a, c, z) is negligible.
4. *Special honest-verifier zero-Knowledge (SHVZK)*: There exists a PPT simulator $\mathcal{S}$ which inputs $x \in L$ and $c \in \mathcal{C}$, outputs the transcript (a, c, z). For any PPT distinguisher $\mathcal{D}$, the advantage that $\mathcal{D}$ can distinguish between transcript generated by the real protocol and transcript generated by $\mathcal{S}$ is negligible concerning probability $\mathsf{negl}(\lambda)$.

2.4 System Model of MDKG

The roles involved in our MDKG scheme are K independent participant entities. We make the assumption that a secure communication channel exists between each pair of participants, ensuring that no external party can access or manipulate the transmitted information. Additionally, we assume the presence of a reliable broadcast channel capable of delivering identical messages simultaneously to all participants. The system model of our scheme closely follows the discrete logarithm-based DKG scheme outlined by Gennaro et al. [22] as below.

1. **VSS generation.** For each $i \in \{1, \ldots, K\}$, participant $\mathcal{P}_i$ performs Shamir secret sharing to share his secret $\mathbf{s}_i$ and generate the secret share $\mathbf{s}_i^{(j)}$ for each participant $\mathcal{P}_j$ where $j \in \{1, \ldots, K\}$. Note that $\mathcal{P}_i$ will also generate a secret share $\mathbf{s}_i^{(i)}$ for himself. Then, $\mathcal{P}_i$ sends $\mathbf{s}_i^{(j)}$ to $\mathcal{P}_j$.
2. **VSS Verification.** After receiving $\mathcal{P}_i$'s secret share $\mathbf{s}_i^{(j)}$, participant $\mathcal{P}_j$ checks whether the received share $\mathbf{s}_i^{(j)}$ is consistent by using a method similar to Pedersen VSS [32]. If the verification fails, $\mathcal{P}_j$ broadcasts a complaint against $\mathcal{P}_i$. Then, $\mathcal{P}_i$ resends the secret share to $\mathcal{P}_j$. In the meantime, $\mathcal{P}_i$ generates the corresponding NIZK proofs and makes them public. Each participant can verify them and confirm whether $\mathcal{P}_i$ is an honest participant. Finally, the honest participants will reach a consensus on a set $\mathcal{Q}$, consisting of qualified participants who successfully generated the VSS.
3. **Share computation.** Each qualified $\mathcal{P}_j$ can compute $\mathbf{s}^{(j)} := \sum_{i \in \mathcal{Q}} \mathbf{s}_i^{(j)}$ to get his partial share, and the secret key is implicitly represented as $\sum_{i \in \mathcal{Q}} \mathbf{s}_i$.
4. **Common public key computation.** Each qualified participant $\mathcal{P}_i$ publishes a public key $\mathbf{b}_{i0}$ corresponding to his secret $\mathbf{s}_i$. And the public key $\mathbf{b}_{i0}$'s correctness can be verified by each $\mathbf{s}_i^{(j)}$. If verification fails, the honest participants will publish their public keys of shares $\mathbf{s}_i^{(j)}$ to reconstruct $\mathbf{b}_{i0}$ publicly. Finally, the common public key is computed as $\mathbf{b} := \sum_{i \in \mathcal{Q}} \mathbf{b}_{i0}$.

3 Non-interactive Zero-Knowledge Proof from MSIS

In this section, we present a NIZK proof scheme derived from the MSIS problem, building on the foundational work outlined in [29].

3.1 Module-Lattice-Based NIZK Proof

For an odd prime q, consider a matrix with n rows and m columns, along with two boundary parameters η and γ, where $\gamma > 1$. The secret key is defined as $\mathbf{s} \in R_\eta^m$, while the corresponding public key consists of $\mathbf{A} \in R_q^{n \times m}$ and $\mathbf{t} = \mathbf{A}\mathbf{s} \in R_q^n$ where $\mathbf{A} \xleftarrow{\$} R_q^{n \times m}$. Our NIZK proof scheme contains Algorithm 1 and Algorithm 2. Among them, B_τ is defined as the set of elements in the polynomial ring R where exactly τ coefficients are either 1 or -1, while all other coefficients are zero. The hash function is defined as $\mathsf{H} : \{0,1\}^* \to B_\tau$, which can be implemented as described in [16].

3.2 Security Proof

Our scheme can be interpreted as a Fiat-Shamir transformation [20] applied to a Σ-protocol over module-lattice. Consequently, if the interactive variant of the scheme meets the security criteria outlined in Sect. 2.3, the Fiat-Shamir transformation can be applied to ensure that our scheme retains the zero-knowledge proof properties under non-interactive conditions [6]. Thus, we will demonstrate that the interactive variant of our scheme fulfills the requirements of *completeness*, *soundness*, and *special honest-verifier zero-knowledge (SHVZK)*.

Algorithm 1. Module-lattice-based NIZK proof

Require: Private key s and public key $(\mathbf{A}, \mathbf{t})$
Ensure: NIZK proof π
1: $\mathbf{z} := \perp$
2: **while** $\mathbf{z} = \perp$ **do**
3:    $\mathbf{y} \leftarrow D_s^m$
4:    $\mathbf{a} := \mathbf{A}\mathbf{y}$
5:    $c \in B_\tau := \mathsf{H}(\mathbf{A}, \mathbf{t}, \mathbf{a})$
6:    $\mathbf{z} := \mathbf{y} + c\mathbf{s}$
7:    **if** $\|\mathbf{z}\|_\infty \geq \gamma$ **then**
8:       $\mathbf{z} := \perp$
9:    **else**
10:      Set $\mathbf{z} := \perp$ with probability $1 - \min\left(\frac{D_s^m(\mathbf{z})}{MD_{c\mathbf{s},s}^m(\mathbf{z})}, 1\right)$
11:    **end if**
12: **end while**
13: **return** $\pi := \mathsf{Encode}(\mathbf{z}, c)$

Algorithm 2. Module-lattice-based NIZK verification

Require: Public key $(\mathbf{A}, \mathbf{t})$ and NIZK proof π
Ensure: Bits 1 or 0 corresponding to Accept/Reject
1: $(\mathbf{z}, c) := \mathsf{Decode}(\pi)$
2: **if** $\|\mathbf{z}\|_\infty \geq \gamma$ **then**
3:    **return** 0
4: **else**
5:    **return** $[c = \mathsf{H}(\mathbf{A}, \mathbf{t}, \mathbf{A}\mathbf{z} - c\mathbf{t})]$
6: **end if**

Correctness. The proof is considered correct provided that Algorithm 2 consistently produces an output of 1. This can be derived from Algorithm 1: Since π is generated by Algorithm 1, it must satisfy $\|\mathbf{z}\|_\infty \geq \gamma$, and $\mathbf{A}\mathbf{z} - c\mathbf{t} = \mathbf{A}(\mathbf{y} + c\mathbf{s}) - c\mathbf{A}\mathbf{s} = \mathbf{A}\mathbf{y} + c\mathbf{A}\mathbf{s} - c\mathbf{A}\mathbf{s} = \mathbf{A}\mathbf{y} = \mathbf{a}$. Thus, $\mathsf{H}(\mathbf{A}, \mathbf{t}, \mathbf{A}\mathbf{z} - c\mathbf{t}) = \mathsf{H}(\mathbf{A}, \mathbf{t}, \mathbf{a}) = c$ holds, Algorithm 2 will always output 1, which means the proof is correct.

Completeness. Given that our proof scheme adheres to the correctness requirement, it follows that the scheme also satisfies the completeness requirement.

Soundness. The soundness of our scheme can be established by relying on the computational hardness of the MSIS problem. We assume the existence of a PPT adversary $\mathcal{A}$ capable of generating a valid transcript $(\mathbf{a}, c_0, \mathbf{z}_0)$ with non-negligible probability. Given the same $\mathbf{w}$, there exists a knowledge extractor $\mathcal{E}$ capable of rewinding the process. Then, the extractor $\mathcal{E}$ obtains a valid transcript $(\mathbf{a}, c_1, \mathbf{z}_1)$ such that $\mathbf{A}\mathbf{z}_1 - c_1\mathbf{t} = \mathbf{a}$ with $c_1 \neq c_0$ and $\mathbf{z}_1 \neq \mathbf{z}_0$. We set $c' := c_0 - c_1$ and $\mathbf{z}' := \mathbf{z}_0 - \mathbf{z}_1$. Then, we have $\mathbf{A}\mathbf{z}' - \mathbf{t}c' = \mathbf{0}$ i.e. $[\mathbf{A}|\mathbf{t}]\begin{bmatrix}\mathbf{z}'\\c'\end{bmatrix} = \mathbf{0}$. Since all coefficients of c are 0 or ± 1, $\|c\|_\infty \leq 1 < \gamma$. Thus, $\|c'\|_\infty < 2\gamma$ and

$\|\mathbf{z}'\|_\infty < 2\gamma$ hold which means that we have found a short integer solution to the MSIS problem s-MSIS$_s$ $(m+1, n, q, 2\gamma)$ with a non-negligible probability. However, since such an MSIS problem is hard, $\mathcal{A}$ cannot generate a valid transcript with a non-negligible probability, and our proof scheme is sound.

SHVZK. With the input public key $(\mathbf{A}, \mathbf{t}) \in L$ and the challenge $c \in \mathcal{C}$, the simulator $\mathcal{S}$ can firstly sample $\mathbf{z} \leftarrow D_s^m$ where $\|\mathbf{z}\|_\infty < \gamma$. Then, $\mathcal{S}$ computes

$$\mathbf{a} := \mathbf{A}\mathbf{z} - c\mathbf{t} \tag{2}$$

Finally, $\mathcal{S}$ produces the transcript $(\mathbf{a}, c, \mathbf{z})$ and sends it to the honest verifier $\mathcal{V}$. From the perspective of $\mathcal{V}$, since $c \xleftarrow{\$} B_\tau$, $\mathbf{a}$ satisfies the uniform distribution on R_q^n. Thus, due to the hardness of the Decision-MSIS problem, the probability that $\mathcal{V}$ can distinguish whether $\mathbf{a}$ is generated by Eq. (2) or computed by $\mathbf{A}\mathbf{y}$ where $\mathbf{y} \leftarrow D_s^m$ is negligible. Furthermore, since Eq. (2) holds and the distribution of $\mathbf{z} \leftarrow D_s^m$ is indistinguishable from that in the real execution of the protocol, $\mathcal{V}$ will always accept the transcript $(\mathbf{a}, c, \mathbf{z})$.

4 Concrete Construction of MDKG

In this section, we introduce the specific construction of our MDKG scheme. Let $q > 2$ be a prime integer, $\mathbf{A} \in R_q^{n \times m}$ and $\widetilde{\mathbf{A}} \in R_q^{n \times m}$ be two public matrices where n and m are defined as the dimension of matrix. Additionally, let N be a positive integer representing the degree of a polynomial, η be a small positive integer which represents a bound value. Define a hash function H : $\{0,1\}^* \to B_\tau$, and let S_η be the set of integers in the range $[-\eta, \eta]$. Let T and K be two positive integers, where T represents the threshold and K denotes the total number of participants, respectively. The structure of the MDKG scheme is composed of four phases, denoted as VSSGen, VSSVerify, SSCom, CPKCom as follows:

1. **VSS Generation** VSSGen. For each $i \in \{1, \ldots, K\}$, participant $\mathcal{P}_i$ performs Algorithm 3 m times to generate m module-lattice-based Lagrange interpolation polynomials $F_h(x, X)$ over $R_q[x]$ and m polynomials r_k^h for $h \in \{1, \ldots, m\}$, and sets $\mathbf{r}_k := \left(r_k^1, r_k^2, \ldots, r_k^m\right)^T$ for $k \in \{0, \ldots, T-1\}$.

Denote $\mathbf{r}_0$ by $\mathbf{s}_i$, each $\mathcal{P}_i$ computes his secret share for every $\mathcal{P}_j$ as $\mathbf{s}_i^{(j)} := (F_1(j, X), \ldots, F_m(j, X))^T \in R_q^m$. For $\mathbf{r}_0$, $\mathcal{P}_i$ samples an error $\mathbf{e}_i \in D_{R_q^n, s}$. Based on these, $\mathcal{P}_i$ computes $\mathbf{b}_{i0} := \mathbf{A}\mathbf{r}_0 + \mathbf{e}_i = \mathbf{A}\mathbf{s}_i + \mathbf{e}_i \in R_q^n$ and $\mathbf{b}_{ik} := \mathbf{A}\mathbf{r}_k \in R_q^n$ for $k \neq 0$. Similarly, $\mathcal{P}_i$ performs Algorithm 3 m times to generate m module-lattice-based Lagrange interpolation polynomials $\widetilde{F}_h(x, X)$ and m polynomials $\widetilde{r}_k^h$, and sets $\widetilde{\mathbf{r}}_k := \left(\widetilde{r}_k^1, \widetilde{r}_k^2, \ldots, \widetilde{r}_k^m\right)^T$. Denote $\widetilde{\mathbf{r}}_0$ by $\widetilde{\mathbf{s}}_i$, each $\mathcal{P}_i$ computes the share $\widetilde{\mathbf{s}}_i^{(j)}$ for every $\mathcal{P}_j$ where $j \in \{1, \ldots, K\}$. For $\widetilde{\mathbf{r}}_0$, $\mathcal{P}_i$ samples an error $\widetilde{\mathbf{e}}_i \in D_{R_q^n, s}$. Using these, $\mathcal{P}_i$ computes $\widetilde{\mathbf{b}}_{i0} := \widetilde{\mathbf{A}}\widetilde{\mathbf{r}}_0 + \widetilde{\mathbf{e}}_i = \widetilde{\mathbf{A}}\widetilde{\mathbf{s}}_i + \widetilde{\mathbf{e}}_i \in R_q^n$, and for $k \in \{1, \ldots, T-1\}$, computes $\widetilde{\mathbf{b}}_{ik} := \widetilde{\mathbf{A}}\widetilde{\mathbf{r}}_k \in R_q^n$.

Algorithm 3. Module-lattice-based Shamir secret sharing

Require: Security parameter λ, bound value η
Ensure: A module-lattice-based Lagrange interpolation polynomial $F(x, X)$ and polynomials r_k for $k \in \{0, \ldots, T-1\}$
1: Sample $N \cdot T$ elements $r_{ik} \xleftarrow{\$} S_\eta$ for $i \in \{0, \ldots, N-1\}$ and $k \in \{0, \ldots, T-1\}$
2: Set $F(x, X) := (1, X, \ldots, X^{N-1}) \begin{pmatrix} r_{0,0} & r_{0,1} & \cdots & r_{0,T-1} \\ r_{1,0} & r_{1,1} & \cdots & r_{1,T-1} \\ \vdots & \vdots & \ddots & \vdots \\ r_{N-1,0} & r_{N-1,1} & \cdots & r_{N-1,T-1} \end{pmatrix} \begin{pmatrix} 1 \\ x \\ \vdots \\ x^{T-1} \end{pmatrix}$
3: Record $r_k := \sum_{i=0}^{N-1} r_{ik} X^i \in R_q$ and $F(x, X) = \sum_{k=0}^{T-1} r_k x^k$ holds
4: **return** $F(x, X)$ and r_k for $k \in \{0, \ldots, T-1\}$

Finally, $\mathcal{P}_i$ computes all $\widehat{\mathbf{b}}_{ik} := \mathbf{b}_{ik} + \widetilde{\mathbf{b}}_{ik}$ and broadcasts them for $k \in \{0, \ldots, T-1\}$. Finally, $\mathcal{P}_i$ sends $\mathbf{s}_i^{(j)}$ and $\widetilde{\mathbf{s}}_i^{(j)}$ to participants $\mathcal{P}_j$.

Remark 1. Equation (3) holds. This relationship is also applicable between $\widetilde{\mathbf{s}}_i^{(j)}$ and $\{\widetilde{\mathbf{r}}_k\}_{k \in \{0, \ldots, T-1\}}$ for $j \in \{1, \ldots, K\}$.

$$(\mathbf{r}_0, \mathbf{r}_1, \ldots, \mathbf{r}_{T-1}) \begin{pmatrix} 1 & 1 & \cdots & 1 \\ 1 & 2 & \cdots & K \\ \vdots & \vdots & \ddots & \vdots \\ 1 & 2^{T-1} & \cdots & K^{T-1} \end{pmatrix} = \begin{pmatrix} r_0^1 & r_1^1 & \cdots & r_{T-1}^1 \\ r_0^2 & r_1^2 & \cdots & r_{T-1}^2 \\ \vdots & \vdots & \ddots & \vdots \\ r_0^m & r_1^m & \cdots & r_{T-1}^m \end{pmatrix} \begin{pmatrix} 1 & 1 & \cdots & 1 \\ 1 & 2 & \cdots & K \\ \vdots & \vdots & \ddots & \vdots \\ 1 & 2^{T-1} & \cdots & K^{T-1} \end{pmatrix}$$
$$= \begin{pmatrix} F_1(1, X) & F_1(2, X) & \cdots & F_1(K, X) \\ F_2(1, X) & F_2(2, X) & \cdots & F_2(K, X) \\ \vdots & \vdots & \ddots & \vdots \\ F_m(1, X) & F_m(2, X) & \cdots & F_m(K, X) \end{pmatrix} = \left(\mathbf{s}_i^{(1)}, \mathbf{s}_i^{(2)}, \ldots, \mathbf{s}_i^{(K)}\right) \quad (3)$$

Remark 2. $\mathbf{s}_i = \mathbf{r}_0 = (F_1(0, X), \ldots, F_m(0, X))^T = \sum_{j \in S} \mathbf{s}_i^{(j)} L_j^S$ holds for a subset S of at least T participants because Eq. (4) holds.

$$\sum_{j \in S} F_h(j, X) L_j^S = (1, X, \ldots, X^{N-1}) \begin{pmatrix} \sum_{j \in S} L_j^S \sum_{k=0}^{T-1} r_{0,k} j^k \\ \sum_{j \in S} L_j^S \sum_{k=0}^{T-1} r_{1,k} j^k \\ \vdots \\ \sum_{j \in S} L_j^S \sum_{k=0}^{T-1} r_{N-1,k} j^k \end{pmatrix} = \sum_{i=0}^{N-1} r_{i,0} X^i = F_h(0, X)$$
(4)

2. **VSS Verification** VSSVerify. Each $\mathcal{P}_j$ verifies the proof associated with the statements involving $\mathbf{s}_i^{(j)}$ and $\widetilde{\mathbf{s}}_i^{(j)}$: $\mathcal{P}_j$ verifies whether Eq. (5) holds.

$$\mathbf{A}\mathbf{s}_i^{(j)} + \widetilde{\mathbf{A}}\widetilde{\mathbf{s}}_i^{(j)} \approx_q \sum_{k=0}^{T-1} \widehat{\mathbf{b}}_{ik} j^k \quad (5)$$

"$\approx_q$" requires that the infinite norm of the difference between the two sides modulo q is less than $2\sqrt{s \cdot n \cdot N}$. If the verification does not succeed, $\mathcal{P}_j$

issues a complaint about $\mathcal{P}_i$. Conversely, if the verification is successful, $\mathcal{P}_j$ accepts the shares. If $\mathcal{P}_i$ receives at least T valid complaints, he is disqualified. In the event that $\mathcal{P}_j$ reports a failure in the verification of $\mathcal{P}_i$'s proofs, $\mathcal{P}_i$ is required to regenerate the secret shares $\mathbf{s}_i^{(j)}$ and $\widetilde{\mathbf{s}}_i^{(j)}$ in accordance with Eq. (5) and subsequently sends them to $\mathcal{P}_j$. Then, $\mathcal{P}_i$ computes $len_\eta := \left\lfloor \log_{(2\eta+1)} q \right\rfloor$ and perform Algorithm 4 to decomposes them into $\mathbf{s}_{il}^{(j)}, \widetilde{\mathbf{s}}_{il}^{(j)}$, such that $\sum_{l=0}^{len_\eta} \mathbf{s}_{il}^{(j)} (2\eta+1)^k = \mathbf{s}_i^{(j)}$, $\sum_{l=0}^{len_\eta} \widetilde{\mathbf{s}}_{il}^{(j)} (2\eta+1)^k = \mathbf{s}_i^{(j)}$ hold.

Algorithm 4. Polynomial vector decompose

Require: Polynomial vector $\mathbf{s} = \left(s^1, s^2, \ldots, s^m\right)^T$ where each $s^h = s_0^h + s_1^h X + \cdots + s_{N-1}^h X^{N-1}$ for $h \in \{1, 2, \ldots, m\}$, bound value η, size value len_η
Ensure: A set of polynomial vectors $\mathbf{s}_l$ in $R_{[-\eta, \eta]}^m$
1: $st := \mathbf{s}$
2: **for** l from 0 to len_η **do**
3:   **for** h from 1 to m **do**
4:     **for** i from 0 to $N-1$ **do**
5:       Set $s_{li}^h := st_i^h \mod (2\eta+1)$ and $st_i^h := \lfloor st_i^h / (2\eta+1) \rfloor$
6:     **end for**
7:     Set $s_l^h := s_{l,0}^h + s_{l,1}^h X + \cdots + s_{l,N-1}^h X^{N-1}$
8:   **end for**
9:   Set $\mathbf{s}_l := \left(s_l^1, s_l^2, \ldots, s_l^m\right)^T$
10: **end for**
11: **return** $\{\mathbf{s}_l\}_{l \in \{0, 1, \ldots, len_\eta\}}$

Finally, $\mathcal{P}_i$ computes the public keys $\mathbf{t}_{il}^{(j)} := \mathbf{A}\mathbf{s}_{il}^{(j)}$, $\widetilde{\mathbf{t}}_{il}^{(j)} := \widetilde{\mathbf{A}}\widetilde{\mathbf{s}}_{il}^{(j)}$ and generates the NIZK proofs $\pi_{il}^{(j)}$ and $\widetilde{\pi}_{il}^{(j)}$ of each share $\mathbf{s}_{il}^{(j)}$ and $\widetilde{\mathbf{s}}_{il}^{(j)}$ by running Algorithm 1 for $l \in \{0, \ldots, len_\eta\}$.

In terms of public channels, $\mathcal{P}_i$ responds to the complaint by revealing the public keys and NIZK proofs $\left\{\mathbf{t}_{il}^{(j)}, \widetilde{\mathbf{t}}_{il}^{(j)}, \pi_{il}^{(j)} \widetilde{\pi}_{il}^{(j)}\right\}_{l \in \{0, \ldots, len_\eta\}}$ for $\mathcal{P}_j$ so that everyone can verify the proofs by running Algorithm 2 and Eq. (6). If this verification succeeds, all participants can confirm that $\mathcal{P}_i$ holds the correct secret shares for $\mathcal{P}_j$, and the scheme continues as normal.

$$\sum_{l=0}^{len_\eta} \widetilde{\mathbf{t}}_{il}^{(j)} (2\eta+1)^k + \sum_{l=0}^{len_\eta} \widetilde{\mathbf{t}}_{il}^{(j)} (2\eta+1)^k \approx_q \sum_{k=0}^{T-1} \widehat{\mathbf{b}}_{ik} j^k \qquad (6)$$

Additionally, if any of the resent secret shares from $\mathcal{P}_i$ can not satisfy $\mathbf{t}_{il}^{(j)} = \mathbf{A}\mathbf{s}_{il}^{(j)}$ and $\widetilde{\mathbf{t}}_{il}^{(j)} = \widetilde{\mathbf{A}}\widetilde{\mathbf{s}}_{il}^{(j)}$ where $l \in \{0, \ldots, len_\eta\}$, $\mathcal{P}_j$ can directly broadcast them, and $\mathcal{P}_i$ will be disqualified. Ultimately, all honest participants will reach a consensus on the same set of qualified participants, denoted as $\mathcal{Q}$.

Remark 3. Equation (6) is equivalent to Eq. (5) because $\sum_{l=0}^{len_\eta} \widetilde{\mathbf{t}}_{il}^{(j)} (2\eta+1)^k = \sum_{l=0}^{len_\eta} \mathbf{A}\mathbf{s}_{il}^{(j)} (2\eta+1)^k = \mathbf{A}\mathbf{s}_i^{(j)}$ and $\sum_{l=0}^{len_\eta} \widetilde{\mathbf{t}}_{il}^{(j)} (2\eta+1)^k = \mathbf{A}\mathbf{s}_i^{(j)}$ hold.

3. **Secret Share Computation** SSCom. The secret key is implicitly expressed as $\mathbf{s} := \sum_{i \in \mathcal{Q}} \mathbf{s}_i$. Each participant $\mathcal{P}_j$ computes their individual secret share of $\mathbf{s}$ as $\mathbf{s}^{(j)} := \sum_{i \in \mathcal{Q}} \mathbf{s}_i^{(j)}$. Because $\mathbf{s}_i = \sum_{j \in S} \mathbf{s}_i^{(j)} L_j^S$ holds for a subset S of at least T participants (proven in Remark 2), $\mathbf{s} = \sum_{i \in \mathcal{Q}} \mathbf{s}_i = \sum_{i \in \mathcal{Q}} \sum_{j \in S} \mathbf{s}_i^{(j)} L_j^S = \sum_{j \in S} L_j^S \sum_{i \in \mathcal{Q}} \mathbf{s}_i^{(j)} = \sum_{j \in S} \mathbf{s}^{(j)} L_j^S$ also holds.

4. **Common public key computation** CPKCom. Each qualified participant $\mathcal{P}_i$ broadcasts $\{\mathbf{b}_{ik}\}_{k \in \{0,\ldots,T-1\}}$ where $i \in \mathcal{Q}$. Subsequently, each qualified participant $\mathcal{P}_j$, where $j \in \mathcal{Q}$, verifies the values broadcast by the other qualified participants by checking whether Eq. (7) holds.

$$\mathbf{A}\mathbf{s}_i^{(j)} \approx_q \sum_{k=0}^{T-1} \mathbf{b}_{ik} j^k \quad (7)$$

If Eq. (7) does not hold for $\mathcal{P}_i$, then $\mathcal{P}_j$ will lodge a complaint against $\mathcal{P}_i$ by broadcasting the public key $\{\mathbf{t}_{il}^{(j)}, \widetilde{\mathbf{t}}_{il}^{(j)}\}_{l \in \{0,\ldots,len_\eta\}}$. This key satisfies Eq. (5), yet fails to satisfy Eq. (7). And the common public key share $\mathbf{b}_{i0}$, of $\mathcal{P}_i$, who has received at least one valid complaint, can be reconstructed using the Lagrange interpolation formula. Finally, the common public key can be computed by $\mathbf{b} := \sum_{i \in \mathcal{Q}} \mathbf{b}_{i0}$ for all qualified participants $\{\mathcal{P}_i\}_{i \in \mathcal{Q}}$.

Remark 4. $\mathbf{b}$ is a public key corresponding to all secret shares $\mathbf{s}^{(j)}$ because $\mathbf{b} = \sum_{i \in \mathcal{Q}} \mathbf{b}_{i0} = \sum_{i \in \mathcal{Q}} (\mathbf{A}\mathbf{s}_i + \mathbf{e}_i) = \mathbf{A} \sum_{i \in \mathcal{Q}} \mathbf{s}_i + \sum_{i \in \mathcal{Q}} \mathbf{e}_i = \mathbf{A} \sum_{j \in S} \mathbf{s}^{(j)} L_j^S + \sum_{i \in \mathcal{Q}} \mathbf{e}_i$ holds for a subset S of at least T qualified participants.

5 Security Analysis

The security of our MDKG scheme is characterized by two key properties: robust correctness and secrecy, introduced by Gennaro et al. [22] as below.

Robust Correctness. For K participants, where there are at most $T-1$ malicious participants and at least T honest participants, correctness ensures that each honest participant $\mathcal{P}_j$ can generate a secret share $\mathbf{s}^{(j)}$ of the secret key. Robust correctness requires [22] that even if at most $T-1$ malicious participants submit incorrect secret shared values, there is still an effective algorithm to reconstruct the secret. We refer to [22] and introduce the definition of robust correctness in module-lattice as below.

Definition 5 (Robust correctness). *Our MDKG protocol for K participants $\{\mathcal{P}_i\}_{i \in \{1,\ldots,K\}}$ is considered robustly correct if, for any positive integer $T \leq K$, any subset $\mathcal{I} \subseteq \{1,\ldots,K\}$ with $|\mathcal{I}| \geq T$ and $K - |\mathcal{I}| < T$, and any probabilistic polynomial-time (PPT) adversary $\mathcal{A}$, the following conditions hold: All subsets* of at least T honest participants in $\mathcal{I}$ will derive an equivalent secret sharing **s**; all honest participants in $\mathcal{I}$ will agree on a unique common public key **b**; even if at most $T - 1$ of the submitted secret shares are incorrect (due to participants controlled by $\mathcal{A}$), the common public key **b** can still be computed effectively.

Secrecy. For the secret **s**, secrecy ensures that no information about **s** can be obtained by the adversary through the MDKG protocol, except for the value of the public key **b**. Following the approach in [9], we introduce a simulator Sim that, given a public key **b** chosen uniformly and randomly, simulates the behavior of honest participants and interacts with the adversary $\mathcal{A}$.

Definition 6 (Secrecy). *Our MDKG protocol is said to have secrecy for K participants $\{\mathcal{P}_i\}_{i \in \{1,\ldots,K\}}$ if, for any positive integer $T \leq K$, any subset of honest participants $\mathcal{I}$ with $|\mathcal{I}| \geq T$ and $K - |\mathcal{I}| < T$, and any PPT adversary $\mathcal{A}$ that corrupts at most $T - 1$ participants, there exists a simulator Sim. Given the public key **b** as input, the output distribution of Sim simulating the MDKG scheme must be indistinguishable from the output distribution of the real MDKG scheme, as observed by $\mathcal{A}$, where the output consists of the public key **b**.*

We propose Theorem 1 and Theorem 2 and provide relevant proofs in Appendix A and Appendix B. And the soundness and zero-knowledge property of the NIZK proof can be reduced to the hardness of the MSIS problem.

Theorem 1 (Robust correctness). *Given the modulus $q \geq 2$, the dimensions m and n of the matrix, and an error distribution, our MDKG scheme satisfies the requirement of robust correctness in Definition 5, if the NIZK proof is sound.*

Theorem 2 (Secrecy). *Given the modulus $q \geq 2$, the dimensions m and n of the matrix, and an error distribution, our MDKG scheme satisfies the requirement of secrecy of Definition 6, if the NIZK proof is zero-knowledge, and if the decision-MLWE problem d-MLWE$_\mathbf{s}\left(m, n, q, D_{R_q^n, s}\right)$ and the decision-MSIS problem d-MSIS$_\mathbf{s}(m, n, q, \eta)$ are hard.*

6 Performance Analysis

In this section, we compare the performance of our scheme with two other lattice-based DKG schemes [18,35] in terms of running time and parameter size. Our MDKG scheme is more lightweight than theirs. It not only does not use additional encryption [18] and multi-party computing [35] algorithms, but also maintains robust correctness and secrecy.

Parameters Sets. We select the parameters for the security level κ of our MDKG scheme, as presented in Table 1, based on the lattice estimator [3]. The bytesizes of secret sharing, common public key, and NIZK proof are computed as $|ss| := mN\lceil \log_2 q \rceil$, $|pk| := nN\lceil \log_2 q \rceil$, and $|\pi| :=$

$2N \left(len_\eta + 1\right) \left(m \left\lceil \log_2 \gamma \right\rceil + 1\right)$ respectively because $\|c\|_\infty < 1$ and $\|\mathbf{z}\|_\infty < \gamma$. We compute the byte sizes of VSSGen, SSCom, and CPKCom as $2|ss|$, $|ss|$, and $|pk|$, respectively. Considering that the zero-knowledge proof is only generated when a complaint is generated, we do not add the byte size of the NIZK proof to the calculation of the size of each stage. Compared with the other two lattice-based DKG schemes [18,35] in Table 2, our MDKG scheme reduces the secret sharing size by 74% and has obvious size advantages in most processes because it does not completely require NIZK proof and is designed based on module-lattice.

Table 1. Parameters and sizes of MDKG scheme

κ	q	m	n	N	s	η	γ	M	MLWE C/Q	MSIS C/Q	Secret share	Common public key	NIZK proof
128	8380417	4	4	256	$\sqrt{2}$	3	2^{17}	3.38	123/112	123/112	3 KB	3 KB	39 KB

Table 2. Size and runtime comparison of lattice-based DKG scheme

	VSSGen	SSCom	CPKCom	Runtime
[18]	56 KB	14 KB	12.8 KB	-
[35]	23.5 KB	0.78 KB	18.75 KB	455 ms
MDKG	6 KB	3 KB	3 KB	0.45 ms

Running Time. We implemented our MDKG scheme using the official reference code of the Dilithium algorithm [16]. The running time of each phase was measured on an Ubuntu 20.04 (64-bit) operating system with an Intel(R) Core(TM) i5-13600K 3.50 GHz processor and 32 GB of RAM.

In our scheme, the time costs for NIZK proof generation and verification are fixed at 0.31 ms and 0.09 ms. And the total time cost of the NIZK proof for each complaint is $(len_\eta + 1)(0.31 + 0.09) = 3.6$ ms. We break down the time costs of our MDKG scheme into common parameter generation (CommonParamGen), VSSGen, VSSVerify stage (FirstVerify), SSCom, verification before the CPKCom stage (SecondVerify), and CPKCom. Since NIZK is only generated when there are complaints, we do not include their time overhead in the various stages of the scheme runtime calculation. We fix $K = 100$ and $T = 25$ respectively, as T and K increase, we show the results in Fig. 1. The results show that the time consumption of VSSGen is significantly affected by the T and K, and the single execution time is also longer; the single execution of other steps is only affected by T, and the single execution time is very short.

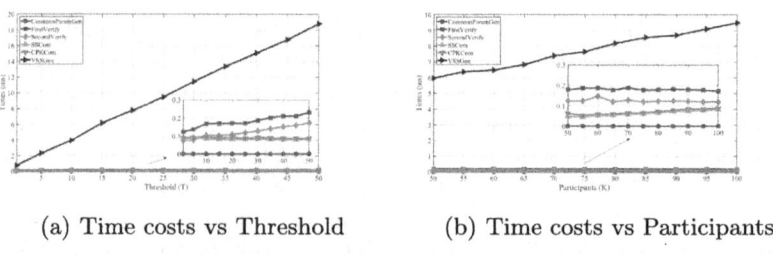

(a) Time costs vs Threshold (b) Time costs vs Participants

Fig. 1. Time costs vs Threshold and Participants

We compare the runtime of our scheme with that of Tang et al.'s scheme [35] for the specific case where $T = 2$ and $K = 3$. The results of this comparison, as shown in Table 2, demonstrate that our MDKG scheme reduces the time cost by at least 90% compared to Tang et al.'s scheme. This improvement is primarily due to the fact that our scheme avoids the need for complex multi-party computation and MAC calculations, which are required in their approach. Additionally, the performance of our scheme may be affected by the NIZK proof, but experiments show that as long as the number of complaints generated is relatively small (less than 120 in the above setting), our scheme still has less time cost.

7 Conclusion

Distributed key generation can effectively overcome a single-point failure problem and securely generate secret shares. In this paper, we propose a module-lattice-based distributed key generation scheme called MDKG. Our scheme is resistant to quantum attacks and combines the NIZK proof to minimize the possibility of exposing secret shares when complaints occur. In addition, we prove the robust correctness and secrecy which are introduced by Gennaro et al. [22]. We also analyze its performance and computation consumption. The results show that our MDKG scheme has high performance and provides a distributed key generation method for lattice-based threshold schemes. Our MDKG scheme reduces the secret sharing size by 74% and reduces the time cost by at least 90% compared to the existing lattice-based DKG scheme. Future work includes research on higher performance and smaller size MDKG, and the combination of MDKG and threshold schemes.

Acknowledgments. The work was supported by the Major Program (JD) of Hubei Province (No. 2023BAA027), R&D Program of Hubei Province (2023BAB171), the National Natural Science Foundation of China (Nos. 62325209, U23A20302, 62202490, 62272350), the New 20 Project of Higher Education of Jinan (No. 202228017) and the Fundamental Research Funds for the Central Universities (Nos. 2042023KF0203, 2042024kf1013).

Disclosure of Interests. The authors have no competing interests to declare that are relevant to the content of this article.

A Proof of Robust Correctness

Given K participants, T is defined as the threshold. Consider an index set $\mathcal{I}$ representing honest parties, where $|\mathcal{I}| \geq T$ and $K - |\mathcal{I}| < T$, we consider the presence of a PPT adversary $\mathcal{A}$ that attempts to compromise the robust correctness of the MDKG scheme. We construct a PPT adversary $\mathcal{B}^{\mathcal{A}}$ against the soundness of the module-lattice-based NIZK proof which has a black box access to $\mathcal{A}$. Then $\mathcal{B}^{\mathcal{A}}$ will break the soundness of the NIZK proof.

During the execution, if a complaint is publicly broadcast, $\mathcal{B}^{\mathcal{A}}$ emulates the behavior of the honest participants $\{\mathcal{P}_i\}_{i \in \mathcal{I}}$. It then engages in an interaction with $\mathcal{A}$, simulating the responses and actions of the honest parties. Finally, $\mathcal{B}^{\mathcal{A}}$ selects an index $i \in \mathcal{Q} \setminus \mathcal{I}$ and generates the statement $\left\{\left(\mathbf{t}_{i,0}^{(j)}, \ldots, \mathbf{t}_{i,len_n}^{(j)}\right)\right\}_{j \in \mathcal{I}}$, along with the corresponding proof components $\left\{\left(\pi_{i,0}^{(j)}, \ldots, \pi_{i,len_n}^{(j)}\right)\right\}_{j \in \mathcal{I}}$.

Let $\mathsf{Break}_{\mathsf{NIZK}}$ represent the event in which at least one of the NIZK proofs submitted by a participant within the qualified set $\mathcal{Q}$, controlled by $\mathcal{A}$, constitutes a valid proof for a statement that is invalid. Its advantage is denoted as $\mathsf{Adv}_{\mathsf{NIZK}}$. Then, if $\mathsf{Break}_{\mathsf{NIZK}}$ occurs, $\mathcal{B}^{\mathcal{A}}$ will pick that proof and output it with probability $1/|\mathcal{Q} \setminus \mathcal{I}| > 1/(K - |\mathcal{I}|)$. Thus, the advantage of $\mathcal{B}^{\mathcal{A}}$ against the soundness of module-lattice-based NIZK proof is at least $\mathsf{Adv}_{\mathsf{NIZK}}/(K - |\mathcal{I}|)$.

If $\mathsf{Break}_{\mathsf{NIZK}}$ does not occur, each participant $\mathcal{P}_i$ in the set $\mathcal{Q}$ has honestly executed the scheme, which means that all revealed zero-knowledge proofs generated by $\mathcal{P}_i$ for $\mathcal{P}_j$ can be verified (see the proof in Sect. 3.2) where $i \in \mathcal{Q}$ and $j \in \{1, \ldots, K\}$. We can deduce that $\sum_{k=0}^{T-1} \mathbf{b}_{ik} j^k = \mathbf{A} \sum_{k=0}^{T-1} \mathbf{r}_{ik} j^k + \mathbf{e}_i = \mathbf{A} \mathbf{s}_i^{(j)} + \mathbf{e}_i$. Similarly, $\sum_{k=0}^{T-1} \widetilde{\mathbf{b}}_{ik} j^k = \widetilde{\mathbf{A}} \widetilde{\mathbf{s}}_i^{(j)} + \widetilde{\mathbf{e}}_i$ holds as well. Thus, Eq. (8) holds.

$$\sum_{k=0}^{T-1} \widehat{\mathbf{b}}_{ik} j^k = \sum_{k=0}^{T-1} \left(\mathbf{b}_{ik} + \widetilde{\mathbf{b}}_{ik}\right) j^k = \mathbf{A} \mathbf{s}_i^{(j)} + \widetilde{\mathbf{A}} \widetilde{\mathbf{s}}_i^{(j)} + \mathbf{e}_i + \widetilde{\mathbf{e}}_i \tag{8}$$

Since $\mathbf{e}_i, \widetilde{\mathbf{e}}_i \in R_q^n$ are sampled from the distribution $D_{R_q^n, s}$. We can deduce that $\mathbf{e}_i + \widetilde{\mathbf{e}}_i < 2\sqrt{s \cdot n \cdot N}$ with an whelming probability from Lemma 1. And it can be judged from Eq. (8) that Eq. (5) and Eq. (7) hold with an overwhelming probability. Consequently, the advantage of $\mathcal{A}$ in compromising the robust correctness of the MDKG scheme is bounded by $\mathsf{Adv}_{correctness} \leq \frac{1}{K-|\mathcal{I}|} \cdot \mathsf{Adv}_{\mathsf{NIZK}} + \mathsf{negl}(\lambda)$, which is negligible. Furthermore, we note that $\mathcal{A}$ may generate a false but valid $\mathbf{s}_i^{'(j)}$ instead of $\mathbf{s}_i^{(j)}$ for $\mathcal{P}_j$, which means $\mathbf{s}_i^{'(j)} \neq \mathbf{s}_i^{(j)}$ but $\mathcal{A}$ honestly generates the proof for $\mathbf{s}_i^{'(j)}$ satisfies Eq. (9) which refers from Eq. (7).

$$\mathbf{A} \mathbf{s}_i^{'(j)} \approx_q \sum_{k=0}^{T-1} \mathbf{b}_{ik} j^k \tag{9}$$

It means that there exists an error $\mathbf{e}_i' \in R_q^n$ which all coefficients of each polynomial element is in $D_{R_q^n, s}$ such that $\mathbf{A} \mathbf{s}_i^{'(j)} + \mathbf{e}_i' = \mathbf{A} \mathbf{s}_i^{(j)} + \mathbf{e}_i$. There are currently no hard problems that indicate that it would be hard to find such a collision. Fortunately, although the adversary $\mathcal{A}$ generates an incorrect but valid $\mathbf{s}_i'^{(j)}$, such $\mathbf{s}_i'^{(j)}$ will not affect the correctness of the scheme at all. It can replace the correct $\mathbf{s}_i^{(j)}$ as the secret share sent by $\mathcal{P}_i$ to $\mathcal{P}_j$.

B Proof of Secrecy

Given K participants, T is defined as the threshold. We posit the existence of a PPT adversary $\mathcal{A}$ that can compromise up to $T-1$ participants. The set of participants who remain uncorrupted is defined as $\mathcal{I}$, where $\mathcal{I}$ is a subset of $\{1,\ldots,K\}$. We design a simulator $\mathsf{Sim} = (\mathsf{Sim}_1, \mathsf{Sim}_2)$ to emulate the behavior of the honest participants $\{\mathcal{P}_i\}_{i\in\mathcal{I}}$ and the random oracle. The simulation is constructed in such a way that it becomes indistinguishable from the actual execution of the MDKG scheme. We design the game between the real world and simulated environment to prove the secrecy of the MDKG scheme as follows:

Game 0. In this game, the simulator Sim faithfully simulates the honest participants as in the real world. It consists of two parts, where Sim_1 faithfully simulates the honest participants, and Sim_2 simulates a random oracle by maintaining a query list.

Game 1. This game closely resembles Game 0, with the key distinction being that the common public key $\mathbf{b}$ is provided beforehand. Sim_1 simulates the process of computing the public key as follows:

- Fix a participant $\mathcal{P}_{fix}$ where $fix \in \mathcal{Q}$.
- Broadcast $\mathbf{b}_{i0} = \mathbf{A}\mathbf{r}_{i0} + \mathbf{e}_i$ and $\{\mathbf{b}_{ik} = \mathbf{A}\mathbf{r}_{ik}\}_{k\in\{1,\ldots,T-1\}}$ for all $i \in \mathcal{Q}\setminus\{fix\}$.
- For the participant $\mathcal{P}_{fix}$ who is simulated by Sim_1, sample an error $\mathbf{e}_{fix}^* \in D_{R_q^n,s}$, compute

$$\mathbf{b}_{fix,0}^* := \mathbf{b} - \sum_{i\in\mathcal{Q}\setminus\{fix\}} \mathbf{b}_{i0} \qquad (10)$$

- We assume that $\mathcal{A}$ corrupts a subset of qualified participants, denoted as $\mathcal{C}$, where $\mathcal{C}$ is a proper subset of the qualified set $\mathcal{Q}$. In addition, we define $\mathcal{C}' := \mathcal{C} \cup \{0\}$. We assume that $fix \notin \mathcal{C}$, then, Sim_1 reconstruct the public shares polynomials for $\mathcal{P}_{fix}$ by computing $\mathbf{t}_{fix}^{(x)} = \mathbf{A}\mathbf{s}_{fix}^{(x)} := L_{x,0}^{\mathcal{C}'} \cdot \mathbf{b}_{fix,0}^* + \sum_{i\in\mathcal{C}} L_{x,i}^{\mathcal{C}'} \cdot \mathbf{A}\mathbf{s}_i^{(fix)}$ where $x \in \mathcal{C}$ and $L_{x,i}^{\mathcal{C}'}$ represents the Lagrange interpolation parameter, computed by Eq. (11). The coefficient vector of $\mathbf{t}_{fix}^{(x)}$ corresponding to the term "x^k" is represented as $\mathbf{b}_{fix,k}^*$ for each $k \in \mathcal{Q}$.

$$L_{x,i}^{\mathcal{C}'} := \prod_{j\in\mathcal{C}'\setminus\{i\}} \frac{x-j}{i-j} \mod q \qquad (11)$$

- Broadcast $\mathbf{b}_{ik}$ and $\mathbf{b}_{fix,k}^*$ for all $i \in \mathcal{Q}\setminus\{fix\}$ and $k \in \mathcal{C}$.
- Verify if Eq. (7) holds for all corrupted participants $i \in \mathcal{C}$. If the there is any verification fails for $i \in \mathcal{C}$ and $j \in \mathcal{Q}$, Sim_1 broadcasts a complaint against $\mathcal{P}_i$ and reveals the public key $\left\{\mathbf{t}_{il}^{(j)}, \widetilde{\mathbf{t}}_{il}^{(j)}\right\}_{l\in\{0,\ldots,len_\eta\}}$.

- Reconstruct the public key formula of the complained participant through Lagrange interpolation.

Game 2. This game resembles **Game 1**, with the key distinction that in this case, Sim_2 does not generate π_{sim} honestly for a legitimate participant $\mathcal{P}_{sim}$, where sim is an element of $\mathcal{I}$. Instead, Sim generates proof π_{sim} by calling $\left\{\pi_{sim}^{(j)}\right\}_{j\in\mathcal{C}} \leftarrow \mathsf{Sim}_1^{\mathsf{Sim}_2}\left(\mathbf{A}, \left\{\mathbf{s}_{sim}^{(j)}\right\}_{j\in\mathcal{C}}\right)$. Sim_2 simulates a random oracle by keeping a query list but it will forward the list of the oracle queries to Sim_1.

We firstly prove that **Game 1** is indistinguishable from **Game 0** if the MLWE problem d-MLWE$_{\mathbf{s}}(m, n, q, \chi)$ is hard. In **Game 0** and **Game 1**, the simulator Sim all perform the VSS generation and VSS verification process honestly. Therefore, we can confirm that the sets of valid participants generated by **Game 1** and **Game 0** in the VSSGen phase are the same. And we can make the following inferences:

- In $\mathcal{A}$'s view, he knows the polynomials $\left\{F_{hi}(x,X), \widetilde{F}_{hi}(x,X)\right\}$, secret shares for $h \in \{1, \ldots, m\}$ and $i \in \mathcal{C}$. And $\mathcal{A}$ also knows all corrupted secret shares $\left\{\mathbf{s}_{il}^{(j)}, \widetilde{\mathbf{s}}_{il}^{(j)}\right\}$ for $i \in \mathcal{Q}, j \in \mathcal{C}$, and $l \in \{0, \ldots, len_\eta\}$. In addition, $\mathcal{A}$ can get all the public values $\widehat{\mathbf{b}}_{ik}$ for $i \in \mathcal{Q}$ and $k \in \{1, \ldots, T-1\}$.
- In Sim's view, he knows all polynomials $\left\{F_{hi}(x,X), \widetilde{F}_{hi}(x,X)\right\}$, secret shares $\left\{\mathbf{s}_{il}^{(j)}, \widetilde{\mathbf{s}}_{il}^{(j)}\right\}$, and public values $\widehat{\mathbf{b}}_{ik}$ for $h \in \{1, \ldots, m\}, i \in \mathcal{Q}, j \in \mathcal{C}, l \in \{0, \ldots, len_\eta\}$ and $k \in \{1, \ldots, T-1\}$ because Sim simulates the honest participants $\{\mathcal{P}_i\}_{i\in\mathcal{I}}$ and can computes all the corrupted polynomials by Lagrange interpolation with enough secret shares sent from $\mathcal{A}$.

It means that the simulator Sim can compute all public key $\mathbf{b}_{i,0}$ for $i \in \mathcal{Q}$ which includes the public key corrupted by $\mathcal{A}$. Then in the CPKCom phase, Sim picks a simulated participant $\mathcal{P}_{fix}$ for $fix \in \mathcal{I}$ and modifies the value that he should broadcast, so that the common public key computed from $\mathcal{A}$'s perspective is equal to the common public key $\mathbf{b}$ given initially. The only difference between **Game 1** and **Game 0** is that the value of $\mathbf{b}_{fix,k}$ where $k \in \{0, \ldots, T-1\}$.

Since the public shares polynomials $\mathbf{t}_{fix}^{(x)}$ of $\mathcal{P}_{fix}$ reconstructed by Sim is generated by Lagrange interpolation based on $\mathbf{b}_{fix,0}^*$ where $x \in \mathcal{Q}$ (the corresponding secret shares polynomials are implicitly reconstructed). And all the secret shares sent to $\mathcal{P}_k$ corrupted by $\mathcal{A}$ where $k \in \mathcal{C}$, the secret share value for each $k \in \mathcal{C}$ does not change with the change of the secret shares polynomials of $\mathcal{P}_{fix}$. Since $\left\{\widehat{\mathbf{b}}_{ik}\right\}_{k\in\{0,\ldots,T-1\}}$ broadcast in the VSSGen phase have not changed, from the adversary $\mathcal{A}$'s view, $\mathbf{b}_{fix,k}^*$ for $k \in \mathcal{Q}$ broadcast by $\mathcal{P}_{fix}$ and the secret shares sent to all $\mathcal{P}_i$ can satisfy both Eq. (5) and Eq. (7).

In **Game 0**, each coefficient of $\mathcal{P}_{fix}$'s secret shares polynomials is sampled uniformly at random which means that the public shares $\mathbf{b}_{fix,k}$ and the secret shares $\mathbf{s}_{fix}^{(j)}$ follows a uniform distribution for $j \in \mathcal{Q}, k \in \{0, \ldots, T-1\}$. Thus, the distribution of $\mathcal{P}_{fix}$'s public shares is indistinguishable from the uniform distribution because the hardness of the MLWE problem d-MLWE$_{\mathbf{s}}\left(m, n, q, D_{R_q^n, s}\right)$

and the MSIS problem d-MSIS$_s$ (m, n, q, η). And in Game 1, the first column coefficients of $\mathcal{P}_{fix}$'s public shares are computed by Eq. (10), the other column coefficients of $\mathcal{P}_{fix}$'s public shares are computed from the Lagrange interpolation. Since **b** is initially sampled uniformly at random, the distribution of the first column coefficients of $\mathcal{P}_{fix}$ follow the uniform distribution. Furthermore, since the values of $\mathbf{s}_{fix}^{(j)}$ do not change where $j \in \mathcal{C}$, that is, they also follow the uniform distribution, the coefficients of each column of the public shares polynomials reconstructed by Lagrange interpolation follow the uniform distribution i.e. $\mathbf{b}^*_{fix,k} \sim \mathcal{U}(R_q^n)$ for $k \in \{0, \ldots, T-1\}$. Thus, Game 1 is indistinguishable from Game 0 with an overwhelming probability.

In Game 2, Sim$_2$ pick an honest participant $\mathcal{P}_{sim}$ in VSSGen phase where $sim \in \mathcal{I}$. When $\mathcal{P}_j$ broadcasts a complain against $\mathcal{P}_{sim}$ where $j \in \mathcal{C}$, Sim generates proof π_{sim} by calling $\left\{\pi_{sim}^{(j)}\right\}_{j \in \mathcal{C}} \leftarrow \mathsf{Sim}_1^{\mathsf{Sim}_2}\left(\mathbf{A}, \left\{\mathbf{s}_{sim}^{(j)}\right\}_{j \in \mathcal{C}}\right)$. Sim$_2$ simulates a random oracle by keeping a query list but it will forward the list of the oracle queries to Sim$_1$. Then if $\mathcal{A}$ can distinguish Game 2 with a non-negligible advantage, Sim can break the zero-knowledge of NIZK proof. Thus, Game 2 is indistinguishable from Game 1. We complete the proof.

References

1. Adida, B.: Helios: web-based open-audit voting. In: Proceedings of the 17th Conference on Security Symposium, SS 2008, pp. 335–348. USENIX Association, USA (2008). https://doi.org/10.5555/1496711.1496734
2. Ajtai, M.: Generating hard instances of lattice problems. In: Proceedings of the Twenty-Eighth Annual ACM Symposium on Theory of Computing, pp. 99–108 (1996). https://doi.org/10.1145/237814.237838
3. Albrecht, M.R., Player, R., Scott, S.: On the concrete hardness of learning with errors. J. Math. Cryptol. **9**(3), 169–203 (2015). https://doi.org/10.1515/jmc-2015-0016
4. Atapoor, S., Baghery, K., Cozzo, D., Pedersen, R.: Practical robust DKG protocols for CSIDH. In: Tibouchi, M., Wang, X. (eds.) Applied Cryptography and Network Security, pp. 219–247. Springer, Cham (2023). https://doi.org/10.1007/978-3-031-33491-7_9
5. Bellare, M., Goldreich, O.: On defining proofs of knowledge. In: Brickell, E.F. (ed.) CRYPTO 1992. LNCS, vol. 740, pp. 390–420. Springer, Heidelberg (1993). https://doi.org/10.1007/3-540-48071-4_28
6. Bellare, M., Rogaway, P.: Random oracles are practical: a paradigm for designing efficient protocols. In: Proceedings of the 1st ACM Conference on Computer and Communications Security, CCS 1993, pp. 62–73. Association for Computing Machinery, New York (1993). https://doi.org/10.1145/168588.168596
7. Benhamouda, F., Camenisch, J., Krenn, S., Lyubashevsky, V., Neven, G.: Better zero-knowledge proofs for lattice encryption and their application to group signatures. In: Sarkar, P., Iwata, T. (eds.) ASIACRYPT 2014. LNCS, vol. 8873, pp. 551–572. Springer, Heidelberg (2014). https://doi.org/10.1007/978-3-662-45611-8_29

8. Benhamouda, F., Krenn, S., Lyubashevsky, V., Pietrzak, K.: Efficient zero-knowledge proofs for commitments from learning with errors over rings. In: Pernul, G., Ryan, P.Y.A., Weippl, E. (eds.) ESORICS 2015. LNCS, vol. 9326, pp. 305–325. Springer, Cham (2015). https://doi.org/10.1007/978-3-319-24174-6_16
9. Beullens, W., Disson, L., Pedersen, R., Vercauteren, F.: CSI-RAShi: distributed key generation for CSIDH. In: Cheon, J.H., Tillich, J.-P. (eds.) PQCrypto 2021 2021. LNCS, vol. 12841, pp. 257–276. Springer, Cham (2021). https://doi.org/10.1007/978-3-030-81293-5_14
10. Boldyreva, A.: Threshold signatures, multisignatures and blind signatures based on the gap-Diffie-Hellman-group signature scheme. In: Desmedt, Y.G. (ed.) PKC 2003. LNCS, vol. 2567, pp. 31–46. Springer, Heidelberg (2003). https://doi.org/10.1007/3-540-36288-6_3
11. Brakerski, Z., Gentry, C., Vaikuntanathan, V.: (leveled) fully homomorphic encryption without bootstrapping. ACM Trans. Comput. Theory (TOCT) **6**(3), 1–36 (2014). https://doi.org/10.1145/2633600
12. Brakerski, Z., Langlois, A., Peikert, C., Regev, O., Stehlé, D.: Classical hardness of learning with errors. In: Proceedings of the Forty-Fifth Annual ACM Symposium on Theory of Computing, STOC 2013, pp. 575–584. Association for Computing Machinery, New York (2013). https://doi.org/10.1145/2488608.2488680
13. Cascudo, I., David, B.: Publicly verifiable secret sharing over class groups and applications to DKG and YOSO. In: Joye, M., Leander, G. (eds.) Advances in Cryptology – EUROCRYPT 2024, pp. 216–248. Springer, Cham (2024). https://doi.org/10.1007/978-3-031-58740-5_8
14. De Santis, A., Desmedt, Y., Frankel, Y., Yung, M.: How to share a function securely. In: Proceedings of the Twenty-Sixth Annual ACM Symposium on Theory of Computing, STOC 1994, pp. 522–533. Association for Computing Machinery, New York (1994). https://doi.org/10.1145/195058.195405
15. Desmedt, Y.: Threshold cryptosystems. In: Seberry, J., Zheng, Y. (eds.) AUSCRYPT 1992. LNCS, vol. 718, pp. 1–14. Springer, Heidelberg (1993). https://doi.org/10.1007/3-540-57220-1_47
16. Ducas, L., et al.: Crystals-dilithium: a lattice-based digital signature scheme. IACR Trans. Cryptogr. Hardw. Embed. Syst. **2018**(1), 238–268 (2018). https://doi.org/10.13154/tches.v2018.i1.238-268
17. Esgin, M.F., Steinfeld, R., Sakzad, A., Liu, J.K., Liu, D.: Short lattice-based one-out-of-many proofs and applications to ring signatures. In: Deng, R.H., Gauthier-Umaña, V., Ochoa, M., Yung, M. (eds.) ACNS 2019. LNCS, vol. 11464, pp. 67–88. Springer, Cham (2019). https://doi.org/10.1007/978-3-030-21568-2_4
18. Espitau, T., Niot, G., Prest, T.: Flood and submerse: distributed key generation and robust threshold signature from lattices. In: Reyzin, L., Stebila, D. (eds.) Advances in Cryptology – CRYPTO 2024, pp. 425–458. Springer, Cham (2024). https://doi.org/10.1007/978-3-031-68394-7_14
19. Feldman, P.: A practical scheme for non-interactive verifiable secret sharing. In: 28th Annual Symposium on Foundations of Computer Science (SFCS 1987), pp. 427–438 (1987). https://doi.org/10.1109/SFCS.1987.4
20. Fiat, A., Shamir, A.: How to prove yourself: practical solutions to identification and signature problems. In: Odlyzko, A.M. (ed.) CRYPTO 1986. LNCS, vol. 263, pp. 186–194. Springer, Heidelberg (1987). https://doi.org/10.1007/3-540-47721-7_12
21. Fouque, P.-A., Stern, J.: One round threshold discrete-log key generation without private channels. In: Kim, K. (ed.) PKC 2001. LNCS, vol. 1992, pp. 300–316. Springer, Heidelberg (2001). https://doi.org/10.1007/3-540-44586-2_22

22. Gennaro, R., Jarecki, S., Krawczyk, H., Rabin, T.: Secure distributed key generation for discrete-log based cryptosystems. In: Stern, J. (ed.) EUROCRYPT 1999. LNCS, vol. 1592, pp. 295–310. Springer, Heidelberg (1999). https://doi.org/10.1007/3-540-48910-X_21
23. Goldreich, O., Micali, S., Wigderson, A.: How to play any mental game, or a completeness theorem for protocols with honest majority, pp. 307–328. Association for Computing Machinery, New York (2019). https://doi.org/10.1145/3335741.3335755
24. Groth, J.: Non-interactive distributed key generation and key resharing. IACR Cryptol. ePrint Arch. 339 (2021). https://eprint.iacr.org/2021/339
25. Gurkan, K., Jovanovic, P., Maller, M., Meiklejohn, S., Stern, G., Tomescu, A.: Aggregatable distributed key generation. In: Canteaut, A., Standaert, F.-X. (eds.) EUROCRYPT 2021. LNCS, vol. 12696, pp. 147–176. Springer, Cham (2021). https://doi.org/10.1007/978-3-030-77870-5_6
26. Katz, J.: Round optimal robust distributed key generation. IACR Cryptol. ePrint Arch. 1094 (2023). https://eprint.iacr.org/2023/1094
27. Kokoris Kogias, E., Alp, E.C., Gasser, L., Jovanovic, P.S., Syta, E., Ford, B.A.: Calypso: private data management for decentralized ledgers. Proc. VLDB Endow. **14**(4), 586–599 (2021). https://doi.org/10.14778/3436905.3436917
28. Langlois, A., Stehlé, D.: Worst-case to average-case reductions for module lattices. Des. Codes Crypt. **75**(3), 565–599 (2014). https://doi.org/10.1007/s10623-014-9938-4
29. Lyubashevsky, V.: Lattice signatures without trapdoors. In: Pointcheval, D., Johansson, T. (eds.) EUROCRYPT 2012. LNCS, vol. 7237, pp. 738–755. Springer, Heidelberg (2012). https://doi.org/10.1007/978-3-642-29011-4_43
30. Micciancio, D., Regev, O.: Worst-case to average-case reductions based on gaussian measures. SIAM J. Comput. **37**(1), 267–302 (2007). https://doi.org/10.1137/S0097539705447360
31. Pedersen, T.P.: A threshold cryptosystem without a trusted party. In: Davies, D.W. (ed.) EUROCRYPT 1991. LNCS, vol. 547, pp. 522–526. Springer, Heidelberg (1991). https://doi.org/10.1007/3-540-46416-6_47
32. Pedersen, T.P.: Non-interactive and information-theoretic secure verifiable secret sharing. In: Feigenbaum, J. (ed.) CRYPTO 1991. LNCS, vol. 576, pp. 129–140. Springer, Heidelberg (1992). https://doi.org/10.1007/3-540-46766-1_9
33. Regev, O.: On lattices, learning with errors, random linear codes, and cryptography. J. ACM (JACM) **56**(6), 1–40 (2009). https://doi.org/10.1145/1568318.1568324
34. Shamir, A.: How to share a secret. Commun. ACM **22**(11), 612–613 (1979). https://doi.org/10.1145/359168.359176
35. Tang, G., Pang, B., Chen, L., Zhang, Z.: Efficient lattice-based threshold signatures with functional interchangeability. IEEE Trans. Inf. Forensics Secur. **18**, 4173–4187 (2023). https://doi.org/10.1109/TIFS.2023.3293408
36. Zheng, Z., Xie, S., Dai, H., Chen, X., Wang, H.: Blockchain challenges and opportunities: a survey. Int. J. Web Grid Serv. **14**(4), 352–375 (2018). https://doi.org/10.1504/IJWGS.2018.10016848

Turtle Wins Rabbit Again: Faster Modulus Reduction for RNS-CKKS

Lianglin Yan[1,2](✉), Pengfei Zeng[1,2], and Mingsheng Wang[1]

[1] Key Laboratory of Cyberspace Security Defense, Institute of Information Engineering, CAS, Beijing, China
wangmingsheng@iie.ac.cn
[2] School of Cyber Security, University of Chinese Academy of Sciences, Beijing, China
{yanlianglin20,zengpengfei20}@mails.ucas.ac.cn

Abstract. As a famous fully homomorphic encryption scheme that supports floating-point/real numbers, CKKS is widely applied in many real-world scenarios, such as privacy-preserving machine learning and data analysis. These applications demand high efficiency and fast implementation, making the variant scheme based on the residue number system (RNS) prevalent.

In this paper, we present an exact modulus reduction algorithm for RNS-CKKS. Instead of using the fast basis conversion method, our algorithm employs iterative rescaling to achieve the desired result. Compared to the state-of-the-art, the algorithm originally proposed by Cheon et al. (SAC'18) and optimized by Halevi et al. (CT-RSA'19), our algorithm notably eliminates all floating-point computations. Furthermore, it reduces the number of modular multiplications and additions when $\ell > (k^2 - k - 2)/2$, thereby achieving better efficiency, where ℓ is the level of ciphertexts and k is the number of auxiliary moduli. To demonstrate its practicability, we apply the proposed algorithm to CKKS bootstrapping. Experimental results show that our approach accelerates the modulus reduction algorithm in bootstrapping by a factor of $15.2\% \sim 20.3\%$ for different ciphertext levels with a fixed $k = 5$.

Keywords: Fully homomorphic encryption · CKKS · Residue number system · Modulus reduction · Rescaling

1 Introduction

In recent years, fully homomorphic encryption (FHE) has gradually become an important tool in the realm of privacy preservation. It allows one to perform computations on the encrypted data without knowing the secret. To date, there are three kinds of mainstream encryption schemes in FHE, i.e., BGV/BFV [6,7,12] for integers over finite fields, FHEW/TFHE [10,11] for bit data and CKKS [9] for complex/real numbers. Notably, the CKKS scheme offers an efficient solution for scenarios demanding secure computations involving decimals or floating-point numbers, such as privacy-preserving machine learning [20,21] and secure genome-wide association studies (GWAS) [3,4]. However, with the development of data analysis, models are becoming increasingly complex, such as deep neural networks (DNNs) in machine learning [15,19], which require efficient implementations of homomorphic operations.

Improving efficiency can be approached from various aspects. One effective strategy is to optimize the algorithm design, particularly focusing on the most time-consuming operations. This often includes the optimization of homomorphic multiplication and key-switching [13,16,17], as they are the most computationally intensive steps in homomorphic computations.

The other perspective is the representation of polynomials and large numbers within the machine's memory and processing units. FHE schemes are built upon the cyclotomic polynomial ring modulo a prime Q (very large), thus the ciphertexts are high-degree polynomials with large coefficients. Representation and modular arithmetic of these polynomials are complicated because of the word size limitation (64-bit). To promote efficiency, Gentry et al. [13] proposed the double-CRT representation, which is derived from the Chinese Remainder Theorem (CRT). The first CRT utilizes the Residue Number System (RNS) to decompose a polynomial into multiple polynomials with small moduli. Then, the second CRT converts each of the small polynomials into an integer vector using the Number Theoretic Transform (NTT). As a result, all polynomials are represented as matrices with small moduli (within 64-bit) and polynomial addition/multiplication corresponds to the component-wise addition/multiplication of these matrices.

Besides polynomial additions and multiplications, the CKKS scheme also includes some non-arithmetic computations (division of polynomials by an integer and rounding) that cannot be directly implemented in RNS representation. These non-arithmetic computations mainly occur during the modulus switching steps, which are integral components of many homomorphic operations such as homomorphic multiplications, rescaling operations, rotations and conjugations, thereby greatly affecting the implementation efficiency. To address this, Cheon et al. [8] introduced several modulus switching algorithms that are compatible with RNS representation, thus obtained a full RNS variant of CKKS scheme. These algorithms contain approximate modulus raising, approximate modulus reduction and rescaling operations. The constructions of their approximate modulus raising/reduction rely on the fast basis conversion technique, proposed by Bajard et al. [2], which is not an exact algorithm[1] and requires a considerable amount of computational overhead.

1.1 Our Contributions

In this paper, we develop the methodology of modulus switching and design an exact and faster modulus reduction algorithm in RNS representation.

---

[1] It can be corrected to the exact version but requires additional floating-point divisions and rounding operations, see [14]. In this work, we consider the exact version as the baseline.

- Following the idea of Cheon et al. [8], we notice that the modulus switching algorithm is not unique, we can, therefore, design new modulus switching algorithms, provided that the resulting ciphertext remains a valid encryption of the same plaintext as in the original algorithm. We theoretically explore the commonalities between modulus reduction and rescaling operations, and find that the modulus reduction can be achieved through a procedure analogous to that in rescaling.
- We design a novel *exact* modulus reduction algorithm in RNS by employing the rescaling iteratively. We prove the correctness of the algorithm through a theorem. Additionally, we provide a complexity comparison showing that, compared to the state-of-the-art, our new algorithm (i) entirely eliminates all the floating-point operations and (ii) requires fewer modular arithmetic operations (multiplications/additions) when $\ell > (k^2 - k - 2)/2$, where ℓ denotes ciphertext level and k is the number of auxiliary moduli.
- To demonstrate the practicability, we implement our algorithm using the Lattigo library [18] and apply it to the CKKS bootstrapping procedure. Experimental results show that the new modulus reduction algorithm achieves a speed-up of $1.15 \sim 1.20$ times over the previous method, when $10 \le \ell \le 24$ and $k = 5$ (the typical parameters in bootstrapping).

1.2 Related Works

The double-CRT technique, initially used in BGV/BFV schemes [13], represents all polynomials in the NTT form for fast implementation. However, some operations, like the decryption in FV [12] scheme, involve the coefficient-wise division and rounding that are difficult to handle in RNS. The conventional approach translates the polynomial back to its standard representation by using CRT-composition, and then recovers the RNS representation after the coefficient-wise division and rounding operations. This CRT-reconstruction is computationally expensive, and the intermediate values are not in RNS, thus requiring multi-precision arithmetic.

To achieve a full RNS variant of FV scheme, Bajard et al. [2] proposed a fast basis conversion that converts polynomials from a modulus basis to another modulus basis while keeping their RNS representation. However, their conversion is not exact and the result polynomials contain a small additional error. Based on the fast basis conversion, Cheon et al. [8] constructed several approximate modulus switching algorithms to achieve the full RNS variant of the CKKS scheme. Due to inherent approximation of the CKKS scheme, the small error introduced by fast basis conversion does not affect correctness, rather, it merely reduces the precision.

Halevi, Polyakov and Shoup [14] introduced a correction term to eliminate the additional error in the fast basis conversion, thereby refining Bajard et al.'s scheme with a lower amount of noise. But the correction term involves floating-point division and rounding. In this work, we use integer arithmetic operations in RNS to achieve exact modulus reduction.

1.3 Road-Map

In Sect. 2, we briefly describe the CKKS scheme and its RNS representation, we also elaborate the technique of fast basis conversion. Section 3 presents the existing modulus switching algorithms within the context of RNS-CKKS. In Sect. 4, we detail our optimizations, accompanied by a proof of correctness and a complexity comparison. Section 5 gives the implementation and performance while Sect. 6 summarizes the paper.

2 Background

In this section, we provide the basic notations of this paper, and present the CKKS scheme and its RNS-representation as preliminaries.

2.1 Notations

Let N be a power-of-two integer and $\mathcal{R} = \mathbb{Z}[X]/(X^N+1)$ be the $2N$-th cyclotomic ring. $q_0, q_1, \cdots, q_L, p_0, \cdots, p_{k-1}$ are prime integers in 64-bit, $Q_\ell = \prod_{j=0}^{\ell} q_j$ for $0 \le \ell \le L$ and $P = \prod_{i=0}^{k-1} p_i$. For simplicity, we also write Q_L as Q. The residue ring modulo Q is $\mathcal{R}_Q = \mathcal{R}/(Q)$. Moreover, we enforce $q_j \equiv 1 \mod 2N$ (as well as p_i) so that $\mathcal{R}_{q_j}$ is NTT-friendly for $0 \le j \le L$. $\mathbb{Z}_M^* = \{x \in \mathbb{Z}_M : \gcd(x, M) = 1\}$ is the multiplicative group of units in $\mathbb{Z}_M$. $\langle \cdot, \cdot \rangle$ denotes the inner-product of two vectors. $\lfloor \cdot \rceil$ is the nearest rounding operation with ties rounded towards $+\infty$. $[x]_Q \in [-Q/2, Q/2)$ denotes the reduction of x modulo Q. In RNS, we call a finite ordered set $\mathcal{B} = \{q_0, q_1, \cdots, q_\ell\}$ a basis if all elements are pairwise coprime, and write $[a]_\mathcal{B} = (a_0, a_1, \cdots, a_\ell)$ where $a_j = [a]_{q_j}$ for $0 \le j \le \ell$.

2.2 The CKKS Scheme

The CKKS scheme, proposed by Cheon et al. [9], is built upon the Ring Learning with Error (RLWE) problem. It encrypts a vector of complex numbers while the errors generated by homomorphic operations are regarded as the computational errors in the approximate computations.

To achieve a SIMD (Single Instruction Multiple Data) manner, CKKS encodes multiple messages into a polynomial over $\mathcal{R}$ via a variant of the canonical embedding. Let $\mathcal{K} = \mathbb{Q}[X]/(X^N+1)$ be the $2N$-th cyclotomic field and $\zeta = \exp(\pi i/N)$ be a primitive $2N$-th root of unity. The canonical embedding of $\mathcal{K}$ is defined as $a(X) \mapsto (a(\zeta), a(\zeta^3), \cdots, a(\zeta^{2N-1}))$. Notice that $a(\zeta^j) = \overline{a(\zeta^{2N-j})}$ for $j \in \mathbb{Z}_{2N}^*$, so we define a variant of the canonical embedding $\tau : \mathcal{K} \to \mathbb{C}^{N/2}$ as

$$a(X) \mapsto (a(\zeta), a(\zeta^5), \cdots, a(\zeta^{4j+1}), \cdots, a(\zeta^{2N-3}))_{0 \le j < N/2}.$$

In decoding, the mapping τ transforms a plaintext $m(X) \in \mathcal{R}$ into a complex vector while preserving the homomorphism properties of both addition and multiplication. And the inverse function of τ is employed in encoding function that map a complex vector into a polynomial in $\mathcal{R}$.

The ciphertext of CKKS relies on the ring $\mathcal{R}_{Q_\ell}$ with a chain of moduli $\{Q_\ell\}_{0\le \ell < L}$. Each ciphertext is constructed based on RLWE problem and is represented as $\mathtt{ct} \in \mathcal{R}^2_{Q_\ell}$ at a level ℓ. For plaintext $\mathtt{pt} \in \mathcal{R}$ and secret key $\mathtt{sk}$, the ciphertext $\mathtt{ct}$ (at level ℓ) is a valid encryption of $\mathtt{pt}$ if $[\langle \mathtt{ct},\mathtt{sk}\rangle]_{Q_\ell} \approx \mathtt{pt}$ where $\approx$ indicates a bounded error. Let $\mathtt{pt}_1$ and $\mathtt{pt}_2$ be two plaintexts. Homomorphic addition and multiplication at level ℓ would output ciphertexts $\mathtt{ct}_{\mathtt{add}}$ and $\mathtt{ct}_{\mathtt{mult}}$ respectively, satisfying $[\langle \mathtt{ct}_{\mathtt{add}},\mathtt{sk}\rangle]_{Q_\ell} \approx \mathtt{pt}_1 + \mathtt{pt}_2$ and $[\langle \mathtt{ct}_{\mathtt{mult}},\mathtt{sk}\rangle]_{Q_\ell} \approx \mathtt{pt}_1 \times \mathtt{pt}_2$.

Homomorphic operations in the CKKS scheme encounter the challenge of noise growth, which can degrade the correctness of computations. To address this issue, Cheon et al. [9] introduced the Rescale operation which scales a ciphertext $\mathtt{ct} \in \mathcal{R}^2_{Q_\ell}$ by a scaling factor $1/q_\ell$ and outputs ciphertext $\mathtt{ct}' \in \mathcal{R}^2_{Q_{\ell-1}}$ with reduced noise. This operation involve division and rounding, which must be meticulously handled in RNS.

Additionally, key-switching is a core technique in homomorphic operations (e.g., homomorphic multiplication, rotation and conjugation). It requires to temporarily lift the ciphertext modulus from Q to PQ using an auxiliary modulus P (called ModUp), and reduce the modulus back to Q at the final stage (called ModDown). Specifically, for a ciphertext $\mathtt{ct} \in \mathcal{R}^2_{Q_\ell}$, ModUp outputs the same $\mathtt{ct}$ but with ciphertext modulus PQ_ℓ, while ModDown takes a ciphertext $\mathtt{ct}' \in \mathcal{R}^2_{PQ_\ell}$ as input and outputs $\lfloor \frac{1}{P}\mathtt{ct}'\rceil \in \mathcal{R}^2_{Q_\ell}$. These two procedures are easy to implement in standard representation, but difficult in RNS (we elaborate this in Sect. 3).

2.3 The RNS Representation

In the Residue Number System (RNS), the large coefficients of ciphertext polynomials are decomposed into small integers within 64-bit, enabling all operations in the implementation to be performed using native word-sized operations without requiring multi-precision computations. For example, a coefficient $a \in \mathbb{Z}_{Q_\ell}$ is represented as $(a_0, a_1, \cdots, a_\ell)$ where $a_j = [a]_{q_j}$ for $0 \le j \le \ell$. As mentioned before, polynomial additions and multiplications can be performed efficiently in the RNS because the polynomials are represented in the NTT form, and what need to be discussed is the modulus switching procedures. A powerful technique for modulus switching in RNS is fast basis conversion, introduced by Bajard et al. [2].

Fast Basis Conversion. Let $\mathcal{B} = \{q_0, q_1, \cdots, q_\ell\}$ and $\mathcal{C} = \{p_0, p_1, \cdots, p_{k-1}\}$ be two bases where q_j and p_i are all prime integers within 64-bit for $0 \le j \le \ell$ and $0 \le i < k$. The fast basis conversion can convert a polynomial from the basis $\mathcal{B}$ to the basis $\mathcal{C}$ without involving any computations over large numbers.

In RNS, a polynomial in $\mathcal{R}_{Q_\ell}$ is represented as a set of small polynomials in $\prod_{j=0}^{\ell} \mathcal{R}_{q_j}$ by independently decomposing the large coefficients into small numbers using the basis $\mathcal{B}$. For easy description, we only consider a coefficient

$a \in \mathbb{Z}_{Q_\ell}$. For $[a]_\mathcal{B} = (a_0, \cdots, a_\ell) \in \prod_{j=0}^{\ell} \mathbb{Z}_{q_j}$, the fast basis conversion takes it as input and outputs an element in $\prod_{i=0}^{k-1} \mathbb{Z}_{p_i}$ by computing

$$\mathsf{Conv}_{\mathcal{B} \to \mathcal{C}}([a]_\mathcal{B}) = \left(\sum_{j=0}^{\ell} \left[a_j \cdot q_j^* \right]_{q_j} \cdot \hat{q}_j \pmod{p_i} \right)_{0 \le i < k} \quad (1)$$

where $\hat{q}_j = \prod_{j' \ne j} q_{j'}$ and $q_j^* = \left[\hat{q}_j^{-1} \right]_{q_j}$. Note that $\sum_{j=0}^{\ell} \left[a_j \cdot q_j^* \right]_{q_j} \cdot \hat{q}_j = a + e Q_\ell$ where e is a small integer satisfying $|a + e Q_\ell| \le (\ell+1) Q_\ell / 2$. Therefore, the output of Eq. (1) is $[a + e Q_\ell]_\mathcal{C}$ instead of exactly $[a]_\mathcal{C}$. For CKKS, the additional error e has no effect on correctness, but increases the noise.

Halevi et al. [14] corrected the fast basis conversion by refining Eq. (1) as

$$\mathsf{Conv}_{\mathcal{B} \to \mathcal{C}}([a]_\mathcal{B}) = \left(\sum_{j=0}^{\ell} \left[a_j \cdot q_j^* \right]_{q_j} \cdot \hat{q}_j - v \cdot [Q_\ell]_{p_i} \pmod{p_i} \right)_{0 \le i < k} \quad (2)$$

where $v = \left\lfloor \sum_{j=0}^{\ell} \frac{[a_j \cdot q_j^*]_{q_j}}{q_j} \right\rceil$. The refined formula outputs exactly $[a]_\mathcal{C}$ but the computation of v involves additional floating-point division and rounding operations. As the state-of-the-art, the refined fast basis conversion had been implemented in various homomorphic encryption libraries [1,18,23]. For the sake of uniformity, in the subsequent sections, fast basis conversion refers to the refined version.

3 Existing Modulus Switching in RNS-CKKS

Here we discuss the approximate modulus switching algorithms proposed by Cheon et al. [8]. Their constructions rely on the fast basis conversion. More importantly, they offer a innovative methodology when designing the modulus reduction algorithm. We also provide the complexity analysis of Cheon et al.'s algorithms for the comparison in the next section. Although Cheon et al.'s original paper adopted the approximate fast basis conversion from [2], we use the refined version from [14] to achieve exact algorithms.

3.1 Modulus Raising

In key-switching procedure, modulus raising is a function that lifts ciphertext modulus from Q_ℓ to PQ_ℓ. Let $\mathcal{B} = \{q_0, q_1, \cdots, q_\ell\}$, $\mathcal{C} = \{p_0, p_1, \cdots, p_{k-1}\}$ and $\mathcal{D} = \mathcal{B} \cup \mathcal{C}$ be three bases. The modulus raising algorithm takes $[a]_\mathcal{B}$ (the RNS representation of $a \in \mathbb{Z}_{Q_\ell}$) as input, and outputs $[a]_\mathcal{D}$. More specifically, it employs the (refined) fast basis conversion algorithm to generate $[a]_\mathcal{C}$, which is then concatenated with $[a]_\mathcal{B}$ to get $[a]_\mathcal{D}$. See Algorithm 1 for the details.

Algorithm 1. Modulus Raising: $\mathsf{ModUp}_{\mathcal{B}\to\mathcal{C}}([a]_{\mathcal{B}})$

Input $[a]_{\mathcal{B}} = (a_0, a_1, \cdots, a_\ell)$
Output $[a]_{\mathcal{D}} = (a_0, \cdots, a_\ell, a_{\ell+1}, \cdots, a_{\ell+k})$
1: Compute $(a_{\ell+1}, \cdots, a_{\ell+k}) = \mathsf{Conv}_{\mathcal{B}\to\mathcal{C}}([a]_{\mathcal{B}})$ using Eq. (2).
2: **return** $(a_0, \cdots, a_\ell, a_{\ell+1}, \cdots, a_{\ell+k})$.

Complexity Analysis. The complexity of Algorithm 1 is actually the complexity of the fast basis conversion algorithm. According to Eq. (2), fast basis conversion involves (integer) modular multiplication, (integer) modular addition, floating-point division, floating-point addition and rounding operations. We count these operations as follow.

1. The computation of $x_j = \left[a_j \cdot q_j^*\right]_{q_j}$ for $0 \leq j \leq \ell$ requires $\ell + 1$ modular multiplications.
2. $y_i = \sum_{j=0}^{\ell} x_j \cdot \hat{q}_j \mod p_i$ for $0 \leq i < k, 0 \leq j \leq \ell$ requires $k(\ell+1)$ modular multiplications and $k\ell$ modular additions.
3. Computing $v = \left\lfloor \sum_{j=0}^{\ell} \frac{x_j}{q_j} \right\rceil$ requires $\ell + 1$ floating-point divisions, ℓ floating-point additions and one rounding.
4. $y_i - v \cdot [Q_\ell]_{p_i} \pmod{p_i}$ require k modular multiplications and k modular additions (we unify subtraction into modular addition).

Totally, the complexity of Algorithm 1 is: $k\ell + \ell + 2k + 1$ modular multiplications, $k\ell + k$ modular additions, $\ell + 1$ floating-point divisions, ℓ floating-point additions and 1 rounding.

3.2 Modulus Reduction

Let $\mathcal{B} = \{q_0, q_1, \cdots, q_\ell\}$, $\mathcal{C} = \{p_0, p_1, \cdots, p_{k-1}\}$ and $\mathcal{D} = \mathcal{B} \cup \mathcal{C}$ be three bases. Modulus reduction is also a component in key-switching procedure. For a large integer $a \in \mathbb{Z}_{PQ_\ell}$, it takes $[a]_{\mathcal{D}}$ as input and returns $\left[\lfloor \frac{a}{P} \rceil\right]_{\mathcal{B}}$. Since all numbers are represented in RNS, it is challenging to perform the division $\frac{a}{P}$.

Cheon et al.'s algorithm aims to compute the result of $\left[\lfloor \frac{a}{P} \rfloor\right]_{\mathcal{B}}$ instead of directly computing $\frac{a}{P}$. Using the division theorem (see Lemma 1), we can write

$$a = t \cdot P + r \qquad (3)$$

where $r = [a]_P \in [-P/2, P/2)$ and $t = \lfloor \frac{a}{P} \rfloor$. Then, we need to compute $[t]_{q_j}$ for each q_j in $\mathcal{B}$. Rewriting Eq. (3) by modulo q_j, we have

$$[a]_{q_j} = [t]_{q_j} \cdot [P]_{q_j} + [[a]_P]_{q_j} \pmod{q_j}.$$

Then, $[t]_{q_j} = P^{-1}([a]_{q_j} - [[a]_P]_{q_j}) \pmod{q_j}$, where $P^{-1} \pmod{q_j} = (\prod_{i=0}^{k-1} p_i)^{-1}$ $\pmod{q_j}$ can be pre-computed, $[a]_{q_j}$ is provided in $[a]_{\mathcal{D}}$ and $[[a]_P]_{q_j}$ is essentially the fast basis conversion of $[a]_{\mathcal{C}}$ from $\mathcal{C}$ to $\mathcal{B}$ (note that $[a]_{\mathcal{C}} = [[a]_P]_{\mathcal{C}}$).

Algorithm 2. Modulus Reduction: $\mathsf{ModDown}_{\mathcal{D}\to\mathcal{B}}([a]_\mathcal{D})$

Input $[a]_\mathcal{D} = (a_0, \cdots, a_\ell, a_{\ell+1}, \cdots, a_{\ell+k})$
Output $\left[\lfloor \frac{a}{P} \rfloor\right]_\mathcal{B}$
1: Compute $(b_0, \cdots, b_\ell) = \mathsf{Conv}_{\mathcal{C}\to\mathcal{B}}(a_{\ell+1}, \cdots, a_{\ell+k})$ using Eq. (2).
2: **for** $j = 0, 1, \cdots, \ell$ **do**
3: $\quad a'_j = (\prod_{i=0}^{k-1} p_i)^{-1} \cdot (a_j - b_j) \pmod{q_j}$.
4: **end for**
5: **return** $(a'_0, a'_1, \cdots, a'_\ell)$.

Complexity Analysis. Algorithm 2 contains a fast basis conversion and a loop. Hence its complexity is:

1. Similar to Algorithm 1, the fast basis conversion from $\mathcal{C}$ to $\mathcal{B}$ requires $k\ell + \ell + 2k + 1$ modular multiplications, $k\ell + k$ modular additions, k floating-point divisions, $k - 1$ floating-point additions and 1 rounding.
2. The loop in step 2–4 requires $\ell + 1$ modular multiplications and $\ell + 1$ modular additions.

Totally, the complexity of Algorithm 2 is: $k\ell + 2\ell + 2k + 2$ modular multiplications, $k\ell + \ell + k + 1$ modular additions, k floating-point divisions, $k - 1$ floating-point additions and 1 rounding.

3.3 Rescaling

Rescaling is also a modulus switching operation in CKKS. It scales the coefficients of ciphertext polynomials from $\mathbb{Z}_{Q_\ell}$ to $\mathbb{Z}_{Q_{\ell-1}}$. Let $\mathcal{B} = \{q_0, q_1, \cdots, q_\ell\}$ and $\mathcal{B}' = \{q_0, q_1, \cdots, q_{\ell-1}\}$. For a large coefficient $a \in \mathbb{Z}_{Q_\ell}$, rescaling takes $[a]_\mathcal{B}$ as input and outputs $\left[\lfloor \frac{a}{q_\ell} \rfloor\right]_{\mathcal{B}'}$. From this perspective, rescaling achieves the similar functionality to ModDown, but its algorithm is easier as q_ℓ is a small integer in RNS.

Similar to Eq. (3), we write $a = t' \cdot q_\ell + r'$ with $r' = [a]_{q_\ell}$ and $t' = \lfloor \frac{a}{q_\ell} \rfloor$. By modulo q_j for each q_j in $\mathcal{B}'$, we have $[t']_{q_j} = q_\ell^{-1} \cdot ([a]_{q_j} - [a]_{q_\ell}) \pmod{q_j}$, where $q_\ell^{-1} \pmod{q_j}$ is pre-computed, $[a]_{q_j}$ and $[a]_{q_\ell}$ are both included in $[a]_\mathcal{B}$. Therefore, rescaling algorithm only involves the modular multiplications and modular additions (we unify subtraction into modular addition).

Algorithm 3. Rescaling: $\mathsf{Rescale}_{\mathcal{B}\to\mathcal{B}'}([a]_\mathcal{B})$

Input $[a]_\mathcal{B} = (a_0, \cdots, a_\ell)$
Output $\left[\lfloor \frac{a}{q_\ell} \rfloor\right]_{\mathcal{B}'}$
1: **for** $j = 0, 1, \cdots, \ell - 1$ **do**
2: $\quad a'_j = q_\ell^{-1} \cdot (a_j - a_\ell) \pmod{q_j}$.
3: **end for**
4: **return** $(a'_0, a'_1, \cdots, a'_{\ell-1})$.

Complexity Analysis. According to Algorithm 3, the complexity of rescaling is: ℓ modular multiplications and ℓ modular additions (we unify subtraction into modular addition).

4 Our Optimizations

In this section, we propose a novel *exact* modulus reduction algorithm which does not require any floating-point operations. Furthermore, when the ciphertext is at a high level ℓ and the number of auxiliary moduli k is not very large, our algorithm involves less modular multiplications and additions than Algorithm 2. All notations remain consistent with previous sections.

4.1 Iterative Rescaling for Modulus Reduction

Our Motivation. As mentioned before, modulus reduction takes $[a]_\mathcal{D}$ as input and returns $\left[\lfloor \frac{a}{P} \rceil\right]_\mathcal{B}$, and rescaling takes $[a]_\mathcal{B}$ as input and outputs $\left[\lfloor \frac{a}{q_\ell} \rceil\right]_{\mathcal{B}'}$. It seems that modulus reduction is a "giant step" (reducing multiple primes $p_0, \cdots, p_k$), whereas rescaling is a "baby step" (only reducing prime q_ℓ). However, complexity analysis reveals that the cost of the rescaling operation is much lower than the modulus reduction algorithm. So we think: it seems possible to construct a modulus reduction algorithm with reduced time complexity by employing rescaling iteratively.

The Improved Algorithm. Let $\mathcal{B} = \{q_0, q_1, \cdots, q_\ell\}$, $\mathcal{C} = \{p_0, p_1, \cdots, p_{k-1}\}$ and $\mathcal{D} = \mathcal{B} \cup \mathcal{C}$ be three bases, $a \in \mathbb{Z}_{PQ_\ell}$ be a large integer. For rescaling operation, if the input is $[a]_\mathcal{D}$, then the output would be $\lfloor \frac{a}{p_{k-1}} \rceil$ over basis $\{q_0, \cdots, q_\ell, p_0, \cdots, p_{k-2}\}$. Iteratively, if we input $\lfloor \frac{a}{p_{k-1}} \rceil$ over basis $\{q_0, \cdots, q_\ell, p_0, \cdots, p_{k-2}\}$, the output would be $\lfloor \frac{1}{p_{k-2}} \cdot \lfloor \frac{a}{p_{k-1}} \rceil \rceil$ over basis $\{q_0, \cdots, q_\ell, p_0, \cdots, p_{k-3}\}$. Rerun rescaling operation k times, we obtain $\lfloor \frac{1}{p_0} \cdot \lfloor \frac{1}{p_1} \cdots \lfloor \frac{a}{p_{k-1}} \rceil \rceil \rceil$ over basis $\mathcal{B}$. Correctness proof in the next subsection will demonstrate that the final output is exactly $\left[\lfloor \frac{a}{P} \rceil\right]_\mathcal{B}$.

Algorithm 4. Improved Modulus Reduction

Input $[a]_\mathcal{D} = (a_0, \cdots, a_\ell, a_{\ell+1}, \cdots, a_{\ell+k})$
Output $\left[\lfloor \frac{a}{P} \rceil\right]_\mathcal{B}$
1: **for** $i = 0, 1, \cdots, k-1$ **do**
2:     **for** $j = 0, 1, \cdots, \ell+k-i-1$ **do**
3:        $a_j = q_{\ell+k-i}^{-1} \cdot (a_j - a_{\ell+k-i}) \pmod{q_j}$.
4:     **end for**
5: **end for**
6: **return** $(a_0, a_1, \cdots, a_\ell)$

4.2 Correctness Proof

Before proving the correctness of Algorithm 4, we first give the well-known division theorem whose proof is omitted.

Lemma 1 (Division Theorem). *For integers a, b where $b \neq 0$, there exist unique integers q, r such that $a = q \cdot b + r$ and $-\frac{|b|}{2} \leq r < \frac{|b|}{2}$.*

In fact, we know that the quotient $q = \lfloor \frac{a}{b} \rceil$ where $\lfloor \cdot \rceil$ is the nearest rounding operation with ties rounded towards $+\infty$. Then, we have Theorem 1 that ensures the correctness of Algorithm 4.

Theorem 1. *Assume $k \geq 1$, $p_0, p_1, \cdots, p_{k-1}$ are prime integers and $P = \prod_{i=0}^{k-1} p_i$. Given an integer a, we have $\lfloor \frac{1}{p_0} \cdot \lfloor \frac{1}{p_1} \cdots \lfloor \frac{a}{p_{k-1}} \rceil \rceil \rceil = \lfloor \frac{a}{P} \rceil$.*

Proof. According to Lemma 1, we write $a = t \cdot P + r$ with $t = \lfloor \frac{a}{P} \rceil$ and $|r| \leq P/2$. Moreover, $|r| \leq (P-1)/2$ as P is odd. Let $P_j = \prod_{i=0}^{j} p_i$ where $0 \leq j < k$. Similarly, we have

$$a = t_1 \cdot p_{k-1} + r_1 \text{ with } t_1 = \lfloor \frac{a}{p_{k-1}} \rceil \text{ and } |r_1| \leq \frac{p_{k-1} - 1}{2},$$

$$t_1 = t_2 \cdot p_{k-2} + r_2 \text{ with } t_2 = \lfloor \frac{t_1}{p_{k-2}} \rceil \text{ and } |r_2| \leq \frac{p_{k-2} - 1}{2},$$

$$\vdots$$

$$t_{k-1} = t_k \cdot p_0 + r_k \text{ with } t_k = \lfloor \frac{t_{k-1}}{p_0} \rceil \text{ and } |r_k| \leq \frac{p_0 - 1}{2}.$$

We need to prove $t_k = t$. Putting all these equations together, we have

$$a = ((t_k \cdot p_0 + r_k) \cdot p_1 + r_{k-1}) \cdots p_{k-1} + r_1$$
$$= t_k p_0 p_1 \cdots p_{k-1} + r_k p_1 p_2 \cdots p_{k-1} + \cdots + r_2 p_{k-1} + r_1$$
$$= t_k P + R$$

where $R = r_k \prod_{i=1}^{k-1} p_i + \cdots + r_2 p_{k-1} + r_1$. Additionally, we have

$$\left| r_k \prod_{i=1}^{k-1} p_i \right| \leq \frac{p_0 - 1}{2} \prod_{i=1}^{k-1} p_i = \frac{P}{2} - \frac{\prod_{i=1}^{k-1} p_i}{2},$$

$$\left| r_{k-1} \prod_{i=2}^{k-1} p_i \right| \leq \frac{p_1 - 1}{2} \prod_{i=2}^{k-1} p_i = \frac{\prod_{i=1}^{k-1} p_i}{2} - \frac{\prod_{i=2}^{k-1} p_i}{2},$$

$$\vdots$$

$$|r_2 p_{k-1}| \leq \frac{p_{k-2} - 1}{2} p_{k-1} = \frac{p_{k-2} p_{k-1}}{2} - \frac{p_{k-1}}{2},$$

$$|r_1| \leq \frac{p_{k-1} - 1}{2}.$$

Summing these inequalities, we have $|R| \leq \frac{P-1}{2}$. According to the uniqueness of the remainder in Lemma 1, we have $r = R$. Then, $t_k = t$. $\square$

We also remark that this theorem can be proven via a mathematical induction.

4.3 Complexity Comparison

Algorithm 4 involves two layers of loops. For each i, the inner-loop requires $\ell + k - i$ modular multiplications and $\ell + k - i$ modular additions (we unify subtraction into modular addition). Since i varies from 0 to $k - 1$, the total number of modular multiplications (same as modular additions) is

$$\sum_{i=0}^{k-1}(\ell + k - i) = \frac{k(\ell + k + \ell + 1)}{2} = k\ell + \frac{k(k+1)}{2}.$$

Table 1 summarizes the computational complexity of Algorithm 2 versus the proposed Algorithm 4. Notably, our improved algorithm eliminates floating-point operations entirely. Furthermore, it reduces the number of modular multiplications and additions compared to the previous method when $\ell+k+1 > k(k+1)/2$, i.e., $\ell > (k^2 - k - 2)/2$. For the parameter setting of CKKS scheme in practice, the parameter k is typically set to approximately 5, while the maximal ℓ attains values up to 27. For example, in [5], the highest level L and the number of p_i in the recommended four parameter sets are respectively $24/23/21/27$ and $5/4/5/6$. This implies that our improved modulus reduction algorithm can be applied when the ciphertext is at a high level. And when the level drops to a critical value during homomorphic computations, the original algorithm (Algorithm 2) can be used again.

Table 1. Complexity comparison of previous modulus reduction algorithm (Algorithm 2) and ours (Algorithm 4). "FP" represents "Floating-Point". All operations are word-sized in a 64-bit machine.

Algorithm	# Modular Mult	# Modular Add	#FP div	#FP Add	# round
Algorithm 2	$k\ell + 2\ell + 2k + 2$	$k\ell + \ell + k + 1$	k	$k - 1$	1
Algorithm 4	$k\ell + k(k+1)/2$	$k\ell + k(k+1)/2$	-	-	-

5 Implementation and Performance

In this section, we conduct several experiments to demonstrate the practicability of our proposed algorithm. Firstly, we provide a benchmark test which includes the running time (CPU cycles) of previous modulus reduction algorithm (Algorithm 2) and our improved algorithm (Algorithm 4) for different values of ℓ and k. Then, we apply our method to the CKKS bootstrapping procedure because, as a built-in circuit of CKKS scheme, bootstrapping is performed at a relatively high level. We develop the code based on the Lattigo library [18] which implements CKKS in a full-RNS manner. The source code is available at https://github.com/Moohee-yll/FasterModulusReductionInCKKS. All experiments were performed on a machine with Intel(R) Xeon(R) Gold 6230R 2.10GHz and 252GB memory, running on a single-threaded mode. The environment information includes Go version 1.21.1, GOARCH=amd64, GOOS=linux.

5.1 Benchmark Test

Previous theoretical complexity analysis shows that our proposed algorithm achieves a better performance than the existing algorithm when the ciphertext level ℓ and the number of auxiliary moduli k satisfy $\ell > (k^2 - k - 2)/2$. Here, we perform a benchmark test to present the actual runtime of the two modulus reduction algorithms under varying ℓ and k. Specifically, for $4 \leq k \leq 7$ and $0 \leq \ell \leq 30$ (which covers the common used values), we run the Algorithm 2 and Algorithm 4 and record the CPU cycles. All values are the average number of 100 times executions.

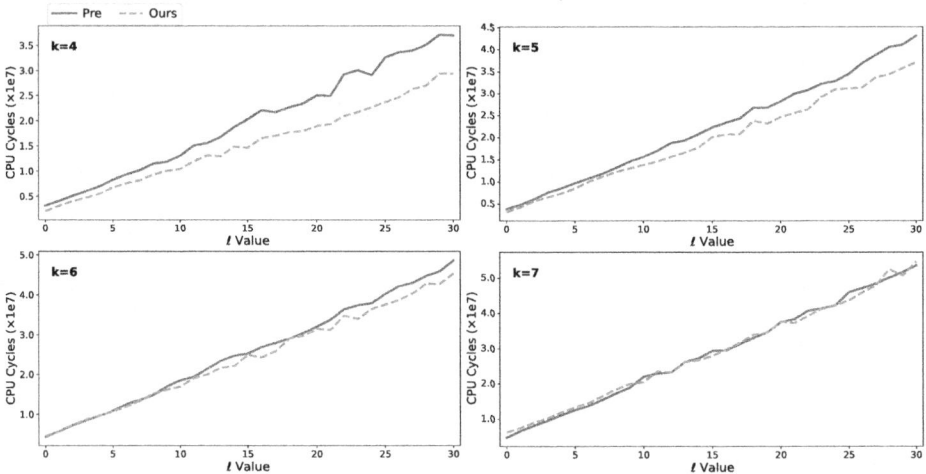

Fig. 1. CPU cycles of Algorithm 2 (Pre) and Algorithm 4 (Ours) for $4 \leq k \leq 7$ and $0 \leq \ell \leq 30$ ($N = 2^{16}$).

As shown in Fig. 1, our improved algorithm demonstrates progressive performance gains with increasing ℓ values when $k \leq 6$. For $k = 7$, both algorithms exhibit comparable efficiency. Notably, in the case of $k = 4$ or 5, the novel algorithm still outperforms the previous algorithm when $\ell \leq (k^2 - k - 2)/2$ (e.g., $\ell \leq 4$). This is because we avoid the floating-point operations.

5.2 Applying to Bootstrapping

CKKS bootstrapping inherently elevates the ciphertext modulus level, thereby the homomorphic operations in bootstrapping are performed at relatively higher levels. To show the applicability of our algorithm, we apply the Algorithm 4 to the bootstrapping procedure.

Table 2. Overview of parameter set I in [5]. h is the Hamming weight of secret key. "**Left**" represents the remaining moduli after bootstrapping.

$\log(PQ)$	N	h	L	k
1546	2^{16}	192	24	5

$\log(q_j)$				$\log(p_i)$
Left	StC	Sine	CtS	
$60 + 9 \cdot 40$	$3 \cdot 39$	$8 \cdot 60$	$4 \cdot 56$	$5 \cdot 61$

Table 2 describes the parameter set I suggested in [5], where bootstrapping is performed when $\ell \geq 10$. We apply Algorithm 4 to bootstrapping under this parameter set. The bootstrapping duration should not be considered a meaningful metric for evaluating algorithmic improvement, as our algorithm only saves a few tenths of a second while the total time of bootstrapping procedure is around 15–20 s. To capture the improvements of our new algorithm, we execute Algorithm 2 and Algorithm 4 for $10 \leq \ell \leq 24$ with a fixed $k = 5$, and measure their running time (values averaged over 200 trials). Then, we also count the number

Table 3. The runtime of the previous (Prev) and the new (Ours) modulus reduction operations in CKKS bootstrapping with parameter set I of [5] ($N = 2^{16}, L = 24, k = 5$). "#Calls" denotes the number of times the ModDown function is called at each level, while each time value represents one call. All timing values are the average of 200 runs.

ℓ		24	23	22	21	20	19	18	17
#Calls		7	9	9	5	6	8	12	8
Timing	Prev	15.9	15.3	14.7	14.0	13.3	12.8	12.1	11.4
(ms)	Ours	13.5	12.9	12.5	12.0	11.4	10.9	10.4	9.9
Speed-up		1.178	1.186	1.176	1.167	1.167	1.174	1.163	1.152

ℓ		16	15	14	13	12	11	10	**Total**
#Calls		4	4	4	4	9	9	9	107
Timing	Prev	11.0	10.5	9.8	9.3	8.9	8.1	7.7	1254.6
(ms)	Ours	9.4	9.0	8.4	7.8	7.4	6.9	6.4	1067.4
Speed-up		1.170	1.167	1.167	1.192	1.203	1.174	1.203	1.175

of times the ModDown function is called at each level of bootstrapping. The results in Table 3 show that our method accelerates the ModDown algorithm by a factor of $15.2\% \sim 20.3\%$.

6 Conclusion

This work introduces a novel modulus reduction algorithm which achieves a better efficiency compared to the state-of-the-art algorithm when the ciphertext level is high. Specifically, we proposed the theorem 1 which states that the modulus reduction can be achieved by employing rescaling operation iteratively. And the complexity analysis demonstrates that our improved algorithm avoids the floating-point operation in implementations. When the ciphertext level ℓ and the number of auxiliary prime moduli k satisfy $\ell > (k^2 - k - 2)/2$, our method reduces the number of modular multiplications and additions. Therefore, the proposed algorithm can be applied to the bootstrapping procedure since the ciphertext level is raised in bootstrapping. Experiments demonstrates that our method achieves a $1.15\times \sim 1.20\times$ acceleration in the modulus reduction algorithm when evaluated with a typical bootstrapping parameter set.

Additionally, our proposed algorithm does not involve floating-point operations, which makes it suitable for hardware-constrained deployments or environments where floating-point operations are expensive.

Acknowledgements. We would like to thank the anonymous reviewers for their helpful suggestions. This work was supported by the National Key Research and Development Program of China (No. 2020YFA0712303), and the "Climbing Program" Special Project for Basic and Frontier Research of the Institute of Information Engineering, Chinese Academy of Sciences (No. 2024000051).

References

1. Badawi, A.A., et al.: OpenFHE: open-source fully homomorphic encryption library. Cryptology ePrint Archive, Paper 2022/915 (2022). https://eprint.iacr.org/2022/915
2. Bajard, J.-C., Eynard, J., Hasan, M.A., Zucca, V.: A full RNS variant of FV like somewhat homomorphic encryption schemes. In: Avanzi, R., Heys, H. (eds.) SAC 2016. LNCS, vol. 10532, pp. 423–442. Springer, Cham (2017). https://doi.org/10.1007/978-3-319-69453-5_23
3. Blatt, M., Gusev, A., Polyakov, Y., Goldwasser, S.: Secure large-scale genome-wide association studies using homomorphic encryption. Proc. Natl. Acad. Sci. **117**(21), 11608–11613 (2020). https://doi.org/10.1073/pnas.1918257117
4. Blatt, M., Gusev, A., Polyakov, Y., Rohloff, K., Vaikuntanathan, V.: Optimized homomorphic encryption solution for secure genome-wide association studies. BMC Med. Genomics **13**(7), 83 (2020). https://doi.org/10.1186/s12920-020-0719-9
5. Bossuat, J.-P., Mouchet, C., Troncoso-Pastoriza, J., Hubaux, J.-P.: Efficient bootstrapping for approximate homomorphic encryption with non-sparse keys. In: Canteaut, A., Standaert, F.-X. (eds.) EUROCRYPT 2021. LNCS, vol. 12696, pp. 587–617. Springer, Cham (2021). https://doi.org/10.1007/978-3-030-77870-5_21

6. Brakerski, Z.: Fully homomorphic encryption without modulus switching from classical GapSVP. In: Safavi-Naini and Canetti [22], pp. 868–886. https://doi.org/10.1007/978-3-642-32009-5_50
7. Brakerski, Z., Gentry, C., Vaikuntanathan, V.: (Leveled) fully homomorphic encryption without bootstrapping. In: Goldwasser, S. (ed.) ITCS 2012, pp. 309–325. ACM (2012). https://doi.org/10.1145/2090236.2090262
8. Cheon, J.H., Han, K., Kim, A., Kim, M., Song, Y.: A full RNS variant of approximate homomorphic encryption. In: Cid, C., Jacobson, Jr., M.J. (eds.) SAC 2018. LNCS, vol. 11349, pp. 347–368. Springer, Cham (2019). https://doi.org/10.1007/978-3-030-10970-7_16
9. Cheon, J.H., Kim, A., Kim, M., Song, Y.: Homomorphic encryption for arithmetic of approximate numbers. In: Takagi, T., Peyrin, T. (eds.) ASIACRYPT 2017. LNCS, vol. 10624, pp. 409–437. Springer, Cham (2017). https://doi.org/10.1007/978-3-319-70694-8_15
10. Chillotti, I., Gama, N., Georgieva, M., Izabachène, M.: TFHE: fast fully homomorphic encryption over the torus. J. Cryptol. **33**(1), 34–91 (2019). https://doi.org/10.1007/s00145-019-09319-x
11. Ducas, L., Micciancio, D.: FHEW: bootstrapping homomorphic encryption in less than a second. In: Oswald, E., Fischlin, M. (eds.) EUROCRYPT 2015. LNCS, vol. 9056, pp. 617–640. Springer, Heidelberg (2015). https://doi.org/10.1007/978-3-662-46800-5_24
12. Fan, J., Vercauteren, F.: Somewhat practical fully homomorphic encryption. Cryptology ePrint Archive, Report 2012/144 (2012). https://eprint.iacr.org/2012/144
13. Gentry, C., Halevi, S., Smart, N.P.: Homomorphic evaluation of the AES circuit. In: Safavi-Naini, R., Canetti, R. (eds.) CRYPTO 2012. LNCS, vol. 7417, pp. 850–867. Springer, Heidelberg (2012). https://doi.org/10.1007/978-3-642-32009-5_49
14. Halevi, S., Polyakov, Y., Shoup, V.: An improved RNS variant of the BFV homomorphic encryption scheme. In: Matsui, M. (ed.) CT-RSA 2019. LNCS, vol. 11405, pp. 83–105. Springer, Cham (2019). https://doi.org/10.1007/978-3-030-12612-4_5
15. He, K., Zhang, X., Ren, S., Sun, J.: Deep residual learning for image recognition. In: 2016 IEEE Conference on Computer Vision and Pattern Recognition (CVPR), pp. 770–778 (2016). https://doi.org/10.1109/CVPR.2016.90
16. Kim, A., Polyakov, Y., Zucca, V.: Revisiting homomorphic encryption schemes for finite fields. In: Tibouchi, M., Wang, H. (eds.) ASIACRYPT 2021. LNCS, vol. 13092, pp. 608–639. Springer, Cham (2021). https://doi.org/10.1007/978-3-030-92078-4_21
17. Kim, M., Lee, D., Seo, J., Song, Y.: Accelerating HE operations from key decomposition technique. In: Handschuh, H., Lysyanskaya, A. (eds.) CRYPTO 2023, Part IV. LNCS, vol. 14084, pp. 70–92. Springer, Cham (2023). https://doi.org/10.1007/978-3-031-38551-3_3
18. Lattigo v6 (2024). https://github.com/tuneinsight/lattigo. ePFL-LDS, Tune Insight SA
19. LeCun, Y., Bengio, Y., Hinton, G.: Deep learning. Nature **521**, 436–444 (2015). https://doi.org/10.1038/nature14539
20. Lee, E., et al.: Low-complexity deep convolutional neural networks on fully homomorphic encryption using multiplexed parallel convolutions. In: Chaudhuri, K., Jegelka, S., Song, L., Szepesvari, C., Niu, G., Sabato, S. (eds.) Proceedings of the 39th International Conference on Machine Learning. Proceedings of Machine Learning Research, vol. 162, pp. 12403–12422. PMLR (2022). https://proceedings.mlr.press/v162/lee22e.html

21. Lee, J.W., et al.: Privacy-preserving machine learning with fully homomorphic encryption for deep neural network. IEEE Access **10**, 30039–30054 (2022). https://doi.org/10.1109/ACCESS.2022.3159694
22. Safavi-Naini, R., Canetti, R. (eds.): CRYPTO 2012. LNCS, vol. 7417. Springer, Heidelberg (2012)
23. Microsoft SEAL (release 4.1) (2023). https://github.com/Microsoft/SEAL. Microsoft Research, Redmond, WA

A BGV-Subroutined CKKS Bootstrapping Algorithm Without Sine Approximation

Jingjing Fan[1], Chi Zhang[1], Zejiu Tan[2], Zoe Lin Jiang[2], Man Ho Au[3](✉), and Siu Ming Yiu[1]

[1] University of Hong Kong, Pokfulam, Hong Kong
{jjfan,smyiu}@cs.hku.hk, zhchi@connect.hku.hk
[2] Harbin Institute of Technology, Shenzhen, Shenzhen, China
tanzejiu@stu.hit.edu.cn, zoeljiang@hit.edu.cn
[3] The Hong Kong Polytechnic University, Kowloon, Hong Kong
mhaau@polyu.edu.hk

Abstract. Bootstrapping plays a critical role in fully homomorphic encryption (FHE) systems like BGV, BFV, and CKKS, enabling versatile leveled multiplication. While BGV and BFV utilize a digit extraction-based approach for bootstrapping, CKKS employs a scaled sine function to approximate the modular reduction operation, followed by homomorphic evaluation of sine functions using polynomials. This method introduces a fixed lower bound on approximation error.

Rather than building on existing CKKS bootstrapping approaches, we propose an innovative alternative pathway that eliminates the need for sine function fitting in modular reduction approximation. Our method leverages BGV bootstrapping as a subroutine and introduces a ciphertext transformation technique to transform CKKS ciphertexts into BGV-compatible format, thereby enabling the integration of the BGV bootstrapping framework within the CKKS process. Such incorporation complements the existing result that BGV and BFV bootstrapping algorithms are equivalent and CKKS bootstrapping algorithm can serve as a subroutine in BFV bootstrapping algorithm. In addition, compared with the trivial way to construct a CKKS-subroutined BGV bootstrapping containing an intermediate transformation from BGV to BFV then to CKKS, we design a CKKS-subroutined BGV bootstrapping with a direct transformation. As a result, we point out the transformation among BGV, BFV and CKKS bootstrapping algorithms, enabling the optimization of each algorithm's strengths and advantages.

Keywords: Fully Homomorphic Encryption · bootstrapping · BGV · CKKS · bootstrapping approximation

1 Introduction

Homomorphic encryption (HE) is a powerful cryptographic technique that allows clients to encrypt their data and send the encrypted data to an external party for computation without revealing the underlying data [19]. By enabling secure data processing in the cloud [20], facilitating e-voting [9], and supporting privacy-preserving machine learning [4], homomorphic encryption offers a versatile solution for protecting sensitive information while allowing for valuable computations to be performed on encrypted data.

One of the key components that propelled the advancement of FHE schemes is bootstrapping. It is an operation that enables unbounded number of multiplications to be performed on encrypted data [13]. The main idea of bootstrapping is to run the decryption algorithm homomorphically, using an encryption of the secret key. The result is a new ciphertext encrypting the same message with a smaller error. In the decryption algorithm, generally, calculation of $\mathsf{Enc}(\mathbf{m} \bmod t)$ for some integer t is needed and we call it the homomorphic modulus operation.

BGV, BFV and CKKS are three mainstream FHE schemes, and all of them incorporate bootstrapping as a crucial component. All three schemes use encoding to allow for the encryption of plaintext data into a format that can be operated on homomorphically. The main idea of BGV and BFV bootstrapping is to use digit extraction method for homomorphic modulus computation [6,14]. While the form of the plaintext supported is limited, no error is introduced during the homomorphic modulus operation [14].

CKKS bootstrapping uses sine functions to approximate the modulus function, then uses polynomials to approximate sine functions for bootstrapping [7]. During this process, the restriction of plaintext is eliminated but bootstrapping error with fixed lower bound will occur [7]. And we call it bootstrapping error.

Existing work showed that BGV and BFV bootstrapping algorithms are equivalent [11], and CKKS bootstrapping can work as a subroutine of BFV bootstrapping [17]. However, it is still unknown if BGV/BFV can be a subroutine in CKKS bootstrapping so that we can complete the "BGV-BFV-CKKS transformation triangle".

In this paper, we propose an innovative method for CKKS bootstrapping that eliminates the need for sine function fitting in modular reduction approximation. Our method leverages BGV bootstrapping as a subroutine and introduces a ciphertext transform technique to convert complex vectors in CKKS ciphertext to integer vectors, thus incorporating the BGV subroutine in CKKS bootstrapping. Such incorporation also complements the side from BGV to CKKS of the BGV-BFV-CKKS transformation triangle. In addition, compared with the trivial way to construct a CKKS-subroutined BGV bootstrapping containing an intermediate transformation from BGV to BFV before transforming to CKKS, we design a CKKS-subroutined BGV bootstrapping with a direct transformation.

Subsequently, we come up with a BGV-BFV-CKKS transformation triangle. In [11], Geenlen et al. provided a transformation between BGV and BFV bootstrapping algorithms. In [17], Kim et al. proposed a CKKS-subroutined BFV bootstrapping algorithm, thus constructed the transformation from BFV bootstrapping to CKKS bootstrapping. In our work, we complement the side from

BGV to CKKS and from CKKS to BGV in the BGV-BFV-CKKS transformation triangle.

1.1 Related Works

Homomorphic encryption (HE) was initially introduced by Rivest et al. in 1978 [19]. A groundbreaking advancement in 2009 by Gentry [12] marked the emergence of the first feasible fully homomorphic encryption (FHE) scheme. Since then, FHE has seen significant progress towards practical implementation, with BFV, BGV, and CKKS requiring bootstrapping after a fixed number of multiplication operations, known as depth.

The BGV bootstrapping technique, initially proposed by Gentry in 2009 [12] and later enhanced by Alperin-Sherriff et al. in 2013 [1], saw the introduction of the digit extraction method by Halevi et al. in 2015 [14]. This method, widely utilized in both BGV and BFV bootstrapping algorithms, has become a standard approach. In 2018, Chen et al. introduced a bootstrapping procedure for BFV, asserting its superior efficiency over BGV [6]. However, this claim was later revised by Geelen et al. in 2023 [11], who demonstrated the equivalence in hardness between BGV and BFV bootstrapping. The CKKS bootstrapping process diverges from BGV/BFV methods by utilizing sine functions to approximate homomorphic modulus operations and fitting polynomials into these sine functions [7]. Subsequent refinements to the CKKS bootstrapping procedure have been proposed in follow-up works [2,3,15,16,18].

While the BGV and BFV bootstrapping methods require fewer computation operations but the efficiency is constrained by the plaintext modulus, the CKKS bootstrapping approach gets rid of this constraint at the cost of potential errors from approximations In a recent study by Kim et al. [17], a BFV bootstrapping algorithm incorporating a CKKS bootstrapping subroutine was proposed, resulting in a BFV bootstrapping method with increased flexibility in plaintext modulus.

1.2 Our Contribution

In this paper, we provide a novel CKKS bootstrapping method by utilizing a BGV bootstrapping algorithm as a subroutine. Then we further propose a BGV bootstrapping method by utilizing a CKKS bootstrapping algorithm as a subroutine. After that, we prove the transformation among these bootstrapping algorithms.

Incorporating BGV Bootstrapping into CKKS and Coefficient Transformation Method. We observe that after the ModRaise operation in CKKS bootstrapping, the ciphertext $\mathbf{c_0}, \mathbf{c_1}$ satisfies $\mathbf{c_0} + \mathbf{c_1} \cdot \mathbf{s} = q\mathbf{I} + \mathbf{m} \mod Q$. At this stage, the CKKS ciphertext has a similar structure as BGV ciphertext. When treating $q\mathbf{I}$ as an error term needed to be annihilated and $\mathbf{m}$ as a plaintext, we would like to bootstrapp the CKKS ciphertext using the BGV bootstrapping algorithm. With proper prime p and integer Q coprime with p, we can finde s.t. $p^r\mathbf{e} = q\mathbf{I}$. Then the error term $p^r\mathbf{e}$ can be annihilated and encrypted message $\mathbf{m}$ can be obtained by the digit extraction method in BGV bootstrapping.

A Direct Construction CKKS-Subroutined BGV Bootstrapping and Transformation Among Bootstrapping Algorithms. In [1], Alperin et al. showed BGV scheme can be transformed to BFV via a scalar multiplication. In [17], Kim et al. gave a CKKS-subroutined BFV bootstrapping. So a natural way to construct a CKKS-subroutined BGV bootstrapping is to transform the BGV sample to a BFV sample, apply the CKKS-subroutined BFV bootstrapping algorithm, and transform back to a BGV sample. In this paper, we provide a more straightforward method construct a CKKS-subroutined BGV bootstrapping algorithm. In our construction, we get rid of the transformation between BGV and BFV ciphertexts.

Based on the previous results, we prove that the bootstrapping methods of BGV, BFV, and CKKS are transformable. The probability of transformation among different bootstrapping methods offers the advantage of adapting to varying computational requirements and optimizing efficiency.

1.3 Paper Organization

In Sect. 2, we cover the necessary background and preliminaries. Then we exhibit our design of bootstrapping algorithms in Sect. 3. In Sect. 4, we prove the transformation among different bootstrapping methods. Finally, we conclude in Sect. 5.

2 Preliminary

In this section, we introduce the notation and definition used in the rest part of the paper.

2.1 Notation

For N, a fixed power of 2, let $R = \mathbb{Z}[X]/(X^N + 1)$ $R_q = \mathbb{Z}_q[X]/(X^N + 1)$ be the cyclotomic ring over the integers modulo q with coefficients in $[-\lfloor \frac{q}{2} \rfloor, \lfloor \frac{q}{2} \rfloor)$. Define $Y = X^{N/2n}$ for n to be some power of 2. All ring elements of R are shown in lower case letters (e.g. $a \in R$) or explicit as polynomials (e.g. $a(X) \in R$). We use $\mathbf{a}$ to denote a vector and $\mathbf{a}[i]$ to denote the i-th bit of the vector. The infinity norm of $\mathbf{a}$ is denoted as $||\mathbf{a}||_\infty$.

We use $[\cdot]_q$ to denote the central reduction modulo q, $\lfloor \cdot \rfloor$, $\lfloor \cdot \rceil$ to be the rounding to the previous integer and the closest integer, respectively. For an integer z, we denote its p-digit decomposition modulo $z = \sum_k z\langle k \rangle p^k$, where $z\langle k \rangle \in [-p/2, p/2)$ and i-through-j middle digits as $z\langle j, \ldots, i\rangle = \sum_{k=i}^{j} z\langle k \rangle p^k$ for $i < j$.

2.2 Ring Learning with Errors

Let χ and ψ be distributions over R. The ring learning with errors assumption with respect to the parameters (R, q, χ, ψ) goes as follows: given $a, b \leftarrow \mathcal{U}(R_q)$, $s \leftarrow \chi$, $e \leftarrow \psi$, a RLWE sample $(a, a \cdot s + e)$ is indistinguishable from (a, b). The security of the lattice-based fully homomorphic encryption schemes mentioned in this paper (BFV [10], BGV [5], CKKS [8]) relies on this assumption.

2.3 Approximated Homomorphic Encryption (CKKS)

The CKKS scheme [8] is a homomorphic encryption scheme that is specifically designed for handling computations on real-number data. It supports approximate arithmetic operations over the complex numbers $\mathbb{C}$. The CKKS scheme is as follows:

CKKS Scheme

- CKKS.setup(1^λ): Given a security parameter λ, outputs public parameters $\mathsf{pp} = (R, q, \chi, \psi)$ where q is a positive integer and χ, ψ are distributions over R.
- CKKS.Enc(m, pk): Given a message $m \in R$ and a public key $\mathsf{pk} \in R_q^2$, sample $z \leftarrow \chi$, $e_0, e_1 \leftarrow \psi$. Output a ciphertext $\mathsf{ct} = z \cdot \mathsf{pk} + (e_0 + m, e_1)$.
- CKKS.Dec (ct, sk): Given a secret key $\mathsf{sk} = s$, and a ciphertext $\mathsf{ct} = (c_0, c_1)$, output $m' = c_0 + c_1 \cdot s$.

Here, since CKKS is the approximated homomorphic encryption, m and m' are not necessarily the same, but with similar values, i.e. $m \approx m'$.

Encoding and Decoding in CKKS. The encoding and decoding algorithm allows for efficient encoding and manipulation of multiple plaintext values into a single ciphertext. In the encoding algorithm, a message vector $\mathbf{m} \in \mathbb{C}^{n/2}$ is mapped to a polynomial $m(X) \in R$. The decoding algorithm works vice versa.

Remark 1 (Slot-encoded and Coefficient-encoded Cipertexts Representations). In CKKS, the ciphertexts are in the coefficient-encoded state before bootstrapping. For example, for a ciphertext ct with a decryption structure $t(X) = \langle \mathsf{ct}, \mathsf{sk} \rangle$, with coefficients $t_0, \cdots, t_{N-1}$, we denote the ring element $t(X)$ by its corresponding coefficient vector $\mathbf{t} = (t_0, t_1, \cdots, t_{N-1})$ and call it in the coefficient-encoded state. In the CoeffToSlot procedure, the coefficients of $t(X)$ are put into the plaintext slots to be evaluated coefficient-wisely. After this step, we say that the ciphertexts are in the slot-encoded state and denote the ring element $t(X)$ by t.

CKKS Bootstrapping. The bootstrapping method is an important part in CKKS scheme. It enables versatile leveled multiplication. The first CKKS bootstrapping algorithm was proposed by Cheon et al. [7]. The main idea of CKKS bootstrapping method is as follows:

- **ModRaise**: Given a ciphertext ct = $\text{Enc}(\mathbf{m} \mod q)$, set $\mathbf{t} = \langle \text{ct} \cdot \text{sk} \rangle (\mod X^N + 1)$. Here $\mathbf{t} = q\mathbf{I} + \mathbf{m}$, where $\mathbf{I} \in R$ and $||\mathbf{I}||_\infty < K$ for some small constant K. So ct itself can be viewed as an encryption of $t(X) = t_0 + t_1 X + \cdots + t_{N-1} X^{N-1}$ in a large modulus $Q > q$ due to $[\langle \text{ct} \cdot \text{sk} \rangle]_Q = \mathbf{t}$. The bootstrapping procedure of CKKS aims to evaluate the reduction modulus q, $F(\mathbf{t}) = [\mathbf{t}]_q$ using arithmetic operations. Thus we can calculate the encryption of the original message $\mathbf{m} = [\mathbf{t}]_q$ with a ciphertext modulus $Q > q$.
- **CoeffToSlot**: Given a ciphertext ct with a decyprion structure $t(X) = \langle \text{ct} \cdot \text{sk} \rangle$, put the coefficients $t_0, \cdots, t_{N-1}$ into plaintext slots to facilitate a faster evaluation of the modulus reduction function $F(\mathbf{t}) = [\mathbf{t}]_q$.
- **EvalMod**: Perform homomorphic modulus operation on $t(X) = \text{Enc}(m + qI \mod Q)$ and get $\text{ct}' = \text{Enc}(m \mod Q')$
- **SlotToCoeff**: The final step of the CKKS bootstrapping procedure is to transform the ciphertext in the slot-encoded state to coefficient-encoded state.

In this paper, we will present the traditional CKKS bootstrapping algorithm by CKKS.boot$(c_0, c_1, q_{in}, q_{out})$ where (c_0, c_1) is a CKKS ciphertext, q_{in} refers to the modulus of the input ciphertext and q_{out} refers to the modulus of the output ciphertext. The functionality of the CKKS can be parameterized as the tuple $(q_{in}, q_{out}, B_{out})$. Here q_{in} and q_{out} are defined as above and B_{out} refers to the upperbound for the differences between input and output plaintext (Fig. 1).

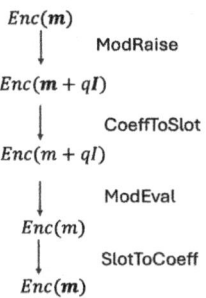

Fig. 1. General CKKS bootstrapping procedure, adapted from [7]

2.4 Homomorphic Encryption (BGV/BFV)

The BGV and BFV homomorphic encryption have a similar encryption and bootstrapping construction. In this paper, we mainly focus on the BGV homomorphic encryption scheme.

BGV Scheme. We will first present the BFV scheme proposed in [5].

- BGV.Setup(1^λ): Given the security parameter λ, output the public parameter $\mathsf{pp} = (R, p, r, \chi, \psi)$, where χ, ψ are distributions over R and p is a prime and r an integer.
- BGV.KeyGen(pp): Given the public parameter pp, sample $s \leftarrow \chi$, $a \leftarrow \mathcal{U}(R_q)$ and $e \leftarrow \psi$. Output a secret key $\mathsf{sk} = (1, s) \in R^2$ and a public key $\mathsf{pk} = (b, a)$, where $b = a \cdot s + p^r e \mod q$.
- BGV.Enc(pk, m): Given a public key $\mathsf{pk} \in R_q^2$, message $m \in R_{p^r}$. Sample $z \in \chi$, $e_0, e_1 \in \psi$. Output $\mathsf{ct} = z \cdot \mathsf{pk} + (p^r e_0 + m, p^r e_1) \mod q$
- BGV.Dec(sk, ct): Given a ciphertext $\mathsf{ct} = (c_0, c_1)$ and a secret key $\mathsf{sk} = (1, s)$, output a message $m = c_0 + c_1 \cdot s \mod p^r$.

Slot and Coefficient Transformation in BGV. In BGV, the coefficient-to-slot and the slot-to-coefficient transformations are also referred to as linear transformation and inverse linear transformation [11]. And we will follow the notations as in Remark 1.

BGV Bootsrapping. In this paper, we mainly follow the BGV bootstrapping method in [11]. We first give a brief idea of the BGV bootstrapping. First, p and q are chosen to be coprime. Set $\mathbf{c}'_\mathbf{i} = p^{e-r} c_i$ for $i \in \{0, 1\}$. So set $\mathbf{u} = \mathbf{c_0}' + \mathbf{c_1}' \cdot \mathbf{s} = p^{e-r}\mathbf{m} + p^e e \mod q$. There exists $\mathbf{g}$ s.t. $\mathbf{w} = [\mathbf{u}]_{p^e} = [p^{e-r}\mathbf{m} + \mathbf{g}]_{p^e}$. Then perform homomorphic modulus and get $\mathbf{m} = [\lfloor \mathbf{w}/p^{e-r} \rceil]_{p^r}$. To ensure the correctness of the homomorphic decryption, we require $\|\mathbf{g}\|_\infty < p^{e-r}/2$. The BGV bootstrapping algorithm goes as follows:

- **Inner product evaluation**: At this stage, the ciphertexts are in the coefficient-encoded stage, so we present the ring elements using their coefficient vector. Given a ciphertext $\mathsf{Enc}(\mathbf{m})$, calculate $\mathsf{Enc}(p^{e-r}\mathbf{m} + \mathbf{e})$ by scalar multiplication and homomorphic evaluation.
- **Linear transformation**: Move the noisy terms into the slots, encoding one coefficient per slot.
- **Digit extraction**: Perform a slot-wise rounding procedure on each of the ciphertexts and decrease the modulus from p^e to p^r
- **Inverse linear transformation**: Move the noise-free slots back into coefficients

Here we list some useful theorems in BGV bootstrapping procedure (Fig. 2).

Theorem 1 ([14]). *Let $p > 1$, $r \geq 1$ and $q = p^r + 1$ be integers with p being an odd prime. Let z be an integer s.t. $|z/q| + |[z]_q| < (q-1)/2$. Then $[z]_q = z\langle 0\rangle - z\langle r\rangle \mod p$.*

Theorem 2 ([6]). *Let p be a prime, $v < e$ be positive integers, u be an integer input modulo p^e with digits $u = u\langle e-1, \ldots, 0\rangle = \sum_{k=0}^{e-1} u_k p^k$. Then there is an algorithm with $\sqrt{2pev}$ multiplications and arithmetic depth $v \log p + \log e$ which returns $u\langle e_1, \ldots, v\rangle = \sum_{k=0}^{e-1} u\langle k\rangle p^k$.*

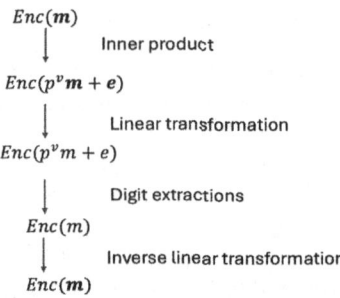

Fig. 2. General BGV bootstrapping procedure, adapted from [6]

In this paper, we will present the traditional BGV bootstrapping algorithm by BGV.boot(c_0, c_1, q, Q) where (c_0, c_1) is a BGV ciphertext, q refers to the plaintext modulus and Q refers to the ciphertext modulus. The functionality of the CKKS can be parameterized as the tuple (q, Q, B_{out}). Here q and Q are defined as above and B_{out} refers to the upper bound for the differences between input and output plaintext.

3 New CKKS Bootstrapping and BGV Bootstrapping

In this section, we present our new construction of BGV-subroutined CKKS bootstrapping and CKKS-subroutined BGV bootstrapping.

3.1 New CKKS Bootstrapping

Here we would like to modify the BGV bootstrapping method to finish the CKKS bootstrapping procedure.

We will first give a brief illustration of our idea. We offer a BGV-subroutined CKKS bootstrapping algorithm. In CKKS scheme, before bootstrapping, the ciphertext Enc(**m**) is in the coefficient-encoded state. Then we perform the ModRaise on the ciphertext. After that, the encoded message is removed from the least significant bit to the most significant bit. At this step, we denote the ciphertext by c_0', c_1'. Then we proceed bootstrapping using BGV.boot. We employ the inner product procedure and calculate $Enc(p^{e-r}\mathbf{m} + \mathbf{e})$.

Next, perform the digit extraction operation on the ciphertext. Finally, use the inverse linear transformation and get the ciphertext back to the coefficient-encoded state.

Before proving the correctness of Algorithm 1, we first illustrate a theorem used in the proof (Fig. 3).

Theorem 3 (Bootstrapping Failure Probability [2]). *Let h be the Hamming weight of the secret* **s** *and a constant $K < h$. $\|\tilde{I}\|_\infty \in [-h, h]$. Given that*

Algorithm 1. NewCKKS.Boot

Input: a CKKS ciphertext c_0, c_1.
Output: a CKKS ciphertext c_0', c_1'.
- $Q = p^e + 1$ for some prime p.
- $(c_0', c_1') \xleftarrow{\text{ModRaise}} (c_0, c_1, Q)$
- $(\tilde{c_0'}, \tilde{c_1'}) \leftarrow$ BGV.boot(c_0', c_1')
- **Return** $(\tilde{c_0'}, \tilde{c_1'})$.

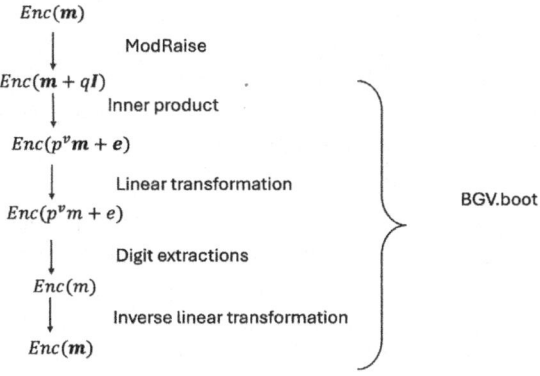

Fig. 3. Our CKKS bootstrapping pipeline.

the ciphertext encrypts a message $m(Y)$ with $Y = X^{N/2n}$, the exact probability $f(K, h, n) = \Pr[||\tilde{I}||_\infty > K]$ can be computed by

$$1 - (\frac{2}{(h+1)!}(\sum_{i=0}^{\lfloor K + 0.5(h+1) \rfloor} (-1)^i \binom{h+1}{i}(K + 0.5(h+1) - i)^{h+1} - 1)))^{2n}$$

Table 1 shows the choice of K when achieving a failure probability of $\approx 2^{-15}$ and $n = 2^{15}$.

Table 1. $\Pr[||\tilde{I}(Y)|| > K] \approx 2^{-16}$ for $n = 2^{15}$ and variable h. [2]

$\log_2(h)$	6	7	8	9	10	11	12	13	14	15
K	14	20	29	41	58	82	116	163	232	328
$\log_2(\Pr[||\tilde{I}(Y)|| > K])$	−14.6	−14.6	−15.7	−15.6	−15.5	−15.4	−15.4	−15.4	−15.4	−15.4
$K/\sqrt{h}$	1.75	1.76	1.81	1.81	1.81	1.81	1.81	1.81	1.81	1.81

Theorem 4. *Suppose that the algorithm* BGV.boot *in Algorithm 1 is a BGV bootstrapping algorithm with functionality* $(q, p^e + 1, B_{out})$, *there exists a CKKS bootstrapping algorithm with functionality* $(q, p^e + 1, B_{out})$.

Proof. We first prove that this algorithm works, and then show that the algorithm is an effective bootstrapping algorithm. Given a CKKS example c_0, c_1, where $c_0 + c_1 \cdot s = m \mod q$. First apply the ModRaise in the CKKS bootstrapping procedure in [7]. The CKKS example, whose modulus is q is expressed in the modulus $p^e + 1 > q \geq p^r \geq ||m||_\infty$. This gives rise to a ciphertext c'_0, c'_1, where $c'_0 + c'_1 \cdot s = qI + m \mod p^e + 1$. Here, we require q and $p^e + 1$ to be coprime.

After the ModRaise step, we perform the coefficient transformation method on the CKKS ciphertext so that it can be encoded using the BGV linear transformation.

Next, we will prove that the ciphertext satisfies the requirement of the digit extraction method in BGV bootstrapping with small failure probability. Then we set $e = \frac{q}{p^r}I \mod (p^e + 1)$. Here, e exists because q and $p^e + 1$ are coprime. Then we set $\tilde{c_i}' = c_i' p^{e-r}$ We have $\tilde{c_0}' + \tilde{c_1}' \cdot s = p^e e + p^{e-r} m = -e + p^{e-r} m \mod (p^e + 1)$. Set $K = \frac{p^{e-r}}{2}$, by Theorem 3, with overwhelming probability, $||I||_\infty < K$. So $||e||_\infty = \frac{q}{p^r}||\tilde{I}||_\infty \leq p^{e-r}/2$ with overwhelming probability. So, the ciphertext $(\tilde{c_0}', \tilde{c_1}')$ can be correctly decrypted in the homomorphic decryption procedure of BGV [14]. After that, perform the linear transformation, digit extractions, and inverse linear transformation. So the algorithm is a valid bootstrapping algorithm.

After BGV.boot, we can retrieve $(\tilde{c_0}, \tilde{c_1}')$, satisfying $\tilde{c_0}' + \tilde{c_1}' \cdot s = m + e$. Denote $m + e$ by m'. We have $||m' - m||_\infty \leq ||e||_\infty \leq B_{out}$. So this algorithm is an efficient CKKS bootstrapping algorithm.

3.2 New BGV Bootstrapping

In the BGV Bootstrapping method, we borrow the idea of [17]. We first transform the BGV ciphertext to BFV ciphertext, then apply the BFV bootstrapping method in [17], which uses a CKKS bootstrapping as a subroutine.

Algorithm 2. NewBGV.Boot

Input: a BGV ciphertext c_0, c_1; integers q, the base modulus of the plaintext; integer Q, the base modulus of the ciphertext.
Output: a BGV ciphertext c_0''', c_1''' integers q, Q.
- $c_0 + c_1 = qe + m \mod Q$
- $(c_0', c_1') \leftarrow (\lfloor c_0/q \rfloor, \lfloor c_1/q \rfloor) \mod Q/q$, where $\lfloor c_0/q \rfloor$ represents the coefficient-wise rounding
- $(c_0'', c_1'') \leftarrow \mathsf{CKKS.boot}(c_0'', c_1'', Q/q, Q)$
- $c_0''' = c_0 - qc_0''$, $c_1''' = c_1 - qc_1''$

Theorem 5. *Suppose that the algorithm* CKKS.boot *in Algorithm 2 is a CKKS bootstrapping algorithm with functionality* (q, Q, B_{out}), *there exists a BGV bootstrapping algorithm with functionality* (q, Q, B_{out}).

Proof. Let $(\mathbf{c_0}, \mathbf{c_1})$ be a BGV ciphertext with $\mathbf{c_0} + \mathbf{c_1} \cdot \mathbf{s} = q\mathbf{e} + \mathbf{m} \mod (Q)$. Since q is an integer, we have $\mathbf{c_0}' + \mathbf{c_1}' \cdot \mathbf{s} = \mathbf{e} \mod Q/q$. Note that if we regard $(\mathbf{c_0}', \mathbf{c_1}')$ as a ciphertext, then $\mathbf{e}$ becomes the plaintext. Now we perform the CKKS bootstrapping algorithm CKKS.boot with functionality CKKS.boot $(Q/q, Q, B_{out})$, we can get a CKKS ciphertext $(\mathbf{c_0}'', \mathbf{c_1}'')$ s.t. $\mathbf{c_0}'' + \mathbf{c_1}'' \cdot \mathbf{s} = \mathbf{e}' \mod Q$. So the output $(\mathbf{c_0}''', \mathbf{c_1}''')$ satisfies $\mathbf{c_0}''' + \mathbf{c_1}''' \cdot \mathbf{s} = q(\mathbf{e} - \mathbf{e}') + \mathbf{m} \mod Q$. And we can see that the new error $||(\mathbf{e} - \mathbf{e}')||_\infty = \leq B_{out}$. Hence this is a valid and efficient BGV bootstrapping algorithm.

4 Analysis

In the previous section, we give a BGV-subroutined CKKS bootstrapping algorithm, and a CKKS-subroutined BGV bootstrapping algorithm. The two algorithms prove the transformation between BGV and CKKS bootstrapping algorithm. Together with the previous works, which offer a transformation between BGV and BFV bootstrapping algorithm [11] and a CKKS-subroutined BFV bootstrapping algorithm [17], we can complete the BGV-BFV-CKKS transformation triangle.

This triangle exhibits the possibility of transforming among BGV, BFV, and CKKS bootstrapping algorithms. This can leverage the strengths of different algorithms, optimizing performance and overcoming limitations present in individual bootstrapping techniques. Additionally, switching between algorithms can provide flexibility in addressing specific use cases or accommodating varying computational requirements. Overall, the ability to transform among different bootstrapping algorithms enables homomorphic encryption systems to achieve enhanced functionality and performance across a wide range of applications. According to [11], the digit extraction method in both BGV and BFV bootstrapping algorithms will not introduce new error to $\mathsf{Enc}(\mathbf{m})$. Hence, our BGV-subroutined CKKS bootstrapping algorithm will be free from bootstrapping error with a proper choice of parameters. With parameters $p^e > q \geq p^r \geq ||m||_\infty$, the BGV-subroutined CKKS bootstrapping algorithm will eliminate the bootstrapping noise.

5 Conclusion

This paper has advanced the theoretical foundations of fully homomorphic encryption by establishing direct bootstrapping transformations between the BGV and CKKS schemes. Our key contribution is a novel CKKS bootstrapping method that eliminates the need for traditional sine-function approximations by leveraging BGV's exact digit extraction procedure. This approach not only removes approximation errors but also demonstrates the feasibility of using BGV as a subroutine within CKKS.

Conversely, we have designed a CKKS-subroutined BGV bootstrapping algorithm, enabling transformations in the opposite direction. Together with prior work on BGV-BFV and BFV-CKKS bootstrapping, our results complete the

BGV-BFV-CKKS transformation triangle, proving that these three major FHE schemes are mutually convertible. This framework allows practitioners to flexibly switch between schemes based on computational requirements—opting for BGV/BFV when exact arithmetic is critical, or CKKS when approximate efficiency is preferred.

The implications of this work extend beyond theory. By unifying these bootstrapping methods, we enable hybrid use cases, such as combining CKKS's fast approximate computations with BGV's precise decryption in privacy-preserving machine learning pipelines. Future research could explore optimizations for real-world implementations, including parameter tuning and hardware acceleration, to further bridge the gap between theoretical feasibility and practical deployment.

Acknowledgment. This work is partially supported by the Research Grant Council of Hong Kong (Project No.: C1029-22G), The Hong Kong Polytechnic University (Project No.: P0046340), National Natural Science Foundation of China (62272131), Shenzhen Science and Technology Major Special Project (KJZD20230923114608017).

References

1. Alperin-Sheriff, J., Peikert, C.: Practical bootstrapping in quasilinear time. In: Canetti, R., Garay, J.A. (eds.) CRYPTO 2013. LNCS, vol. 8042, pp. 1–20. Springer, Heidelberg (2013). https://doi.org/10.1007/978-3-642-40041-4_1
2. Bossuat, J.-P., Mouchet, C., Troncoso-Pastoriza, J., Hubaux, J.-P.: Efficient bootstrapping for approximate homomorphic encryption with non-sparse keys. In: Canteaut, A., Standaert, F.-X. (eds.) EUROCRYPT 2021. LNCS, vol. 12696, pp. 587–617. Springer, Cham (2021). https://doi.org/10.1007/978-3-030-77870-5_21
3. Bossuat, J., Troncoso-Pastoriza, J.R., Hubaux, J.: Bootstrapping for approximate homomorphic encryption with negligible failure-probability by using sparse-secret encapsulation. In: Ateniese, G., Venturi, D. (eds.) Applied Cryptography and Network Security - 20th International Conference, ACNS 2022, Rome, Italy, 20–23 June 2022, Proceedings. Lecture Notes in Computer Science, vol. 13269, pp. 521–541. Springer (2022). https://doi.org/10.1007/978-3-031-09234-3_26
4. Boura, C., Gama, N., Georgieva, M., Jetchev, D.: Simulating homomorphic evaluation of deep learning predictions. In: Dolev, S., Hendler, D., Lodha, S., Yung, M. (eds.) CSCML 2019. LNCS, vol. 11527, pp. 212–230. Springer, Cham (2019). https://doi.org/10.1007/978-3-030-20951-3_20
5. Brakerski, Z.: Fully homomorphic encryption without modulus switching from classical GapSVP. In: Safavi-Naini, R., Canetti, R. (eds.) CRYPTO 2012. LNCS, vol. 7417, pp. 868–886. Springer, Heidelberg (2012). https://doi.org/10.1007/978-3-642-32009-5_50
6. Chen, H., Han, K.: Homomorphic lower digits removal and improved FHE bootstrapping. In: Nielsen, J.B., Rijmen, V. (eds.) EUROCRYPT 2018. LNCS, vol. 10820, pp. 315–337. Springer, Cham (2018). https://doi.org/10.1007/978-3-319-78381-9_12

7. Cheon, J.H., Han, K., Kim, A., Kim, M., Song, Y.: Bootstrapping for approximate homomorphic encryption. In: Nielsen, J.B., Rijmen, V. (eds.) EUROCRYPT 2018. LNCS, vol. 10820, pp. 360–384. Springer, Cham (2018). https://doi.org/10.1007/978-3-319-78381-9_14
8. Cheon, J.H., Kim, A., Kim, M., Song, Y.: Homomorphic encryption for arithmetic of approximate numbers. In: Takagi, T., Peyrin, T. (eds.) ASIACRYPT 2017. LNCS, vol. 10624, pp. 409–437. Springer, Cham (2017). https://doi.org/10.1007/978-3-319-70694-8_15
9. Cramer, R., Gennaro, R., Schoenmakers, B.: A secure and optimally efficient multi-authority election scheme. Eur. Trans. Telecommun. **8**(5), 481–490 (1997). https://doi.org/10.1002/ETT.4460080506
10. Fan, J., Vercauteren, F.: Somewhat practical fully homomorphic encryption. IACR Cryptol. ePrint Arch. 144 (2012). http://eprint.iacr.org/2012/144
11. Geelen, R., Vercauteren, F.: Bootstrapping for BGV and BFV revisited. J. Cryptol. **36**(2), 12 (2023). https://doi.org/10.1007/S00145-023-09454-6
12. Gentry, C.: Fully homomorphic encryption using ideal lattices. In: Mitzenmacher, M. (ed.) Proceedings of the 41st Annual ACM Symposium on Theory of Computing, STOC 2009, Bethesda, MD, USA, 31 May–2 June 2009, pp. 169–178. ACM (2009). https://doi.org/10.1145/1536414.1536440
13. Gentry, C., Halevi, S.: Implementing gentry's fully-homomorphic encryption scheme. In: Paterson, K.G. (ed.) EUROCRYPT 2011. LNCS, vol. 6632, pp. 129–148. Springer, Heidelberg (2011). https://doi.org/10.1007/978-3-642-20465-4_9
14. Halevi, S., Shoup, V.: Bootstrapping for HElib. In: Oswald, E., Fischlin, M. (eds.) EUROCRYPT 2015. LNCS, vol. 9056, pp. 641–670. Springer, Heidelberg (2015). https://doi.org/10.1007/978-3-662-46800-5_25
15. Jutla, C.S., Manohar, N.: Sine series approximation of the mod function for bootstrapping of approximate HE. In: Dunkelman, O., Dziembowski, S. (eds.) Advances in Cryptology - EUROCRYPT 2022 - 41st Annual International Conference on the Theory and Applications of Cryptographic Techniques, Trondheim, Norway, 30 May–3 June 2022, Proceedings, Part I. Lecture Notes in Computer Science, vol. 13275, pp. 491–520. Springer (2022). https://doi.org/10.1007/978-3-031-06944-4_17
16. Kim, A., et al.: General bootstrapping approach for RLWE-based homomorphic encryption. IEEE Trans. Comput. **73**(1), 86–96 (2024). https://doi.org/10.1109/TC.2023.3318405
17. Kim, J., Seo, J., Song, Y.: Simpler and faster BFV bootstrapping for arbitrary plaintext modulus from CKKS. In: Luo, B., Liao, X., Xu, J., Kirda, E., Lie, D. (eds.) Proceedings of the 2024 on ACM SIGSAC Conference on Computer and Communications Security, CCS 2024, Salt Lake City, UT, USA, 14–18 October 2024, pp. 2535–2546. ACM (2024). https://doi.org/10.1145/3658644.3670302
18. Lee, J.-W., Lee, E., Lee, Y., Kim, Y.-S., No, J.-S.: High-precision bootstrapping of RNS-CKKS homomorphic encryption using optimal minimax polynomial approximation and inverse sine function. In: Canteaut, A., Standaert, F.-X. (eds.) EUROCRYPT 2021. LNCS, vol. 12696, pp. 618–647. Springer, Cham (2021). https://doi.org/10.1007/978-3-030-77870-5_22
19. Rivest, R.L., Adleman, L., Dertouzos, M.L., et al.: On data banks and privacy homomorphisms. Found. Secure Comput. **4**(11), 169–180 (1978)
20. Wang, Y., Malluhi, Q.M.: Privacy preserving computation in cloud using noise-free fully homomorphic encryption (FHE) schemes. In: Askoxylakis, I., Ioannidis, S., Katsikas, S., Meadows, C. (eds.) ESORICS 2016. LNCS, vol. 9878, pp. 301–323. Springer, Cham (2016). https://doi.org/10.1007/978-3-319-45744-4_15

PolarKyber: Polished Kyber with Smaller Ciphertexts, Greater Security Redundancy, and Lower Decryption Failure Rate

Chen An[1,2], Ziyao Liu[1,2], Xianhui Lu[1,2(✉)], and Jingnan He[1,2]

[1] State Key Laboratory of Cyberspace Security Defense, Institute of Information Engineering, CAS, Beijing, China
{anchen,liuziyao,luxianhui,hejingnan}@iie.ac.cn
[2] School of Cyber Security, University of Chinese Academy of Sciences, Beijing, China

Abstract. Kyber is one of the most representative lattice-based public key encryption schemes and has been standardized by the National Institute of Standards and Technology (NIST) as a post-quantum cryptographic standard. However, several aspects remain suboptimal. First, its ciphertext size is considerably larger than that of traditional public-key encryption schemes, posing significant challenges for deployment and migration. Second, its security redundancy is insufficient, as reflected in the decision of the International Organization for Standardization (ISO) and the International Electrotechnical Commission (IEC) to exclude the NIST Level I parameters from their standards. Third, the decryption failure rates (DFR) for NIST Levels III and V do not adequately align with their intended security levels. To address these issues, we propose PolarKyber, an enhanced version of Kyber that incorporates polar codes. PolarKyber effectively reduces ciphertext size, strengthens security redundancy, and lowers the DFR. Specifically, our enhancements achieve a ciphertext size reduction of 8.82%–18.37%, an effective security improvement of 9–20 bits, and a DFR reduction that ensures compliance with security requirements. These benefits come at the cost of a 17.85%–39.77% increase in computational overhead, primarily during decryption. However, this trade-off is reasonable, as lattice-based schemes are efficient, with ciphertext size being the main deployment bottleneck.

Keywords: Post-quantum cryptography · Lattice-based cryptography · Module learning with errors · Channel capacity · Polar codes

---

This work was supported in part by the National Key Research and Development Program of China under Grant 2022YFB2702701, and in part by the Chinese Academy of Sciences (CAS) Project for Young Scientists in Basic Research under Grant YSBR-035.

© The Author(s), under exclusive license to Springer Nature Singapore Pte Ltd. 2026
J. Han et al. (Eds.): ICICS 2025, LNCS 16217, pp. 286–305, 2026.
https://doi.org/10.1007/978-981-95-3540-8_16

1 Introduction

Lattice-based cryptography [24] is one of the most competitive candidates for post-quantum cryptography. Its security relies on the inherent hardness of fundamental lattice problems, which are widely believed to be intractable for both classical and quantum computers. Key assumptions such as learning with errors (LWE) [28], ring learning with errors (RLWE) [22], and module learning with errors (MLWE) [16] serve as the foundation for numerous cryptographic protocols, making lattice-based cryptography a leading approach in the development of quantum-resistant cryptographic standards. Kyber [10] is a public key encryption (PKE) scheme based on the MLWE problem, offering indistinguishability under chosen-plaintext attacks (IND-CPA) security. When transformed using the Fujisaki-Okamoto (FO) framework, it becomes a key encapsulation mechanism (KEM) that achieves indistinguishability under chosen-ciphertext attacks (IND-CCA) security. In its SP 800-227 draft, NIST states that IND-CPA security is sufficient to meet the requirements of many practical applications, such as ephemeral key establishment [2]. As a result of its strong security guarantees and efficient performance [25], Kyber was selected as one of the first post-quantum cryptographic algorithms in the NIST standardization process. It supports three security levels—I, III, and V—corresponding to Kyber512, Kyber768, and Kyber1024, respectively, ensuring its suitability for a wide range of security-critical applications.

However, certain aspects of Kyber still remain suboptimal with respect to ciphertext size, security redundancy, and DFR. The ciphertext size of Kyber is significantly larger than that of traditional public-key encryption schemes, posing substantial challenges for migration. For example, in transport layer security (TLS) deployments, larger ciphertexts result in higher handshake latency. As demonstrated by the experimental results in [5], Kyber incurs greater handshake delays than elliptic curve X25519. Additionally, the security redundancy of Kyber512 is insufficient, which is why the ISO/IEC did not select the Level I parameters during the standardization process [13]. Furthermore, Kyber's DFR exhibits certain flaws, particularly at Level III and Level V, where the DFR does not align with the expected security level. Specifically, the DFR of Kyber768, which is 2^{-164}, fails to meet the required threshold of 2^{-192}, and the DFR of Kyber1024, at 2^{-174}, falls short of the standard requirement of 2^{-256}.

The ciphertext size, security redundancy, and DFR are interrelated and mutually constrained. In lattice-based schemes, security is primarily determined by the polynomial dimension and the noise-to-modulus ratio; higher values of both contribute to stronger security. To improve security, one can increase the polynomial dimension, but this naturally increases the ciphertext size. Alternatively, increasing the noise-to-modulus ratio requires either enlarging the noise or reducing the modulus, both of which lead to an elevated DFR. To address the issue of security redundancy without expanding the ciphertext size, it is necessary to maintain the polynomial dimension while adjusting only the noise-to-modulus ratio. However, this approach results in a higher DFR, which can be mitigated by employing more robust error correction codes (ECC). This presents a promising direction for future research.

Polar codes are the first explicitly constructed codes proven to achieve channel capacity [6]. As demonstrated by Wang and Ling [34], both the encoding and decoding algorithms of polar codes operate in constant time, making them an attractive option for enhancing the Kyber scheme. In 2023, Papadopoulos and Wang [26] incorporated a rate-1/2 polar code into Kyber to reduce its DFR. However, a significant challenge arises from the fixed message length of 256 bits, which leaves no space for the inclusion of parity check bits for error correction. In their approach, Papadopoulos and Wang sacrificed ciphertext size by halving the message length to accommodate the error correction code, effectively doubling the ciphertext size in order to transmit the full 256-bit message. This trade-off in size is often impractical in real-world applications of Kyber. Consequently, addressing the three critical issues of Kyber—ciphertext size, security redundancy, and decryption failure rate—without increasing the ciphertext size presents a substantial challenge.

1.1 Our Contribution

For lattice-based PKE schemes like Kyber, the decryption process can be abstracted as $v - \mathbf{s}^T\mathbf{u} = \lceil q/2 \rceil \cdot m + w$, where $\lceil q/2 \rceil \cdot m$ represents the modulation of the message, and the noise term w is interpreted as Gaussian white noise in the channel. This enables the decryption process to be modeled as 256 parallel, independent additive white Gaussian noise (AWGN) channels. Based on Kyber's parameter configuration, the capacity of each channel exceeds 3 bits, meaning that Kyber's usable message space is greater than 256×3 bits. However, the currently standardized version of Kyber transmits only 1 bit per channel, resulting in a message space of just 256 bits. In this paper, we propose the PolarKyber scheme, which modulates 2 bits per coefficient to address the challenge of using polar codes without increasing the ciphertext size. The main technical contributions of this paper are as follows:

Modeling PolarKyber as a 2-Bit AWGN Channel Model to Address Message Space Limitations. We model the PolarKyber scheme as 256 parallel, independent two-bit channels, establishing a direct correspondence between its encryption and decryption processes and the communication flow. By modulating 2 bits per coefficient—i.e., transmitting two bits per channel—we effectively expand the message space. However, when modulating 2 bits per coefficient, the bits exhibit correlation, which may lead to error propagation. To mitigate this, we employ Gray coding, ensuring that adjacent modulation symbols differ by only a single bit. This approach minimizes error propagation, guaranteeing that errors between adjacent symbols affect at most one bit, thereby enhancing the robustness of the scheme.

Constructing a Relaxed Channel Model to Estimate the DFR. To address the issue that the PolarKyber channel model does not satisfy the memoryless condition required for polar codes, we construct a Relaxed Channel model consisting of multiple parallel, independent single-bit channels. By analyzing the reliability relationship between these two models, we leverage the upper bound on the DFR of the Relaxed Channel model to estimate the upper bound on the DFR of the PolarKyber scheme. Additionally, the DFR required in communication is typically around 2^{-30}, while cryptographic standards require a DFR of at least 2^{-128}. This discrepancy makes simulation experiments infeasible. Therefore, we employ a degrading construction algorithm [18] to theoretically derive an upper bound for the DFR.

The PolarKyber scheme integrates the two aforementioned techniques and provides two versions. Both versions ensure that the DFR meets the security level requirements for all three security levels. The versions differ in their priorities: one emphasizes optimizing ciphertext size, while the other focuses on enhancing security strength. Specifically, the two versions are as follows:

Reducing Ciphertext Size Without Compromising Security. By preserving the original Kyber scheme's security and distributions, the PolarKyber scheme achieves ciphertext size reductions of 12.50%, 8.82%, and 16.33% at the three corresponding security levels. When the distribution in the PolarKyber scheme is allowed to differ from that of the Kyber scheme, the ciphertext size at security level V is further reduced to 1280 bytes, resulting in an 18.37% reduction.

Enhancing Security and Reducing Ciphertext Size Simultaneously. For classical security, at security level I, using an identical distribution, security is increased by 9 bits while the ciphertext size is reduced by 4.17%. With a variable distribution, a 6-bit security increase corresponds to an 8.33% reduction in ciphertext size, and a 2-bit security increase leads to a 12.50% reduction in size. At security level III, an identical distribution yields a 10-bit increase in security and a 5.88% reduction in ciphertext size. With a variable distribution, an 18-bit security increase results in an 8.82% reduction in ciphertext size. At security level V, an identical distribution provides a 14-bit increase in security and an 8.16% reduction in ciphertext size. When using a variable distribution, a 20-bit security increase corresponds to a 12.24% reduction, an 18-bit security increase results in a 14.29% reduction, and a 15-bit security increase achieves a 16.33% reduction in ciphertext size.

1.2 Related Work

Many lattice-based PKE schemes have incorporated ECC to enhance performance. For instance, the NewHope-Simple scheme [4] utilizes repetition codes, while the LAC scheme [20] leverages BCH codes to reduce modulus size. Recent research has modeled the decryption process in LWE/RLWE/MLWE-based encryption schemes as a decoding problem over a noisy channel, where ECC are employed to lower DFR, reduce ciphertext size, or improve security.

In 2019, Lee et al. [17] introduced BCH codes into the LWE-based FrodoKEM scheme to enhance its security. Fritzmann et al. [12] applied BCH and LDPC codes, as well as their combinations, to the RLWE-based NewHope Simple key exchange scheme to improve security. Maringer et al. [23] evaluated the theoretical capacity of RLWE/MLWE channels and suggested using BCH codes to increase transmission rates and reduce DFR. Bocharova et al. [9] combined BCH codes with modulo Q pulse amplitude modulation (PAM) and vector dequantization techniques to reduce DFR in RLWE and MLWE-based encryption schemes.

Progress has also been made in applying lattice codes to PKE schemes. In 2016, van Poppelen [27] introduced a Leech lattice-based scheme. In 2022, Saliba et al. [30] proposed an E8-lattice-based FrodoKEM scheme. More recently, in 2024, Lyu et al. [21] employed Barnes-Wall lattice codes in FrodoPKE, achieving enhanced security and reduced ciphertext sizes. Liu and Sakzad [19] applied Barnes-Wall and Leech lattice codes to the Kyber scheme, resulting in significant reductions in ciphertext size and DFR.

Regarding polar codes, Wang and Ling [33,34] applied polar codes to the NewHope-Simple scheme in 2021 and 2023 to improve security. In 2023, Papadopoulos and Wang [26] utilized a rate-1/2 polar code in Kyber to reduce its DFR.

2 Preliminary

2.1 Notation

Ring. Let $\mathcal{R} = \mathbb{Z}[X]/(X^n+1)$ and $\mathcal{R}_q = \mathbb{Z}_q[X]/(X^n+1)$, where n is the degree of the monic polynomial. In Kyber, n is fixed at 256.

Matrices and Vectors. Matrices and vectors are denoted by bold uppercase and lowercase letters, respectively. The transpose of a vector $\mathbf{v}$ is written as $\mathbf{v}^T$, and the infinity norm $\|\mathbf{v}\|_\infty$ represents the maximum absolute value among its entries.

Sampling and Distribution. For a set S, the notation $s \leftarrow S$ indicates that s is selected uniformly at random from S. If S is a probability distribution, this means s is sampled according to the distribution S. For a polynomial $f(x) \in \mathcal{R}_q$ or a vector of such polynomials, this operation is applied to each coefficient independently. When x is a bit string and S is a distribution that takes x as input, $y \sim S := \text{Sam}(x)$ represents the output y produced by S with input x, which can be extended to any desired length. The centered binomial distribution (CBD) over $\mathbb{Z}$ is defined as $\psi_\eta = B(2\eta, 0.5) - \eta$, where the corresponding variance is $\eta/2$. When we write $f(x) \sim \psi_\eta$, with $f(x) \in \mathcal{R}_q$, we mean that each coefficient of the polynomial is independently sampled from ψ_η. The n-ary CBD with fixed Hamming weight is denoted as $\psi_{n,h}$, where $h \in (0, n/2)$ is even. For a random variable distributed according to this distribution, its Hamming weight is fixed at h, with $h/2$ instances of both 1's and -1's, and $n - h$ instances of 0. The continuous normal distribution over $\mathbb{R}$, with mean μ and variance σ^2, is denoted as $\mathcal{N}(\mu, \sigma^2)$. The discrete uniform distribution over $\mathbb{Z}$, with minimum $a \in \mathbb{Z}$ and maximum $b \in \mathbb{Z}$, is denoted as $\mathcal{U}(a, b)$.

Compress and Decompress. Let $x \in \mathbb{R}$ be a real number, then $\lceil x \rfloor$ represents rounding to the closest integer with ties rounded up. Let $x \in \mathbb{Z}_q$ and $d \in \mathbb{Z}$ such that $2^d < q$. We define

$$\text{Compress}_q(x, d) = \lceil (2^d/q) \cdot x \rfloor \mod 2^d,$$
$$\text{Decompress}_q(x, d) = \lceil (q/2^d) \cdot x \rfloor.$$

2.2 Kyber Scheme

Define the message space as $\mathbb{Z}_2^n = \{0, 1\}^n$, where each message $m \in \mathbb{Z}_2^n$ can be represented as a polynomial in $\mathcal{R}$ with coefficients from $\{0, 1\}$. We consider the PKE scheme Kyber = (KeyGen, Enc, Dec), as detailed in Algorithms 1 to 3 [7]. Let k, d_u, and d_v be positive integer parameters, as specified in Table 1. The security level presented in the table is estimated using the lattice-estimator tool [3], covering both classical and quantum bit-security strength. These estimates reflect the cost under optimal attack models. For the classical security analysis, we employ the cost model proposed in [1], while for the quantum case, we use the model from [15]. Encap and Decap represent the computational performance of the encapsulation and decapsulation algorithms in the Kyber scheme, respectively. Both were evaluated based on the C reference implementation of Kyber [8]. All performance measurements were conducted on a system running Ubuntu 18.04, equipped with an AMD Ryzen 7 PRO 4750U processor (base frequency 1.70 GHz) and 14.9 GB of memory.

Algorithm 1. KeyGen()

1: $\rho, \sigma \leftarrow \{0, 1\}^{256}$
2: $\mathbf{A} \sim \mathcal{R}_q^{k \times k} := \text{Sam}(\rho)$
3: $(\mathbf{s}, \mathbf{e}) \sim \psi_{\eta_1}^k \times \psi_{\eta_1}^k := \text{Sam}(\sigma)$
4: $\mathbf{t} := \mathbf{As} + \mathbf{e}$
5: **return** $(pk := (\mathbf{t}, \rho), sk := \mathbf{s})$

The ciphertext size, denoted as $\text{Size}(ct)$, is measured in bytes and is computed using the following formula:

$$\text{Size}(ct) = \frac{256 \cdot k \cdot d_u + 256 \cdot d_v}{8} \text{ bytes}.$$

Algorithm 2. $\text{Enc}(pk = (\mathbf{t}, \rho), m \in \mathbb{Z}_2^n)$

1: $r \leftarrow \{0,1\}^{256}$
2: $\mathbf{A} \sim \mathcal{R}_q^{k \times k} := \text{Sam}(\rho)$
3: $(\mathbf{r}, \mathbf{e}_1, e_2) \sim \psi_{\eta_1}^k \times \psi_{\eta_2}^k \times \psi_{\eta_2} := \text{Sam}(r)$
4: $\mathbf{u} := \text{Compress}_q(\mathbf{A}^T \mathbf{r} + \mathbf{e}_1, d_u)$
5: $v := \text{Compress}_q(\mathbf{t}^T \mathbf{r} + e_2 + \lceil q/2 \rfloor \cdot m, d_v)$
6: **return** $ct := (\mathbf{u}, v)$

Algorithm 3. $\text{Dec}(sk = \mathbf{s}, ct = (\mathbf{u}, v))$

1: $\mathbf{u} := \text{Decompress}_q(\mathbf{u}, d_u)$
2: $v := \text{Decompress}_q(v, d_v)$
3: **return** $\text{Compress}_q(v - \mathbf{s}^T \mathbf{u}, 1)$

Table 1. Kyber parameters.

Scheme	n	k	q	η_1	η_2	d_u	d_v	Size(ct)/B	DFR	Security Level	Encap/cycle	Decap/cycle
Kyber512	256	2	3329	3	2	10	4	768	2^{-139}	140 131Q	61406	80851
Kyber768	256	3	3329	2	2	10	4	1088	2^{-164}	201 189Q	99299	127441
Kyber1024	256	4	3329	2	2	11	5	1568	2^{-174}	270 255Q	144235	180092

The DFR is defined as the probability that the decrypted message $\hat{m}$ does not match the original message m. Formally, it is expressed as $\text{DFR} \triangleq \Pr(\hat{m} \neq m)$. In the Kyber scheme, the DFR can be calculated as follows [10]:

$$\text{DFR} := \Pr(\|w\|_\infty \geq \lceil q/4 \rfloor),$$

where w represents the decryption noise. The decryption noise w is given by:

$$w = v - \mathbf{s}^T \mathbf{u} - \lceil q/2 \rfloor \cdot m = \mathbf{e}^T \mathbf{r} + e_2 + c_v - \mathbf{s}^T(\mathbf{e}_1 + \mathbf{c}_u),$$

where $c_v \leftarrow \psi_{d_v}$, $\mathbf{c}_u \leftarrow \psi_{d_u}^k$ are rounding noises generated due to the compression operation.

2.3 Multi-bit Modulation

We describe a scheme that modulates B bits into a coefficient, where $B \geq 1$ [9]. Let $Q = 2^B$ and $h = \lceil q/2^B \rfloor$. In order to reduce the amount of bit errors, we will introduce Q-PAM, using the B-bit Gray code.

Gray: $\mathbb{Z}_Q \to \mathbb{Z}_Q$

$$\sum_{i=0}^{B-1} b_i 2^i \mapsto b_{B-1} 2^{B-1} + \sum_{i=0}^{B-2} (b_{i+1} \oplus b_i) 2^i,$$

where $\oplus$ denotes addition modulo 2. We extend the Gray map on vectors over $\mathbb{Z}_Q$ by applying it on each coordinate.

Let $\mathcal{C} \in \mathbb{Z}_2^N$ be an $[N, K]$ binary linear code, where $N = nB$ denotes the length of the codeword and $K = n$ denotes the length of the message. The codeword $c \in \mathcal{C}$ is then partitioned into blocks of B bits, resulting in the following expression for b:

$$b = \left(\sum_{j=0}^{B-1} 2^j c_j, \sum_{j=0}^{B-1} 2^j c_{B+j}, \ldots, \sum_{j=0}^{B-1} 2^j c_{(n-1)B+j} \right) \in \mathbb{Z}_Q^n.$$

Next, we apply the B-bit Gray code to remap the Q-ary symbols b, obtaining $a = \text{Gray}(b)$. Then, a scaling factor h is applied to modulate a into the transmitted symbol $x = h \cdot a$. This sequence of operations is referred to as Gray coded Q-PAM. Notably, all computations are conducted over $\mathbb{Z}_q$, ensuring that each symbol remains within the valid range defined by q. The detailed process can be found in Algorithm 4.

Algorithm 4. Multi-Bit Modulation ($c \in \mathbb{Z}_2^N$)

1: $b = \left(\sum_{j=0}^{B-1} 2^j c_j, \sum_{j=0}^{B-1} 2^j c_{B+j}, \ldots, \sum_{j=0}^{B-1} 2^j c_{(n-1)B+j} \right)$
2: $a = \text{Gray}(b)$
3: $x = h \cdot a$
4: **return** x

2.4 Polar Codes

Polar codes, introduced by Arikan [6], are the first error-correcting codes that provably achieve the capacity for any discrete memoryless channel. They leverage a property called polarization, which divides the bit channels into two groups: one approaching a capacity of 1 for reliable information transmission, and the other nearing a capacity of 0 for transmitting frozen bits, typically set to 0.

In the binary field, a vector u of length N is transformed using the matrix $G_N = \begin{bmatrix} 1 & 0 \\ 1 & 1 \end{bmatrix}^{\otimes l}$, where $N = 2^l$ and $\otimes$ is the Kronecker product, yielding the codeword $c = uG_N$. In non-systematic polar encoding with a code length of N and a message length of K, the vector $u = \{u_A, u_{A^c}\}$ where $A \subset \{1, \cdots, N\}$, u_A represents the message m of length K, and u_{A^c} represents the frozen bits of length $(N - K)$ that have been pre-agreed upon by both legitimate parties. The elements in the set A correspond to the indices of K reliable bit channels following channel polarization, whereas the elements in the set A^c correspond to the indices of $(N - K)$ unreliable bit channels. The codeword c can be expressed as $c = u_A G_A + u_{A^c} G_{A^c}$, where G_A and G_{A^c} denote the submatrices of G_N formed by the rows with indices in A and A^c, respectively.

A typical polar decoding algorithm is the successive cancellation (SC) decoding algorithm [6]. For a binary discrete memoryless channel (BDMC) W, the block error rate (BLER) of polar codes satisfies BLER $\leq \sum_{i \in A} Z(W_N^i)$ [6], where $Z(W_N^i)$ is the Bhattacharyya parameter of the i-th polarized channel [32].

3 The AWGN Channel Model for Kyber

Many studies have modeled lattice-based public-key encryption schemes as message transmission over noisy channels [9,23]. The encryption and decryption processes of Kyber can be interpreted as the transmission of symbols over an AWGN channel. Specifically, the Kyber decryption problem can be modeled as:

$$y = v - \mathbf{s}^T\mathbf{u} = \lceil q/2 \rfloor \cdot m + \mathbf{e}^T\mathbf{r} + e_2 + c_v - \mathbf{s}^T(\mathbf{e}_1 + \mathbf{c}_u) = x + w \in \mathcal{R}_q.$$

$m \in \mathbb{Z}_2^n$ denotes the message to be transmitted, where each bit of m corresponds to a binary message. The term $x = \lceil q/2 \rfloor \cdot m$ refers to 1-bit high-part encoding, which corresponds to the modulation stage in the communication process. Therefore, it can be said that each coefficient modulates a single bit of the message. The error term, $w = \mathbf{e}^T\mathbf{r} + e_2 + c_v - \mathbf{s}^T(\mathbf{e}_1 + \mathbf{c}_u)$, represents the noise in the AWGN channel. The corresponding channel model is depicted in Fig. 1, which illustrates the structure of the communication system in the context of the Kyber scheme.

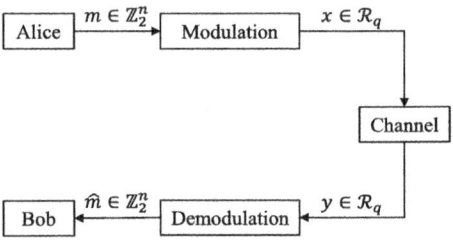

Fig. 1. AWGN channel model for Kyber.

The capacity of the AWGN channel is given by $C = \frac{1}{2}\log_2(1 + \text{SNR})$ [31], where SNR quantifies the ratio between signal power and noise power, computed as $\text{SNR} = E_s/\sigma_w^2$. Here, $E_s = (\lceil q/4 \rfloor)^2$ denotes the average signal energy per transmitted symbol, and σ_w^2 is the variance of the noise term w. Table 2 presents the capacity of the Kyber channel. It is clear that the capacity of the Kyber channel exceeds 3 bits per channel use (bpcu), meaning that each coefficient in Kyber has the potential to transmit up to 3 bits of message. However, the currently standardized Kyber scheme transmits only 1 bit per coefficient, resulting in suboptimal utilization of the channel's capacity. To fully exploit this potential, the number of bits modulated per coefficient can be increased. This enhancement, however, leads to an increase in the DFR, necessitating the use of advanced channel coding techniques to ensure that the DFR remains within acceptable bounds.

In this paper, we adopt polar codes as a representative class of $[N, K]$ binary linear codes within the multi-bit modulation framework. Polar codes are the

Table 2. Capacity of the Kyber channel.

Scheme	σ_w^2	SNR	C/bpcu
Kyber512	6252.2	110.7834	3.4023
Kyber768	5868.2	118.0328	3.4476
Kyber1024	3356.9	206.3332	3.8479

first class of explicitly constructed codes proven to achieve channel capacity [6]. Furthermore, as shown by Wang and Ling [34], both the encoding and decoding algorithms of polar codes can be implemented in constant time. Given that polar code lengths are typically powers of two, and that the capacity of each Kyber channel is approximately 3 bits, we employ a 2-bit modulation technique to the Kyber scheme, as described in Sect. 2.3, where $B = 2$, $Q = 4$, $h = \lceil q/4 \rceil$, $K = n$, and $N = 2n$.

4 Polar Code-Enhanced Kyber Scheme

In this section, we first introduce the PolarKyber scheme, an enhanced variant of Kyber that incorporates polar coding and 2-bit modulation. We then establish a channel model for PolarKyber, representing it as a set of parallel, independent 2-bit AWGN channels, and construct a Relaxed Channel model to analyze its DFR. Finally, we present experimental results and comparisons, including four parameter configurations for PolarKyber, which are compared with the performance of the original Kyber scheme, along with an analysis of the computational overhead introduced by integrating polar codes.

4.1 PolarKyber Scheme

Our PolarKyber scheme consists of three algorithms:

- PolarKyber.KeyGen: the key generation algorithm, as illustrated in Algorithm 5.
- PolarKyber.Enc: the encryption algorithm, as illustrated in Algorithm 6.
- PolarKyber.Dec: the decryption algorithm, as illustrated in Algorithm 7.

The algorithm PolarKyber.KeyGen randomly generates a public-key and secret-key pair (pk, sk). In this algorithm, $\psi_\mathbf{s}^k$ and $\psi_\mathbf{e}^k$ denote the distributions from which $\mathbf{s}$ and $\mathbf{e}$ are sampled, respectively.

The algorithm PolarKyber.Enc takes the public key pk and the message m as inputs, encrypting the message m into the ciphertext ct. In this algorithm, $\psi_\mathbf{r}^k$, $\psi_{\mathbf{e}_1}^k$, and ψ_{e_2} represent the distributions from which $\mathbf{r}$, $\mathbf{e}_1$, and e_2 are sampled, respectively. The subroutine PolarEnc implements the encoding algorithm of polar codes, which transforms the n-bit message m into the N-bit codeword c.

The algorithm PolarKyber.Dec takes the private key sk and the ciphertext ct as inputs, recovering the corresponding message. The subroutine PolarDec refers

Algorithm 5. PolarKyber.KeyGen()

1: $\rho, \sigma \leftarrow \{0,1\}^{256}$
2: $\mathbf{A} \sim \mathcal{R}_q^{k \times k} := \mathrm{Sam}(\rho)$
3: $(\mathbf{s}, \mathbf{e}) \sim \psi_{\mathbf{s}}^k \times \psi_{\mathbf{e}}^k := \mathrm{Sam}(\sigma)$
4: $\mathbf{t} := \mathbf{As} + \mathbf{e}$
5: **return** $(pk := (\mathbf{t}, \rho), sk := \mathbf{s})$

Algorithm 6. PolarKyber.Enc($pk = (\mathbf{t}, \rho), m \in \mathbb{Z}_2^n$)

1: $r \leftarrow \{0,1\}^{256}$
2: $\mathbf{A} \sim \mathcal{R}_q^{k \times k} := \mathrm{Sam}(\rho)$
3: $(\mathbf{r}, \mathbf{e}_1, e_2) \sim \psi_{\mathbf{r}}^k \times \psi_{\mathbf{e}_1}^k \times \psi_{e_2} := \mathrm{Sam}(r)$
4: $\mathbf{u} := \mathrm{Compress}_q(\mathbf{A}^T \mathbf{r} + \mathbf{e}_1, d_u)$
5: $c := \mathrm{PolarEnc}(m)$
6: $b = \left(\sum_{j=0}^{1} 2^j c_j, \sum_{j=0}^{1} 2^j c_{2+j}, \ldots, \sum_{j=0}^{1} 2^j c_{N-2+j} \right)$
7: $a = \mathrm{Gray}(b)$
8: $x = \lceil q/4 \rfloor \cdot a$
9: $v := \mathrm{Compress}_q(\mathbf{t}^T \mathbf{r} + e_2 + x, d_v)$
10: **return** $ct := (\mathbf{u}, v)$

to the decoding algorithm of polar codes, which decodes the received signal y into the message estimate $\hat{m}$. If $m = \hat{m}$, decryption is successful; otherwise, decryption fails.

Algorithm 7. PolarKyber.Dec($sk = \mathbf{s}, ct = (\mathbf{u}, v)$)

1: $\mathbf{u} := \mathrm{Decompress}_q(\mathbf{u}, d_u)$
2: $v := \mathrm{Decompress}_q(v, d_v)$
3: $y := v - \mathbf{s}^T \mathbf{u}$
4: $\hat{m} := \mathrm{PolarDec}(y)$
5: **return** $\hat{m}$

4.2 The 2-Bit AWGN Channel Model for PolarKyber

The channel model for the PolarKyber scheme, incorporating polar coding and 2-bit modulation, is referred to as the PolarKyber-2bit Channel, as shown in Fig. 2. Modulation corresponds to the 2-bit high-part encoding technique described in lines 6 to 8 of Algorithm 6. It is crucial to note that in the AWGN channel, the decoding algorithm for polar codes inherently includes the demodulation step.

The PolarKyber-2bit Channel consists of n mutually parallel, independent two-bit channels. The output is represented as:

$$y = v - \mathbf{s}^T \mathbf{u} = \lceil q/4 \rfloor \cdot \mathrm{Gray}(b) + \mathbf{e}^T \mathbf{r} + e_2 + c_v - \mathbf{s}^T(\mathbf{e}_1 + \mathbf{c}_u) = x + w \in \mathcal{R}_q.$$

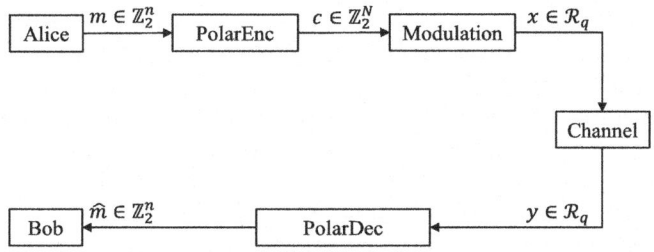

Fig. 2. 2-bit AWGN channel model for PolarKyber.

Let $x = \lceil q/4 \rfloor \cdot \text{Gray}(b)$ and $w = \mathbf{e}^T \mathbf{r} + e_2 + c_v - \mathbf{s}^T(\mathbf{e}_1 + \mathbf{c}_u)$. Then, the output of a single channel in the PolarKyber-2bit Channel can be expressed as:

$$y_i = x_i + w_i \in \mathbb{Z}_q, i = 0, 1, \ldots, n-1,$$

where the generation process of the channel input x_i is detailed in Table 3. The noise term w_i, which follows a Gaussian distribution with mean 0 and standard deviation σ_w, represents the channel noise. Thus, each polynomial coefficient of $v - \mathbf{s}^T \mathbf{u} \in \mathcal{R}_q$ is modeled as an AWGN channel, where each channel transmits a 2-bit codeword.

Table 3. Modulation mapping of a single channel.

	00	01	10	11
$(c_{2i}, c_{2i+1}) \in \mathbb{Z}_2^2$				
$b_i = 2c_{2i} + c_{2i+1}$	0	1	2	3
Gray-code mapping: $a_i = \text{Gray}(b_i)$	0	1	3	2
4-PAM: $x_i = h \cdot a_i \in \mathbb{Z}_q, h = \lceil q/4 \rfloor$	0	832	2497	1665

4.3 DFR Analysis of PolarKyber

Polar codes are typically designed under the assumption that the channel is a BDMC. However, it is evident that the PolarKyber-2bit Channel does not conform to this assumption. To address this issue, we construct a new AWGN channel with worse reliability to estimate the DFR of the PolarKyber scheme. This new channel is referred to as the Relaxed Channel. Specifically, the DFR of PolarKyber corresponds to the BLER after SC decoding of the polar code. We use the BLER over the Relaxed Channel as an upper-bound estimate for the BLER of the original PolarKyber-2bit Channel.

The Relaxed Channel consists of $2n$ parallel, independent single-bit channels, each satisfying the BDMC condition required for polar codes. The output of each individual channel in the Relaxed Channel is given by:

$$y'_j = x'_j + w'_j \in \mathbb{Z}_{q'}, q' = \lceil q/2 \rfloor, j = 0, 1, \ldots, 2n-1,$$

where $x'_j = \lceil q'/2 \rceil \cdot c_j$, and w'_j follows a Gaussian distribution with mean 0 and standard deviation σ_w. Each channel in the Relaxed Channel transmits 1 bit of codeword information.

The relationship between the PolarKyber-2bit Channel and the Relaxed Channel is illustrated in Fig. 3, where the $\succ$ symbol indicates that the PolarKyber-2bit Channel has higher reliability than the Relaxed Channel. In other words, the channel quality of the PolarKyber-2bit Channel is superior to that of the Relaxed Channel. The justification for this relationship will be provided below.

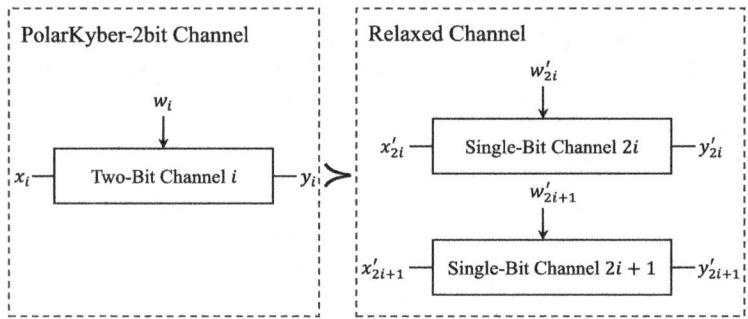

Fig. 3. Reliability comparison of PolarKyber-2bit Channel and Relaxed Channel.

Remark 1. The reliability of an AWGN channel is commonly characterized by the SNR. The higher the SNR, the higher the channel reliability and the lower the BLER. We calculate the SNR of a single channel in both PolarKyber-2bit Channel and Relaxed Channel using the formula SNR $= E_s/\sigma^2$. In the PolarKyber-2bit Channel, the 2-bit high-part encoding can be viewed as a Gray coded 4-PAM, with the symbol set $\{0, \lceil q/4 \rceil, \lceil q/2 \rceil, \lceil 3q/4 \rceil\}$. The SNR of a single channel in the PolarKyber-2bit Channel is given by:

$$\text{SNR}_{\text{polar}} = \frac{\left(\lceil \frac{q}{4} \rceil\right)^2 + \left(\lceil \frac{q}{2} \rceil\right)^2 + \left(\lceil \frac{3q}{4} \rceil\right)^2}{4\sigma_w^2} = \frac{7q^2}{32\sigma_w^2}.$$

In the Relaxed Channel, the 1-bit high-part encoding corresponds to a 2-PAM, with the symbol set $\{0, \lceil q'/2 \rceil\}$. The SNR of a single channel in the Relaxed Channel is given by:

$$\text{SNR}_{\text{relaxed}} = \frac{\left(\lceil \frac{q'}{2} \rceil\right)^2}{2\sigma_w^2} = \frac{q^2}{32\sigma_w^2} < \frac{7q^2}{32\sigma_w^2} = \text{SNR}_{\text{polar}}.$$

This analysis suggests that the PolarKyber-2bit Channel has a higher reliability than the Relaxed Channel.

Table 4. Maximum tolerable noise standard deviation for polar code at various security levels.

ECC	DFR	max_σ_w
polar(512, 256, 1/2)	$2^{-128.1953}$	119
polar(512, 256, 1/2)	$2^{-192.8579}$	103.6
polar(512, 256, 1/2)	$2^{-256.8761}$	91.5

Therefore, in analyzing the DFR of PolarKyber, which is equivalent to its BLER, we estimate its upper bound using the BLER upper bound of the Relaxed Channel. This bound is computed using polar code construction algorithms, such as the degrading construction [18].

4.4 Experimental Results and Comparisons

In this section, we analyze the maximum noise tolerance of polar codes and their application in three distributions: the original Kyber distribution, identical distribution, and variable distribution. We also discuss the computational overhead introduced by polar codes.

Maximum Noise Tolerance of Polar Codes. As shown in Table 4, ECC is a polar code with a code length of 512, a message length of 256, and a code rate of 1/2. To meet the DFR requirements for security levels I, III and V, which correspond to a BLER of less than 2^{-128}, 2^{-192}, and 2^{-256} respectively, the table provides the maximum noise standard deviation that the polar code can tolerate, denoted as max_σ_w. This value represents the maximum noise intensity the code can withstand at different security levels. Therefore, when adjusting the Kyber parameters, it is essential to ensure that the standard deviation of the noise w remains below max_σ_w. If the noise exceeds this threshold, the polar code will not be able to reduce the DFR to the required security level.

Polar Codes Applied to the Original Kyber Distribution. The PolarKyber$_{\text{Orig}}$ scheme represents the integration of polar codes into the original Kyber scheme. In this enhanced scheme, the distributions of the parameters s, e, r, e_1, and e_2 remain consistent with those in the original Kyber scheme, ensuring that the bit-security strength is preserved. The integration of polar codes facilitates greater noise tolerance, enabling ciphertext compression. Additionally, the DFR of PolarKyber$_{\text{Orig}}$ is lower compared to that of the original Kyber scheme, enhancing overall performance without compromising security.

As shown in Table 5, the polar code configurations for three security levels are presented, along with the corresponding ciphertext sizes and decryption failure rates. Compared to the original Kyber scheme, the ciphertext size of the PolarKyber$_{\text{Orig}}$ scheme is reduced by 12.50%, 8.82%, and 16.33%, respectively.

Table 5. Parameters of the PolarKyber$_{\text{Orig}}$.

Scheme	ECC	(d_u, d_v)	Size(ct)/B	DFR	Security Level$_{pk}$	Security Level$_u$
PolarKyber512$_{\text{Orig}}$	polar(512, 256, 1/2)	(8, 5)	672	$2^{-140.6135}$	140 131Q	162 152Q
PolarKyber768$_{\text{Orig}}$	polar(512, 256, 1/2)	(9, 4)	992	$2^{-272.1128}$	201 189Q	219 207Q
PolarKyber1024$_{\text{Orig}}$	polar(512, 256, 1/2)	(9, 5)	1312	$2^{-323.1981}$	270 255Q	295 278Q

Polar Codes Applied to Identical Distribution. As presented in Table 1, the parameter distributions for Kyber768 and Kyber1024 are identical, with $\eta_1 = \eta_2 = 2$. In contrast, Kyber512 employs non-identical distributions, where $\eta_1 = 3$ and $\eta_2 = 2$. Through detailed analysis, it becomes evident that this distinction arises from a balance between security and correctness in the Kyber512 scheme. Specifically, when $\eta_1 = \eta_2 = 2$, the classical bit-security strength is 133, which provides only a slight increase over 128, resulting in insufficient security redundancy. Conversely, setting $\eta_1 = \eta_2 = 3$ leads to a DFR of 2^{-121}, which is higher than the 2^{-128} threshold, thereby failing to meet the security level I requirement.

It is clear that the original design intent of the Kyber scheme was to achieve identical distributions, as this simplifies both the implementation and analysis. In this section, we explore the application of polar codes to the Kyber scheme, assuming identical distributions for all relevant parameters. Specifically, **s**, **e**, **r**, $\mathbf{e}_1$, and e_2 are all modeled as following a CBD, ψ_η, with variance $\eta/2$. This modified scheme is referred to as PolarKyber$_{\text{Id}}$.

Table 6 presents the minimum ciphertext size achievable by the PolarKyber$_{\text{Id}}$ scheme, ensuring that the bit-security strength is at least as high as that of the original Kyber scheme. It also illustrates the maximum bit-security strength attainable with a ciphertext size smaller than that of the original Kyber scheme. Specifically, when the bit-security strength matches that of the Kyber scheme, the minimum ciphertext size compressed by the PolarKyber$_{\text{Id}}$ scheme is the same as that of the PolarKyber$_{\text{Orig}}$ scheme, as shown in Table 5. Notably, at security level I, the value of η_2 for the PolarKyber512$_{\text{Id}}$ scheme is set to 3, while for the PolarKyber512$_{\text{Orig}}$ scheme, η_2 is 2. As a result, the DFR for the former is slightly higher. For classical bit-security strength, the PolarKyber512$_{\text{Id}}$ scheme achieves a 4.17% reduction in ciphertext size while increasing the bit-security by 9 bits. Similarly, the PolarKyber768$_{\text{Id}}$ scheme reduces the ciphertext size by 5.88% with a 10-bit increase in bit-security. Furthermore, the PolarKyber1024$_{\text{Id}}$ scheme results in an 8.16% reduction in ciphertext size while enhancing bit-security by 14 bits.

Polar Codes Applied to Variable Distribution. Due to ciphertext compression, the bit-security strength of **u** is higher than that of pk, but the overall security of the scheme is determined by the lower security of pk. In other words, the security mentioned earlier all refer to the security of pk. Table 5 includes the security levels of pk and **u** in the PolarKyber$_{\text{Orig}}$ scheme.

Table 6. Parameters of the PolarKyber$_{\text{Id}}$.

Scheme	ECC	η	(d_u, d_v)	Size(ct)/B	DFR	Security Level
PolarKyber512$_{\text{Id}}$	polar(512, 256, 1/2)	3	(8, 5)	672	$2^{-134.1109}$	140 131Q
		5	(9, 5)	736	$2^{-167.1944}$	149 140Q
PolarKyber768$_{\text{Id}}$	polar(512, 256, 1/2)	2	(9, 4)	992	$2^{-272.1128}$	201 189Q
		3	(9, 5)	1024	$2^{-252.0343}$	211 198Q
PolarKyber1024$_{\text{Id}}$	polar(512, 256, 1/2)	2	(9, 5)	1312	$2^{-323.1981}$	270 255Q
		3	(10, 5)	1440	$2^{-310.6946}$	284 268Q

Therefore, we can adjust the distributions of pk and $\mathbf{u}$, which can be divided into two main approaches. The first approach is to maintain the bit-security strength of pk by keeping the distributions of $\mathbf{s}$ and $\mathbf{e}$ unchanged. The goal is to reduce the bit-security of $\mathbf{u}$ to match that of pk, which involves decreasing the variance of the distributions of $\mathbf{r}$ and $\mathbf{e}_1$ in order to further compress the ciphertext size of PolarKyber$_{\text{Orig}}$ scheme. We refer to this new scheme as PolarKyber$_{\text{Var-one}}$. It is worth noting that when the variance of the distribution is reduced to $\eta < 1$, we use a CBD with fixed Hamming weight h. The second approach is to maintain the bit-security strength of $\mathbf{u}$ by keeping the distributions of $\mathbf{r}$ and $\mathbf{e}_1$ and the value of d_u unchanged. The aim is to enhance the bit-security of pk by increasing the variance of the distributions of $\mathbf{s}$ and $\mathbf{e}$, aligning its bit-security strength closer to that of $\mathbf{u}$, while ensuring that the ciphertext size remains smaller than that of the original Kyber scheme, thereby improving the overall security. We denote this new scheme as PolarKyber$_{\text{Var-two}}$.

PolarKyber$_{Var\text{-}one}$: Maintaining Security with Further Ciphertext Size Compression. Through an exhaustive search of the distributions of $\mathbf{r}$ and $\mathbf{e}_1$, as well as the optimal values for d_u and d_v, as shown in Table 7, we found that at security level V, when the bit-security strength of $\mathbf{u}$ is reduced to match that of pk, the PolarKyber1024$_{\text{Var-one}}$ scheme can achieve further ciphertext compression, reducing the size to 1280 bytes—a reduction of 18.37%. However, further compression is not feasible at security levels I and III.

Table 7. Parameters of the PolarKyber$_{\text{Var-one}}$.

Scheme	ECC	$(h_\mathbf{r}, h_{\mathbf{e}_1})$	(d_u, d_v)	Size(ct)/B	DFR	Security Level$_{pk}$	Security Level$_u$
PolarKyber1024$_{\text{Var-one}}$	polar(512, 256, 1/2)	(82, 110)	(9, 4)	1280	$2^{-266.6267}$	270 255Q	270 255Q

PolarKyber$_{Var\text{-}two}$: Enhancing Security with Ciphertext Size Smaller than the Original Scheme. Through an exhaustive search of the distributions of $\mathbf{s}$ and $\mathbf{e}$, as well as the value of d_v, as detailed in Table 8, we have successfully enhanced the security of pk while ensuring that the ciphertext size remains smaller than that of the original Kyber scheme. In other words, the bit-security strength of the PolarKyber$_{\text{Var-two}}$ scheme exceeds that of the original Kyber scheme, while simultaneously achieving a reduction in ciphertext size. Specifically, for classical bit-security, at security level I, bit-security is increased by 6 bits with an 8.33% reduction in ciphertext size, and by 2 bits with a 12.50% reduction in ciphertext size. At security level III, bit-security is increased by 18 bits, accompanied by an 8.82% reduction in ciphertext size. At security level V, bit-security is increased by 20 bits with a 12.24% reduction in ciphertext size, by 18 bits with a 14.29% reduction in ciphertext size, and by 15 bits with a 16.33% reduction in ciphertext size.

Table 8. Parameters of the PolarKyber$_{\text{Var-two}}$.

Scheme	ECC	(η_s,η_e)	(d_u,d_v)	Size(ct)/B	DFR	Security Level$_{pk}$	Security Level$_u$
PolarKyber512$_{\text{Var-two}}$	polar(512, 256, 1/2)	(3, 6)	(8, 6)	704	$2^{-129.7649}$	146 137Q	162 152Q
		(3, 4)	(8, 5)	672	$2^{-131.1887}$	142 133Q	
PolarKyber768$_{\text{Var-two}}$	polar(512, 256, 1/2)	(2, 9)	(9, 4)	992	$2^{-195.7158}$	219 206Q	219 207Q
PolarKyber1024$_{\text{Var-two}}$	polar(512, 256, 1/2)	(2, 7)	(9, 7)	1376	$2^{-256.9896}$	290 274Q	295 278Q
		(2, 6)	(9, 6)	1344	$2^{-268.4749}$	288 271Q	
		(2, 5)	(9, 5)	1312	$2^{-262.8978}$	285 269Q	

Computational Overhead Introduced by Polar Codes. In our proposed scheme, for different security levels, we utilize polar codes with a code length of 512 bits, a message length of 256 bits, and a code rate of 1/2. For the decoding algorithm of the polar codes, we employ the SC decoding algorithm. The computation of the log likelihood ratio (LLR) using multi-bit modulation technique is provided in [9].

The computational performance of the polar codes is presented in Table 9, where PolarEnc and PolarDec denote the computational performance of the encoding and decoding algorithms, respectively. All results were obtained using our own optimized C implementations, which reflect the best and fastest versions we have developed to date. Performance testing was carried out on a system running Ubuntu 18.04, equipped with an AMD Ryzen 7 PRO 4750U processor clocked at 1.70 GHz and 14.9 GB of memory. According to the performance data, the introduction of polar codes increases the computational cost of the encapsulation process in the Kyber scheme by 0.13%, 0.08%, and 0.06% for different security levels, respectively. For the decapsulation process, the corresponding increases in computational cost are 39.77%, 25.23%, and 17.85%. Notably, the computational overhead introduced by polar codes is relatively lower in scenarios with higher error correction performance requirements.

Note: Compared to traditional public-key encryption schemes such as RSA [29] and elliptic curve cryptography [14], the Kyber scheme offers superior computational performance but suffers from larger ciphertext sizes [11]. Therefore, trading computational cost for reduced ciphertext size represents a reasonable balance.

Table 9. Computational performance of polar codes.

ECC	PolarEnc/cycle	PolarDec/cycle
polar(512, 256, 1/2)	80	32073

5 Conclusion

Our PolarKyber scheme integrates polar coding and the 2-bit modulation technique into the original Kyber scheme to address the issue of limited message space. This scheme is modeled as the PolarKyber-2bit Channel. To analyze the DFR of our scheme, we construct the Relaxed Channel model. Through comparison, we demonstrate that the PolarKyber scheme significantly enhances the DFR, reduces ciphertext size, and increases the security. Specifically, it meets the DFR requirements for all three security levels. The PolarKyber$_{Orig}$ scheme reduces the ciphertext size while maintaining the same bit-security strength as the original Kyber scheme. Furthermore, the PolarKyber$_{Id}$ and PolarKyber$_{Var}$ schemes provide flexibility, allowing for either ciphertext size reduction or an increase in bit-security strength. However, the integration of polar codes into the Kyber scheme introduces computational overhead. In the future, we will focus on optimizing the implementation of the polar code decoding algorithm to mitigate this issue.

6 Discussion

This paper investigates the integration of polar codes into the IND-CPA-secure Kyber public key encryption scheme. The proposed enhancement preserves the original security proofs, concrete security estimates, and DFR analysis. These improvements are also applicable to the corresponding key encapsulation mechanism. Under the IND-CCA security model, the incorporation of polar codes does not affect the established security proofs or concrete security estimates; however, it may impact the estimation of the DFR. Specifically, under the IND-CPA security model, the noise coefficients are typically assumed to be independent and identically distributed (i.i.d.). In contrast, in CCA attacks that exploit decryption failures, this independence assumption no longer holds, making the estimation of the DFR an open problem. Therefore, further investigation into DFR estimation under the IND-CCA security model is necessary to comprehensively assess the overall security implications of integrating polar codes.

References

1. Report on the security of lwe: Improved dual lattice attack (2022). https://api.semanticscholar.org/CorpusID:251600824
2. Recommendations for key-encapsulation mechanisms (initial public draft). Technical report 800-227, National Institute of Standards and Technology (2025). https://csrc.nist.gov/pubs/sp/800/227/ipd. Initial Public Draft

3. Albrecht, M.R., Player, R., Scott, S.: On the concrete hardness of learning with errors. J. Math. Cryptol. **9**(3), 169–203 (2015)
4. Alkim, E., Ducas, L., Pöppelmann, T., Schwabe, P.: Newhope without reconciliation. Cryptology ePrint Archive (2016)
5. Alnahawi, N., Müller, J., Oupický, J., Wiesmaier, A.: SoK: post-quantum TLS handshake. Cryptology ePrint Archive (2023)
6. Arikan, E.: Channel polarization: a method for constructing capacity-achieving codes for symmetric binary-input memoryless channels. IEEE Trans. Inf. Theory **55**(7), 3051–3073 (2009)
7. Avanzi, R., et al.: Algorithm specifications and supporting documentation (version 3.0). Technical report, Submission to the NIST postquantum project (2020)
8. Avanzi, R., et al.: CRYSTALS-Kyber: submission to the NIST post-quantum cryptography standardization project (round 3). https://csrc.nist.gov/Projects/post-quantum-cryptography/post-quantum-cryptography-standardization/round-3-submissions (2020)
9. Bocharova, I.E., Hollmann, H.D., Khathuria, K., Kudryashov, B.D., Skachek, V.: Coding with cyclic pam and vector quantization for the RLWE/MLWE channel. In: 2022 IEEE International Symposium on Information Theory (ISIT), pp. 666–671. IEEE (2022)
10. Bos, J., et al.: Crystals-kyber: a CCA-secure module-lattice-based KEM. In: 2018 IEEE European Symposium on Security and Privacy (EuroS&P), pp. 353–367. IEEE (2018)
11. Demir, E.D., Bilgin, B., Onbasli, M.C.: Performance analysis and industry deployment of post-quantum cryptography algorithms. arXiv preprint arXiv:2503.12952 (2025)
12. Fritzmann, T., Pöppelmann, T., Sepulveda, J.: Analysis of error-correcting codes for lattice-based key exchange. In: Cid, C., Jacobson Jr, M. (eds.) SAC 2018. LNCS, vol. 11349, pp. 369–390. Springer, Cham (2019). https://doi.org/10.1007/978-3-030-10970-7_17
13. ISO/IEC: ISO/IEC 18033-2:2006/DAMD 2. information technology – security techniques – encryption algorithms – part 2: Asymmetric ciphers – amendment 2 (2025)
14. Koblitz, N.: Elliptic curve cryptosystems. Math. Comput. **48**(177), 203–209 (1987)
15. Laarhoven, T., Mosca, M., Van De Pol, J.: Finding shortest lattice vectors faster using quantum search. Des. Codes Crypt. **77**(2), 375–400 (2015)
16. Langlois, A., Stehlé, D.: Worst-case to average-case reductions for module lattices. Des. Codes Crypt. **75**(3), 565–599 (2015)
17. Lee, E., Kim, Y.S., No, J.S., Song, M., Shin, D.J.: Modification of FrodoKEM using gray and error-correcting codes. IEEE Access **7**, 179564–179574 (2019)
18. Liu, L., Yan, Y., Ling, C., Wu, X.: Construction of capacity-achieving lattice codes: polar lattices. IEEE Trans. Commun. **67**(2), 915–928 (2018)
19. Liu, S., Sakzad, A.: Lattice codes for crystals-kyber. arXiv preprint arXiv:2308.13981 (2023)
20. Lu, X., et al.: LAC: practical ring-LWE based public-key encryption with byte-level modulus. Cryptology ePrint Archive (2018)
21. Lyu, S., Liu, L., Ling, C., Lai, J., Chen, H.: Lattice codes for lattice-based PKE. Des. Codes Crypt. **92**(4), 917–939 (2024)
22. Lyubashevsky, V., Peikert, C., Regev, O.: On ideal lattices and learning with errors over rings. In: Gilbert, H. (ed.) EUROCRYPT 2010. LNCS, vol. 6110, pp. 1–23. Springer, Heidelberg (2010). https://doi.org/10.1007/978-3-642-13190-5_1

23. Maringer, G., Puchinger, S., Wachter-Zeh, A.: Information-and coding-theoretic analysis of the RLWE/MLWE channel. IEEE Trans. Inf. Forensics Secur. **18**, 549–564 (2022)
24. Micciancio, D., Regev, O.: Lattice-based cryptography. In: Bernstein, D.J., Buchmann, J., Dahmen, E. (eds.) Post-Quantum Cryptography, pp. 147–191. Springer, Heidelberg (2009). https://doi.org/10.1007/978-3-540-88702-7_5
25. NIST: Module-lattice-based key encapsulation mechanism standard. Federal Information Processing Standards Publication (FIPS) NIST FIPS 203 IPD (2023)
26. Papadopoulos, I., Wang, J.: Polar codes for module-LWE public key encryption: the case of kyber. Cryptography **7**(1), 2 (2023)
27. van Poppelen, A.: Cryptographic decoding of the Leech lattice. Master's thesis (2016)
28. Regev, O.: On lattices, learning with errors, random linear codes, and cryptography. J. ACM (JACM) **56**(6), 1–40 (2009)
29. Rivest, R.L., Shamir, A., Adleman, L.: A method for obtaining digital signatures and public-key cryptosystems. Commun. ACM **21**(2), 120–126 (1978)
30. Saliba, C., Luzzi, L., Ling, C.: Error correction for FrodoKEM using the Gosset lattice. arXiv preprint arXiv:2110.01740 (2021)
31. Shannon, C.E.: A mathematical theory of communication. Bell Syst. Tech. J. **27**(3), 379–423 (1948)
32. Tal, I., Vardy, A.: How to construct polar codes. IEEE Trans. Inf. Theory **59**(10), 6562–6582 (2013)
33. Wang, J., Ling, C.: How to construct polar codes for ring-LWE-based public key encryption. Entropy **23**(8), 938 (2021)
34. Wang, J., Ling, C.: Polar coding for ring-LWE-based public key encryption. Cryptogr. Commun. **15**(2), 397–431 (2023)

Lion: A New Ring Signature Construction from Lattice Gadget

Yanting Li[1], Pingbin Luo[1], Xinjian Chen[1], and Qiong Huang[1,2(✉)]

[1] South China Agricultural University, Guangzhou, China
{liyant,robin2022,xchen}@stu.scau.edu.cn, qhuang@scau.edu.cn
[2] Guangdong University of Finance, Guangzhou, China

Abstract. Ring signature schemes are being used in many scenarios that require both accreditation and privacy of the user's identity. For example, ring confidential transaction protocol in Monero uses ring signature as an underlying scheme. Almost all ring signatures deployed in practice are based on number-theoretic assumptions which are vulnerable to quantum computers. Existing efficient ring signature schemes rely on NTRU lattices or structured lattices which are more susceptible to attacks than generic lattices. We propose a new generic construction of ring signature based on generic lattices and give an instantiation called LION, which has the best security and strikes the best balance between computational efficiency and storage overheads. Compared with RAPTOR, which is almost as practical as number-theoretic-based ring signatures, LION has comparable computational efficiency and smaller signature size(roughly 1.07KB per user). Based on standard hardness assumptions, LION follows the strongest security definitions of ring signature in random oracle model. In addition, LION does not need to rely on trusted third parties.

Keywords: Ring signature · Generic lattices · Lattice gadget

1 Introduction

Ring signature is originally proposed to cater for the scenario of whistle blowing [1]. A ring signature on a message can convince an arbitrary verifier that the message was signed by one of the ring members but does not reveal the signer's identity. Thus ring signatures can protect the identity of the signer while giving credence to the message. This implies two properties of ring signatures: anonymity and unforgeability (see Appendix A). Ring signature schemes are being used in many scenarios that require both accreditation and privacy of the user's identity. The most classic scenarios are electronic voting systems and cryptocurrency systems. For example, in ring confidential transaction protocol used in Monero, ring signature is used as an underlying scheme [2].

Y. Li and P. Luo—Equal contribution.

Almost all ring signatures deployed in practice are based on number-theoretic assumptions which are vulnerable to quantum computing attacks [3,4]. Since breaking a random instance of a lattice problem is as hard as solving the worst-case instance, lattice-based cryptography has become one of the most promising candidates for the quantum apocalypse. In order to deploy the lattice-based ring signatures in practice, efficiency and security are important factors to consider. To date, there exist a number of lattice-based ring signature schemes, with the efficiency progressively improved.

1.1 Related Work

We first recall the most classical two ring signature schemes that inspired the design of subsequent schemes. In 2001, Rivest, Shamir and Tauman [1] introduced a generic construction(RST) of ring signatures. RST construction is based on one-way trapdoor permutations, where each user's secret key is the trapdoor corresponding to his public key. Any user can choose any set of possible signers including himself to form a ring, and use his secret key and the others' public keys to sign any message, without getting other ring member's approval or assistance. The signer uses his secret key to invert his permutation, and uses the inverse to construct the ring signature. In 2002, Abe, Ohkubo and Suzuku [5] proposed a new generic construction(AOS) which can make use of both hash-and-sign signature and three-move sigma-protocol-based signature. It translates 1-out-of-n proof into a ring signature. AOS yields shorter signatures than RST by setting computation in a circular chain. Both RST and AOS are secure in random oracle model and the signature size are linear to the ring size.

Among lattice based ring signature schemes, RAPTOR [6], ESLL19 [7], and DUALRING [8] have the most efficient running time. These three schemes represent three different ways of ring signature construction. Firstly, ESLL19 has signature size logarithmic with the ring size, by adapting its efficient one-out-of-many proof to a ring signature. But the one-out-of-many proof relies on zero-knowledge proof technique, which is not practically efficient. Zero-knowledge proofs are always intricate and thus imply huge computational overhead. As far as we know, almost all logarithmic size ring signature schemes [9–11] are based on zero-knowledge proof technique, and therefore not efficient enough in practical applications. Secondly, DUALRING [8] has linear signature size to the ring size (although the DL-based setting can be compressed into logarithmic size via an sum argument of knowledge system, the lattice-based setting has linear size). Inspired by AOS, DUALRING contains two circles: a circle of commitments and a circle of challenges, and the ring signature is composed of n challenges and a single response. The signature size can be significantly shortened by being instantiated with cryptosystems that have a small challenge size and a large response size. But DUALRING's construction needs to use a common V_1 function in the Verification algorithm, which may pose a security threat: if the V_1 function has a trapdoor, then DUALRING will be insecure. Lastly, RAPTOR [6] has linear signature size to the ring size and has low running time overhead (verification takes around 32 ms and signature generation takes 40 ms when the ring size is 50). RAPTOR's running time is almost as efficient as RST ring signatures. Inspired by RST, RAPTOR proposes a new primitive to make the RST structure compatible with lattices. The new primitive is called Chameleon Hash Plus (CH^+) which has a trapdoor. With the trapdoor, it is easy to find collisions for any hash value. By using FALCON [12] to instantiate CH^+, RAPTOR obtains a small ring signature size (roughly 1.3 KB per user). But FALCON is based on NTRU lattices that are more susceptible to attacks than generic lattices. There exist a series of approaches on attacking NTRU lattices, such as search attack [13], dimension-reduction attack [14], subfield attack [15] and Hybrid attack [16]. In addition, there are also some attacks directly targeting FALCON [17,18]. Although RAPTOR is resistant to search attacks, it may be vulnerable to the other attacks mentioned above. ESLL19 and DUALRING are based on module lattices(M-SIS/M-LWE), whose algebraic structure brings better efficiency but more potential risks [19] than generic lattices. Therefore, it is necessary to construct an efficient ring signature scheme over generic lattices for better security. And for practicality, the building blocks used in a ring signature scheme should be easy to implement.

1.2 Our Contributions

We present a new generic construction of ring signature from lattice gadget, and give an instantiation based on generic lattices. We call our instantiation LION because it is more secure but not less efficient. The main building blocks of our generic construction are *trapdoor* and *preimage sampling*, which can be efficiently implemented with gadget in practice. The preimage sampling algorithm can be divided into two phases: online phase and offline phase. The offline phase can be computed in advance to improve efficiency. To reduce the public key size, we make the random matrix $\bar{\mathbf{A}}$ be used commonly in public key generation. The matrix $\bar{\mathbf{A}}$ is expanded from a random *seed* by a pseudorandom generator PRG, which allows us to remove the trusted third party from the setup phase. This further enhances the security of ring signature. The comparison of signature size of our LION and other lattice-based ring signature schemes mentioned above is shown in Table 2. The comparison of running time is shown in Table 3. The signature size of LION is smaller than RAPTOR and smaller than ESLL19 for ring size ≤ 385. LION provides an optimal balance between computational efficiency and storage overheads. Based on generic lattices, instead of NTRU lattices or structured lattices, LION has the best security. For ring size ≤ 385, LION is the optimal ring signature scheme.

Technical Overview: Inspired by the construction of trapdoors proposed by Micciancio and Peikert [20], we propose a new generic construction of ring signature based on gadget matrix $\mathbf{G}$. The public matrix with trapdoor $\mathbf{R}$ is $\mathbf{A} = [\bar{\mathbf{A}}|\mathbf{G} - \bar{\mathbf{A}}\mathbf{R}]$. For a ring member i, it uses $\mathbf{R}_i$ as secret key, and $\mathbf{A}_i$ as public key. For l ring members, the key pairs are $\{(\mathbf{R}_i, \mathbf{A}_i = [\bar{\mathbf{A}}_i|\mathbf{G} - \bar{\mathbf{A}}_i\mathbf{R}_i])\}_1^l$. To reduce the size of public key, we use the $\bar{\mathbf{A}}$ as a common parameter for all ring members, and the public keys become $\{\widehat{\mathbf{A}}_i = \mathbf{G} - \bar{\mathbf{A}}\mathbf{R}_i\}_1^l$ that are still random while $\mathbf{R}_i$'s entries are small and random. In this way, the dimension of the public key is reduced. $\bar{\mathbf{A}}$ should be generated by a trusted third party in the setup phase. What if the third party becomes untrustworthy? To remove the trusted third party from the setup phase, we use a pseudorandom generator PRG to generate $\bar{\mathbf{A}}$ to avoid $\bar{\mathbf{A}}$ having a trapdoor. Then in the setup phase, only a common random seed $seed_{\bar{\mathbf{A}}}$ and a safe PRG are needed. Generated from a random seed and a secure PRG which is widely used in practice (such as SHAKE256), the probability of $\bar{\mathbf{A}}$ having a trapdoor is negligible. Thus, the third party can hardly embed a trapdoor in $\bar{\mathbf{A}}$.

To sign a message, the signer firstly generates fake "signatures" $\{\mathbf{x}_i\}_{i \neq \pi}$ for other ring members, and uses secret key to sample a preimage $\mathbf{x}_\pi$. For security, the distributions of all the $\{\mathbf{x}_i\}_1^l$ should be indistinguishable. The signer computes $\mathbf{u}_\pi := \mathsf{H}(salt, M, L_{pk}) - \sum_{i \neq \pi} \widehat{\mathbf{A}}_i \mathbf{x}_i \mod q$, and uses secret key $\mathbf{R}_\pi$ to sample a preimage $\mathbf{x}$ such that $[\bar{\mathbf{A}}|\widehat{\mathbf{A}}_\pi]\mathbf{x} = \mathbf{u}_\pi \mod q$. Then parse $\mathbf{x}$ as $\mathbf{x}_0$ and $\mathbf{x}_\pi$ to make the dimension of $\mathbf{x}_\pi$ the same as column dimension of $\widehat{\mathbf{A}}_\pi$. Thus the signature size is shortened because the dimension of $\widehat{\mathbf{A}}_\pi$ has been reduced. The ring signature of the ring $L_{pk} = \{\widehat{\mathbf{A}}_i\}_1^l$ on message M is $\sigma = (salt, \{\mathbf{x}_0; \mathbf{x}_1, ..., \mathbf{x}_l\})$.

To verify a ring signature, the verifier firstly checks whether $\mathbf{x}_0$ and $\{\mathbf{x}_i\}_1^l$ are all short, and then checks whether $\mathbf{x}_0$ and $\{\mathbf{x}_i\}_1^l$ satisfy the equation $\mathsf{H}(salt, M, L_{pk}) = \bar{\mathbf{A}}\mathbf{x}_0 + \sum_{i=1}^l \widehat{\mathbf{A}}_i \mathbf{x}_i \mod q$. Note that without the secret key, it is hard to find such a ring signature that satisfies the above verification conditions.

Our new generic construction is more secure because it is based on generic lattice and does not require a trusted third party. Besides, our construction is efficient. All the building blocks of our construction have efficient implementations in practice, which ensures the practicality. Since $\bar{\mathbf{A}}$ is common and is expanded from a seed, the public keys size of ring members is reduced, and the ring signature size becomes smaller correspondingly.

2 Preliminaries

2.1 Notations

Let λ denote the security parameter. Let $\mathbb{Z}_q$ be the integers in $[-\frac{q}{2}, \frac{q}{2})$. $\{pk_i\}_1^l$ denotes the set $\{pk_1, ..., pk_l\}$. $[l]$ denotes the set $\{1, ..., l\}$. H denotes hash function $\{0,1\}^* \rightarrow \mathbb{Z}_q^n$. For a discrete Gaussian distribution D, let $x \leftarrow D$ represents the sample x drawn from D. Let $a \xleftarrow{\$} S$ denote the sample $a \leftarrow U(S)$.

Lower-case bold letters (e.g. $\mathbf{x} = (x_1, x_2, ..., x_n)$) denotes the column vectors. Matrices are denoted by upper-case bold letters (e.g. $\mathbf{A}$). By default we use ℓ_2-norm of a vector. The largest singular value of matrix $\mathbf{M}$ is denoted as $s_1(\mathbf{M}) = \max_{\mathbf{x} \neq \mathbf{0}} \frac{\|\mathbf{M}\mathbf{x}\|}{\|\mathbf{x}\|}$. $\Sigma \succ 0$ denotes that Σ is positive definite.

2.2 Lattice and Hardness Assumptions

Lattice. An n-dimension lattice Λ of rank $k \leq n$ is a discrete additive subgroup of $\mathbb{R}^n$. Given k linearly independent basis vectors $\mathbf{B} = \{\mathbf{b}_1, ..., \mathbf{b}_k \in \mathbb{R}^n\}$, the lattice generated by $\mathbf{B}$ is

$$\Lambda(\mathbf{B}) = \Lambda(\mathbf{b}_1, ..., \mathbf{b}_k) = \left\{ \sum_{i=1}^{k} x_i \cdot \mathbf{b}_i | x_i \in \mathbb{Z} \right\} = \left\{ \mathbf{Bx} | \mathbf{x} \in \mathbb{Z}^k \right\}.$$

For $n, m \in \mathbb{N}$, a modulus $q \geq 2$, given an arbitrary matrix $\mathbf{A} \in \mathbf{Z}_q^{n \times m}$, we often use q-ary lattice $\Lambda_q^{\perp}(\mathbf{A})$ and its coset $\Lambda_{\mathbf{u}}^{\perp}(\mathbf{A})$, defined as follows:

$$\Lambda_q^{\perp}(\mathbf{A}) = \{\mathbf{x} \in \mathbb{Z}^m : \mathbf{A} \cdot \mathbf{x} = \mathbf{0} \bmod q\},$$

$$\Lambda_{\mathbf{u},q}^{\perp}(\mathbf{A}) = \{\mathbf{x} \in \mathbb{Z}^m : \mathbf{A} \cdot \mathbf{x} = \mathbf{u} \bmod q\} = \Lambda_q^{\perp}(\mathbf{A}) + \mathbf{c},$$

where $\mathbf{c}$ is an integral solution such that $\mathbf{Ac} = \mathbf{u} \bmod q$.

Hardness Assumptions. Short Integer Solution(SIS) problem [21] and Learning With Errors(LWE) problem [22] are two hard problems frequently used in lattice-based cryptography constructions. We recall the SIS problem and its Hermite normal form(HNF) as well as the LWE problem below. The security of our scheme relies on the hardness of these problems.

Definition 1 (SIS). *For any $n, m, q \in \mathbb{Z}$ and $\beta \in \mathbb{R}$, define the short integer solution problem $\mathsf{SIS}_{n,m,q,\beta}$ as follows: Given $\mathbf{A} \xleftarrow{\$} \mathbb{Z}_q^{n \times m}$, find a non-zero vector $\mathbf{x} \in \mathbb{Z}^m$ such that $\|\mathbf{x}\| \leq \beta$, and $\mathbf{Ax} = \mathbf{0} \bmod q$.*

Lemma 1 (Hardness of SIS [21]). *For any $m = poly(n)$, any $\beta = poly(n) > 0$, if $q \geq \beta \cdot \omega(\sqrt{n \log n})$, then $\mathsf{SIS}_{n,m,q,\beta}$ and $\mathsf{ISIS}_{n,m,q,\beta}$ are at least as hard as the standard worst-case lattice problem GapSVP_γ (Shortest Vector Problem) and SIVP_γ (Shortest Independent Vector Problem), for some approximation factor $\gamma = \beta \cdot poly(n)$.*

Distribution of SIS Samples is statistically close to $U(\mathbb{Z}_q^{n \times m} \times \mathbb{Z}_q^n)$ when $\beta \gg q^{n/m}$ [23]. In Hermite normal form (HNF), the public matrix A is of the form $[\mathbf{I}_n | \mathbf{A}']$. The HNF version of SIS are denoted as HNF.SIS, which is as hard as the standard version.

Definition 2 (LWE). *Let $n, m \in \mathbb{N}$ and $q \geq 2$, χ_s, χ_e be the distribution over $\mathbb{Z}$, define the learning with errors problem $\mathsf{LWE}_{n,m,q,\chi_s,\chi_e}$ as fallows: Given $\mathbf{A} \xleftarrow{\$} \mathbb{Z}_q^{n \times m}, \mathbf{s} \leftarrow \chi_s, \mathbf{e} \leftarrow \chi_e, \mathbf{y} = \mathbf{As} + \mathbf{e} \in \mathbb{Z}_q^n$, finding $(\mathbf{s}, \mathbf{e})$.*

Lemma 2 (Hardness of LWE [22]). *For any $m = poly(n), q \leq 2^{poly(n)}$, let $\chi_s = U(\mathbb{Z}_q), \chi_e = D_{\mathbb{Z},s}$ where $s \geq 2\sqrt{n}$, then $\mathsf{LWE}_{n,m,q,\chi_s,\chi_e}$ are at least as hard as the standard worst-case lattice problem SIVP_γ, for some approximation factor $\gamma = O(nq/s)$.*

The distribution of LWE sample $(\mathbf{A}, \mathbf{y} := \mathbf{As} + \mathbf{e})$ is computationally indistinguishable from $U(\mathbb{Z}_q^{n \times m} \times \mathbb{Z}_q^n)$ [22]. According to [24], sampling $\mathbf{s}$ from the error distribution does not change the hardness of LWE.

2.3 Discrete Gaussian on Lattices

Smoothing Parameter. The smoothing parameter $\eta_\epsilon(\Lambda)$ is the smallest real $s > 0$ such that $\rho_{1/s}(\Lambda^* \setminus \{\mathbf{0}\}) \leq \epsilon$. Above the smoothing parameter, the basic shape of the Gaussian distribution does not change from lattice Λ to its coset $\Lambda + \mathbf{c}$. Besides, in integer lattice, the Gaussian distribution over it is statistically close to the uniform distribution over integer space. The upper bound of smoothing parameter is $\omega(\sqrt{\log(2n(1+1/\epsilon))/\pi})$.

Lemma 3 ([25], Lemma 4.2). *For any $s \geq \eta_\epsilon(\Lambda_q^\perp(\mathbf{G}))$, $\mathbf{G} = \mathbf{I}_n \otimes \mathbf{g}^t$ with $\mathbf{g}^t = (1, b, ..., b^{k-1}), k = \lceil \log_b(q) \rceil, b \geq 2$. The following two distributions are statistically close:*
  1. *first sample $\mathbf{u} \leftarrow U(\mathbb{Z}_q^n)$, then sample $\mathbf{x} \leftarrow D_{\Lambda_{\mathbf{u},q}^\perp(\mathbf{G}),s}$, output $(\mathbf{x}, \mathbf{u})$;*
  2. *first sample $\mathbf{x} \leftarrow D_{\mathbb{Z}^{nk},s}$, then compute $\mathbf{u} = \mathbf{Gx} \mod q$, output $(\mathbf{x}, \mathbf{u})$.*

Linear Transformation of Discrete Gaussians. Regarding the linear transformation $\mathbf{T}$ of a discrete Gaussian, as long as the original discrete Gaussian(over an integer space) is smooth enough in $\Lambda_q^\perp(\mathbf{T})$, the transformed distribution is statistically close to another discrete Gaussian over transformed integer space.

Lemma 4 ([25], Lemma 2.8). *Let $\mathbf{T} \in \mathbb{Z}^{n \times m}$ such that $\mathbf{T}\mathbb{Z}^m = \mathbb{Z}^n$, $\Lambda_q^\perp(\mathbf{T}) = \{\mathbf{x} \in \mathbb{Z}^m : \mathbf{Tx} = \mathbf{0} \mod q\}$, $\Sigma = \mathbf{TT}^t$. For $\epsilon \in (0, 1/2)$, $\hat{\epsilon} = \epsilon + O(\epsilon^2)$, $r \geq \eta_\epsilon(\Lambda_q^\perp(\mathbf{T}))$, the max-log distance between $\mathbf{T} \cdot D_{\mathbb{Z}^m, r}$ and $D_{\mathbb{Z}^n, r\sqrt{\Sigma}}$ is at most $4\hat{\epsilon}$.*

3 Our Construction

We present a new generic construction of ring signature scheme in this section. Inspired by the idea of gadget matrix in [20], our ring signature scheme is based on generic lattice, instead of NTRU lattice or structured lattice. This makes our scheme more secure. We first review the trapdoor construction and preimage sampling with gadget matrix in [20]. Then we describe our new generic construction of ring signatures and finally analyse its security.

3.1 Trapdoor and Preimage Sampling with Gadget Matrix

Preimage sampling is finding a preimage of a one-way trapdoor function. For example, given $(\mathbf{A}, \mathbf{u})$, preimage sampling is using the trapdoor of $\mathbf{A}$ to find a preimage $\mathbf{x}$, such that $\mathbf{Ax} = \mathbf{u}$. The algorithm used to implement preimage sampling is called SamplePre. Gadget matrix has a simple form $\mathbf{G} = \mathbf{I}_n \otimes \mathbf{g}^t$ with $\mathbf{g}^t = (1, b, ..., b^{k-1}), k = \lceil \log_b q \rceil, b \geq 2$, which can simplify the preimage sampling. Micciancio and Peikert [20] converted finding $\mathbf{x}$ for $\mathbf{A}\mathbf{x} = \mathbf{u}$ into finding $\mathbf{x}'$ for $\mathbf{G}\mathbf{x}' = \mathbf{u}$, which significantly improves the efficiency of preimage sampling. The linear transformation matrix $\mathbf{T}$ from $\mathbf{A}$ to $\mathbf{G}$ (i.e. $\mathbf{AT} = \mathbf{G}$) is the key to constructing $\mathbf{x}$ from $\mathbf{x}'$.

For any matrix $\bar{\mathbf{A}}$ with randomly chosen entries and matrix $\mathbf{R}$ with randomly chosen small entries such that $\bar{\mathbf{A}}\mathbf{R}$ satisfies the hardness of SIS problem, Micciancio and Peikert generated a public matrix $\mathbf{A} = [\bar{\mathbf{A}}|\mathbf{G} - \bar{\mathbf{A}}\mathbf{R}]$. $\mathbf{T} := \begin{bmatrix}\mathbf{R}\\\mathbf{I}\end{bmatrix}$ is the linear transformation matrix such that $\mathbf{AT} = \mathbf{G}$. Thus $\mathbf{R}$ is the corresponding trapdoor of $\mathbf{A}$. Preimage sampling algorithm SampPre requires only $(\mathbf{R}, r \geq \eta_\epsilon(\Lambda_q^\perp(\mathbf{G})), s \geq r \cdot s_1(\mathbf{T}), \mathbf{u})$ as inputs, and efficiently outputs a preimage $\mathbf{x}$ such that $\mathbf{A}\mathbf{x} = \mathbf{u} \mod q$ as follows: $\mathbf{p} \leftarrow D_{\mathbb{Z}^m, \sqrt{\Sigma_\mathbf{p}}}$ with $\Sigma_\mathbf{p} := s^2\mathbf{I}_m - r^2\mathbf{T}\mathbf{T}^t$, $\mathbf{x}' \leftarrow D_{\Lambda_{\mathbf{u}',q}^\perp(\mathbf{G}),r}$ with $\mathbf{u}' = \mathbf{u} - \mathbf{A}\mathbf{p}$, $\mathbf{x} := \mathbf{p} + \mathbf{T}\mathbf{x}' \mod q$. The distribution of $\mathbf{x} := \mathbf{p} + \mathbf{T}\mathbf{x}' \mod q$ is statistically close to spherical discrete Gaussian distribution $D_{\mathbb{Z}^m,s}$.

3.2 New Generic Construction

We describe our new generic construction of ring signature based on gadget matrix in this section. Let $L_{pk} = \{pk_1, ...pk_l\} = \{\mathbf{A}_i\}_1^l$ be a ring, the signer is denoted by the subscript π, $\pi \in [l]$. Analogously to [20], we can use the preimage as a signature. In the ring L_{pk}, each member can obtain a preimage $\mathbf{x}$ by using his secret key sk. The signer can obtain a "real" preimage using algorithm SampPre, and simulate "confusing" preimage by randomly choosing vectors from $D_{\mathbb{Z}^m,s}$. The signing process can be constructed as follows:

1: (*simulate the "confusing" preimages*) For all $i \in [l], i \neq \pi$, choose $\mathbf{x}_i \leftarrow D_{\mathbb{Z}_q^m,s}$ such that $\|\mathbf{x}_i\| \leq \beta$, and compute $\mathbf{u}_i = \mathbf{A}_i\mathbf{x}_i \mod q$.
2: (*generate the signer's "real" preimage*) $salt \xleftarrow{\$} \mathbb{Z}^n$, $\mathbf{u}_\pi := \mathsf{H}(salt, M, L_{pk}) - \sum_{i \neq \pi} \mathbf{u}_i \mod q$, sample $\mathbf{x}_\pi \leftarrow \mathsf{SampPre}(\mathbf{A}_\pi, \mathbf{R}_\pi, \mathbf{u}_\pi, r, s)$.
3: Return $\sigma = (salt, L_{pk}, \{\mathbf{x}_i\}_1^l)$ as the ring signature on the message M.

To verify a ring signature σ on the message M, the verification process is as follows:

1: Parse $\sigma = (salt, L_{pk}, \{\mathbf{x}_i\}_1^l)$.
2: Check whether for all $\{\mathbf{x}_i\}_1^l$, $\|\mathbf{x}_i\| \leq \beta$; return $reject$ if not.
3: Check whether $\sum_{i=1}^l \mathbf{A}_i\mathbf{x}_i = \mathsf{H}(salt, M, L_{pk})$; return $reject$ if not.
4: Return $accept$.

In step 3 of the above verification process, we can observe that

$$\mathsf{H}(salt, M, L_{pk}) := \sum_{i=1}^l \mathbf{A}_i\mathbf{x}_i$$

$$= \sum_{i=1}^{l} [\bar{\mathbf{A}}_i | \mathbf{G} - \bar{\mathbf{A}}_i \mathbf{R}_i] \mathbf{x}_i$$

$$= \sum_{i=1}^{l} [\bar{\mathbf{A}}_i | \mathbf{G} - \bar{\mathbf{A}}_i \mathbf{R}_i] \begin{bmatrix} \mathbf{x}_{i0} \\ \mathbf{x}_{i1} \end{bmatrix} \ (where \ \mathbf{x}_i = \begin{bmatrix} \mathbf{x}_{i0} \\ \mathbf{x}_{i1} \end{bmatrix})$$

$$= \sum_{i=1}^{l} \bar{\mathbf{A}}_i \mathbf{x}_{i0} + \sum_{i=1}^{l} (\mathbf{G} - \bar{\mathbf{A}}_i \mathbf{R}_i) \mathbf{x}_{i1}$$

$$= \sum_{i=1}^{l} \bar{\mathbf{A}}_i \mathbf{x}_{i0} + \sum_{i=1}^{l} \widehat{\mathbf{A}}_i \mathbf{x}_{i1}, \ where \ \widehat{\mathbf{A}}_i = \mathbf{G} - \bar{\mathbf{A}}_i \mathbf{R}_i$$

What if we make all the $\bar{\mathbf{A}}_i$ be a common $\bar{\mathbf{A}}$? According to Lemma 1, for randomly chosen matrices $\mathbf{R}_i$ with small enough entries, it is hard to find any $\mathbf{R}_i$ in the SIS samples $\{(\bar{\mathbf{A}}, \bar{\mathbf{A}}\mathbf{R}_i)\}_1^l$. The distribution of $\{(\bar{\mathbf{A}}, \bar{\mathbf{A}}\mathbf{R}_i)\}_1^l$ is statistically close to random samples from $U(\mathbb{Z}_q^{n \times \bar{m}} \times \mathbb{Z}_q^{n \times nk})$. The distribution of $\{\widehat{\mathbf{A}}_i = \mathbf{G} - \bar{\mathbf{A}}\mathbf{R}_i\}_1^l$ is statistically close to $U(\mathbb{Z}_q^{n \times nk})$. Then $\{\widehat{\mathbf{A}}_i\}_1^l$ are indistinguishable and simulatable without secret key. So we can let the matrix $\bar{\mathbf{A}}$ be common to all members of the ring and add matrix $\bar{\mathbf{A}}$ to the public parameter pp. Thus the public keys of ring members become matrices $\{\widehat{\mathbf{A}}_i = \mathbf{G} - \bar{\mathbf{A}}\mathbf{R}_i\}_1^l$. In this way, we reduce the size of public keys, and the dimension of signature correspondingly.

Then the step 3 of above verification process becomes:

$$\mathsf{H}(salt, M, L_{pk}) := \sum_{i=1}^{l} \bar{\mathbf{A}} \mathbf{x}_{i0} + \sum_{i=1}^{l} \widehat{\mathbf{A}}_i \mathbf{x}_{i1}.$$

However, what if the common matrix $\bar{\mathbf{A}}$ also has a trapdoor? The ring signature scheme would be insecure. With the trapdoor of $\bar{\mathbf{A}}$, one can easily forge a valid ring signature although he's not a ring member. To avoid the above situation, our $\bar{\mathbf{A}}$ is expanded from a random seed by a pseudorandom generator PRG, such as AES256 and SHAKE256 [26]. In this way, the probability of $\bar{\mathbf{A}}$ being maliciously embedded in a trapdoor is negligible.

We now introduce our generic ring signature scheme formally. In Setup (Algorithm 1), ring signature system takes security parameter as input, and generates some public parameters, also called system parameters, which will be used as implicit inputs in other algorithms. Public parameter pp includes a seed $seed_{\bar{\mathbf{A}}}$ and pseudorandom generator PRG used to generated $\bar{\mathbf{A}}$, a gadget matrix $\mathbf{G}$ used to simplify key generation and preimage sampling, a *collision resistant* hash function H used to hash a message to a number field, and r, s used as Gaussian parameters in preimage sampling. Let $\max\{s_1(\mathbf{T})\}$ denotes the maximum $s_1(\mathbf{T})$ over all eligible $\mathbf{T} := [\mathbf{R}^t | \mathbf{I}]^t$.

Algorithm 1. Setup (1^λ)

input: λ is a security parameter
output: public parameters $pp = (seed_{\bar{\mathbf{A}}}, q, \mathbf{G}, \mathsf{PRG}, \mathsf{H})$
1: $seed_{\bar{\mathbf{A}}} \xleftarrow{\$} \{0,1\}^{256}$
2: integer modulus q depends on security parameter
3: choose a gadget matrix $\mathbf{G} \in \mathbb{Z}_q^{n \times \widehat{m}}$
4: $\mathsf{PRG}(seed) \to \mathbb{Z}_q^{n \times \bar{m}}, \mathsf{H}: \{0,1\}^* \to \mathbb{Z}_q^n$
5: $r \geq \eta_\epsilon(\Lambda_q^\perp(\mathbf{G})), s \geq r \cdot \max\{s_1(\mathbf{T})\}$
6: return $pp = (seed_{\bar{\mathbf{A}}}, q, r, s, \mathbf{G}, \mathsf{PRG}, \mathsf{H})$

Running KeyGen (Algorithm 2), each ring member obtains his key pair (sk, pk). The distributions of public keys $\{\widehat{\mathbf{A}}_i\}_1^l$ are indistinguishable by randomly choosing secret key $\mathbf{R}_i$ with small entries. According to the hardness of SIS problem, it is hard to retrieve any secret key from public keys.

Algorithm 2. KeyGen (pp)

input: pp is the public parameters
output: a key pair (sk, pk), secret key $sk = \mathbf{R} \in \mathbb{Z}_q^{\bar{m} \times \widehat{m}}$, public key $pk = \widehat{\mathbf{A}} \in \mathbb{Z}_q^{n \times \widehat{m}}$
1: $\bar{\mathbf{A}} \leftarrow \mathsf{PRG}(seed_{\bar{\mathbf{A}}})$, $\mathbf{R} \xleftarrow{\$} \mathbb{Z}_q^{\bar{m} \times \widehat{m}}$ with small entries, $m = \bar{m} + \widehat{m}$
2: $\widehat{\mathbf{A}} := \mathbf{G} - \bar{\mathbf{A}}\mathbf{R} \mod q$
3: return $(\widehat{\mathbf{A}}, \mathbf{R})$

In algorithm Sign (Algorithm 3), taking its own secret key sk and ring members' public keys L_{pk} as inputs, a ring member can generate a ring signature σ on message M. According to Lemma 3 and Lemma 4, the distribution of $\mathbf{x}$ is statistically close to $D_{\mathbb{Z}_q^m, s}$, and the distribution of $\mathbf{x}_\pi$ is statistically close to $D_{\mathbb{Z}_q^{\widehat{m}}, s}$. For the sake of unforgeability, all the $\mathbf{x}_i, i \in \{0, 1, ..., l\}$ should be small enough. The analysis of unforgeability is shown in Sect. 3.3.

In algorithm Verify (Algorithm 4), first make sure that all the $\mathbf{x}_i, i \in \{0, 1, ..., l\}$ are small enough. Then, since $\bar{\mathbf{A}}$ is generated by a pseudorandom generator, $\bar{\mathbf{A}}$ is statistically random and has a negligible probability for having a trapdoor. Thus only the member in the ring L_{pk} can generate a valid ring signature. If the equation in step 4 is true, the ring signature σ for message M was signed by a member of the ring L_{pk}.

3.3 Security Analysis

We now firstly analyze the correctness of our new generic construction of ring signature from lattice gadget, and then prove the unforgeability and anonymity of it. We follow the strongest security definitions presented by Bender et al. [27] in 2006: anonymity against full key exposure(full anonymity) and unforgeability with insider corruption.

Algorithm 3. Sign $(pk_\pi, sk_\pi, M, L_{pk})$

    input: $L_{pk} = \{\widehat{\mathbf{A}}_i\}_1^l$, $pk_\pi = \widehat{\mathbf{A}}_\pi \in L_{pk}$, $sk_\pi = \mathbf{R}_\pi$, message M
    output: a ring signature $\sigma = (salt, \{\mathbf{x}_0; \mathbf{x}_1, \mathbf{x}_2, ..., \mathbf{x}_l\})$
1: for all $i \in [l]$, $i \neq \pi$, choose $\mathbf{x}_i \leftarrow D_{\mathbb{Z}_q^{\widehat{m}}, s}$ such that $\|\mathbf{x}_i\| \leq \beta_2$, and compute $\mathbf{u}_i = \widehat{\mathbf{A}}_i \mathbf{x}_i$
2: $salt \xleftarrow{\$} \{0,1\}^{320}$, $\mathbf{u}_\pi := \mathsf{H}(salt, M, L_{pk}) - \sum_{i \neq \pi} \mathbf{u}_i \mod q$
3: sample $\mathbf{x} \leftarrow \mathsf{SampPre}([\bar{\mathbf{A}}|\widehat{\mathbf{A}}_\pi], \mathbf{R}_\pi, \mathbf{u}_\pi, r, s)$
4: parse $\mathbf{x}$ as $\begin{pmatrix} \mathbf{x}_0^{\bar{m}} \\ \mathbf{x}_\pi^{\widehat{m}} \end{pmatrix}$
5: **if** $\|\mathbf{x}_0\| > \beta_1$ or $\|\mathbf{x}_\pi\| > \beta_2$ **then**
6:     restart
7: **end if**
8: return $(salt, \{\mathbf{x}_0; \mathbf{x}_1, \mathbf{x}_2, ..., \mathbf{x}_l\})$

Algorithm 4. Verify (σ, M, L_{pk})

    input: $\sigma = (salt, \{\mathbf{x}_0; \mathbf{x}_1, \mathbf{x}_2, ..., \mathbf{x}_l\})$ is the ring signature on the message M, L_{pk} is the public key list of the ring
    output: $accept/reject$
1: parse $\sigma = (salt, \{\mathbf{x}_0; \mathbf{x}_1, \mathbf{x}_2, ..., \mathbf{x}_l\})$
2: for all $i \in \{1, ..., l\}$, check whether $\|\mathbf{x}_i\| \leq \beta_2$
3: check whether $\|\mathbf{x}_0\| \leq \beta_1$
4: check whether $\mathsf{H}(salt, M, L_{pk}) = \bar{\mathbf{A}}\mathbf{x}_0 + \sum_{i=1}^{l} \widehat{\mathbf{A}}_i \mathbf{x}_i \mod q$
5: if all pass, return $accept$. Otherwise, return $reject$.

Correctness. The signature generated from algorithm **Sign** can inherently passes the verification algorithm **Verify**. Since the signature σ generated from algorithm **Sign** contains a $\mathbf{x}_j, j \in [l]$ such that $\begin{pmatrix} \mathbf{x}_0 \\ \mathbf{x}_j \end{pmatrix}$ satisfies

$$\bar{\mathbf{A}}\mathbf{x}_0 + \widehat{\mathbf{A}}_j \mathbf{x}_j = \mathbf{u}_j = \mathsf{H}(salt, M, L_{pk}) - \sum_{i \neq j} \mathbf{u}_i \mod q.$$

Hence σ satisfies the check in step 4 in algorithm **Verify**. Besides, all the $\mathbf{x}_i, i \in \{1, ..., l\}$ satisfy $\|\mathbf{x}_i\| \leq \beta_2$ and $\|\mathbf{x}_0\| \leq \beta_1$, which satisfies the checks in step 2 and step 3 in algorithm **Verify**. Consequently, σ will be accepted.

Theorem 1 (Unforgeability). *Our new generic construction of ring signature is unforgeable with insider corruption in the random oracle model, assuming* $\mathsf{SIS}_{n, \bar{m}+l\widehat{m}, q, 2\sqrt{\beta_1^2 + l\beta_2^2}}$ *is hard.*

Proof. Suppose there exists a PPT adversary $\mathcal{A}$ who can break the unforgeability of our generic construction of ring signature scheme, we can construct a challenger $\mathcal{C}$ to solve $\mathsf{SIS}_{n, \bar{m}+l\widehat{m}, q, 2\sqrt{\beta_1^2 + l\beta_2^2}}$ with overwhelming probability. $\mathcal{C}$ interacts with $\mathcal{A}$ as follows:

- *Setup.* Let q_e be the total number of extraction queries that can be issued by adversary $\mathcal{A}$. Challenger $\mathcal{C}$ simulates Setup and KeyGen. $\mathcal{C}$ maintains a list $Klist$ for KeyGen. $\mathcal{C}$ simulates hash function H with random oracle $\mathcal{HO}$ and maintains a list $Hlist$. $Klist$ and $Hlist$ are initialized to be empty. Given a SIS instance $\mathbf{A} \in \mathbb{Z}_q^{n \times (\bar{m}+l\widehat{m})}$, $\mathcal{C}$ parses it as $\mathbf{A} = [\bar{\mathbf{A}}|\mathbf{A}_{s1}|\mathbf{A}_{s2}|\cdots|\mathbf{A}_{sl}]$. Then $\mathcal{C}$ randomly chooses an index set $\{u1,...,ul\} \in [q_e]$ and sets $\widehat{\mathbf{A}}_{ui} = \mathbf{A}_{si}$. For $i \in [q_e]\backslash\{ui\}_1^l$, $\mathcal{C}$ randomly chooses $\mathbf{R}_i$ as in KeyGen, computes $\widehat{\mathbf{A}}_i := \mathbf{G} - \bar{\mathbf{A}}\mathbf{R}_i$ and stores $\langle \mathbf{R}_i, \widehat{\mathbf{A}}_i \rangle$ in $Klist$. $\mathcal{C}$ sends $\bar{\mathbf{A}}$ and $\{\widehat{\mathbf{A}}_i\}_1^{q_e}$ to $\mathcal{A}$.

- *Query.* Adversary $\mathcal{A}$ queries $\mathcal{CO}$ for corruption secret key of $\widehat{\mathbf{A}}_j, j \in [q_e]$. If $j \in \{ui\}_1^l$, $\mathcal{C}$ aborts; otherwise $\mathcal{C}$ searches $Klist$ to find $(\widehat{\mathbf{A}}_j, \mathbf{R}_j)$ and returns $\mathbf{R}_j$ to $\mathcal{A}$. In signature queries, $\mathcal{A}$ queries $(M^{(j)}, L_{pk}^{(j)}, pk_i^{(j)})$ to $\mathcal{SO}$ in an adaptive manner. $\mathcal{C}$ first chooses a $salt^{(j)}$, and simulates $\mathcal{HO}(salt^{(j)}, M^{(j)}, L_{pk}^{(j)})$. $\mathcal{C}$ chooses sufficiently short $\mathbf{x}_0^{(j)} \leftarrow D_{\mathbb{Z}_q^{\bar{m}}, s}$, $\mathbf{x}_i^{(j)} \leftarrow D_{\mathbb{Z}_q^{\widehat{m}}, s}$, and computes $\mathbf{y}^{(j)} := \bar{\mathbf{A}}\mathbf{x}_0^{(j)} + \sum_{i=1}^{lj} \widehat{\mathbf{A}}_i^{(j)} \mathbf{x}_i^{(j)} \mod q$, with $\widehat{\mathbf{A}}_i^{(j)} \in L_{pk}^{(j)}$. Then $\mathcal{C}$ returns $\mathbf{y}^{(j)}$ as the value of $\mathcal{HO}(salt^{(j)}, M^{(j)}, L_{pk}^{(j)})$ and stores the tuple $\langle (salt^{(j)}, M^{(j)}, L_{pk}^{(j)}), \mathbf{x}_0^{(j)}, \{\mathbf{x}_i^{(j)}\}_1^{lj} \rangle$ in $Hlist$. Finally $\mathcal{C}$ returns $\sigma^{(j)} := (salt^{(j)}, \mathbf{x}_0^{(j)}, \{\mathbf{x}_i^{(j)}\}_1^{lj})$ to $\mathcal{A}$ as the return of $\mathcal{SO}(M^{(j)}, L_{pk}^{(j)}, pk_i^{(j)})$.

- *Output.* $\mathcal{A}$ outputs a forgery $(M^*, L_{pk}^*, \sigma^* := (salt^*, \mathbf{x}_0^*, \{\mathbf{x}_i^*\}_1^{l*}))$.

If the forgery $(M^*, L_{pk}^*, \sigma^* := (salt^*, \mathbf{x}_0^*, \{\mathbf{x}_i^*\}_1^{l*}))$ satisfies the conditions in Definition 4 and $l* \leq l$, $\mathcal{C}$ can construct a solution for $SIS_{n, \bar{m}+l\widehat{m}, q, 2\sqrt{\beta_1^2 + l\beta_2^2}}$ with a probability of $[\frac{l(q_e-l)}{q_e^2} \cdot \frac{C_l^{l*}}{C_{q_e}^{l*}}]$. Without loss of generality, it can be assumed that $\mathcal{A}$ has made a hash query on $(salt^*, M^*, L_{pk}^*)$ before the signature forgery, since if not $\mathcal{C}$ can immediately simulate $\mathcal{HO}$ $(salt^*, M^*, L_{pk}^*)$. $\mathcal{C}$ searches the tuple $\langle (salt^*, M^*, L_{pk}^*), \mathbf{x}_0^{**}, \{\mathbf{x}_i^{**}\}_1^{l*} \rangle$ in $Hlist$. If in L_{pk}^* all the $\widehat{\mathbf{A}}_i \in \{\widehat{\mathbf{A}}_{ui}\}_1^l = \{\mathbf{A}_{si}\}_1^l$, it can be assumed that $\widehat{\mathbf{A}}_i$s in L_{pk}^* are $\{\mathbf{A}_{si}\}_1^{l*}$ without loss of generality. $\mathcal{C}$ computes $(\mathbf{x}_i^{**} - \mathbf{x}_i^*)$ for $\widehat{\mathbf{A}}_i \in L_{pk}^*$, also $\in \{\mathbf{A}_{si}\}_1^{l*}$, and sets $\mathbf{x}_j = \mathbf{0}$ for $\{\mathbf{A}_{sj}\}_{l*+1}^l$. Then $\begin{pmatrix} (\mathbf{x}_0^{**} - \mathbf{x}_0^*) \\ (\mathbf{x}_1^{**} - \mathbf{x}_1^*) \\ \cdots \\ (\mathbf{x}_{l*}^{**} - \mathbf{x}_{l*}^*) \\ \mathbf{0} \end{pmatrix}$ is a solution for SIS instance $\mathbf{A} = [\bar{\mathbf{A}}|\mathbf{A}_{s1}|\cdots|\mathbf{A}_{sl*}|\mathbf{A}_{sl*+1}|\cdots|\mathbf{A}_{sl}]$.

Theorem 2 (Anonymity). *Our new generic construction of ring signature is full anonymous in the random oracle model, according to the statistical property of our ring signature.*

Proof. Assume there exists a PPT adversary $\mathcal{A}$ playing the game Game$_{\text{anon}}$ described in the security model (see Appendix A). $\mathcal{C}$ interacts with $\mathcal{A}$ as follows:

- *Setup.* Let q_e be the total number of extraction queries that can be issued by adversary $\mathcal{A}$. Challenger $\mathcal{C}$ runs Setup and KeyGen. $\mathcal{C}$ maintains a list $Klist$ for

KeyGen. $\mathcal{C}$ simulates hash function H with random oracle $\mathcal{HO}$ and maintains a list $Hlist$. $Klist$ and $Hlist$ are initialized to be empty. $\mathcal{C}$ runs KeyGen(pp) to generate key pairs $(\mathbf{R}_1, \widehat{\mathbf{A}}_1), ..., (\mathbf{R}_{q_e}, \widehat{\mathbf{A}}_{q_e})$ and stores them in $Klist$. Then the public keys $L_{q_e} = \{\widehat{\mathbf{A}}_i\}_1^{q_e}$ are sent to $\mathcal{A}$.

- *Query.* Adversary $\mathcal{A}$ queries $\mathcal{CO}$ in an adaptive manner. When $\mathcal{A}$ queries $\widehat{\mathbf{A}}_j \in L_{q_e}$, $\mathcal{C}$ finds the tuple $(\mathbf{R}_j, \widehat{\mathbf{A}}_j)$ in $Klist$ and returns $\mathbf{R}_j$ to $\mathcal{A}$. $\mathcal{A}$ picks a list of user public keys $L_{pk} = \{\widehat{\mathbf{A}}_i\}_1^l$, a message M and two public keys $\widehat{\mathbf{A}}_{i_0}, \widehat{\mathbf{A}}_{i_1} \in L_{pk}$. $\mathcal{A}$ sends $(L_{pk}, M, (\widehat{\mathbf{A}}_{i_0}, \widehat{\mathbf{A}}_{i_1}))$ to $\mathcal{C}$. $\mathcal{C}$ randomly chooses $b \in \{0,1\}$ and runs Sign($pp, \mathbf{R}_{i_b}, L_{pk}, M$) to obtain the ring signature $\sigma = (salt, \mathbf{x}_0, \{\mathbf{x}_i\}_1^l)$. $\mathcal{C}$ sends σ to $\mathcal{A}$.
- *Output.* $\mathcal{A}$ outputs a guess $b^* \in \{0,1\}$.

The probability of $b^* = b$ is negligibly close to $1/2$, by the statistical property of ring signature. Suppose that $\sigma^{(0)} = (salt^{(0)}, \mathbf{x}_0^{(0)}, \{\mathbf{x}_i^{(0)}\}_1^l)$ is generated by Sign($\widehat{\mathbf{A}}_{i_0}, \mathbf{R}_{i_0}, L_{pk}, M$), and $\sigma^{(1)} = (salt^{(1)}, \mathbf{x}_0^{(1)}, \{\mathbf{x}_i^{(1)}\}_1^l)$ is generated by Sign($\widehat{\mathbf{A}}_{i_1}, \mathbf{R}_{i_1}, L_{pk}, M$). For $i \in [l], j \in \{0,1\}$, all the $\begin{pmatrix} \mathbf{x}_0^{(j)} \\ \mathbf{x}_i^{(j)} \end{pmatrix}$ are statistically close to $D_{\mathbb{Z}_q^m, s}$ and satisfy the equation $\mathsf{H}(salt, M, L_{pk}) = \bar{\mathbf{A}} \mathbf{x}_0^{(j)} + \sum_{i=1}^l \widehat{\mathbf{A}}_i \mathbf{x}_i^{(j)}$ mod q. This implies that adversary $\mathcal{A}$ cannot get additional information from the ring signature $\sigma^{(0)}$ and $\sigma^{(1)}$. Besides, both $salt^{(0)}$ and $salt^{(1)}$ are randomly chosen from $\{0,1\}^{320}$, and all of $\{\mathbf{x}_i^{(0)}\}$ and $\{\mathbf{x}_i^{(1)}\}$ are $\widehat{m}$-dimensional vectors obeying discrete Gaussian distribution. $\sigma^{(0)}$ and $\sigma^{(1)}$ are indistinguishable. Thus the probability of $\mathcal{A}$ guessing correctly is negligibly close to $1/2$.

4 Instantiation with Compact Gadget

In this section, we give an instantiation of our new generic construction from compact gadget [28]. We call our instantiation LION, because it is more secure but not less efficient. Compact gadget is a square matrix $\mathbf{P}$ which works with a specialized preimage sampler, called *semi-random sampler*. We first review the compact gadget and semi-random sampler, and then give our instantiation in detail.

4.1 Compact Gadget and Semi-random Sampler

Compact gadget is a square matrix $\mathbf{P} \in \mathbb{Z}^{n \times n}$ along with $\mathbf{Q} \in \mathbb{Z}^{n \times n}$ such that $\mathbf{P}\mathbf{Q} = Q \cdot \mathbf{I}_n$. For public matrix $\mathbf{A} \in \mathbb{Z}_Q^{n \times m}$, the trapdoor $\mathbf{T} \in \mathbb{Z}_Q^{m \times n}$ for $\mathbf{A}$ satisfies $\mathbf{AT} = \mathbf{P}$ mod Q. A public matrix $\mathbf{A}$ can be generated as $\mathbf{A} := [\mathbf{I}|\bar{\mathbf{A}}|\mathbf{P} + \bar{\mathbf{A}}\mathbf{S} + \mathbf{E}]$ and the corresponding trapdoor is $\mathbf{T} = \begin{bmatrix} -\mathbf{E} \\ -\mathbf{S} \\ \mathbf{I} \end{bmatrix}$. According to LWE problem(see Definition 2), it is hard to recover the trapdoor from the public matrix. And the public matrix is simulatable without trapdoor since its distribution is indistinguishable from uniform distribution.

Semi-random sampler can computes a short approximate preimage $\mathbf{x} \in \mathbb{Z}_Q^m$ for $\mathbf{u} \in \mathbb{Z}_Q^n$ such that $\mathbf{Ax} = \mathbf{u} - \mathbf{e} \mod Q$, where $\mathbf{e}$ is a small error. It proceeds in two steps: (1)*deterministic error decoding*: using lattice decoding algorithm LatticeDecoder to compute an error $\mathbf{e}$ such that $\mathbf{u} - \mathbf{e} = \mathbf{Pc} \in \Lambda(\mathbf{P})$ with deterministic lattice decoding; (2)*random preimage sampling*: the improvement of preimage sampling in compact gadget is that $\mathbf{x}'$ can be easily chosen from the coset $\Lambda(\mathbf{Q}) + \mathbf{c}$. A preimage is computed as $\mathbf{x} := \mathbf{p} + \mathbf{Tx}' \mod Q$.

4.2 Instantiation and Security Analysis

We describe our instantiation with compact gadget in this section and then analyze the security of it. For simplicity, we can use diagonal matrices $(\mathbf{P}, \mathbf{Q}) = (p\mathbf{I}_n, q\mathbf{I}_n)$. According to LWE problem, $\widehat{\mathbf{A}}$ can avoid key recovery attacks. B_1 is the centered binomial distribution. Our instantiation is from Algorithm 5 to Algorithm 8. To improve efficiency, the perturbation vector $\mathbf{p}$ is precomputed in offline phase.

Security Analysis. Full anonymity of our LION is straightforward from the statistical property of preimage sampling, and can be proved in the same way as the generic construction. The proof of unforgeability is similar to the generic construction. We sketch the proof below. If there exists a PPT adversary $\mathcal{A}$ who can break the unforgeability of LION, we can construct a challenger $\mathcal{C}$ to solve $\mathsf{HNF.SIS}_{n,n+\bar{m}+ln,Q,2\sqrt{(\beta_1)^2+(\beta_2)^2+l(\beta_3)^2}}$ with non-negligible probability as follows: given a $\mathsf{HNF.SIS}_{n,n+\bar{m}+ln,Q,2\sqrt{(\beta_1)^2+(\beta_2)^2+l(\beta_3)^2}}$ instance $\mathbf{A}$, challenger $\mathcal{C}$ parses it as $\mathbf{A} := [\mathbf{I}_n|\bar{\mathbf{A}}|\mathbf{A}_{s1}|\cdots|\mathbf{A}_{sl}]$, and simulates the value of $\mathcal{HO}(salt, M, L_{pk})$ as $\mathbf{y} := \mathbf{x}_{00} + \mathbf{e} + \bar{\mathbf{A}}\mathbf{x}_{01} + \sum_{i=1}^{lj} \widehat{\mathbf{A}}_i \mathbf{x}_i \mod Q$. If adversary $\mathcal{A}$ can return a valid forgery $(M^*, L_{pk}^*, \sigma^* := (salt^*, \mathbf{x}_{01}^*, \{\mathbf{x}_i^*\}_1^{l*}))$, challenger $\mathcal{C}$ searches the tuple $\left\langle (salt^*, M^*, L_{pk}^*), \mathbf{x}_{01}^{**}, \{\mathbf{x}_i^{**}\}_1^{l*}, \mathbf{x}_{00}^{**} + \mathbf{e}^{**} \right\rangle$ in $Hlist$. Denote $\mathbf{x}_{00}^* + \mathbf{e}^* := \mathsf{H}(salt^*, M^*, L_{pk}^*) - \bar{\mathbf{A}}\mathbf{x}_{01}^* - \sum_{i=1}^{l*} \widehat{\mathbf{A}}_i \mathbf{x}_i \mod Q$. If in L_{pk}^* all the $\widehat{\mathbf{A}}_i \in \{\widehat{\mathbf{A}}_{ui}\}_1^l = \{\mathbf{A}_{si}\}_1^l$, it can be assumed that $\widehat{\mathbf{A}}_i$s in L_{pk}^* are $\{\mathbf{A}_{si}\}_1^{l*}$ without loss of generality. $\mathcal{C}$ computes $(\mathbf{x}_i^{**} - \mathbf{x}_i^*)$ for $\widehat{\mathbf{A}}_i \in L_{pk}^*$, also $\in \{\mathbf{A}_{si}\}_1^{l*}$, and sets $\mathbf{x}_j = \mathbf{0}$ for $\{\mathbf{A}_{sj}\}_{l*+1}^l$. Then
$$\begin{pmatrix} (\mathbf{x}_{00}^{**} + \mathbf{e}^{**}) - (\mathbf{x}_{00}^* + \mathbf{e}^*) \\ (\mathbf{x}_{01}^{**} - \mathbf{x}_{01}^*) \\ (\mathbf{x}_1^{**} - \mathbf{x}_1^*) \\ \cdots \\ (\mathbf{x}_{l*}^{**} - \mathbf{x}_{l*}^*) \\ \mathbf{0} \end{pmatrix}$$
is a solution for SIS instance $\mathbf{A} = [\mathbf{I}_n|\bar{\mathbf{A}}|\mathbf{A}_{s1}|\cdots|\mathbf{A}_{sl*}|\mathbf{A}_{sl*+1}|\cdots|\mathbf{A}_{sl}]$.

Lion: A New Ring Signature Construction from Lattice Gadget

Algorithm 5. LION.Setup (1^λ)

input: λ is a security parameter
output: public parameters $pp = (seed_{\bar{\mathbf{A}}}, Q, \mathbf{P}, \mathbf{Q}, \mathsf{PRG}, \mathsf{H})$, where $\mathbf{PQ} = Q\mathbf{I}_n$
1: choose two integers p, q depended on security parameter, $Q := pq$
2: $\mathbf{P} := p\mathbf{I}_n, \mathbf{Q} := q\mathbf{I}_n$
3: $seed_{\bar{\mathbf{A}}} \xleftarrow{\$} \{0,1\}^{256}$
4: $\mathsf{PRG}(seed) \to \mathbb{Z}_q^{n \times \bar{m}}, \mathsf{H} : \{0,1\}^* \to \mathbb{Z}_q^n$
5: $r \geq \eta_\epsilon(\Lambda_q^\perp(\mathbf{G})), s \geq r \cdot \max\{s_1(\mathbf{T})\}$
6: return $pp = (seed_{\bar{\mathbf{A}}}, Q, r, s, \mathbf{P}, \mathbf{Q}, \mathsf{PRG}, \mathsf{H})$

Algorithm 6. LION.KeyGen (pp)

input: pp is the common parameters
output: a key pair (sk, pk), secret key $sk = \mathbf{R} \in \mathbb{Z}_q^{(\bar{m}+n) \times n}$, public key $pk = \widehat{\mathbf{A}} \in \mathbb{Z}_q^{n \times n}$
1: $\bar{\mathbf{A}} \leftarrow \mathsf{PRG}(seed_{\bar{\mathbf{A}}}), \mathbf{S} \leftarrow \chi_s^{\bar{m} \times n}, \mathbf{E} \leftarrow \chi_e^{n \times n}, \chi_s = \chi_e = B_1$
2: $\widehat{\mathbf{A}} := \mathbf{P} + \bar{\mathbf{A}}\mathbf{S} + \mathbf{E} \mod Q, \mathbf{R} := \begin{bmatrix} -\mathbf{E} \\ -\mathbf{S} \end{bmatrix}$
3: return $(\widehat{\mathbf{A}}, \mathbf{R})$

Algorithm 7. LION.Sign $(pk_\pi, sk_\pi, M, L_{pk})$

input: $L_{pk} = \{\widehat{\mathbf{A}}_i\}_1^l, pk_\pi = \widehat{\mathbf{A}}_\pi \in L_{pk}, sk_\pi = \mathbf{R}_\pi, M$ is a message
output: a ring signature $\sigma = (salt, \{\mathbf{x}_{01}; \mathbf{x}_1, \mathbf{x}_2, ..., \mathbf{x}_l\})$
Offline phase:
1: $\mathbf{T}_\pi := \begin{bmatrix} \mathbf{R}_\pi \\ \mathbf{I}_n \end{bmatrix}, \Sigma_\mathbf{p} := s^2 \mathbf{I}_m - r^2 \mathbf{T}\mathbf{T}^t$
2: sample $\mathbf{p} \leftarrow D_{\mathbb{Z}^m, \sqrt{\Sigma_\mathbf{p}}}$
Online phase:
3: For all $i \in [l], i \neq \pi$, choose $\mathbf{x}_i \leftarrow D_{\mathbb{Z}_q^n, s}$ such that $\|\mathbf{x}_i\| \leq \beta_3$, and compute $\mathbf{u}_i = \widehat{\mathbf{A}}_i \mathbf{x}_i$
4: $salt \xleftarrow{\$} \{0,1\}^{320}, \mathbf{u}_\pi := \mathsf{H}(salt, M, L_{pk}) - \sum_{i \neq \pi} \mathbf{u}_i \mod Q, \mathbf{u}' = \mathbf{u}_\pi - \bar{\mathbf{A}}\mathbf{p} \mod Q$
5: $(\mathbf{c}, \mathbf{e}) \leftarrow \text{LatticeDecoder}(\mathbf{u}', \mathbf{P})$ such that $\mathbf{c} \in \mathbb{Z}^n$ and $\mathbf{u}' - \mathbf{e} = \mathbf{Pc}$
6: sample $\mathbf{x}' \leftarrow D_{\Lambda(\mathbf{Q})+\mathbf{c}, r}$
7: compute $\mathbf{x} := \mathbf{p} + \mathbf{T}\mathbf{x}' \mod Q$
8: parse $\mathbf{x}$ as $\begin{pmatrix} \mathbf{x}_{00} \\ \mathbf{x}_{01} \\ \mathbf{x}_\pi \end{pmatrix}$
9: $\mathbf{e} := \mathbf{u}_\pi - \mathbf{I}_n \mathbf{x}_{00} - \bar{\mathbf{A}}\mathbf{x}_{01} - \widehat{\mathbf{A}}_\pi \mathbf{x}_\pi \mod Q$
10: **if** $\|\mathbf{e} + \mathbf{x}_{00}\| > \beta_1$ or $\|\mathbf{x}_{01}\| > \beta_2$ or $\|\mathbf{x}_\pi\| > \beta_3$ **then**
11:     restart
12: **end if**
13: return $(salt, \{\mathbf{x}_{01}; \mathbf{x}_1, \mathbf{x}_2, ..., \mathbf{x}_l\})$

Algorithm 8. LION.Verify (σ, M, L_{pk})

input: $\sigma = (salt, \{\mathbf{x}_{01}; \mathbf{x}_1, \mathbf{x}_2, ..., \mathbf{x}_l\})$ is the ring signature on the message M, L_{pk} is the public key list of the ring
output: $accept/reject$
1: parse $\sigma = (salt, \{\mathbf{x}_{01}; \mathbf{x}_1, \mathbf{x}_2, ..., \mathbf{x}_l\})$
2: For all $i \in \{1, ..., l\}$, check whether $\|\mathbf{x}_i\| \leq \beta_3$
3: check whether $\|\mathbf{x}_{01}\| \leq \beta_2$
4: check whether $\|\mathsf{H}(salt, M, L_{pk}) - \sum_{i\in[l]} \widehat{\mathbf{A}}_i \mathbf{x}_i - \bar{\mathbf{A}}\mathbf{x}_{01}\| \leq \beta_1$
5: If all pass, return $accept$. Otherwise, return $reject$.

4.3 Concrete Parameters and Comparisons

In this section, we give some recommended parameters for LION in Table 1. We analyze the cost of known lattice attacks and use BKZ algorithm to estimate the bit-security following the core-SVP methodology.[1] The cost of BKZ with blocksize b is estimated as $2^{0.292b}$ [29] in the classical setting and $2^{0.265b}$ [30] in the quantum setting. Let Gaussian parameter $r = \eta_\epsilon(\Lambda(\mathbf{Q})) = q \cdot \eta_\epsilon(\mathbb{Z}^n) = q\sqrt{\ln(2n(1+1/\epsilon))/\pi}, \epsilon = 2^{-36}$, the standard deviation of the preimage is $s = 1.05 \cdot (r/\sqrt{2\pi}) \cdot \sqrt{(1+q^2)/(2q^2)} \cdot (\sqrt{n} + \sqrt{\bar{m}+n})$ and acceptance boundaries $\beta_1 = \sqrt{ns^2 + \frac{n(p^2-1)}{12}}, \beta_2 = \sqrt{\bar{m}} \cdot s, \beta_3 = \sqrt{n} \cdot s$. Let l denote the ring size. The signature size is $320 + (\bar{m} + nl)(\log_2 s + 3)$ bits. Public key size is $n^2 \cdot \log_2 Q$ bits.

Table 1. Recommended parameters

security level	100bit	128bit	192bit	256bit
Dimensions $(n, \bar{m})$	(682,718)	(828,870)	(1150,1320)	(1464,1664)
Modulus Q	2^{17}	2^{17}	2^{19}	2^{19}
Gadget parameters (p, q)	$(2^{13}, 2^4)$	$(2^{13}, 2^4)$	$(2^{14}, 2^5)$	$(2^{14}, 2^5)$
Gaussian parameter r	51.20	51.35	103.23	103.61
Standard deviation s	965.39	1066.60	2557.80	2892.08
Acceptance bound β_1	66705.52	74649.01	182342.48	212118.48
Acceptance bound β_2	25868.08	31460.20	92929.63	117974.38
Acceptance bound β_3	25211.24	30691.43	86739.35	110657.68
Key recovery:				
$\quad$ BKZ blocksize b	344	442	660	879
$\quad$ Classical core $-$ SVP security	100	129	193	257
$\quad$ Quantum core $-$ SVP security	91	117	175	233
Forgery:				
$\quad$ BKZ blocksize b	342	439	657	876
$\quad$ Classical core $-$ SVP security	100	128	192	256
$\quad$ Quantum core $-$ SVP security	90	116	174	232

---

[1] We estimate the BKZ blocksize by python script provided at https://github.com/pq-crystals/security-estimates

Table 2 shows the comparison of signature size of Lion with the practical linear size ring signature scheme Raptor [6], the smallest linear size ring signature scheme DualRing [8], and the most practical logarithmic size ring signature scheme ESLL19 [7]. Table 3 shows the comparison of running time and security basis of these schemes. Based on zero-knowledge proof technique, the running time of ESLL19 is around 150 s and 22 s for Sign and Verify respectively with 50 ring members [8], which is not practical enough. In contrast, being based on chameleon hash plus preimage sampling technique, Raptor only consumes 40ms for Sign and 32ms for Verify. Referring to the implementation of digital signature based on compact gadget in HuFu [26], with additional $(l+1)n^2 + n\bar{m}$ times integer multiplication operations and $n[(l+1)(n-1) + (\bar{m}-1)] + l + 3$ times addition operations, the running time of our Lion is expected to be around 42ms for Sign and 38ms for Verify with 50 ring members and around 187ms for Sign and 183ms for Verify with 250 ring members. Thus, Lion can strike an optimal balance between computational efficiency and storage overheads in practice. Raptor, ESLL19 and DualRing are based on NTRU lattice or module lattice, which are more susceptible to attacks than generic lattices. Therefore our Lion has the best security basis.

Table 2. Comparison of signature size at 100bit-security level

ring size l	8	64	256	385	512	growth
ESLL19 [7]	76.97 KB	129.92 KB	276.84 KB	415.40 KB	425.78 KB	logarithmic
DualRing-LB [8]	4.63 KB	6.05 KB	10.93 KB	14.21 KB	17.43 KB	linear
Raptor [6]	9.89 KB	80.6 KB	332.6 KB	475.99 KB	633.0 KB	linear
Lion	9.77 KB	69.98 KB	276.42 KB	415.12 KB	551.67 KB	linear

Table 3. Comparison of running time

	ring size $l=10$		ring size $l=50$		ring size $l=250$		security basis
	Sign	Verify	Sign	Verify	Sign	Verify	
ESLL19 [7]	120 s	14 s	150 s	22 s	–	–	M-SIS&M-LWE
DualRing-LB [8]	5 s	1 s	10 s	3 s	30 s	12 s	M-SIS&M-LWE
Raptor [6]	9.5 ms	6.5 ms	40 ms	32 ms	–	–	NTRU
Lion	12.7ms	8.5 ms	42 ms	38ms	187 ms	183 ms	SIS&LWE

5 Conclusion

We develop a new ring signature generic construction from lattice gadget and our instantiation Lion obtains the optimal balance between computational efficiency and signature size. Based on generic lattices, LION is the most secure. We reuse the random matrix $\bar{\mathbf{A}}$ in key generation to reduce the public key size of the ring member. And $\bar{\mathbf{A}}$ is expanded from a seed to avoid the risk of having a trapdoor. Thus LION can work without trusted setup.

Acknowledgments. This work is supported by the Major Program of Guangdong Basic and Applied Research (2019B030302008), the National Natural Science Foundation of China (62272174), and the Science and Technology Program of Guangzhou (2024A04J6542).

A Ring Signature

A ring signature scheme is composed of a tuple of probabilistic algorithms (Setup, KeyGen, Sign, Verify) described as follows:

- Setup(1^λ) $\to$ pp: Taking as input the security parameter 1^λ, the algorithm outputs the public parameters pp.
- KeyGen(pp) $\to$ (sk, pk): Taking as input the public parameter pp, the algorithm outputs a key pair (sk, pk), where pk is public key and sk is the corresponding secret key.
- Sign(pp, sk, L_{pk}, M) $\to$ σ: The algorithm takes as input the public parameter pp, the signer's secret key sk, a list of public keys L_{pk} of the ring(the signer's public key $\in L_{pk}$) and a message M to be signed, and outputs a ring signature σ on message M w.r.t the ring L_{pk}.
- Verify(pp, L_{pk}, M, σ) $\to$ "$accept/reject$": The algorithm takes as input the public parameter pp, the list of public keys L_{pk} of the ring, a message M and the ring signature σ on M, and outputs "$accept$" if the ring signature is valid or "$reject$" otherwise.

For simplicity, pp is usually used as an implicit input in the algorithms KeyGen, Sign, Verify.

Security Model. The security of a ring signature scheme consists of two requirements, namely *Anonymity* and *Unforgeability*. We follow the strongest security definitions presented by Bender et al. [27] in 2006: anonymity against full key exposure(full anonymity) and unforgeability with insider corruption. Full anonymity means that the adversary $\mathcal{A}$ cannot identify the actual signer even though all of the secret keys in the ring are exposed. Unforgeability with insider corruption means that one cannot forge a valid ring signature when he's not a member of the ring. We first introduce the following oracles which will be used in the security model of ring signature scheme:

- Corruption Oracle $\mathcal{CO}(pk_i) \to sk_i$: On input a user public key pk_i, $\mathcal{CO}$ returns the corresponding secret key sk_i.
- Sign Oracle $\mathcal{SO}(M, L_{pk}, pk_\pi) \to \sigma$: On input a message M, a list of public keys L_{pk} and the public key of the signer $pk_\pi \in L_{pk}$, $\mathcal{SO}$ returns a valid ring signature σ on M and L_{pk}.

- Hash Oracle $\mathcal{HO}(salt, M, L_{pk}) \to \mathbf{u}$: On input a message M, a random $salt$ and the public key list L_{pk}, $\mathcal{HO}$ returns a hash value over integer lattice.

Definition 3 (Full Anonymity). *For a ring signature scheme (*Setup, KeyGen, Sign, Verify*), full anonymity is defined by a game* Game$_{anon}$ *between a PPT adversary $\mathcal{A}$ and a challenger $\mathcal{C}$:*

- *Setup. Challenger $\mathcal{C}$ runs* Setup(1^λ) *and sends the public parameter pp to adversary $\mathcal{A}$. $\mathcal{C}$ runs* KeyGen(pp) *to generate key pairs $(sk_1, pk_1), ..., (sk_u, pk_u)$, where u is a game parameter. Send the public keys $L_u = \{pk_i\}_1^u$ to $\mathcal{A}$.*
- *Query. Adversary $\mathcal{A}$ queries $\mathcal{CO}$ in an adaptive manner. $\mathcal{A}$ picks a list of user public keys $L_{pk} = \{pk_i\}_1^l = \{pk_1, pk_2, ..., pk_l\}$, a message M and two public keys $pk_{i_0}, pk_{i_1} \in L_{pk}$. $\mathcal{A}$ sends $(L_{pk}, M, (pk_{i_0}, pk_{i_1}))$ to $\mathcal{C}$. $\mathcal{C}$ randomly chooses $b \in \{0, 1\}$ and runs* Sign(pp, sk_{i_b}, L_{pk}, M) *to obtain the ring signature σ. $\mathcal{C}$ sends σ to $\mathcal{A}$.*
- *Output. $\mathcal{A}$ outputs a guess $b^* \in \{0, 1\}$, and succeeds if $b^* = b$.*

If the success probability of $\mathcal{A}$ is negligibly close to $\frac{1}{2}$, the ring signature scheme achieves full anonymity.

Definition 4 (Unforgeability). *For a ring signature scheme (*Setup, KeyGen, Sign, Verify*), unforgeability is defined by a game* Game$_{forge}$ *between a PPT adversary $\mathcal{A}$ and a challenger $\mathcal{C}$:*

- *Setup. Challenger $\mathcal{C}$ runs* Setup(1^λ) *and sends the public parameter pp to adversary $\mathcal{A}$. $\mathcal{C}$ runs* KeyGen(pp) *to generate key pairs $(sk_1, pk_1), ..., (sk_u, pk_u)$, where u is a game parameter. Then the public keys $L_u = \{pk_i\}_1^u$ are sent to $\mathcal{A}$.*
- *Query. Adversary $\mathcal{A}$ queries $\mathcal{CO}$ and $\mathcal{SO}$ for a polynomial bounded number of times in an adaptive manner.*
- *Output. $\mathcal{A}$ outputs a forgery $(M^*, L_{pk}^*, \sigma^*)$.*

$\mathcal{A}$ succeeds if all of the following conditions are satisfied:

(1) $L_{pk}^* \subseteq L_l \backslash C$, where C is the set of all the corrupted users;
(2) (M^*, L_{pk}^*) has not been submitted to $\mathcal{SO}$;
(3) Verify(L_{pk}^*, M^*, σ^*) $=$ *Accept.*

If the success probability of $\mathcal{A}$ is negligible, the ring signature scheme achieves unforgeability.

References

1. Rivest, R.L., Shamir, A., Tauman, Y.: How to leak a secret. In: Boyd, C. (ed.) ASIACRYPT 2001. LNCS, vol. 2248, pp. 552–565. Springer, Heidelberg (2001). https://doi.org/10.1007/3-540-45682-1_32

2. Yuen, T.H., et al.: RingCT 3.0 for blockchain confidential transaction: shorter size and stronger security. In: Bonneau, J., Heninger, N. (eds.) FC 2020. LNCS, vol. 12059, pp. 464–483. Springer, Cham (2020). https://doi.org/10.1007/978-3-030-51280-4_25
3. Shor, P.: Algorithms for quantum computation: discrete logarithms and factoring. In: Proceedings 35th Annual Symposium on Foundations of Computer Science, pp. 124–134 (1994). https://doi.org/10.1109/SFCS.1994.365700
4. Regev, O.: An efficient quantum factoring algorithm. J. ACM **72**(1) (2025). https://doi.org/10.1145/3708471
5. Abe, M., Ohkubo, M., Suzuki, K.: 1-out-of-n signatures from a variety of keys. In: Zheng, Y. (ed.) ASIACRYPT 2002. LNCS, vol. 2501, pp. 415–432. Springer, Heidelberg (2002). https://doi.org/10.1007/3-540-36178-2_26
6. Lu, X., Au, M.H., Zhang, Z.: Raptor: a practical lattice-based (linkable) ring signature. In: Deng, R.H., Gauthier-Umaña, V., Ochoa, M., Yung, M. (eds.) ACNS 2019. LNCS, vol. 11464, pp. 110–130. Springer, Cham (2019). https://doi.org/10.1007/978-3-030-21568-2_6
7. Esgin, M.F., Steinfeld, R., Liu, J.K., Liu, D.: Lattice-based zero-knowledge proofs: new techniques for shorter and faster constructions and applications. In: Boldyreva, A., Micciancio, D. (eds.) CRYPTO 2019. LNCS, vol. 11692, pp. 115–146. Springer, Cham (2019). https://doi.org/10.1007/978-3-030-26948-7_5
8. Yuen, T.H., Esgin, M.F., Liu, J.K., Au, M.H., Ding, Z.: *DualRing*: generic construction of ring signatures with efficient instantiations. In: Malkin, T., Peikert, C. (eds.) CRYPTO 2021. LNCS, vol. 12825, pp. 251–281. Springer, Cham (2021). https://doi.org/10.1007/978-3-030-84242-0_10
9. Esgin, M.F., Steinfeld, R., Sakzad, A., Liu, J.K., Liu, D.: Short lattice-based one-out-of-many proofs and applications to ring signatures. In: Deng, R.H., Gauthier-Umaña, V., Ochoa, M., Yung, M. (eds.) ACNS 2019. LNCS, vol. 11464, pp. 67–88. Springer, Cham (2019). https://doi.org/10.1007/978-3-030-21568-2_4
10. Lyubashevsky, V., Nguyen, N.K., Seiler, G.: SMILE: set membership from ideal lattices with applications to ring signatures and confidential transactions. In: Malkin, T., Peikert, C. (eds.) CRYPTO 2021. LNCS, vol. 12826, pp. 611–640. Springer, Cham (2021). https://doi.org/10.1007/978-3-030-84245-1_21
11. Lyubashevsky, V., Nguyen, N.K.: BLOOM: bimodal lattice one-out-of-many proofs and applications. In: Agrawal, S., Lin, D. (eds.) ASIACRYPT 2022. LNCS, vol. 13794, pp. 95–125. Springer, Cham (2022). https://doi.org/10.1007/978-3-031-22972-5_4
12. Fouque, P.A., et al.: Falcon: Fast-fourier lattice-based compact signatures over NTRU. Submission NIST's Post-quantum Cryptogr. Standard. Process **36**(5), 1–75 (2018)
13. Grover, L.K.: A fast quantum mechanical algorithm for database search. In: Proceedings of the Twenty-Eighth Annual ACM Symposium on Theory of Computing, pp. 212–219 (1996). https://doi.org/10.1145/237814.237866
14. Yang, Z., Fu, S., Qu, L., Li, C.: A lower dimension lattice attack on NTRU. SCI. CHINA Inf. Sci. **61**(5), 1–3 (2017). https://doi.org/10.1007/s11432-017-9175-y
15. Albrecht, M., Bai, S., Ducas, L.: A subfield lattice attack on overstretched NTRU assumptions. In: Robshaw, M., Katz, J. (eds.) CRYPTO 2016. LNCS, vol. 9814, pp. 153–178. Springer, Heidelberg (2016). https://doi.org/10.1007/978-3-662-53018-4_6
16. Howgrave-Graham, N.: A hybrid lattice-reduction and meet-in-the-middle attack against NTRU. In: Menezes, A. (ed.) CRYPTO 2007. LNCS, vol. 4622, pp. 150–169. Springer, Heidelberg (2007). https://doi.org/10.1007/978-3-540-74143-5_9

17. Zhang, S., Lin, X., Yu, Y., Wang, W.: Improved power analysis attacks on falcon. In: Hazay, C., Stam, M. (eds.) EUROCRYPT 2023. LNCS, vol. 14007, pp. 565–595. Springer, Cham (2023). https://doi.org/10.1007/978-3-031-30634-1_19
18. Lin, X., et al.: Thorough power analysis on falcon gaussian samplers and practical countermeasure. Cryptology ePrint Archive, Paper 2025/351 (2025). https://eprint.iacr.org/2025/351
19. Ducas, L., Plançon, M., Wesolowski, B.: On the shortness of vectors to be found by the ideal-SVP quantum algorithm. In: Boldyreva, A., Micciancio, D. (eds.) CRYPTO 2019. LNCS, vol. 11692, pp. 322–351. Springer, Cham (2019). https://doi.org/10.1007/978-3-030-26948-7_12
20. Micciancio, D., Peikert, C.: Trapdoors for lattices: simpler, tighter, faster, smaller. In: Pointcheval, D., Johansson, T. (eds.) EUROCRYPT 2012. LNCS, vol. 7237, pp. 700–718. Springer, Heidelberg (2012). https://doi.org/10.1007/978-3-642-29011-4_41
21. Ajtai, M.: Generating hard instances of lattice problems (extended abstract). In: Proceedings of the Twenty-Eighth Annual ACM Symposium on Theory of Computing, STOC 1996, pp. 99–108. Association for Computing Machinery, New York (1996). https://doi.org/10.1145/237814.237838
22. Regev, O.: On lattices, learning with errors, random linear codes, and cryptography. J. ACM **56**(6) (2009). https://doi.org/10.1145/1568318.1568324
23. Lyubashevsky, V.: Lattice signatures without trapdoors. In: Pointcheval, D., Johansson, T. (eds.) EUROCRYPT 2012. LNCS, vol. 7237, pp. 738–755. Springer, Heidelberg (2012). https://doi.org/10.1007/978-3-642-29011-4_43
24. Brakerski, Z., Langlois, A., Peikert, C., Regev, O., Stehlé, D.: Classical hardness of learning with errors. In: Proceedings of the Forty-Fifth Annual ACM Symposium on Theory of Computing, STOC 2013, pp. 575–584. Association for Computing Machinery, New York (2013). https://doi.org/10.1145/2488608.2488680
25. Chen, Y., Genise, N., Mukherjee, P.: Approximate trapdoors for lattices and smaller hash-and-sign signatures. In: Galbraith, S.D., Moriai, S. (eds.) ASIACRYPT 2019. LNCS, vol. 11923, pp. 3–32. Springer, Cham (2019). https://doi.org/10.1007/978-3-030-34618-8_1
26. Yu, Y., et al.: HuFu: hash-and-sign signatures from powerful gadgets (2023)
27. Bender, A., Katz, J., Morselli, R.: Ring signatures: stronger definitions, and constructions without random oracles. In: Halevi, S., Rabin, T. (eds.) TCC 2006. LNCS, vol. 3876, pp. 60–79. Springer, Heidelberg (2006). https://doi.org/10.1007/11681878_4
28. Yu, Y., Jia, H., Wang, X.: Compact lattice gadget and its applications to hash-and-sign signatures. In: Handschuh, H., Lysyanskaya, A. (eds.) CRYPTO 2023. LNCS, vol. 14085, pp. 390–420. Springer, Cham (2023). https://doi.org/10.1007/978-3-031-38554-4_13
29. Becker, A., Ducas, L., Gama, N., Laarhoven, T.: New directions in nearest neighbor searching with applications to lattice sieving. In: Proceedings of the Twenty-Seventh Annual ACM-SIAM Symposium on Discrete Algorithms, SODA 2016, pp. 10–24. Society for Industrial and Applied Mathematics, USA (2016). https://doi.org/10.5555/2884435.2884437
30. Laarhoven, T.: Search problems in cryptography: from fingerprinting to lattice sieving (2016)

Anonymity and Privacy

MagWatch: Exposing Privacy Risks in Smartwatches Through Electromagnetic Signals

Haowen Xu[1], Tianya Zhao[2], Xuyu Wang[2], Jun Dai[1(✉)], and Xiaoyan Sun[1(✉)]

[1] Worcester Polytechnic Institute, Worcester, MA 01609, USA
{hxu4,jdai,xsun7}@wpi.edu
[2] Florida International University, Miami, FL 33199, USA
{tzhao010,xuywang}@fiu.edu

Abstract. As smartwatches become increasingly integrated into daily life, their electromagnetic (EM) emissions introduce a significant yet overlooked privacy risk. This study systematically examines how EM leakage from smartwatches can be exploited to infer user interactions and behavioral patterns. We propose *MagWatch*, a novel non-intrusive attack that applies wavelet transform for signal processing and leverages a CNN-LSTM model to identify applications and in-app activities, achieving up to 90% accuracy across multiple smartwatch models. Our findings reveal a critical security vulnerability, demonstrating that attackers can passively monitor EM emissions to reconstruct user interactions, exposing sensitive information such as communication habits and app usage patterns. This research highlights the urgent need for privacy-preserving countermeasures in wearable technology and establishes a foundation for future studies on EM side-channel security risks.

Keywords: Side Channel Attack · Privacy Leakage · Electromagnetic Signal · Smartwatch

1 Introduction

The widespread adoption of smartwatches has made them an integral part of users' daily lives. According to market research, the global smartwatch market is projected to reach 253 million units by 2025 [21]. Users increasingly rely on smartwatches not only for payments, communication, navigation, and remote control but also for continuous health monitoring, including heart rate tracking, blood oxygen measurement, and sleep analysis [5,10,11,20]. As these devices become more autonomous, many users interact with them independently of their smartphones [3], making them attractive targets for security threats.

Although extensive research has been conducted on the security of mobile devices and the Internet of Things (IoT), the security of smartwatches remains relatively underexplored. Previous studies have identified various vulnerabilities in wearable devices, such as motion sensor-based attacks [25], weaknesses in authentication mechanisms, and security risks associated with third-party applications [12]. However, most existing research has focused mainly on software-based vulnerabilities, largely overlooking side-channel threats in smartwatches.

In this study, we identify and analyze for the first time a significant electromagnetic (EM) leakage issue in smartwatches, particularly when operating on cellular networks. Unlike Bluetooth-only smartwatches, cellular-enabled smartwatches generate stronger EM emissions due to their higher power consumption and continuous network communication. These emissions originate from various hardware components, including wireless communication modules (4G/5G), processors, and sensors, all of which contribute to a unique EM side-channel footprint. When users interact with their smartwatches—such as receiving notifications, making calls, or synchronizing data—the wireless transmission and computational workload induce distinguishable EM patterns. Similarly, different in-app activities trigger specific processing states, leading to identifiable EM leakage signatures that an attacker can exploit. We have designed and implemented MagWatch to demonstrate the feasibility of leveraging our reported electromagnetic side-channel leakage to launch a contactless, fine-grained, and scalable attack on smartwatches for the first time.

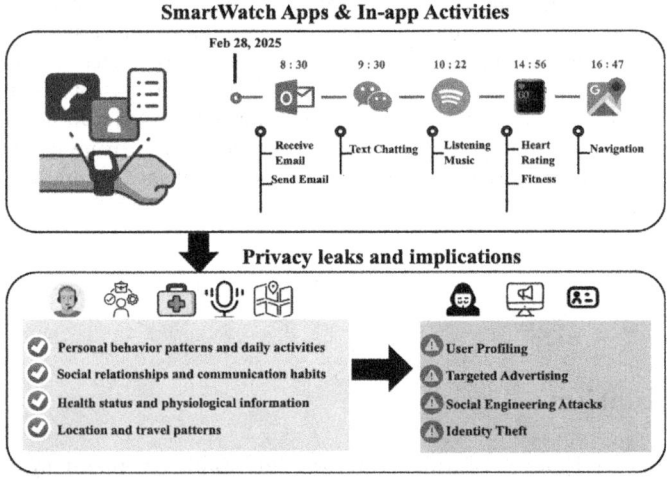

Fig. 1. Smartwatch Privacy Leaks and Their Consequences

EM side-channel attacks pose a particularly severe security risk as they leverage unintended physical emissions rather than software vulnerabilities [7,16,17], allowing adversaries to infer user activities without requiring malware installation or direct access to the device. Unlike cryptographic attacks that target encryption algorithms, EM-based techniques passively extract sensitive information from seemingly benign system operations, making them difficult to detect and mitigate. Moreover, these attacks can be conducted remotely, enabling an adversary to monitor smartwatch EM emissions from a distance without physical access, significantly increasing their feasibility as a scalable attack vector.

As shown in Fig. 1, beyond merely recognizing the applications in use (e.g., social media, music, or productivity apps), EM side-channel attacks can also identify fine-grained in-app activities, exposing detailed user behavior patterns. This means that even without direct access to the smartwatch, an adversary could infer whether a user is engaging in communication, media consumption, or work-related tasks, raising serious privacy concerns. More critically, the combination of app identification and in-app activity recognition enables adversaries to construct long-term behavioral profiles, revealing user preferences, app usage habits, daily routines, social interaction frequency, and even work intensity [22]. In certain cases, this could extend to exposing personal life patterns or professional confidentiality [19], demonstrating that EM side-channel attacks have far-reaching implications beyond simple app recognition.

We propose and demonstrate MagWatch, a contactless EM side-channel attack, and evaluate its effectiveness, in uncovering three key aspects of user privacy: app launching, in-app activity recognition, and behavioral inference. Our experimentation, conducted on multiple smartwatch models, demonstrates that MagWatch achieves high classification performance in different scenarios. Specifically, for *application recognition*, MagWatch successfully classifies 16 smartwatch applications, achieving an average accuracy above 90%, with only navigation and heart-rate monitoring apps showing moderate mis-classification due to similar sensor usage. Regarding *in-app activity recognition*, across multiple applications, MagWatch can differentiate specific in-app activities, with classification accuracy reaching over 85% in apps such as WeChat, Outlook, and Spotify. The attack exhibits exceptional effectiveness at close range, maintaining high accuracy. While accuracy gradually declines beyond 7.5 cm and drops below 20% at 12.5 cm, this aligns with the inherent range characteristics of EM-based attacks, which are optimized for short-distance precision, as consistently observed in the literature, such as [16,17].

This paper presents the following key contributions:

- This paper conducts systematic analysis of electromagnetic leakage in smartwatches, evaluating its impact across various applications and user activities.
- This paper introduces MagWatch, a novel contactless side-channel attack model that utilizes EM leakage to infer user interactions without requiring software exploitation.
- This paper explores countermeasures against EM-based privacy threats, proposing effective mitigation strategies while highlighting previously underestimated security risks in smartwatches.

By addressing this emerging security challenge, we aim to provide new insights into smartwatch privacy risks and contribute to the broader field of wearable device security and electromagnetic side-channel analysis.

2 Background and Related Work

2.1 Background

Electromagnetic (EM) signals arise from the movement of electric charges, as described by Maxwell's equations. When an electric current flows through a conductor, it generates both electric and magnetic fields. According to Ampère's Law with Maxwell's correction [15]:

$$\nabla \times \mathbf{B} = \mu_0 \mathbf{J} + \mu_0 \varepsilon_0 \frac{\partial \mathbf{E}}{\partial t} \tag{1}$$

where $\mathbf{B}$ is the magnetic field, $\mathbf{J}$ is the current density, μ_0 is the permeability of free space, and ε_0 is the permittivity of free space. This equation indicates that both electric currents and time-varying electric fields contribute to the formation of magnetic fields. In electronic devices, rapid switching of transistors, varying clock speeds, and fluctuating power consumption introduce dynamic electromagnetic emissions. These emissions, often categorized as electromagnetic interference (EMI), are a byproduct of hardware activity and can serve as a side channel for information leakage. In smartwatches, just like in smartphones [7], various hardware components contribute to the generation of EM signals. CPU and memory operations induce fluctuating electrical currents as processes are executed, leading to distinct magnetic field variations. Power management circuits dynamically regulate voltage and current, producing low-frequency magnetic fluctuations. Display drivers and touchscreen circuits generate periodic electromagnetic variations as screens refresh or register user interactions. Since different applications invoke different hardware modules upon launching, they induce unique electromagnetic patterns, as illustrated in Fig. 2.

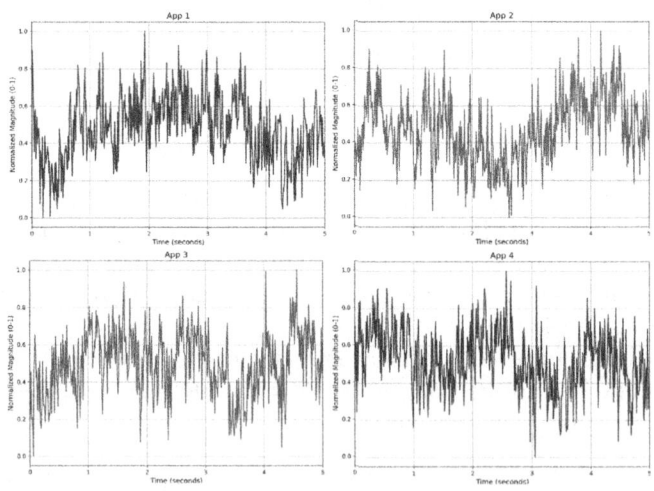

Fig. 2. Different EM Patterns of Four Apps (as Illustration)

2.2 Related Work

Most prior research on smartwatch privacy has focused on data collection and user perception rather than side-channel threats. Emmanuel Sebastian Udoh [23] and HongSuk Yoon [27] have both conducted user research on smartwatch privacy risks. Yoon's qualitative study focuses on users' perspectives regarding information tailoring and data privacy through surveys and interviews, while Udoh's exploratory study assesses American college students' privacy awareness and attitudes toward smartwatch-related privacy issues. A key concern raised in previous studies [12,24] is the "Privacy Paradox", the phenomenon where users express concerns about privacy but fail to take adequate protective measures. Many smartwatch users underestimate the extent of personal data their devices collect and share, including GPS location, health metrics, and communication logs [14]. Furthermore, the integration of third-party applications exacerbates privacy risks, as user data can be shared beyond their control, leading to potential security breaches. These researches rely on user surveys and behavioral studies rather than direct hardware-based security assessments.

Although privacy concerns regarding smartwatch applications have been studied, electromagnetic (EM) side-channel risks in smartwatches remain largely unexplored. Most existing EM side-channel attack research has been conducted on smartphones and computers, leaving wearable devices unexamined. Zhu et al. [28] first utilized mobile phone magnetometers to analyze electromagnetic radiation footprint from nearby computers, enabling app and webpage inference. Tao et al. [16,17] investigated electromagnetic side-channel leakage in smartphones during wireless charging, revealing how sensitive information can be inferred from EM emissions. Similarly, Yushicheng et al. [4] studied EM-based privacy leakage from computers, exposing vulnerabilities in cryptographic implementations and system operations. Furthermore, Yongjian Fu et al. [7] demonstrated how EM emissions from smartphones can be exploited to infer user activities and extract sensitive data. Additionally, prior work has demonstrated EM-based fingerprinting of USB devices [8] and profiling of IoT device activities through side-channel emissions [1]. These studies highlight the risks associated with EM-based side channels in traditional computing devices, yet the potential privacy threats posed by EM emissions from smartwatches remain largely unexamined. Given the compact design, reliance on multiple sensors, and frequent connectivity with smartphones and other IoT devices, smartwatches may exhibit distinct EM leakage characteristics, warranting further investigation.

3 Threat Model

We consider a realistic threat scenario where a victim wears a smartwatch and engages in daily activities such as making payments, answering calls, and monitoring heart rate (Fig. 3). These interactions generate distinct electromagnetic (EM) emissions, which an attacker can exploit to infer sensitive behaviors. This attack can occur in public or semi-public environments such as *cafes, offices, and public transportation hubs (e.g., airport, train station, or subway compartments)*, where users frequently interact with their smartwatches near tables and desks.

We assume that the attacker can be in close proximity to the victim, either by being physically present in the same space or by strategically placing hidden EM sensing devices in frequently visited locations. The adversary may be an unauthorized third party, such as a cybercriminal, a corporate espionage agent, or a surveillance entity seeking to extract private information. The attacker can discreetly deploy malicious EM sensing devices, such as software-defined radios (SDRs) or concealed antennas, under tables or desks to passively capture smartwatch-generated EM signals. Due to their small size, these sensing devices can be disguised as common objects such as earphones, chargers, power banks, or wireless mice, making them difficult to detect. Once deployed, these devices enable continuous and covert data collection. Beyond hardware attacks, the attacker may also exploit a compromised smartphone app running on the victim's paired device, using side channels to extract EM data in the background and upload it to a remote server for further analysis. More sophisticated attackers may deploy multiple hidden sensors across an environment, such as different tables in a shared workspace or a cafe, allowing them to aggregate signals from various perspectives to enhance tracking accuracy and improve the reconstruction of user activities. By analyzing these captured EM emissions, the attacker can infer financial transactions, call and messaging patterns, health monitoring behaviors, and authentication gestures, posing significant privacy risks. This threat model demonstrates that EM side channels can be practically leveraged to infer smartwatch activities without requiring direct access to the device, highlighting the feasibility and stealthiness of such attacks.

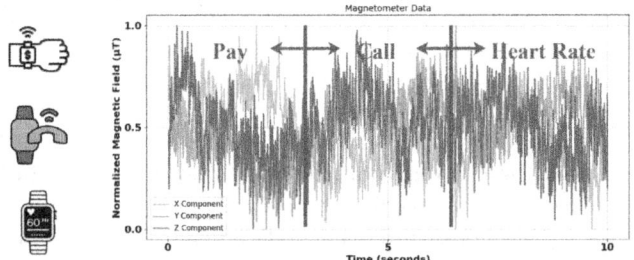

Fig. 3. Motivating Example Scenario: A user wearing a smartwatch sequentially makes a payment, receives a call, and checks their heart rate. Each action activates the smartwatch's corresponding functions, generating distinct electromagnetic signals throughout the process.

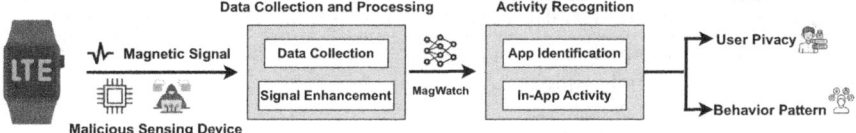

Fig. 4. Overview of MagWatch

4 Approach Overview

In this section, we provide a detailed overview of MagWatch, followed by an in-depth discussion of each stage in the pipeline. As illustrated in Fig. 4, the proposed system begins with data collection, where magnetic signals emitted by the smartwatch are captured using a malicious sensing device. These signals undergo enhancement and processing to improve their quality before being analyzed by the MagWatch model. The recognition phase leverages machine learning techniques to infer in-app activities and app launching events. Ultimately, this extracted information can be used to deduce user privacy risks and uncover behavioral patterns.

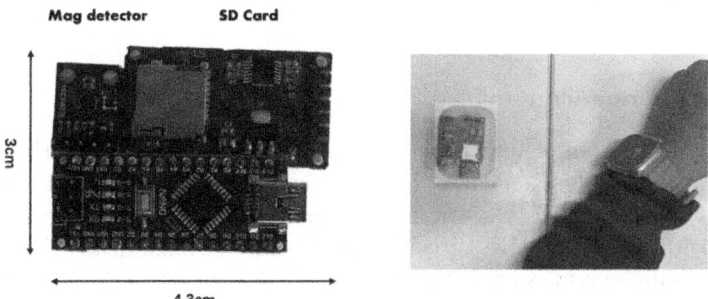

Fig. 5. Attack Device and Attack Scenarios

4.1 Data Collection

Data collection involves discreetly capturing EM signals in public or semi-public environments. Specifically, to record electromagnetic emissions from an iWatch 11.2 and iWatch 10 under various operating conditions, we developed a covert attack device designed for unobtrusive data collection, as shown in Fig. 5. This device is equipped with a QMC5883L sensor and an Arduino Nano, which processes the captured electromagnetic signals. Its compact and concealed design enhances effectiveness for covert data collection.

Algorithm 1. Electromagnetic Signal Enhancement via Wavelet Analysis

Require: EM signal sequence $X = \{x_1, x_2, \ldots, x_n\}$ with axes $\{x, y, z\}$, window size w
Ensure: Enhanced signal sequence $X_{enhanced}$
1: Initialize empty list $X_{enhanced}$
2: **while** number of samples in $X \geq w$ **do**
3:     Fetch subsequence X' of length w from X
4:     **for** each axis $a \in \{x, y, z, magnitude\}$ **do**
5:        $S_a \leftarrow$ signal values of axis a from X'
6:        $S_a \leftarrow S_a - \text{mean}(S_a)$ ▷ Remove DC component
7:        $S_a \leftarrow S_a / \text{std}(S_a)$ ▷ Normalize signal
8:        $C_a \leftarrow \text{WaveletDecompose}(S_a, \text{'db4'}, \text{level} = 4)$
9:        **for** each detail coefficient level $i \in \{1, 2, 3, 4\}$ **do**
10:           $T_i \leftarrow 0.5 \times \text{std}(C_a[i])$ ▷ Adaptive threshold
11:           $C_a[i] \leftarrow \text{SoftThreshold}(C_a[i], T_i)$
12:           $C_a[i] \leftarrow \text{sign}(C_a[i]) \times |C_a[i]|^{0.75}$ ▷ Non-linear enhancement
13:        **end for**
14:        $S'_a \leftarrow \text{WaveletReconstruct}(C_a, \text{'db4'})$
15:        $S'_a \leftarrow S'_a[0:w]$ ▷ Trim to original length
16:     **end for**
17:     Create enhanced frame $F_{enhanced}$ by combining all axes S'_a
18:     Append $F_{enhanced}$ to $X_{enhanced}$
19: **end while**
20: **return** $X_{enhanced}$

4.2 Signal Enhancement

This paper proposes an electromagnetic signal enhancement algorithm based on wavelet transform [6], specifically designed for processing multi-axis electromagnetic signals with subtle variations. The proposed method integrates wavelet decomposition with adaptive thresholding to enhance significant signal features while preserving overall signal integrity.

For wavelet selection, the Daubechies-4 (db4) wavelet is employed as the basis function due to its optimal balance between smoothness and localization, making it particularly suitable for electromagnetic signal analysis. The algorithm utilizes a multi-level wavelet decomposition strategy, where the signal is decomposed into approximation and detail coefficients across multiple frequency bands, enabling localized feature extraction.

To effectively suppress noise while retaining essential signal structures, adaptive thresholding is applied based on the statistical properties of wavelet coefficients, as shown in Algorithm 1. Specifically, the threshold value at each decomposition level is computed dynamically as:

$$T_j = 0.5 \times \sigma(C_a[i]) \tag{2}$$

where $\sigma(C_a[i])$ represents the standard deviation of the wavelet coefficients at level j. After thresholding, the detail coefficients undergo a non-linear enhancement transformation formulated as:

$$\hat{d}_{j,k} = \text{sign}(d_{j,k})\left((|d_{j,k}| - T_j)_+ \cdot |d_{j,k}|^{0.75}\right) \quad (3)$$

where $(x)_+$ denotes the positive part of x, ensuring that only coefficients exceeding the threshold contribute to the enhanced signal. The power-law transformation with an exponent of 0.75 amplifies relevant features while suppressing noise interference.

The proposed approach effectively enhances both global and localized signal variations, ensuring robust performance across diverse signal conditions. By processing each spatial axis independently and subsequently integrating them into a coherent enhanced signal, the method preserves both temporal and spatial relationships within the data.

This methodology is particularly advantageous for electromagnetic signal processing, where traditional enhancement techniques may struggle to capture subtle but meaningful variations. The combination of wavelet-based decomposition, adaptive thresholding, and non-linear enhancement makes this approach especially suitable for applications requiring high-precision signal analysis, such as pattern recognition in electromagnetic data.

4.3 Activity Recognition

Convolutional Neural Networks (CNNs) [26] are extensively used across different domains and demonstrate strong performance in feature extraction and classification tasks. Once we obtain the enhanced EM signal sequence, we first use a CNN to extract key features for subsequent activity recognition. To balance efficiency and performance, we opt for a simple two-layer CNN in this study, ensuring feature extraction without imposing a heavy computational load.

The first convolutional block consists of a convolution layer with 64 filters of size 7 and a stride of 1, followed by a max-pooling layer with stride 2, a ReLU activation [9], and a batch normalization layer. The second convolutional block has a similar structure as the previous one, except that the convolution layer is different. The second convolutional layer has a kernel size of 5, a stride of 1, and 128 output channels. After extracting the EM features, we then introduce two Long Short-Term Memory (LSTM) [13] layers to model the temporal dependencies. The first LSTM layer, with 128 units, processes the output features from the convolutional layers and transfers the information to the subsequent layer. To prevent overfitting and enhance generalization, a dropout layer with a rate of 0.3 is applied after the first LSTM layer. The second LSTM layer, consisting of 64 units, further refines the learned sequential features and outputs a fixed-length representation of the data, capturing the underlying temporal patterns. Last, a fully connected layer with 64 units and a ReLU activation function is added to process the output from the LSTM layers for the final activity prediction. A dropout layer follows to help prevent overfitting.

To validate the effectiveness of these hyperparameter settings, we evaluate the impact of different LSTM unit sizes and dropout rates on model accuracy in Sect. 5.3. The results indicate that a configuration of 128 LSTM units with a dropout rate of 0.3 achieves the highest validation accuracy of 93%.

5 Evaluation

5.1 Experiment Setup

Our evaluation involves two sets of equipment: an attack device for data collection and a computing device for processing, training, and testing the collected data. We can use the attack device we designed to discreetly collect the electromagnetic signals, as introduced in Sect. 4.1. We can also opt to use a smartphone's built-in electromagnetic sensor, as demonstrated in this experimental setup, for data collection instead. The experiment was carried out by collecting electromagnetic (EM) signals from the victim's device at a sampling rate of 10 milliseconds per data point.

The environment was intentionally non-isolated, with electromagnetic interference from other electronic devices to simulate a realistic setting. To capture the side-channel signals covertly, the receiver was placed underneath the victim's desk, directly below the target device. This setup replicates a realistic attack scenario where the attacker collects EM emissions without direct physical access to the victim's device, a method commonly used in practice, as documented in the literature.

The Sensor Logger app [2] was used to record electromagnetic signals, ensuring a controlled and systematic data acquisition process. To maintain consistency, the battery level of the smartwatch was kept between 60% and 80%, preventing extreme power states from affecting the collected signals. Additionally, all background applications on the smartwatch were forcefully closed, ensuring that only the target application was running during the experiments, allowing for consistent and comparable data.

5.2 Datasets

We construct four datasets using commodity smartwatches under different conditions to evaluate their effectiveness in Sect. 5.4. These datasets are collected from four smartwatch models (Apple Watch Ultra 2, Apple Watch 10, Xiaomi Watch S3, and Huawei Watch 4) to train various models and assess their performance in app recognition, in-app activity classification, signal analysis across different distances, and cross-device application.

Table 1. App Categories and Corresponding App List

Category	App List (on Both Android and iOS)
Video & Music	Youtube, Spotify, Apple Music
Social	Wechat, Message, Outlook, QQ Email, WhatsApp
Navigation	Apple Map, Google Map, Baidu Map, Weather
Health	Heart Rate, Fitness
Pay	BofA, Alipay

- $\mathcal{D}_{\mathbf{app}}$: We selected 16 commonly used smartwatch apps listed in Table 1 from categories such as health, music, communication, and navigation. For each app in the categories, we recorded 3 s of EM signals and repeated the collection 100 times.
- $\mathcal{D}_{\mathbf{act}}$: We selected 5 representative apps from the collected set and recorded EM signals for 5 different in-app activities. Each activity was recorded for 3 s and repeated 100 times.
- $\mathcal{D}_{\mathbf{dis}}$: We collected EM signals at different distances from the smartwatch, spanning from 0 cm to 20 cm, with increments of 2.5 cm, to analyze the signal variations across a range of proximities. Each recording lasted 10 s and was repeated 100 times per distance.
- $\mathcal{D}_{\mathbf{dev}}$: We collected usage data from four smartwatch devices while connected to a cellular network. Each device's dataset includes three key components: app recognition, in-app activity classification, and signal analysis at different distances.

These datasets, gathered from Apple Watch Ultra 2, Apple Watch 10, Xiaomi Watch S3, and Huawei Watch 4, are used to train models and evaluate their performance in distinguishing apps, identifying specific activities within apps, and assessing the impact of distance on signal variations. This comprehensive dataset enables cross-device comparisons and enhances the robustness of our analysis.

5.3 Hyperparameter Evaluation

In our hyperparameter optimization experiments for the CNN-LSTM architecture, we conducted a grid search across LSTM units (50–250) and dropout rates (0.1–0.5). Figure 6 illustrates two cross-sectional analyses of the hyperparameter tuning results: the left graph examines the effect of varying LSTM units with the dropout rate fixed at 0.3, while the right graph explores different dropout rates with the LSTM unit count fixed at 128, evaluating validation accuracy across different settings.

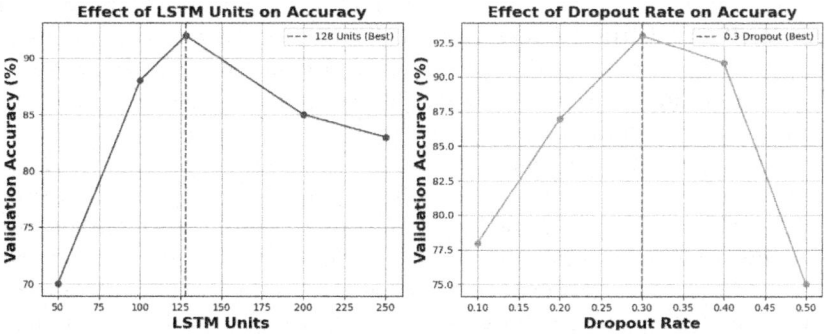

Fig. 6. Effect of Different Units and Dropout Rates

Our results indicated that 128 LSTM units yielded the highest validation accuracy, as additional units increased complexity without further improving performance, leading to a decline in validation accuracy. Similarly, a dropout rate of 0.3 achieved the best trade-off between regularization and performance, resulting in 93% validation accuracy. Higher dropout rates (0.4–0.5) led to excessive information loss, whereas lower dropout rates (0.1–0.2) provided insufficient regularization, reducing the model's generalization ability.

5.4 Effectiveness

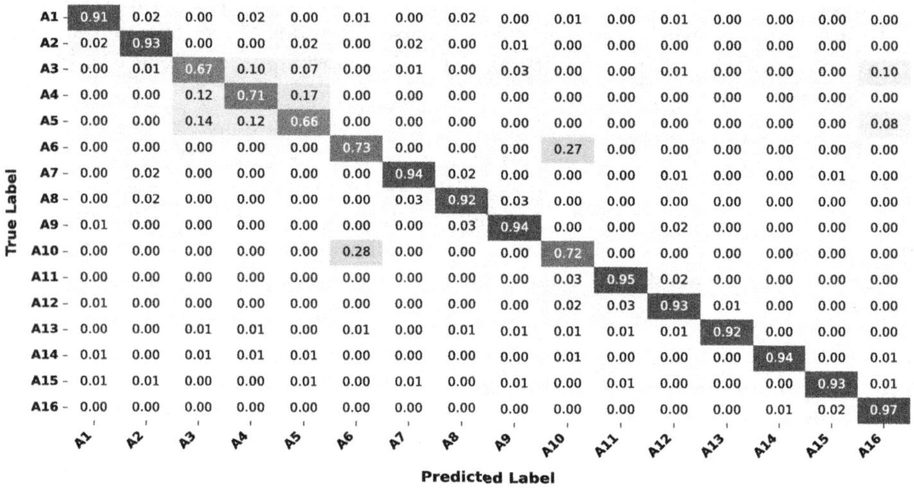

Fig. 7. The Classification Performance of MagWatch on iWatch

Effectiveness of App Launching Recognition. The confusion matrix in Fig. 7 presents the effectiveness of our method in recognizing 16 different applications on iWatch Ultra 2 (used here for illustration) based on EM signals. To evaluate the classification performance, we utilize 80% of each application from the $\mathcal{D}_{\mathbf{app}}$ for training and assess the model using the remaining 20% data.

Overall, the recognition model achieves an average accuracy exceeding 90% across most applications, demonstrating its robustness in distinguishing different app usage scenarios. However, specific application groups, such as A3, A4, and A5 (navigation and mapping apps) and A6 and A10 (heart rate and fitness apps), exhibit lower classification accuracy. This is likely due to their reliance on the same underlying sensors (GPS for navigation apps and heart rate sensors for fitness-related apps), resulting in similar EM signal patterns that increase misclassification rates. Despite this challenge, our model is still capable of effectively distinguishing applications based on the type of sensors they utilize. This indicates that even when fine-grained app differentiation is difficult, EM signal analysis remains effective in *identifying the broader sensor usage patterns*, such as whether an app primarily interacts with GPS, heart rate sensors, or other components.

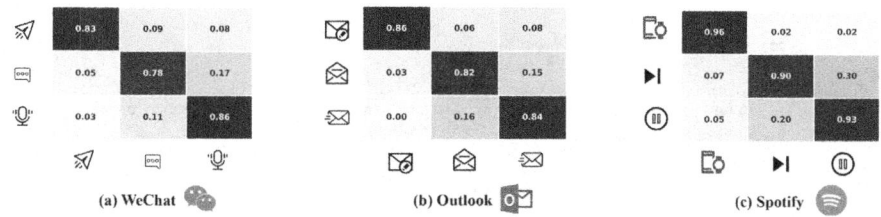

Fig. 8. Accuracy of In-App Activity Recognition Across Different Applications

Effectiveness of In-App Activity Recognition. To further validate the effectiveness of MagWatch, we conduct additional experiments by using $\mathcal{D}_{act}$ of three widely used applications—Spotify, WeChat, and Outlook—each involving three distinct activities. We then implement a CNN-based classification model to assess the feasibility of distinguishing fine-grained in-app activities based on their EM signatures.

Specifically, as shown in Fig. 8(a)-Fig. 8(c), in Spotify, we evaluate the recognition of skipping a song, pausing playback, and switching devices. In WeChat, we analyze the ability to classify receiving a message, sending a message, and listening to a voice message. Similarly, for Outlook, we investigate the recognition of receiving an email, sending an email, and editing an email.

The results show MagWatch's classification accuracy of 83.0%, 78.0%, and 86.0% for WeChat, 82.0%, 84.0%, and 86.0% for Outlook, and 90.0%, 93.0%, and 96.0% for Spotify, in recognizing the in-app activities elaborated above. These results validate the effectiveness of MagWatch, which not only identifies the launched application but also accurately recognizes fine-grained in-app activities.

Impact of Position and Distance. Figure 9 illustrates the impact of distance on the classification accuracy of app launching recognition. In practice, an attacker could place a disguised device near the target smartwatch at various distances to capture EM emissions. To evaluate this, we conducted experiments by positioning the attacking device at distances ranging from 0 cm to 20 cm at increment of 2.5 cm.

The results indicate that while the model maintains high accuracy at close range, the performance gradually declines as the distance increases. Beyond 7.5 cm, the accuracy drops significantly, and at 12.5 cm, it falls below 20%, making app recognition nearly ineffective. This decline, as consistently observed in the literature such as [16,17], suggests that as the distance increases, the EM

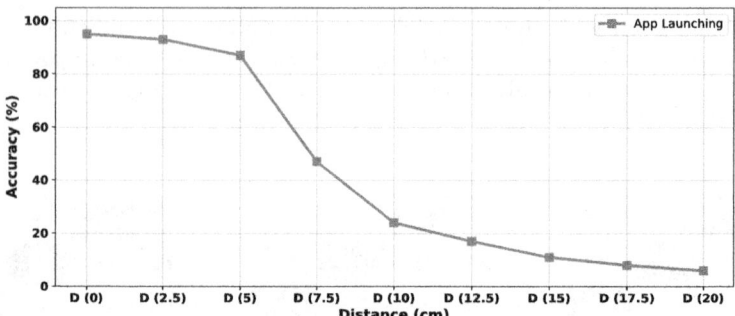

Fig. 9. Impact of Distance

signal disturbances become too weak to be reliably captured, limiting the effectiveness of remote attacks. This outcome highlights the practical constraints of EM-based side-channel attacks, where attackers must be in close proximity to achieve high classification accuracy. The results suggest that physical distance and shielding could serve as natural countermeasures to mitigate such privacy risks.

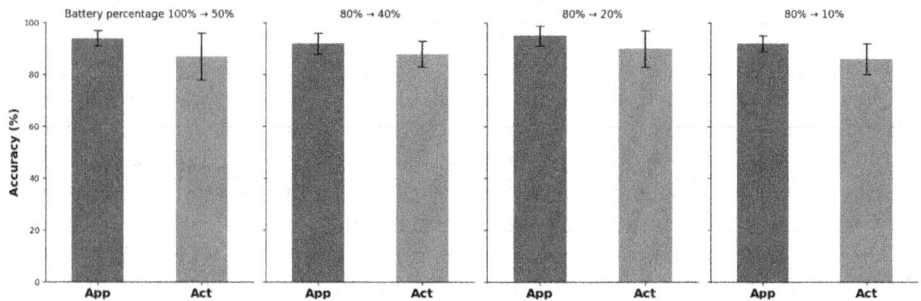

Fig. 10. Impact of Battery Level

Impact of Battery Level. To evaluate the impact of smartwatch battery levels on the performance of MagWatch, we conducted experiments under four different battery conditions: 60%, 40%, 20%, and 10%. The dataset for these experiments was collected from the Apple Watch Ultra 2 (used here for example), including both app usage data (denoted as "App" data in Fig. 10) and in-app activity data (denoted as "Act" data). The classification accuracy, as shown in Fig. 10, represents the mean accuracy across multiple trials and remains largely unaffected by battery fluctuations. One key factor contributing to this stability is the power-efficient architecture of modern smartwatches, which ensures consistent computational performance regardless of battery level. These findings underscore MagWatch's strong generalization capability across different battery conditions, reinforcing its robustness and reliability for electromagnetic-based inference. This ensures that MagWatch can be robustly operated under varying real-world scenarios without concerns about performance degradation due to battery fluctuations.

5.5 Analysis of Cold/Hot Start

Table 2. Classification Results of EM Signals Generated During COLD/HOT-Start

	kNN	SVM	CNN	CNN-LSTM
Cold	67%	75%	83%	93%
Hot	6%	9%	13%	16%

As stated in [18], cold start and hot start exhibit distinct characteristics during the application launch process. Cold start refers to launching an application from scratch, requiring CPU initialization, memory allocation, and data loading. This process generates more prominent EM signal characteristics, making it easier to identify. In contrast, a hot start occurs when an application is partially loaded in the background, allowing the system to restore data from cache with reduced CPU activity. This results in weaker EM signal characteristics, which makes identification more challenging.

This phenomenon is not limited to smartphones; it also applies to smartwatches and other wearable devices, as demonstrated in Table 2. The table presents the results of applying various classification models (kNN, SVM, CNN, CNN-LSTM) to smartwatch EM emission data collected from cold start versus hot start scenarios. It clearly shows that classification accuracy is high in cold start scenarios, particularly with CNN-LSTM, while accuracy is significantly lower in hot start scenarios across all models. Given that smartwatches typically use more aggressive power management strategies, the result in Table 2 is expected, as the differences in resource scheduling between cold and hot starts are even more pronounced in smartwatch devices.

6 Discussion

In this section, we discuss the limitations of MagWatch, the potential defense countermeasure, and the future work.

Limitations. We have implemented MagWatch to demonstrate the feasibility of electromagnetic (EM) side-channel inference for smartwatch applications. While the results are promising, several limitations remain in the current work.

First, MagWatch is evaluated in controlled experimental settings where the smartwatch remains stationary during app interactions. However, we have not fully explored its performance in dynamic scenarios, such as users wearing the smartwatch while walking or performing other activities. Theoretically, MagWatch could still capture meaningful EM traces in these cases by refining signal preprocessing techniques and adapting feature extraction methods. However, movement introduces additional noise and variability, making real-time inference more challenging.

Second, our experiments are conducted under relatively close-range and controlled environmental conditions to validate the feasibility of EM-based inference. The accuracy of MagWatch may degrade as the distance between the sensor and the smartwatch increases due to EM signal attenuation. To extend its applicability, further studies are required to refine the model to accommodate varying distances and external interference.

Countermeasures. Several countermeasures can help mitigate EM information leakage from smartwatches. One approach is physical shielding with ferromagnetic materials to reduce magnetometer interference, though this is often overlooked due to design constraints prioritizing size and weight. Increasing physical distance(in Sect. 5.4) can naturally reduce privacy risks. Detecting and excluding nearby devices that may capture EM signals could further enhance protection. Lastly, current EM emission regulations may be too lax; enforcing stricter standards could prevent adversaries from inferring user behavior through app usage patterns.

Future Work. For future work, we plan to explore real-world deployment scenarios, considering factors such as user motion, device orientation, and environmental EM noise. Additionally, we aim to improve the generalization of MagWatch by integrating adaptive learning techniques to enhance robustness across different smartwatch models and operating conditions.

7 Conclusion

This paper explores and demonstrates the feasibility of EM-based side-channel attacks towards smartwatches, where EM emissions generated during smartwatch operations can be captured to reveal both app usage and in-app activities without requiring direct device access. We have designed and implemented MagWatch, a novel non-intrusive attack that utilizes wavelet transform for signal processing and a CNN-LSTM model to identify applications and in-app activities, achieving up to 90% accuracy across multiple smartwatch models. To the best of our knowledge, this is the first attack targeting smartwatch EM emissions to infer user privacy and behavior patterns related to app interaction/navigation.

Acknowledgment. Drs. Xiaoyan Sun and Jun Dai are supported by NSF DGE-2409851. Dr. Jun Dai is also supported by NSF DGE-1934285/2403603.

References

1. Amodei, A., et al.: Experimental analysis of side-channel emissions for IOT devices activities' profiling. In: 2023 IEEE International Workshop on Metrology for Industry 4.0 & IoT (MetroInd4.0&IoT), pp. 42–47 (2023). https://doi.org/10.1109/MetroInd4.0IoT57462.2023.10180188
2. App Store: Sensor Logger. https://apps.apple.com/us/app/sensorlogger-csv-export/id15052035. Accessed 22 Feb 2025
3. Chen, X., Chen, W., Liu, K., Chen, C., Li, L.: A comparative study of smartphone and smartwatch apps. In: Proceedings of the 36th Annual ACM Symposium on Applied Computing, pp. 1484–1493 (2021)
4. Cheng, Y., et al.: Magattack: guessing application launching and operation via smartphone. In: Proceedings of the 2019 ACM Asia Conference on Computer and Communications Security, pp. 283–294 (2019)
5. Chuah, S.H.W., Rauschnabel, P.A., Krey, N., Nguyen, B., Ramayah, T., Lade, S.: Wearable technologies: the role of usefulness and visibility in smartwatch adoption. Comput. Hum. Behav. **65**, 276–284 (2016)
6. Farge, M., et al.: Wavelet transforms and their applications to turbulence. Annu. Rev. Fluid Mech. **24**(1), 395–458 (1992)
7. Fu, Y., Yang, L., Pan, H., Chen, Y.C., Xue, G., Ren, J.: Magspy: Revealing user privacy leakage via magnetometer on mobile devices. IEEE Trans. Mob. Comput. (2024)
8. Ibrahim, O.A., Sciancalepore, S., Oligeri, G., Pietro, R.D.: Magneto: fingerprinting USB flash drives via unintentional magnetic emissions. ACM Trans. Embed. Comput. Syst. **20**(1) (2020). https://doi.org/10.1145/3422308
9. Ioffe, S., Szegedy, C.: Batch normalization: accelerating deep network training by reducing internal covariate shift. In: International Conference on Machine Learning, pp. 448–456. PMLR (2015)
10. Isakadze, N., Martin, S.S.: How useful is the smartwatch ECG? Trends Cardiovasc. Med. **30**(7), 442–448 (2020)
11. Jat, A.S., Grønli, T.M.: Smart watch for smart health monitoring: a literature review. In: International Work-Conference on Bioinformatics and Biomedical Engineering, pp. 256–268. Springer (2022)
12. Kang, H., Jung, E.H.: The smart wearables-privacy paradox: a cluster analysis of smartwatch users. Behav. Inf. Technol. **40**(16), 1755–1768 (2021)
13. Karim, F., Majumdar, S., Darabi, H., Chen, S.: LSTM fully convolutional networks for time series classification. IEEE Access **6**, 1662–1669 (2017)
14. Kim, J.W., Lim, J.H., Moon, S.M., Jang, B.: Collecting health lifelog data from smartwatch users in a privacy-preserving manner. IEEE Trans. Consum. Electron. **65**(3), 369–378 (2019)
15. Maxwell, J.C.: The Scientific Papers of James Clerk Maxwell..., vol. 2. University Press, Cambridge (1890)
16. Ni, T., et al.: Exploiting contactless side channels in wireless charging power banks for user privacy inference via few-shot learning. In: Proceedings of the 29th Annual International Conference on Mobile Computing and Networking, pp. 1–15 (2023)
17. Ni, T., et al.: Uncovering user interactions on smartphones via contactless wireless charging side channels. In: 2023 IEEE Symposium on Security and Privacy (SP), pp. 3399–3415. IEEE (2023)
18. Pan, H., et al.: Magthief: stealing private app usage data on mobile devices via built-in magnetometer. In: 2021 18th Annual IEEE International Conference on Sensing, Communication, and Networking (SECON), pp. 1–9. IEEE (2021)

19. Qiao, Y., Zhao, X., Yang, J., Liu, J.: Mobile big-data-driven rating framework: measuring the relationship between human mobility and app usage behavior. IEEE Network **30**(3), 14–21 (2016)
20. Reeder, B., David, A.: Health at hand: a systematic review of smart watch uses for health and wellness. J. Biomed. Inform. **63**, 269–276 (2016)
21. StraitsResearch: Report: Smartwatch market trends, growth, and forecast 2033. https://straitsresearch.com/report/smartwatch-market. Accessed 06 Jan 2025
22. Tu, Z., et al.: Your apps give you away: distinguishing mobile users by their app usage fingerprints. Proc. ACM Interact. Mob. Wearable Ubiquit. Technol. **2**(3), 1–23 (2018)
23. Udoh, E.S., Alkharashi, A.: Privacy risk awareness and the behavior of smartwatch users: A case study of indiana university students. In: 2016 Future Technologies Conference (FTC), pp. 926–931 (2016). https://doi.org/10.1109/FTC.2016.7821714
24. Udoh, E.S., Alkharashi, A.: Privacy risk awareness and the behavior of smartwatch users: a case study of Indiana university students. In: 2016 Future Technologies Conference (FTC), pp. 926–931. IEEE (2016)
25. Williams, M., Nurse, J.R., Creese, S.: (smart) watch out! encouraging privacy-protective behavior through interactive games. Int. J. Hum Comput Stud. **132**, 121–137 (2019)
26. Wu, J.: Introduction to convolutional neural networks. national Key Lab for Novel Software Technology. Nanjing University. China **5**(23), 495 (2017)
27. Yoon, H.S., Shin, D.-H., Kim, H.: Health information tailoring and data privacy in a smart watch as a preventive health tool. In: Kurosu, M. (ed.) HCI 2015. LNCS, vol. 9171, pp. 537–548. Springer, Cham (2015). https://doi.org/10.1007/978-3-319-21006-3_51
28. Zhu, Z., Pan, H., Chen, Y.C., Ji, X., Zhang, F., You, C.W.: Magattack: remote app sensing with your phone. In: Proceedings of the 2016 ACM International Joint Conference on Pervasive and Ubiquitous Computing: Adjunct, pp. 241–244 (2016)

Privacy-Preserving, Secure and Certificate-Based Integrity Auditing for Cloud Storage

Wenhao Wang[1], Yu Li[1], Yinxia Sun[2(✉)], Yuan Zhang[1,3(✉)], and Sheng Zhong[1,3]

[1] School of Computer Science, Nanjing University, Nanjing 210023, China
{wenhao.wang,liyu}@smail.nju.edu.cn, {zhangyuan,zhongsheng}@nju.edu.cn
[2] School of Computer Science and Electronic Information,
Nanjing Normal University, Nanjing 210023, China
[3] State Key Laboratory for Novel Software Technology, Nanjing University,
Nanjing 210023, China

Abstract. Cloud data security and privacy have garnered significant attention in recent years, with data integrity being a key concern. Most existing auditing protocols overlook the unforgeability of integrity proofs against Man-in-the-Middle (MitM) attacks. Specifically, MitM attackers can forge integrity proofs to conceal tampering with the data, misleading the data owner (DO) into believing the data is unchanged. This could lead to more severe consequences when the DO unknowingly uses compromised data. In this paper, we propose PSCIA, a privacy-preserving, secure and certificate-based integrity auditing protocol for cloud storage. Firstly, PSCIA provides unforgeability for the integrity proofs to resist MitM attackers. Secondly, we introduce a novel privacy attack during auditing that can reveal whether DOs have modified the cloud data, and then propose a defense method. Thirdly, by employing the certificate-based cryptosystem, PSCIA eliminates the need for secure channels, resolves the key escrow problem, and reduces certificate management burden. Furthermore, we formally prove that PSCIA is secure in the random oracle model. Finally, theoretical analysis and experiments demonstrate that PSCIA improves computational efficiency over related schemes, reducing the number of bilinear pairings computed by the DO and CSP from linear to constant complexity.

Keywords: Integrity Auditing · Privacy-preserving · Cloud Storage · Certificate-based Cryptosystem

1 Introduction

As clouds offer on-demand and high-quality storage services, they have become indispensable in our daily lives. Data owners (DOs), such as resource-limited devices, can reduce their local storage and data management costs by outsourcing their large-scale data to cloud servers owned by cloud service providers (CSPs) like Amazon and Google [7,20]. However, due to the transfer of data management from DOs to CSP, DOs face high-risk security threats [10,27]. Generally speaking, personal data often includes private and sensitive information, such as business secrets and health reports. If this information is modified, destroyed, or deleted in cloud servers, it would be disastrous for DOs. Therefore, it is crucial for DOs to regularly audit the integrity of their cloud data [21]. To facilitate this, hiring a third-party auditor (TPA) is an effective and common method to alleviate the computational and communication burden on DOs [12].

As an efficient way to guarantee the integrity of cloud data, remote auditing enables DOs to audit data without downloading the entire dataset. In remote auditing, the CSP addresses a random challenge from the TPA by providing an integrity proof. The TPA then verifies this proof and communicates the audit results back to the DO. However, typical remote auditing protocols allow anyone with access to the data to create a valid integrity proof, posing hidden risks. For instance, a Man-in-the-Middle (MitM) attacker [3] could seize the CSP-TPA communication channel. If the MitM attacker breaches the cloud database, obtains the target dataset, and alters some data blocks, they could delay or prevent the DO from discovering these changes by replacing the proof generated by the CSP with a forged proof. Should the unknowing DO download the modified data during this period, it could lead to greater damage. Thus, against MitM attackers, ensuring the unforgeability of integrity proofs is essential.

Additionally, curious TPAs may threaten DO's data modification privacy. For example, sensitive financial data, like account balances or credit scores, is typically stored in specific dataset locations. When the DO modifies this data, the associated tags update, altering the integrity proof if the challenge remains unchanged. In this scenario, a curious TPA could repeatedly issue the same challenge to some targeted data blocks, not only detecting potential modifications to them based on changes in the integrity proof but also inferring the approximate time of modification. We refer to this privacy attack as challenge replay attacks (CRAs). Most TPA-assisted remote auditing protocols overlook CRAs, necessitating a resistance mechanism in such systems.

Many current remote auditing protocols depend on the public key infrastructure (PKI) [8]. Although widely used in practice, PKI introduces considerable certificate management overhead, such as certificate issuance and revocation [2,19]. To address this, the identity-based cryptosystem (IBC) [19] has been explored. In IBC, a user's public key is their identity, and the private key is generated by a trusted authority known as the private key generator (PKG). However, the private key must be securely delivered, requiring secure channels. Moreover, IBC suffers from the key escrow problem, since the PKG can generate users' private keys at will, posing a significant trust risk. Certificateless cryptography (CLC) was introduced as a compromise between PKI and IBC. In CLC, the user's private key is derived from a combination of a user-generated secret and a partial private key issued by the key generation center (KGC). This design avoids key escrow and reduces certificate management. However, it still requires a secure channel to transmit the partial private key. To resolve these issues, Gentry [6] proposed the certificate-based cryptosystem (CBC). In CBC, only users with valid certificates possess legitimate identities, allowing them to generate valid signatures and decrypt ciphertexts. Unlike traditional PKI, CBC embeds user identity information in the certificate, simplifying verification and reducing management overhead. Moreover, users generate and retain their own private keys, ensuring key escrow freeness, and certificates can be transmitted over public channels without compromising security.

1.1 Our Contributions

Our main contributions are summarized as follows:

1. We propose PSCIA, a certificate-based integrity auditing protocol that ensures integrity proof unforgeability, neutralizing MitM attacker threats. Built on CBC, PSCIA reduces certificate management overhead, eliminates key escrow risks, and removes the requirement for secure channels.
2. To counter the challenge replay attacks (CRAs), we introduce a novel challenge generation method that safeguards DOs's data modification privacy.
3. We formally analyze the security of PSCIA and prove its security under the random oracle model.
4. We conduct both theoretical and experimental analyses of PSCIA, demonstrating its high computational efficiency. Specifically, the DO and CSP in PSCIA are required to compute a constant number of bilinear pairings, which is better than other relative schemes.

1.2 Related Work

To alleviate the local storage burden, DOs typically do not retain local copies of their outsourced dataset after uploading, retaining only its hash value or metadata label [25]. In this context, Deswarte et al. [4] proposed the first remote auditing scheme to verify the integrity of the outsourced dataset. Their scheme requires the TPA to download the entire dataset from the CSP, compute its hash value, and compare the computed hash value with the originally stored one to audit data integrity. However, the efficiency of their scheme is generally regarded as unacceptable.

To address this issue, Ateniese et al. [1] introduced the provable data possession (PDP) model, eliminating the need to download the entire outsourced dataset. They created homomorphic verifiable tags (HVTs) to audit the integrity of the outsourced dataset. Subsequently, Wang et al. [23] proposed a publicly verifiable PDP scheme that utilizes a fully trusted TPA, aiming to alleviate the auditing workload for the DO.

These schemes rely on the public key infrastructure, where certificate management complexity hinders scalability of cloud storage systems [2,19]. To overcome this, identity-based remote auditing protocols [10,15,25] have emerged. Li et al. [10] proposed an efficient identity-based remote auditing scheme, which improves the auditing efficiency by batch auditing. Liu et al. [15] introduced a scheme using identity-based sanitized signatures to protect sensitive data privacy. Zhang et al. [25] proposed a protocol leveraging smart contracts for integrity auditing, reducing reliance on fully trusted TPAs. However, a common limitation of identity-based remote auditing schemes is the key escrow problem and the requirement for secure channels, which also restrict their scalability in large cloud systems.

Recently, Li et al. [11] proposed CBDIAP, a certificate-based remote auditing protocol. By employing CBC, CBDIAP alleviates the certificate management burden on the certificate authority, dispenses with secure channel requirements, and ensures key escrow freeness. However, Zhang et al. [26] demonstrated that the HVTs in CBDIAP can be forged by the CSP and proposed a provably secure HVT construction.

As cloud storage privacy concerns grow, privacy has become a critical factor in remote auditing protocol design [5,9,15,24,27]. For instance, Gai et al. [5] proposed PPADT, a privacy-preserving remote auditing scheme protecting DO identity from TPAs. Li et al. [9] introduced a remote auditing protocol designed for resource-limited DOs, allowing TPAs to verify data integrity while preserving data privacy. Wu et al. [24] proposed a public auditing protocol which protects the privacy of both outsourced data and its upload timestamp. Nevertheless, most current TPA-assisted remote auditing schemes overlook the privacy leakage risks associated with data modifications.

2 Preliminaries

2.1 System Model

The system model of PSCIA, depicted in Fig. 1, comprises four entity types:

- **Data owner (DO)**: A resource-constrained device, the DO uploads its dataset to cloud storage to eliminate the local storage burden.
- **Third-party auditor (TPA)**: Hired by the DO, the TPA issues random integrity challenges to the CSP, verifies the returned integrity proof, and reports the auditing result to the DO.
- **Cloud service provider (CSP)**: With powerful storage and computational capabilities, the CSP provides data storage services to DOs. In this paper, the CSP providing cloud storage services to a DO is referred to as the DO's designated CSP (DCSP).
- **System administrator (SA)**: The SA issues certificates to assign legal identities to other entities, granting them access to the cloud storage system only if they possess their valid certificates.

Definition 1. *PSCIA is an integrity remote auditing protocol that consists of four processes: System Setup, Registration, Outsourcing, and Integrity Auditing.*

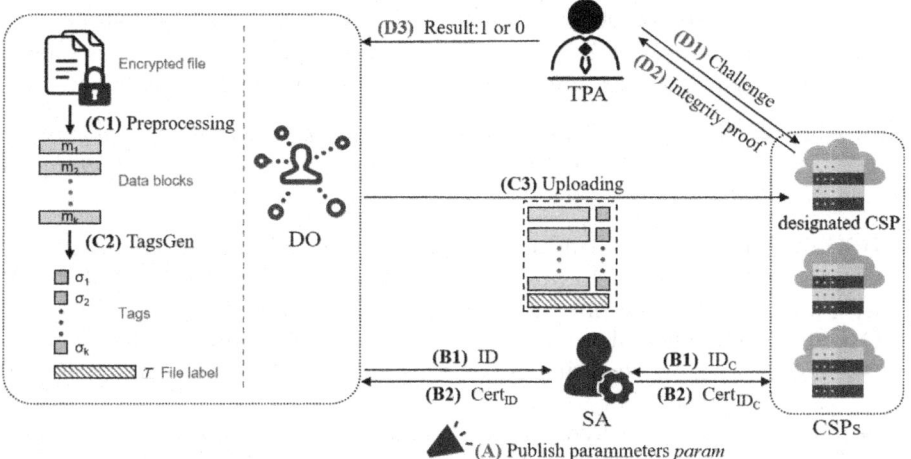

Fig. 1. The system model in PSCIA: (A) System Setup, (B) Registration, (C) Outsourcing, (D) Integrity Auditing.

- **System Setup**: The SA generates system parameters $params$ and a master private key msk, then publishes $params$.
- **Registration**: Each cloud user, including CSPs and DOs, generates their private/public key pair (usk_{ID}, upk_{ID}), submits their identity ID and public key upk_{ID} to the SA, and receives a verifiable certificate $Cert_{ID}$.
- **Outsourcing**: The DO encrypts and partitions the original file into a dataset F, generates its label τ and tags for all data blocks of F, and uploads τ, the data blocks and the tags to the DCSP.
- **Integrity Auditing**: The TPA sends a random challenge $chal$ to the DCSP. The DCSP responds with an integrity proof PF, which the TPA verifies and reports the result to the DO.

2.2 Threat Model

The system administrator (SA), representing the strongest MitM attacker, is modeled as the adversary $\mathcal{A}$. Therefore, we allow $\mathcal{A}$ access to the master private key msk but prohibit if from altering the DCSP's public key. $\mathcal{A}$ can perform following attacks:

- *Forgery attacks*: $\mathcal{A}$ spoofs the TPA by forging a valid integrity proof for the latest challenge.
- *Replay attacks*: $\mathcal{A}$ reuses a prior valid proof to mislead the TPA for a new challenge.
- *Replace attacks*: $\mathcal{A}$ attempts to forge a valid integrity proof by replacing challenged data blocks with others.

The unforgeability of integrity proofs, denoted as UF_p, ensures that, after the DO uploads the dataset and deletes the local copy, only the DCSP can generate a valid integrity proof. Accordingly, we define PSCIA's security as follows:

Definition 2. Security. *If the probability that any probabilistic polynomial-time (PPT) adversary $\mathcal{A}$ can win the following interactive game is negligible, we consider PSCIA to be secure.*

- **Initialization**: The challenger $\mathcal{C}$ generates public parameters $params$, a master private key msk, and a private/public key pair (usk_{ID}, upk_{ID}) of the DO with identity ID, then provides $\mathcal{A}$ with $params$, msk, ID, and upk_{ID}.
- **Query on Oracles**: $\mathcal{A}$ can adaptively query the following oracles:
  - *KeyGen.* Upon receiving a DCSP's identity ID_C, $\mathcal{C}$ returns its private/public key pair to $\mathcal{A}$.
  - *TagsGen.* Upon receiving a DCSP's identity ID_C and a dataset F named $fname$, $\mathcal{C}$ returns the label τ of $fname$ and tags $\{\sigma_i\}_{i=1}^k$, where k is the amount of data blocks in the dataset $fname$.
  - *ProofGen.* Upon receiving a DCSP's identity ID_C, a challenge $chal = \{fname, \{i, cnum_i\}_{i \in I}\}$, and data blocks with tags $\{m_i, \sigma_i\}_{i=1}^k$, $\mathcal{C}$ returns the integrity proof PF to $\mathcal{A}$.
- **Output**: $\mathcal{A}$ outputs a valid integrity proof PF^* for a new DCSP with identity ID_{C^*} based on a fresh challenge from $\mathcal{C}$.

2.3 Privacy Model

In PSCIA, TPAs are semi-honest, meaning they perform auditing tasks faithfully but seek additional information through challenge replay attacks (CRAs). By repeatedly auditing sensitive data blocks over time, TPAs can detect modifications to them by the DO. This compromises the DO's data modification privacy, adversely affecting the DO.

2.4 Definitions

Definition 3. Admissible Bilinear Map *[17]. Let $\mathbb{G}_1$ and $\mathbb{G}_T$ be two multiplicative cyclic groups of prime order q. An admissible bilinear map $e: \mathbb{G}_1 \times \mathbb{G}_1 \to \mathbb{G}_T$ should satisfy the following three properties:*

(1) *Bilinear: For all $a, b \in \mathbb{Z}_q$ and $g, h \in \mathbb{G}_1$, the equation $e(g^a, h^b) = e(g, h)^{ab}$ holds.*
(2) *Non-degenerate: If g is a generator of $\mathbb{G}_1$, $e(g, g)$ is a generator of $\mathbb{G}_T$.*
(3) *Computable: For all $g, h \in \mathbb{G}_1$, always an algorithm exists to compute $e(g, h)$ efficiently.*

Definition 4. Collision-resistant Hash Function *[18]. For input space $\mathbb{M}$ and output space $\mathbb{N}$, a collision-resistant function is a mapping $H: \mathbb{M} \to \mathbb{N}$ that maps a message in $\mathbb{M}$ into a message digest with fix-size in $\mathbb{N}$. The mapping satisfies the collision-resistant property, that is, for a hash function H, it is computationally infeasible to find two distinct messages (m, m') such that $H(m) = H(m')$.*

Definition 5. *Bilinear Diffie-Hellman (BDH) Problem Assumption* *[2]. For a, b, c which are randomly picked and a generator $g \in \mathbb{G}_1$. Given g^a, g^b and g^c, it is hard for any PPT algorithm to output $e(g, g)^{abc}$.*

3 Our Proposed Protocol

Based on the existing homomorphic verifiable tag (HVT) constructions in [23, 26], we design a novel HVT construction and propose the PSCIA protocol. Before detailing the protocol, we outline the key improvements driving its design.

3.1 Main Idea

First, we identify that MitM attacks succeed due to the homomorphic property of the HVTs, which allows any entity with access to all HVTs to forge integrity proofs. To counter this, we temporarily disable the homomorphic property of the HVTs, and restore it when generating integrity proofs. Specifically, inspired by Lin et al. [14], we allow the DO to generate HVTs by embedding the identity information of the DCSP. After the DO deletes local backups, only the DCSP can restore the homomorphic property and generate valid integrity proofs. In this way, we introduce the unforgeability property to the integrity proofs, thereby preventing MitM attacks.

Additionally, to resist challenge replay attacks (CRAs) from semi-honest TPAs, we propose a novel challenge generation method called re-randomization challenge. The foundation reason why CRAs succeed is that the choice of data blocks to challenge and the challenge coefficients are entirely determined by the TPA. However, allowing the DCSP to control the challenged data blocks is not feasible, as it may intentionally select intact data blocks. Our approach is for both the DCSP and the TPA to co-determine the challenge coefficients. In this case, by altering the challenge coefficients, the CSP returns a distinct integrity proof even for identical challenges.

3.2 System Setup

Given a security parameter λ, the System Administrator (SA) does:

1. Generate a bilinear group $(\mathbb{G}_1, \mathbb{G}_T, q, e, g)$ of a prime order q, where $q > 2^\lambda$. Here, g is a generator of $\mathbb{G}_1$, and $e : \mathbb{G}_1 \times \mathbb{G}_1 \to \mathbb{G}_T$ is the bilinear map.
2. Randomly choose $s \in \mathbb{Z}_q^*$ as the master private key msk. The master public key is $mpk = g^s$.
3. Let $H_1 : \{0,1\}^* \to \mathbb{G}_1$, $H_2 : \{0,1\}^* \to \mathbb{Z}_q$, $H_3 : \{0,1\}^* \to \mathbb{G}_1$, $H_4 : \{0,1\}^* \to \mathbb{Z}_q$, $H_5 : \{0,1\}^* \times \mathbb{G}_T \to \mathbb{G}_1$, $H : \{0,1\}^* \times \mathbb{G}_T \to \mathbb{Z}_q^*$, $H_0 : \{0,1\}^* \to \mathbb{G}_1$ be collision-resistant hash functions.
4. Let $\phi : \mathbb{Z}_q^* \times \mathbb{Z}_q^* \to \mathbb{Z}_q^*$ be a pseudo-random function.
5. Publish the system parameters $params = \{q, \mathbb{G}_1, \mathbb{G}_T, e, g, mpk, H_1, H_2, H_3, H_4, H_5, H, H_0, \phi\}$ and keep msk secret.

3.3 Registration

The *Registration* process comprises three phases: *KeyGen*, *CertGen*, and *UserAuth*.

- **Phase 1 (KeyGen)**: Given $params$ and an identity $ID \in \{0,1\}^N$, where N is the maximum bit length of the DO's identity, the DO randomly chooses $x_{ID} \in \mathbb{Z}_q^*$ as its private key usk_{ID} and computes its public key $upk_{ID} = g^{x_{ID}}$.
- **Phase 2 (CertGen)**: Using $params$, the master private key msk, the public key upk_{ID}, and the identity ID, the SA generates and sends a certificate $Cert_{ID} = H_1(ID, upk_{ID})^{msk}$ to the DO.
- **Phase 3 (UserAuth)**: Upon receiving the certificate $Cert_{ID}$, the DO verifies it using the equation $e(Cert_{ID}, g) = e(H_1(ID, upk_{ID}), mpk)$. If the equation holds, registration is successful.

The CSP with identity ID_C also follows the same process to obtain its private/public key pair ($usk_{ID_C} = x_{ID_C} \in \mathbb{Z}q^*, upk_{ID_C} = g^{x_{ID_C}}$) and a valid certificate $Cert_{ID_C} = H_1(ID_C, upk_{ID_C})^{msk}$, thereby completing the registration.

3.4 Outsourcing

The *Outsourcing* process comprises three phases: *Preprocessing*, *TagsGen* and *Uploading*.

- **Phase 1 (Preprocessing)**: Let $F \in \{0,1\}^*$ be an encrypted dataset named $fname$. The DO partitions F into k data blocks $m_1, ..., m_k$, each containing n sectors, represented as $F = (m_{ij})_{k \times n}$, where $m_{ij} \in \mathbb{Z}_q$.
- **Phase 2 (TagsGen)**:
  1. The DO randomly selects $r, p \in \mathbb{Z}_q^*$ and calculates $R = g^r, P = g^p$ and $\tau_0 = Cert_{ID}^{\frac{1}{usk_{ID}+H_2(ID,fname,k,n,R,P)}}$. Then, the DO generates $2n$ group elements $\alpha_j = g^{H_4(ID,j,usk_{ID})}, \beta_j = \alpha_j^r, j \in [1,n]$. The DO sets $\tau = \{\tau_0, R, P, \{\alpha_j, \beta_j\}_{j=1}^n, fname, t, k, n\}$ as a label of the dataset $fname$, where t is a timestamp.
  2. Given the identity ID_C and public key upk_{ID_C}, for each $m_i, i \in [1,k]$, the DO generates tags $\sigma_i, i \in [1,k]$ with

$$\sigma_i = (H_3(ID, i, \tau) \cdot g^{\sum_{j=1}^n H_4(ID,j,usk_{ID})m_{ij}})^r \cdot H_5(m_i, E^p), \quad (1)$$

where

$$E = e(mpk, H_1(ID_C, upk_{ID_C})) \cdot e(upk_{ID_C}, H_0(ID_C)). \quad (2)$$

- **Phase 3 (Uploading)**: The DO uploads the label τ, data blocks $m_1, ..., m_k$ and tags $\sigma_1, ..., \sigma_k$ to the DCSP, retaining only τ locally, and deleting the blocks and tags from local storage.

3.5 Integrity Auditing

The *Integrity Auditing* process comprises three phases: *Challenge*, *ProofGen* and *ProofVer*.

- **Phase 1 (Challenge)**: Given the label τ of the dataset owned by the DO with identity ID, the TPA selects a nonempty subset $I \subseteq [1, k]$ and distinct random elements $cnum_i \in \mathbb{Z}_q^*$ for each $i \in I$. The TPA then sends the challenge $chal = \{ID, fname, \{i, cnum_i\}_{i \in I}\}$ to the DCSP.
- **Phase 2 (ProofGen)**: Upon receiving $chal$, the DCSP verifies that all $cnum_i$ are unique. If true, it randomly chooses $rnum, tnum \in \mathbb{Z}_q^*$, computes $\gamma_j = \beta_j^{rnum}$ for $j \in [1, n]$, and calculates $c_i = \phi(cnum_i, tnum)$ for $i \in I$, then proceeds to compute:

$$\psi_j = \sum_{i \in I} c_i m_{ij} + rnum, \quad j \in [1, n], \tag{3}$$

$$\sigma = \prod_{i \in I} \left(\sigma_i \cdot H_5(m_i, E')^{-1}\right)^{c_i}, \quad E' = e\left(P, Cert_{ID_C} \cdot H_0(ID_C)^{usk_{ID_C}}\right). \tag{4}$$

Next, the DCSP randomly chooses $u \in \mathbb{Z}_q^*$ and calculates:

$$U = e(g, g)^u, \quad V = H(\psi_1, ..., \psi_n, U), \quad W = \left(\sigma \cdot \prod_{j=1}^{n} \gamma_j\right)^V \cdot g^u. \tag{5}$$

Finally, the DCSP returns an integrity proof $PF = \{\tau, V, W, \{\psi_j\}_{j=1}^n, tnum\}$ to the TPA.

- **Phase 3 (ProofVer)**: Upon receiving PF, the TPA first verifies the label τ of dataset F using:

$$e\left(\tau_0, upk_{ID} \cdot g^{H_2(ID, fname, k, n, R, P)}\right) = e\left(H_1(ID, upk_{ID}), mpk\right). \tag{6}$$

If satisfied, the TPA calculates $c_i = \phi(cnum_i, tnum)$ for $i \in I$, $\sigma' = \prod_{i \in I} H_3(ID, i, \tau)^{c_i} \cdot \prod_{j=1}^n \alpha_j^{\psi_j}$, and $U' = e(\sigma', R^{-1})^V \cdot e(W, g)$, then checks:

$$V = H(\psi_1, ..., \psi_n, U') \tag{7}$$

If this holds, the TPA confirms that the dataset F on the DCSP is intact and returns 1 to the DO; otherwise, it returns 0, indicating that tampering has occurred.

4 Property Analysis

4.1 Correctness

Theorem 1. *If the SA, DCSP, DO and TPA act honestly throughout the described processes, the integrity proof will pass verification.*

Proof. The correctness of Eqs. 6 and 7 in the ProofVer phase follows from properties of bilinear pairings. For simplicity, we denote $H_2(ID, fname, k, n, R, P)$ as h_2.

$$e\left(\tau_0, upk_{ID} \cdot g^{h_2}\right) = e\left(Cert_{ID}^{\frac{1}{usk_{ID}+h_2}}, g^{usk_{ID}+h_2}\right) \\ = e\left(H_1(ID, upk_{ID}), mpk\right) \tag{8}$$

The Eq. 7 holds, i.e., $V = H(\psi_1, ..., \psi_n, U') = H(\psi_1, ..., \psi_n, U)$, which means $U = U'$.

$$\begin{aligned} U' = e(\sigma', R^{-1})^V \cdot e(W, g) &= e\left(\sigma', R\right)^{-V} \cdot e\left(\left(\sigma \cdot \prod_{j=1}^{n} \gamma_j\right)^V \cdot g^u, g\right) \\ &= e(\sigma', R)^{-V} \cdot e(\sigma \cdot \prod_{j=1}^{n} \gamma_j, g)^V \cdot e(g, g)^u \\ &= \left(e(\sigma', R)^{-1} \cdot e\left(\sigma \cdot \prod_{j=1}^{n} \gamma_j, g\right)\right)^V \cdot e(g, g)^u \\ &= e(g, g)^u = U, \end{aligned} \tag{9}$$

where

$$\begin{aligned} &e\left(\sigma \cdot \prod_{j=1}^{n} \gamma_j, g\right) \\ &= e\left(\prod_{i \in I}\left(\sigma_i \cdot (H_5(m_i, E'))^{-1}\right)^{c_i} \cdot \prod_{j=1}^{n} \gamma_j, g\right) \\ &= e(\prod_{i \in I}\left(H_3(ID, i, \tau) \cdot g^{\sum_{j=1}^{n} H_4(ID, j, usk_{ID}) \cdot m_{ij}}\right)^{c_i r} \\ &\quad \cdot \prod_{i \in I}(H_5(m_i, E^p) \cdot (H_5(m_i, E'))^{-1})^{c_i} \cdot \prod_{j=1}^{n} \alpha_j^{rnum \cdot r}, g) \\ &= e\left(\prod_{i \in I} H_3(ID, i, \tau)^{c_i} \cdot \prod_{i \in I}\left(g^{\sum_{j=1}^{n} H_4(ID, j, usk_{ID}) \cdot m_{ij}}\right)^{c_i} \cdot \prod_{j=1}^{n} \alpha_j^{rnum}, R\right) \\ &= e\left(\left(\prod_{i \in I} H_3(ID, i, \tau)^{c_i}\right) \cdot \prod_{i \in I} \prod_{j=1}^{n} \alpha_j^{c_i m_{ij}} \cdot \prod_{j=1}^{n} \alpha_j^{rnum}, R\right) \\ &= e\left(\left(\prod_{i \in I} H_3(ID, i, \tau)^{c_i}\right) \cdot \prod_{j=1}^{n} \alpha_j^{\psi_j}, R\right) = e(\sigma', R), \end{aligned} \tag{10}$$

and

$$E^p = e(mpk, H_1(ID_C, upk_{ID_C}))^p \cdot e(upk_{ID_C}, H_0(ID_C))^p$$
$$= e(g^p, H_1(ID_C, upk_{ID_C})^{msk}) \cdot e(g^p, H_0(ID_C)^{usk_{ID_C}}) \quad (11)$$
$$= e(P, Cert_{ID_C} \cdot H_0(ID_C)^{usk_{ID_C}}) = E'.$$

4.2 Security

According to our threat model shown in Sect. 2.2, we consider a probabilistic polynomial time (PPT) adversary $\mathcal{A}$ and a PPT algorithm $\mathcal{B}$ to accomplish the game. The following theorem establishes the security guarantees of PSCIA.

Theorem 2. *If the BDH problem is hard on $\mathbb{G}_1$, PSCIA satisfies UF_p.*

Proof. Assume there exists a adversary $\mathcal{A}$ capable of producing a valid integrity proof with a non-negligible advantage $\epsilon_{\mathcal{A}}$. A PPT algorithm $\mathcal{B}$, acting as a simulator, solves the BDH problem with an advantage $\epsilon'_{\mathcal{A}}$.

Initialization. Let $params = (q, G_1, G_T, e, g, mpk = g^{msk}, H_1, H_2, H_3, H_4, H)$ represents the public parameters. The simulator $\mathcal{B}$ provides both $params$ and msk to $\mathcal{A}$. The hash functions H_0 and H_5 are modeled as random oracles. $\mathcal{B}$ simulates a DO with identity ID and obtains its key pair (usk_{ID}, upk_{ID}) using the $KeyGen$ algorithm. It then delivers ID and upk_{ID} to $\mathcal{A}$. For simplicity, assume that all datasets have the same structure, i.e., $F = (m_{ij})_{k \times n}, m_{ij} \in \mathbb{Z}_q$

Query on Oracles. $\mathcal{A}$ can adaptively query the following oracles:

Hash. Before any hash queries, $\mathcal{B}$ randomly chooses $d^* \in [1, q_{H_0}]$, where q_{H_0} is the maximum number of H_0 hash queries. $\mathcal{B}$ then initializes two empty hash lists to record queries and responses:

- H_0: For the d-th query on ID_{C_d}, if ID_{C_d} is already contained in L_{H_0}, $\mathcal{B}$ returns the stored value. Otherwise, it randomly selects $\omega_d \in \mathbb{Z}_q^*$ and sets $H_0(IDC_d) = g^b$ when $d = d^*$, or $H_0(ID_{C_d}) = g^{\omega_d}$ otherwise. Next, $\mathcal{B}$ adds $H_0(ID_{C_d})$ to L_{H_0}.
- H_5: Let the d-th hash query to H_5 be y_d. If y_d exists in L_{H_5}, $\mathcal{B}$ response to this query according to L_{H_5}. Otherwise, it randomly chooses $\{Y_{i,d}\}_{i=1}^n \in \mathbb{G}_1$, sets $H_5(y_d) = \{Y_{i,d}\}_{i=1}^n$, and adds $(y_d, H_5(y_d))$ to L_{H_5}.

KeyGen. $\mathcal{B}$ maintains an initially empty list L_C consisting of three tuples $(ID_{C_d}, usk_{ID_{C_d}}, upk_{ID_{C_d}})$. Upon receiving a query on ID_{C_d}, if the identifier exists in L_C, $\mathcal{B}$ returns the corresponding tuple. Otherwise, it generates the private/public key pair as $(usk_{ID_{C_d}}, upk_{ID_{C_d}}) = (\Delta, g^a)$ when $d = d^*$, or (x_d, g^{x_d}) otherwise, where $x_d \in \mathbb{Z}_q^*$ is chosen uniformly at random. Here, a is unknown to both $\mathcal{B}$ and $\mathcal{A}$. The resulting pair $(usk_{ID_{C_d}}, upk_{ID_{C_d}})$ is returned to the adversary and recorded in L_C.

TagsGen. On receiving a preprocessed dataset F named $fname$ and a CSP's identity ID_{C_d}, $\mathcal{B}$ generates tags by invoking the $TagsGen$ algorithm. The only difference in generating tags of data blocks is $\sigma_i = \tilde{\sigma}_i \cdot Y_{i,d}$, where

$\tilde{\sigma}_i = (H_3(ID, i, \tau_d) \cdot g^{\sum_{j=1}^{n} H_4(ID,j,usk_{ID})m_{ij}})^r$. If $d \neq d^*$, $\mathcal{B}$ returns $\{\tau_d, \{\sigma_i\}_{i=1}^{k}\}$; otherwise, it returns $\{\tau_{d^*}/\{P_{d^*}\}, P_{d^*} = g^c, \{\sigma_i\}_{i=1}^{k}\}$.

ProofGen. Upon receiving a query on $(ID_{C_d}, \{\sigma_i\}_{i=1}^{k}, chal = \{fname, \{i, cnum_i\}_{i \in I}\})$, $\mathcal{B}$ confirms if $fname$ was queried in $TagsGen$ Oracle. If so, it validates the tags, then proceeds as follows:

$\mathcal{B}$ randomly selects $rnum, tnum \in \mathbb{Z}_q^*, i \in I$ and computes $\gamma_j = \beta_j^{rnum}$ and $c_i = \phi(cnum_i, tnum)$ for $i \in I$. Using the data blocks and tags, $\mathcal{B}$ calculates $\psi_j = \sum_{i \in I} c_i m_{ij} + rnum$ and $\sigma = \prod_{i \in I} (\sigma_i \cdot (Y_{i,d})^{-1})^{c_i}$. $\mathcal{B}$ randomly chooses $u \in \mathbb{Z}_q^*$ and calculates U, V, W using Eq. 5. Finally, $\mathcal{B}$ delivers $PF_d = \{\tau_d, V, W, \{\psi_j\}_{j=1}^{n}, tnum\}$ to $\mathcal{A}$.

Output. $\mathcal{B}$ selects a novel challenge $chal^* = \{fname^*, \{i, cnum_i\}_{i \in I}\}$, where $fname^*$ has been queried in $TagsGen$ Oracle and sends it to $\mathcal{A}$. $\mathcal{A}$ outputs $\{ID_C^*, chal^*, PF^* = \{\tau^*, V^*, W^*, \{\psi_j\}_{j=1}^{n}, tnum\}$. The challenge H_5 query is defined as $\{m_i, Q^*\}$. $\mathcal{A}$ wins if:

1. The proof PF^* is generated by $\mathcal{A}$, not the DCSP.
2. The proof PF^* is valid, satisfying:

$$\begin{aligned}
e\left(\sigma^* \cdot \prod_{j=1}^{n} \gamma_j, g\right) &= e\left(\prod_{i \in I}(\sigma_i \cdot H_5(m_i, Q^*)^{-1})^{c_i} \cdot \prod_{j=1}^{n} \gamma_j, g\right) \\
&= e(\prod_{i \in I}\left(H_3(ID, i, \tau) \cdot g^{\sum_{j=1}^{n} H_4(ID, j, usk_{ID}) \cdot m_{ij}}\right)^{c_i r} \\
&\quad \cdot \prod_{i \in I}((Y_{i,d^*}) \cdot H_5(m_i, Q^*)^{-1})^{c_i} \cdot \prod_{j=1}^{n} \alpha_j^{rnum \cdot r}, g) \\
&= e\left(\left(\prod_{i \in I} H_3(ID, i, \tau)^{c_i}\right) \cdot \prod_{j=1}^{n} \alpha_j^{\psi_j}, R\right)
\end{aligned} \quad (12)$$

Due to the randomness of $\{\{Y_{i,d}\}_{i=1}^{k}, x_d, \omega_d\}, d = [1, q_{H_0}]$, $\mathcal{A}$ cannot distinguish between the game and a real attack scenario, thereby ensuring indistinguishability.

If $Y_{i,d^*} = H_5(m_i, Q^*)$, where Q^* is the challenge hash query, the integrity proof PF^* is valid. Therefore, $\mathcal{A}$ outputs a forgery $Q^* = E^c = e(g, Cert_{ID_{C_{d^*}}})^c \cdot e(g^a, g^b)^c$. $\mathcal{B}$ then solves the BDH problem by calculating $e(g,g)^{abc} = \frac{Q^*}{e(g^c, Cert_{ID_{C_{d^*}}})}$.

Probability Analysis. If the following conditions hold, $\mathcal{B}$ can get a solution of the BDH problem. S_1: $\mathcal{A}$ succeeds in the game. S_2: $ID_C^* = ID_{C_{d^*}}$.

1. S_1 happens means that Q^* has been queried to H_5 query. In this case, the probability is $\frac{1}{q_{H_5}} \epsilon_{\mathcal{A}}$, where q_{H_5} denotes the maximum number of H_5 hash queries.

2. If S_2 happens, the probability is $\frac{1}{q_{H_0}}$.

Thus, $\mathcal{B}$'s success probability is $\epsilon'_\mathcal{A} \geq (\frac{1}{q_{H_0} \cdot q_{H_5}}) \epsilon_\mathcal{A}$. Given the computational hardness of the BDH problem in $\mathbb{G}_1$, this probability is negligible. Consequently, the probability that $\mathcal{A}$ can successfully forge a valid integrity proof is negligible. Above all, Theorem 2 is proven.

4.3 Privacy

In PSCIA, the TPA initially generates a preliminary challenge $chal = \{ID, fname, \{i, cnum_i\}_{i \in I}\}$, which is not yet the final challenge that DCSP needs to address. The coefficients $\{c_i\}_{i \in I}$ of the final challenge are derived using a pseudo-random function $\phi(cnum_i, tnum), i \in I$, where $tnum \in \mathbb{Z}_q^*$ is randomly selected by the DCSP. This ensures that the probability of producing identical integrity proofs for repeated challenges is negligible, effectively safeguarding the DO's data modification privacy.

5 Evaluations

5.1 Theoretical Analysis

In this section, we compare our scheme with LHSDC [14] and LHSDE [13], as both of these protocols are capable of defending against Man-in-the-Middle (MitM) attacks. Besides, LHSDC and LHSDE do not involve an explicit tags generation operation. Instead, these protocols rely on the generation of data signatures, which serve a function analogous to that of tag generation in other schemes. Consequently, we conducted a comparative analysis to evaluate their performance in this context. Similarly, in the subsequent sections, we extend this comparative approach by examining the signature combination process alongside proof generation, as well as signature verification alongside proof verification. This analysis enables us to evaluate the efficiency and security trade-offs between these essential operations across different protocols.

In this paper, we disregard addition and inverse operation in $\mathbb{Z}_q^*$, as well as the hash mapping to $\mathbb{Z}_q^*$, due to their negligible costs. Moreover, we introduce some notations in Table. 1 to facilitate the discussion of the later evaluations.

It is worth noting that the bilinear paring in PSCIA is symmetric and only $\mathbb{G}_1$ and $\mathbb{G}_T$ are involved. In contrast, the bilinear pairing in LHSDC and LHSDE is asymmetric, incorporating an additional multiplicative cyclic group $\mathbb{G}_2$. In the interest of objectivity, we follow the original schemes and assume that the symmetric and asymmetric bilinear pairings have the same computational cost.

5.1.1 Computation Costs The computation costs of the *KeyGen* and *TagsGen* algorithms are shown in Table 2. Generally speaking, PSCIA has the most computation costs due to the necessity of generating the certificate for DO, but the extra operations ($T_{E_1} + T_H$) are only completed by SA. *TagsGen* algorithm

Table 1. Notations of Evaluations

Notations	Meanings
T_H	Map-to-point hash operation
$T_{M_{1/2/T}}$	Multiplicative operation in $\mathbb{G}_{1/2/T}$
$T_{E_{1/2/T}}$	Exponential operation in $\mathbb{G}_{1/2/T}$
T_P	Bilinear pairing operation
c	The amount of challenged data blocks
k	The amount of data blocks in a dataset
n	The amount of sectors in each data block
$|\mathbb{G}_{1/2/T}|$	The bit length of an element in $\mathbb{G}_{1/2/T}$
$|\mathbb{Z}_q|$	The bit length of an element in $\mathbb{Z}_q$

is used to generate tags, which is performed by the DO. Note that, calculating bilinear pairings is extremely time-consuming. Note that, in LHSDE and LHSDC, the number of bilinear pairings that DO needs to calculate is linear, which is unacceptable for DO with low computing capacity. But in PSCIA, no matter how many tags DO needs to generate, DO only needs to calculate two bilinear pairings. The computation costs of PSCIA of *TagsGen* are $2T_P + (2k+4)T_{E_1} + 2kT_{M_1} + T_{M_T} + (2k+2)T_H$. We reduce the number of bilinear pairings that the DO needs to compute from linear to constant.

For proof generation, the computation cost in computing bilinear pairing of PSCIA is constant with the number of challenged data blocks c, while those of LHSDE and LHSDC are linear with $(c+1)$. Thus, the PSCIA have low computation costs to generate proof. In terms of proof verification, DCSP in PSCIA costs $4T_P + (c+2)T_{E_1} + (c+2)T_{M_1} + (c+1)T_H$ to verify the integrity proof, which is significantly lower than LHSDC and LHSDE with $T_P + (k+n)T_{E_1} + T_{E_T} + kT_H$ and $2T_P + (k+n)T_{E_1} + T_{E_2} + T_{E_T} + (k+n+1)T_M + kT_H$ respectively. Table 3 provides the detailed comparisons of the computational overheads of the *Integrity Auditing* process.

Table 2. Comparisons of Computation Costs of *KeyGen* and *TagsGen* Phases

Scheme	KeyGen	Generation costs of k tags
LHSDE	T_{E_2}	$kT_P + k(n+k+1)T_{E_1} + kT_{E_T} + k(k+n+2)T_{M_1} + 3kT_H$
LHSDC	T_{E_2}	$kT_P + k(n+k+1)T_{E_1} + kT_{E_T} + k(k+n+2)T_{M_1} + 3kT_H$
PSCIA	$2T_{E_1} + T_H$	$2T_P + (2n+3k+3)T_{E_1} + 2kT_{M_1} + (2k+2)T_H + T_{M_T}$

5.1.2 Communication Costs We introduce the communication costs in each of the three processes: *Registration*, *Outsourcing* and *Integrity Auditing*. Note

Table 3. Comparisons of Computation Costs in Integrity Auditing Process

Scheme	Operation	Computation costs
LHSDE	Combine	$(c+1)T_P + 2cT_{E_1} + cT_{E_T} + 2cT_{M_1} + 2cT_H$
	Signature verification	$T_P + (k+n)T_{E_1} + T_{E_T} + kT_H$
LHSDC	Combine	$(c+1)T_P + (2c+2)T_{E_1} + (c+1)T_{E_T}$ $+(c+1)T_{M_1} + (2c+1)T_H$
	Signature verification	$2T_P + (k+n)T_{E_1} + 2T_{E_2} + T_{E_T}$ $+(k+n+1)T_{M_1} + T_{m_T} + kT_H$
PSCIA	Proof generation	$2T_P + (n+2c+3)T_{E_1} + (n+c+1)T_{M_1} + (c+2)T_H + T_{E_T}$
	Proof verification	$4T_P + (n+c+2)T_{E_1} + 2T_{M_1} + (c+2)T_H + T_{E_T}$

that, LHSDE and LHSDC are basic linearly homomorphic signatures, so we simulate DO, DCSP and TPA to perform auditing tasks. In order to avoid the additional overhead of converting LHS to a remote auditing protocol, we only call the original algorithm in LHSDE and LHSDC during the simulation.

In the *Registration* process, since SA needs to generate certificate for DO, the communication costs of PSCIA are $2|\mathbb{G}_1| + N$ which is obviously higher than those of others, where N is the maximum number of bits for user's identity. The communication costs of LHSDE and LHSDC are $|\mathbb{G}_1| + N$.

During the *Outsourcing* process, DO performs *Tag Generation* phase to generate tags, and then uploads both data blocks and their corresponding tags to the DCSP. In this case, the communication costs of PSCIA are $(k+2n+3)|\mathbb{G}_1| + kn|\mathbb{Z}_q|$. And the communication costs of LHSDE and LHSDC are $k|\mathbb{G}_1| + kn|\mathbb{Z}_q|$.

During the *Integrity Auditing* process, due to the different challenge generation strategies, we ignore the communication costs related to the challenge. The communication overhead between the CSP and the TPA in PSCIA is $(2n+4)|\mathbb{G}_1| + (n+c+2)|\mathbb{Z}_q|$ bits, consisting of $c|\mathbb{Z}_q|$ bits for the challenge and the remainder for the response. The communication overheads of LHSDE and LHSDC are $|\mathbb{G}_T| + (n+k)|\mathbb{Z}_q|$ and $|\mathbb{G}_1| + |\mathbb{G}_2| + (n+k)|\mathbb{Z}_q|$, respectively.

In summary, in terms of communication costs, PSCIA has no advantage, because in order to conduct data integrity auditing with greater security and privacy, PSCIA needs to increase the number of parameters to be transmitted.

5.2 Experiment Analysis

In this section, we present the experimental performance of PSCIA. We use C for the non-cryptographic computations and the Pairing-based Cryptography library [16] for the cryptographic computations. In order to facilitate a fair comparison of the performance among PSCIA, LHSDC, and LHSDE, we have chosen to uniformly initialize pairing with parameter **a.param**. This type of pairings is based on the curve $y^2 = x^3 + x$, which embeds degree 2 over a finite field $GF(q)$, where q is a large prime with $q = 3 \pmod 4$ [22]. We execute our programs on a ubuntu 22.04 device, i7-13700f CPU @2.10GHz, 4GB RAM. In our experiments, the test file size is 1 MB (1048576 bytes), divided into m blocks, with each block consisting of n sectors. Each sector corresponds to a 160-bit element in $\mathbb{Z}_q$, resulting in mn elements representing the file. Therefore, mn needs to satisfy $(m-1)n \leq \frac{1048576B}{160} \leq mn$. To minimize the impact of randomness, we conduct the experiment 100 times and average the results.

Tags Generation Costs: We vary the number of sectors between 10 and 100 and compare the performance of our protocol with LHSDC and LHSDE in this experiment. As shown in Fig. 2, while tag generation costs decreases with the increase of sectors for all three protocols, PSCIA demonstrates a significant performance advantage compared to LHSDC and LHSDE. When we split a 1MB file to 525 data blocks (each of them have 100 sectors), PSCIA only requires 2.82 s, which is significantly lower compared to LHSDC's 38.10 s and LHSDE's 38.05 s. Figure 3 shows that even as the dataset grows synchronously, the growth rate of PSCIA is much slower than that of LHSDE and LHSDC, approximately one-quarter of the latter two.

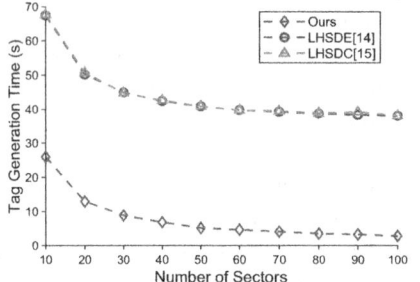

Fig. 2. Comparison of tags generation costs for different number of sectors.

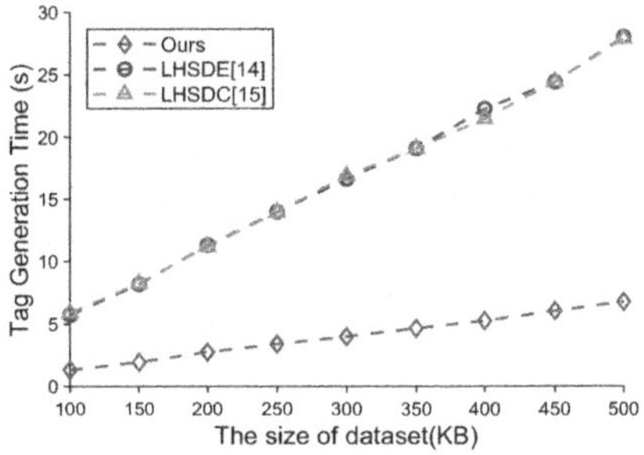

Fig. 3. Comparison of tags generation costs for different size of datasets.

Proof Generation Costs: We compare the costs of proof generation by setting sectors to 20, which means we have 2622 data blocks in total. We set c varies from 20 to 160. As is shown in Fig. 4, time increases roughly linearly as c increases for all three protocols. There is almost no difference between LHSDE and LHSDC, but PSCIA's performance is much better than both.

Proof Verification Costs: Using same setting with proof generation, we play the role of TPA to compare the costs of proof verification. From Fig. 5, we can easily seen the performance of LHSDE is slightly better than that of LHSDC. Both are approximately a straight line parallel to the horizontal axis. This is because the proof verification complexity of LHSDE and LHSDC is independent of c, which is consistent with our theoretical analysis in Sect. 5.1.1.

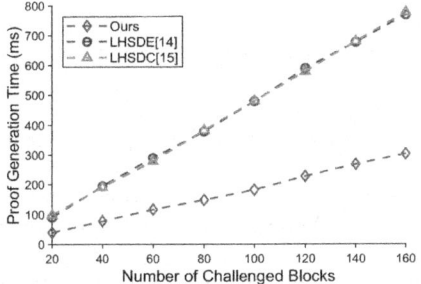
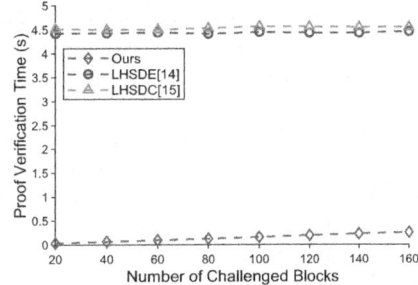

Fig. 4. Comparison of proof generation costs.

Fig. 5. Comparison of proof verification costs.

6 Conclusion

Throughout this paper, we first present a novel MitM attack targeting remote auditing schemes and design a novel HVT construction specifically to resist it. Then, incorporating the HVT design, we propose PSCIA, a privacy-preserving, secure, and certificate-based integrity auditing protocol for cloud storage. By employing the certificate-based cryptosystem, PSCIA makes secure channels unnecessary, resolves the key escrow problem, and reduces the burden of certificate management, which enhances system scalability and reduces overhead. Furthermore, to protect data modification privacy of DO, we design a novel challenge generation method to resist the challenge replay attacks initiated by semi-honest TPAs. The formal security analysis proves that PSCIA is secure under the random oracle model. Finally, theoretical analysis and experimental results show that PSCIA provides high computational efficiency compared with state-of-the-art schemes. In particular, we reduce the number of bilinear pairings that DO and CSP need to compute from linear to constant levels.

Acknowledgments. Thanks to the anonymous reviewers for their constructive suggestions. This work was supported in part by the Natural Science Foundation on Frontier Leading Technology Basic Research Project of Jiangsu under Grant BK20222001, NSFC Grants No.62272215, No.62272222, No.62372177, and the Fundamental Research Funds for the Central Universities (No. 2024300401).

References

1. Ateniese, G., et al.: Provable data possession at untrusted stores. In: Proceedings of the 14th ACM conference on Computer and communications security, pp. 598–609 (2007)
2. Boneh, D., Franklin, M.: Identity-based encryption from the weil pairing. In: Annual international cryptology conference, pp. 213–229. Springer (2001)
3. Callegati, F., Cerroni, W., Ramilli, M.: Man-in-the-middle attack to the https protocol. IEEE Secur. Privacy **7**(1), 78–81 (2009)

4. Deswarte, Y., Quisquater, J.J., Saïdane, A.: Remote integrity checking. In: Working conference on integrity and internal control in information systems, pp. 1–11. Springer (2003)
5. Gai, C., Shen, W., Yang, M., Yu, J.: Ppadt: privacy-preserving identity-based public auditing with efficient data transfer for cloud-based IOT data. IEEE Internet Things J. **10**(22), 20065–20079 (2023)
6. Gentry, C.: Certificate-based encryption and the certificate revocation problem. In: International Conference on the Theory and Applications of Cryptographic Techniques, pp. 272–293. Springer (2003)
7. Gu, K., Zhang, W., Wang, X., Li, X., Jia, W.: Dual attribute-based auditing scheme for fog computing-based data dynamic storage with distributed collaborative verification. IEEE Trans. Netw. Serv. Manag. (2023). https://ieeexplore.ieee.org/abstract/document/10102665
8. Hellman, M.: New directions in cryptography. IEEE Trans. Inf. Theory **22**(6), 644–654 (1976)
9. Li, J., Zhang, L., Liu, J.K., Qian, H., Dong, Z.: Privacy-preserving public auditing protocol for low-performance end devices in cloud. IEEE Trans. Inf. Forensics Secur. **11**(11), 2572–2583 (2016)
10. Li, X., et al.: An identity-based data integrity auditing scheme for cloud-based maritime transportation systems. IEEE Trans. Intell. Transp. Syst. **24**(2), 2556–2567 (2022)
11. Li, Y., Zhang, F.: An efficient certificate-based data integrity auditing protocol for cloud-assisted wbans. IEEE Internet Things J. **9**(13), 11513–11523 (2021)
12. Li, Y., Zhang, F.: Remote data auditing for cloud-assisted wbans with pay-as-you-go business model. Chin. J. Electron. **32**(2), 248–261 (2023)
13. Lin, C.-J., Huang, X., Li, S., Wu, W., Yang, S.-J.: Linearly Homomorphic Signatures with Designated Entities. In: Liu, J.K., Samarati, P. (eds) ISPEC 2017. LNCS, vol. 10701, pp. 375–390. Springer, Cham (2017). https://doi.org/10.1007/978-3-319-72359-4_22
14. Lin, C., Xue, R., Huang, X.: Linearly homomorphic signatures with designated combiner. In: International Conference on Provable Security, pp. 327–345. Springer (2021)
15. Liu, Z., Ren, L., Li, R., Liu, Q., Zhao, Y.: Id-based sanitizable signature data integrity auditing scheme with privacy-preserving. Comput. Secur. **121**, 102858 (2022)
16. Lynn, B., et al.: The pairing-based cryptography library. Internet: crypto. stanford. edu/pbc/ (2006)
17. Menezes, A., Vanstone, S., Okamoto, T.: Reducing elliptic curve logarithms to logarithms in a finite field. In: Proceedings of the twenty-third annual ACM symposium on Theory of computing, pp. 80–89 (1991)
18. Rogaway, P., Shrimpton, T.: Cryptographic Hash-Function Basics: Definitions, Implications, and Separations for Preimage Resistance, Second-Preimage Resistance, and Collision Resistance. In: Roy, B., Meier, W. (eds.) FSE 2004. LNCS, vol. 3017, pp. 371–388. Springer, Heidelberg (2004). https://doi.org/10.1007/978-3-540-25937-4_24
19. Shamir, A.: Identity-based cryptosystems and signature schemes. In: Advances in Cryptology: Proceedings of CRYPTO 84 4, pp. 47–53. Springer (1985)
20. Shen, W., Qin, J., Yu, J., Hao, R., Hu, J., Ma, J.: Data integrity auditing without private key storage for secure cloud storage. IEEE Trans. Cloud Comput. **9**(4), 1408–1421 (2019)

21. Song, M., Hua, Z., Zheng, Y., Huang, H., Jia, X.: Blockchain-based deduplication and integrity auditing over encrypted cloud storage. IEEE Trans. Dependable and Secure Comput. (2023)
22. Su, Y., Li, Y., Yang, B., Ding, Y.: Decentralized self-auditing scheme with errors localization for multi-cloud storage. IEEE Trans. Dependable Secure Comput. **19**(4), 2838–2850 (2021)
23. Wang, Q., Wang, C., Ren, K., Lou, W., Li, J.: Enabling public auditability and data dynamics for storage security in cloud computing. IEEE Trans. Parallel Distrib. Syst. **22**(5), 847–859 (2010)
24. Wu, T., Yang, G., Mu, Y., Guo, F., Deng, R.H.: Privacy-preserving proof of storage for the pay-as-you-go business model. IEEE Trans. Dependable Secure Comput. **18**(2), 563–575 (2019)
25. Zhang, Q., Qian, S., Cui, J., Zhong, H., Wang, F., He, D.: Blockchain-based privacy-preserving deduplication and integrity auditing in cloud storage. IEEE Trans. Comput. (2025)
26. Zhang, Y., Chang, J., Chen, Y., Xu, R.: Certificate-based remote auditing protocol with privacy protection and deduplication functions for cloud-assisted applications. Comput. Netw. p. 111083 (2025)
27. Zhou, L., Fu, A., Yang, G., Wang, H., Zhang, Y.: Efficient certificateless multi-copy integrity auditing scheme supporting data dynamics. IEEE Trans. Dependable Secure Comput. **19**(2), 1118–1132 (2020)

Unbalanced Private Computation on Set Intersection with Reduced Computation and Communication

Zelin Tang[1], Hua Guo[1,2(✉)], Yewei Guan[1], and Kaijie Yang[1]

[1] School of Cyber Science and Technology, Beihang University, Beijing, China
{by2339204,hguo,ame_reiori,kaijieyang}@buaa.edu.cn
[2] Key Laboratory of Data and Intelligent System Security (NKU), Ministry of Education, Tianjin, China

Abstract. Private computation on set intersection (PCSI) allows two parties to compute fine-grained functions on set intersection without revealing anything else. Although several efficient PCSI protocols are available in the literature, only recently have techniques been applied to improve their performance in an unbalanced setting, where the set size of one party is much smaller than the other.

In this work, we focus on improving the efficiency of unbalanced PCSI based on the framework comprising shared characteristic (SC) and permuted private equality test (p-PEQT). Firstly, our key insight is that SC can be efficiently constructed from batch keyword private information retrieval (PIR), requiring only a single keyword PIR query instead of multiple PIR queries. We further provide a novel construction of batch keyword PIR that performs remarkably for small entries, and then propose a communication-efficient SC based on the batch keyword PIR. Secondly, previous p-PEQT protocols depend on public-key operations, leading to substantial computation overhead. We consider symmetric-key operations to efficiently construct p-PEQT. Specifically, our p-PEQT construction is based on distributed point function (DPF) and additionally improves online performance thanks to preprocessing. Finally, we obtain a series of unbalanced PCSI protocols with reduced computation and communication from our SC and p-PEQT, and these protocols are secure against semi-honest adversaries in the standard model.

We implement our protocols and compare them with the state-of-the-art works. The experiments show that our PCSI-card achieves a 2.1× speedup under LAN setting and a 1.9× speedup under WAN setting. It also reduces communication by $1.1 - 1.8\times$, depending on the set sizes.

Keywords: Private set intersection · Private computation on set intersection · Batch keyword private information retrieval · Distributed point function

1 Introduction

Private set intersection (PSI) [7] allows two parties to compute the intersection of their data sets without revealing anything else. Although PSI has a wide range of applications such as genome testing and contact discovery, there are also certain real-world scenarios requiring only partial/aggregate information about the intersection to be revealed. Google [13] presented private intersection-sum for measuring ad conversion rates. And Facebook [2] proposed private secret shared set intersection as an extension of PSI. After that, Garimella et al. [8] introduced private computation on set intersection (PCSI), which is a series of cryptographic protocols to securely compute fine-grained functions on set intersection.

Most of works in PCSI [6,8,10,20] focus on a balanced setting, but in practical applications, such as client-server cases, the server frequently manages considerably larger datasets compared to the client. Directly applying the existing PCSI protocols in the balanced setting leads to significant overhead, as the communication complexity scales linearly with the size of the larger set. Therefore, PCSI protocols with sublinear communication complexity for large datasets are necessary to handle unbalanced scenarios.

Only a few existing works have been conducted to consider PCSI in the unbalanced setting. Recently, Tu et al. [17] designed an unbalanced PCSI framework including shared characteristic (SC) and permuted matrix private equality test (pm-PEQT). Considering the client-server case, if an element of the client is in the intersection, SC shares 0 (the server and the client obtain the same shared characteristic), otherwise SC shares a random value. After that, pm-PEQT tests whether the shared characteristics are equal and returns the permuted result to the server, which directly implies the cardinality of intersection. For SC, [17] proposed a concrete construction from fully homomorphic encryption (FHE) with logarithmic communication in the large set. Unfortunately, their SC's asymptotic complexity exhibits a constant term due to the bin-partition technique. The constant term predominates when the client's set is relatively small, leading not only to inefficient SC but also to the need for a pm-PEQT instantiation with large parameter to construct PCSI.

To remove this constant term, Zhang et al. [19] considered batch private information retrieval (PIR) and oblivious key-value store (OKVS) to construct oblivious key-value retrieval (OKVR), which can be viewed as SC. In addition, they proposed a generalization of pm-PEQT: permuted private equality test (p-PEQT), and provided two constructions. However, their SC construction requires a constant number (e.g., 2 − 3) of PIR queries for each element in the client's set, resulting in a considerable number of queries in total. Furthermore, both of their p-PEQT constructions are based on public-key operations, leading to substantial computation cost.

In this work, we are interested in efficient constructions of SC and p-PEQT in the unbalanced PCSI framework, with the aim of achieving concrete improvements in both communication and computation.

1.1 Contributions

Our core idea involves designing SC by employing batch keyword PIR rather than batch PIR and constructing p-PEQT through symmetric key operations. The main contributions are outlined below:

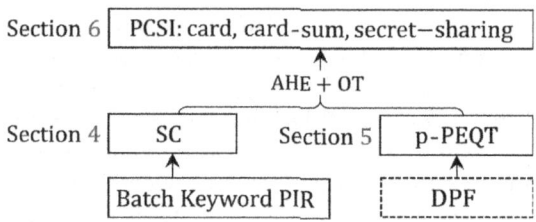

Fig. 1. Overview of unbalanced PCSI framework. The rectangles with solid lines are contributions of our work.

- **Efficient constructions of SC and p-PEQT.** We first provide a novel construction of batch keyword PIR that demonstrates remarkable performance for small entries, and then propose a communication-efficient SC based on our batch keyword PIR. After that, we give a computationally efficient construction of p-PEQT utilizing symmetric-key operations. Specifically, our p-PEQT construction builds upon distributed point function (DPF), and further optimizes online performance through preprocessing.
- **Unbalanced PCSI protocols with reduced communication and computation.** By leveraging the efficient constructions of SC and p-PEQT, we present a series of efficient unbalanced PCSI protocols, including PCSI-card for intersection cardinality, PCSI-card-sum for intersection cardinality and sum, and PCSI-secret-sharing for secret shared intersection. Our protocols achieve sublinear communication in the size of the large set and are secure against semi-honest adversaries in the standard model.
- **Evaluations.** We implement our unbalanced PCSI protocols and compare them with the state-of-the-art protocols. The experiments show that our PCSI-card achieves a 2.1× speedup in running time under LAN setting and a 1.9× speedup under WAN setting. And the communication of our PCSI-card is 1.1 − 1.8× better depending on set sizes.

1.2 Related Works

There are various methodologies available to securely compute PCSI-card in the unbalanced two-party setting. Lv et al. [12] presented an unbalanced PCSI-card protocol utilizing Bloom filter. Subsequent to [12], Li and Gao [11] considered Cuckoo filter and trivial PIR to construct a PCSI-card protocol in the unbalanced setting. However, both of their works involve communication overhead that scales linearly with the size of the large set, rendering it impractical. After that, Wu and Yuen [18] proposed an efficient unbalanced PCSI-card protocol with sublinear communication complexity based on the unbalanced PSI scheme [5]. Unfortunately, their work has a large constant term in the asymptotic complexity when the sender's set is relatively small, due to the bin partition technique.

Several works also provided unified PCSI frameworks for specific functionalities including PCSI-card, PCSI-card-sum, PCSI-secret-sharing, etc. Chen et al. [4] proposed the first unbalanced PCSI framework based on general circuit constructions. Their protocols obtain secret shares of the intersection utilizing their labeled PSI, and then securely compute functionalities from the corresponding circuits. However, it is difficult to conduct performance experiments of their protocols due to their theoretical circuit constructions. Afterwards, Chen et al. [6] presented an unbalanced PCSI* framework that leaks the cardinality to the sender. Recent work [17] designed the framework including SC and p-PEQT. Since their concrete SC construction is based on the same unbalanced PSI scheme [5] as [18], the asymptotic complexity of their PCSI also includes a large constant term, leading to inefficiency in an extremely unbalanced setting.

2 Technical Overview

We first introduce the framework of unbalanced PCSI and then give our efficient constructions of SC and p-PEQT. For convenience, we denote the parties involved in PCSI as the sender $\mathcal{S}$ and the receiver $\mathcal{R}$, and their respective input sets $X = \{x_i\}_{i \in [m]}$ and $Y = \{y_j\}_{j \in [n]}$ with $m \ll n$. Let $\phi : [m] \to [m_c]$ ($m_c \geq m$) denote the injective function of $\mathcal{S}$.

2.1 Unbalanced PCSI Protocols From SC and P-PEQT

We propose unbalanced PCSI protocols based on the framework in Fig. 1, and review the PSCI functionalities in Fig. 2.

Parameters:

- Two parties: the sender $\mathcal{S}$ and the receiver $\mathcal{R}$, set size m and n, bit length of set elements l.

Functionality:

- Wait for input $X = \{x_i\}_{i \in [m]} \in \{\{0,1\}^l\}^m$ from $\mathcal{S}$.
- Wait for input $Y = \{y_j\}_{j \in [n]} \in \{\{0,1\}^l\}^n$ from $\mathcal{R}$.
- [**card**]: $\mathcal{R}$ gets $card = |X \cap Y|$.
- [**card-sum**]: $\mathcal{R}$ gets $card = |X \cap Y|$ and $sum = \sum_{x_i \in Y} x_i$.
- [**secret-shares**]: $\mathcal{S}$ gets $\mathbf{z_1}$, $\mathcal{R}$ gets $\mathbf{z_2}$, where $\{\mathbf{z_1}[i] + \mathbf{z_2}[i]\}_{i \in [card]}$ is a random permutation of $X \cap Y$.

Fig. 2. Private Computing on Set Intersection Functionality $\mathcal{F}_{\mathsf{pcsi}}$

In the framework, $\mathcal{S}$ and $\mathcal{R}$ first play as P_1 and P_2 to invoke SC and p-PEQT. SC shares different characteristics to P_1 and P_2 based on whether $x_i \in Y$. Specifically, P_1 receives $\mathbf{r}' = \{r'_k\}_{k \in [m_c]}$ and P_2 receives $\mathbf{r} = \{r_k\}_{k \in [m_c]}$. And $r'_k = r_k$ if there exists $i \in [m]$ such that $\phi(i) = k$ and $x_i \in Y$, otherwise $r'_k \neq r_k$. Then P_1 and P_2 test whether $r'_k = r_k$ for $k \in [m_c]$ by p-PEQT, and P_2 (i.e., $\mathcal{R}$) receives a permuted indication vector $\mathbf{b} = \{b_k\}_{k \in [m_c]}$. For PCSI-card, $\mathcal{R}$ computes the Hamming weight of $\mathbf{b}$ as the intersection cardinality. Based on PCSI-card, PCSI-card-sum and PCSI-secret-sharing can be constructed combining with additively homomorphic encryption (AHE) and oblivious transfer (OT).

We observe that PCSI-card constitutes the majority of the framework's overhead by experiment. Therefore, it is crucial to provide efficient constructions of SC and p-PEQT to further improve the performance of PCSI-card, which leads to a series of PCSI protocols with reduced communication and computation.

2.2 Communication-Efficient SC

As shown in Fig. 3, SC shares 0 (which means that P_1 and P_2 obtain the same shared value) if and only if $x \in Y$. In the construction of SC proposed by Zhang et al. [19], P_2 first encodes Y into the data structure $D = D_0 || D_1$ with multi-point oblivious pseudorandom function (mpOPRF) and sparse OKVS, where D_0 represents the sparse part, D_1 represents the dense part and $|D_1| = o(|D_0|)$. Subsequently, for each $x \in X$, P_1 retrieves an arbitrary number of elements from D_1 and α ($\alpha = 2 - 3$) elements from D_0, and then sums all these elements to obtain the final result r'. For the dense part D_1 with $O(logn)$ size, P_2 can send D_1 directly to P_1. For the sparse part D_0, P_1 retrieves α elements separately by batch PIR. The dense part D_1 and the α elements retrieved from the sparse part D_0 are both random for P_1 due to the double obliviousness property of OKVS. Since P_1 requires α PIR queries for each $x \in X$, their construction requires $\alpha \times m_c$ PIR queries in total, leading to significant communication overhead.

Parameters:

- Two parties: P_1 and P_2, set sizes m and n, bit length of set elements l.

Functionality:

- Wait for input $X = \{x_i\}_{i \in [m]} \in \{\{0,1\}^l\}^m$ and injective function $\phi : [m] \to [m_c]$ ($m_c \geq m$) from P_1.
- Wait for input $Y = \{y_j\}_{j \in [n]} \in \{\{0,1\}^l\}^n$ from P_2.
- Give $\mathbf{r}' = \{r'_k\}_{k \in [m_c]}$ to P_1, and give $\mathbf{r} = \{r_k\}_{k \in [m_c]}$ to P_2. And $r'_k = r_k$ if there exists $i \in [m]$ such that $\phi(i) = k$ and $x_i \in Y$, otherwise $r'_k \neq r_k$.

Fig. 3. Shared Characteristic Functionality $\mathcal{F}_{\mathsf{sc}}$

We aim to compress α queries into a single query. And our key insight lies in that batch keyword private information retrieval (PIR) requires only a single keyword PIR query instead of α PIR queries for each $x_i \in X$, achieving the desired compression. We further note that SC only needs to share small-sized characteristics (e.g., 8 byte) to satisfy the false-positive probability, whereas existing efficient batch keyword PIR schemes are optimized for large-sized characteristics (e.g., 8K byte). Hence, we first provide a novel construction of batch keyword PIR that performs well for small entries by adopting vectorized batch PIR [14], and then propose communication-efficient SC based on our batch keyword PIR.

The general idea of our SC is as follows. P_2 encodes Y into D by the key-value filter [3]. For $x \in X$, P_1 retrieves an arbitrary number of elements from D in a single keyword PIR query. Then P_2 computes the response based on the query and D. The elements in the response need to be summed to reconstruct l' (the corresponding value obtained from query). To achieve this, P_2 uses the response merging algorithm of batch keyword PIR to add all elements in the response, instead of sending them separately to P_1. After that, P_2 homomorphically adds a random value r and l', and sends the result to P_1. Finally, P_1 decrypts the result and subtracts l (the actual corresponding value) to obtain r'.

In the end, we analyze the efficiency. Firstly, our SC is communication-efficient because only a single keyword PIR query is required for each $x \in X$. Secondly, our SC does not directly reveal elements of D to P_1, so the double obliviousness property of OKVS is unnecessary to ensure security. Thus, we remove mpOPRF and adopt the key-value filter with a high encoding rate, which reduces the size of D, resulting in computational improvement.

2.3 Computation-Efficient P-PEQT

p-PEQT checks if $r'_k = r_k$ for $k \in [m_c]$ and gives the permuted result to P_2. Different from Tu et al.'s p-PEQT protocols [17] which rely on expensive public-key operations, we consider only symmetric-key operations to construct a more efficient p-PEQT. In addition, we introduce preprocessing to generate and distribute secret keys, which further reduces the online overhead (Fig. 4).

Parameters:

- Two parties: P_1 and P_2, set size m_c, bit length of set elements σ.

Functionality:

- Wait for input $\mathbf{r}' = \{r'_k\}_{k \in [m_c]} \in \{\{0,1\}^\sigma\}^{m_c}$ from P_1.
- Wait for input $\mathbf{r} = \{r_k\}_{k \in [m_c]} \in \{\{0,1\}^\sigma\}^{m_c}$ from P_2.
- Select a random permutation π over $[m_c]$. For $k \in [m_c]$, set $b_k = 1$ if $r'_{\pi(k)} = r_{\pi(k)}$, and $b_k = 0$ otherwise.
- Give π to P_1, and give $\mathbf{b} = \{b_k\}_{k \in [m_c]}$ to P_2.

Fig. 4. Permuted Private Equality Test Functionality $\mathcal{F}_{\text{p-peqt}}$

In the preprocessing phase, we first select random point functions $\{f_k(x) = 1 \text{ if } x = \alpha_k, \text{ and } 0 \text{ otherwise}\}_{k \in [m_c]}$, and then generate the corresponding DPF keys $\{(\mathbf{k_1}[k], \mathbf{k_2}[k])\}_{k \in [m_c]}$. After that, we pick a random permutation π over $[m_c]$ and compute $\{\alpha_k\}_{k \in [m_c]}$'s secret shares $\mathbf{s_1}, \mathbf{s_2}$, where $\mathbf{s_1}[k] \oplus \mathbf{s_2}[k] = \alpha_k$. Finally,

$(\mathbf{k_1}, \mathbf{s_1})$ and π are distributed to P_1, and $(\{\mathbf{k_2}[\pi(k)]\}_{k \in [m_c]}, \mathbf{s_2})$ are distributed to P_2.

In the online phase, P_2 computes $v_k := \mathbf{s_2}[k] \oplus r_k$ for $k \in [m_c]$, and sends them to P_1. For $k \in [m_c]$, P_1 computes $\alpha'_k := \mathbf{s_1}[\pi(k)] \oplus r'_{\pi(k)} \oplus v_{\pi(k)}$ and $\mathbf{b_1}[k] := \mathsf{Eval}(\mathbf{k_1}[\pi(k)], \alpha'_k)$ utilizing the DPF evaluation algorithm. Subsequently, P_1 sends $\{\alpha'_k\}_{k \in [m_c]}$ and $\mathbf{b_1}$ to P_2, and considers π as output. In the end, P_2 computes $\mathbf{b_2}[k] := \mathsf{Eval}(\mathbf{k_2}[\pi(k)], \alpha'_k)$ and $b_k := \mathbf{b_1}[k] + \mathbf{b_2}[k]$ for $k \in [m_c]$, and obtains $\mathbf{b} = \{b_k\}_{k \in [m_c]}$ as output. Note that $\alpha'_k = \mathbf{s_1}[\pi(k)] \oplus \mathbf{s_2}[\pi(k)] \oplus r'_{\pi(k)} \oplus r_{\pi(k)} = \mathbf{s_1}[\pi(k)] \oplus \mathbf{s_2}[\pi(k)]$ if $r_{\pi(k)} = r'_{\pi(k)}$, therefore $b_k = 1$ by the correctness of DPF. Similarly, $b_k = 0$ if $r_{\pi(k)} \neq r'_{\pi(k)}$.

3 Preliminaries

3.1 Notation

We use κ and λ as the computational and statistical parameters, respectively. $[n]$ denotes the set $\{1, \ldots, n\}$. $\{\{0,1\}^l\}^n$ denotes all sets consisting of n l-bit strings. We say that a function f is negligible in λ if it vanishes faster than the inverse of any polynomial in κ, and denote $f(\lambda) = \mathsf{negl}(\lambda)$. We use the abbreviation PPT to denote probabilistic polynomial time in λ.

3.2 Cuckoo Hashing

Cuckoo hashing maps n items into $m = (1 + \epsilon) \cdot n$ bins and a stash using w random hash functions $h_1, \ldots, h_w$, where $\epsilon > 0$ is a constant. It allows only one item to be stored in a bin. By adopting the corresponding parameters of [15], we reduce the stash size to 0 while achieving a hashing failure probability of $2^{-\lambda}$.

3.3 Key-Value Filter

Key-value filter [3] is a form of storage that allows mapping a set of keys to corresponding values. Its formal definition is illustrated in Definition 1.

Definition 1. *A key-value filter, parameterized by the key domain $\mathcal{K}$ and the value domain $\mathcal{V}$, consists of three algorithms with the following syntax:*

- *KVF.Setup$(\epsilon) \to (\mathsf{H}, \mathsf{fpt}_\epsilon)$: on input the false-positive probability ϵ, outputs a set of hash functions $\mathsf{H} = \{h_1, \ldots, h_t\}$ for $t \in \mathbb{N}$ and a fingerprint function $\mathsf{fpt}_\epsilon : \mathcal{K} \times \mathcal{V} \to \{0,1\}^\mu$ which is parametrized by the false-positive probability via the polynomial $\mu = \mu(\epsilon)$.*
- *KVF.Encode$(\mathsf{D}, \mathsf{H}, \mathsf{fpt}_\epsilon) \to (b, \mathsf{F})$: on input key-value pairs $\mathsf{D} \in (\mathcal{K} \times \mathcal{V})^n$, the set of hash functions H and the fingerprint function fpt_ϵ, outputs $b \in \{0,1\}$ and a filter F. If $b \neq 1$, it may be necessary to regenerate the filter.*
- *KVF.Decode$(x, \mathsf{H}, \mathsf{F}) \to y'$: on input a key x, the set of hash functions H and the filter F, outputs a value $y' \in \mathcal{V}$.*

3.4 Batch (Keyword) PIR

We give the formal definition of batch keyword PIR in Definition 2. And we can obtain batch PIR if we restrict the key domain $\mathcal{K} = [|\mathsf{D}|]$.

Definition 2. *A batch (keyword) PIR scheme, parameterized by the key domain $\mathcal{K}$ and the value domain $\mathcal{V}$, consists of the following algorithms:*

- PIR.Set(D, sp) → $\tilde{\mathsf{D}}$: *on input database* D *and system parameters* sp, *outputs* $\tilde{\mathsf{D}}$ *which are plaintexts of merged buckets.*
- PIR.Query(I, sp) → Q: *on input client's query batch* $I = \{k_i\}_{i \in [b]}$ *and the system parameters* sp, *outputs* Q *which are encrypted queries of merged buckets.*
- PIR.Resp($Q, \tilde{\mathsf{D}}, \mathsf{sp}$) → R: *on input the encrypted queries of merged buckets* Q, *the plaintexts of merged buckets* $\tilde{\mathsf{D}}$ *and the system parameters* sp, *outputs a list of responses* R.
- PIR.RMerge(R) → T: *on input the responses* R, *outputs ciphertexts of merged responses* T.
- PIR.Rec(T, sp) → A: *on input the ciphertexts of merged responses* T *and the system parameters* sp, *outputs responses* A.

3.5 Distributed Point Function

DPF is for secretly sharing a vector of 2^n elements in which only a single element is non-zero, and was first introduced by Boyle and Giboa in [1] to reduce secret sharing 2^n to $O(n)$. The formal definition of DPF is illustrated in Definition 3.

Definition 3. *A standard DPF scheme, parameterized by a finite field $\mathbb{F}$, consists of the following algorithms:*

- Gen(α, β) → (k_1, k_2): *on input a string $\alpha \in \{0,1\}^n$ and value $\beta \in \mathbb{F}$, output two DPF keys representing secret shares of a dimension-2^n vector that has value β only at the α-th position and is zero everywhere else.*
- Eval$_i(k_i, x)$ → y_i: *on input DPF key k_i and $x \in \{0,1\}^n$, output $y_i \in \mathbb{F}$, the secret-shared value of the dimension-2^n vector at the position indexed by x.*

4 Shared Characteristic from Batch Keyword PIR

In this section, we construct a novel batch keyword PIR: vectorized batch keyword PIR [14], demonstrating remarkable performance for small entries. Then we design a communication-efficient SC from vectorized batch keyword PIR.

4.1 Vectorized Batch Keyword PIR

Overall Process. We provide a general description to help build intuition. At first, the client inserts the query batch I into B buckets using Cuckoo hashing and the server inserts the database D into B buckets using simple hashing.

Figure 5 illustrates the next steps for each bucket. We set the size of the first two dimensions $N1_B = 3$, the size of the last dimension $N2_B = 3$, the plaintext vector size $n = 3$, and the client's query is consist of $d = 3$ query ciphertexts.

The server uses hash functions F_1, F_2 to place key-value pairs at different positions in the matrix. To obtain a d-dimensional hypercube, the server then encodes key-value pairs at the same position using the key-value filter and arranges the results $\{\mathsf{KVFij}\}_{i \in [N1_B], j \in [N1_B]}$ in different slices in order. After that, the server rotates and encodes slice0, slice1 and slice2 into $N1_B \times N2_B$ plaintexts in step $\boxed{1}$. Due to rotations in the server's computation, the positions of desired entries shift, but these shifts are public and can be easily calculated by the client. Thus, the client rotates and encrypts query1 in step $\boxed{2}$, and the rotation and encryption results of query2 and query3 are shown in step $\boxed{5}$ and step $\boxed{7}$, respectively.

Table 1. System Parameters sp

Notations	Descriptions
b	Size of client query batch I
B	Number of buckets, usually set to $1.5b$
$\{h_i\}_{i \in [w]}$	Hash functions for Cuckoo hashing and simple hashing
N_B	Size of the largest bucket
d-dimensional hypercube	Server's hypercube generated by hashing functions and the database D
$\{F_i\}_{i \in [d-1]}$	Hash functions for d-dimensional hypercube generation
KVF	A key-value filter
$N1_B$	Size of the first two dimensions
$N2_B$	Size of the last dimensions
n	Size of slots per ciphertext
$g_B = n/N1_B$	Number of buckets that a ciphertext can hold

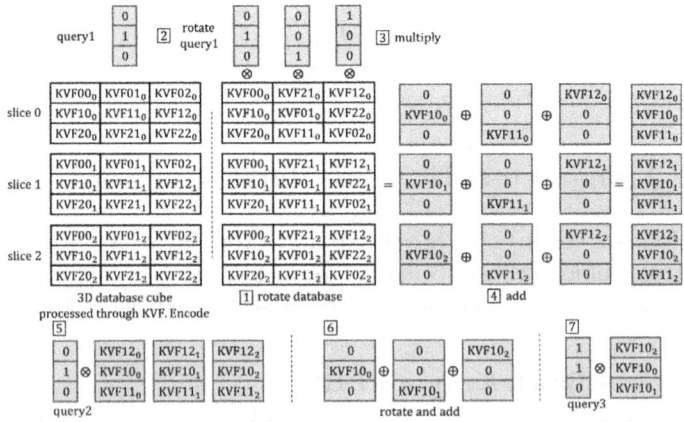

Fig. 5. General process of vectorized batch keyword PIR for each bucket.

In step $\boxed{3}$, $\boxed{4}$, $\boxed{5}$, $\boxed{6}$ and $\boxed{7}$, the server generates a response for each bucket through ciphertext addition, multiplication, and rotation. To merge responses from different buckets, the server utilizes PIR.RMerge algorithm. As shown in Fig. 6, the server automatically reconstructs the corresponding values during the responses merging process.

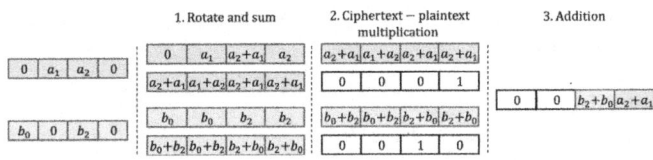

Fig. 6. Merging of responses from different buckets.

Algorithm Details. Compare to vectorized batch PIR, our modifications are limited to PIR.Set and PIR.Query. To simplify description, we use algorithms in vectorized batch PIR including VecPIRSetup (hypercube process in each bucket), PtsRotate (plaintext rotation) and EncodeToPt (plaintext encoding).

Algorithm 2. PIR.Query algorithm

Input:
  $I = \{k_1, \ldots, k_b\}$, Client query batch
  System parameters sp in Table 1
Output: $\tilde{Q}_0, \ldots, \tilde{Q}_{\lceil B/g_B \rceil - 1}$
1: Set $\epsilon = \lambda + \lceil log(B) \rceil$, and let $(\mathsf{H}, \mathsf{fpt}_\epsilon) \leftarrow$ KVF.Setup(ϵ).
2: Apply Cuckoo hashing to I using $h_1, \ldots, h_w$
3: **for** $i = 0 : B - 1$ **do**
4:     Represent k as key in the i-th bucket, $(j_1, \ldots, j_{d-1})$ as $(d-1)$-dim coordinates
5:     **for** $l = 1 : d - 1$ **do**
6:         $j_l = F_l(k)$
7:         $j'_l = \sum_{m=1}^{l} j_m$
8:         Generate vector v_l as one-hot encoding of j'_l
9:         $Q_i[l] = \mathsf{EncodeToPt}(v_l)$ ▷ Plaintext encoding.
10:     **end for**
11:     Represent $\{j_d | j_d \in \mathsf{H}(k)\}$ as entries for the last dim
12:     $\{j'_d\} = \{j_d + \sum_{m=1}^{d-1} j_m | j_d \in \mathsf{H}(k)\}$
13:     $v_d = \{0\}^{N_{2B}}$, set the x-th position to 1 if $x \in \{j'_d\}$
14:     $Q_i[d] = \mathsf{EncodeToPt}(v_d)$
15: **end for**
16: **for** $i = 0 : \lceil B/g_B \rceil - 1$ **do**
17:     $Q'_i = \sum_{k=0}^{g_B-1} \mathsf{PtsRotate}(Q_{i \cdot g_B + k}, k)$ ▷ Merge the plaintexts of different buckets.
18: **end for**
19: Encrypt $Q'_0, \ldots, Q'_{\lceil B/g_B \rceil - 1}$ to get $\tilde{Q}_0, \ldots, \tilde{Q}_{\lceil B/g_B \rceil - 1}$

Algorithm 1. PIR.Set algorithm

Input:
    D, the server's database with N key-value pairs
    System parameters sp in Table 1

Output:
    $\tilde{D}_0, \ldots, \tilde{D}_{\lceil B/g_B \rceil - 1}$, merged plaintext buckets

1: For each key-value pair $(k, v) \in$ D, copy it to buckets $B_{h_1(k)}, \ldots, B_{h_w(k)}$
2: Set $\epsilon = \lambda + \lceil log(B) \rceil$, and let $(H, fpt_\epsilon) \leftarrow$ KVF.Setup(ϵ)
3: **for** $i = 0 : B - 1$ **do**
4:     Initialize $P \leftarrow \emptyset, F \leftarrow \emptyset$
5:     **for** $(k, v) \in B_i$ **do**
6:         Set $ords_{d-1} = (F_1(k), \ldots, F_{d-1}(k))$
7:         $P[ords_{d-1}] \leftarrow P[ords_{d-1}] \cup (k, v)$
8:     **end for**
9:     **for** $P[ords_{d-1}] \in P$ **do**
10:       $F[ords_{d-1}] \leftarrow$ KVF.Encode$(P[ords_{d-1}], H, fpt_\epsilon)$
11:       Pad $F[ords_{d-1}]$ to size $N2_B$ with dummy value
12:     **end for**
13:     Initialize F' with $N2_B \cdot \prod_{l=1}^{d-1} F_l^{max}$ dummy entries, and F_l^{max} is F_l's maximum
14:     $F'[j_1 + \sum_{k=2}^{d} j_k \cdot \prod_{l=1}^{k-1} F_l^{max}] = F[j_1, \ldots, j_k]$
15:     $B'_i \leftarrow$ VecPIRSetup(F') ▷ Rotate and encode F' of each bucket into plaintexts.
16: **end for**
17: **for** $i = 0 : \lceil B/g_B \rceil - 1$ **do**
18:     $\tilde{D}_i = \sum_{k=0}^{g_B - 1}$ PtsRotate(B', k) ▷ Merge the plaintexts of different buckets.
19: **end for**
20: Output $\tilde{D}_0, \ldots, \tilde{D}_{\lceil B/g_B \rceil - 1}$

PIR.Set. In Algorithm 1, the server inserts D into B buckets by simple hashing. From line 4 to line 14, the server represents key-value pairs of each bucket as a d-dimensional hypercube using $\{F_i\}_{i \in [d-1]}$ and KVF. After that, the server uses VecPIRSetup to rotate and encode the hypercube of each bucket into plaintexts. Finally, the server uses PtsRotate to merge the plaintexts of different buckets.

PIR.Query. In Algorithm 2, the client places the query batch I into B buckets applying the Cuckoo hashing. From line 4 to line 10, the client computes and rotates the first $d - 1$ queries using $\{F_i\}_{i \in [d-1]}$ and encodes them into plaintexts by EncodeToPt. From line 13 to line 15, the client computes and rotates the last query of each bucket by KVF and encodes it into plaintext. In the end, the client merges the plaintexts of different buckets and encrypts them into ciphertexts.

We construct our batch keyword PIR scheme based on vectorized batch PIR [14] using key-value filters [3], thus the efficiency of our scheme can be inferred from the performance of vectorized batch PIR. As shown in the experiments of [14], vectorized batch PIR outperforms other existing batch PIR schemes in terms of communication and computation when handling smaller entry size.

4.2 Protocol Details

We show the formal description of our SC protocol in Fig. 7.

Parameters:

- Two parties: P_1 and P_2, set size m and n, where $m \ll n$.
- System parameters sp for batch keyword PIR.
- A key-value filter scheme KVF with key space $\mathcal{K} = \{0,1\}^*$ and value space $\mathcal{V} = \mathbb{F}$ where $\mathbb{F}$ is a finite field. Set ϵ based on the statistical parameter λ, and let $(\mathsf{H}, \mathsf{fpt}_\epsilon) \leftarrow \mathsf{KVF.Setup}(\epsilon)$.

Input of P_1: $X = \{x_i\}_{i \in [m]} \in \{\{0,1\}^*\}^m$, injective function $\phi : [m] \to [m_c]$
Input of P_2: $Y = \{y_j\}_{j \in [n]} \in \{\{0,1\}^*\}^n$
Protocol:

1. P_2 defines database $\mathsf{D} := \{(y_j, 0)\}_{j \in [n]}$, and computes the merged plaintext buckets $\tilde{\mathsf{D}} \leftarrow \mathsf{PIR.Set}(\mathsf{D}, \mathsf{sp})$.
2. P_2 defines the simple hashing table of Y in PIR.Set as $Y^* := \{Y_k^*\}_{k \in [m_c]}$. Then, P_2 selects random values $\mathbf{r} := \{r_k\}_{k \in [m_c]}$ for each bucket where $r_k \in \mathbb{F}$, and encrypts $\mathbf{r}$ based on sp to obtain the ciphertexts T_R.
3. P_1 defines the Cuckoo hashing table of X in PIR.Query as $X^* := \{x_k^*\}_{k \in [m_c]}$. And $\{x_{\phi(i)}^*\}_{i \in [m]}$ are the non-dummy item bins of X^*. Then, P_1 computes a list of encrypted queries $\tilde{Q} \leftarrow \mathsf{PIR.Query}(X^*, \mathsf{sp})$, and sends $\tilde{Q}$ to P_2.
4. P_2 computes the ciphertexts of responses $R \leftarrow \mathsf{PIR.Resp}(\tilde{Q}, \tilde{\mathsf{D}}, \mathsf{sp})$.
5. P_2 computes the ciphertexts of merged responses $T \leftarrow \mathsf{PIR.RMerge}(R, \mathsf{sp})$.
6. P_2 homomorphically computes $T' \leftarrow T + T_R$, and sends T' to P_1.
7. P_1 computes $\{u_k\}_{k \in [m_c]} \leftarrow \mathsf{PIR.Rec}(T', \mathsf{sp})$.
8. P_1 computes $r_k' := u_k - \mathsf{fpt}_\epsilon(x_k^*, 0)$ for $k \in [m_c]$, and outputs $\mathbf{r}' = \{r_k'\}_{k \in [m_c]}$. P_2 outputs $\mathbf{r} = \{r_k\}_{k \in [m_c]}$.

Fig. 7. Shared Characteristic Protocol from Vectorized Batch Keyword PIR

Correctness. We give the correctness analysis of our SC in Appendix 8.

Security. We prove the security of our SC in Theorem 1.

Theorem 1. *Given a batch keyword PIR scheme with query privacy, the protocol in Fig. 7 securely computes $\mathcal{F}_{\mathsf{sc}}$ against semi-honest adversaries.*

Proof. We provide the proof of Theorem 1 in Appendix 8.

5 Permuted Private Equality Test from DPF

We design our computation-efficient p-PEQT based on DPF [1], as detailed in Fig. 8. For key generation and distribution, we adopt an offline trusted third-party in our concrete p-PEQT construction. Alternatively, secret keys can be generated and distributed by garbled circuits during the preprocessing phase.

Parameters:

- Party P_1, party P_2, an offline trusted dealer $\mathcal{D}$.
- Set size m_c, bit length of set elements σ.
- A DPF scheme in Definition 3 with a finite field $\mathbb{F}$.

Input of P_1: $\mathbf{r}' = \{r'_1, \ldots, r'_{m_c}\} \in \{\{0,1\}^\sigma\}^{m_c}$
Input of P_2: $\mathbf{r} = \{r_1, \ldots, r_{m_c}\} \in \{\{0,1\}^\sigma\}^{m_c}$

Protocol:

1. The dealer $\mathcal{D}$ randomly selects $\{\alpha_k\}_{k \in [m_c]} \in \{\{0,1\}^\sigma\}^{m_c}$, and computes 2-party additive secret shares $\{\mathbf{s_1}[k], \mathbf{s_2}[k]\}_{k \in [m_c]}$ of $\{\alpha_k\}_{k \in [m_c]}$.
   $\mathcal{D}$ computes DPF keys $(\mathbf{k_1}[k], \mathbf{k_2}[k]) \leftarrow \mathsf{Gen}(\alpha_k, 1)$ for $k \in [m_c]$ and samples a random permutation π over $[m_c]$. $\mathcal{D}$ distributes $(\mathbf{k_1}, \mathbf{s_1})$ and π to P_1, and distributes $(\{\mathbf{k_2}[\pi(k)]\}_{k \in [m_c]}, \mathbf{s_2})$ to P_2.
2. P_2 computes $v_k := \mathbf{s_2}[k] \oplus r_k$ for $k \in [m_c]$, and sends $\{v_k\}_{k \in [m_c]}$ to P_1.
3. P_1 computes $\alpha'_k := \mathbf{s_1}[\pi(k)] \oplus r'_{\pi(k)} \oplus v_{\pi(k)}$ and $\mathbf{b_1}[k] := \mathsf{Eval}(\mathbf{k_1}[\pi(k)], \alpha'_k)$ for $k \in [m_c]$, then sends $\{\alpha'_k\}_{k \in [m_c]}$ and $\mathbf{b_1}$ to P_2.
4. P_2 computes $\mathbf{b_2}[k] := \mathsf{Eval}(\mathbf{k_2}[\pi(k)], \alpha'_k)$ and $b_k := \mathbf{b_1}[k] + \mathbf{b_2}[k]$ for $k \in [m_c]$.
5. P_1 outputs π, P_2 outputs $\mathbf{b} = \{b_k\}_{k \in [m_c]}$.

Fig. 8. Permuted Private Equality Test Protocol from DPF

5.1 Correctness

For all $k \in [m_c]$, we consider the following two cases: (1) If $r'_k = r_k$, we have $\alpha'_k = \mathbf{s_1}[\pi(k)] \oplus \mathbf{s_2}[\pi(k)]$. Therefore, we have $b_k = \mathbf{b_1}[k] + \mathbf{b_2}[k] = 1$ according to the correctness of DPF. (2) if $r'_k \neq r_k$, we have $\alpha'_k = \mathbf{s_1}[\pi(k)] \oplus \mathbf{s_2}[\pi(k)] \oplus r'_k \oplus r_k = \alpha_{\pi(k)} \oplus r'_k \oplus r_k$. Thus, we have $b_k = \mathbf{b_1}[k] + \mathbf{b_2}[k] = 0$ by the DPF's correctness.

5.2 Security

Theorem 2. *The construction of Fig. 8 securely implements functionality 4 based on DPF against semi-honest adversaries.*

Proof. We provide the proof of Theorem 2 in Appendix 8.

6 Private Computation on Set Intersection

In this section, we propose a series of PCSI protocols based on our communication-efficient SC and computation-efficient p-PEQT. The formal description is given in Fig. 9.

Our unbalanced PCSI protocols are obtained according to the framework in [17], and the correctness analysis and security proof are the same as [17], therefore we do not provide them here.

Parameters:

- Two parties: sender $\mathcal{S}$ and receiver $\mathcal{R}$, set size m and n where $n \gg m$, bit length of set elements l.
- Ideal $\mathcal{F}_{sc}$, $\mathcal{F}_{p\text{-}peqt}$ in Figure 3, 4 respectively. 1-out-of-2 OT functionality [16].

Input of $\mathcal{S}$: $X = \{x_i\}_{i \in [m]} \in \{\{0,1\}^l\}^m$
Input of $\mathcal{R}$: $Y = \{y_j\}_{j \in [n]} \in \{\{0,1\}^l\}^n$
Protocols:

1) $\mathcal{S}$ sets injective function $\phi : [m] \to [m_c]$ based on Cuckoo hashing. $\mathcal{S}$ and $\mathcal{R}$ play as P_1 and P_2 to invoke $\mathcal{F}_{sc}$: $\mathcal{S}$ inputs X and ϕ, and $\mathcal{R}$ inputs Y. As a result, $\mathcal{S}$ outputs $\mathbf{r'} = \{r'_i\}_{i \in [m_c]}$, and $\mathcal{R}$ outputs $\mathbf{r} = \{r_i\}_{i \in [m_c]}$, where m_c denotes the size of $\mathcal{S}$'s Cuckoo hashing table $\{x_i^*\}_{i \in [m_c]}$.
2) $\mathcal{S}$ and $\mathcal{R}$ play as P_1 and P_2 to invoke $\mathcal{F}_{p\text{-}peqt}$: $\mathcal{S}$ inputs $\mathbf{r'}$, and $\mathcal{R}$ inputs $\mathbf{r}$. As a result, $\mathcal{S}$ outputs a permutation π over $[m_c]$, and $\mathcal{R}$ outputs $\mathbf{b} = \{b_i\}_{i \in [m_c]}$, where $b_i = 1$ if $r_{\pi(i)} = r'_{\pi(i)}$, and $b_i = 0$ otherwise.

1. **PCSI-card.**

3) $\mathcal{R}$ computes and outputs $card = \sum_{i=1}^{m_c} b_i$.

2. **PCSI-card-sum.**

3) $\mathcal{S}$ chooses m_c random values $\{e_i\}_{i \in [m_c]}$ such that $\sum_{i=1}^{m_c} e_i = 0$. $\mathcal{S}$ computes $v_i^0 := e_i$ and $v_i^1 := x_i^* + e_i$ for $i \in [m_c]$.
4) $\mathcal{S}$ and $\mathcal{R}$ run 1-out-of-2 OT: for $i \in [m_c]$, $\mathcal{S}$ inputs (v_i^0, v_i^1), and $\mathcal{R}$ inputs b_i. Thus $\mathcal{R}$ obtains $u_i = v_i^1$ if $b_i = 1$, and $u_i = v_i^0$ otherwise. Then $\mathcal{R}$ computes and outputs $sum = \sum_{i=1}^{m_c} u_i$.

3. **PCSI-secret-sharing.**

3) $\mathcal{S}$ encrypts x_i^* to get c_i for $i \in [m_c]$, and sends $\mathbf{c} = \{c_i\}_{i \in [m_c]}$ to $\mathcal{R}$.
4) $\mathcal{R}$ removes c_i if $b_i = 0$, and obtains $\{c'_j\}_{j \in [card]}$. After that, $\mathcal{R}$ selects a random permutation π' over $[card]$, and obtains $\{c_j^*\}_{j \in [card]}$ where $c_j^* = c'_{\pi'(j)}$. Finally, $\mathcal{R}$ randomly selects $\mathbf{s'} = \{s'_j\}_{j \in [card]}$, then homomorphically computes $\{c_j^* - s'_j\}_{j \in [card]}$ and sends all the ciphertexts to $\mathcal{S}$.
5) $\mathcal{S}$ decrypts the ciphertexts, and outputs $\mathbf{s} = \{s_j\}_{j \in [card]}$. $\mathcal{R}$ outputs $\mathbf{s'}$.

Fig. 9. Unbalanced PCSI Protocols from SC and p-PEQT

7 Implementation and Performance

In this section, we experimentally evaluate our PCSI protocols and compare them with the state-of-the-art works in terms of communication and runtime.

7.1 Experimental Setup

We conduct our experiments on a single Intel Core i9-10900X at 3.70 GHz and 128GB RAM. All experiments are executed with 8 threads. To evaluate the efficiency of the protocols in different network setting, we use two simulated network settings, including LAN (10Gbps bandwidth and 0.05ms RTT latency) and WAN (100Mbps, 10Mbps and 1Mbps bandwidth and 80ms RTT latency).

7.2 Implementation Details

The used dataset is 128-bit random strings. We set the computational security parameter $\kappa = 128$ and the statistical parameter $\lambda = 40$. We set the parameters of batch PIR following [14]. Bit-length μ of key-value filter's decoding result is set to 64 to guarantee the correctness. Output domain of DPF is $\{0, 1\}$. Our implementation is written in C++. We choose binary fuse filter [9] when the hypercube dimension $d = 2$ and random matrix when $d = 3$.

To compare with other works [17,19], we use their open source code mpc4j (https://github.com/alibaba-edu/mpc4j). For [19], we exclude the OT phase from unbalanced PSU to evaluate the performance of unbalanced PCSI-card. In addition, [17,18] proposed unbalanced PCSI-card based on the same unbalanced PSI scheme [5]. Although the protocol in [18] handles the long item issue, its communication overhead and computation overhead are essentially the same as in [17] for our settings (the sender's set is relatively small). Furthermore, there is no open source code for [18]. Thus, we only evaluate [17].

7.3 Performance Evaluation

We compare our PCSI-card protocol with [17] (Tu-ddh) and [19] (Zhang-ddh, Zhang-pk). We focus on PCSI-card from DDH assumption based p-PEQT, as it shows better performance than Permute+Share based p-PEQT in our experiments. A detailed benchmark for small set sizes $m \in \{2^4, 2^6, 2^8, 2^{10}\}$ and large set sizes $n \in \{2^{18}, 2^{20}, 2^{22}\}$ is presented in Table 2. And Fig. 10 illustrates the variation of communication/running time with small set size/bandwidth. All other PCSI protocols including PCSI-card sum and PCSI-secret-sharing, are built upon PCSI-card based on the same framework. Thus, it is enough to evaluate PCSI-card to compare the performance of a series of PCSI protocols.

In addition, since SC contributes the main overhead of PCSI-card, the performance of PCSI-card protocol inherently reflects the efficiency of SC.

At last, we compare our p-PEQT with p-PEQT protocols from DDH assumption and Permute + Share [19]. Figure 11 illustrates the optimal online performance of our DPF based p-PEQT.

Communication Comparison. As shown in Table 2 and the top left figure in Fig. 10, our protocol achieves the lowest communication among all protocols when the size of small set is less than 2^8. Compared to the best result, the communication of our PCSI-card is $1.1 - 1.9$ better. For Tu-ddh, each ciphertext contains 1683 plaintext slots, and as long as the sender's elements do not fill all slots (corresponding to a set size of approximately 2^{10}), the communication

Table 2. Communication cost (in MB) and running time (in seconds) of our PCSI-card protocol to the state-of-the-art works. The best result is marked in ▓▓▓▓.

Param.		Protocol	Comm.(MB)		Running time(s)							
					LAN		100Mbps		10Mbps		1Mbps	
n	m		Setup	Online	Setup	Online	Setup	Online	Setup	Online	Setup	Online
2^{18}	2^4	Tu-ddh[17]	0.77	1.33	3.98	1.54	4.20	2.05	4.76	3.01	10.30	12.58
		Zhang-ddh[19]	18.00	0.63	16.44	1.84	18.04	2.29	30.97	2.74	160.62	7.28
		Zhang-pk[19]	18.00	0.65	60.95	12.13	62.55	12.42	75.51	12.89	205.11	17.57
		Ours	19.31	**0.38**	**2.48**	**1.37**	**4.78**	**1.64**	**19.16**	**1.91**	**164.89**	**4.65**
	2^6	Tu-ddh[17]	0.77	1.33	3.50	1.54	3.72	2.05	4.28	3.01	9.82	12.58
		Zhang-ddh[19]	18.00	0.63	20.10	1.31	21.70	1.76	34.66	2.24	164.26	6.99
		Zhang-pk[19]	18.00	0.66	51.47	3.67	53.07	3.96	66.03	4.44	195.63	9.19
		Ours	19.47	**0.63**	**2.20**	**0.97**	**4.71**	**1.26**	**19.12**	**1.71**	**165.21**	**6.25**
	2^8	Tu-ddh[17]	0.77	1.34	3.43	1.51	3.65	2.02	4.21	2.98	9.75	12.63
		Zhang-ddh[19]	18.00	1.26	28.57	1.21	30.17	1.71	43.13	2.62	172.73	11.69
		Zhang-pk[19]	18.00	1.46	81.87	3.10	83.47	3.62	96.43	4.67	226.03	15.18
		Ours	20.25	**1.15**	**2.29**	**0.59**	**4.92**	**0.92**	**21.39**	**1.75**	**171.27**	**10.03**
	2^{10}	Tu-ddh[17]	0.77	**1.39**	3.57	1.54	3.79	2.05	4.35	3.05	9.89	13.06
		Zhang-ddh[19]	18.00	4.99	32.40	1.36	34.00	2.16	46.96	5.75	176.56	41.68
		Zhang-pk[19]	18.00	3.29	185.23	3.74	186.83	4.40	199.79	6.77	329.39	30.46
		Ours	21.51	2.42	**4.25**	**1.17**	**7.23**	**1.60**	**22.13**	**3.35**	**181.83**	**20.77**
2^{20}	2^4	Tu-ddh[17]	0.77	1.86	12.72	2.06	12.94	2.61	13.50	3.95	19.04	17.34
		Zhang-ddh[19]	18.00	0.67	63.62	5.62	65.22	6.07	78.18	6.56	207.78	11.38
		Zhang-pk[19]	18.00	0.80	137.51	16.66	139.11	17.12	152.07	17.70	281.67	23.46
		Ours	19.31	**0.38**	**3.69**	**2.81**	**6.01**	**3.08**	**20.39**	**3.35**	**166.04**	**6.09**
	2^6	Tu-ddh[17]	0.77	1.87	12.65	2.14	12.87	2.69	13.43	4.04	18.97	17.50
		Zhang-ddh[19]	18.00	1.28	63.08	2.00	64.68	2.50	77.64	3.42	207.24	12.64
		Zhang-pk[19]	18.00	**0.80**	203.91	12.25	205.51	12.66	218.47	12.71	348.07	13.29
		Ours	19.45	0.88	**3.12**	**1.98**	**5.54**	**2.29**	**19.97**	**2.92**	**166.13**	**9.26**
	2^8	Tu-ddh[17]	0.77	**1.88**	12.57	2.05	12.79	2.60	13.35	3.95	18.89	17.49
		Zhang-ddh[19]	18.00	2.52	79.17	1.60	80.77	2.20	93.73	4.02	223.33	22.16
		Zhang-pk[19]	18.00	2.63	188.75	4.03	190.35	4.64	203.31	6.53	332.91	25.47
		Ours	20.15	1.89	**3.45**	**1.07**	**6.10**	**1.46**	**20.58**	**2.82**	**170.22**	**16.43**
	2^{10}	Tu-ddh[17]	0.77	**1.92**	12.96	2.10	13.18	2.65	13.74	4.04	19.28	17.86
		Zhang-ddh[19]	18.00	5.05	126.02	1.93	127.86	2.49	140.82	6.13	270.42	42.49
		Zhang-pk[19]	18.00	5.85	336.45	4.53	338.05	5.40	351.01	9.61	480.61	51.73
		Ours	40.98	1.84	**26.30**	**1.54**	**30.34**	**1.93**	**60.10**	**3.25**	**360.04**	**16.50**
2^{22}	2^4	Tu-ddh[17]	0.77	3.20	49.25	2.74	49.47	3.40	50.03	5.70	55.57	28.74
		Zhang-ddh[19]	18.00	**0.86**	260.30	13.02	261.90	13.49	274.86	14.11	404.46	20.30
		Zhang-pk[19]	18.00	1.38	418.14	27.08	419.74	27.59	432.70	28.58	562.30	38.52
		Ours	19.44	1.38	**6.53**	**3.85**	**9.04**	**4.20**	**23.40**	**5.19**	**169.54**	**15.13**
	2^6	Tu-ddh[17]	0.77	3.20	490.80	**2.70**	491.02	3.36	491.58	5.66	497.12	28.70
		Zhang-ddh[19]	18.00	1.47	302.77	8.62	304.37	9.14	317.33	10.20	446.93	20.78
		Zhang-pk[19]	18.00	1.99	501.30	18.06	503.14	18.62	516.10	20.05	645.70	34.38
		Ours	19.44	**1.38**	**6.53**	3.85	**9.04**	**4.20**	**23.40**	**5.19**	**169.54**	**15.13**
	2^8	Tu-ddh[17]	0.77	3.21	49.38	**2.78**	49.60	**3.44**	50.16	5.75	55.70	28.86
		Zhang-ddh[19]	18.00	5.12	258.43	2.74	260.03	3.55	272.99	7.24	402.59	44.10
		Zhang-pk[19]	18.00	3.21	742.86	13.80	744.46	14.46	757.42	16.77	887.02	39.88
		Ours	20.00	**2.89**	**7.92**	3.94	**10.69**	4.41	**25.24**	**6.36**	**173.49**	**27.19**
	2^{10}	Tu-ddh[17]	0.77	3.26	48.26	**2.71**	48.72	**3.37**	49.28	**5.72**	54.82	29.19
		Zhang-ddh[19]	18.00	9.46	336.50	3.69	338.10	4.85	351.06	11.66	480.66	79.77
		Zhang-pk[19]	18.00	9.90	775.72	9.74	777.32	10.93	790.28	18.06	919.88	89.34
		Ours	40.72	**2.96**	**105.39**	3.13	**109.49**	3.61	**140.17**	5.74	**438.38**	**27.05**

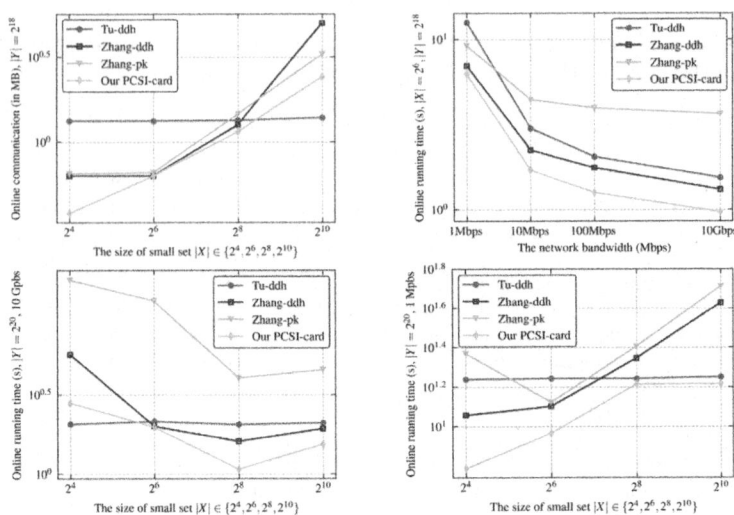

Fig. 10. Communication cost (in MB) and runtime (in secs) comparing to Tu-ddh [17], Zhang-ddh and Zhang-pk [19]. Both x and $y-$axis are in log scale. The top left figure shows the communication increases as the small set size increases. The top right figure shows the runtime decreases as the bandwidth increases. The bottom two figures show the runtime increases as the small set size increases in different bandwidths.

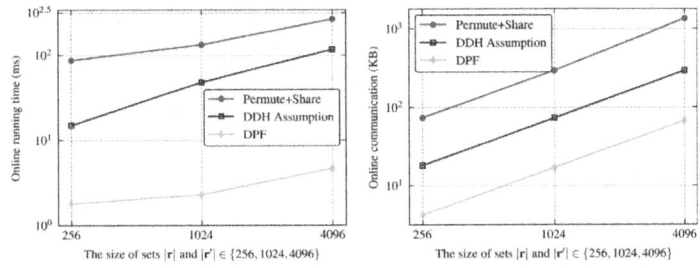

Fig. 11. Communication cost (in KB) and running time (in ms) comparing to p-PEQT protocols from DDH assumption and Permute + Share [19]. Both x and $y-$axis are in log scale.

cost remains constant. Therefore, Zhang-ddh, Zhang-pk and our PCSI-card perform better in the extremely unbalanced setting ($m \leq 2^8$). For Zhang-ddh and Zhang-pk, they incur relatively high communication overhead, as each element in the sender's set requires 2 PIR queries. Although our PCSI-card requires only a single keyword PIR query for each element, it still presents certain issues. First, our batch keyword PIR requires encoding the data into hypercubes, with sizes of the first two dimensions being a power of 2. Second, we use hash functions to map the data into the hypercubes. The two factors mentioned above result in vacant positions within the hypercubes, preventing a 2 times reduction in communication. We find that there are also vacant positions in the hypercubes of Zhang-ddh and Zhang-pk. But when $m = 2^6$, $n = 2^{20}$ and $m = 2^4$, $n = 2^{22}$, there are relatively more vacant positions in our work, resulting in a higher communication overhead compared to Zhang-ddh and Zhang-pk. Nevertheless, our protocol consistently outperforms the others in the vast majority of cases, as illustrated in Table 2.

Running Time Comparison. The experimental results in Table 2 and Fig. 10 indicate that our protocols demonstrate optimal computational performance in most settings. Specifically, the running time of our PCSI-card is $1.1 - 2.1$ times shorter than the best result from the state-of-the-art works in different bandwidth settings. For $m = 2^4, n = 2^{20}$, in 1Mbps bandwidth, our PCSI-card requires 6.09 seconds, while Zhang-pk requires 23.46 seconds, achieving a $3.9\times$ improvement, Zhang-ddh requires 11.38 seconds, achieving a $1.9\times$ improvement, and Tu-ddh requires 17.34 seconds, achieving a $2.8\times$ improvement.

8 Conclusion

In our work, we designed efficient constructions of SC and p-PEQT. For SC, we first provided a novel batch keyword PIR construction that performs well for small entries, and then proposed a communication-efficient SC from our batch keyword PIR. For p-PEQT, we gave a computationally efficient construction utilizing symmetric-key operations. Specifically, our construction builds upon DPF, and has further optimized online performance through preprocessing. In the end, we presented a series of unbalanced PCSI protocols with reduced communication and computation by combining the efficient constructions of SC and p-PEQT. While our protocol achieves improvements in efficiency, it still has certain limitations and can not effectively defend against malicious adversaries. Future work could focus on addressing this issue to further enhance the security and robustness of unbalanced PCSI.

Acknowledgments. This work is supported by the Beijing Natural Science Foundation (4242022) and the CCF-NSFOCUS "Kunpeng" Research Fund (CCF-NSFOCUS 2023006).

A Correctness Analysis of SC

For all $k \in [m_c]$, we consider the following two cases:

(1) If $x_k^* \in Y$, there is an $Y_k^* \in Y, y_j \in Y_k^*$ s.t $x_k^* = y_j$. Then we have $u_k = \mathsf{KVF.Decode}(x_k^*, \mathsf{H}, \mathsf{F}) + r_k = \mathsf{KVF.Decode}(y_j, \mathsf{H}, \mathsf{F}) + r_k$, where F is the key-value filter corresponding to the k-th bin. Since $\mathsf{KVF.Decode}(y_j, \mathsf{H}, \mathsf{F}) = \mathsf{fpt}_\epsilon(y_j, 0)$ by the inclusion correctness of key-value filter, $r_k' = u_k - \mathsf{fpt}_\epsilon(x_k^*, 0) = \mathsf{fpt}_\epsilon(y_j, 0) - \mathsf{fpt}_\epsilon(x_k^*, 0) + r_k = r_k$.

(2) If $x_k^* \notin Y$, a collision occurs if $\mathsf{KVF.Decode}(x_k^*, \mathsf{H}, \mathsf{F}) = \mathsf{fpt}_\epsilon(x_k^*, 0)$, $r_k' = u_k - \mathsf{fpt}_\epsilon(x_k^*, 0) = r_k$. To quantify the probability of collision, we consider two possible types of events:

(2.1) $\mathsf{H}(x_k^*) = \mathsf{H}(y_j)$ for some $y_j \in Y_k^*$. Then we have $\mathsf{KVF.Decode}(x_k^*, \mathsf{H}, \mathsf{F}) = \mathsf{fpt}_\epsilon(y_j, 0)$. Since $\mathsf{fpt}_\epsilon(x_k^*, 0) = hash(x_k^*)\|0$ is a random value in $\{0,1\}^\mu$, the probability that it exactly equals to $\mathsf{fpt}_\epsilon(y_j, 0)$ is $2^{-\mu}$.

(2.2) $\mathsf{H}(x_k^*) \neq \mathsf{H}(y_j)$ for all $y_j \in Y_k^*$. Then we have $\mathsf{KVF.Decode}(x_k^*, \mathsf{H} = \{h_1, \cdots, h_t\}, \mathsf{F}) = \mathsf{F}[h_1(x_k^*)] + \cdots + \mathsf{F}[h_t(x_k^*)] = l_k$. And the probability that $\mathsf{fpt}_\epsilon(x_k^*, 0) = hash(x_k^*)\|0$ equals to l_k is $2^{-\mu}$.

Setting $\mu = \lambda + \lceil log_2(m_c) \rceil$, our protocol is correct with $1 - 2^{-\lambda}$ probability, where λ is the statistical security parameter.

B Proof of Theorem 1

Proof. We exhibit simulators Sim_{P_1} and Sim_{P_2} for simulating corrupt P_1 and P_2 respectively, and argue the indistinguishability of the produced transcript from the real execution.

Corrupt P_1 : $\mathsf{Sim}_{P_1}(X, \phi, \mathbf{r}')$ simulates the view of corrupt semi-honest P_1 as follows. Sim_{P_1} randomly selects T' in step 6. Since $\{r_k\}_{k \in [m_c]}$ are random for P_1, $\{u_k = r_k + \mathsf{fpt}(x_k^*, 0)\}_{k \in [m_c]}$ are uniformly distributed, and the view in simulation is computationally indistinguishable from the real view.

Corrupt P_2: $\mathsf{Sim}_{P_2}(Y, \mathbf{r})$ simulates the view of corrupt semi-honest P_2 as follows. Sim_{P_2} randomly selects X, and computes the encrypted queries $\tilde{Q}$. The query privacy of batch keyword PIR scheme guarantees the view generated by Sim_{P_2} is computationally indistinguishable from the real view.

C Proof of Theorem 2

Proof. We exhibit simulators Sim_{P_1} and Sim_{P_2} for simulating corrupt P_1 and P_2 respectively, and argue the indistinguishability of the produced transcript from the real execution.

Corrupt P_1: $\mathsf{Sim}_{P_1}(\mathbf{r}', \pi)$ simulates the view of corrupt semi-honest P_1.

1. Sim_{P_1} selects m_c random DPF keys $\mathbf{k_1}$ and m_c random shares $\mathbf{s_1}$, then appends them to the view.
2. Sim_{P_1} selects m_c random values $\{v_k\}_{k \in [m_c]}$, and appends them to the view.

Now we argue that the view output by Sim_{P_1} is indistinguishable from the real one. We define four hybrid transcripts T_0, T_1, T_2, T_3 where T_0 is real view of P_1, and T_3 is the output of Sim_{P_1}.

- Hybrid$_0$. The first hybrid is the real interaction described in Fig. 8. Let T_0 denote the real view of P_1.
- Hybrid$_1$. Let T_1 be the same as T_0, except that $\mathbf{k_1}$ are replaced by m_c random DPF keys. Since $\{\alpha_k\}_{k \in [m_c]}$ are uniformly distributed, this hybrid is computationally indistinguishable from T_0 by the secrecy of DPF.
- Hybrid$_2$. Let T_2 be the same as T_1, except that $\mathbf{s_1}$ are replaced by m_c random shares. Now that $\mathbf{s_1}$ are additive secret shares of $\{\alpha_1, \ldots, \alpha_{m_c}\}$, they are uniformly distributed. Therefore, T_1 and T_2 are statistically indistinguishable.

– Hybrid$_3$. Let T_3 be the same as T_2, except that $\{v_k\}_{k\in[m_c]}$ are replaced by m_c random values. Since $\mathbf{s_2}$ are random for P_1, $\{v_k = \mathbf{s_2}[k] \oplus r_k\}_{k\in[m_c]}$ are consequently uniformly distributed, ensuring the view is indistinguishable from real execution.

Corrupt P_2: $\text{Sim}_{P_2}(\mathbf{r}, \mathbf{b})$ simulates the view of corrupt semi-honest P_2. It executes as follows:

1. Sim_{P_2} selects m_c random DPF keys $\mathbf{k_2}$ and m_c random shares $\mathbf{s_2}$, then appends them to the view.
2. Sim_{P_2} selects m_c random values $\{\alpha'_k\}_{k\in[m_c]}$, and computes DPF evaluation results $\mathbf{b_1}$ base on $\{\alpha'_k\}_{k\in[m_c]}$, then appends them to the view.

Now we argue that the view output by Sim_{P_2} is indistinguishable from the real one. We define four hybrid transcripts T_0, T_1, T_2, T_3 where T_0 is real view of P_2, and T_3 is the output of Sim_{P_2}.

– Hybrid$_0$. The first hybrid is the real interaction described in Fig. 8. Let T_0 denote the real view of P_2.
– Hybrid$_1$. Let T_1 be the same as T_0, except that $\{\mathbf{k_2}[\pi(k)]\}_{k\in[m_c]}$ are replaced by m_c random DPF keys. Firstly, Sim_{P_2} randomly selects π and $\{\alpha_k\}_{k\in[m_c]}$, then computes $(\mathbf{k_1}[\pi(k)], \mathbf{k_2}[\pi(k)]) \leftarrow \text{Gen}(\alpha_{\pi(k)}, 1)$ for $k \in [m_c]$. Since $\{\alpha_{\pi(k)}\}_{k\in[m_c]}$ are uniformly distributed, this hybrid is computationally indistinguishable from T_0 by the secrecy of DPF.
– Hybrid$_2$. Let T_2 be the same as T_1, except that $\mathbf{s_2}$ are replaced by m_c random shares. Now that $\mathbf{s_2}$ are additive secret shares of $\{\alpha_1, \ldots, \alpha_{m_c}\}$, they are uniformly distributed. Therefore, T_1 and T_2 are statistically indistinguishable.
– Hybrid$_3$. Let T_3 be the same as T_2, except that $\{\alpha'_k\}_{k\in[m_c]}$ are replaced by m_c random values and $\mathbf{b_1}$ are computed based on $\{\alpha'_k\}_{k\in[m_c]}$. Specifically, Sim_{P_2} selects $\alpha'_k := \alpha_{\pi(k)}$ if $b_k = 1$, and selects random value for α'_k if $b_k = 0$. After that, Sim_{P_2} computes $\mathbf{b_1}[k] := \text{Eval}(\mathbf{k_1}[\pi(k)], \alpha'_k)$ for $k \in [m_c]$. Since $\{\alpha'_k\}_{k\in[m_c]}$ are uniformly distributed, and the values of $\{\alpha'_k\}_{k\in[m_c]}$ and $\mathbf{b_1}$ ensure the correctness, this hybrid is indistinguishable from real execution.

References

1. Boyle, E., Gilboa, N., Ishai, Y.: Function secret sharing: improvements and extensions. In: Proceedings of the 2016 ACM SIGSAC Conference on Computer and Communications Security, pp. 1292–1303 (2016)
2. Buddhavarapu, P., Knox, A., Mohassel, P., Sengupta, S., Taubeneck, E., Vlaskin, V.: Private matching for compute. Cryptol. ePrint Arch. (2020)
3. Celi, S., Davidson, A.: Call me by my name: simple, practical private information retrieval for keyword queries. In: Proceedings of the 2024 on ACM SIGSAC Conference on Computer and Communications Security, pp. 4107–4121 (2024)
4. Chen, H., Huang, Z., Laine, K., Rindal, P.: Labeled PSI from fully homomorphic encryption with malicious security. In: Proceedings of the 2018 ACM SIGSAC Conference on Computer and Communications Security, pp. 1223–1237 (2018)

5. Chen, H., Laine, K., Rindal, P.: Fast private set intersection from homomorphic encryption. In: Proceedings of the 2017 ACM SIGSAC Conference on Computer and Communications Security, pp. 1243–1255 (2017)
6. Chen, Y., Zhang, M., Zhang, C., Dong, M., Liu, W.: Private set operations from multi-query reverse private membership test. In: IACR International Conference on Public-Key Cryptography, pp. 387–416. Springer (2024)
7. Freedman, M.J., Nissim, K., Pinkas, B.: Efficient private matching and set intersection. In: International Conference on the Theory and Applications of Cryptographic Techniques, pp. 1–19. Springer (2004)
8. Garimella, G., Mohassel, P., Rosulek, M., Sadeghian, S., Singh, J.: Private set operations from oblivious switching. In: IACR International Conference on Public-Key Cryptography, pp. 591–617. Springer (2021)
9. Graf, T.M., Lemire, D.: Binary fuse filters: fast and smaller than XoR filters. J. Exper. Algorithmics (JEA) **27**(1), 1–15 (2022)
10. Ion, M., et al.: On deploying secure computing: private intersection-sum-with-cardinality. In: 2020 IEEE European Symposium on Security and Privacy (EuroS&P), pp. 370–389. IEEE (2020)
11. Li, H., Gao, Y.: Efficient private set intersection cardinality protocol in the reverse unbalanced setting. In: International Conference on Information Security, pp. 20–39. Springer (2022)
12. Lv, S., et al.: Unbalanced private set intersection cardinality protocol with low communication cost. Futur. Gener. Comput. Syst. **102**, 1054–1061 (2020)
13. Miao, P., Patel, S., Raykova, M., Seth, K., Yung, M.: Two-sided malicious security for private intersection-sum with cardinality. In: Annual International Cryptology Conference, pp. 3–33. Springer (2020)
14. Mughees, M.H., Ren, L.: Vectorized batch private information retrieval. In: 2023 IEEE Symposium on Security and Privacy (SP), pp. 437–452 (2023)
15. Pinkas, B., Schneider, T., Zohner, M.: Scalable private set intersection based on OT extension. ACM Trans. Priv. Secur. (TOPS) **21**(2), 1–35 (2018)
16. Rabin, M.O.: How to exchange secrets with oblivious transfer. Cryptol. ePrint Arch. (2005)
17. Tu, B., Zhang, X., Bai, Y., Chen, Y.: Fast unbalanced private computing on (labeled) set intersection with cardinality. Cryptol. ePrint Arch. (2023)
18. Wu, M., Yuen, T.H.: Efficient unbalanced private set intersection cardinality and user-friendly privacy-preserving contact tracing. In: 32nd USENIX Security Symposium (USENIX Security 23), pp. 283–300 (2023)
19. Zhang, C., et al.: Unbalanced private set union with reduced computation and communication. In: Proceedings of the 2024 on ACM SIGSAC Conference on Computer and Communications Security, pp. 1434–1447 (2024)
20. Zhang, C., Chen, Y., Liu, W., Zhang, M., Lin, D.: Linear private set union from {Multi-Query} reverse private membership test. In: 32nd USENIX Security Symposium (USENIX Security 23), pp. 337–354 (2023)

Artemis: Decentralized, Secure, and Efficient Safety Monitoring with Dynamic Trajectories

Meng Li[1,2], Zhuangwei Li[1,2], Yifei Chen[1,2], Yan Qiao[1,2(✉)], and Mauro Conti[3,4]

[1] Key Laboratory of Knowledge Engineering with Big Data (Hefei University of Technology, HFUT), Ministry of Education, Hefei, China
[2] School of Computer Science and Information Engineering, HFUT, Hefei, China
qiaoyan@hfut.edu.cn
[3] Department of Mathematics and HIT Center, University of Padua, Padua, Italy
[4] Örebro University, Örebro, Sweden

Abstract. Secure safety monitoring is an intriguing and practical feature of current ride-hailing services. It enhances the riders' safety after they are onboard a driver's vehicle while not violating location privacy. Existing work suffer from the problems of unnecessary trajectory uploading, centralized trajectory monitoring, redundant trajectory monitoring, and assuming static trajectories. In this work, we propose a decentralized, secure, and efficient safety monitoring scheme Artemis to guarantee rider safety while supporting dynamic trajectories, i.e., permitted trajectory deviations. In specific, we design a 2-out-of-n private threshold signature protocol to achieve trajectory authenticity, design a decentralized RHS platform and a secure trajectory similarity computation protocol to guarantee both efficiency and location privacy, and design a secure three-party computation protocol to ensure deviation authenticity. Formal security experiments and proofs are provided. Experimental results from extensive experiments based on Ethereum and Intel SGX2 demonstrate that Artemis is highly efficient.

Keywords: Ride-hailing services · Safety monitoring · Security · Location Privacy · Efficiency

1 Introduction

1.1 Background

During the past fifteen years, Ride-Hailing Services (RHSs) run by commercial Ride-Hailing Service Providers (RHSPs) have been providing convenient transportation to riders via recruiting voluntary drivers, optimizing matching algorithms, and upgrading Ride-Hailing Apps (RHAs). Such services contribute to alleviating traffic pressure, creating employment positions, and stimulating sharing economy. Two representative RHSPs are Uber and DiDi.

Despite significant progress, the security problem of current RHSs remain a giant obstacle to shaping the RHSs toward perfection. The root of the problem comes from the untrustworthy RHSPs that are (1) leaking users's data due to system malfunction, program bugs, and hacker attacks [1], (2) prying into users' privacy (e.g., identity and location) for some employee being curious or the platform conducting user profiling [2], and (3) sharing users's privacy when they are malicious [3]. To cope with these issues, a plethora of privacy-preserving schemes have been proposed, such as privacy-preserving ride-hailing with accountability [4], privacy-preserving ride-matching with prediction [5], privacy-preserving and fog-based ride-hailing [6], and privacy-preserving ride matching against collusion attack [7].

Recently, a new ride-hailing feature has attracted public attention, i.e., Safety Monitoring. Given that there have been reports of harassment, assault and robbing riders in RHSs [8–10], a growing interest in the development of such a feature is obvious. It allows the RHSP monitors the trajectories of drivers, who are transporting riders en route, via real-time GPS tracking to detect any suspicious behavior when a trajectory deviation occurs [11–13]. However, safety monitoring poses privacy threats to users because it requires them to upload real-time trajectories to the RHSP. These privacy concerns have given rise to privacy-preserving safety monitoring that achieves secure trajectory similarity measure. One typical example is pSafety [13], which made use of the RHSP and a Crypto Service Provider (CSP) to compute trajectory similarity over encrypted trajectories.

1.2 Problems and Motivations

After digging into existing work, we have successfully identified four problems described as follows. **P1. Unnecessary trajectory uploading**. The riders have to upload their encrypted trajectory to the RHSP periodically during the ride, resulting in extra communication, i.e., electricity consumption for their smartphones. **P2. Centralized trajectory monitoring**. A centralized RHSP has to perform the monitoring task for all riders, which obviously not only incurs too much burden for the RHSP, but also constitutes a larger security risk for users. **P3. Redundant trajectory monitoring**. For each ongoing RHS order, the RHSP has to launch a trajectory deviation monitoring process for both the rider and the driver, simultaneously and respectively. Moreover, the RHSP has to communicate with the CSP to compare encrypted trajectories, which involves heavy computation and multi-round communication. **P4. Assuming static trajectories**. All planned trajectories that are compared with for safety monitoring are assumed to be static, which overlook multiple real-life scenarios and limits the practicability of traffic monitoring (Fig. 1).

Correspondingly, we are enlightened to present our four motivations as below. **M1. Minimum trajectory uploading**. Considering that most RHSPs including the RHS magnate DiDi, monitor the real-time locations of drivers in process [?] while not forcing access to riders' locations, we only need the drivers to

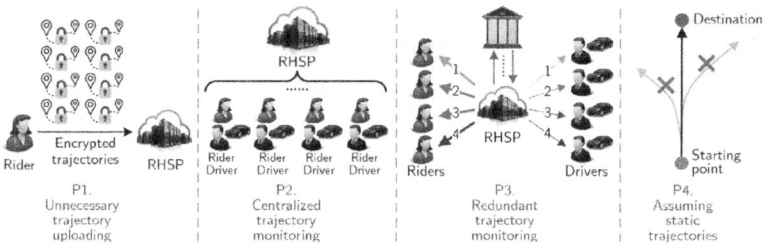

Fig. 1. Problems of Existing Work and Motivations of Artemis.

upload their encrypted trajectory to the RHSP periodically, setting the riders free from unnecessary trajectory uploading as well as accompanying costs. **M2. Decentralized trajectory monitoring.** Inspired by the advancement of blockchain [14,15], which is decentralized, transparent, and immutable, we divide the RHS region into a set of small service areas and assign the monitoring task in an area to an Roadside Unit (RSU). All RSUs combined will form a decentralized service architecture. **M3. Minimum trajectory monitoring.** Similar to M1, given than riders no longer need to upload trajectories periodically besides the original trajectory, the RHSP only needs to monitor the trajectory of the driver during a ride. Further, the RHSP does not have to communicate with another cloud server to compute trajectory similarity. **M4. Supporting dynamic trajectories.** Not all planned trajectories are the actually traversed ones since there are possible scenarios leading to an inevitable trajectory diversion, such as traffic jam, blocked roads (due to road construction and accidents), and taking a wrong turn (due to driver's negligence). Therefore, after a ride begins, we allow the users to deviate from their original trajectories and choose a new route.

1.3 Possible Approaches and Technical Challenges

Given M1, M2, and M3, it is not difficult to come up with solutions. Regarding M4, there are two possible approaches to realize such a new feature. First, we can continue to adopt pSafety's approach. When there is a reasonable trajectory deviation, the rider and driver plan a new route from the current location to the destination. However, such an approach would inevitably incur substantial computational and communication costs. Second, we can ask the users to plan a new route and then co-sign the route.

Nevertheless, there are three technical challenges to be tackled. TC1. How to balance privacy and efficiency for riders in safety monitoring? Encrypting-and-submitting trajectories to the RHSP facilitates trajectory monitoring, but it incurs too many costs. Although appointing the driver in service to submit trajectories is convenient (M1, M3), this will leave the initiative to submit trajectories to the untrustworthy driver. TC2. How to maintain good efficiency for decentralized monitoring while not sacrificing users' location privacy? Decentralizing the monitoring tasks into a set of RSUs (M2) can reduce computation and communication pressure of the RHSP, but it expands the scope of privacy leakages by designating RSUs to handle risers' trajectories, especially when the RSUs are not fully trusted. TC3. How to defend against collusion attack from an RSU and a rider toward a fake permitted deviation? When the RHSP colludes with a rider that is not on board to maliciously deviate from the planned trajectory (M4) under the radar, they may try to forge a permitted deviated trajectory by asking the colluding rider to sign the deviated trajectory on behalf of the real rider without his/her knowledge.

1.4 Solutions and Proposed Scheme

For TC1, we design a 2-out-of-n private threshold signature protocol to achieve trajectory authenticity. We only require the drivers to upload encrypted trajectories to the RHS platform (not necessarily the RHSP, to be explained later) and we assume that each driver holds a tamper-resistant On-Board Unit (OBU) to process trajectories and be responsible for communication. Meanwhile, we harness Accountable Threshold Signature [16] to enable the rider and the driver, i.e., 2 signers from a group of n signers, to co-sign the trajectories. We install on each RSU a Trusted Execution Environment (TEE), i.e., Intel SGX2 [17,18], to privately aggregate and verify the trajectory. In this way, we retain the control of trajectories in the hands of the riders and achieve trajectory authenticity. *For TC2, we design a decentralized RHS platform and a secure trajectory similarity computation protocol to guarantee both efficiency and location privacy.* First, we deploy a set of RSUs to assist the RHSP and a CSP in building a decentralized RHS platform based on blockchain [19,20]. Each RSU will handle a subset of ride matching tasks and trajectory monitoring tasks, reducing the RHSP's operating pressure. Second, we embrace the Longest Common Subsequence (LCSS) [21] to measure trajectory similarity while secure the locations by utilizing Cheon-Kim-Kim-Song (CKKS) [22] scheme. Third, to reduce privacy leakage during signature verification and trajectory comparison, we make use of the TEE while avoiding multi-round communication with CSP and improving comparison efficiency. *For TC3, we design a secure three-party computation protocol to ensure deviation authenticity.* We exploit the Group Signature (GS) with selective linkability [23] to allow the riders to send an anonymous identity and a new trajectory to a nearby RSU. The RSU will faithfully check whether the trajectory is permitted by the original rider by interacting with TEE, which blinds and encrypts anonymous identities and sends back two encrypted anonymous identities for final comparison. By doing so, identity consistency is ensured and the collusion attack is thwarted.

1.5 Contributions

To address and resolve these technical challenges, **we propose a decentralized, secure, and efficient safety monitoring scheme Artemis to guarantee rider safety while supporting dynamic trajectories, i.e., permit-**

ted trajectory deviations. Our contributions of this work is summarized as follows.

- We design the first ride-monitoring scheme that facilitates trajectory authenticity, location privacy, and collusion resistance.
- We formally define and prove the privacy and security of Artemis.
- We build a prototype of Artemis and conduct extensive experiments to evaluate its performance while comparing with existing work.

2 Related Work

In Table 1, we compare Artemis with existing work. Chaudhry et al. [11] focused on the passenger (rider) safety in RHSs. They emphasize a dire need of launching security measures to guarantee that riders' safety is ensured from the starting point to their destination. The authors have introduced the current situation of RHSs and RHSPs, such as Uber and DiDi, then discussed some possible solutions to ensure passenger safety during and after the ride. Such solutions are Dash Cam and Watchdog Network, Distress Alarm, and Miscellaneous Ones. Yu et al. [13] proposed pSafety, a privacy-preserving safety monitoring scheme for RHSs. Specifically, they design two secure trajectory similarity computation algorithms based on somewhat homomorphic encryption [22] to plan an original route $\mathcal{T}^{\mathrm{ori}}$ and measure trajectory deviation. Riders and drivers have to periodically upload their real-time trajectories T_r and T_d to the RHSP after encryption. Two cloud servers, i.e., RHSP and CSP, are used to collaboratively compute trajectory similarity for both $(T^r, \mathcal{T}^{\mathrm{ori}})$ and $(T^d, \mathcal{T}^{\mathrm{ori}})$. During this process, even though the locations are protected from the semi-honest RHSP, the minimum of two trajectories is still leaked. Han et al. [24] proposed a privacy-preserving travel recommendation scheme based on stay points over raw encrypted trajectory data. Given that user trajectories and travel routes have different stay points with space-time properties, they design an adapted LCSS computation algorithm to measure the similarity of two trajectories. To protect location privacy, they also develop some secure two-party computation primitives based on the Paillier cryptosystem [25]. Wang et al. [26] proposed TraSQ, an encrypted trajectory similarity query for mobile e-health systems. They employ XZ* index and bijective function to produce the unique trajectory encoding value, pre-filtering out some trajectories. Then, they utilize Paillier algorithm, two clouds, and secure computing protocols to protect the true trajectory similarity from cloud servers.

3 Problem Statement

3.1 System Model

The system model of Artemis consists of RHSP, CSP, RSU, Rider, Driver, and consortium blockchain as portrayed in Fig. 2.

Table 1. The Comparison Among Existing Work and Artemis

Scheme	Feature				
	System Model	Security Model	Collusion Attack	Location Privacy	Dynamic Trajectory
Chaudhry et al. [11]	Centralized	Honest	-	-	-
Yu et al. [13]	Centralized	Semi-honest	-	√	-
Han et al. [24]	Centralized	Semi-honest	-	√	-
Wang et al. [26]	Centralized	Semi-honest	-	√	-
Artemis	Decentralized	Malicious	√	√	√

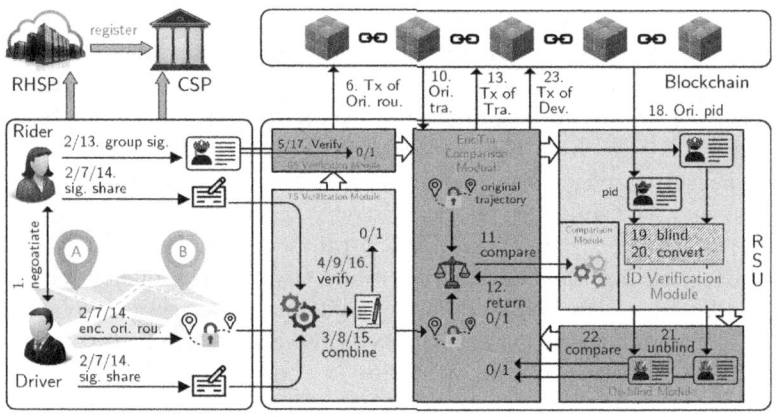

Fig. 2. System Model of Artemis.

RHSP is a ride-hailing service platform providing rides to riders who needs convenient transportation and job opportunities to drivers who are willing to utilize idle vehicles. It manages the basic information of all users when they register. It maintains a consortium blockchain $\mathcal{BC}$ with the CSP and RSUs to receive and track ride requests, ride responses, and ride information. Its main task in this work is to coordinate with the CSP and all RSUs to conduct trajectory monitoring for riders.

CSP is a reinforced cloud server responsible for generating all public parameters of underlying techniques including CKKS, ATS, GS, and TEE. It also issues keys to registered users and RSUs. It assists the RHSP and RSUs to facilitate trajectory monitoring.

RSU is an edge node deployed by the network operator or the RHSP with certain computation and communication capabilities [7]. It receives ride requests from riders and ride responses from drivers nearby, then assigns appropriate drivers to riders using a matching algorithm, which is not the focus on this work. Each RSU is equipped with TEE that processes sensitive user information during trajectory monitoring. An RSU also registers to the CSP and communicates with the $\mathcal{BC}$ as a blockchain node.

Rider is the user who needs a ride from the RHS platform. Once submitting a ride request to a nearby RSU, they await an assigned driver and gets on the driver's vehicle. The rider negotiates with the driver an original trajectory, which is encrypted and sent to the $\mathcal{BC}$ by the RSU.

Driver is the user who provides transportation services to RHS rides. Once submitting a ride response to a nearby RSU, they wait for an assigned rider and pick up the rider. While the vehicle is in motion, the driver's OBU periodically uploads real-time trajectories to the nearest RSU.

Consortium Blockchain is the underlying infrastructure for communication between RSUs and partial storage of ride order information. It receives the encrypted original trajectories from RSUs and replies to RSUs' requests for information of an original trajectory. The choice of a consortium blockchain over a private blockchain emphasizes the fact that the RHSP does not own the RSUs.

3.2 Security Model

We assume that the CSP is fully trusted and it is not possible for an adversary to compromise the CSP. We adopt the common honest-but-curious assumption [5,27,28] for the internal entities including riders, most of drivers, most of RSUs, and RHSP. While adhering strictly to their prescribed operational guidelines, they will simultaneously try to extract riders' locations by monitoring and analyzing transmitted communication packets, i.e., violating riders' location privacy. Besides, we consider the collusion attack between RSU and driver, where the colluding parties will share information and try to forge a valid trajectory deviation on behalf of an honest rider on board.

3.3 Design Objectives

Informally, we have four design objectives described as follows. We will give formal definitions and proofs in Sect. 5. **Trajectory authenticity**. The original trajectory submitted by a driver can be verified by the RSU that communicates with them without knowing the users' identities. **Location privacy**. During the process of trajectory monitoring, the locations of honest users including riders and driver, are protected from external adversaries and internal adversaries. **Deviation authenticity**. When an RSU colludes with a driver, they cannot forge a valid trajectory deviation for the honest rider on board. **Efficiency**. Artemis should realize good efficiency regarding computational costs and communication overhead of system entities.

4 The Proposed Scheme Artemis

4.1 Overview

Artemis consists of six phases: System Initialization, Entity Registration, Route Planning, Secure Trajectory Uploading, Secure Deviation, and Secure Trajectory

Monitoring. It has nine building blocks: CKKS $\Gamma = $ (HGen, HEnc, HDec, HAdd, HMul), ATS $\Delta = $ (AGen, ASign, ACombine, AVerify, ATrace), GS $\Lambda = $ (GGen, GJoin, GSign, GVerify, GBlind, GConvert, GUnblind), NIZK $\Pi = $ (P, V), PKE $\Sigma = $ (EGen, Enc, Dec), SIG $\Upsilon = $ (SGen, Sign, Verify), and two secure hash functions $\mathcal{H}_1, \mathcal{H}_2$.

4.2 System Initialization

The CSP initializes the system by running generation functions of the underlying techniques. First, it runs $\Gamma.\mathsf{HGen}(\lambda)$: given a security parameter λ, chooses a power-of-two $M = M(\lambda, q_L)$, an integer $h = h(\lambda, q_L)$, an integer $P = P(\lambda, q_L)$, and a real value $\sigma = \sigma(\lambda, q_L)$; samples $s \leftarrow \mathsf{HWT}(h)$, $a \leftarrow \mathcal{R}_{q_L}$, and $e \leftarrow \mathcal{DG}(\sigma^2)$; samples $a' \leftarrow \mathcal{R}_{P \cdot q_L}$ and $e' \leftarrow \mathcal{DG}(\sigma^2)$; outputs a secret key $sk = (1, s)$, a public key $pk = (b, a) \in \mathcal{R}_{q_L}^2$ where $b = -as + e$, and an evaluation key $evk = (b', a') \in \mathcal{R}_{P \cdot q_L}^2$ where $b' = -a's + e' + Ps^2 \bmod P \cdot q_L$.

Then, it runs $\Delta.\mathsf{AKGen}(1^\lambda, n, t) \to (pp^{\mathrm{ATS}}, \{ask_i\}_{i=1}^n, apk)$: given λ, a number of signers n, and a threshold t, outputs public parameters pp^{ATS}, a set of private keys $\{ask_i\}_{i=1}^n$ and a public key apk, where $ask_1, \cdots, ask_n \leftarrow \mathbb{Z}_q$ and $apk_i = g_i^{ask_i}$ for $i \in [n]$, $n = n_1 + n_2$.

Next, it runs $\Lambda.\mathsf{GSetup}(\lambda)$: given a security parameter λ, invokes $\mathcal{G}(1^\lambda) \to (p, \mathbb{G}_1, \mathbb{G}_2, \mathbb{G}_T, e, g_1, g_2)$ and $_{\$}\mathbb{G}_1 \to (g, h, h_1, h_2)$, randomly chooses three secret keys $isk, csk, bsk \leftarrow_{\$} \mathbb{Z}_p^*$, computes three pubic keys $ipk = g_2^{isk}$, $cpk = g^{csk}$, $bpk = g^{bsk}$, sets $gpk = (\mathbb{G}_1, \mathbb{G}_2, \mathbb{G}_T, p, e, g_1, g_2, g, h, h_1, h_2, ipk, cpk)$, and outputs $(gpk, bpk, isk, csk, bsk)$.

Collaborating with the RHSP and RSUs, the CSP initializes and maintains a consortium blockchain $\mathcal{CB}$ that runs atop a consensus algorithm. Finally, the CSP releases all public parameters.

4.3 Entity Registration

Riders register to the CSP to obtain two secret keys ask^r, $gsk^r = (A, x, y, s)$. Drivers register to the CSP to obtain a secret key ask^d. TEEs register to the CSP to obtain evk, csk, and a pair of private key and public key $(sk_i^{\mathrm{tee}}, pk_i^{\mathrm{tee}})$.

4.4 Route Planning

When a rider id_i^r hails a ride by using an RHA to communicate with a nearby RSU, we assume that the RHSP and the RSU assign the most matched driver id_i^d to id_i^r. Then, id_i^r and id_i^d plan an original route $\mathcal{T}^{\mathrm{ori}} = ((l_{1x}, l_{1y}), \cdots, (l_{vx}, l_{vy}))$ with a length of v by using a starting point S, a destination D, downloaded real-time traffic, and a navigation algorithm (e.g., Dijkstra).

4.5 Secure Trajectory Uploading

For the original trajectory, the driver id_i^d encrypts $\mathcal{T}^{\text{ori}} = ((l_{1x}, l_{1y}), \cdots, (l_{ux}, l_{uy}))$ by invoking $\Gamma.\mathsf{HGen}(pk, \cdot)$ and computes a hash value of $h^{\text{ori}} = \mathcal{H}_1(\mathcal{E}\mathcal{T}^{\text{ori}})$:

$$\mathcal{E}\mathcal{T}^{\text{ori}} = \{(El_{ix}, El_{iy})\}, i \in [u]$$

$El_{ix} = v \cdot pk + (l_{ix} + e_0, e_1) \bmod q_L$, $El_{iy} = v \cdot pk + (l_{iy} + e_0, e_1) \bmod q_L$.

Then, both rider id_i^d and driver id_i^r sign h^{ori} by using $\Delta.\mathsf{ASign}(ask_i^r, h^{\text{ori}}, \mathcal{S})$ and $\Delta.\mathsf{ASign}(ask_i^d, h^{\text{ori}}, \mathcal{S})$.

$$r_i^r \xleftarrow{\$} \mathbb{Z}_q, \; R_i^r = g^{r_i^r}, r_i^d \xleftarrow{\$} \mathbb{Z}_q, \; R_i^d = g^{r_i^d}, \; R = R_i^r \cdot R_i^d, c \leftarrow \mathcal{H}_2(pk, R, h^{\text{ori}}) \in \mathbb{Z}_q,$$
$$z_i^r \leftarrow r_i^r + ask_i^r \cdot c \in \mathbb{Z}_q, \; z_i^d \leftarrow r_i^d + ask_i^d \cdot c \in \mathbb{Z}_q, \; ss_i^r = (R_i^r, z_i^r), ss_i^d = (R_i^d, z_i^d),$$

where $\mathcal{S} \subseteq [n]$ is a set of size 2 of participating signers.

Next, id_i^r signs h^{ori} with $\Lambda.\mathsf{GSign}(gpk, gsk_i^r, h^{\text{ori}})$ to get a group signature σ_i^r:

$$\alpha \xleftarrow{\$} \mathbb{Z}_p^*, \; pid_{i1}^r = g^\alpha, pid_{i2} = cpk^\alpha h^y, \; r_1, r_2, r_3 \xleftarrow{\$} \mathbb{Z}_p^*,$$
$$A' = A^{r_1}, \; \hat{A} = A'^x(g_1 h_1^y h_2^s)^{r_1}, \; d = (g_1 h_1^y h_2^s)^{r_1} h_2^{-r_2}, \; r_3 = r_1^{-1}, \; s' = s - r_2 r_3,$$
$$\tau \leftarrow \mathsf{SPK}\{(x, y, r_2, r_3, s', \alpha) : \; nym_1 = g^\alpha \land nym_2 = cpk^\alpha h^y$$
$$\land \hat{A}/d = A'^{-x} h_2^{r_2} \land g_1 h_1^y = d^{r_3} h_2^{-s'}\}(h^{\text{ori}}),$$
$$gs_i^r \leftarrow (A', \hat{A}, d, \pi), \; pid_i^r = (pid_{i1}^r, pid_{i2}^r), \; \sigma_i^r = (gs_i^r, pid_i^r).$$

Afterward, id_i^d encrypts $\mathcal{S}$ with $\Sigma.\mathsf{Enc}$ and pk^{tee} to get $\mathcal{E}\mathcal{S}$. Finally, id_i^d uploads $(\mathcal{E}\mathcal{T}^{\text{ori}}, h^{\text{ori}}, \mathcal{E}\mathcal{S}, ss_i^r, ss_i^d, \sigma_i^r)$ to the RSU. We note that for the trajectories excluding original ones and deviated ones, the rider does not generate σ_i^r.

4.6 Secure Deviation

When the rider in transit decides to deviate from $\mathcal{T}^{\text{ori}}$, id_i^r and id_i^d proceeds the same as they did in Route Planning except with a current location CL. Besides, the new route is considered as an original route from CL to D, and the uploading process refers to 4.5 where the deviation message is $(\mathcal{E}\mathcal{T}^{\text{dev}}, h^{\text{dev}}, \mathcal{E}\mathcal{S}, ss_i^r, ss_i^d, \sigma_i^r)$.

4.7 Secure Trajectory Monitoring

4.7.1 Without Permitted Trajectory Deviation

(1) **For an original message** $(\mathcal{E}\mathcal{T}^{\text{ori}}, h^{\text{ori}}, \mathcal{E}\mathcal{S}, ss_i^r, ss_i^d, \sigma_i^r)$, the RSU invokes its local TEE to combine the two signature shares with $\mathsf{ACombine}(apk, h^{\text{ori}}, \mathcal{S}, \{ss_j\}_{S_j \in \mathcal{S}})$: $\mathcal{S} = \Sigma.\mathsf{Dec}(sk^{\text{tee}}, \mathcal{E}\mathcal{S})$, $z = z_i^r + z_i^d \in \mathbb{Z}_q$, $R = R_i^r R_i^d$, $cs = (R, z, \mathcal{S})$. Then, the TEE verifies cs with $\Delta.\mathsf{AVerify}(apk, h^{\text{ori}}, cs)$ by computing

$pk_{\mathcal{S}} = \prod_{id_i^d \in \mathcal{S}} apk_i^d$ and checking check $g^z \stackrel{?}{=} pk_{\mathcal{S}} \cdot R$. If the ATS signature is valid, the TEE will notify the RSU, which verifies σ_i^r by checking whether $A' \neq 1_{\mathbb{G}_1}$ and $e(A', ipk) = e(\hat{A}, g_2)$. If the group signature is valid, the RSU uploads an original transaction $\text{Tx}^{\text{ori}} = (\text{"Original"}, h^{\text{ori}}, \sigma_i^r)$ with a signature.

(2) **For a normal message** $(\mathcal{ET}^{\text{tra}}, h^{\text{tra}}, \mathcal{ES}, ss_i^r, ss_i^d, h^{\text{ori}})$, the RSU first retrieves the original route $\mathcal{ET}^{\text{ori}}$ by searching the local storage or retrieving from other RSUs with h^{ori}. Say the RSU now has $\mathcal{ET}^{\text{tra}} = \{(El_{ix}, El_{iy})\}, i \in [v]$ and $\mathcal{ET}^{\text{ori}} = \{(El_{ix}, El_{iy})\}, i \in [u]$. The RSU now checks trajectory similarity.

1. Set LCSS = 0, $\mathcal{ET}^{\text{left}} = \mathcal{ET}^{\text{tra}}_{(1,v)}$, $\mathcal{ET}^{\text{Right}} = \mathcal{ET}^{\text{ori}}_{(1,u)}$, and a threshold ϵ.
2. If $\mathcal{ET}^{\text{left}} = \emptyset$ or $\mathcal{ET}^{\text{right}} = \emptyset$, $\text{LCSS}(\mathcal{ET}^{\text{left}}, \mathcal{ET}^{\text{right}}) = 0$.
3. $Edist = (El_{1x}^{\text{left}} \ominus El_{1x}^{\text{right}})^2 \oplus (El_{1y}^{\text{left}} \ominus El_{1y}^{\text{right}})^2$, send $dist$ to TEE.
   3.1 TEE returns $dist = \text{HDec}(sk, dist)$.
   3.2 If $dist < \epsilon^2$, LCSS = $1 + \text{LCSS}(ET_{(1,v-1)}, ET_{(1,u-1)})$.
4. Else, LCSS = $\max\{\text{LCSS}(ET^{\text{left}}_{1(1,v-1)}, ET^{\text{right}}), \text{LCSS}(ET^{\text{left}}, ET^{\text{right}}_{(1,u-1)})\}$.
5. Sim = LCSS/v.

4.7.2 With Permitted Trajectory Deviation

For a deviation message $(\mathcal{ET}^{\text{dev}}, h^{\text{dev}}, \mathcal{ES}, ss_i^r, ss_i^d, \sigma_i^r)$, the recipient RSU and TEE verify the ATS and GS as in 4.7.1, then verify the new route as follows.

1. TEE retrieves the original pid^{ori} from $\mathcal{CB}$.
2. TEE blinds pid^{ori} and pid_i^r as follows.
   2.1 $\alpha, \beta, \gamma \leftarrow_\$ \mathbb{Z}_p^*$, $\alpha', \beta', \gamma' \leftarrow_\$ \mathbb{Z}_p^*$, $Npid_1^{\text{ori}} = pid_1^{\text{ori}} g^\beta$, $Npid_2^{\text{ori}} = g^\alpha$, $Npid_3^{\text{ori}} = pid_2^{\text{ori}} cpk^\beta bpk^\alpha$, $t_1^{\text{ori}} = g^\gamma$, $t_2^{\text{ori}} = bpk^\gamma h^{\text{ori}}$.
   2.4 $Npid^{\text{ori}} = (Npid_1^{\text{ori}}, Npid_2^{\text{ori}}, Npid_3^{\text{ori}})$, $t^{\text{ori}} = (t_1^{\text{ori}}, t_2^{\text{ori}})$.
   2.5 compute $Npid^{\text{dev}}$ and t^{dev} similarly to 2.1–2.4.
3. TEE converts $(Npid^{\text{ori}}, t^{\text{ori}})$ and $(Npid^{\text{dev}}, t^{\text{dev}})$ as follows.
   3.1 $r \leftarrow_\$ \mathbb{Z}_p^*$, $Npid_1^{\text{ori}'} = (Npid_2^{\text{ori}})^r$, $Npid_2^{\text{ori}'} = (Npid_3^{\text{ori}} (Npid_1^{\text{ori}})^{-csk})^r$.
   3.2 $r_1, r_2 \leftarrow_\$ \mathbb{Z}_p^*$, $Npid_1^{\text{ori}''} = Npid_1^{\text{ori}'} g^{r_1}$, $Npid_2^{\text{ori}''} = Npid_2^{\text{ori}'} bpk^{r_1}$.
   3.3 $t_1^{\text{ori}'} = t_1^{\text{ori}} g^{r_2}$, $t_2^{\text{ori}'} = t_2^{\text{ori}} bpk^{r_2}$.
   3.4 $Npid^{\text{ori}'} = (Npid_1^{\text{ori}''}, Npid_2^{\text{ori}''})$, $t^{\text{ori}'} = (t_1^{\text{ori}'}, t_2^{\text{ori}'})$.
   3.5 compute $(Npid^{\text{dev}''}, t^{\text{dev}'})$ similarly to 3.1–3.4.
   3.6 randomly permute $(Npid^{\text{ori}''}, t^{\text{ori}'})$ and $(Npid^{\text{dev}''}, t^{\text{dev}'})$.
4. RSU unblinds and compares as follows.
   4.1 $\overline{Npid^{\text{ori}}} = Npid_2^{\text{ori}''} (Npid_1^{\text{ori}''})^{-bsk}$, $\overline{t^{\text{ori}}} = t_2^{\text{ori}'} (t_1^{\text{ori}'})^{-bsk}$.
   4.2 $\overline{Npid^{\text{dev}}} = Npid_2^{\text{dev}''} (Npid_1^{\text{dev}''})^{-bsk}$, $\overline{t^{\text{dev}}} = t_2^{\text{dev}'} (t_1^{\text{dev}'})^{-bsk}$.
   4.3 check $\overline{Npid^{\text{ori}}} \stackrel{?}{=} \overline{Npid^{\text{dev}}}$ and $\overline{t^{\text{ori}}} \stackrel{?}{=} \overline{t^{\text{dev}}}$.

If the two verification are valid, the RSU acknowledges the new route and uploads a deviation transaction $\text{Tx}^{\text{dev}} = (\text{"Deviation"}, h^{\text{dev}}, \sigma_i^r)$ with a signature.

5 Security Analysis

We first define security experiments corresponding to the three objectives, namely trajectory authenticity, location privacy, and deviation authenticity.

The first one is experiment of trajectory authenticity (Fig. 3). The adversary attempts to forge a valid ATS and a valid group signature. In step 4, the gsk is not shared with the adversary $\mathcal{A}_1$ because it does not contribute to the attack.

---

1. $(n, \mathcal{S}) \xleftarrow{\$} \mathcal{A}_0(1^\lambda)$ where $\mathcal{S} \subseteq [n]$ and $|\mathcal{S}| = 2$ $\quad\quad\quad\quad\quad\quad\quad\quad\quad$ **Exp**^{unf}
2. $(sk, pk, evk, n, t, \{ask_i\}_{i=1}^n, apk, (ipk, isk), (cpk, csk), (bpk, bsk), pp) \leftarrow \mathsf{Setup}(1^\lambda)$
3. $(\{ask_i^d, gsk_i^r\}_{i=1}^{n_1}, \{ask_i\}_{i=1}^{n_2}, \{sk_i^{\text{tee}}, pk_i^{\text{tee}}\}_{i=1}^{n_3}) \leftarrow \mathsf{Register}(\{id_i^r\}, \{id_i^d\}, \{id_i^u\})$
4. $(\mathcal{T}', \gamma', \sigma') \leftarrow \mathcal{A}_1^{\mathcal{O}_0(\cdots), \mathcal{O}_1(\cdots)}(pk, n, t, ipk, cpk, bpk, pp, ask_i^r, ask_i^d)$

$\mathcal{O}_0(\mathcal{S}_j, \mathcal{T}_j)$ returns signature shares $\{\Delta.\mathsf{ASign}(ask_j, \mathcal{T}_j, \mathcal{S}_i)\}_{id_j \in \mathcal{S}_i}$ by invoking the $\mathsf{UploadOri}(\mathcal{T}_i, pk, ask_i^r, ask_i^d, gpk, gsk_i^r, pk_i^{\text{tee}})$, $\mathcal{S}_i \subseteq [n]$ and $|\mathcal{S}_i| = 2$

$\mathcal{O}_1(\mathcal{S}_j, \mathcal{T}_j)$ returns a group signature $\Lambda.\mathsf{GSign}(gpk, gsk_j, \mathcal{T}_j)$ by invoking the $\mathsf{UploadOri}(\mathcal{T}_i, pk, ask_i^r, ask_i^d, gpk, gsk_i^r, pk_i^{\text{tee}})$, $id_j \in \mathcal{S}_j$, $\mathcal{S}_i \subseteq [n]$ and $|\mathcal{S}_i| = 2$

Winning condition:

Let $(\mathcal{S}_1, \mathcal{T}_1), (\mathcal{S}_2, \mathcal{T}_2), \cdots$ be $\mathcal{A}_1$'s queries to $\mathcal{O}_0$, let $\mathcal{S}' \leftarrow \cup \mathcal{S}_i$, union over queries to $\mathcal{O}_0(\mathcal{S}_i, \mathcal{T})$, if no $\mathcal{O}_0$ on $\mathcal{T}'$, set $\mathcal{S}'$ as $\emptyset$, let $\mathcal{S}_t \leftarrow \Delta.\mathsf{ATrace}(pk, \mathcal{T}', \gamma')$

Output 1 if $\Delta.\mathsf{AVerify}(pk, \mathcal{T}', \sigma') = 1$ and either $\mathcal{S}_t \not\subseteq \mathcal{S} \cup \mathcal{S}'$ or if $\mathcal{S}_t = \bot$ and $\Lambda.\mathsf{GVerify}(gpk, \mathcal{T}', \sigma') = 1$

Fig. 3. Experiment of Trajectory Authenticity.

---

1. $b \xleftarrow{\$} \{0, 1\}$ $\quad\quad\quad\quad\quad\quad\quad\quad\quad\quad\quad\quad\quad\quad\quad\quad\quad\quad\quad\quad\quad\quad\quad\quad\quad\quad$ **Exp**^{pPub}
2. $((\mathcal{T}_0, \mathcal{S}_0), (\mathcal{T}_1, \mathcal{S}_1)) \leftarrow_\$ \mathcal{A}_2(1^\lambda)$
3. $(sk, pk, evk, n, t, \{ask_i\}_{i=1}^n, apk, (ipk, isk), (cpk, csk), (bpk, bsk), pp) \leftarrow \mathsf{Setup}(1^\lambda)$
4. $(\{ask_i^d, gsk_i^r\}_{i=1}^{n_1}, \{ask_i\}_{i=1}^{n_2}, \{sk_i^{\text{tee}}, pk_i^{\text{tee}}\}_{i=1}^{n_3}) \leftarrow \mathsf{Register}(\{id_i^r\}, \{id_i^d\}, \{id_i^u\})$
5. $b' \leftarrow \mathcal{A}_3^{\mathcal{O}_2(\cdots), \mathcal{O}_3(\cdots), \mathcal{O}_4(\cdots)}(pk, n, t, ipk, cpk, bpk, pp)$
6. Output $(b' = b)$

$(\mathcal{ET}_b, \mathcal{ES}_b) \leftarrow \mathcal{O}_2(\mathcal{T}_b)$ by invoking $\mathsf{UploadOri}(\mathcal{T}_b, pk, ask_i^r, ask_i^d, gpk, gsk_i^r, pk_i^{\text{tee}})$
$(\mathcal{ET}_b, \mathcal{ES}_b) \leftarrow \mathcal{O}_3(\mathcal{T}_b)$ by invoking $\mathsf{UploadTra}(\mathcal{T}_b, pk, ask_i^r, ask_i^d, gpk, gsk_i^r, pk_i^{\text{tee}})$
$(\mathcal{ET}_b, \mathcal{ES}_b) \leftarrow \mathcal{O}_4(\mathcal{T}_b)$ by invoking $\mathsf{Deviate}(\mathcal{T}_b, pk, ask_i^r, ask_i^d, pk_i^{\text{tee}})$

Fig. 4. Experiment of Privacy against the Public.

The second one and the third one are experiment of privacy against the public (Fig. 4) and experiment of privacy against the RSU (Fig. 5), respectively. The attack agenda is to decide which of the two chosen trajectories is encrypted given a challenge ciphertext. The difference between Exp 2 and Exp 3 is that the adversary in Fig. 5 can access the driver's secret key.

The fourth one is the experiment of deviation authenticity (Fig. 6), where the adversary has to forge two valid signatures and make the forged group signature linkable to the previous group signature generated by the honest rider.

---

1. $b \xleftarrow{\$} \{0,1\}$ $\quad$ **Exp^{pRSU}**
2. $((\mathcal{T}_0, \mathcal{S}_0), (\mathcal{T}_1, \mathcal{S}_1)) \leftarrow_\$ \mathcal{A}_4$ // $\mathcal{A}_2$ generates two trajectories
3. $(sk, pk, evk, n, t, \{ask_i\}_{i=1}^n, apk, (ipk, isk), (cpk, csk), (bpk, bsk), pp) \leftarrow \mathsf{Setup}(1^\lambda)$
4. $(\{ask_i^d, gsk_i^r\}_{i=1}^{n_1}, \{ask_i\}_{i=1}^{n_2}, \{sk_i^{\text{tee}}, pk_i^{\text{tee}}\}_{i=1}^{n_3}) \leftarrow \mathsf{Register}(\{id_i^r\}, \{id_i^d\}, \{id_i^u\})$
5. $b' \leftarrow \mathcal{A}_5^{\mathcal{O}_5(\cdots), \mathcal{O}_6(\cdots), \mathcal{O}_7(\cdots)}(pk, n, t, ipk, cpk, bpk, pp, ask_i^r, ask_i^d)$
6. Output $(b' = b)$

$(\mathcal{ET}_b, \mathcal{ES}_b) \leftarrow \mathcal{O}_5(\mathcal{T}_b)$ by invoking $\mathsf{UploadOri}(\mathcal{T}_b, pk, ask_i^r, ask_i^d, gpk, gsk_i^r, pk_i^{\text{tee}})$
$(\mathcal{ET}_b, \mathcal{ES}_b) \leftarrow \mathcal{O}_6(\mathcal{T}_b)$ by invoking $\mathsf{UploadTra}(\mathcal{T}_b, pk, ask_i^r, ask_i^d, gpk, gsk_i^r, pk_i^{\text{tee}})$
$(\mathcal{ET}_b, \mathcal{ES}_b) \leftarrow \mathcal{O}_7(\mathcal{T}_b)$ by invoking $\mathsf{Deviate}(\mathcal{T}_b, pk, ask_i^r, ask_i^d, pk_i^{\text{tee}})$

Fig. 5. Experiment of Privacy against the RSU.

---

1. $(n, \mathcal{S}, \mathcal{T}^{\text{ori}}) \xleftarrow{\$} \mathcal{A}_6(1^\lambda)$ where $\mathcal{S} \subseteq [n]$ and $|\mathcal{S}| = 2$ $\quad$ **Exp^{dev}**
2. $(sk, pk, evk, n, t, \{ask_i\}_{i=1}^n, apk, (ipk, isk), (cpk, csk), (bpk, bsk), pp) \leftarrow \mathsf{Setup}(1^\lambda)$
3. $(\{ask_i^d, gsk_i^r\}_{i=1}^{n_1}, \{ask_i\}_{i=1}^{n_2}, \{sk_i^{\text{tee}}, pk_i^{\text{tee}}\}_{i=1}^{n_3}) \leftarrow \mathsf{Register}(\{id_i^r\}, \{id_i^d\}, \{id_i^u\})$
4. $(\mathcal{T}', \gamma', \sigma') \leftarrow \mathcal{A}_7^{\mathcal{O}_8(\cdots), \mathcal{O}_9(\cdots)}(pk, n, t, ipk, cpk, bpk, pp, ask_i^d)$

$\mathcal{O}_8(\mathcal{S}_j, \mathcal{T}_j)$ returns signature shares $\{\Delta.\mathsf{ASign}(ask_j, \mathcal{T}_j, \mathcal{S}_i)\}_{id_j \in \mathcal{S}_i}$ by invoking the $\mathsf{UploadDev}(\mathcal{T}_j, pk, ask_i^r, ask_i^d, gpk, gsk_i^r, pk_i^{\text{tee}})$, $\mathcal{S}_i \subseteq [n]$ and $|\mathcal{S}_i| = 2$
$\mathcal{O}_9(\mathcal{S}_j, \mathcal{T}_j)$ returns a group signature $\Lambda.\mathsf{GSign}(gpk, gsk_j, \mathcal{T}_j)$ by invoking the $\mathsf{UploadDev}(\mathcal{T}_j, pk, ask_i^r, ask_i^d, gpk, gsk_i^r, pk_i^{\text{tee}})$, $id_j \in \mathcal{S}_j$, $\mathcal{S}_i \subseteq [n]$ and $|\mathcal{S}_i| = 2$

Winning condition:
$\quad$ Let $(\mathcal{S}_1, \mathcal{T}_1), (\mathcal{S}_2, \mathcal{T}_2), \cdots$ be $\mathcal{A}_1$'s queries to $\mathcal{O}_0$, let $\mathcal{S}' \leftarrow \cup \mathcal{S}_i$, union over queries to $\mathcal{O}_0(\mathcal{S}_i, \mathcal{T}')$, if no $\mathcal{O}_0$ on $\mathcal{T}'$, set $\mathcal{S}'$ as $\emptyset$, let $\mathcal{S}_t \leftarrow \Delta.\mathsf{ATrace}(pk, \mathcal{T}', \gamma')$

$\quad$ Output 1 if
$\quad\quad \Delta.\mathsf{AVerify}(pk, \mathcal{T}', \sigma') = 1$ and either $\mathcal{S}_t \not\subseteq \mathcal{S} \cup \mathcal{S}'$ or if $\mathcal{S}_t = \bot$ and
$\quad\quad \Lambda.\mathsf{GVerify}(gpk, \mathcal{T}', \sigma') = 1$ and
$\quad\quad \Lambda.\mathsf{Unblind}(bsk, (Npid_1, t_1)) = \Lambda.\mathsf{Unblind}(bsk, (Npid_2, t_2))$ where
$((Npid_1, t_1), (Npid_2, t_2)) = \Lambda.\mathsf{Convert}(gpk, csk, bpk, (Npid^{\text{ori}}, t^{\text{ori}}), (Npid^{\text{dev}}, t^{\text{dev}}))$,
$(Npid^{\text{ori}}, t^{\text{ori}}) = \Lambda.\mathsf{Blind}(gpk, bpk, pid^{\text{ori}}, \mathcal{T}^{\text{ori}})$
$(Npid^{\text{dev}}, t^{\text{dev}}) = \Lambda.\mathsf{Blind}(gpk, bpk, pid^{\text{dev}}, \mathcal{T}^{\text{dev}})$.

Fig. 6. Experiment of Deviation Authenticity.

Theorem 1. *The Artemis scheme is trajectory authentic, location privacy-preserving, and deviation authentic, assuming that the ATS scheme Δ is unforgeable, the GS scheme Λ is unforgeable, the CKKS scheme Γ is secure, the TEE is confidentiality-preserving.*

We now prove Theorem 1. The proof is captured in three lemmas.

Lemma 1. *The Artemis scheme is trajectory authentic, assuming that the ATS scheme Δ is unforgeable and the GS scheme Λ is unforgeable.*

Proof. Let Ω be the Artemis scheme, and let $\mathcal{A} = (\mathcal{A}_0, \mathcal{A}_1)$ be the Probabilistic Polynomial-Time (PPT) adversary attacking the scheme. We construct the following PPT adversary $\mathcal{A}'$ breaking the underlying ATS scheme Δ.

Adversary $\mathcal{A}'$:
$\mathcal{A}'$ is given (pp^{ATS}, apk) as inputs.
  1. Run $\mathcal{A}$ as a subroutine and answer as follows
  1.1 When $\mathcal{A}$ request signature shares on $(\mathcal{S}_j, \mathcal{T}_j)$:

- Query $\mathcal{O}_0(\cdots)$ to obtain $(\mathcal{ET}^{\text{ori}}, h^{\text{ori}}, \mathcal{ES}, ss_i^r, ss_i^d, \mathcal{ES}, \sigma_i^r)$ of an honest execution of the UploadOri$(\cdots)$, return the signature shares (ss_j^r, ss_j^d).

  1.2 When $\mathcal{A}$ requests a group signature on $(\mathcal{S}_j, \mathcal{T}_j)$:

- Query $\mathcal{O}_1(\cdots)$ to obtain $(\mathcal{ET}^{\text{ori}}, h^{\text{ori}}, \mathcal{ES}, ss_i^r, ss_i^d, \mathcal{ES}, \sigma_i^r)$ of an honest execution of the UploadOri$(\cdots)$, return the group signature σ_i^r.

  2. Outputs whatever $\mathcal{A}$ forges.

The view of $\mathcal{A}$ when run as a subroutine by $\mathcal{A}'$ is identical to the view of $\mathcal{A}$ in experiment $\mathbf{Exp}^{\text{unf}}$. Therefore, when $\mathcal{A}$ outputs both two valid signature shares (ss_j^r, ss_j^d) on $(\mathcal{S}_j, \mathcal{T}_j)$ and a group signature σ_j on $(\mathcal{S}_j, \mathcal{T}_j)$, $\mathcal{A}$ successfully breaks the underlying ATS scheme and the GS scheme. Since both $\Pr[\text{forge}_{\mathcal{A}',\Delta}(\lambda) = 1]$ and $\Pr[\text{forge}_{\mathcal{A}',\Lambda}(\lambda) = 1]$ are negligible, we have that breaking the trajectory authenticity of Artemis is negligible, i.e., $\text{Adv}^{\text{unf}}_{\mathcal{A},\Omega} \leqslant \text{negl}(\lambda)$. □

Lemma 2. *The Artemis scheme is location privacy-preserving, assuming that the CKKS scheme Γ is secure and the TEE is confidentiality-preserving.*

Proof. The proof of location privacy preservation is two-folds. The first one relies on the security of the CKKS scheme that encrypts the locations in a trajectory. The second one depends on the TEE assisting in the trajectory similarity checking. Due to the page limit, we only give a proof sketch here.

Let $\mathcal{A} = (\mathcal{A}_2, \mathcal{A}_3, \mathcal{A}_4)$ be the PPT adversary attacking the scheme. We can construct a PPT adversary $\mathcal{A}'$ breaking the underlying CKKS scheme Γ. $\mathcal{A}'$ will invoke $\mathcal{A}$ as a subroutine and output what $\mathcal{A}$ outputs. If $\mathcal{A}$ can tell whether $\mathcal{T}_0$ or $\mathcal{T}_1$ is encrypted with an advantage that is not negligible, then $\mathcal{A}'$ has a non-negligible advantage in $\mathbf{Exp}^{\text{adv}}_{\mathcal{A}',\Gamma}(\lambda)$, which contradicts the claimed security of CKKS, i.e., $\text{Adv}^{\text{pPub}}_{\mathcal{A},\Omega} \leqslant \text{negl}(\lambda)$. The proof also applies to $\text{Adv}^{\text{pRSU}}_{\mathcal{A},\Omega} \leqslant \text{negl}(\lambda)$. In the trajectory similarity checking, the TEE only decrypt the distance between two encrypted trajectories, i.e., $dist = \text{HDec}(sk, dist)$. Such a value will not help violate the location privacy since the adversary only holds a distance and there are infinite combinations of two trajectories sharing the same distance. □

Lemma 3. *The Artemis scheme is deviation authentic, assuming that the GS scheme Λ is unforgeable.*

Proof. The proof is similar to the one in Lemma 1 with one difference being relying on the security of the GS scheme. Due to the page limit, we omit the detailed proof here. □

6 Performance Analysis

6.1 Experimental Settings

Dataset. We employ a real-world ride-hailing dataset made publicly available by Didi Chuxing. Specifically, we utilize the trajectory and trip records collected in Chengdu's Second Ring Road area over the course of November 2016. The dataset includes full-sample GPS trajectories and anonymized trip-level information, where each trajectory is sampled at 2–4 second intervals. Each data point contains a pseudonymized driver ID, an anonymized order ID, a Unix timestamp, and corresponding GPS coordinates in the GCJ-02 system. All trip data has been pre-processed to ensure privacy, and identifiers have been irreversibly desensitized.

Key Parameters. For trajectory data encryption, Artemis adopts the CKKS approximate homomorphic encryption scheme, implemented via the HEAAN library. In our experiments, we set the polynomial modulus degree to 8192 and select a scaling factor of 240, following the recommended security guidelines of HEAAN and the CKKS design. The total coefficient modulus is chosen to support adequate computational depth for encrypted similarity computations while ensuring a 128-bit security level. For trajectory authenticity and non-repudiation, Artemis adopts a 2-out-of-n Accountable Threshold Signature (ATS) protocol implemented over elliptic curve cryptography. The protocol is instantiated with a 128-bit prime order group. The group signature setup is instantiated over a pairing-friendly elliptic curve (Type-3 pairing, 128-bit prime order). The similarity threshold for trajectory comparison, denoted as ϵ, is empirically set to 20 m, which allows for limited natural deviations in route while reliably detecting unauthorized detours or malicious behavior.

Metrics. We evaluate the computational cost and communication overhead of all entities in all phases.

Setup. Our experiments were conducted in a simulated environment equipped with Intel SGX support. Specifically, we used a workstation running Ubuntu 20.04.6 LTS with an Intel Core^{TM} i7-9750H CPU 2.60 GHz and 8 GB of RAM. The trusted execution environment was provided by Intel SGX, supporting secure enclave execution for all critical Artemis components, including trajectory encryption, LCSS-based similarity computation, and signature validation. Homomorphic encryption operations adopted the CKKS scheme, implemented via the HEAAN library [29]. Threshold signature protocols were instantiated using elliptic curve cryptographic primitives. All modules, such as secure trajectory generation, 2-out-of-n threshold signature aggregation, and encrypted similarity computation within the TEE, were deployed and benchmarked in this controlled setting to ensure replicability and comprehensive performance profiling.

6.2 Computational Costs

To evaluate the computational efficiency of the Artemis framework, we conducted extensive simulations within an Intel SGX-based trusted execution environment.

The performance metrics reflect average runtimes under a secure and realistic operating environment as shown in Table 2. Encryption costs on average 83.3 ms, while decryption within SGX costs 54.5 ms. This is acceptable due to the periodic upload model in RHSs. Secure trajectory comparison using LCSS logic within the SGX enclave takes 283 ms per 100-point trace. Generating a partial signature by the rider or driver takes 0.5 ms, signature aggregation takes 0.1 ms, and verification within SGX takes 0.24 ms, totaling under 1 ms-making this component lightweight and efficient. Generating and verifying anonymous group signatures used for detour approval both take under 0.5 ms, which is negligible in the total processing pipeline. Zero-knowledge proof generation and verification each require less than 0.25 ms. Overall, the most computationally demanding task is the LCSS comparison for encrypted trajectories. Importantly, Artemis leverages secure trajectory similarity computation using LCSS in a TEE and avoids multi-round communication to ensure efficient and near real-time performance.

Table 2. Average Computation Time of Main Operations in Artemis

Algorithm	Function	Meaning	Average Time (ms)
CKKS	$\Gamma.\text{HEnc}$	Encryption	83.3
CKKS	$\Gamma.\text{HDec}$	Decryption	54.5
ATS	$\Delta.\text{ASign}$	Signature generation	0.5
ATS	$\Delta.\text{ACombine}$	Signature aggregation	0.1
ATS	$\Delta.\text{AVerify}$	Signature verification	0.24
Group Signature	$\Lambda.\text{GSign}$	Signing	0.4
Group Signature	$\Lambda.\text{GVerify}$	Verification	0.5
NIZK	$\Pi.\text{P}$	Proof generation	0.12
NIZK	$\Pi.\text{V}$	Proof verification	0.21
LCSS	LCSS	Similarity computation	283

6.3 Communication Overhead

Communication costs in Artemis arise from encrypted data uploads, signature transmissions, and encrypted result deliveries as shown in Table 3. A trajectory comprising 100 GPS points encrypted under CKKS, results in a ciphertext of approximately 21.8 KB. Threshold signatures and group signatures are compact, typically resulting in a combined payload of 0.86 KB per session. This is negligible relative to trajectory data. The encrypted comparison result are returned to the client. This includes a similarity score, occupying merely 32 bytes. The communication between the RSU and SGX enclaves involves ciphertext-based distance calculations for 100 GPS points, with a data size of 2.7 KB, and remains confined to the local RSU environment. In sum, the dominant communication cost lies in trajectory ciphertexts. However, drivers upload encrypted trajectories periodically, and such data can be efficiently handled in batches, the overall overhead is trivial over current 4G/5G vehicular networks.

Table 3. Storage Size of Different Components per 100 Locations

Component	Size per 100 locations
Encrypted GPS Trajectory	21.8 KB
Threshold/Group Signature	0.86 KB
Encrypted Similarity Score	32 bytes
Encrypted Distances	2.7 KB

7 Conclusions

In this paper, we propose a decentralized, secure, and efficient safety monitoring scheme Artemis. We design a 2-out-of-n private threshold signature protocol to achieve trajectory authenticity We design a decentralized RHS platform and a secure trajectory similarity computation protocol to guarantee both efficiency and location privacy. We design a secure three-party computation protocol to ensure deviation authenticity. Both theoretical analysis and extensive experiments confirm Artemis' security and efficiency. We believe there are more interesting research directions in the area of secure safety monitoring.

Acknowledgments. It is supported by the National Natural Science Foundation of China (NSFC) under the grant No. 62372149, No. U23A20303, and No. 62572168. It is supported by the Anhui Provincial Natural Science Foundation 2508085MF151. This research is supported by the Key Laboratory of Knowledge Engineering with Big Data (the Ministry of Education of China), under grant number No. BigKEOpen2025-04. It is partially supported by EU LOCARD Project under Grant H2020-SU-SEC-2018-832735.

References

1. Black & White: Uber covered up data breach, paid off hackers £75,000. https://blackandwhiteinsurance.co.uk/news-and-resources/uber-data-breach/
2. CIO Africa: Study reveals ride-hailing apps as the most data-hungry (2022). https://cioafrica.co/study-reveals-ride-hailing-apps-as-the-most-data-hungry/
3. POLITICO: Uber fined €290 million for sending drivers' data outside Europe (2024). https://www.politico.eu/article/uber-fined-e290-million-for-sending-drivers-data-outside-europe/

4. Pham, A., Dacosta, I., Endignoux, G., Troncoso-Pastoriza, J.R., Huguenin, K., Hubaux, J.-P.: ORide: a privacy-preserving yet accountable ride-hailing service. In: Proceedings of the 26th USENIX Security Symposium, pp. 1235–1252. Vancouver, Canada (2017)
5. Huang, J., Luo, Y., Fu, S., Xu, M., Hu, B.: pRide: privacy-preserving online ride hailing matching system with prediction. IEEE Trans. Veh. Technol. (TVT) **70**(8), 7413–7425 (2021)
6. Sun, J., Xu, G., Zhang, T., Alazab, M., Deng, R.H.: A practical fog-based privacy-preserving online car-hailing service system. IEEE Trans. Inf. Forensics Secur. (TIFS) **17**, 2862–2877 (2022)
7. Li, M., Chen, Y., Lal, C., Conti, M., Martinelli, F., Alazab, M.: Nereus: anonymous and secure ride-hailing service based on private smart contracts. IEEE Trans. Dependable Secure Comput. (TDSC) **20**(4), 2849–2866 (2023). https://doi.org/10.1109/TDSC.2022.3192367.
8. CNN Business: Uber releases safety data: 998 sexual assault incidents including 141 rape reports in 2020 (2022). https://www.cnn.com/2022/06/30/tech/uber-safety-report/index.html
9. Uber driver accused of kidnapping says rider made the whole thing up (2021). https://www.fox13news.com/news/uber-driver-accused-of-kidnapping-says-rider-made-the-whole-thing-up
10. United States Government Accountability Office: Ridesharing and taxi safety information on assaults against drivers and passengers (2024). https://www.gao.gov/assets/d24106742.pdf
11. Chaudhry, B., Yasar, A.-U.-H., El-Amine, S., Shakshuki, E.: Passenger safety in ride-sharing services. Procedia Comput. Sci. **130**, 1044–1050 (2018)
12. Ahmed, S.A., Dogra, D.P., Kar, S., Roy, P.P.: Trajectory-based surveillance analysis: a survey. IEEE Trans. Circ. Syst. Video Technol. (TCSVT) **29**(7), 1985–1997 (2019)
13. Yu, H., Zhang, H., Jia, X., Chen, X., Yu, X.: pSafety: privacy-preserving safety monitoring in online ride hailing services. IEEE Trans. Dependable Secure Comput. (TDSC) **20**(1), 209–224 (2023)
14. Nakamoto, S.: Bitcoin: a peer-to-peer electronic cash system (2009). https://bitcoin.org/bitcoin.pdf
15. Ali, M.S., Vecchio, M., Pincheira, M., Dolui, K., Antonelli, F., Rehmani, M.H.: Applications of blockchains in the internet of things: a comprehensive survey. IEEE Commun. Surv. Tutorials (COMST) **21**(2), 1676–1717 (2019)
16. Boneh, D., Komlo, C.: Threshold signatures with private accountability. In: Proceedings of the 42nd Annual International Cryptology Conference (CRYPTO), pp. 551–581. Santa Barbara, USA (2022). https://doi.org/10.1007/978-3-031-15985-5_19
17. McKeen, F., et al.: Intelc software guard extensions (IntelIntel® SGX) support for dynamic memory management inside an enclave". In: Proceedings 5th International Workshop on Hardware and Architectural Support for Security and Privacy (HASP), pp. 1–9, Seoul, South Korea (2016)
18. Intel: Which platforms support intel® software guard extensions (Intel® SGX) SGX2?. https://www.intel.com/content/www/us/en/support/articles/000058764/software/intel-security-products.html
19. Wood, G.: Ethereum: a secure decentralised generalised transaction ledger (2014). https://ethereum.github.io/yellowpaper/paper.pdf
20. Ethereum. https://ethereum.org

21. Su, H., Liu, S., Zheng, B., Zhou, X., Zheng, K.: A survey of trajectory distance measures and performance evaluation. VLDB J. **29**(1), 3–32 (2020)
22. Cheon, J.H., Kim, A., Kim, M., Song, Y.: Homomorphic encryption for arithmetic of approximate numbers. In: Proceedings 23rd International Conference on the Theory and Applications of Cryptology and Information Security, (ASIACRYPT), pp. 409–437, Hong Kong, China (2017)
23. Garms, L., Lehmann, A.: Group signatures with selective linkability. In: Proceedings of the 22nd IACR International Conference on Practice and Theory of Public-Key Cryptography (PKC), pp. 190–220, Beijing, China (2019)
24. Han, L., Luo, W., Lu, R., Zheng, Y., Yang, A., Lai, J.: Privacy-preserving travel recommendation based on stay points over outsourced spatio-temporal data. IEEE Trans. Intell. Transport. Syst. (TITS) **25**(10), 12999–13013 (2024)
25. Paillier, P.: Public-key cryptosystems based on composite degree residuosity classes. In: Proceedings of the 17th International Conference on the Theory and Application of Cryptographic Techniques (EUROCRYPT), pp. 223–238, Prague, Czech Republic (1999)
26. Wang, X., Miao, Y., Zhang, S., Li, Q., Zhang, J., Ding, K.: Efficient encrypted trajectory similarity query over mobile e-health cloud. IEEE Internet Things J. (IoTJ) **12**, 12748–12760 (2024)
27. Luo, Y., Jia, X., Fu, S., Xu, M.: pRide: privacy-preserving ride matching over road networks for online ride-hailing service. IEEE Trans. Inf. Forensics Secur. (TIFS) **14**(7), 1791–1802 (2019)
28. Yu, H., Jia, X., Zhang, H., Shu, J.: Efficient and privacy-preserving ride matching using exact road distance in online ride hailing services. IEEE Trans. Serv. Comput. **15**(4), 1841–1854 (2022)
29. HEAAN-Python. https://github.com/Huelse/HEAAN-Python?tab=readme-ov-file

Privacy-Preserving Framework for k-Modes Clustering Based on Personalized Local Differential Privacy

Yuling Luo[1,2], Zhangrui Wang[1,2], Xue Ouyang[1,2(✉)], Siyuan Zu[1,2], Qiang Fu[1,2], Sheng Qin[1,2], and Junxiu Liu[1,2]

[1] Guangxi Key Lab of Brain-inspired Computing and Intelligent Chips, School of Electronic and Information Engineering, Guangxi Normal University, Guilin 541004, China
{yuling0616,ouyangxue,qiangfu,qinsheng}@gxnu.edu.cn,
{wangzhangrui,zusiyuan}@stu.gxnu.edu.cn, j.liu@ieee.org
[2] Education Department of Guangxi Zhuang Autonomous Region, Key Laboratory of Nonlinear Circuits and Optical Communications (Guangxi Normal University), Guilin 541004, China

Abstract. As artificial intelligence advances, data clustering faces significant privacy challenges. Existing Differential Privacy (DP) k-modes methods typically rely on fully trusted data collectors, while Local Differential Privacy (LDP) underperforms compared to DP; neither approach adequately addresses personalized privacy requirements. This paper introduces a novel Personalized Local Differential Privacy k-modes (PLDP k-modes) algorithm. This method utilizes a PK-RR mechanism to perturb data locally at the user side and employs server-coordinated iterative centroid updates, thereby protecting users' real data and respecting personalized privacy needs. A key contribution includes the incorporation of a weight function into the perturbation algorithm to enhance utility, alongside a centroid perturbation method to counteract inference attacks. Notably, this marks the first application of PLDP in k-modes clustering. Theoretical analysis and experimental results confirm the algorithm's privacy guarantees and demonstrate a superior privacy-utility tradeoff, with its clustering utility surpassing that of state-of-the-art DP k-modes algorithms. The proposed method effectively addresses the privacy-utility challenge in k-modes clustering.

Keywords: Cluster · k-modes · Differential privacy · Personalized local differential privacy

1 Introduction

The extensive application of smart devices has significantly enhanced the data collection capability of distributed computing systems [1]. The development of cloud computing has laid the foundation for the large-scale analysis of data, and the information extracted from massive amounts of data possesses tremendous value [2]–[3]. As a classical unsupervised data analysis method, the k-means algorithm has become a commonly used approach for clustering numerical data due to its efficiency and simplicity [4]. In order to eliminate the limitation that the k-means algorithm is only applicable to numerical data, the k-modes algorithm has been proposed to cluster categorical data [5]. However, k-modes are often used in a non-private manner. Despite the great potential of clustering, there is a risk of user privacy leakage. Sensitive information such as financial, location, and medical information may lead to privacy leakage. Traditional anonymization methods cannot resist re-identification and background knowledge attacks, and attackers can associate and identify users' private information. Differential privacy (DP), on the other hand, can address such privacy issues.

Differential Privacy (DP) [6], as a reliable privacy-preserving algorithm, is widely applied in clustering analysis [7]. Among these, the DP k-means algorithm is extensively utilized [8] due to its efficiency and privacy-preserving capabilities. The privacy-aware data clustering (PADC) k-means algorithm, introduced in [9], enhances the precision of cluster partitioning by identifying outliers based on data density and assessing similarity using the weighted relative distance. Significant improvements in clustering utility are achieved by the novel Differential Privacy Kernel Clustering (DP-KCCM) algorithm, which employs cluster merging and adaptive noise mechanisms [10]. However, research on DP algorithms for k-modes remains limited. The recently proposed DP-Modes-Lloyd algorithm [11] achieves efficient clustering while preserving privacy by updating noisy modes through the computation of noisy counts. Although effective clustering can be achieved by existing DP-based approaches, the practical absence of a fully trusted server leads to significant challenges. Difficulties in user data collection and privacy breaches due to data leakage are among these issues, hindering the widespread application of these algorithms.

Local Differential Privacy (LDP) [12,13] can solve the problem of untrusted servers by moving the perturbation algorithm from the server side to the user side, making the user private data only owned by the user. LDP has also been applied to practical cases to create feasible solutions [14]–[16]. LDP mainly uses random responses [17] as a basic perturbation mechanism. However, researchers have not used it to solve privacy problems in k-modes due to the excessive noise added by LDP, facing the challenge of reducing the utility of clustering [18]. The influence of noise is further amplified in the clustering iterations. At the same time, the LDP cannot satisfy the personalized privacy requirements of different users and data in practical applications, leading to difficulties in collecting data and unnecessary utility loss caused by uniform privacy protection strength. Based on the above discussion, the main issue to address is how to achieve an excellent privacy-utility tradeoff in the k-modes model. Reference [19] introduced the concept of Personalized Local Differential Privacy (PLDP), which allows each user to independently set the privacy level of their data, thereby providing an effective solution to this problem. Building on the PLDP approach, we propose a clustering framework based on the K-Modes algorithm, ensuring data utility while protecting user privacy.

An improved perturbation mechanism PK-RR is proposed in this framework to perturb the user data at the local side and send it to the server for k-modes clustering. A good utility of clustering is obtained while satisfying PLDP. In addition, a centroid perturbation algorithm is proposed, which prevents privacy leakage caused by inference attacks by perturbing the centroids in the iterative process. The effect of the perturbation mechanism on the utility of clustering is reduced by a variant of the PK-RR mechanism in this algorithm. The main contributions of this paper are as follows.

- A clustering framework based on the PLDP k-modes algorithm is proposed. To the best of our knowledge, this paper is the first attempt at adopting PLDP in k-means clustering. This framework ensures that user privacy is not leaked while obtaining high-quality clustering through the PK-RR mechanism. Meanwhile, the user's personalized privacy requirements are also satisfied.
- Centroid perturbation algorithm is proposed to address the potential leakage of private information during iteration. It prevents inference attacks and further protects user privacy while reducing the impact of perturbations on the utility of clustering.
- Theoretical analysis demonstrates the privacy protection capability of the proposed mechanism, and experimental results on real datasets show that the algorithm proposed by this paper achieves a superior privacy-utility trade-off, and the utility of clustering outperforms the state-of-the-art DP k-modes algorithm.

The rest of this paper is organized as follows. The basic concepts required for this framework and the related technical foundation are introduced in 2. The proposed approach is presented in Sect. 3. The experimental results and analysis are illustrated in Sect. 4. Finally, the paper is concluded in Sect. 5.

2 Background

2.1 Personalized Local Differential Privacy

Definition 1 ((G_i, ε_u) - PLDP). Given a set of privacy requirements (G_i, ε_u) to one user $i \in [1, n]$, a randomized mechanism $F : D \to R$ satisfies (G_i, ε_u) – PLDP if and only if for any possible output result $t^* (t^* \subseteq R)$ on any two records t and $t'(t, t' \subseteq G_i)$ satisfying $\Pr[F(t) = t^*] \le e^{\varepsilon_u} \times \Pr\left[F\left(t'\right) = t^*\right]$, when G_i is set to the domain D and all users are unified ε, PLDP is equivalent to LDP.

Differential privacy has two combinatorial properties: sequential and parallel combinatorial properties. These properties will be applied in Sect. 3.3 for the privacy analysis of the proposed framework.

2.2 DP k-Modes

DP-Modes-Lloyd was proposed in [17] and is the first DP k-modes algorithm. DP-Modes-Lloyd satisfies differential privacy by adding noise to the clustering centroids through the Laplace mechanism. The precondition of this algorithm is that the third-party data collector is trusted. The main steps of DP-Modes-Lloyd are as follows.

1. Randomly select k points as the initial centroid in the data domain. ε is the privacy budget, and T is the number of iterations.
2. The Hamming distance $HamDis(X, C)$ between each data point and all initial centroids is calculated, and the data points are assigned to the class represented by the nearest clustering centroid.

Definition 2 (Hamming distance). *[21]* Suppose X is a data point and C is a cluster centroid, denoted $X = \{x_1, \ldots, x_d\}$ and $C = \{c_1, \ldots, c_d\}$, x_j and c_j are the j-th attribute of X and C, respectively. The Hamming distance in the clustering algorithm is defined as follows.

$$HamDis(X,C) = \sum_{j=1}^{d} x_j \oplus c_j.$$

1. In each cluster, calculating the count of all values taken for each attribute is $\mathbf{SUM} = \{sum_{vj}(1), \ldots, sum_{vj}(t)\}, 1 \leq v \leq k, 1 \leq j \leq d$, where k represents the number of clusters, d represents the number of attributes of the data, and t represents the v-th cluster of data with t different values in the j-th attribute. $sum_{vj}(t)$ is the count of the t-th value of the j-th dimensional attribute of the data in the v-th cluster.
2. The Laplace noise satisfying differential privacy is added to $\mathbf{SUM}$ to obtain $\mathbf{SUM}^{'} = \mathbf{SUM} + Laplace\left(e^{-\varepsilon/(d+T)}\right) = \{sum^{'}_{vj}(1), \ldots, sum^{'}_{vj}(t)\}$, and the j-th attribute of the the v-th cluster centroid is updated to the maximum count value $\max_{y \in \{1,\ldots,t\}} sum^{'}_{vj}(y)$ in $\mathbf{SUM}^{'}$.
3. Continue iterating unless a set threshold or a specified number of iterations is reached.

3 Proposed Approach

In this section, the PLDP k-modes clustering algorithm is proposed, and its privacy [22] is demonstrated. The privacy issues in clustering and the attack model are analyzed first. The overall flow of the proposed framework is then described, and the corresponding design of the perturbation mechanism based on PLDP is given. Finally, the privacy guarantee of the proposed overall framework is proved theoretically.

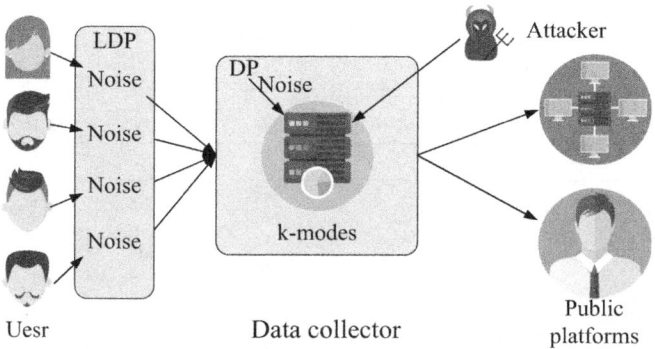

Fig. 1. DP and LDP k-modes clustering model.

3.1 Overview

The DP and LDP k-modes clustering model is shown in Fig. 1. The privacy issues faced by this model and the solutions are analyzed in this subsection. A third-party data collector collects sensitive data (e.g., location, income, cases, etc.) from many users, processes it using the k-modes algorithm, and shares or publishes the model results to partners or public platforms [23]. Many users, as data generators, want to be able to protect their data to prevent their privacy from being compromised. When users are faced with third-party collectors (e.g., service providers, etc.) managing their data, protecting their privacy becomes an issue that must be addressed.

As shown in Fig. 1, when data leaves the user's local side and is uploaded to a data collector, the user's privacy can be stolen by an attacker. DP is considered an effective solution to this problem by perturbing the user's data on a third-party [24] collector so that neither the attacker nor the subsequent release can cause a leakage of the user's privacy. However, attackers may be external, or data collectors may be honest and curious, also known as semi-honest collectors, following service protocols but trying to analyze personal private information from real user data. Although it is possible to defend against semi-honest collectors by perturbing user data locally through LDP before uploading it to third-party collectors, it still faces the challenge of reduced clustering utility due to the limitations and excessive noise of LDP. Meanwhile, the risk of privacy leakage cannot be completely avoided by simply perturbing the data in clustering. So the problem is to design a model that achieves a better utility of clustering without the impact of semi-honest collectors.

A clustering framework based on the PLDP k-modes algorithm is proposed in this paper to address the issues mentioned above. A randomized perturbation algorithm satisfying PLDP is used to perturb the user's local data, eliminating the risk of semi-honest collectors while satisfying personalized privacy requirements and enhancing the utility of clustering. Meanwhile, an iterative clustering

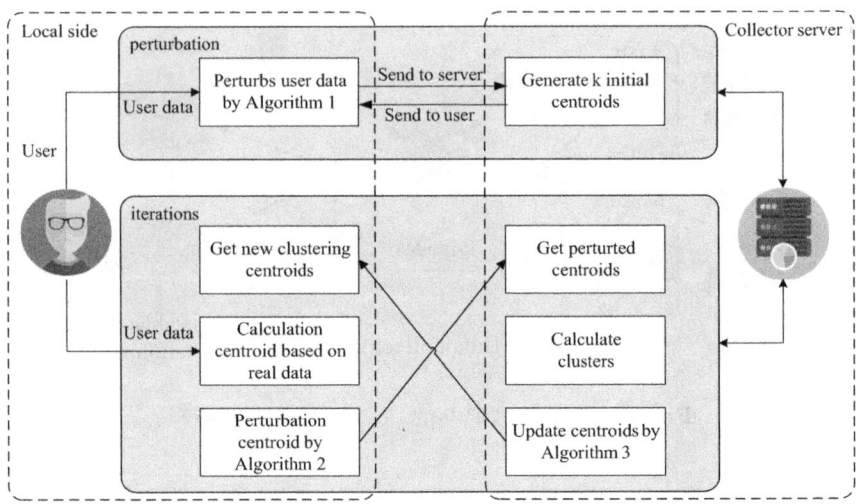

Fig. 2. Cluster privacy-preserving framework based on PLDP k-modes.

centroid perturbation algorithm perturbs the real clustering information locally to prevent privacy leakage due to inference attacks [25].

3.2 Proposed Framework

A framework based on PLDP k-modes that can solve the above problem is proposed, and its overall framework is shown in Fig. 2. The clustering model has a user set $U = \{u_1, u_2, \ldots, u_i, \ldots, u_n\}$, an attribute set $A = \{a_1, a_2, \ldots, a_j, \ldots, a_d\}$. User u_i has a d-dimensional data vector $S_i = \{s_1, s_2, \ldots, s_j, \ldots, s_d\}$. Where a_j is the set of values in the range of the j-th attribute, $s_j \in a_j$ and $d = |S_i|$ is the number of attributes. The target of k-modes is to classify the user data into k cluster centroids $C = \{c_1, c_2, \ldots, c_k\}$. Table 1 describes the important symbols and their descriptions used in this paper.

Figure 2 illustrates a proposed PLDP-based privacy-preserving clustering framework operating between a local user side and a collector server. Initially, user data S_i is perturbed locally to S_i^* via Algorithm 1 and transmitted to the server [26]. The server then provides initial centroids and manages the iterative k-modes clustering process. In each iteration, users identify their nearest centroid c_i using their true data S_i, send a perturbed version c_i^* generated by Algorithm 2 to the server. The server subsequently updates the global centroids based on the aggregated perturbed centroids C^* and the initially perturbed user data S_i^*, repeating this cycle until convergence. Further procedural details are elaborated in Sects. 3.2 and 3.2

Perturbation Method. As shown in Fig. 2, the perturbation part consists of two components [27], user data perturbation at the local side and initial

Table 1. Summary of Some Significant Symbol Notations

Symbol	Definition
U	user set
A	attribute set
d	data dimension
S_i	user u_i data vector
S_i^*	user u_i data vector after perturbation
C	cluster centroid set
C^*	centroid after perturbation
ε_u	user u_i privacy budget
G_i	user u_i security range
R	perturbation candidate set
$L(\cdot)$	weight function
$L_C(\cdot)$	index function
B_i	acceptable maximum Hamming distance
B_i^{index}	acceptable maximum sequential index
Sum_C	the count of all values taken for each attribute in each cluster

centroid selection on the server. The perturbation part perturbs the user data at the local side by the proposed PK-RR mechanism. Both prevented attacks from semi-honest collectors and obtained smaller errors and better utility through the PK-RR, an improved K-RR mechanism.

User Data Perturbation. In contrast to continuous data, discrete data is encoded by sorting all possible values of each feature a_j of attribute set A. The user's features are transformed into corresponding ordinal numbers, i.e., $a_j \in \{1, 2, \ldots, K\}$. Transforming the raw information into integers reduces the communication cost and decreases the algorithmic coupling. Unlike DP k-modes, since users cannot trust semi-honest third-party data collectors, the data are perturbed at the local side in this paper to ensure that the users' real data are known only to the users themselves. The perturbation method being used is based on the K-RR mechanism. However, the direct use of the LDP-based K-RR mechanism to perturb user data leads to the following four problems.

1. When the size of A is large and the privacy budget ε is small, the K-RR mechanism will lead to large statistical errors and loss of clustering utility.
2. Since the privacy budget ε has to be allocated to each dimension of the data, when the data dimension is large leads to a smaller privacy budget for each dimension to the extent that the usability of the data decreases greatly.
3. For the general K-RR mechanism, when the output is not equal to the input, the user data is usually replaced by anyone perturbed data that is not equal to the user data. It means that the probability of outputting any perturbed data is equal. The similarity between the perturbation data and the user data

greatly impacts the utility of clustering, and the similarity between different perturbation data and the user data varies. Therefore, the direct use of K-RR causes a decrease in the usability of data and leads to a decrease in the utility of clustering.

4. According to the definition of LDP, the privacy budget ε is uniform for each user (user data). However, different users and data usually have different privacy requirements. A uniformized privacy budget ε cannot satisfy the actual privacy requirements of users.

To solve the above problem, the PLDP-based personalized K-RR (PK-RR) mechanism is proposed in this paper to implement the PLDP k-modes algorithm. To address the first problem above, the K-RR mechanism assumes that each data needs to be indistinguishable within the global scope of the data [28]distribution, which is usually unnecessary. For example, suppose a user's location is within the university campus. In that case, the user usually allows the data collector to know their location's approximate range (e.g., street or district) rather than requiring the location data to be indistinguishable within a city or larger area. The PK-RR uses a safe range G_i and only selects perturbation outputs within the safe range when perturbing [29]. It avoids excessive privacy protection and ensures the utility of clustering. For the second problem, the PK-RR mechanism chooses to perturb multidimensional data as a whole, thus avoiding the problem of reduced data availability due to privacy budget allocation. For example, a two-dimensional binary data $\{1,1\}$, the perturbation candidate set R of PK-RR have $\{1,0\},\{0,1\},\{0,0\}$. It can be found that the Hamming distances between the three perturbation outputs and the inputs are different. But in the K-RR mechanism, these three perturbation outputs are selected with the same probability. And the similarity between perturbation data and user data has a great impact on the utility of clustering. Therefore, for the third problem, PK-RR improves the K-RR mechanism by assigning weights to each perturbed output, determined by the Hamming distance between the input user data and the perturbed output. Specifically, given the input data S_i, perturbation candidate set $R = \{r_1, \ldots, r_n\}$. Let $L(S_i, r_x)$ be the weight function. Finally, to satisfy the different privacy requirements of users, PK-RR allows users to choose their security range G_i and privacy budget ε_u. PK-RR is proven to satisfy PLDP. PK-RR is specifically described as follows.

S_i is the user data. S_i^* is the data perturbed by the PK-RR algorithm. G_i and ε_u are the security range and privacy budget chosen by the user according to their privacy requirements. B_i represents the acceptable maximum Hamming distance between the S_i^* and S_i. $L(S_i, r_x)$ is a weights function based on Hamming distance, where $L(S_i, S_i) = 1$ and $G_i = \{r_x \mid r_x \in R, L(S_i, r_x) \leq B_i\}$ is the security candidate set of S_i. r_x is any one of the centroids in R. All r_x in G_i are sorted by the Hamming distance between them and S_i in ascending order. The formula for the PK-RR is shown as follows.

$$P(S_i^* \mid S_i) = \begin{cases} \frac{\left(e^{\varepsilon_u/(B_i-1)}\right)^{-L(S_i, S_i^*)}}{\sum_{r_x \in G_i} \left(e^{\varepsilon_u/(B_i-1)}\right)^{-L(S_i, r_x)}}, & \text{if } L(S_i, S_i^*) \leq B_i, \\ 0, & \text{if } L(S_i, S_i^*) > B_i, \end{cases}$$

Algorithm 1. PK-RR Perturbation Data

Input: privacy budget ε_u, acceptable maximum Hamming distance B_i, user u_i data S_i, perturbation candidate set R
Output: user u_i data S_i after perturbation S_i^*
1: $G_i = \{r_x \mid r_x \in R, L(S_i, r_x) \leq B_i\}$
2: $m = \text{size}(G_i)$
3: $list = [], dis = 0$
4: **for** $x \leftarrow 0$ to $m - 1$ **do**
5: $\quad list[x] = \dfrac{\left(e^{\varepsilon_u/(B_i-1)}\right)^{-L(S_i, r_x)}}{\sum_{r_x \in G_i}\left(e^{\varepsilon_u/(B_i-1)}\right)^{-L(S_i, r_x)}}$
6: **end for**
7: $z \leftarrow$ Take a random value from $(0, 1)$
8: **for** $j \leftarrow 0$ to $m - 1$ **do**
9: $\quad$ **if** $\sum_{x=0}^{j} list[x] \geq z$ **then**
10: $\quad\quad dis = j$
11: $\quad\quad$ BREAK
12: $\quad$ **end if**
13: **end for**
14: $S_i^* \leftarrow$ Randomly select a data from G_i whose Hamming S distance to S_i is dis
15: RETURN S_i^*

where R represents all possible candidate data. G_i is the set of security candidates corresponding to the data S_i of user u_i, where the Hamming distance of all candidate data to S_i is less than B_i. In contrast to the K-RR mechanism, which outputs all perturbed data with the same probability, PK-RR mechanism constrains and grades the output range and probability based on the Hamming distance between the perturbed data and the user data. PK-RR mechanism ensures that the output is more similar to the user data with higher probability, thus reducing the error caused by perturbation and improving data usability.

The general process of perturbation of user data is shown in Algorithm 1, where S_i^* is obtained according to the perturbation of Eq. 3.2 and then sent to the collector server. The privacy proof of Algorithm 1 is shown in Sect. 4.

Initial Centroid Selection. The server randomly generates k d-dimensional initial centroids C based on the S_i^* and sent them to the user.

The PK-RR mechanism perturbs the user data at the local side, which allows the data to leave the local side and be analyzed by the data collector safely. Although the data collector cannot obtain the user's privacy information from the perturbed data, it can still infer it from the information in the clustering iteration. The next subsection will analyze the privacy issues in clustering iterations and propose a solution.

Clustering Iteration Method. As shown in Fig. 2, the local side enters the iterative process after the user data perturbation is completed. The centroids from the server are first accepted, then new centroids are calculated based on the real user data. Although the server cannot infer privacy information from

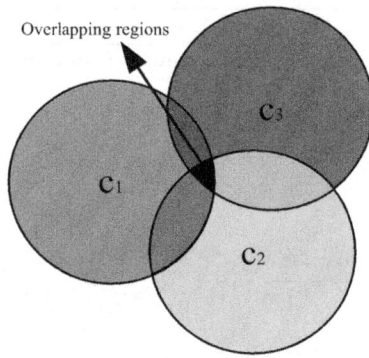

Fig. 3. Privacy leakage from iterative centroids.

the user data, the clustering information of the user belonging to that cluster, i.e., the iteration centroids sent to the server in each iteration, may reveal user privacy. Because the clusters to which users belong are calculated from real data over multiple iterations, the server can infer the approximate distribution or exact value of the user data as the clusters to which users belong change, and the iteration centroids are updated. For example, assuming the user data is two-dimensional location data, the user can be positioned in a circular region in each iteration. In multiple iterations, overlapping these circular regions will help the server locate the user's exact location or exact location range. As shown in Fig. 3, the user's location will be positioned within the black area after three iterations.

Centroid Perturbation. Cluster centroid set $C = \{c_1, c_2, \ldots, c_k\}$ $(1 \leq x \leq k)$, c_x is any one of the centroids in C, c_i is the nearest centroid to S_i. c_i^* is the c_i perturbed by the PK-RR algorithm. G_i and ε_u are the security range and privacy budget chosen by the user according to their privacy requirements. B_i^{index} represents the acceptable maximum sequential index between the c_i and c_i^*. All c_x in C are sorted by the Hamming distance between them and c_i in ascending order. Let $L_C(c_i, c_x)$ denote the index of c_x in the sorted list, where $L_C(c_i, c_i)$ = 1 and $G_i = \{c_x \mid c_x \in C, L_C(c_i, c_x) \leq B_i^{index}\}$ is the security candidate set of c_i. The formula is shown as follows.

$$P(c_i^* \mid c_i) = \begin{cases} \dfrac{\left(e^{\varepsilon_u/(B_i^{index}-1)}\right)^{-L_C(c_i, c_i^*)}}{\sum_{c_x \in G_i} \left(e^{\varepsilon_u/(B_i^{index}-1)}\right)^{-L_C(c_i, c_x)}} & \text{if } L_C(c_i, c_i^*) \leq B_i^{index}, \\ 0 & \text{if } L_C(c_i, c_i^*) > B_i^{index}. \end{cases}$$

From Eq. 3.2, it can be seen that the centroids which are outside the security range G_i will not be used as the output of the perturbation mechanism, and the

Algorithm 2. PK-RR Perturbation Centroid

Input: privacy budget ε_u, acceptable maximum sequential index B_i^{index}, user u_i data S_i, cluster centroid set C
Output: perturbed centroid c_i^*

1: $G_i = \{c_x \mid c_x \in C, L_C(c_i, c_x) \leq B_i^{index}\}$
2: $c_i \leftarrow$ the nearest centroid to S_i
3: $m = \text{size}(G_i)$
4: $list = [\,], index = 0$
5: **for** $x \leftarrow 1$ to m **do**
6: $\quad list[x] = \dfrac{\left(e^{\varepsilon_u/(B_i^{index}-1)}\right)^{-L_C(c_i,c_x)}}{\sum_{c_x \in G_i}\left(e^{\varepsilon_u/(B_i^{index}-1)}\right)^{-L_C(c_i,c_x)}}$
7: **end for**
8: $z \leftarrow$ Take a random value from $(0,1)$
9: **for** $j \leftarrow 1$ to m **do**
10: $\quad$ **if** $\sum_{x=1}^{j} list[x] \geq z$ **then**
11: $\quad\quad index = j$
12: $\quad\quad$ BREAK
13: $\quad$ **end if**
14: **end for**
15: $c_i^* \leftarrow$ Select the centroid in G_i with the serial number $index$
16: RETURN c_i^*

centroids which are within the security range G_i will be output with different probabilities according to the Hamming distance from c_i. The detailed procedure for the centroid perturbation is described in Algorithm 2. After completing the centroid perturbation, the local side sends the perturbed centroid c_i^* to the collector server for the next round of clustering iterations.

Send the new centroid to the local side after the calculation is completed. Clustering iterations are performed as described above until the clustering is complete.

3.3 Privacy Analysis

This section proves that Algorithm 1 and Algorithm 2 satisfy the definition of differential privacy and further prove that the overall framework satisfies the definition of differential privacy.

Theorem 1. Algorithm 1 provides (G_i, ε_u) – PLDP for each user u_i with (G_i, ε_u). For any two values $S_i, S_i', S_i^* \in G_i$, according to Eq. 3.2, there is

$$\frac{\Pr[S_i^* \mid S_i]}{\Pr[S_i^* \mid S_i']} = \frac{\dfrac{\left(e^{\varepsilon_u/(B_i-1)}\right)^{-L(S_i, S_i^*)}}{\sum_{r_x \in G_i}\left(e^{\varepsilon_u/(B_i-1)}\right)^{-L(S_i, r_x)}}}{\dfrac{\left(e^{\varepsilon_u/(B_i-1)}\right)^{-L(S_i', S_i^*)}}{\sum_{r_x \in G_i}\left(e^{\varepsilon_u/(B_i-1)}\right)^{-L(S_i', r_x)}}} = e^{\varepsilon_u\left(\frac{L\left(S_i', S_i^*\right) - L(S_i, S_i^*)}{B_i - 1}\right)} \leq e^{\varepsilon_u}.$$

Therefore, the probability ratio of the output S_i^* is less than e^{ε_u} for any two values S_i, S_i' within the security range G_i.

Theorem 2. Algorithm 2 provides (G_i, ε_u) – PLDP for centroid with (G_i, ε_u).

Proof. For any two centroids $c_i, c_i', c_i^* \in G_i$, there is

$$\frac{\Pr\left[c_i^* \mid c_i\right]}{\Pr\left[c_i^* \mid c_i'\right]} = \frac{\left(e^{\varepsilon_u/\left(B_i^{index}-1\right)}\right)^{-L_C(c_i,c_i^*)}}{\sum_{c_x \in G_i} \left(e^{\varepsilon_u/\left(B_i^{index}-1\right)}\right)^{-L_C(c_i,c_x)}} = e^{\frac{\varepsilon_u\left(L_C(c_i',c_i^*)-L_C(c_i,c_i^*)\right)}{B_i-1}} \leq e^{\varepsilon_u}.$$

From Eq. 3.3, the probability ratio of the output c_i^* is less than e^{ε_u} for any two centroids c_i, c_i' within the security range G_i. According to Eq. 3.2 of the definition of PLDP, it is clear that Algorithm 2 satisfies (G_i, ε_u) – PLDP. □

It is worth noting that the traditional LDP-based clustering solutions require the privacy budget to be allocated to each dimension, whether the data is perturbed directly or after coding. In contrast, the proposed scheme is designed to perturb the multidimensional data as a whole, so there is no requirement to allocate the privacy budget, which greatly improves the utility of the data. Specifically, suppose there is a ten-dimensional user data with a privacy budget of ε. According to Property 1 in Sect. 2.1, the traditional scheme obtains the privacy budget $\frac{\varepsilon}{10}$ for each dimension, leading to excessive noise. In contrast, the scheme proposed by this paper only requires a single perturbation with the privacy budget ε.

4 Experimental Evaluation

In this section, experiments are designed to investigate the improvements in the proposed framework compared to the existing DP k-modes algorithm and how the relevant parameters influence the utility of the proposed framework.

4.1 Experimental Environment and Datasets

The hardware platform for this experiment uses Windows 10 operating system, Intel(R) Core(TM) i7-11700 CPU @ 2.50GHz, and 32.00 GB RAM. The experimental platform uses python 3.7. Eight categorical datasets from the UCI database were used for the experiments in this paper. Table 2 shows the dataset names, the amount of data N, the number of attributes A and the number of clusters K.

Table 2. Dataset Information

	Vote	Zoo	Mushroom	Lung-Cancer	Soybean	Breast-Cancer	Chess	Adult
N	435	101	8124	32	47	699	3196	48842
A	16	16	22	56	21	9	36	8
K	2	7	2	3	4	2	2	2

4.2 Experimental Setup and Evaluation Metrics

This paper focuses on three aspects of experimenting with the proposed framework.

1. Compare the utility of clustering with state-of-the-art algorithms for uniform k values under different ε. The PLDP k-modes algorithm proposed in this paper is compared with the DP-Modes-Lloyd algorithm [17]. The DP-Modes-Lloyd algorithm serves as our main comparison baseline, and it is the first proposed, best and representative algorithm for k-Modes pattern clustering under the center difference privacy model compared to other baselines. This comparison effectively highlights the practical advantages of our PLDP-based approach over existing DP methods. The DP-Modes-Lloyd updates the noisy mode of the subset by computing noisy counts of all attribute values and taking, for each attribute, the value whose count is maximum. It is worth noting that the DP-Modes-Lloyd algorithm is based on differential privacy and does not prevent attacks by semi-honest servers.
2. Compare the effects of different setting changes on the utility of clustering. Comparative experiments were established to understand the impact of the key mechanisms on the clustering utility. To understand the impact of the security range mechanism on the utility of clustering, an experimental analysis of the utility of clustering of the algorithm in different security range cases was performed.

The utility of clustering is assessed using the Normalised Intra-Cluster Variance (NICV) [17]. The essential goal of the k-modes algorithm is to divide the data into k clusters based on minimizing the error function, with Hamming distance as the evaluation metric. Therefore NICV can directly reflect the utility of clustering, while the NICV value can also reasonably reflect the impact of privacy protection mechanisms on the utility of clustering. NICV is defined as

$$NICV = \frac{1}{N}\sum_{i=1}^{k}\sum_{S_i \in c_i} HamDis\left(S_i, c_i\right),$$

where N represents the total number of users, k represents the number of centroids, S_i represents the user data, and $S_i \in c_i$ represents the centroid c_i is the closest centroid to S_i. $HamDis\left(S_i, c_i\right)$ is the calculation of the Hamming distance between data and centroid.

4.3 Experimental Analysis

A privacy analysis of the proposed algorithm is presented in Sect. 3.3, which demonstrates that the proposed algorithm can effectively protect user privacy. In this subsection, the clustering utility of the PLDP k-modes algorithm will be analyzed by experiments. The results of the experiments are shown below. The PLDP k-modes algorithm is compared with the DP-Modes-Lloyd algorithm [17] for NICV values on eight datasets in the first experiment. ε_u is set at $[0, 2]$, considering the utility of clustering for two cases of iteration rounds $T = 1$ and $T = 5$, respectively. Also, two extreme situations for each dataset illustrate the experimental results better. $\varepsilon_u = \infty$ represents the best utility of clustering without privacy mechanisms, and $\varepsilon_u = 0$ represents the utility of clustering when adding infinite noise with random output, which can be interpreted as the worst utility of clustering.

Figure 4 shows the performance of PLDP k-modes. It can be seen that the privacy budget ε_u increases, the NICV of both algorithms decreases on each dataset, which is as expected. According to the definition of differential privacy, as ε_u increases, the less noise is added, therefore, the better the utility of clustering. Secondly, PLDP k-modes have significantly lower NICV than DP-Modes-Lloyd in most cases, i.e., PLDP k-modes have better clustering utility. This proves that the proposed method in this paper is reasonable and effective. Finally, the NICV values of $T = 5$ are higher than those at $T = 1$ in most cases. The reason is that the centroid perturbation adds additional noise to the clustering during the iterative process, which leads to a poorer clustering utility.

The effect of the security range is next examined. The user can select the appropriate security range G_i by specifying the value of B_i or B_i^{index}. The role of B_i or B_i^{index} in the mechanism is similar. The effect of B_i and B_i^{index} on the utility of clustering is mainly explored in the experimental analysis.

The effect of B_i on the utility of clustering was first explored in the experiment. ε_u is set to 1 and $T = 5$. Since the security region chosen by different users is usually not equal in practical applications, five cases of B_i are set to explore its effect on the clustering utility. Min represents the minimum value that B_i can take, which is set to 2, and Max represents the maximum value that can be taken, i.e., the dimensionality of the data. Med is the median value, which is set to $\frac{Max+Min}{2}$. Two other cases in practical applications are considered, assuming The distribution of B_i is uniform or normal, the mean of both distributions is set to Med, and the standard deviation of the normal distribution is set to $\frac{Max-Min}{6}$.

Figure 5 shows the utility of clustering for different B_i on the eight datasets. It can be seen as B_i increases from Min to Max, the NICV value becomes larger, and the utility of clustering decreases on all datasets. It shows that B_i is negatively correlated with the utility of clustering. As illustrated in Fig. 5, A larger security range of user data leads to the poorer utility of clustering.

As shown in Fig. 6, the effect of B_i on the clustering utility is consistent with the results in Fig. 5. A larger B_i leads to poorer clustering utility, and a normal distribution has a better clustering utility than a uniform distribution. The effect

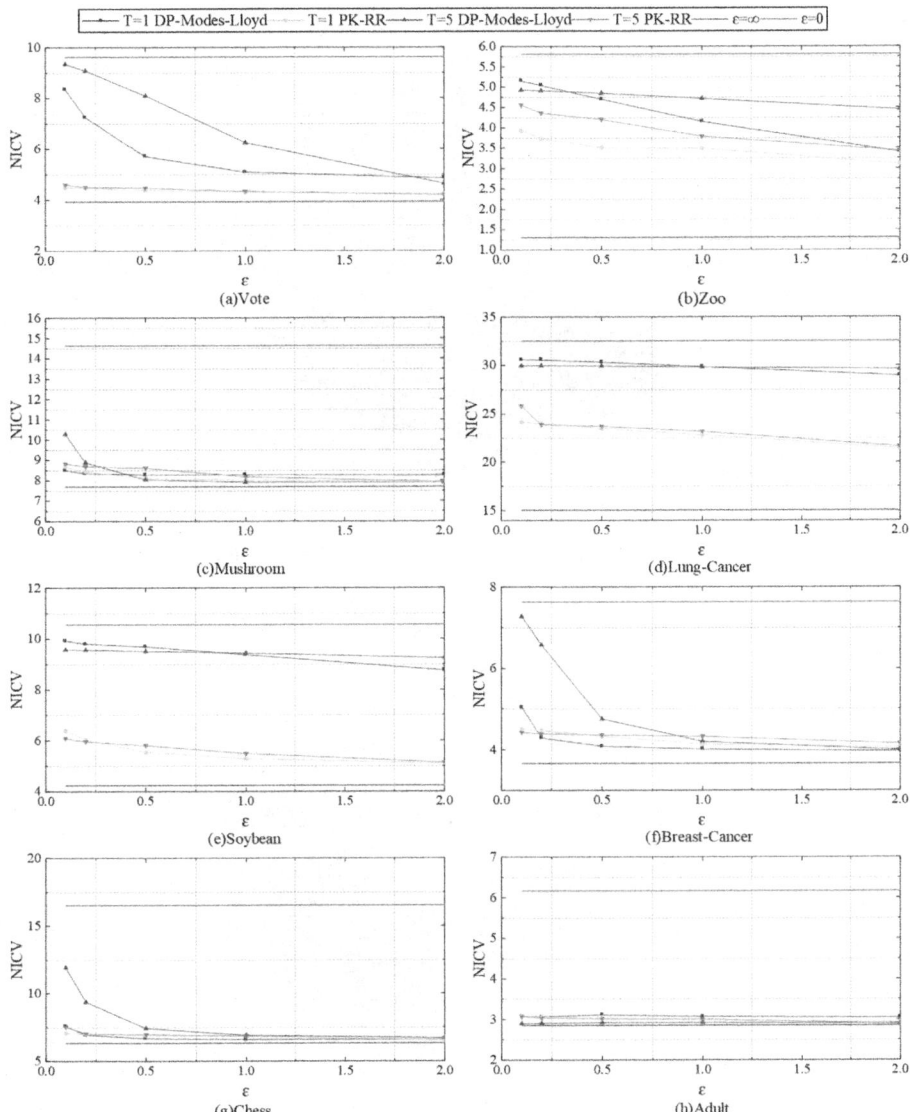

Fig. 4. Comparative experiments with respect to ε.

of B_i^{index} is next explored. It can be seen that as B_i^{index} increases from Min to Med and then to Max, the NICV value increases and the utility of clustering decreases. This indicates that B_i^{index} is negatively correlated with the utility of clustering. As the safety range of the user's iteration centroids increases, it will lead to the poorer utility of clustering. Also, it can be found that the results of uniform distribution and Med are similar, and the utility of clustering is

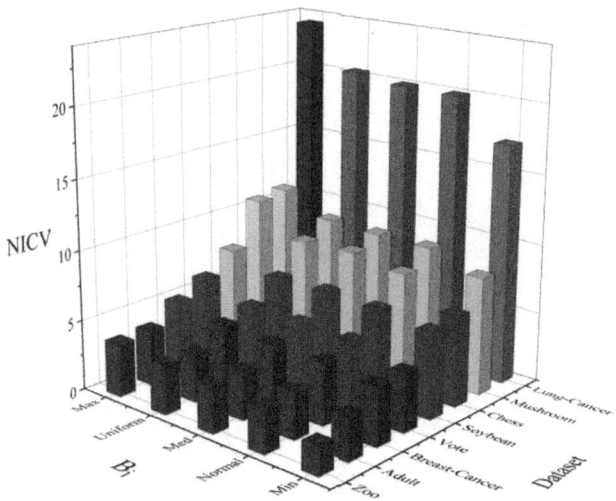

Fig. 5. Performance under the different B_i and the different dataset.

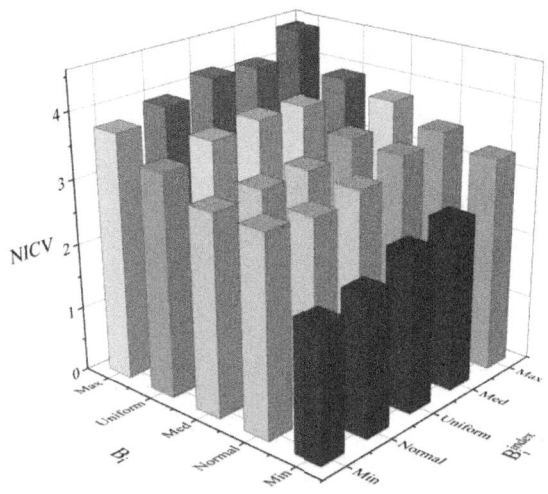

Fig. 6. Performance under the different B_i and the different B_i^{index} in Zoo dataset.

better for normal distribution than uniform distribution. In general, the overall regularity of B_i^{index} is similar to B_i.

Based on the experimental analysis above, the proposed algorithm in this paper improves the utility of clustering while ensuring the strength of privacy protection, and the experiments illustrate that the desired effect is achieved.

5 Conclusion

Clustering models are vital for data analysis but risk user privacy without proper safeguards. Current methods struggle with the privacy-utility balance: DP approaches rely on trusted servers, while LDP offers insufficient utility and neither address personalized privacy needs or the perturbed-real data relationship. This paper proposes a PLDP k-modes clustering framework using a novel PK-RR perturbation mechanism for user data and centroids, targeting an efficient and secure k-modes algorithm.

Theoretical analysis and real-dataset experiments confirm our framework's superior privacy-utility trade-off, outperforming state-of-the-art DP-based k-modes algorithms in clustering utility. It protects private information, enhances model utility, and meets personalized privacy demands. Future work will focus on optimizing privacy budget allocation through adaptive algorithms, potentially using reinforcement learning or designing new evaluation metrics based on data characteristics, to further develop efficient and secure DP clustering algorithms. Our future research will focus on optimizing communication efficiency in ultra-large-scale distributed environments through techniques like model update compression, reduced communication rounds, and asynchronous updates, while preserving privacy and model utility.

Acknowledgments. This research was supported by the National Natural Science Foundation of China under Grant 62462009, Guangxi Natural Science Foundation under Grant 2022GXNSFFA035028, the Guangxi Science and Technology Projects under Grant GuiKeAD24010047, Guangxi Young Elite Scientist Sponsorship Program under Grant GXYESS2025144.

References

1. Wang, X., Yang, L.T., Song, L., Wang, H., Ren, L., Deen, M.J.: A tensor-based multiattributes visual feature recognition method for industrial intelligence. IEEE Trans. Ind. Inform. **17**(3), 2231–2241 (2021)
2. Liu, Q., Peng, Y., Pei, S., Wu, J., Peng, T., Wang, G.: Prime inner product encoding for effective wildcard-based multi-keyword fuzzy search. IEEE Trans. Serv. Comput., 1–13 (2020)
3. Liu, Q., Tian, Y., Wu, J., Peng, T., Wang, G.: Enabling verifiable and dynamic ranked search over outsourced data. IEEE Trans. Serv. Comput. **15**(1), 69–82 (2022)
4. Wang, S., Sun, Y., Bao, Z.: On the efficiency of k-means clustering: evaluation, optimization, and algorithm selection. In: Proceedings of the VLDB Endowment, pp. 163–175 (2020)
5. Huang, Z.: Extensions to the k-Means algorithm for clustering large data sets with categorical values. Data Min. Knowl. Discov. **2**(3), 283–304 (1998)
6. Dwork, C.: Differential privacy. In: Information Security and Cryptography, Languages and Programming (ICALP), pp. 1–12 (2006)
7. Nguyen, H.H.: Privacy-preserving mechanisms for k-modes clustering. Comput. Secur. **78**(1), 60–75 (2018)

8. Xiao, Y., Xiong, L.: Protecting locations with differential privacy under temporal correlations. In: Proceedings of the ACM Conference on Computer and Communications Security, pp. 1298–1309 (2015)
9. Xiong, J., et al.: Enhancing privacy and availability for data clustering in intelligent electrical service of IoT. IEEE Internet Things J. **6**(2), 1530–1540 (2019)
10. Ni, T., Qiao, M., Chen, Z., Zhang, S., Zhong, H.: Utility-efficient differentially private k-means clustering based on cluster merging. Neurocomputing **424**(1), 205–214 (2021)
11. Huang, L., Wu, J., Shi, D., Dey, S., Shi, L.: Differential privacy in distributed optimization with gradient tracking. IEEE Trans. Autom. Contr. **69**(9), 5727–5742 (2024)
12. Shin, H., Kim, S., Shin, J., Xiao, X.: Privacy enhanced matrix factorization for recommendation with local differential privacy. IEEE Trans. Knowl. Data Eng. **30**(9), 1770–1782 (2018)
13. Yang, M., Guo, T., Zhu, T., Tjuawinata, I., Zhao, J., Lam, K.Y.: Local differential privacy and its applications: a comprehensive survey. Elsevier B.V (2024)
14. Erlingsson, Ú., Pihur, V., Korolova, A.: RAPPOR: randomized aggregatable privacy-preserving ordinal response. In: Proceedings of the ACM Conference on Computer and Communications Security, pp. 1054–1067 (2014)
15. Ding, B., Kulkarni, J., Yekhanin, S.: Collecting telemetry data privately. In: Advances in Neural Information Processing Systems, pp. 3572–3581 (2017)
16. Zhang, X., Chen, Z., Chen, G., Feng, X., Shen, Q., Wu, Z.: RPPFL: robust and privacy-preserving federated learning via trusted execution environments. In: ICASSP 2025 - 2025 IEEE International Conference on Acoustics, Speech and Signal Processing (ICASSP), pp. 1–5. IEEE (2025)
17. Warner, S.L.: Randomized response: a survey technique for eliminating evasive answer bias. J. Am. Stat. Assoc. **60**(309), 63–66 (1965)
18. Gu, X., Li, M., Xiong, L., Cao, Y.: Providing input-discriminative protection for local differential privacy. In: International Conference on Data Engineering, pp. 505–516 (2020)
19. Chen, R., Li, H., Qin, A.K., Kasiviswanathan, S.P., Jin, H.: Private spatial data aggregation in the local setting. In: 2016 IEEE 32nd International Conference on Data Engineering, ICDE 2016, pp. 289–300 (2016)
20. Li, N., Lyu, M., Su, D., Yang, W.: Differential privacy: from theory to practice. In: Synthesis Lectures on Information Security, Privacy, and Trust, pp. 1–138 (2016)
21. Huang, Z.: Extensions to the k-means algorithm for clustering large data sets with categorical values. Data Min. Knowl. Discov. **2**(3), 283–304 (1998)
22. Ye, Q.Q., Meng, X.F., Zhu, M.J., Huo, Z.: Survey on local differential privacy. J. Softw. **29**(7), 1981–2005 (2018)
23. Song, H., et al.: APLDP: adaptive personalized local differential privacy data collection in mobile crowdsensing. Comput. Secur. **136** (2024)
24. Liu, J., et al.: Enhancing the robustness of random Boolean networks by epigenetic regulation. IEEE Trans. Comput. Biol. Bioinform. **22**(1), 261–270 (2025)
25. Zhang, E.Z., Guan, Y., Lu, R., Zhang, H.: A communication-efficient conjunctive query scheme under local differential privacy. In: GLOBECOM 2024 - 2024 IEEE Global Communications Conference, pp. 31–36. IEEE (2024)
26. Nguyen, H.H.: Privacy-preserving mechanisms for k-modes clustering. Comput. Secur. **78**(1), 60–75 (2018)
27. Luo, Y., et al.: DPO-Face: differential privacy obfuscation for facial sensitive regions. Comput. Secur. **154** (2025)

28. Chen, Y., Gan, W., Huang, G., Wu, Y., Yu, P.S.: Privacy-preserving federated discovery of DNA motifs with differential privacy. Expert Syst. Appl. **249** (2024)
29. Luo, Y., Li, Y., Qin, S., Fu, Q., Liu, J.: Copyright protection framework for federated learning models against collusion attacks. Inf. Sci. (NY) **680** (2024)

AnoST: An Anonymous Optimistic Verification System Based on Off-Chain State Transition

Qiyuan Gao[1], Qianhong Wu[1(✉)], Junxiang Nong[2], and Qi Liu[1]

[1] Beihang University, Beijing, China
{gaoqy,qianhong.wu}@buaa.edu.cn
[2] Renmin University of China, Beijing, China
fpdqwq@yeah.net

Abstract. Current blockchain optimistic verification schemes typically face critical challenges such as inadequate verifier workload verification, verifier identity exposure risks, and low transaction acceptance rates due to batch rollbacks, significantly impacting off-chain verification efficiency and security. To address these issues, we propose AnoST, an efficient transaction verification mechanism with enhanced verifier anonymity protection. First, we introduce an anonymous staking-based verifier election mechanism to preserve verifier privacy and mitigate identity exposure risks. Second, we propose a State Transition Commitment (STC) mechanism, enabling direct blockchain verification of verifier workloads, effectively resolving the Verifier's Dilemma (Luu et al., CCS 2015). Furthermore, we develop a partial rollback mechanism leveraging State Transition Trees (STT), significantly improving transaction acceptance rates and system efficiency. Experimental results demonstrate that AnoST achieves notable advantages in security, anonymity protection, and transaction approval rates, highlighting its promising practicality and scalability.

Keywords: Blockchain Rollup · Optimistic Verification · Verifier's Dilemma

1 Introduction

In recent years, with the rapid development of distributed computing technologies, third-party applications have increasingly required reliable and efficient verification of computation results [3,7,12,17,25,31]. Blockchain [5,6,32], characterized by decentralization, transparency, and tamper-resistance, has emerged as a mainstream solution for computational verification, finding widespread application in edge computing [10,28,30], federated learning [19,22,26], decentralized finance (DeFi) [24], and other domains.

However, the traditional verification model of blockchain, constrained by factors such as independent node computation, limited block capacity [9,18], and consensus mechanisms [2,4,14,29], exhibits significant performance bottlenecks, failing to meet the growing demand for efficient verification. To mitigate these issues, verification schemes represented by Optimistic Rollup [15,27] have been proposed. Optimistic Rollup transfers complex computational tasks off-chain, submitting only the final state commitment on-chain and employing fraud-proof mechanisms when disputes arise, greatly enhancing on-chain verification efficiency and transaction throughput.

Despite these advantages, current Optimistic Rollup implementations still face several critical challenges in practical deployments. First, insufficient economic incentives for verifiers conducting off-chain validations often lead to the *Verifier's Dilemma* [13,16], where rational verifiers tend to avoid actively performing validations, thereby compromising system security. Second, the absence of anonymity protection mechanisms for verifiers makes them vulnerable targets for malicious attacks, significantly increasing security risks. Furthermore, the prevailing batch rollback mechanisms reject entire transaction batches due to individual faulty transactions, severely impacting the approval rate of valid transactions.

1.1 Problem Statement

In this section, we will analyze and distill the issues identified above, in order to clearly define the design objectives for our solution.

Verifier Identity Exposure and Security Risks. Optimistic Rollup relies on verifiers to monitor off-chain transactions and submit fraud proofs to prevent fraudulent states from being committed on-chain. However, blockchain Publicity leads to exposure of verifier identities [20,23], making them susceptible to bribery, targeted attacks, or even internal collusion. Such risks severely undermine the security and reliability of the verification process, ultimately threatening the overall system stability.

Verifier's Dilemma. The Verifier's Dilemma arises when rational verifiers choose not to actively verify transactions due to cost considerations, especially when the majority of transactions are expected to be correct. Verifiers tend to rely on others to perform verification tasks, creating collective inaction. If all verifiers adopt this strategy, the system may lack adequate oversight, allowing error transactions to successfully enter the blockchain and significantly diminishing the system's security and integrity.

Excessive Verification Period and Batch Rollback Issues. Current Optimistic Rollup schemes typically enforce prolonged verification dispute periods, involving multiple interactive rounds on-chain to identify disputed transactions. This results in significant delays in final transaction confirmations, substantially affecting system throughput and verification efficiency. Additionally, if a transaction batch contains erroneous transactions, the entire batch is rejected, dramatically reducing the throughput efficiency of valid transactions and negatively impacting overall system performance.

Therefore, designing an efficient off-chain transaction verification mechanism that ensures verifier anonymity and effectively incentivizes active participation from verifiers has become a critical and urgent issue for blockchain-based Optimistic Rollup solutions.

1.2 Our Contributions

In summary, designing an efficient verification mechanism that ensures verifier anonymity while incentivizing proactive verifier participation has emerged as a critical challenge in off-chain scaling solutions. To address this, we propose the AnoST with the following major contributions:

1. We introduce an anonymous staking-based verifier election mechanism, which ensures verifier identity anonymity throughout the verification process, fundamentally mitigating bribery, collusion, and targeted attack risks associated with verifier identity exposure.
2. We propose an on-chain verification mechanism leveraging the State Transition Commitment (STC). Through this mechanism, the blockchain can directly verify the actual workload performed by off-chain verifiers by comparing state commitment hashes, effectively resolving the verifier's dilemma.
3. Building upon the STC mechanism, we further design a refined partial confirmation and rollback scheme using the State Transition Tree (STT). With this scheme, when erroneous transactions are detected in a batch, the system precisely rolls back only the incorrect states and their dependent transactions, thus significantly enhancing the overall transaction throughput and verification efficiency.
4. We conduct comprehensive experimental analyses on AnoST. The results demonstrate that our mechanism achieves acceptable on-chain computational and storage overhead, while providing highly efficient transaction verification performance, confirming its practical effectiveness.

2 Preliminaries

2.1 Optimistic Rollup

Optimistic Rollup [8] is a widely adopted blockchain scaling solution that improves throughput by executing transactions off-chain and submitting only state commitments on-chain. It operates under the assumption that transactions are valid unless challenged, triggering on-chain fraud proofs only when disputes arise. This design significantly reduces on-chain computation while preserving correctness.

Table 1 compares several representative protocols. Most systems expose verifier identities and lack mechanisms to enforce honest verification, leading to the

Verifier's Dilemma. Arbitrum introduces interactive dispute resolution with logarithmic complexity, while others, including AnoST, support one-shot challenges. Only AnoST ensures verifier anonymity, proves verification effort via STC, and supports precise transaction-level rollback for efficient error localization.

2.2 Aztec Private Transfers

Aztec [1] enables confidential asset transfers on Ethereum by representing each asset as a note: $N = \mathsf{Commit}(v, \rho, pk)$, where v is the value, ρ is randomness, and pk is the recipient's public key. A private transfer consumes input notes $\{N_i\}$ and creates output notes $\{N'_j\}$, preserving value: $\sum_i v_i = \sum_j v'_j$. A zero-knowledge proof π is generated to prove correctness and ownership without revealing v, ρ, or pk. This construction guarantees value privacy, sender/receiver anonymity, and public verifiability.

Table 1. Comparison of Optimistic Verification Protocols

Feature	Optimism [8]	Arbitrum [11]	Cartesi [21]	AnoST
Verifier Identity	Public Addr	Public Addr	Public Addr	Anony ID
Verifier's Dilemma	✗	✓	✗	✓
Challenge Round	O(1)	O(log n)	O(1)	O(1)
Proof Method	Tx Re-exe	VM-step	Vm-Segment	Tx Re-exe
Rollback	✗	✗	✗	✓

3 Definitions and System Model

We consider a set of verifiers in the system, denoted as $\mathcal{V} = \{V_1, V_2, \ldots, V_m\}$. Each verifier $V_i \in \mathcal{V}$ is assumed to be rational, meaning they always make decisions aimed at maximizing their own economic benefit. A verification task $\mathcal{T}$ (hereinafter referred to as transaction set) is defined as a set of transactions to be verified, i.e., $\mathcal{T} = \{\tau_1, \tau_2, \ldots, \tau_n\}$, where each transaction τ_i triggers an update to the system state upon execution. Moreover, we assume that economic incentives provided by the system fully cover the costs incurred by honest executors and verifiers.

To clearly define the analytical scope and preconditions of this paper, we introduce the following two fundamental assumptions:

Assumption 1 (Blockchain Trust Assumption): We assume the consensus mechanism of the underlying blockchain is secure and trustworthy, meaning attackers cannot modify or revoke states that have been finalized by the blockchain's consensus protocol. Consequently, any transaction result confirmed by the blockchain consensus is considered correct and trustworthy by the verifiers. In addition, as a trusted arbitrator, the blockchain can correctly handle challenges initiated by verifiers and correctly allocate incentives to each participant.

Assumption 2 (Honest Majority Assumption): We assume there exists at least one honest verifier within the verifier set $\mathcal{V}$. Honest verifiers strictly follow the protocol rules when performing transaction verification and promptly submit valid fraud proofs upon detecting incorrect transactions. Moreover, we assume that these fraud proofs are guaranteed to be submitted within the predetermined challenge window period. Combined with the blockchain trust assumption, this ensures the correctness and reliability of the transaction verification outcomes.

Definition 1 (State Transition Tree, STT). *A State Transition Tree (STT) is a DAG-based structure that explicitly records the execution order of transactions and their corresponding system states. Each node represents either a transaction or a resulting system state, clearly linking transaction inputs to their outputs. The root hash of the STT summarizes the entire batch execution result.*

Definition 2 (State Transition Commitment, STC). *A State Transition Commitment (STC) is an on-chain commitment of the STT root hash submitted by executors or verifiers. It represents a cryptographic summary of off-chain executed transactions and serves both as proof of correctness and verification workload.*

Definition 3 (Anonymous Staking Identifier). *An Anonymous Staking Identifier for a verifier V_i is defined based on an anonymous staking protocol as an elliptic curve commitment:*

$$C_i = r_i \mathbf{G} + v \mathbf{H}$$

where v represents the staking amount, r_i is a randomly generated blinding factor, and $\mathbf{G}, \mathbf{H}$ are publicly known base points on the elliptic curve. Each verifier is required to use this anonymous staking identifier when participating in verification tasks, ensuring that the link between the verifier's real identity and staking amount cannot be externally traced. This effectively mitigates risks of bribery, attacks, or collusion by preserving verifier anonymity and security.

We abstract the functional modules of the blockchain-based computing verification model in Fig. 1, which includes three types of entities: Blockchain, Executor, Verifiers.

- **Blockchain**: The blockchain acts as the final arbitrator, handling transaction challenges initiated by verifiers, distributing incentives, and managing verifiers' anonymous identities.
- **Executor**: The executor executes transactions off-chain and submits state commitments to the blockchain for verification by verifiers.
- **Verifiers**: Verifiers independently validate transaction executions submitted by the executor, initiating on-chain challenges when discrepancies are detected to ensure authenticity and correctness.

The entire process consists of four stages?the workflow can be summarized in the following steps.

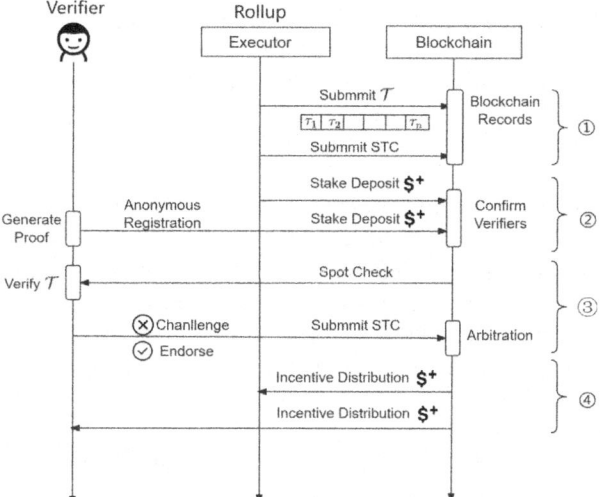

Fig. 1. AnoST Workflow

- ***Step 1 (Submitting Transactions):*** The executor completes the execution of the transaction set $\mathcal{T}$ off-chain and submits the corresponding state transition commitment to the blockchain, which records and stores this commitment.
- ***Step 2 (Staking Collateral):*** Both executors and verifiers are required to stake a certain amount of collateral to the blockchain to incentivize honest participation. The blockchain deducts this collateral if dishonesty or malicious behavior is later identified.
- ***Step 3 (Verifying Transactions):*** The system enters a verification window period, during which the blockchain randomly and anonymously schedules a set of verifiers $\mathcal{V}$ to independently verify the transaction set $\mathcal{T}$. Based on their independent verification outcomes, verifiers interact with the blockchain by either initiating explicit on-chain challenges if incorrect transactions are detected or submitting state commitments consistent with the executor.
- ***Step 4 (Distributing Incentives):*** Upon conclusion of the verification window, the blockchain aggregates the verification results submitted by executors and verifiers, subsequently executing collateral deductions and incentive distributions accordingly, thereby incentivizing honest behaviors and penalizing dishonest actions.

4 Anonymous Verifier Election

This section introduces the workflow of the Anonymous Verifier Election (AVE) mechanism, a critical component ensuring system security, verifier anonymity, and protection against targeted malicious attacks.

Intuition. The AVE protocol of AnoST effectively protects verifier identity privacy. Specifically, each user registering as a verifier submits anonymous stake commitments along with verifiable anonymous identities. Verifiers register with randomized identity tags to ensure both identity confidentiality and unpredictability. When interacting with the blockchain, verifiers must provide these anonymous identity proofs along with zero-knowledge membership proofs.

After staking, the system enters the verification phase, during which verifiers generate zk-SNARK proofs to demonstrate correctness of their staked node property information. In particular, when creating anonymous stake commitments, each verifier node generates a unique identifier C_{stake}, concealing real identities and serving as credentials for system interactions.

To verify the correctness of staked amounts, we utilize Pedersen commitments to conceal exact staking values. Leveraging the additive homomorphic property of Pedersen commitments, we construct a balancing term ΔC to ensure value conservation. Additionally, we introduce a unique staking identifier N_i to prevent double-spending attacks, thereby ensuring the uniqueness and security of each staking transaction.

Construction. The AVE protocol is illustrated in Fig. 2, where the underlined segments indicate specific implementation actions performed by each verifier. These actions will be elaborated in detail in the subsequent section. The AVE protocol comprises the following stages:

- **Initialization Phase**: The system is first initialized, establishing public data structures and public parameters. Participating verifiers then generate their associated cryptographic keys.
- **Staking and Registration Phase**: Verifier nodes intending to participate in anonymous staking apply to join the verification system. During the anonymous staking and registration process, an identity identifier is generated. Successfully staked nodes are then aggregated into a global verifier list, from which verifiers are selected.
- **Verification Phase**: The system assigns a group of verifiers to each task. Assigned verifiers independently perform transaction verification. If discrepancies are detected, verifiers initiate an on-chain challenge. Based on the verification outcomes, the blockchain subsequently enforces rewards and penalties for the involved parties.

4.1 Protocol and Verifiers Interaction Details

The AVE protocol describes the overall process of anonymous verification of tasks. Next, we will give a detailed description of the key functions involved.

$$\underline{\mathsf{AnonyReg}(1^\lambda, pp) \rightarrow (C_i, Ukey_i, \pi_{\text{mix}})}$$

In the anonymous registration phase, each verifier needs to complete key generation, anonymous recharge, and system registration.

Protocol Anonymous Verification Mechanism about AnoST
For each verifier V_i to verify the $\mathcal{T} : \{\tau_0, \tau_1, \ldots, \tau_n\}$

★ **Initialization phase**
◇ **System Intilize para and state**
- Initialize an empty Merkle tree $\mathcal{MKT} = \{\}$ for verifier membership proofs
- For each $\mathcal{T}$, initialize verifier list $\mathcal{VL}_\tau = \{\}$
- Initialize a mapping $\mathcal{DS}_\tau = \{\}$ to prevent double-spending
- Select elliptic curve base points $\mathbf{G}$, $\mathbf{H}$
- set basic stake amount v_{stake} about τ
- set $\mathcal{A}_\tau$ represents the stake address about $\mathcal{T}$
- call NIZK.KeyGen($1^\lambda, C_{N_i}, C_{\text{mkt}}, C_{\text{link}}$) → (zpk, zvk)
- set $pp \leftarrow (\mathcal{A}_\tau, zpk, zvk, v_{\text{stake}})$, set $st \leftarrow (\mathcal{MKT}, \mathcal{VL}_\tau, \mathcal{DS}_\tau)$

★ **Staking and Registration phase**
◇ **Each verifier V_i generates system parameters**
- V_i get the value from anonymous mixers
- AnonyReg($1^\lambda, pp$) → $(C_i, Ukey_i, \pi_{\text{mix}})$
- $1 \leftarrow$ NIZK.Vf(pp, π_{mix}), Ensure the Note C_i is correct
- If Verification passed: $\mathcal{VL}_\tau.append(C_i)$

◇ **Each verifier V_i stakes to participate in the election**
- AnonyStake($C_i, pp, UKey_i$) → $(\pi_{N_i}, C_{\text{change, stake, receipt}}, \Delta C)$
- Check $N_i \notin \mathcal{DS}_\tau$, if it exists, exit the process directly
    // *In order to avoid double-spending attacks*
- $1 \leftarrow$ NIZK.Vf(pp, π_{N_i}), If the verification fails, exit directly
- $\mathcal{DS}_\tau.append(N_i)$
- Check $C_i + \Delta C == C_{\text{stake}} + C_{\text{change}}$ or $C_i == C_{\text{stake}} + C_{\text{change}} + \Delta C$
    // *Checking the conservation of amount*
- Check $C_{\text{stake}} == C_{receipt} + v_{stake}\mathbf{H}$
- The pledged funds are deposited into the system fund pool
- $\mathcal{VL}_\tau.append(C_{\text{stake}})$, Regard the C_{stake} as new C_i

★ **Verification and Scheduling phase**
◇ **Schedule verifiers to verify computation tasks τ**
- Sample R from random beacon, set R as the verifier number
- $\mathcal{VL}'_\tau \leftarrow Filter(\mathcal{VL}_\tau, R)$, Compute Merkle tree commiment as $\mathcal{H}_{root}$

◇ **For each verifier V_i in $\mathcal{VL}'_\tau$ to verify tasks τ**
- Endorse($pp, st, C_{\text{stake}}, \mathcal{H}_{root}$) → Res
- Authenticate the identity of anonymous verifiers
- Parse evidence parameter $Res.\pi$ as (π_{path}, π_{link})
- Verify merkle-tree path: $1 \leftarrow$ NIZK.Vf(pp, π_{path})
- Verify linkability proof: $1 \leftarrow$ NIZK.Vf(pp, π_{link})
- Output 1 if and only if the above conditions is true
- Otherwise, output 0 // *The authentication is incorrect*

Fig. 2. Anonymous Verification Election Mechanism AVE

(1) **Key Generation.** New verifiers are required to complete identity registration before joining the system. Each verifier generates a personal key pair

and a random scalar r_i, which is used to construct a Pedersen commitment and also serves as part of the verifier's identity.

(2) **Anonymous Registration.** To prevent Sybil attacks, the system requires verifiers to deposit a certain amount of funds before they can participate. Verifiers anonymously deposit their funds into the system via a mixer, which breaks the traceability between the fund source and the verifier's identity. During this process, the verifier generates a commitment $C_i = r_i \mathbf{G} + v_i \mathbf{H}$ along with a zero-knowledge proof of the committed amount $\pi_{\text{mix}} = \mathsf{NIZK}(\exists v_i = \text{mix amount}, \text{s.t. } C_i = r_i \mathbf{G} + v_i \mathbf{H})$. Upon successful verification, C_i is bound to verifier V_i as their identity representation.

$$\text{AnonyStake}(C_i, pp, UKey_i) \rightarrow (\pi_{N_i}, C_{\text{change, stake}}, \Delta \text{ C})$$

(1) **Stake Preparation.** Verifier selects input note C_i, which representing the current balance and identity state. Based on the staking parameters, they compute two new output notes: the staking note $C_{\text{stake}} = r_{\text{stake}} \mathbf{G} + v_{\text{stake}} \mathbf{H}$ and the change note $C_{\text{change}} = r_{\text{change}} \mathbf{G} + v_{\text{change}} \mathbf{H}$, ensuring value conservation and uniqueness of the commitments.

(2) **Proof Generation.** The verifier first computes the nullifier to prevent double-spending $N_i = \mathsf{H}_1(r_i, C_i, index)$. Then, a zero-knowledge proof $\pi_{N_i} = \mathsf{NIZK}(\exists r_i \text{ s.t. } N_i = \mathsf{H}_1(r_i, C_i, index(C_i)))$ is generated to prove ownership of the input note C_i and the correctness of the output amounts without revealing any sensitive information.

(3) **Staking and Broadcasting.** The prover packages the transaction and broadcasts it to the blockchain. The transaction includes the stake note C_{stake}, change note C_{change}, the nullifier N_i, and the corresponding zero-knowledge proof π_{N_i}, ensuring both transaction validity and anonymity. At the same time, a balance item $\Delta C = r' \mathbf{G}$ is calculated to verify the correctness of the pledge.

(4) **Verification and State Update.** The system verifies the validity of the zero-knowledge proof and checks whether the nullifier N_i has been used before. If valid, the staking note is accepted as locked capital, and the system state is updated accordingly.

In summary, the verifier anonymously stakes assets by splitting an input note C_i into a staking note C_{stake} and a change note C_{change}. A nullifier N_i and zero-knowledge proof are generated to ensure correctness and prevent double-spending without revealing private information.

$$\text{Endorse}(pp, st, C_{\text{stake}}, \mathcal{H}_{root}) \rightarrow Res$$

(1) **Identity Proof Generation.** Verifiers generate a Merkle-tree existence proof π_{path} to ensure that only authorized verifiers can participate in verifying the computational task. Subsequently, verifiers produce a linkability proof $\pi_{link} = \mathsf{NIZK}\{\exists r_{stake}, \text{ s.t. } C_{stake} = r_{stake} \mathbf{G} + v_{stake} \mathbf{H}\}$ to confirm that the current C_{stake} was indeed created by themselves. This process effectively prevents impersonation while preserving verifiers' anonymity.

(2) **Task Verification.** Upon receiving a task requiring verification, verifiers immediately initiate an on-chain challenge if they detect any incorrect transactions within the task. The correctness of the challenged transaction is subsequently determined on-chain.

4.2 Protocol Correctness Explanation

Anonymity Explanation. During the execution of the protocol, each verifier generates an identity commitment C_i using a Pedersen commitment, which serves as their anonymous identity tag. At the same time, a network relay is used during blockchain interactions to hide the source address of transactions. When submitting verification results, the verifier uses a zk-SNARK to prove knowledge of the corresponding random scalar r_i, thereby achieving identity linkage. Throughout the entire protocol, no real-world identity information of the verifier is ever exposed.

Staking Correctness Explanation. Verifiers participate in the election by staking funds. According to the principle of value conservation, we have $v_{\text{change}} = v_i + v_{\text{stake}}$. Meanwhile, the verifier computes a balancing term ΔC to ensure that the randomness in the commitments is not reduced. The system verifies the following equation based on the additive homomorphism of Pedersen commitments:

$$C_i + \Delta C = C_{\text{stake}} + C_{\text{change}}$$

This can be further expanded as:

$$(r_i + r')\mathbf{G} + v_i\mathbf{H} = (r_{\text{stake}} + r_{\text{change}})\mathbf{G} + (v_{\text{stake}} + v_{\text{change}})\mathbf{H}$$

To prevent double-spending attacks, each verifier must submit a proof N_i bound to their identity and C_i before initiating a transaction. The system checks for the existence of N_i to ensure it has not been reused.

5 Commitment and Challenge Verification

In this section, we introduced the State Transition Commitment (STC) mechanism, which effectively verifies verifiers' workloads and addresses the verifier's dilemma. Furthermore, based on the State Transition Tree (STT), we proposed a refined challenge and pruning mechanism, enabling precise rollback of erroneous transactions and efficient confirmation of valid ones, significantly enhancing system verification performance.

5.1 Overview

Optimistic verification security heavily relies on verifiers actively checking off-chain transactions. However, since executors typically submit correct transactions due to strict staking penalties, rational verifiers may avoid costly verification, adopting a strategy called *free-riding*. This behavior, termed the *Verifier's*

Dilemma, can severely undermine security if incorrect transactions escape detection.

To resolve this, we propose a *State Transition Commitment (STC)* mechanism based on a *State Transition Tree (STT)*–a DAG structure explicitly recording transaction state transitions. Executors and verifiers independently compute and submit state transition commitment hashes along with random seeds to the blockchain. Commitment consistency is efficiently verified through hash comparison without costly on-chain re-execution, addressing the verifier's dilemma.

Intuition: The dilemma occurs because blockchains economically cannot verify whether verifiers genuinely performed validations. Our STC mechanism requires verifiers explicitly record transaction input-output states, providing verifiable proof of their verification work.

Specifically, executors first submit an STC hash and random seed. During the verification window, verifiers independently validate transactions, submitting their own STC hashes and seeds, and initiating on-chain challenges if discrepancies occur. After revealing seeds, the blockchain resolves inconsistencies; otherwise, the batch is confirmed correct. This ensures genuine verifier participation.

Based on the **Anonymous Verifier Election** mechanism, we assume verifiers independently verify transactions without collusion, ensuring objectivity and simplifying consistency checks.

Thus, the STC framework explicitly records input-output states as execution commitments on-chain, significantly reducing verification overhead. The detailed mechanism is presented subsequently.

5.2 Mechanism Design and Workflow Description

To clearly illustrate the detailed execution process of the STC mechanism, this section elaborates on constructing the State Transition Tree (STT) and the corresponding on-chain verification procedure. Specifically, the STC mechanism precisely records off-chain transaction state changes and employs efficient hash-consistency verification on-chain, significantly reducing verification costs while ensuring transaction correctness.

The STC mechanism comprises three stages: (1) STT construction, (2) independent verifier commitments, and (3) on-chain hash verification and challenges.

– **Stage 1: STT Construction**

  The off-chain executor strictly records input-output states for each transaction in the batch awaiting verification, specifying dependencies and execution order. After executing all transactions, the executor constructs a complete State Transition Tree ($\mathcal{T}_{\text{exec}}$), comprising *State Nodes* and *Transaction Nodes*. Each *State Node* records the resulting state after transaction execution, defined as:
$$H(S_i) = \mathcal{H}(S_i^{\text{data}})$$

Each *Transaction Node* represents individual transaction execution, containing detailed transaction data (e.g., function calls, parameters), explicitly linking input and output state nodes. Its hash is defined as:

$$H(\tau_i) = \mathcal{H}(\tau_i \parallel H(S_i^{\text{in}}) \parallel H(S_i^{\text{out}}))$$

Here, τ_j ensures transaction uniqueness (e.g., transaction signature or hash of the content itself), enabling explicit identification and verification. All nodes collectively form a DAG, constituting the STT. The executor then calculates node hashes from bottom-up, obtaining the STT root hash ($h_{\text{exec}} = \mathcal{R}_{\mathcal{H}}(\mathcal{T}_{\text{exec}})$), submitted to the on-chain smart contract.

- **Stage 2: Independent Verifier Commitments**
  During the verification window, each verifier independently executes off-chain verification tasks to confirm transaction correctness. Each verifier constructs its own STT ($\mathcal{T}_{v_i}$) following the same method as the executor, calculates hashes for nodes, and submits the resulting root hash to the blockchain:

$$h_{v_i} = \mathcal{R}_{\mathcal{H}}(\mathcal{T}_{v_i})$$

Independent root hash submissions explicitly demonstrate verifiers' verification effort, deterring cheating.

- **Stage 3: On-chain Hash Verification and Challenges**
  After the verification window, the on-chain smart contract automatically checks the consistency of STT root hashes submitted by executors and verifiers. If verifier hashes match the executor's hash (i.e., $h_{v_i} = h_{\text{exec}}$), transaction correctness is confirmed without further action. Discrepancies indicate potential transaction errors, triggering an on-chain challenge resolution process. Detailed challenge handling is elaborated in the next section.

5.3 State Pruning and Partial Transaction Confirmation

Traditional optimistic verification mechanisms typically reject an entire transaction batch upon detecting any single erroneous transaction, significantly limiting transaction acceptance rates. To address this, we propose a *state pruning and partial transaction confirmation mechanism* based on the State Transition Tree (STT), allowing selective rollback of erroneous transactions without impacting correct transactions.

Specifically, after the verification window period, the blockchain checks the consistency of STT root hashes submitted by verifiers and executors. If the root hashes match (i.e., $h_{v_i} = h_{\text{exec}}$), the batch correctness is confirmed. If discrepancies arise, indicating potential erroneous transactions, a verifier submits a path proof $\pi(\tau_k)$ targeting the disputed transaction node τ_k. This proof includes hashes along the path from the STT root node to the challenged node, together with sibling nodes at each layer.

Upon receiving the proof, the on-chain smart contract reconstructs the root hash from the challenged node upwards. If the reconstructed root hash matches the executor's submitted root hash, the verifier's challenge is invalid, resulting in stake forfeiture. If not, the executor is confirmed malicious, and the verifier receives the executor's stake as compensation.

Once transaction τ_k is confirmed erroneous, the smart contract initiates the state pruning procedure, removing τ_k and all subsequent dependent nodes from the executor's STT ($\mathcal{T}_{\text{exec}}$). It then recalculates hashes from bottom to top, deriving a new, valid STT root hash h'_{exec}.

The updated root hash h'_{exec} is publicly announced, ensuring all off-chain nodes quickly synchronize to the latest correct state. Our refined on-chain challenge and pruning mechanism selectively rolls back only erroneous transactions, enabling unaffected transactions to be smoothly validated and confirmed. This significantly improves transaction throughput, reduces on-chain validation and storage overhead, and enhances overall system performance and scalability.

5.4 Protocol Correctness Analysis

We now rigorously analyze the correctness and verifiability of the proposed protocol based on previously defined notations and concepts. Given a transaction batch $\mathcal{T} = \{\tau_i\}_{i=1}^n$, the executor constructs the corresponding State Transition Tree $\mathcal{T}_{\text{exec}}$, submitting its root hash commitment $h_{\text{exec}} = \mathcal{R}_{\mathcal{H}}(\mathcal{T}_{\text{exec}})$ to the blockchain. Each verifier v_i independently verifies the transaction batch $\mathcal{T}$, constructing their own State Transition Tree $\mathcal{T}_{v_i}$ and submitting the root hash $h_{v_i} = \mathcal{R}_{\mathcal{H}}(\mathcal{T}_{v_i})$.

As defined previously, the hash of each transaction node explicitly depends on its input and output state node hashes:

$$H(\tau_j) = \mathcal{H}(\tau_j \parallel H(S_j^{\text{in}}) \parallel H(S_j^{\text{out}})),$$

while each state node directly encodes the post-execution state data:

$$H(S_j) = \mathcal{H}(S_j^{\text{data}}).$$

Thus, verifiers must genuinely verify each transaction $\tau_j \in \mathcal{T}$ to correctly derive intermediate states S_j^{data} and compute a consistent root hash h_{v_i}. Failure to perform complete verification inevitably leads to discrepancies between the verifier's root hash and the executor's commitment, efficiently detectable during on-chain consistency checks. Formally:

$$h_{v_i} \neq h_{\text{exec}} \iff \exists \tau_j, S_j^{\text{in}}, S_j^{\text{out}}, (S_j^{\text{in}}, \tau_j, S_j^{\text{out}}) \notin \mathcal{T}_{v_i}.$$

Additionally, our experimental section introduced a *path-inclusion detection mechanism* to further ensure accurate and efficient on-chain challenges. Specifically, transaction batches may contain multiple dependent transaction paths. When a verifier successfully challenges a transaction τ_x within a path $\mathcal{P}(\tau_x) = \{\tau_1, \ldots, \tau_x, \ldots, \tau_m\}$, all subsequent dependent transactions (τ_y, where $x < y \leq m$) on that path are automatically invalidated. Formally, we denote this property as:

$$\forall \tau_x, \tau_y \in \mathcal{P}, x < y, \quad \tau_y \text{ is invalidated.}$$

This means subsequent challenges targeting transactions following an already successfully challenged transaction in the same path will be ignored by the blockchain, preventing redundant verification efforts.

The correctness of this path detection mechanism inherently leverages the STT structure: each transaction node explicitly depends on the output state of its preceding node. Hence, if a preceding node τ_x is identified as erroneous, the output states of subsequent nodes (τ_y with $y > x$) become invalid by definition. Consequently, straightforward on-chain path detection effectively prevents redundant challenges, significantly reducing verification overhead and improving challenge efficiency.

Finally, the anonymous verifier election mechanism introduced earlier inherently restricts effective communication among verifiers, fundamentally preventing collusion risks. Consequently, our mechanism explicitly ensures verifiers' workload verifiability, completely eliminating incentives and possibilities for free-riding or lazy verification, significantly enhancing the security, correctness, and efficiency of the system even under malicious conditions.

6 Security Analysis

In this section, we rigorously analyze the security properties of AnoST, focusing on three essential guarantees: verifier anonymity, anti-collusion (workload verifiability), and game-theoretic resolution of the Verifier's Dilemma.

6.1 Verifier Anonymity

Definition 4 (Verifier Anonymity). *Let $\mathcal{A}$ denote a probabilistic polynomial-time adversary. We define verifier anonymity via the following indistinguishability game:*

1. *The challenger chooses random commitments (C_0, C_1), corresponding to distinct verifiers (V_0, V_1) with commitments $C_i = r_i \mathbf{G} + v\mathbf{H}$.*
2. *The challenger samples a random bit $b \xleftarrow{\$} \{0,1\}$ and provides C_b to $\mathcal{A}$.*
3. *Adversary $\mathcal{A}$ outputs a guess $b' \in \{0,1\}$.*

AnoST achieves verifier anonymity if for all PPT adversaries $\mathcal{A}$,

$$\left| \Pr[b' = b] - \frac{1}{2} \right| \leq negl(\lambda)$$

where λ is the security parameter.

Theorem 1 (Verifier Anonymity). *AnoST guarantees verifier anonymity under the hardness assumption of the elliptic curve discrete logarithm problem (ECDLP).*

Proof. Verifier anonymity in AnoST fundamentally relies on the Pedersen commitment $C = r\mathbf{G}+v\mathbf{H}$, which is known to provide unconditional hiding. To break anonymity, adversary $\mathcal{A}$ must distinguish between two commitments. Since commitments are perfectly hiding and independent of verifier identity, distinguishing C_0 from C_1 requires the adversary to compute r_i given C_i, thus solving the discrete logarithm of elliptic curve group $\mathbb{G}$. Given ECDLP is computationally infeasible, the adversary's advantage must remain negligible. Hence, AnoST preserves verifier anonymity.

6.2 Anti-collusion: Workload Verifiability

Definition 5 (Workload Verifiability). *Consider a transaction set $T = \{\tau_i\}_{i=1}^n$, where each verifier V_j independently computes a State Transition Tree (STT) T_{V_j} with root hash h_{V_j}. The system ensures workload verifiability if:*

$$h_{V_j} = RH(T_{exec}) \iff \forall \tau_i \in T, V_j \text{ executed } \tau_i \text{ correctly}$$

Theorem 2 (Workload Verifiability). *AnoST achieves workload verifiability assuming collision-resistance and pre-image resistance of the cryptographic hash function.*

Proof. To generate a valid STT root hash h_{V_j} identical to the executor's h_{exec}, a verifier V_j must correctly compute every intermediate state S_i^{out} for each transaction $\tau_i \in T$. The STT node hashes are computed as follows:

$$H(\tau_i) = H(\tau_i \| H(S_i^{in}) \| H(S_i^{out}))$$

If a verifier attempts to bypass actual computation, it must guess or reproduce the exact output state S_i^{out} for each transaction. Given the pre-image resistance and collision-resistance of the hash function, the probability of correctly guessing these outputs without actual computation is negligible. Consequently, generating a consistent h_{V_j} without full verification is computationally infeasible. Thus, AnoST guarantees workload verifiability.

6.3 Resolution of the Verifier's Dilemma

To demonstrate AnoST's resolution of the Verifier's Dilemma, we construct a game-theoretic model to analyze verifier incentives.

Definition 6 (Verifier Game). *We define a strategic game $G = (V, \{S_i\}, \{u_i\})$, where:*

- *$V = \{V_1, \ldots, V_n\}$ is the set of verifiers.*
- *Each verifier V_i selects a strategy $s_i \in S_i = \{\text{Verify}, \text{Skip}\}$.*
- *Payoff function $u_i : S \to \mathbb{R}$ is defined as follows:*
  - *$u_i(\text{Verify}) = r - c$, where r is the verification reward, and c is the verification cost.*

- $u_i(\mathsf{Skip}) = r - P_{detect} \cdot p$, where p is the penalty if the skipping is detected and P_{detect} is the detection probability.

Theorem 3 (Verifier's Dilemma Resolution). *If the system parameters satisfy $P_{detect} \cdot p > c$, then Verify is the strictly dominant strategy, and AnoST resolves the Verifier's Dilemma.*

Proof. To prove this, we calculate the expected payoffs explicitly. A verifier selecting Skip obtains an expected payoff:

$$\mathbb{E}[u_i(\mathsf{Skip})] = r - P_{detect} \cdot p$$

A verifier selecting Verify has a deterministic payoff:

$$u_i(\mathsf{Verify}) = r - c$$

The rational verifier will choose Verify if and only if:

$$r - c > r - P_{detect} \cdot p \quad \Rightarrow \quad P_{detect} \cdot p > c$$

Thus, under the given incentive parameters, the dominant strategy equilibrium is all verifiers performing verification. Consequently, AnoST economically enforces honest verification and resolves the Verifier's Dilemma.

Through this rigorous analysis, we demonstrate that AnoST provides robust security guarantees, ensuring verifier anonymity, effective workload verifiability, and economically-driven resolution of the Verifier's Dilemma.

7 Implementation

The primary evaluation objective of AnoST is to demonstrate its core benefits in precise rollback capabilities, verifier anonymity protection, and the resolution of the Verifier's Dilemma. Our experiments are structured around verifying these security and mechanism effectiveness goals, rather than maximizing general performance metrics such as transaction throughput.

We benchmarked the circuit compilation and proof generation overhead of AnoST's zero-knowledge circuits ($\pi_{\mathrm{mix}}, \pi_{N_i}, \pi_{link}$) using an Intel i5-12400H CPU with 32GB RAM, Circom, and snarkJS. Proofs were generated in a single-threaded environment. On-chain verification times were measured on a blockchain simulation node (AMD EPYC 7302P, 64GB RAM). Results summarized in Table 2 confirm that the computational overhead is moderate, demonstrating practical deployability.

We evaluated AnoST's on-chain gas consumption in the Ethereum Virtual Machine (EVM) environment. Figure 3 presents the gas costs breakdown for verifier committee election without zero-knowledge aggregation, scaling the number of verifiers from 50 to 200. Our results illustrate that AnoST maintains efficient

Table 2. Zero-Knowledge Proof Performance Overhead

Proof Task	Gen Time	Prove Time	Set Time
π_{mix}	297 ms	2.4 ms	1.2 s
π_{N_i}	1.43 s	1.3 s	5.2 s
π_{link}	286 ms	2.3 ms	1.1 s

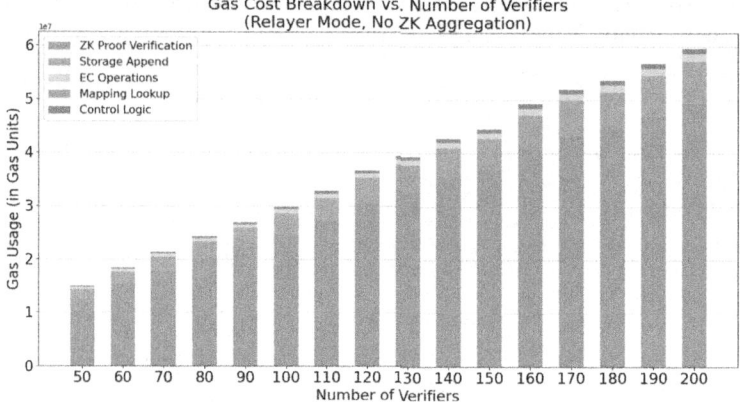

Fig. 3. AnoST Committee Election Gas Cost On Chain

on-chain execution costs, approximately 0.005\$ per execution on Binance Smart Chain (BSC) and 0.0024\$ on Polygon, highlighting its economic feasibility.

To evaluate the effectiveness of the State Transition Tree (STT) mechanism, we randomly selected 50 transaction batches (2000 transactions each) from Ethereum mainnet, simulating varying erroneous transaction ratios. Table 3 summarizes STT construction times, off-chain storage overhead, on-chain pruning times, and transaction acceptance rates.

Table 3. STT Partial Rollback Performance

Error Ratio (%)	Build STT (s)	Storage Cost (KB)	Rollback Time (s)	Pass Rate (%)
0	2.02	148.3	0.00	100.0
5	2.14	152.7	1.72	87.8
10	2.12	157.0	2.41	82.9
15	2.05	162.5	3.07	76.1
20	2.15	168.2	3.88	68.3

The results demonstrate that STT provides efficient, precise rollback capability, significantly maintaining higher acceptance rates compared to full-batch rollback strategies. Despite minor increases in storage overhead and pruning time, the STT mechanism substantially reduces overall on-chain computational load.

We simulated adversarial environments by introducing malicious verifiers into batches of 2000 transactions, testing scenarios with 5%, 10%, and 20% transaction errors. Four defensive mechanisms were evaluated: (1) No Optimization, (2) Path Checking (eliminating redundant challenges via transaction dependency tracking), (3) Blacklist (blocking malicious verifiers), and (4) Combined Optimization (Path Checking and Blacklist).

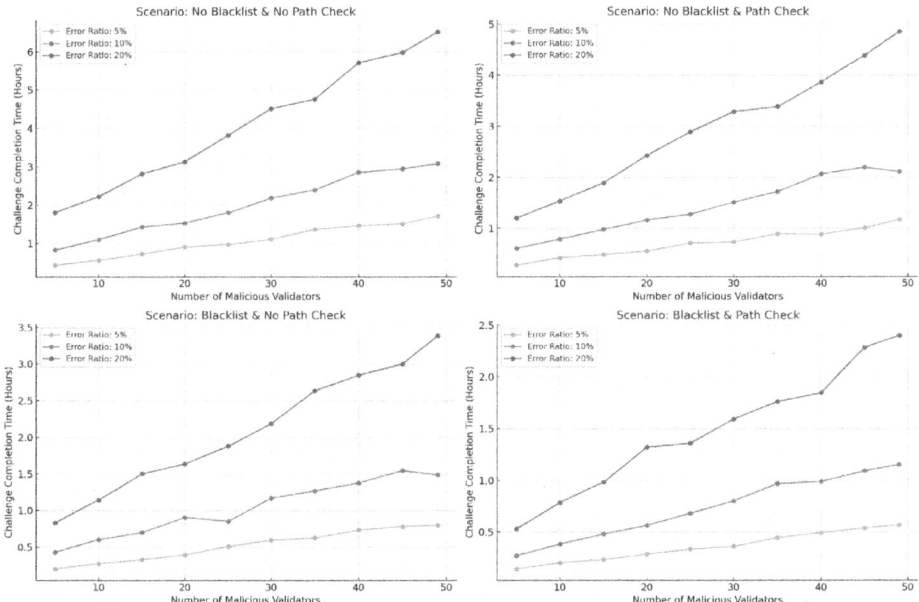

Fig. 4. Challenge Window Duration under Various Optimization Mechanisms

Figure 4 clearly illustrates that the combined optimization strategy effectively reduces the challenge window by up to 65% compared to unoptimized scenarios. This efficiency improvement indicates AnoST's robust defense capability against typical adversarial behaviors encountered in real-world optimistic rollup deployments.

8 Conclusion and Future Work

This paper presents **AnoST**, an anonymous optimistic verification framework for rollup-based blockchains. By integrating anonymous staking-based verifier election with state-centric verification primitives–namely, the State Transition Commitment (STC) and State Transition Tree (STT)–AnoST addresses verifier identity exposure, the Verifier's Dilemma, and coarse-grained rollback limitations. Experimental results confirm its strong anonymity guarantees, enhanced verifier participation, and scalable, fine-grained rollback support.

Future extensions include enhancing AnoST against adaptive verifier collusion, designing robust incentive-compatible reward models, and addressing integration challenges in real-world rollup systems (e.g., Arbitrum, Optimism). These efforts aim to advance AnoST toward scalable and privacy-preserving Layer 2 deployments.

Acknowledgments. This paper is supported by the National Key R&D Program of China through project 2022YFB2702900, the Natural Science Foundation of China through projects U21A20467, U24B20144 and 62272464.

References

1. Aztec Protocol: Aztec protocol: Privacy-first zkrollup on ethereum (2021). https://github.com/AztecProtocol/AZTEC/blob/master/AZTEC.pdf
2. Bach, L.M., Mihaljevic, B., Zagar, M.: Comparative analysis of blockchain consensus algorithms. In: 2018 41st International Convention on Information and Communication Technology, Electronics and Microelectronics (MIPRO), pp. 1545–1550. IEEE (2018)
3. Backes, M., Bugiel, S., Derr, E.: Reliable third-party library detection in android and its security applications. In: Proceedings of the 2016 ACM SIGSAC Conference on Computer and Communications Security (CCS), pp. 356–367. ACM (2016)
4. Baliga, A.: Understanding blockchain consensus models. Persistent **4**(1), 14 (2017)
5. Introducing Ethereum and Solidity. Apress, Berkeley, CA (2017). https://doi.org/10.1007/978-1-4842-2535-6_9
6. Pierro, M.: What is the blockchain? Comput. Sci. Eng. **19**(5), 92–95 (2017)
7. Du, W., Atallah, M.J.: Privacy-preserving cooperative scientific computations. In: Proceedings of the 14th IEEE Computer Security Foundations Workshop (CSFW), pp. 273–282. IEEE (2001)
8. Ethereum Foundation: Optimistic rollups (2025). https://ethereum.org/en/developers/docs/scaling/optimistic-rollups/
9. Göbel, J., Krzesinski, A.E.: Increased block size and bitcoin blockchain dynamics. In: 2017 27th International Telecommunication Networks and Applications Conference (ITNAC), pp. 1–6. IEEE (2017)
10. Guo, S., Hu, X., Guo, S., Qiu, X., Qi, F.: Blockchain meets edge computing: a distributed and trusted authentication system. IEEE Trans. Industr. Inf. **16**(3), 1972–1983 (2020)
11. Kalodner, H., Goldfeder, S., Chen, X., Weinberg, S.M., Felten, E.W.: Arbitrum: scalable, private smart contracts. In: 27th USENIX Security Symposium (USENIX Security 18), pp. 1353–1370 (2018)
12. Kissner, L., Song, D.: Privacy-preserving set operations. In: Shoup, V. (ed.) CRYPTO 2005. LNCS, vol. 3621, pp. 241–257. Springer, Heidelberg (2005). https://doi.org/10.1007/11535218_15
13. Landis, D.: Incentive non-compatibility of optimistic rollups. arXiv preprint arXiv:2312.01549 (2023)

14. Lashkari, B., Musilek, P.: A comprehensive review of blockchain consensus mechanisms. IEEE Access **9**, 43620–43652 (2021)
15. Li, J.: On the security of optimistic blockchain mechanisms. In: Working Paper, SSRN (2023)
16. Luu, L., Teutsch, J., Kulkarni, R., Saxena, P.: Demystifying incentives in the consensus computer. In: Proceedings of the 22nd ACM SIGSAC Conference on Computer and Communications Security (CCS 2015), pp. 706–719. ACM (2015)
17. McAllister, L.K.: Regulation by third-party verification. Boston Coll. Law Rev. **53**, 1–70 (2012)
18. Min, X., Li, Q., Liu, L., Cui, L.: A permissioned blockchain framework for supporting instant transaction and dynamic block size. In: 2016 IEEE Trustcom/BigDataSE/ISPA, pp. 90–96. IEEE (2016)
19. Nguyen, D.C., et al.: Federated learning meets blockchain in edge computing: opportunities and challenges. IEEE Internet Things J. **8**(16), 12806–12825 (2021)
20. Panait, A.E., Olimid, R.F., Stefanescu, A.: Identity management on blockchain – privacy and security aspects. arXiv preprint arXiv:2004.13107 (2020)
21. Project, C.: Cartesi rollups: a modular execution stack for scalable DApps. Whitepaper / technical report, Cartesi Foundation (2022). https://docs.cartesi.io/cartesi-rollups/1.0/
22. Qu, Y., Uddin, M.P., Gan, C., Chen, X., Zhang, L.: Blockchain-enabled federated learning: a survey. ACM Comput. Surv. **55**(4), 1–35 (2022)
23. Salman, T., Zolanvari, M., Erbad, A., Jain, R., Samaka, M.: Security services using blockchains: a state of the art survey. arXiv preprint arXiv:1810.08735 (2018)
24. Schär, F.: Decentralized finance: on blockchain- and smart contract-based financial markets. Fed. Reserve Bank St. Louis Rev. **103**(2), 153–174 (2021)
25. Setty, S., Vu, V., Panpalia, N., Braun, B., Blumberg, A.J., Walfish, M.: Taking proof-based verified computation a few steps closer to practicality. In: 21st USENIX Security Symposium, pp. 253–268. USENIX Association (2012)
26. Shayan, M., Fung, C., Yoon, C.J.M., Beschastnikh, I.: Biscotti: a blockchain system for private and secure federated learning. IEEE Trans. Parallel Distrib. Syst. **32**(7), 1513–1525 (2021)
27. Thibault, L.T., Sarry, T., Hafid, A.S.: Blockchain scaling using rollups: a comprehensive survey. IEEE Access **10**, 93039–93054 (2022)
28. Xiong, Z., Zhang, Y., Niyato, D., Wang, P., Han, Z.: When mobile blockchain meets edge computing. IEEE Commun. Mag. **56**(8), 33–39 (2018)
29. Xu, J., Wang, C., Jia, X.: A survey of blockchain consensus protocols. ACM Comput. Surv. **55**(13s), 1–35 (2023)
30. Yang, R., Yu, F.R., Si, P., Zheng, Z., Zhang, Y.: Integrated blockchain and edge computing systems: a survey, some research issues and challenges. IEEE Commun. Surv. Tutorials **21**(2), 1508–1532 (2019)
31. Yu, X., Yan, Z., Vasilakos, A.V.: A survey of verifiable computation. Mobile Netw. Appl. **22**(3), 438–453 (2017)
32. Zheng, Z., Xie, S., Dai, H.N., Chen, X., Wang, H.: Blockchain challenges and opportunities: a survey. Int. J. Web Grid Serv. **14**(4), 352–375 (2018)

Privacy-Preserving K-Hop Shortest Path Query on Encrypted Graphs Based on Graph Pruning

Ya Gao[1,2], Chao Mu[1,2(✉)], Ming Yang[1,2], and Xiaoming Wu[1,2]

[1] Key Laboratory of Computing Power Network and Information Security, Ministry of Education, Shandong Computer Science Center, Qilu University of Technology (Shandong Academy of Sciences), Jinan 250014, People's Republic of China
[2] Shandong Provincial Key Laboratory of Computer Networks, Shandong Fundamental Research Center for Computer Science, Jinan 250014, People's Republic of China
muchao@sdas.org

Abstract. K-hop Shortest Path Query (KSPQ) is a fundamental graph operation that aims to find the shortest path between two nodes within K hops. As graph data scales, owners outsource storage and computation to cloud servers, using encryption and pruning to ensure privacy and efficiency. However, integrating privacy protection and graph pruning in encrypted environments remains unexplored, and existing methods struggle to balance security and query efficiency. To address this, we propose a Graph Pruning-Based Privacy-Preserving K-hop Shortest Path Query (PP-KSPQ) method, emphasizing three core features: 1. Graph Preprocessing: Simplifies the graph by removing edges and nodes that violate constraints, reducing the search space by up to 75%; 2. Accelerated Indexing: Employs the middle node cut technique to efficiently identify candidates and speed up path construction; 3. Encryption Integration: Advanced encryption techniques and Secure Integer Comparison Protocols are employed to enable efficient path queries while preserving privacy. Experiments validate that PP-KSPQ enhances both privacy protection and computational efficiency, demonstrating its effectiveness across real-world datasets.

Keywords: Cloud computing · Privacy-Preserving · K-hop Shortest Path Query · Graph Pruning · Encrypted Graph

1 Introduction

In the digital age, information security has become a cornerstone of modern data management and distributed computing. Social networks, transportation systems [19], and communication [27] infrastructures generate vast amounts of data, with cloud-based storage and computation serving as essential components of these ecosystems. However, this reliance brings significant challenges, including privacy breaches, cyber threats, and computational inefficiencies. To mitigate these risks, data owners frequently employ encryption techniques [24,30] and outsource storage and processing to cloud servers [15,25]. Despite these measures, executing efficient queries in encrypted environments remains a complex task due to three key obstacles: 1. High index construction costs âĂŞ The complexity of large-scale graph data demands substantial computational resources for index generation [8]. 2. Excessive storage overhead âĂŞ Encrypted queries require additional auxiliary data structures, increasing storage burdens [16]. 3. Privacy leakage risks âĂŞ Despite encryption, query execution may inadvertently reveal graph topology or node associations [3]. Against this backdrop, K-hop Shortest Path Query (KSPQ) [32] plays a pivotal role in information security applications, such as: 1. Privacy-preserving social networks, where restricting query scope minimizes the risk of exposing user relationships. 2. Encrypted transportation systems, where secure shortest path computations optimize navigation within constrained trust environments while maintaining confidentiality.

This work introduces the first-ever Privacy-Preserving K-hop Shortest Path Query (PP-KSPQ) [34] method based on graph pruning, designed to enhance query efficiency while strengthening data privacy protection. By integrating graph pruning techniques and secure computation mechanisms, the proposed method significantly improves performance in encrypted environments. Specifically, it incorporates:

1. By removing edges and nodes that fail to meet hop constraints, a simplified subgraph is created. This process drastically reduces the search space and minimizes the time and resources required for subsequent index construction and query operations.
2. Middle node cut techniques are employed to precisely identify candidate path nodes within the simplified graph. This approach eliminates redundant search operations, significantly improving the efficiency and accuracy of index construction.
3. Using encryption techniques and secure integer comparison protocols, this approach effectively ensures data privacy during the query process while meeting the security requirements of encrypted environments.

The structure of this paper is as follows: We summarize the related work in Sect. 2. We introduce models, definitions and preliminaries involved in our scheme in Sect. 2. Section 4 details the scheme's construction. The construction of our scheme is presented in Sect. 4. We give the security analysis in Sect. 5. In Sect. 6, we provide the performance analysis and evaluate the proposed scheme through experiments. Finally, we conclude this paper in Sect. 7.

2 Related Work

With the development of big data and cloud computing, graph data querying and processing have become research hotspots. Researchers have proposed various optimization algorithms to enhance path query efficiency [14,22,31]. As the scale and complexity of graph data increase, the demand for efficient path query algorithms is growing. Data outsourcing enables organizations to store and process large volumes of data through cloud platforms, reducing costs.

In the context of graph data outsourcing, how to handle encrypted data has become a key issue. Researchers [7]are dedicated to designing efficient algorithms that enable fast querying while ensuring data privacy. For example, Zhang et al. [33] proposed a privacy-preserving graph encryption scheme for accurate constrained shortest distance queries. Liu et al. [12] explored techniques for strongly privacy-preserving shortest distance queries. Additionally, Kumar and Chand [9] introduced a publicly verifiable and efficient privacy-preserving system.

Additionally, K-hop shortest distance and shortest path queries have emerged as critical research directions within this context. For example, Song et al. [26] proposed a privacy-preserving k-hop reachability query method, Cheng et al. [3] examined efficient k-hop reachability query processing methods, and Zhao et al. [35] introduced a privacy-preserving arbitrary hop-covered shortest-distance query method. Lai et al. [10] developed an efficient k-hop constrained s-t simple path enumeration method on FPGA. These techniques aim to find the shortest distance or path in graph data that meets specific constraints (e.g. path length, number of nodes, and total weight) [6, 13, 20, 21, 23].

Despite significant progress in research on constrained shortest paths and shortest distances, the increasing scale and complexity of graph data [2, 28] make further efficiency improvements imperative. Therefore, it is crucial to continue to deepen the research on data encryption and fast index construction technologies [1]. In this context, techniques such as graph pruning and bidirectional index construction are essential for enhancing query efficiency. Graph pruning effectively reduces the search space, thereby accelerating query speed. Additionally, research on dynamic graphs [4,5] has demonstrated its importance. Real-time update and query methods for dynamic graphs are continuously evolving, ensuring efficient path queries and data processing even as graph data changes.

3 Models, Definitions And Preliminaries

3.1 Models

System Model. It is shown in Fig. 1, which consists of four entities: *Data Owner*, *Data User*, and two *Cloud Servers*, S1 and S2.

1. ***Data Owner:*** Safeguards graph data and manages access, generating and distributing authorization tokens and private keys to ensure the security of cloud servers and data users. Upon receiving a data request, they construct a simplified graph, generate secure indexes, and send them to cloud server S1 for shortest path queries.
2. ***Data User:*** Requests queries and decrypts results. They obtain an authorization token, send queries to the Data Owner, and use the key set to decode encrypted shortest path data from the cloud server.
3. ***Cloud Server S1:*** Processes secure indexes and query requests, performing node matching, path concatenation, and weight accumulation. It employs

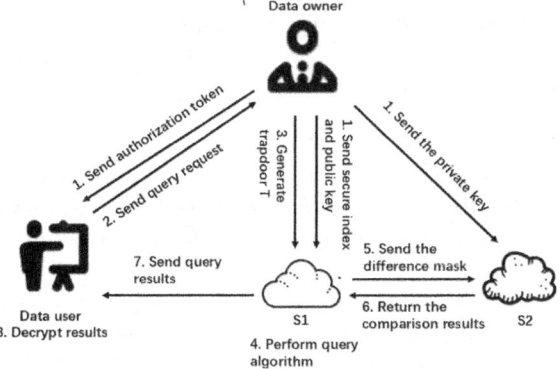

Fig. 1. System Model

a Secure Integer Comparison Protocol to protect weight privacy, integrates S2 comparison results, determines the shortest path, and transmits the final results to the Data User.

4. ***Cloud Server S2:*** Responsible for decrypting and comparing encrypted differences. It holds the Data Owner's private key, utilizes a Secure Integer Comparison Protocol to determine the relationship between c1 and c2, and returns the results to S1.

Threat Model. The system assumes that the Data Owner is trustworthy, securely distributes keys, and is resilient to attacks. Cloud servers S1 and S2 are considered "honest-but-curious" entities, meaning they adhere to protocols but may attempt to infer sensitive information. Our security objective is to safeguard the privacy of the outsourced graph G, ensuring that S1 and S2 cannot derive any information from the secure indexes, queries, or query results.

3.2 Definitions

Scheme Definition. This scheme includes five algorithms, among which:

1. **Key Generation Algorithm:** $(k_1, (pk, sk)) \leftarrow \text{KeyGen}(1^\lambda)$. It takes the security parameter λ as input and outputs the key k_1 and a public/secret key pair (pk, sk).
2. **Graph Encryption Algorithm:** $(I) \leftarrow \text{GraphEnc}(G, k_1, pk)$. It takes the graph G, the key k_1, and the public key pk as inputs, and outputs the secure index set I.
3. **Trapdoor Generation Algorithm:** $(T) \leftarrow \text{TokenGen}(k, s, t, k_1)$. It takes the key k_1 as input and outputs the query trapdoor T.
4. **Result Query Algorithm:** $(\text{Result}) \leftarrow \text{Query}(I, T, pk, sk)$. It takes the trapdoor T, the key pair (pk, sk), and the secure index set I as inputs, and outputs the query Result.
5. **Decryption Algorithm:** $(\text{result}) \leftarrow \text{Decrypt}(k_1, \text{Result})$. It takes the key k_1 as input and outputs the final decrypted result.

Security Definition. In our scheme, we adopt the security definition of Symmetric Searchable Encryption (SSE), which aligns with most privacy-preserving graph query schemes [11,17,33].

Definition 1. *Let $\Pi = (KeyGen, GraphEnc, TokenGen, Query, Decrypt)$ be a scheme for querying the shortest shortest path k hop constrained. The leakage functions are denoted as L_{setup} and L_{query}. Given the security parameter λ, an attacker $\mathcal{A}$, and a simulator $\mathcal{S}$, consider the following real and ideal experiments.*

$Real_\mathcal{A}(\lambda)$: The attacker $\mathcal{A}$ chooses a graph G and sends it to challenger $\mathcal{C}$. $\mathcal{C}$ runs the key generation algorithm KeyGen(1^λ) to generate key. Then, $\mathcal{C}$ constructs a secure index I by running the algorithm GraphEnc(G, k_1, pk) and sends I to $\mathcal{A}$. $\mathcal{A}$ performs a polynomial number of adaptive queries. For each query q, $\mathcal{C}$ generates the query trapdoor T by running the algorithm TokenGen(k, s, t, k_1) and sends T to $\mathcal{A}$. $\mathcal{A}$ executes Query(I, T, pk, sk). Finally, $\mathcal{A}$ outputs a bit $b \in \{0, 1\}$.

$Ideal_{\mathcal{A},\mathcal{S}}(\lambda)$: The attacker $\mathcal{A}$ chooses a graph G and sends it to the challenger $\mathcal{C}$. $\mathcal{C}$ receives the graph G and sends L_{setup} to the simulator $\mathcal{S}$. $\mathcal{S}$ uses L_{setup} to generate a secure index I_1 and sends it to $\mathcal{A}$. $\mathcal{A}$ performs a polynomial number of adaptive queries $q = (k, s, t)$ and sends them to $\mathcal{C}$. $\mathcal{C}$ sends the leakage function L_{query} to $\mathcal{S}$. $\mathcal{S}$ generates simulated query tokens T_1 and sends them to $\mathcal{A}$. $\mathcal{A}$ executes Query(I, T, pk, sk) to obtain the results. Finally, $\mathcal{A}$ outputs a bit $b \in \{0, 1\}$ as the result of the experiment.

We say that Π is (L_{setup}, L_{query})-secure against CQA2 attacks if for all probabilistic polynomial-time (PPT) adversaries A, there exists a PPT simulator S such that:

$$|Pr[Real_\mathcal{A}(\lambda) = 1] - Pr[Ideal_{\mathcal{A},\mathcal{S}}(\lambda) = 1]| \leq \text{negl}(\lambda) \tag{1}$$

where negl(λ) is a negligible function.

3.3 Preliminaries

Paillier. To perform weight calculations on the encrypted graph, we employ the Paillier homomorphic encryption algorithm [18,29] to encrypt distance data. This algorithm exhibits additive homomorphism, allowing addition operations to be conducted directly on ciphertext: $Enc(m_1) \cdot Enc(m_2) = Enc(m_1 + m_2)$. Hence, ensuring computational security and preserving data privacy.

Hash Encryption(Hash), Advanced Encryption Standard(AES). Hash encryption is a process that converts input data of any length into a fixed-length, irreversible hash value using a hash function, characterized by high sensitivity and one-way operation, ensuring that the original data cannot be derived from the output. AES (Advanced Encryption Standard) is a symmetric encryption algorithm that employs 128/192/256-bit keys and block encryption (128-bit blocks), utilizing multiple rounds of substitution and permutation to secure data–using the same key for both encryption and decryption.

Secure Integer Comparison Protocol. S1 generates a random mask and encrypts it using Paillier homomorphic encryption, then performs homomorphic addition with encrypted integers c_1 and c_2, producing masked encrypted values, which are sent to S2. S2 decrypts the received values using its private key, compares the masked numbers to infer the relationship between c_1 and c_2, and returns the comparison result to S1.

Throughout the process, neither S1 nor S2 can access the plaintext of the encrypted integers, ensuring strong privacy protection. Leveraging Paillier homomorphic encryption, this method enhances security and computational efficiency while maintaining accuracy in comparison operations. It is particularly suitable for privacy-preserving queries and secure data analysis scenarios.

4 Scheme Construction

4.1 Graph Pruning

Existing methods are based on full-graph traversal, increasing costs. We propose an optimized algorithm using advanced pruning to efficiently filter candidates, reducing time and computation [34].

Pruning Nodes that Do Not Satisfy the Hop Constraint k. Assuming the original graph G contains 10^6 nodes and 5^9 edges, directly searching for an s-t path that satisfies the hop constraint k across the entire graph results in extremely high computational complexity. By generating a simplified graph G', we can focus only on a portion of the original graph, significantly reducing the computational load and markedly improving the query speed.

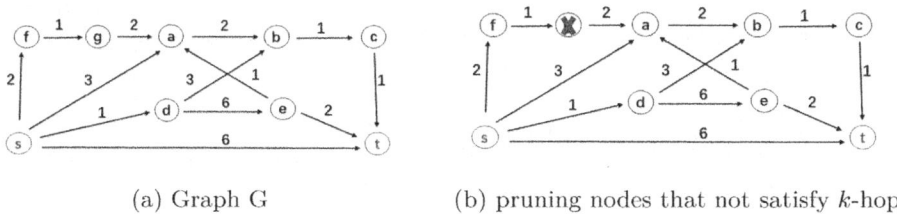

(a) Graph G (b) pruning nodes that not satisfy k-hop

Fig. 2. The original graph G and its simplified version G' after pruning nodes that do not satisfy the k-hop.

Theorem 1. *For the original graph G, calculate two minimum hop path edge count mapping tables: $Dist_s$ and $Dist_t$. $Dist_s$ records the edge count of the minimum path of the hop from the start node s to each nodes u, and similarly for $Dist_t$. Using $Dist_s$ and $Dist_t$ mapping tables, we determine all nodes u involved in the k-step path, i.e., the nodes set V':*

$$V' = \{u \mid u \in V(G) \wedge Dist_s[u] + Dist_t[u] \leq k\} \tag{2}$$

Proof. Suppose a path contains nodes u', but $Dist_s[u'] + Dist_t[u'] > k$. Therefore, the path $p = \{s, \ldots, u', \ldots, t\}$ has $len(p) > k$, which does not satisfy the hop constraint.

Example 1. Given an original graph G (shown in Fig. 2a) and a query $q = (s, t, 5)$, the graph can be pruned. For example, if $Dist_s[g] + Dist_t[g] = 6 > 5$, the vertex g is pruned and is therefore excluded from the vertex set V (as shown in Fig. 2b).

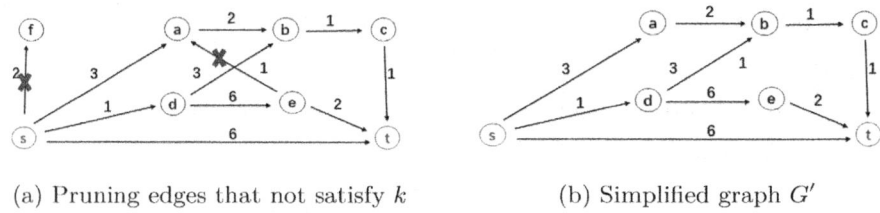

(a) Pruning edges that not satisfy k (b) Simplified graph G'

Fig. 3. Pruning edges that do not satisfy the k-condition from the preliminary simplified graph G', resulting in the final simplified graph G'.

Pruning Pendant Nodes and Edges that Do Not Satisfy the Hop Constraint. After determining the nodes set V' involved in the k-step path, to accelerate the graph retrieval speed and reduce storage space, we introduce the pruning of edges that do not satisfy the hop constraint and pendant nodes with a degree of 1.

Theorem 2. *In the process of building the simplified graph G', if $len_u + Dist_t[v] + 1 \leq k$ for $v \in N_{out}(u)$, then the path is retained; otherwise, the path is removed.*

Theorem 3. *For the simplified graph G', pruning all edges connected to degree-1 pendant nodes that do not form a direct s-t path will not affect the existence of s-t paths that satisfy the hop constraint k.*

Proof. Given a simplified graph G' and a query $q = (s, t, 5)$, the simplified graph G' can be pruned by removing pendant nodes and invalid paths that do not satisfy the hop constraint k. For example, the paths $\{s, f\}$ and $\{s, d, e, a, b, c, t\}$ in Fig. 3a are considered invalid. By pruning, we generate Fig. 3b, eliminating these edges in advance. This does not lead to the loss of paths and avoids a large number of invalid searches.

In the process of constructing a simplified graph, as shown in Algorithm.1, the algorithm aims to prune the graph G to produce a subgraph G' satisfying specific constraints through Breadth-First Search (BFS) and pruning rules. First, BFS is performed from the starting point s and the endpoint t to calculate the minimum hop count for each node(lines 1-2), and based on this, nodes and

---
Algorithm 1: Graph Pruning

Input : G, s, t, k
Output: G'

// Step 1: BFS-based Hop Distance Table from s and t
1    $Dist_s \leftarrow$ BFS_k_hop(G, s, k);
2    $Dist_t \leftarrow$ BFS_k_hop$(\text{reverse}(G), t, k)$;

// Step 2: Initialize pruned graph G'
3    $G' \leftarrow \{\}$;
4   **for** $u \in G$ **do**
5   **if** $u == s$ **or** $u == t$ **or** $(Dist_s[u] + Dist_t[u] \leq k)$ **then**
6        $G'[u] \leftarrow \{\}$;
7   **foreach** $v \in G[u]$ **do**
8   **if** $Dist_s[v] + Dist_t[v] \leq k$ **then**
9            $G'[u][v] \leftarrow G[u][v]$;

10    $q \leftarrow [s]$;
11   dist $\leftarrow \{u : \infty \mid u \in G'\}$;
12   dist$[s] \leftarrow 0$;
13   **while** q *is not empty* **do**
14      $u \leftarrow q.\texttt{pop\_left}()$;
15      $\text{len}_u \leftarrow \text{dist}[u]$;
16   **foreach** $v \in G'[u]$ **do**
17   **if** $\text{len}_u + 1 + Dist_t[v] > k$ **then**
18   Remove edge (u, v) from G';
           // Prune edge
19   dist$[v] \leftarrow \text{len}_u + 1$;
20        $q.\texttt{append}(v)$;

21   **for** $u \in G'$ **do**
22   **if** $\deg(u) = 1$ **and** $u \neq s$ **and** $u \neq t$ **then**
23   Remove u and its edges from G';

24   **return** G'

---

edges that may lie on the *s*-to-*t* path are preliminarily filtered to construct the subgraph G'(lines 4-9). Subsequently, by traversing G' with BFS, edges that could cause the path length to exceed k are removed, ensuring that all paths satisfy the length constraint(lines 10-20). Finally, redundant nodes with a degree of 1 that do not lie on the *s-t* path are deleted to further optimize the graph structure(lines 21-23). The algorithm ultimately outputs a concise and efficient G' that preserves all possible shortest paths.

4.2 Constructing Path Index

In complex networks, we propose an efficient path indexing method based on middle nodes partitioning. By scanning nodes in the simplified graph, we identify the middle nodes partition, significantly optimizing query performance and enhancing indexing efficiency.

Definition 2. *Middle nodes cut (Vc): For a given path $p = \{s,\ldots,vc,\ldots,t\}$, the middle nodes is the $k/2$ node on the path, denoted as vc. The set of middle nodes in all k-step paths that meet this condition is called the middle nodes Partition, represented by Vc.*

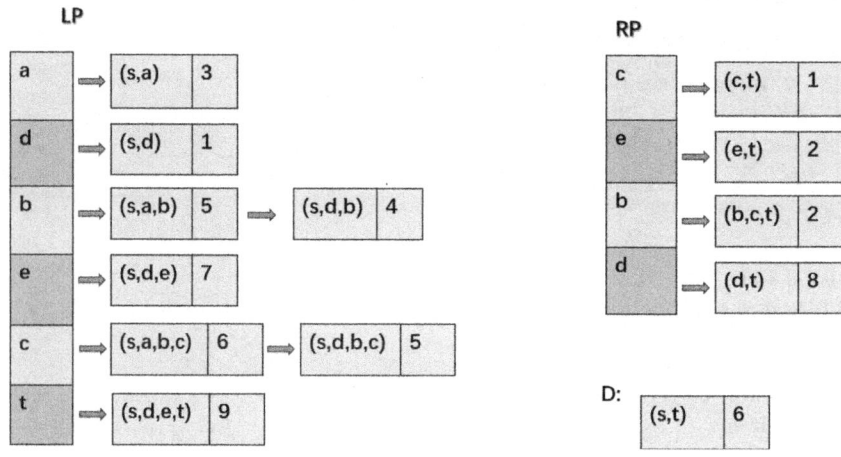

Fig. 4. Path Index

Definition 3. *Stores partial paths left linked to the middle node, satisfying $len(lp) + Dist_t[Vc] \leq k$ and with a length up to $\lceil k/2 \rceil$. It also records the total weight of each path. The Right Path Index (RP) stores right partial paths linked to the middle node, satisfying $len(rp) + Dist_s[Vc] \leq k$ and with a length up to $\lfloor k/2 \rfloor$. It also records the total weight of each path. The Path Set (D) stores paths that go directly from s to t via a single edge, along with their weight.*

Example 2. Given a simplified graph G' and a query $q = (s,t,5)$, the graph is partitioned into left and right segments based on the k-value, ensuring efficient processing of middle nodes, paths(path and weight). Paths(path, weight) where s and t are directly connected via a single edge are stored in D for quick weight checks against paths built from LP and RP in encryption(Fig. 4).

4.3 Scheme Description

In this section, we will describe in detail the construction process of the graph encryption scheme. This scheme includes five algorithms **(KeyGen, GraphEnc, TokenGen, Query, Decrypt)**, with the description of each algorithm as follows:

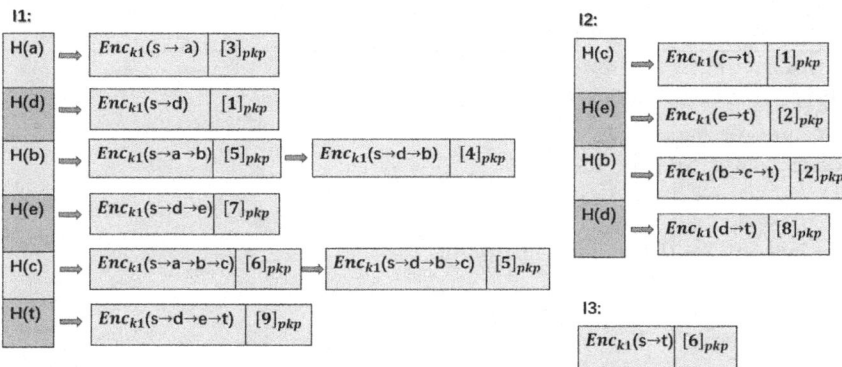

Fig. 5. Secure Index

1) KeyGen: KeyGen generates encryption and decryption keys to ensure data security. Using the security parameter λ, it creates a symmetric key k_1 via AES. Genλ for encrypting data and Paillier encryption key pair (pk, sk) via Paillier. Genλ. The Data Owner sends k_1 to the Data User for path decryption, while pk and sk are assigned to servers S_1 and S_2, respectively, to support the Secure Integer Comparison Protocol, enabling secure computations.

2) GraphEnc: The *Data Owner* encrypts the ordinary LP, RP, D using an encryption Algorithm 2 to create secure indices. Specifically, each middle node Vc in LP and RP is encrypted using the Hash function H(Vc) (line 3). Paths are initialized to store the encrypted path and weight. Each path and weight is encrypted using AES and Paillier homomorphic encryption algorithms to generate encrypted paths P and weights N, which are then inserted into the corresponding lists, and set as the values for $I1$ and I_2 (lines 9 and 16).

Next, the path and weights in D are encrypted. The encrypted path and weight are inserted into the list L_3 and set as the values for I_3 (lines 22-23). Finally, the secure indices containing I_1, I_2, and I_3 (as shown in Fig. 5) is returned to the cloud server $S1$ for subsequent query processing and data access. This encryption mechanism not only ensures the security of graph data on the server.

3) TokenGen: When the Data User requests the shortest path from the starting node s to the destination node t within k hops, the Data Owner generates a query trapdoor T and transmits it to the cloud server. This trapdoor includes the encrypted starting node $Se \leftarrow AES(k_1, s)$, destination node $Te \leftarrow AES(k_1, t)$, and hop count $Ke \leftarrow AES(k_1, k)$, ensuring query privacy. Cloud server $S1$ executes the computation using encrypted parameters without accessing the actual query details, thereby maintaining data security.

4) Query: When the cloud server $S1$ receives the query trapdoor T, it performs a query operation on the secure index, as shown in Algorithm.3. Firstly, the algorithm performs path connections by matching *middle nodes*. Specifically, for each middle nodes in I_1, the hash value $H_1(Vc)$ and its corresponding pathP_1

Algorithm 2: Secure Index Construction

Input : LP, RP, D, keys k_1, and the public key pk
Output: A secure index I

1 initialize dictionaries I_1, I_2, I_3;
2 **foreach** $(Vc, paths) \in LP$ **do**
3 $\quad$ $H_1(Vc) \leftarrow Hash(Vc)$;
4 $\quad$ $L_1 \leftarrow [\,]$;
5 $\quad$ **foreach** $(path, weight) \in paths$ **do**
6 $\quad\quad$ $P_1 \leftarrow AES(k_1, path)$;
7 $\quad\quad$ $N_1 \leftarrow Paillier.Enc(pk, weight)$;
8 $\quad\quad$ insert (P_1, N_1) into L_1;
9 $\quad$ set $I_1[H_1(Vc)] := L_1$;
10 **foreach** $(Vc, paths) \in RP$ **do**
11 $\quad$ $H_2(Vc) \leftarrow Hash(Vc)$;
12 $\quad$ $L_2 \leftarrow [\,]$;
13 $\quad$ **foreach** $(path, weight) \in paths$ **do**
14 $\quad\quad$ $P_2 \leftarrow AES(k_1, path)$;
15 $\quad\quad$ $N_2 \leftarrow Paillier.Enc(pk, weight)$;
16 $\quad\quad$ insert (P_2, N_2) into L_2;
17 $\quad$ set $I_2[H_2(Vc)] := L_2$;
18 $L_3 \leftarrow [\,]$;
19 **foreach** $(path, weight) \in D$ **do**
20 $\quad$ $P_3 \leftarrow AES(k_1, path)$;
21 $\quad$ $N_3 \leftarrow Paillier.Enc(pk, weight)$;
22 $\quad$ insert (P_3, N_3) into L_3;
23 set $I_3 := L_3$;
24 **return** $I = (I_1, I_2, I_3)$;

and weight N_1 are traversed (lines 2–3); the same operation is performed for I_2. If $H_1(Vc)$ matches $H_2(Vc)$ (line 6), the two partial paths are concatenated, and the total weight is calculated (lines 7–8), and the result is added to list $Join_P$. Next, the algorithm identifies the shortest paths. The paths from I_3 and $Join_P$ are merged into All_P. If All_P is empty, an empty list is directly returned. Finally, All_P is traversed, and the shortest path set Min_P is identified using a Secure Integer Comparison Protocol and returned to the Data User.

During the weight comparison process, we employ the Secure Integer Comparison Protocol, leveraging Paillier homomorphic encryption to safeguard data privacy. Server $S1$ generates a random mask enc_mask, computes encrypted weights c_1 and c_2, then sends them to server $S2$. Server $S2$ decrypts the received values using its private key and compares them: If $c_1 > c_2$, it returns $b = 1$. If $c_1 < c_2$, it returns $b = -1$. If $c_1 = c_2$, it returns $b = 0$. This approach ensures both accurate query results and enhanced data privacy protection.

Algorithm 3: Query

Input : $I = (I_1, I_2, I_3), T$
Output: Min_P

1 $Join_P \leftarrow [];$
   //Step 1: Middle nodes (Vc) Matching for Path Joining
2 **foreach** $H_1(Vc) \in I_1$ **do**
3   **foreach** $(P_1, N_1) \in I_1[H_1(Vc)]$ **do**
4   **foreach** $H_2(Vc) \in I_2$ **do**
5   **foreach** $(P_2, N_2) \in I_2[H_2(Vc)]$ **do**
6   **if** $H_1(Vc) == H_2(Vc)$ **then**
7                        $Tol_P \leftarrow P_1 + P_2;$
8                        $Tol_N \leftarrow N_1 + N_2;$
9                        $Join_P.append((Tol_P, Tol_N));$

   //Step 2: Calculate Minimal Paths
10 $Min_N \leftarrow None;$
11 $Min_P \leftarrow [];$
12 $All_P \leftarrow I_3.(P_3, N_3) + Join_P.(Tol_P, Tol_N);$
13 **if** All_P *is empty* **then**
14   **return** $[];$
15 **for** (P, N) *in* All_P **do**
16   **if** Min_N *is None or* $Compare(N, Min_N) == -1$ **then**
17         $Min_P \leftarrow P;$
18         $Min_N \leftarrow N;$
19   **else if** $Compare(N, Min_N) == 0$ **then**
20         $Min_P.append(P);$
21 **return** Min_P

5) Decrypt: When the *Data User* receives the shortest path set Min_P, they use the key k_1 to decrypt each path, and after decryption, remove duplicate nodes from the path to obtain the plaintext *result*.

5 Security Analysis

This section analyzes the proposed scheme's security, using leakage functions L_{setup} and L_{query} to parameterize security definitions.

1. **GraphEnc Leakage:** Leakage function L_{setup} exposes secure index details, including middle nodes, paths from LP, RP and list D.
2. **Query Leakage:** Leakage function L_{query} captures query process leakage, including query and index pattern, exposure from secure integer comparison.

Theorem 4. *If Hash, AES and Paillier scheme are secure, the proposed graph encryption scheme Let $\Pi = (KeyGen, GraphEnc, TokenGen, Query, Decrypt)$ is (L_{setup}, L_{query})-secure under adaptive chosen query attacks (CQA2 security).*

The main idea to prove the above theorem is to construct a simulator S that uses the leakage functions L_{setup} and L_{query} to simulate the secure index I^* and a series of queries Q^*. For any adversary A, if it cannot distinguish between the real game and the ideal game, our scheme achieves CQA2 security.

1. S invokes the key generation algorithm KeyGen(λ) to generate the key set Key $= (k_1, (pk, sk))$.
2. S processes each middle nodes vc in the simplified graph according to the query request and generates a pseudo index $I^* = \{I_1, I_2, I_3\}$ based on the leakage function L_{setup}. The entire index is encrypted using the key.
3. Since Hash, SE, and F are secure, A cannot distinguish the output of Hash, SE, and F from random values. Therefore, A cannot obtain useful information from I^*.
4. Given the leakage function L_{query}, S simulates the query token T^* and sends it to A. Since PE is secure, A cannot obtain any information about the query (k, s, t) from the query token.
5. A receives the query token T^*. Since the path information that meets the hop constraint k in the query is holistic and does not disclose individual information, A cannot infer the value of b from T^* and I^*.

Since the aforementioned cryptographic primitives satisfy security properties, an adversary cannot deduce any information from T^* and I^*. Therefore, we have:

$$|Pr[Real_{\mathcal{A}}(\lambda) = 1] - Pr[Ideal_{\mathcal{A},S}(\lambda) = 1]| \leq \text{negl}(\lambda) \tag{3}$$

where negl(λ) is a negligible function.

Table 1. Properties of Different Graph Encryption Schemes

Property	GRECS	Connor	PGAS	PP-KSPQ(our)
Graph Type	Undirected	Directed	Directed	Directed
Queries	SD	CSD	CSD	CSQ
Accuracy	Approximate	Approximate	Accurate	Accurate
Edge Density	1	1	1	β

6 Performance Analysis and Experiment Evaluation

6.1 Performance Analysis

From Table 1, it is evident that compared to classical methods such as GERCS [17], Connor [24], and PGAS [33], which are designed for constrained shortest distance queries, our PP-KSPQ scheme is specifically tailored for constrained shortest path queries (k-hop). While its construction process shares similarities with these methods, PP-KSPQ optimizes the indexing structure, significantly enhancing query efficiency. Moreover, whereas GERCS and Connor only support approximate queries, PP-KSPQ enables precise queries, improving accuracy.

Additionally, we analyzed the impact of pruning on edge density β. PP-KSPQ incorporates a preprocessing step based on k-hop constraints, systematically pruning the original graph G before index construction. This pruning is performed prior to data upload to the cloud server, eliminating nodes and edges that do not meet the constraints, thereby optimizing storage efficiency and reducing computational overhead. Consequently, the pruned graph exhibits a lower edge density than the original graph, satisfying $\beta \leq 1$. This strategy not only minimizes redundancy but also enhances query performance, making PP-KSPQ highly effective for large-scale graph data processing.

6.2 Experiment Evaluation

We conducted our experiments in a Windows 11 environment, equipped with a 2.3 GHz 12th Gen Intel(R) Core(TM) i7 CPU processor and 16 GB of memory. All experiments were implemented in Python, following the proposed scheme.

We evaluated our method using five real-world graph databases from Stanford's SNAP project, ensuring diverse and representative data. Query requests were generated from each dataset in Table 2, with subgraphs G constructed from 600 randomly selected nodes and their neighbors. Each subgraph produced 200 k-s-t queries to analyze performance across varying hop ranges. This framework highlights the method's robustness and adaptability in complex networks.

Table 2. Databases

Databases	Nodes	Edges	Average clustering coefficient
WikiTalk	2,394,385	4,659,565	0.0526
Slashdot0902	82,168	948,464	0.0603
soc-Epinion1	75,877	508,837	0.1378
soc-LiveJournal1	4,847,571	68,993,773	0.2742
email-Eu-core	1,005	25,571	0.3994

Index Construction Time: The original subgraph G undergoes a pruning process to generate the simplified subgraph G', which removes nodes and edges that do not meet the hop constraint k. This step creates a smaller, more efficient induced subgraph, significantly reducing complexity and improving processing efficiency. Figure 6 illustrates the difference in storage requirements between the original subgraph G and the simplified subgraph G', showing how pruning effectively reduces storage overhead.

Table 3 compares index construction times for the original subgraph G and simplified subgraph G'. G' is built in two stages: graph pruning and index generation. The pruning stage removes nodes and edges violating the k-hop constraint,

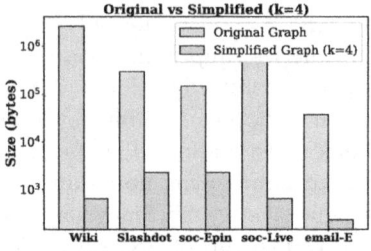
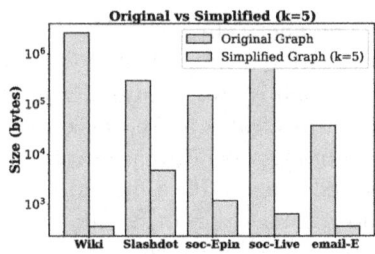

(a) Comparison of Original Graph and Simplified Graph (k=4).

(b) Comparison of Original Graph and Simplified Graph (k=5).

Fig. 6. Comparison of Original Graph and Simplified Graph for k=4 and k=5.

restructured G'. In the index generation stage, the query indices are based on G'. As shown in Table 3, when $k \geq 5$, pruning before index construction significantly reduces time compared to direct index generation on G. Moreover, the time gap increases as k increases, highlighting efficiency and advantages.

Table 3. Comparison of Index Construction Time (G vs G')

Databases	Time (seconds, $\times 10^{-2}$)									
	k=4		k=5		k=6		k=7		k=8	
	G	G'	G	G'	G	G'	G	G'	G	G'
WikiTalk	5.7	5.4↓	4.3	4.2↓	7.9	6.3↓	12.3	6.1↓	23.6	7.5↓
Slashdot0902	0.3	0.7	3.2	0.8↓	4.2	1.2↓	6.4	2.3↓	7.2	2.8↓
soc-Epinion1	0.3	0.5	5.3	0.6↓	3.9	0.7↓	9.4	1.0↓	16.8	1.4↓
soc-LiveJournal1	0.4	0.8	3.3	1.7↓	5.6	1.2↓	8.3	1.3↓	13.6	1.5↓
email-Eu-core	0.3	0.3	5.2	1.2↓	7.0	1.4↓	10.1	1.5↓	91.9	1.9↓

Figure 7a illustrates the construction time of the simplified subgraph G' across various k-hop values. The pruning optimization significantly reduces construction time, with its benefits becoming increasingly pronounced as k-hop values grow. In sparse graphs, the initial pruning phase typically requires more computational resources due to the dispersed edge distribution, which demands greater effort to satisfy k-hop constraints. In contrast, dense graphs benefit from more efficient pruning processes thanks to their concentrated node and edge distributions. As k-hop values increase, performing graph pruning prior to index generation demonstrates notably better time performance compared to directly generating indexes from the original subgraph, further validating the efficiency of the proposed method.

Graph Encryption Time: To evaluate the impact of pruning optimization on encryption efficiency, we measured and compared the encryption times of

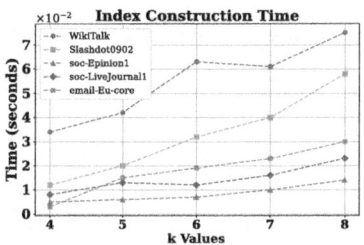 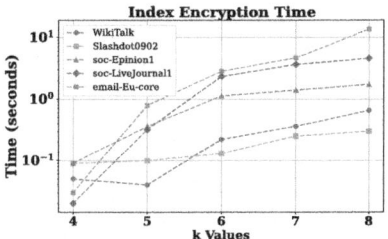

(a) Path Index Construction Time with Varying k Hop Values (b) Path Index Encryption Time with Varying k Hop Values

Fig. 7. Path Index Construction and Encryption Times with Varying k Hop Values

the simplified subgraph G' and the original subgraph G. By eliminating nodes and edges that did not meet the k-hop constraint, pruning significantly reduced the index size, lowering computational complexity and resulting in a substantial reduction in encryption time. Table 4 highlights this difference, demonstrating that the encryption time for G' is markedly shorter than that for G.

Table 4. Comparison of Index Encryption Time (G vs G')

Databases	Time (seconds)									
	k=4		k=5		k=6		k=7		k=8	
	G	G'	G	G'	G	G'	G	G'	G	G'
WikiTalk	6.97	0.05↓	6.31	0.04↓	20.58	0.22↓	79.91	0.36↓	85.64	0.65↓
Slashdot0902	4.65	0.09↓	7.47	0.10↓	9.97	0.13↓	17.23	0.25↓	56.12	0.30↓
soc-Epinion1	0.68	0.09↓	17.02	0.36↓	15.45	1.12↓	41.23	1.39↓	52.13	1.73↓
soc-LiveJournal1	0.56	0.02↓	11.19	0.32↓	11.31	2.31↓	43.65	3.62↓	124.33	2.56↓
email-Eu-core	0.68	0.03↓	19.40	5.19↓	24.11	2.81↓	52.26	4.65↓	173.21	13.56↓

Figure 7b illustrates encryption time trends for G' across varying k-hop values. In dense graphs, pruning retains more indices, leading to slightly higher encryption times. However, the increase remains moderate, underscoring the efficiency and scalability of pruning optimization. Sparse graphs, on the other hand, retain fewer indices after pruning, resulting in a stable encryption time trend. In summary, the simplified subgraph G' showcases significant improvements in encryption efficiency, reducing time overhead and offering a robust solution for processing large-scale network data.

Query Time: Our scheme encrypts query conditions for efficient, private path queries. The shortest path between nodes s and t with a k-hop constraint is computed through interaction between cloud servers $S1$ and $S2$.

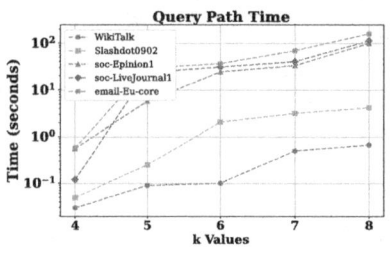 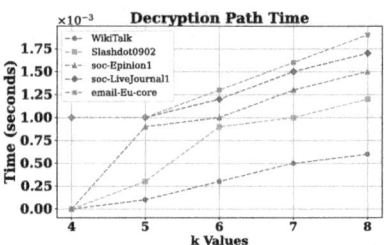

(a) Shortest Path Search Time with Varying k Hop Values

(b) Shortest Path Decryption Time with Varying k Hop Values

Fig. 8. Shortest Path Search and Decryption Times with Varying k Hop Values

Experimental results (Fig. 8a) show notable differences in query time between dense and sparse graphs. Dense graphs, with more connecting edges, require processing more paths, leading to longer query times that increase with k. Sparse graphs show shorter query times and a slower increase with larger k. Overall, simplified graphs offer better stability and scalability in path querying.

Decryption Time: Our decryption technique delivers swift and efficient path processing. Using the authorization token, Data Users can decrypt data to retrieve results. Experimental data in Fig. 8b highlights the longest decryption time as only 0.001 s, underscoring the method's ability to handle large-scale datasets while ensuring secure and efficient path for Data Users.

7 Conclusion

This paper introduces the first-ever Privacy-Preserving K-hop Shortest Path Query (PP-KSPQ) method based on graph pruning, designed to enhance query efficiency while strengthening data privacy protection By eliminating nodes and edges violating K-hop constraints, it reduces computation and storage needs. Secure integer comparison and advanced encryption ensure privacy while optimizing performance.

Experiments confirm PP-KSPQ's effectiveness in privacy protection and efficiency, highlighting its potential in transportation, secure networks, and data analysis. Future work will extend its use to dynamic encrypted graphs, ensuring efficient queries as data evolves.

Acknowledgments. This work was supported in part by the Taishan Scholars Program under Grant tsqn202211203, in part by the Shandong Provincial Innovation Capability Improvement Project for Small and Medium-sized Scientific and Technological Enterprises under Grant 2023TSGC0108, and in part by the QLU/SDAS Pilot Project for Integrated Innovation of Science, Education, and Industry under Grant 2024ZDZX08.

References

1. Akiba, T., Iwata, Y., Yoshida, Y.: Fast exact shortest-path distance queries on large networks by pruned landmark labeling. In: Proceedings of the 2013 ACM SIGMOD International Conference on Management of Data, pp. 349–360 (2013)
2. Anirban, S., Wang, J., Islam, M.S.: Experimental evaluation of indexing techniques for shortest distance queries on road networks. In: 2023 IEEE 39th International Conference on Data Engineering (ICDE), pp. 624–636. IEEE (2023)
3. Cheng, J., Shang, Z., Cheng, H., Wang, H., Yu, J.X.: Efficient processing of k-hop reachability queries. VLDB J. **23**(2), 227–252 (2014)
4. Du, M., Wang, Q., He, M., Weng, J.: Privacy-preserving indexing and query processing for secure dynamic cloud storage. IEEE Trans. Inf. Forensics Secur. **13**(9), 2320–2332 (2018)
5. Du, M., Wu, S., Wang, Q., Chen, D., Jiang, P., Mohaisen, A.: Graphshield: dynamic large graphs for secure queries with forward privacy. IEEE Trans. Knowl. Data Eng. **34**(7), 3295–3308 (2020)
6. D'ascenzo, A., D'emidio, M.: Top-k distance queries on large time-evolving graphs. IEEE Access **11**, 102228–102242 (2023)
7. Guo, J., Sun, J.: Secure shortest distance queries over encrypted graph in cloud computing. Wireless Netw. **30**(4), 2633–2646 (2024)
8. Hu, M., Chen, L., Chen, G., Mu, Y., Deng, R.H.: A pruned pendant vertex based index for shortest distance query under structured encrypted graph. IEEE Trans. Inf. Forensics Secur. (2024)
9. Kumar, M., Chand, S.: A secure and efficient cloud-centric internet-of-medical-things-enabled smart healthcare system with public verifiability. IEEE Internet Things J. **7**(10), 10650–10659 (2020)
10. Lai, Z., Peng, Y., Yang, S., Lin, X., Zhang, W.: PEFP: efficient k-hop constrained st simple path enumeration on FPGA. In: 2021 IEEE 37th International Conference on Data Engineering (ICDE), pp. 1320–1331. IEEE (2021)
11. Li, P., Zhou, F., Xu, Z., Li, Y., Xu, J.: Privacy-preserving top-k nearest keyword search queryies over encrypted graph data. In: 2021 IEEE 6th International Conference on Signal and Image Processing (ICSIP), pp. 531–537. IEEE (2021)
12. Liu, C., Zhu, L., He, X., Chen, J.: Enabling privacy-preserving shortest distance queries on encrypted graph data. IEEE Trans. Dependable Secure Comput. **18**(1), 192–204 (2018)
13. Liu, H., Jin, C., Yang, B., Zhou, A.: Finding top-k shortest paths with diversity. IEEE Trans. Knowl. Data Eng. **30**(3), 488–502 (2017)
14. Liu, T., et al.: Towards indoor temporal-variation aware shortest path query. IEEE Trans. Knowl. Data Eng. **35**(1), 998–1012 (2021)
15. Liu, X., Choo, K.K.R., Deng, R.H., Lu, R., Weng, J.: Efficient and privacy-preserving outsourced calculation of rational numbers. IEEE Trans. Dependable Secure Comput. **15**(1), 27–39 (2016)
16. Lv, C., Wang, J., Sun, S.-F., Wang, Y., Qi, S., Chen, X.: Efficient multi-client order-revealing encryption and its applications. In: Bertino, E., Shulman, H., Waidner, M. (eds.) ESORICS 2021. LNCS, vol. 12973, pp. 44–63. Springer, Cham (2021). https://doi.org/10.1007/978-3-030-88428-4_3
17. Meng, X., Kamara, S., Nissim, K., Kollios, G.: Grecs: graph encryption for approximate shortest distance queries. In: Proceedings of the 22nd ACM SIGSAC Conference on Computer and Communications Security, pp. 504–517 (2015)

18. Paillier, P.: Public-key cryptosystems based on composite degree residuosity classes. In: International Conference on the Theory and Applications of Cryptographic Techniques, pp. 223–238. Springer (1999). https://doi.org/10.1007/3-540-48910-x_16
19. Pan, C., Wang, B.: Prospect of compass navigation system's application in internet of things. J. Telemetry Tracking Command **32**(6), 14–17 (2011)
20. Peng, Y., Lin, X., Zhang, Y., Zhang, W., Qin, L.: Answering reachability and k-reach queries on large graphs with label constraints. The VLDB J. 1–27 (2022)
21. Peng, Y., Zhang, Y., Lin, X., Qin, L., Zhang, W.: Answering billion-scale label-constrained reachability queries within microsecond. Proc. VLDB Endowment **13**(6), 812–825 (2020)
22. Peng, Y., Zhang, Y., Zhang, W., Lin, X., Qin, L.: Efficient probabilistic k-core computation on uncertain graphs. In: 2018 IEEE 34th International Conference on Data Engineering (ICDE), pp. 1192–1203. IEEE (2018)
23. Qiu, X., et al.: Real-time constrained cycle detection in large dynamic graphs. Proc. VLDB Endowment **11**(12), 1876–1888 (2018)
24. Shen, M., Chen, S., Zhu, L., Xiao, R., Xu, K., Du, X.: Privacy-preserving graph encryption for approximate constrained shortest distance queries. In: 2019 IEEE Global Communications Conference (GLOBECOM), pp. 1–6. IEEE (2019)
25. Shen, M., Ma, B., Zhu, L., Mijumbi, R., Du, X., Hu, J.: Cloud-based approximate constrained shortest distance queries over encrypted graphs with privacy protection. IEEE Trans. Inf. Forensics Secur. **13**(4), 940–953 (2017)
26. Song, Y., Ge, X., Yu, J., Hao, R., Yang, M.: Enabling privacy-preserving k-hop reachability query over encrypted graphs. IEEE Trans. Serv. Comput. (2024)
27. Stann, F., Heidemann, J.: RMST: reliable data transport in sensor networks. In: Proceedings of the First IEEE International Workshop on Sensor Network Protocols and Applications, 2003, pp. 102–112. IEEE (2003)
28. Wang, S., Xiao, X., Yang, Y., Lin, W.: Effective indexing for approximate constrained shortest path queries on large road networks (2016)
29. Wang, S., Zheng, Y., Jia, X., Yi, X.: PeGraph: a system for privacy-preserving and efficient search over encrypted social graphs. IEEE Trans. Inf. Forensics Secur. **17**, 3179–3194 (2022)
30. Wang, W., Jia, Z., Xu, M., Li, S.: SPCS: strong privacy-preserving-constrained shortest distance queries on encrypted graphs. IEEE Internet Things J. **9**(22), 22516–22528 (2022)
31. Yu, Z., Yu, X., Koudas, N., Chen, Y., Liu, Y.: A distributed solution for efficient k shortest paths computation over dynamic road networks. IEEE Trans. Knowl. Data Eng. **36**(7), 2759–2773 (2024)
32. Yuan, L., Hao, K., Lin, X., Zhang, W.: Batch hop-constrained st simple path query processing in large graphs. In: 2024 IEEE 40th International Conference on Data Engineering (ICDE), pp. 2557–2569. IEEE (2024)
33. Zhang, C., Zhu, L., Xu, C., Sharif, K., Zhang, C., Liu, X.: PGAS: privacy-preserving graph encryption for accurate constrained shortest distance queries. Inf. Sci. **506**, 325–345 (2020)
34. Zhang, J., Yang, S., Ouyang, D., Zhang, F., Lin, X., Yuan, L.: Hop-constrained st simple path enumeration on large dynamic graphs. In: 2023 IEEE 39th International Conference on Data Engineering (ICDE), pp. 762–775. IEEE (2023)
35. Zhao, X., Wang, M., Jia, Z., Li, S.: Privacy-preserving any-hop cover shortest distance queries on encrypted graphs. IEEE Internet Things J. **11**(9), 16517–16528 (2024)

TA-PDC: Provable Data Contribution with Traceable Anonymous for Group Transactions

Xiaocong Lin[1,2], Weijing You[1,2(✉)], Chenchen Wu[1,2], Wenmao Liu[3], and Qi Gu[3]

[1] College of Computer and Cyber Security, Fujian Normal University, Fuzhou 350117, China
qsx20231396@student.fjnu.edu.cn, youweijing@fjnu.edu.cn
[2] Fujian Provincial Key Laboratory of Network Security and Cryptology, Fujian Normal University, Fuzhou 350117, China
[3] NSFOCUS, Inc., Beijing 100089, China
{liuwenmao,guqi}@nsfocus.com

Abstract. The cloud storage has been increasingly used for convenient data sharing, integration, and transactions among multi-owner groups to promote considerable profits, where the ownership of outsourced data is guaranteed by various provable data possession (PDP) schemes. Existing works mainly focus on ownership confirmation, while overlooking evaluation of distinct data contributions from different group members, which, however, will cause unfair distribution and further commercial dispute. In this paper, by considering strong privacy-preserving demand in public group transactions, we propose **TA-PDC**, the first **P**rovable **D**ata **C**ontribution with **T**raceable **A**nonymous, to resolve absence of evaluation towards data contribution within a multi-owner group. Our key insights are following twofold: 1) designing a novel weighted tag structure based on linkable ring signatures to correctly linking tags with weight embedded to specific owner; 2) using succinct inner product arguments (IPA) to enable a public verifiable commitment about data contribution without disclosing the underlying details. Our TA-PDC supports batch verification to further enhance efficiency and fine-grained contribution querying for greater flexibility. Security analysis and evaluations justify security and practicality of our proposal.

Keywords: Integrity auditing · Multi-ownership · Privacy-preserving · Contribution quantification

1 Introduction

As the collaborative era is coming, cloud storage has been increasingly used for convenient data sharing, integration, and transactions among multi-owner groups. Since data is transferred online, one of the most critical security concerns for group members is losing their ownership of the data. By instantiating ownership as the ability to audit data integrity via provable data possession (PDP), ownership confirmation for multi-owner group has been enabled in several PDP variants for group transactions [1-6].

Specifically, a typical PDP scheme for group transactions involves a challenge-response interaction between the cloud and the multi-owner group. Each owner in the group first generates tags for his/her own data, which will be collected and aggregated for saving storage and bandwidth. Then the group launches an auditing task as a whole through a one-time challenge. Only when the cloud correctly responds to the challenge with valid data-tag pairs could it pass the integrity auditing. Since integrity auditing is initialized and validated by data owners only, the ability to audit data integrity is then regarded as data ownership.

Although existing solutions ensure data ownership respect to each group member, they fail to evaluate distinct contributions from members to the integrated data. However, in practice, each group member always contributes differently in terms of volume and variety, leading to differential contribution and rewards. Taking training a multimodal large language model (MM-LLM) as a quick example, different types of data, e.g. images and texts, from different resources are required, and hence the contribution from each resource should be fairly quantified, which rarely are identical, since different type and volume of data implies different value.

A Straw Scheme. Intuitively, each data owner can sign his/her contribution, including type and volume of data, and upload the signed contribution to the cloud along with tags, which, however, is contradict with the privacy-preserving demand from data owners. More specifically, for data privacy, the details of data contribution are key elements to investment that should not be disclosed to the public. On the other hand, explicitly tracing data contribution back to specific owner in a group violates identity privacy.

In this paper, we enable evaluation towards data contributions with privacy-preserving and propose, to the best of our knowledge, the first **P**rovable **D**ata **C**ontribution with **T**raceable **A**nonymous. Motivated by TA-DPDP [6], which is a PDP scheme with partial anonymity, we adapt linkable ring signatures to provide global anonymity without losing the ability for owners to correctly retrieve their own tags. In addition, we introduce the inner product argument (IPA) to enable a provable variety of data without directly revealing the underlying data weight. Our contributions are summarized as follows:

- We propose TA-PDC for the first time, designing a novel weighted tag structure based on linkable ring signatures to correctly link tags with embedded weights to specific owners, thereby enabling accurate identification of data contributions. In addition, the protocol supports batch verification to improve quantification efficiency.
- We introduce the succinct non-interactive IPA to preserve the privacy of user contribution distributions. Additionally, with user authorization, the system

supports fine-grained queries of contribution values associated with specific data types.
– We provide concrete security definition, security proofs and experimental evaluations, which justify security level and practicality of our proposal. We make the full source code of the experiment publicly available.

2 Related Work

Provable Data Possession. In 2007, Ateniese et al. [7] proposed the first "Provable Data Possession" (PDP) scheme, which enables public verification of outsourced data integrity without retrieving the data, by leveraging homomorphic authenticators to improve efficiency. Concurrently, Juels and Kaliski [8] proposed Proofs of Retrievability (PoR), which further ensures data retrievability by embedding sentinels into error-corrected data. Recognizing the complementary strengths of these two schemes, Shacham and Waters [9] combined PDP and PoR in a unified framework to support unlimited verifications and retrievability.

Provable Data Possession with Privacy Preserving. Auditing protocols that support privacy preservation are also extensively studied. In 2011, Wang et al. [10] employed random masking techniques within homomorphic linear authenticators to enable public verification while protecting data privacy. However, this approach incurs high computational overhead. To address this, Li et al. [11] proposed more efficient constructions and maintain privacy. In 2019, Shen et al. [12] presented a data integrity auditing scheme with sensitive information hiding. Gao et al. [13] proposed a privacy-preserving auditing scheme that allows the TPA to verify keyword-related files without learning sensitive information.

Provable Data Possession of Multi-ownership. In 2014, Wang et al. [14] proposed Oruta, a privacy-preserving scheme based on homomorphic ring signatures that allows public auditing. However, it does not support user revocation. To address this, several schemes [15–17] employed homomorphic authentication combined with proxy re-signatures, enabling the cloud and revoked users to collaboratively convert revoked signatures into valid ones. In 2023, Shen et al. [1] first proposed an accountable auditing mechanism for multi-owner data transactions based on blockchain technology. However, it suffers from key escrow and complex certificate management issues. To overcome these challenges, Zhang et al. [2] proposed a certificateless data auditing mechanism that supports multi-owner transactions. Wang et al. [3–5] leveraged (t, n) threshold signature to enable decentralized multi-owner data transfer. Wu et al. [6] proposed TA-DPDP based on linkable ring signatures to protect member identity privacy in data transaction scenarios. Unfortunately, existing studies have overlooked the critical issue of quantifying member contributions in group data transactions.

3 Preliminaries

3.1 Linkable Ring Signature

In this section, we introduce the linkable ring signature [18–20], which provides anonymity. Given a file $\mathcal{F} = \{m_1, \ldots, m_n\}$, the corresponding signature protocol is defined as:

- $LR.PGen(1^\lambda) \to pp$: Input the security parameter λ, it outputs the system parameter pp.
- $LR.KGen(pp) \to \{(sk_j, pk_j)\}$. Input the system parameter pp: it outputs the key pairs for s signers, where $sk_j = (x_j, y_j)$ and $pk_j \leftarrow g^{x_j} h^{y_j}$.
- $LR.Sign(event, sk_\pi, PK, m_i) \to (\sigma_i, t_i)$: Input the event identifier $event$, the signer's secret key sk_π, the public key set of users $PK = \{pk_1, \ldots, pk_s\}$ and the signed message m_i, output the message tags (σ_i, t_i).
- $LR.Verify(event, \sigma_i, t_i, PK, m_i) \to valid/invalid$: Input the event $event$, the message tags (σ_i, t_i), the public key set of users $PK = \{pk_1, \ldots, pk_s\}$, and the message m_i, it verifies the message signature σ_i and outputs either $valid$ or $invalid$.
- $LR.Link((\sigma_1, t_1), (\sigma_2, t_2)) \to link/unlink$: Input two different tags (σ_1, t_1) and (σ_2, t_2): If $t_1 = t_2$, the algorithm outputs $link$, indicating that the messages m_1 and m_2 were signed by the same user. Otherwise, it outputs $unlink$.

Definition 1 (Discrete Logarithm Assumption). *Given a tuple (g, g^a) for any $a \in_R \mathbb{F}$ as input, output a. Define that the discrete logarithm assumption holds in $\mathbb{G}$ if for any PPT adversary $\mathcal{A}$, $\Pr\left[\mathcal{A}(1^\lambda, g, g^a) = a\right] \leq negl(\lambda)$ holds for arbitrary security parameter λ, where $negl(\cdot)$ is a negligible function.*

3.2 Succinct Non-interactive Inner Product Argument (IPA)

We describe the succinct non-interactive IPA protocol between two vector of field elements [21,22]. Let $\boldsymbol{a}, \boldsymbol{b}$ be the two input vectors of length d. The prover $\mathcal{P}$ wants to convince a verifier $\mathcal{V}$ that $\langle \boldsymbol{a}, \boldsymbol{b} \rangle = \mu$, where $\mathcal{V}$ has μ, and commitments to $\boldsymbol{a}$ and $\boldsymbol{b}$. The IPA protocol is defined as:

- $IPA.Setup(1^\lambda, d) \to (CRS, VK)$: Input the security parameter λ and the vector $\boldsymbol{a}, \boldsymbol{b}$ length d, it outputs the common reference string CRS [23] and the public verification key VK.
- $IPA.Com(CRS, \boldsymbol{a}, \boldsymbol{b}) \to (g_a, g_b)$: Input the CRS and two vectors $(\boldsymbol{a}, \boldsymbol{b})$, it outputs the corresponding KZG commitments [24] g_a, g_b of the two vectors.
- $IPA.ProofGen(CRS, \boldsymbol{a}, \boldsymbol{b}) \to (\mu, \pi)$: Input the CRS and two vectors $(\boldsymbol{a}, \boldsymbol{b})$, it outputs the inner product $\mu = \langle \boldsymbol{a}, \boldsymbol{b} \rangle$ and its IPA proof π.
- $IPA.Verify(CRS, VK, g_a, g_b, \mu, \pi) \to 0/1$: Input the CRS, public verification key VK, commitments g_a, g_b, inner product μ, and IPA proof π. The algorithm outputs 1 if the inner product result μ is computed correctly; otherwise, it outputs 0.

3.3 Polynomial Identity

For any given vector $P \subseteq \mathbb{F}$ and $|P| = d$, we define the Lagrange polynomial $\mathcal{L}_{a,P}(x)$ and Lagrange interpolation polynomial $p(x)$ with respect to P of degree at most $d-1$ as:

$$p(x) = \sum_{a \in P} \mathcal{L}_{a,P}(x) p(a), \text{where } \mathcal{L}_{a,P}(x) = \frac{\prod_{b \in P, b \neq a}(x-b)}{\prod_{b \in P, b \neq a}(a-b)}$$

Our protocol uses the following identity for univariate polynomials.

Lemma 1 (Univariate Sumcheck [25]). *Let $H = \{\omega, \omega^2, \ldots, \omega^d\}$ be a multiplicative subgroup of $\mathbb{F}$ of order d. Given two vector a, b and its Lagrange interpolation polynomial $a(x), b(x) \in \mathbb{F}[x]$ of degree $d-1$ with $a(\omega_i) = a[i], b(\omega_i) = b[i]$, then there exist unique polynomials $q(x)$ and $r(x)$ such that:*

$$a(x)b(x) = q(x)z_H(x) + r(x)x + d^{-1} \cdot \sum_{i \in [d]} a(\omega^i)b(\omega^i).$$

Here $z_H(x)$ is the vanishing polynomial over the set H, i.e., $z_H(\omega^i) = 0$ for each $i \in [d]$. Also, we write $z_H(x)$ as, $z_H(x) = \prod_{i \in [d]}(x - \omega^i) = x^d - 1$.

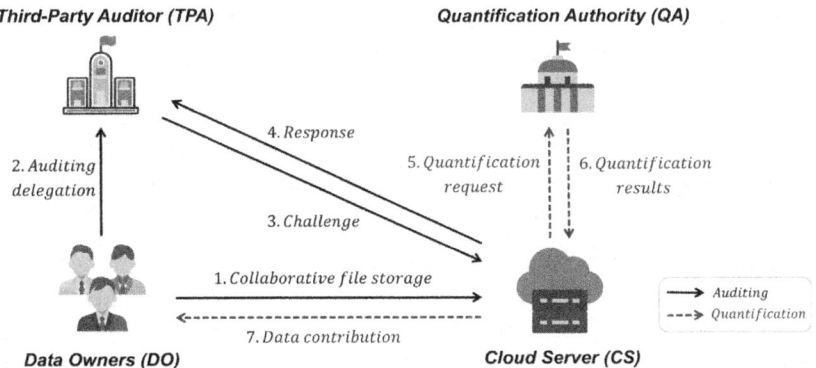

Fig. 1. System Framework

4 System Models and Definitions

4.1 The System Model

As illustrated in Fig. 1, the proposed system consists of four main entities: the cloud server (CS), the data owners (DO), the quantification authority (QA) and the third-party auditor (TPA).

- *The Cloud Server (CS)* is an entity with substantial computational power and storage capacity. It is responsible for storing files uploaded by data owners, responding to integrity audit challenges, and allocating individual user rights based on quantified contributions when the file generates value.
- *The Data Owners (DO)* are a group of users who anonymously collaborate to generate files and store them in the cloud server for shared access. When the files generates value, the system allocates rights to each member based on their quantified contributions.
- *The Quantification Authority (QA)* is an honest but curious entity responsible for quantifying user contributions. However, it may attempt to infer the association between data owners and their contribution.
- *The Third-Party Auditor (TPA)* is a semi-trusted entity with expertise in integrity verification. It is tasked with challenging and verifying the integrity of the data. However, it may attempt to infer the identity of the data owner for each data block using verification metadata.

We briefly describe the system architecture and workflow as follows: DO anonymously collaborate to generate a shared file and then uploaded to the CS for storage. The integrity of file is checked by the TPA with delegations from DO. Upon receiving an audit challenge from the TPA, the CS generates a corresponding integrity proof. The TPA verifies the proof and sends the audit result back to DO. When the file generates value, the CS sends a quantification request to the QA. The QA derives the contribution results of each user based on the linkable information. After verifying the result, the CS allocates corresponding rights to each user according to their quantified contributions.

4.2 The Syntax of TA-PDC

Our protocol consists of six algorithms detailed in the following, with the main notations summarized in Table 1.

- $SysGen(1^\lambda, d) \rightarrow PP$: This is a probabilistic algorithm executed by the system. Input the security parameter λ and the number of data types d, it outputs the system parameter PP.
- $KeyGen(PP) \rightarrow (SK, PK, VK, qsk, qpk)$: This is a probabilistic algorithm executed by DO, the system, and the QA. Input the system parameter PP, DO generates a pair of secret and public keys (SK, PK), the system generates the public verification key VK, and the QA generates a pair of secret and public keys (qsk, qpk).
- $TagGen(PP, m_i, SK, PK, fid, \textbf{\textit{Type}}) \rightarrow (tag_i, g_{c_j}, \overline{td_j})$: This is a deterministic algorithm executed by DO with SK, PK. Input the system parameter PP, data blocks m_i, file identifier fid, and data type vector $\textbf{\textit{Type}}$. It outputs the data blocks tag tag_i, user's contribution commitment g_{c_j}, and user's encrypted trapdoor $\overline{td_j}$.
- $Audit(PP, m_i, tag_i, PK, fid) \rightarrow 1/0$: This is a probabilistic algorithm executed by the CS and TPA in two steps: 1) input a challenge set Q from the

Table 1. Main Notations

Notation	Description
n	The total number of the data blocks in the file F
s	The total number of the data owners
d	The total number of the data types
VK	Public verification key for user contribution
qsk, qpk	The private-public key pair of QA
fid	The identity of the file F
Type	The data type vector
w	The data type weight vector
g_{c_j}	The user j contribution commitment
td_j	The user j linkable trapdoor
Q	The integrity auditing challenge
P	The proof of storage
π	The data owners contribution proof
W_j	The user j contribution
$c_{j\ell}$	The contribution of user j to the ℓ-th data type

TPA, data blocks and tag pairs $(m_i, tag_i)_{i \in Q}$ the CS returns the integrity proof P to the TPA; 2) the TPA verifies the integrity proof P using the DO's PK, file identifier fid, and system parameter PP, while outputs "1" if the verification succeeds, or "0" otherwise.

- $Quantify(PP, t_i, VK, g_{c_j}, \overline{td_j}, \textbf{\textit{Type}}, \textbf{\textit{w}}) \to W_j$: This is a probabilistic algorithm executed by the CS and QA in two steps: 1) input partial tag t_i and the user's encrypted trapdoors $\overline{td_j}$ from the CS, the QA returns user's contribution W_j and proof π to the CS; 2) the CS verifies W_j using the public verification key VK, the contribution commitment g_{c_j}, and the contribution proof π, while obtaining the corresponding contribution W_j for DO.
- $Query(PP, type_\ell, g_{c_j}) \to c_{j\ell}$: This is a deterministic algorithm executed by user and the CS in two steps: 1) upon receiving the queried data type $type_\ell$ from the CS, the user returns the contribution $c_{j\ell}$ along with the proof π_ℓ; 2) the CS verifies $c_{j\ell}$ using the system parameter PP, proof π_ℓ, and contribution commitment g_{c_j}, while obtaining $c_{j\ell}$ corresponding to $type_\ell$.

4.3 The Security Model

The proposed protocol should satisfy the properties of correctness, unforgeability, anonymity, linkability, and nonslanderability, which are defined as follows.

Definition 2 (Correctness). *The proposed protocol is correct if DO and the CS are executed honestly, then for any challenged data blocks m_i, there is always:*

$$Audit(PP, m_i, tag_i, PK, fid) \to 1.$$

Definition 3 (Unforgeability). *The proposed protocol is unforgeable if the data tag tag_i corresponding to block m_i is infeasible to be forged for all adversaries, for any security parameter λ and negligible function $negl(\cdot)$,*

$$\Pr\Big[\; \mathcal{A}^{\mathcal{O}_{TagGen}(SK,PK,\cdot)}(PP) \to (tag'_i) \;\wedge\; (m'_i, tag'_i) \notin \{(m_i, tag_i)\}$$
$$\wedge\; Audit(PP, m'_i, tag'_i, PK, fid) \to 1:$$
$$(SK, PK, \cdot) \leftarrow KeyGen(PP), PP \leftarrow SysGen(1^\lambda, d)\Big] \le negl(\lambda).$$

where $\mathcal{O}_{TagGen}(\cdot, \cdot)$ is the oracle for tag generation queries, and (m_i, tag_i) is the set of pairs that A had queried, the same as in the following.

Definition 4 (Anonymity). *The proposed protocol is anonymous if any adversary cannot distinguish which member generated a given tag_i, for any security parameter λ and negligible function $negl(\cdot)$,*

$$\Pr\Big[\; \mathcal{A}^{\mathcal{O}_{TagGen}(\cdot)}(PP, PK) \to (m'_i, tag'_i = (t'_i, \sigma'_i, K'_i, \widetilde{x'_i}, \widetilde{y'_i}))$$
$$\wedge\; Audit(PP, m'_i, tag'_i, \{pk_j\}_{j\in s, j\neq \pi}, pk_\pi, \cdot) \to 1:$$
$$(SK, PK, \cdot) \leftarrow KeyGen(PP), PP \leftarrow SysGen(1^\lambda, d)\Big] \le negl(\lambda).$$

Definition 5 (Linkability). *The proposed protocol is linkable if any adversary cannot use a single private key sk_π and $type_\ell$ to generate two different tags t, for any security parameter λ and negligible function $negl(\cdot)$,*

$$\Pr\Big[\; \mathcal{A}^{\mathcal{O}_{TagGen}(sk_\pi, type_\ell, \cdot)}(PP, PK) \to (m_i, tag_i = (t_i, \cdot)) \wedge (m'_i, tag'_i = (t'_i, \cdot)) \wedge$$
$$t_i \neq t'_i \wedge Audit(PP, m_i, tag_i, PK, \cdot) \to 1 \wedge Audit(PP, m'_i, tag'_i, PK, \cdot) \to 1:$$
$$(SK, PK, \cdot) \leftarrow KeyGen(PP), PP \leftarrow SysGen(1^\lambda, d)\Big] \le negl(\lambda).$$

Definition 6 (Nonslanderability). *The proposed protocol is nonslanderable if any adversary cannot generate integrity audit tags to falsely accuse others, for any security parameter λ and negligible function $negl(\cdot)$,*

$$\Pr\Big[\; \mathcal{A}^{\mathcal{O}_{TagGen}(sk_\pi, \cdot)}(PP, PK) \to (m_i, tag_i = (t_i, \cdot)) \wedge t_i \neq t_{i'}$$
$$\wedge\; Audit(PP, m_i, tag'_i = (t'_i, \cdot), PK, \cdot) \to 1:$$
$$(SK, PK, \cdot) \leftarrow KeyGen(PP), PP \leftarrow SysGen(1^\lambda, d)\Big] \le negl(\lambda).$$

5 The Proposed Protocol

5.1 Overview

Evaluating data contribution for each member in a multi-owner group is critical for fair distribution. A straightforward solution that appending signature of data contribution to PDP tags is problematic due to privacy issues, i.e., revealing details and owner's identity. To enable a provable data contribution for group transactions with traceable anonymous, we separate data contribution to two aspects, including volume and weight. The first challenge is, *how to obtain the volume of data without disclosing relationship between data and its owner?* Motivated by TA-DPDP [6] that allows partial anonymity for group transactions through linkable ring signatures, we adapt linkable ring signatures again to enable countable and conditionally traceable PDP tags. The *countable* and *conditionally traceable* here mean the volume of data can be counted through tags, and the owner of tags cannot be identified unless the linkable commitment is revealed, respectively.

The next challenge is, *how to manage the linkable commitment and validate data weight without further privacy leakage?* We introduce a semi-trusted Quantification Authority (QA). Besides, we design a novel weighted PDP tag structure that embeds data weight, and leverage a succinct IPA to guarantee public verifiability of embedded data weight without opening weight itself. Then the linkable commitment is stored by QA, and the weighted tag is still managed by the cloud. In this manner, the QA only validates rather than obtaining the data weight belonging to specific owner using linkable commitment and IPA, while the cloud only knows the volume of data from the weighted tag without knowledge of weight and origins of the underlying data. In other words, the data contribution details are separated to two semi-trusted parties, and hence the privacy of data contribution is preserved. In addition, our PDC supports batch verification in ownership confirmation by taking advantage of homomorphic property of the weighted tag.

5.2 The Proposed TA-PDC Protocol

The file F, generated collaboratively by the data owners, is divided into n blocks, denoted $F = (m_1, m_2, ..., m_n)$. Each data block m_i is further partitioned into s sub-blocks, such as $m_i = (m_{i1}, m_{i2}, ..., m_{is})$, where s denotes the number of users in the group. Let $\Omega = \langle Enc, Dec, Sig, Vrf \rangle$ be a hybrid encryption and signature scheme. The encryption algorithm $\Omega.Enc(\cdot)_{qpk}$ denotes encryption under the public key qpk.

- $SysGen(1^\lambda, d) \to PP$: On input the security parameter λ and the number of data types d, the system executes as follows:
  1. Select an elliptic curve group $\mathbb{G}$ with $\mathbb{F}$ as its scalar field.
  2. Pick two generators $g, h \in_R \mathbb{G}$ and a bilinear pairing $e : \mathbb{G} \times \mathbb{G} \to \mathbb{G}_T$.
  3. Choose two hash functions $H_0 : \{0,1\}^* \to \mathbb{G}, H_1 : \{0,1\}^* \to \mathbb{F}$.
  4. Let $H = \{\omega, \omega^2, \ldots, \omega^d\}$ be a multiplicative subgroup of $\mathbb{F}$ of order d.
  5. Sample a random element $\tau \in_R \mathbb{F}$, computes $(g^\tau, h^\tau, g^{\tau^d})$ and common reference string CRS using the Lagrange polynomials defined over H as:

$$\boldsymbol{g}_\mathcal{L} := [g_1, g_2, \cdots, g_d] = [g^{\mathcal{L}_{1,H}(\tau)}, g^{\mathcal{L}_{2,H}(\tau)}, \cdots, g^{\mathcal{L}_{d,H}(\tau)}]$$

$$\boldsymbol{h}_\mathcal{L} := [h_1, h_2, \cdots, h_d] = [h^{\mathcal{L}_{1,H}(\tau)}, h^{\mathcal{L}_{2,H}(\tau)}, \cdots, h^{\mathcal{L}_{d,H}(\tau)}]$$

6. Return system parameter $PP = (\mathbb{G}, \mathbb{F}, e, g, h, H_0, H_1, g^\tau, h^\tau, g^{\tau^d}, CRS)$.

- $KeyGen(PP) \rightarrow (SK, PK, VK, qsk, qpk)$: On input the system parameter PP, DO generates key pairs (PK, SK), the system generates the public verification key VK, and the QA generates key pairs (qsk, qpk) as follows:
  1. The user selects $sk_j = (x_j, y_j) \leftarrow_R \mathbb{F}$ and computers $pk_j = g^{x_j} h^{y_j}$.
  2. The QA select $qsk \leftarrow_R \mathbb{F}$ and computers $qpk = g^{qsk}$.
  3. Determine the data type vector $\boldsymbol{Type} = [type_1, type_2, ..., type_d]$ and its corresponding weight vector $\boldsymbol{w} = [w_1, w_2, ..., w_d]$, while calculating its KZG commitment as $g_w = \prod_{i \in [d]} g_i^{w_i}$.
  4. Calculate the vanishing polynomial commitment as $g_z = g^{\tau^d}/g$.
  5. Return the public verification key $VK = (g_w, g_z, g^{1/d}, h^{1/d})$.

- $TagGen(PP, m_i, SK, PK, fid, \boldsymbol{Type}) \rightarrow (tag_i, g_{c_j}, \overline{td_j})$: Input the PP, file $F = (m_1, m_2, ..., m_n)$, public key set PK, file identifier fid, and data type vector $\boldsymbol{Type}$. The user U_π generates a block tag by sk_π as follows:
  1. Compute $\eta = H_0(fid)$ and $t_i = \eta^{x_\pi H_1(type_\ell)}$, where $type_\ell$ denotes the data type of the block m_i.
  2. Select two random elements $r_{x_i}, r_{y_i} \leftarrow \mathbb{F}$, compute $\varphi = r_{x_i} + x_\pi m_{i\pi}(1 - H_1(type_\ell))$, and calculate:
  $$\sigma_i = g^\varphi h^{r_{y_i}} \cdot H_0(fid||i) \cdot pk_\pi^{H_1(type_\ell) \cdot H_1(r_{x_i})} \prod_{j \neq \pi} pk_j^{m_{ij}},$$
  $$K_i = \eta^{r_{x_i}} \cdot H_0(fid||i) \cdot t_i^{H_1(r_{x_i}) + \sum_{j \neq \pi} m_{ij}}.$$
  3. Compute $\widetilde{x}_i$ and $\widetilde{y}_i$ as:
  $$\widetilde{x}_i = r_{x_i} - x_\pi H_1(type_\ell)(m_{i\pi} - H_1(r_{x_i})),$$
  $$\widetilde{y}_i = r_{y_i} - y_\pi(m_{i\pi} - H_1(type_\ell)H_1(r_{x_i})).$$
  4. After the generation of all block tags, each user obtains contribution vector $\boldsymbol{c}_j = [c_{j1}, c_{j2}, ..., c_{jd}]$ and calculate its commitment $g_{c_j} = \prod_{\ell=1}^{d} g_\ell^{c_{j\ell}}$.
  5. The user computes the linkable trapdoor $td_j = \eta^{x_\pi}$, and encrypt it using $QA's$ public key qpk as $\overline{td_j} = \Omega.Enc(td_j)_{qpk}$.
  6. Let $tag_i = (t_i, \sigma_i, K_i, \widetilde{x}_i, \widetilde{y}_i)$ and upload $\{m_i, tag_i\}_{i \in [1,n]}$ to the CS for remote storage. DO also sends $\{(g_{c_j}, \overline{td_j})\}_{j \in [1,s]}$ to the CS for the contribution verification.

- $Audit(PP, m_i, tag_i, PK, fid) \rightarrow 1/0$: On input the PP, data blocks m_i along with tags tag_i, data owners public key set PK, file identifier fid, The TPA interacts with the CS as follows:
  1. The TPA selects a c-element subset $\mathbb{I} \subseteq [1, n]$ and random values $v_i \in_R \mathbb{F}$ correspond to m_i.
  2. The TPA sends the challenge $Q = \{i, v_i\}_{i \in \mathbb{I}}$ to the CS.
  3. The CS receives the challenge Q, obtains the data blocks m_i and its corresponding tags tag_i, while computing the storage proof P as:
  $$P\sigma = \prod_{i \in \mathbb{I}} (\sigma_i \cdot K_i)^{v_i}, \quad X = \sum_{i \in \mathbb{I}} v_i \widetilde{x}_i, \quad Y = \sum_{i \in \mathbb{I}} v_i \widetilde{y}_i,$$
  $$\mu_j = \sum_{i \in \mathbb{I}} v_i m_{ij} \quad (j \in [1, s]), \quad \Theta = \prod_{i \in \mathbb{I}} t_i^{\sum_{j=1}^{s} v_i m_{ij}}.$$

4. The CS sends the proof $P = (P\sigma, X, Y, \{\mu_j\}_{j\in[1,s]}, \Theta)$ to the TPA.
5. The TPA receives the proof P, and verifies $P\sigma$ validity using the PP and public key PK via the verification equation:

$$P\sigma = (g \cdot \eta)^X \cdot h^Y \cdot \Theta \cdot \prod_{i\in\mathbb{I}} H_0(fid||i)^{2v_i} \cdot \prod_{j=1}^{s} pk_j^{\mu_j}.$$

6. The TPA outputs "1" if the equations hold, indicating that the challenged data blocks are correctly and fully stored. Otherwise, it concludes that at least one of the challenged blocks has been corrupted in the CS.

- $Quantify(PP, t_i, VK, g_{c_j}, \overline{td_j}, \boldsymbol{Type}, \boldsymbol{w}) \to W_j$: Input PP, partial tags t_i, public verification key VK, contribution commitment g_{c_j}, encrypted trapdoor $\overline{td_j}$, type vector $\boldsymbol{Type}$, and weight vector $\boldsymbol{w}$ while the CS interacts with the QA as follows:
  1. The CS sends $\{t_i\}_{i\in[1,n]}$ and $\{\overline{td_j}\}_{j\in[1,s]}$ to the QA.
  2. The QA decrypts the trapdoor as $td_j = \Omega.Dec(\overline{td_j})_{qsk}$.
  3. For each tag t_i, the QA checks whether the equation $td_j^{H_1(type_\ell)} = t_i$ holds. If the equation holds, it set $c_{j\ell} \leftarrow c_{j\ell} + 1$. After processing all tags, the QA obtains the complete set of user contribution vectors $\{\boldsymbol{c}_j = [c_{j1}, c_{j2}, ..., c_{jd}]\}_{j\in[1,s]}$ and only reflects the distribution of contributions without revealing user identities.
  4. The QA verifies whether the total quantity of each data type satisfies the buyer's requirements; otherwise, reject the transaction.
  5. The QA computes commitment $g_{c_j} = \prod_{\ell=1}^{d} g_\ell^{c_{j\ell}}$ and lagrange interpolation polynomial $c_j(x)$ for each contribution vector $\boldsymbol{c}_j$, while calculating the corresponding contribution as $W_j = <\boldsymbol{w}, \boldsymbol{c}_j>$.
  6. The QA computes $\xi_j = H_{FS}(g_{c_j}, g_w, W_j)$, where H_{FS} is a random oracle derived from $H_1(\cdot)$ using domain separation.
  7. The QA computes the lagrange interpolation polynomial $w(x)$ of the weight vector $\boldsymbol{w}$ and the vanishing polynomial $z_H(x) = x^d - 1$, while obtaining three unique polynomial $q_{c_j}(x), r_{c_j}(x), p_{c_j}(x)$ via Lemma 1:

$$c_j(x)w(x) = q_{c_j}(x) \cdot z_H(x) + x \cdot r_{c_j}(x) + W_j \cdot d^{-1},$$
$$p_{c_j}(x) = x \cdot r_{c_j}(x) + W_j \cdot d^{-1}.$$

Then computes the merge polynomials $q_o(x), r_o(x)$ and $p_o(x)$ as:

$$q_o(x) = \xi_1 \cdot q_{c_1}(x) + \xi_2 \cdot q_{c_2}(x) + ... + \xi_s \cdot q_{c_s}(x),$$
$$r_o(x) = \xi_1 \cdot r_{c_1}(x) + \xi_2 \cdot r_{c_2}(x) + ... + \xi_s \cdot r_{c_s}(x),$$
$$p_o(x) = r_o(x) \cdot x + (\xi_1 \cdot W_1 + \xi_2 \cdot W_2 + ... + \xi_s \cdot W_s) \cdot d^{-1}.$$

  8. The QA calculates $g_q = g^{q_o(\tau)}, g_r = g^{r_o(\tau)}, h_p = h^{p_o(\tau)}$ (if $q_o(x) = a_0 + a_1 x + ... + a_{d-1} x^{d-1}, g^{q_o(\tau)} = \prod_{\ell=0}^{d-1} g_\ell^{a_\ell}$) and sets the proof $\pi = (g_q, g_r, h_p)$, and sends $(\{W_j\}_{j\in[1,s]}, \pi, \Omega.Sig(\{W_j\}_{j\in[1,s]}||\pi)_{qsk})$ to CS.

9. The CS recovers $\{W_j\}_{j\in[1,s]}$ and π if $\Omega.Vrf(\{W_j\}_{j\in[1,s]}\|\pi)_{qpk} = 1$. Then, it computes $\xi_j = H_{FS}(g_{c_j}, g_w, W_j)$ and verifies that the following two equations hold using the public verification key VK:

$$e\left(\prod_{j=1}^{s}(g_{c_j})^{\xi_j},\ g_w\right) = e(g_q, g_z) \cdot e(g_r, g^{\tau}) \cdot e\left(\prod_{j=1}^{s} g^{\xi_j W_j},\ g^{1/d}\right),$$

$$e(h_p, g) = e(g_r, h^{\tau}) \cdot e\left(\prod_{j=1}^{s} g^{\xi_j W_j},\ h^{1/d}\right).$$

10. The CS assigns user rights according to the contributions W_j only if both equations hold. Otherwise, it concludes that at least one of the user contributions is incorrect.

- $Query(PP, type_\ell, g_{c_j}) \to c_{j\ell}$: Input PP, queried data type $type_\ell$, and user's contribution commitment g_{c_j}, the CS interacts with the user as follows:
  1. The user computes the polynomial $\psi_\ell(x) = \frac{c_j(x) - c_j(\ell)}{x - \ell}$, derives the proof $\pi_\ell = g^{\psi_\ell(\tau)}$, and sends the queried value $c_{j\ell}$ along with π_ℓ to the CS.
  2. The CS checks the $c_{j\ell}$ and π_ℓ using the user's contribution commitment g_{c_j} via the verification equation:

$$e(g_{c_j}, g) = e(\pi_\ell, g^{\tau}/g^{\ell}) \cdot e(g, g)^{c_{j\ell}}.$$

If the equation holds, then it is confirmed that the user has a contribution of $c_{j\ell}$ under the data type $type_\ell$; otherwise, the claimed contribution is considered invalid.

6 Correctness and Security Analysis

We provide proofs of the following several theorems to demonstrate the achievements of the defined correctness, unforgeability, anonymity, linkability, and non-slanderability defined.

Theorem 1 (Correctness). *The proposed protocol is correct. Concretely, if DO uploads its data honestly and the CS preserves them well, then the proof responded by the CS is valid with overwhelming probability.*

Proof. We prove the correctness of the proposed protocol by showing that the verification equation holds, as it is satisfied only for valid proofs. The correctness follows from the derivation below:

$$P\sigma = \prod_{i\in\mathbb{I}}(\sigma_i \cdot K_i)^{v_i}$$

$$= \prod_{i\in\mathbb{I}}\left(g^{r_{x_i} + x_\pi m_{i\pi}(1 - H_1(type_\ell))} h^{r_{v_i}} H_0(fid\|i) pk_\pi^{H_1(type_\ell) \cdot H_1(r_{x_i})} \prod_{j\neq\pi} pk_j^{m_{ij}}\right)^{v_i}$$

$$\cdot \prod_{i\in\mathbb{I}}\left(\eta^{r_{x_i}}\cdot H_0(fid\|i)\cdot t_i^{H_1(r_{x_i})+\sum_{j\neq\pi}m_{ij}}\right)^{v_i}$$

$$=\prod_{i\in\mathbb{I}}\left(g^{\tilde{x}_i}h^{\tilde{y}_i}\cdot H_0(fid\|i)\cdot\prod_{j=1}^{s}pk_j^{m_{ij}}\right)^{v_i}\cdot\prod_{i\in\mathbb{I}}\left(\eta^{\tilde{x}_i}\cdot H_0(fid\|i)\cdot t_i^{\sum_{j=1}^{s}m_{ij}}\right)^{v_i}$$

$$=(g\cdot\eta)^{\sum_{i\in\mathbb{I}}v_i\tilde{x}_i}\cdot h^{\sum_{i\in\mathbb{I}}v_i\tilde{y}_i}\cdot\prod_{j=1}^{s}pk_j^{\sum_{i\in\mathbb{I}}v_im_{ij}}\cdot\prod_{i\in\mathbb{I}}H_0(fid\|i)^{2v_i}\cdot t_i^{\sum_{j=1}^{s}v_im_{ij}}$$

$$=(g\cdot\eta)^{X}\cdot h^{Y}\cdot\Theta\cdot\prod_{i\in\mathbb{I}}H_0(fid\|i)^{2v_i}\cdot\prod_{j=1}^{s}pk_j^{\mu_j}.$$

Theorem 2 (Unforgeability). *The proposed protocol is unforgeable. Specifically, if DLP is hard, it is computationally infeasible for all adversaries to forge provably valid tags for any data with non-negligible probability in the random oracle model.*

Proof. Suppose that there exists a PPT adversary $\mathcal{A}$ that is capable of breaking the unforgeability of the proposed protocol. We construct a simulator $\mathcal{B}$ to break the DLP assumption by interacting with $\mathcal{A}$ as follows.

- *Setup.* Given n DLP problem instances $(X_1,\ldots,X_s)$ and a generator $g\in\mathbb{G}$, $\mathcal{B}$ chooses $x'\leftarrow_R\mathbb{F}$ and sets $h=g^{x'}$. $\mathcal{B}$ also chooses $y_j\leftarrow_R\mathbb{F}$ and sets $Z_j=X_j\cdot h^{y_j}$ for all $j\in[1,s]$.
- *Hash query.* This phase is for hash queries of H_0 and H_1. The times of queries to $H_0(fid)$ is q_1, to $H_1(r_i)$ is q_2, and to $H_1(r_{x_i})$ is q_3. The queries for the hash $H_0(fid)$, $H_1(r_i)$, and $H_1(r_{x_i})$ are recorded in the records of empty tables T_1,T_2,T_3 generated by $\mathcal{B}$. For queries of the ith block in file with abstract m_i, if they are searchable in tables, $\mathcal{B}$ returns these recorded values; otherwise, it executes as follows:
  - For query $H_0(fid)$, $\mathcal{B}$ selects $a_i\in\mathbb{F}$, sets $H_1(fid)=g^{a_1}$, and responds to the hash query with $H_1(fid)$.
  - For query $H_0(fid\|i)$, $\mathcal{B}$ selects $b_i\in\mathbb{F}$, sets $H_0(fid\|i)=b_i$, records $(fid,i,b_i,\mathcal{A})$ in T_2, and responds to the hash query with $H_0(fid\|i)$.
  - For query $H_1(r_{x_i})$, $\mathcal{B}$ selects $c_i\in\mathbb{F}$, sets $H_1(r_{x_i})=c_i$, records $(r_{x_i},c_i,\mathcal{A})$ in T_3, and responds to the hash query with $H_1(r_{x_i})$.
  - For query $H_1(type_\ell)$, $\mathcal{B}$ selects $d_\ell\in\mathbb{F}$, sets $H_1(type_\ell)=d_\ell$, records $(type_\ell,d_\ell,\mathcal{A})$ in T_4, and responds to the hash query with $H_1(type_\ell)$.
- *Signing query.* On input a signing query for file identifier fid, a set of public keys $\mathcal{PK}=\{Z_1,\ldots,Z_s\}$, the public key for the data owner Z_π, where $\pi\in[1,s]$, the data type $type_l$, and a data block $m_i=\{m_{i1},\ldots,m_{is}\}$, $\mathcal{B}$ simulates as follows:
  - If the query of η_i is signed by Z_π, $\mathcal{B}$ sets $t=X_\pi^{a_i\cdot d_\ell}$.
  - $\mathcal{B}$ randomly chooses $r_{x_i},r_{y_i}\in\mathbb{F}$. $\mathcal{B}$ computes σ_i and K_i by:

$$\sigma_i=g^{r_{x_i}+x_\pi m_{i\pi}(1-d_\ell)}h^{r_{y_i}}b_ipk_\pi^{d_\ell\cdot c_i}\prod_{j\neq\pi}pk_j^{m_{ij}},\ K_i=\eta^{r_{x_i}}\cdot b_i\cdot t_i^{c_i+\sum_{j\neq\pi}m_{ij}}$$

and $\mathcal{B}$ computes $\tilde{x}_i, \tilde{y}_i$ as:

$$\tilde{x}_i = r_{x_i} - x_\pi d_\ell(m_{i\pi} - c_i), \quad \tilde{y}_i = r_{y_i} - y_\pi(m_{i\pi} - d_\ell c_i).$$

- *Forgery.* Eventually, $\mathcal{A}$ successfully forges a valid tag for a data block as $tag'_i = (t'_i, \sigma'_i, K'_i, \tilde{x}'_i, \tilde{y}'_i)$ and $m_i = \{m_{i1}, \ldots, m_{is}\}$ on an fid and a set of public key $\mathcal{PK}$. The corresponding hash responses from $\mathcal{B}$ are $H_1(r_{x_i}) = c_i$. Then there is:

$$\sigma'_i = g^{\tilde{x}'_i} h^{\tilde{y}'_i} \prod_{j=1}^{s} Z_j^{m_{ij}} = g^{(r'_{x_i} - x_\pi d'_\ell(m_{i\pi} - c'_i))} h^{(r'_{y_i} - y_\pi(m_{i\pi} - d'_\ell c'_i))} b_i \prod_{j=1}^{s} Z_j^{m_{ij}}$$

Then, $\mathcal{B}$ can compute

$$\sigma'_i \bigg/ b_i \cdot \prod_{j=1}^{s} Z_j^{m_{ij}} = g^{r'_{x_i} - x_\pi d'_\ell(m_{i\pi} - c'_i)} h^{(r'_{y_i} - y_\pi(m_{i\pi} - d'_\ell c'_i))}.$$

According to the generalized forking lemma, $\mathcal{B}$ can solve the DLP problem by running the forking algorithm to obtain two outputs $tag'_i = (t'_i, \sigma'_i, K'_i, \tilde{x}'_i, \tilde{y}'_i)$ with c'_i and $tag''_i = (t''_i, \sigma''_i, K''_i, \tilde{x}''_i, \tilde{y}''_i)$ with c''_i on the same m_i, where $c'_i \neq c''_i$. Now we have found the solution to the DLP problem:

$$x_\pi = \frac{x' y_\pi (c'_i - c''_i)}{c'_i - c''_i}$$

Theorem 3 (Anonymity). *The proposed protocol is anonymous. Specifically, it is computationally infeasible for any adversary to identify the actual data owner of a data block m_i and its corresponding tag tag_i in the file F.*

Proof. The tag can be generated by any data owner in the ring using their private key, and the auditing process does not reveal any distinguishing information among data owners. We now present an information-theoretic proof showing that adversary $\mathcal{A}$ has zero advantage in identifying the actual data owner. The proof is structured into three parts as follows:

- *Part I.* Let x, ρ, γ be the values such that $t = \eta^{x\rho}$ and $g = h^\gamma$. Furthermore, let $Z_i = h^{z_i}$ for $i = 1$ to n. For each $\pi \in \{1, \ldots, n\}$, consider the values:

$$x_\pi = x \bmod p, \quad y_\pi = z_\pi - x_\pi \gamma \bmod p.$$

Obviously, (x_π, y_π) is a private key corresponding to the public key Z_π (since $Z_\pi = h^{z_\pi} = h^{x_\pi \gamma + y_\pi} = g^{x_\pi} h^{y_\pi}$) and $t = \eta^x = \eta^{x_\pi}$.
- *Part II.* For each possible (x_π, y_π) defined in Part I, consider the values:

$$r_{x_\pi} := \tilde{x} + x_\pi m_{i\pi} \bmod p, \quad r_{y_\pi} := \tilde{y} + y_\pi m_{i\pi} \bmod p.$$

It can be seen that σ is created by the private key (x_π, y_π) using randomness (r_{x_π}, r_{y_π}), for any $\pi \in \{1, \ldots, n\}$.

– **Part III.** It is evident that the distributions of $(x_\pi, y_\pi, r_{x_\pi}, r_{y_\pi})$ for each possible value π are identical and adhere to the distribution of a tag created by the data owner with public key Z_π.

In other words, the tag σ can be generated by any data owner who has the private key (x_π, y_π) for any $\pi \in \{1, \ldots, n\}$, utilizing the randomness (r_{x_π}, r_{y_π}). Even if a adaptive adversary could compute all values $(x_\pi, y_\pi, r_{x_\pi}, r_{y_\pi})$ for all π from 1 to n, it would still be unable to determine the actual identity of data owner.

Theorem 4 (Linkability). *The proposed protocol is linkable. Specifically, if the DLP is hard, it is computationally infeasible for any adversary to produce two valid tags that are unlinked with just one private key.*

Lemma 2. *If an adversary $\mathcal{A}$ knows only one private key $sk_\pi = (x_\pi, y_\pi)$, $\pi \in [1, s]$ and produces a valid tag $tag_i = (t_i, \sigma_i, K_i, \tilde{x}_i, \tilde{y}_i)$ for a file identifier fid, data type $type_\ell$, and $m_i = \{m_{i1}, \ldots, m_{is}\}$, then $t_i = H_0(fid)^{x_\pi H_1(type_\ell)}$, provided that DLP is hard in the random oracle model.*

Proof. We use the standard proof-of-knowledge proving technique in the random oracle model. Suppose $\mathcal{A}$ produces two valid tags $tag_i^1 = (t_i, \sigma_i^1, K_i^1, \tilde{x}_i^1, \tilde{y}_i^1)$ and $tag_i^2 = (t_i, \sigma_i^2, K_i^2, \tilde{x}_i^2, \tilde{y}_i^2)$ in different file blocks $m^1 \neq m^2$, where $t = H_0(fid)^{\hat{x} H_1 type_\ell}$ for $\hat{x} \in \mathbb{Z}_p$. We set $\sigma_i = g^\alpha h^{\alpha'}$ and $K_i = H_0(fid)^\lambda$ while constructing the following equation:

$$m^1 = m_1^1 + \cdots + m_s^1, \tag{1}$$
$$m^2 = m_1^2 + \cdots + m_s^2,$$
$$\alpha = \tilde{x}_i^1 + x_1 m_1^1 + \cdots + x_s m_s^1$$
$$= \tilde{x}_i^2 + x_1 m_1^2 + \cdots + x_s m_s^2,$$

$$\lambda = \tilde{x}_i^1 + \hat{x} m^1 = \tilde{x}^2 + \hat{x} m^2. \tag{2}$$

We evaluate the possible values of $\hat{x}$ for enabling $\mathcal{A}$ to generate two such sequences by having only one private key (x_π, y_π). We consider two cases as follows:

– **Case 1:** Suppose all $m_i^1 = m_i^2$ except when $i = j$ for some $j \in [1, s]$. From (1) and (2), since $m_j^1 \neq m_j^2$, and hence $\tilde{x}_1 \neq \tilde{x}_2$. Then, we have

$$\frac{\tilde{x}_1 - \tilde{x}_2}{m_j^2 - m_j^1} = x_j \quad \text{from (1)},$$

$$\frac{\tilde{x}_1 - \tilde{x}_2}{m_i^2 - m_i^1} = \frac{\tilde{x}_1 - \tilde{x}_2}{m_j^2 - m_j^1} = \hat{x} \quad \text{from (2)}.$$

That is, $x_j = \hat{x}$ which is known by $\mathcal{A}$. Since $\mathcal{A}$ is assumed to know only one private key, we have $j = \pi$.

– **Case 2:** Suppose that all $m_i^1 = m_i^2$ except when $i \in \{j_1, j_2\}$ for some $j_1, j_2 \in [1, s]$. We have $m_{j_1}^1 \neq m_{j_1}^2$, $m_{j_2}^1 \neq m_{j_2}^2$, and $\tilde{x}_1 \neq \tilde{x}_2$. We have

$$\tilde{x}_1 + x_{j_1} m_{j_1}^1 + x_{j_2} m_{j_2}^1 = \tilde{x}_2 + x_{j_1} m_{j_1}^2 + x_{j_2} m_{j_2}^2.$$

If $\pi \in \{j_1, j_2\}$, then $\mathcal{A}$ knows both x_{j_1} and x_{j_2}, which contradicts our assumption that $\mathcal{A}$ knows only one private key. If $\pi \notin \{j_1, j_2\}$, $\mathcal{A}$ obtains the following relation:

$$x_{j_1} + x_{j_2} \phi_2 = \phi_1,$$

where

$$\phi_2 = \frac{m_{j_2}^1 - m_{j_2}^2}{m_{j_1}^1 - m_{j_1}^2} \quad \text{and} \quad \phi_1 = \frac{\tilde{x}_2 - \tilde{x}_1}{m_{j_1}^1 - m_{j_1}^2}.$$

This implies that $\mathcal{A}$ can solve the following problem: Given $g_1, g_2 \in \mathbb{G}$, find $\phi_1, \phi_2 \in \mathbb{F}$ such that $g_1 \cdot g_2^{\phi_2} = g^{\phi_1}$. This is hard if g_1, g_2 are generated independently and randomly. We can show that the problem is computationally equivalent to DLP. We omit the details here. Since we suppose DLP is hard, this case should not exist.

In conclusion, only case 1 is possible. That is, if $\mathcal{A}$ only knows one private key (x_π, y_π), we have $t = H_0(fid)^{x_\pi H_1(type_\ell)}$.

If $\mathcal{A}$ produces two valid tags such that they are unlinkable, that is, $t^1 \neq t^2$, where t^1 and t^2 are the linking tags, we have $t^1 = \eta^{x_{\pi_1} H_1(type_\ell)}$ and $t^2 = \eta^{x_{\pi_2} H_1(type_\ell)}$, where $\eta = H_0(fid)$. From Lemma 1, $\mathcal{A}$ must know x_{π_1} and x_{π_2}. It contradicts the assumption that $\mathcal{A}$ only knows one private key.

Theorem 5 (Nonslanderability). *The proposed protocol is nonslanderable. Specifically, if DLP is hard, it is computationally infeasible for any adversary to forge valid tags on behalf of others with nonnegligible probability in the random oracle model.*

Proof. Suppose $\mathcal{A}$ produces a valid tag $tag_i^* = (t^*, \cdot)$ such that it is not an output from *Signing query* and it is linked to tag_i. Since they are linkable, we have $t^* = t$. From Lemma 2, the data owner must have knowledge of $x_{\pi'}$, where $t^* = t = H_0(fid)^{x_{\pi'} H_1(type_\ell)}$ and $\pi' \in [1, s]$. Since $\mathcal{B}$ uses the private key sk_π to generate σ (as the corresponding linking tag $t^* = H_0(fid)^{x_\pi H_1(type_\ell)}$, we have $\pi = \pi'$. That is, $\mathcal{A}$ must know $sk_{\pi'}$ to generate σ^*, which contradicts the assumption that $\mathcal{A}$ is not allowed to query sk_π.

7 Implementation and Evaluation

This section presents both theoretical and experimental evaluations of our design. For comparison, we select the protocol Oruta [14] as a benchmark because since it support multi-ownership data auditing with anonymous.

7.1 Efficiency Analysis

We implement and evaluate our protocol in golang. Our implementation is publicly available at https://github.com/XiaocongLin01/TA-PDC. We use the BLS12-381 asymmetric pairing based curve implementation from gnark-crypto [26], where each $\mathbb{F}$ element is 256 bits, each $\mathbb{G}_1$ element is 384 bits, and each $\mathbb{G}_2$ element is 768 bits. Simulation experiments are run on Ubuntu 18.04 VMware 17.5 on a laptop running Windows 11 with Intel(R) Core(TM) Ultra 5 125H 3.60 GHz 16 GB RAM. Measurements are performed using Go's built-in benchmarking framework, which adaptively adjusts the number of iterations to produce statistically meaningful results.

Table 2. Communication Cost and Computational Overhead Comparison

Protocols	Oruta [14]	Ours
Tag generation	$ns\|\mathbb{G}_1\|$	$2n\|\mathbb{F}\| + 3n\|\mathbb{G}_1\|$
Proof generation	$k\|\mathbb{F}\| + (k+s)\|\mathbb{G}_1\|$	$(s+2)\|\mathbb{F}\| + 2\|\mathbb{G}_1\|$
Quantification on user side	–	$2\|\mathbb{G}_1\|$
Quantification on QA side	–	$s\|\mathbb{F}\| + 3\|\mathbb{G}_1\|$
Quantification on CS side	–	$(s+n)\|\mathbb{G}_1\|$
User contribution query	–	$\|\mathbb{G}_1\|$
Tag generation	$(k+s)n \cdot \mathbb{G}_1^{Exp} + sn \cdot \mathbb{G}_2^{Exp}$	$(d+s+4)n \cdot \mathbb{G}_1^{Exp}$
Proof generation	$(k+sc) \cdot \mathbb{G}_1^{Exp}$	$3c \cdot \mathbb{G}_1^{Exp}$
Proof verification	$(c+2k) \cdot \mathbb{G}_1^{Exp} + (s+2)P$	$(c+s+2) \cdot \mathbb{G}_1^{Exp}$
Quantification on user side	–	$d \cdot \mathbb{G}_1^{Exp}$
Contribution generation	–	$(s+5d) \cdot \mathbb{G}_1^{Exp}$
Contribution verification	–	$3s \cdot \mathbb{G}_1^{Exp} + 7P$
User contribution query	–	$d \cdot \mathbb{G}_1^{Exp} + \mathbb{G}_2^{Exp} + 3P$

The communication cost analysis is presented in the upper part of Table 2, where k represents the number of sectors [14]. In the tag generation phase, the communication complexity of Oruta is $O(ns)$, while our protocol has a complexity of $O(n)$. In the proof generation phase, our protocol requires only two constant elements in $\mathbb{G}_1$, which is fewer than the $k+s$ elements required by Oruta. In the quantification phase, the communication overhead for the user is constant for two elements in $\mathbb{G}_1$. The QA requires s elements in $\mathbb{F}$ and three elements in $\mathbb{G}_1$, while the CS requires $s+n$ elements in $\mathbb{G}_1$. In the query phase, our protocol requires only one constant element in $\mathbb{G}_1$.

The computational overhead analysis is shown in the lower part of Table 2. Since the time cost of group multiplication is negligible, we only account for exponentiation and pairing operations in the groups. Specifically, exponentiations in $\mathbb{G}_1$ and $\mathbb{G}_2$ are denoted as $\mathbb{G}_1^{Exp}$, $\mathbb{G}_2^{Exp}$ and bilinear pairing as P. According to our benchmarks, a single exponentiation in $\mathbb{G}_1$ takes approximately $82\mu s$, while an exponentiation in $\mathbb{G}_2$ takes $171\mu s$. A single pairing operation requires $537\mu s$. To evaluate the computational overhead of the data auditing phrase, we vary the file size F from 128KB to 8MB, each file being divided into blocks of size 4096 bits. The number of data owners is fixed at 16, and the number of data types is set to 8. Detailed performance results are presented in Fig. 2.

Tag Generation Phrase: The Fig. 2a shows that our protocol requires only 0.998 s per user for an 8MB file. Oruta needs 4.62 s in the same case, which is 4 times slower than ours. As Table 2 shows, although the number of computation operations in our protocol is comparable to Oruta, Oruta performs sn exponentiations in $\mathbb{G}_2$, while all computations in ours are performed in $\mathbb{G}_1$.

Audit Proof Generation Phrase: We set the number of challenged data blocks at 460. As a result, we can see from Fig. 2b that our protocol achieves a constant proof generation time of 0.12 s across different file sizes, which is 5 times faster than Oruta. Referring to Table 2, this is because Oruta requires $k + sc$ group exponentiations, whereas our protocol only performs $3c$ exponentiations.

Audit Proof Verification Phrase: The verification time of our protocol remains constant at $84ms$ for varying file sizes, which is approximately $11ms$ faster than Oruta. This result can be supported by Table 2, Oruta requires an additional $s + 2$ pairing operations, while our protocol does not need pairing operations.

To evaluate the performance of the quantification and query phases, we vary the number of users from 4 to 32, while fixing the number of data types at $\{8, 16, 24\}$ and varying the number of contribution queries from 2 to 16. To enhance computational efficiency, we employ Pippenger's method [27] for multi-exponentiation of group elements. The running times for our protocol are illustrated in Fig. 2.

User Contributions Generation Phrase: According to Fig. 2d, the computational cost for the QA increases with the number of data owners s and the number of data types d. Notably, The QA takes only $127.32ms$ to generate the contribution and the corresponding proofs for 32 users and 24 data types. As Table 2 shows, the QA requires $s + 5d$ exponentiations in $\mathbb{G}_1$ to complete the quantification.

User Contributions Verification Phrase: The Fig. 2e shows that the verification time increases slowly with the number of users s and the number of data types d. Referring to Table 2, this is because the verification process requires only a constant number of 7 pairing operations. The CS completes the verification for 32 users and 24 data types in just $8.14ms$. It is also worth noting that each user only needs to perform d exponentiations in $\mathbb{G}_1$ during the quantification phase.

User Contribution Query Phrase: As shown in Fig. 2f, querying a user's contribution to a specific data type 16 times takes $31.45ms$. The query time increases significantly with the number of queries. Referring to Table 2, this is because each query requires three pairing operations and one exponentiation in $\mathbb{G}_2$.

To evaluate the performance of our protocol in large-scale collaboration scenarios, we extended the number of users s to 1,000 and increased the data types d to 500. The experimental results show that generating the contribution takes $47s$, while verifying the contribution takes only $0.12s$. These results demonstrate that our protocol achieves high efficiency in quantifying user contributions and exhibits strong potential for practical applications.

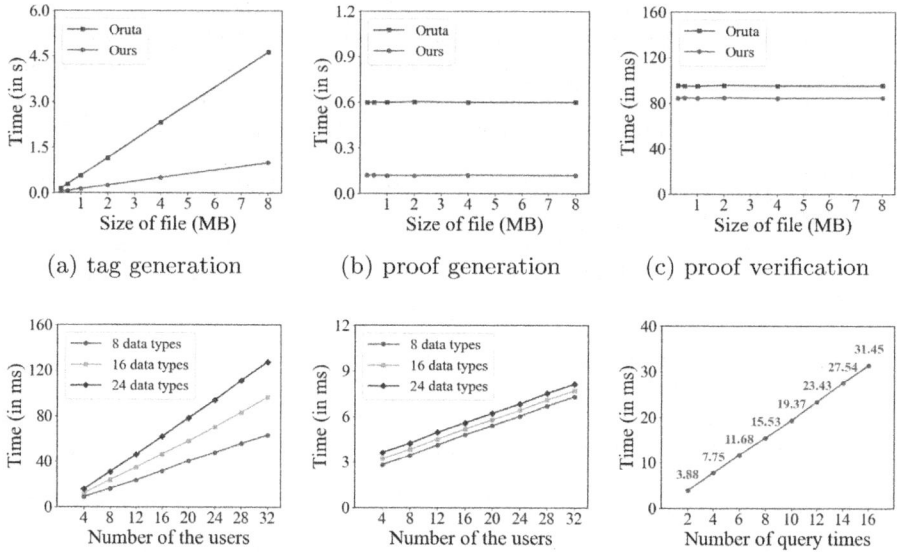

Fig. 2. Time cost of data auditing, quantification and contribution query

8 Conclusion

In this work, we identify that in group transaction scenarios, existing works overlook evaluation of distinct data contributions from different group members. We propose **TA-PDC**, the first **P**rovable **D**ata **C**ontribution with **T**raceable **A**nonymous to addresses this problem. The TA-PDC designs a novel weighted tag structure based on linkable ring signatures and uses IPA to preserve the privacy of data contribution distributions. Security analysis and evaluations justify the security and practicality of our proposal. In future work, we aim to explore more general and practical security models for TA-PDC and further extend it to support richer functional requirements of users.

Acknowledgement. This work was supported in part by the National Natural Science Foundation of China Youth Project (No. 62202102), Scientific and Technological Project of Fujian Province of China (No.2024J08162), and The CCF-NSFOCUS 'Kunpeng' Research Fund (NO.2024004).

References

1. Shen, J., Chen, X., Wei, J., Guo, F., Susilo, W.: Blockchain-based accountable auditing with multi-ownership transfer. IEEE Trans. Cloud Comput. **11**(3), 2711–2724 (2023)
2. Zhang, X., Liu, Q., Liu, B., Zhang, Y., Xue, J.: Dynamic certificateless outsourced data auditing mechanism supporting multi-ownership transfer via blockchain systems. IEEE Trans. Netw. Serv. Manag. **22**(2), 2017–2030 (2025)
3. Wang, Y., You, W., Zhang, Y., Ye, A., Xu, L.: Cloud EMRs auditing with decentralized (t, n)-threshold ownership transfer. Cybersecur. **7**(1), 53 (2024)
4. Wang, Y., Zhang, Y., You, W., Ma, Y., Wang, D.: A fine-grained ownership transfer protocol for cloud EMRs auditing. In: Algorithms and Architectures for Parallel Processing - 24th International Conference, ICA3PP. vol. 15254, pp. 12–21 (2024)
5. Wang, Y., Zhang, Y., Ye, A., Shen, J., Wang, D., Xiang, Y.: Anonymous and efficient (t, n)-threshold ownership transfer for cloud EMRs auditing. IEEE Trans. Inf. Forensics Secur. **20**, 1710–1723 (2025)
6. Wu, C., You, W., Xu, L.: TA-DPDP: dynamic provable data possession for online collaborative system with tractable anonymous. In: Information Security and Cryptology - 20th International Conference, INSCRYPT, pp. 42–62 (2024)
7. Ateniese, G., et al.: Provable data possession at untrusted stores. In: Proceedings of the 14th ACM Conference on Computer and Communications Security, CCS, pp. 598–609 (2007)
8. Juels, A., Jr., B.S.K.: PORs: proofs of retrievability for large files. In: Proceedings of the 2007 ACM Conference on Computer and Communications Security, CCS, pp. 584–597 (2007)
9. Shacham, H., Waters, B.: Compact proofs of retrievability. J. Cryptol. **26**(3), 442–483 (2013)
10. Wang, C., Chow, S.S.M., Wang, Q., Ren, K., Lou, W.: Privacy-preserving public auditing for secure cloud storage. IEEE Trans. Comput. **62**(2), 362–375 (2013)
11. Li, J., Yan, H., Zhang, Y.: Efficient identity-based provable multi-copy data possession in multi-cloud storage. IEEE Trans. Cloud Comput. **10**(1), 356–365 (2022)
12. Shen, W., Qin, J., Yu, J., Hao, R., Hu, J.: Enabling identity-based integrity auditing and data sharing with sensitive information hiding for secure cloud storage. IEEE Trans. Inf. Forensics Secur. **14**(2), 331–346 (2019)
13. Gao, X., Yu, J., Chang, Y., Wang, H., Fan, J.: Checking only when it is necessary: Enabling integrity auditing based on the keyword with sensitive information privacy for encrypted cloud data. IEEE Trans. Dependable Secur. Comput. **19**(6), 3774–3789 (2022)
14. Wang, B., Li, B., Li, H.: Oruta: privacy-preserving public auditing for shared data in the cloud. IEEE Trans. Cloud Comput. **2**(1), 43–56 (2014)
15. Wang, B., Li, B., Li, H.: Public auditing for shared data with efficient user revocation in the cloud. In: Proceedings of the IEEE INFOCOM, pp. 2904–2912 (2013)
16. Wang, B., Li, B., Li, H.: Panda: public auditing for shared data with efficient user revocation in the cloud. IEEE Trans. Serv. Comput. **8**(1), 92–106 (2015)
17. Wang, B., Li, H., Li, M.: Privacy-preserving public auditing for shared cloud data supporting group dynamics. In: Proceedings of IEEE International Conference on Communications, ICC, pp. 1946–1950 (2013)
18. Liu, J.K., Au, M.H., Susilo, W., Zhou, J.: Linkable ring signature with unconditional anonymity. IEEE Trans. Knowl. Data Eng. **26**(1), 157–165 (2013)

19. Liu, J.K., Wei, V.K., Wong, D.S.: Linkable spontaneous anonymous group signature for ad hoc groups. In: Australasian Conference on Information Security and Privacy, ACISP, pp. 325–335. Springer (2004)
20. Tsang, P.P., Wei, V.K.: Short linkable ring signatures for e-voting, e-cash and attestation. In: International Conference on Information Security Practice and Experience, ISPEC, pp. 48–60. Springer (2005)
21. Campanelli, M., Nitulescu, A., Ràfols, C., Zacharakis, A., Zapico, A.: Linear-map vector commitments and their practical applications. In: International Conference on the Theory and Application of Cryptology and Information Security, ASIACRYPT, pp. 189–219. Springer (2022)
22. Das, S., Camacho, P., Xiang, Z., Nieto, J., Bünz, B., Ren, L.: Threshold signatures from inner product argument: succinct, weighted, and multi-threshold. In: Proceedings of the 2023 ACM SIGSAC Conference on Computer and Communications Security, CCS, pp. 356–370 (2023)
23. Campanelli, M., Nitulescu, A., Ràfols, C., Zacharakis, A., Zapico, A.: Linear-map vector commitments and their practical applications. In: 28th International Conference on the Theory and Application of Cryptology and Information Security, ASIACRYPT. vol. 13794, pp. 189–219. Springer (2022)
24. Kate, A., Zaverucha, G.M., Goldberg, I.: Constant-size commitments to polynomials and their applications. In: Abe, M. (ed.) ASIACRYPT 2010. LNCS, vol. 6477, pp. 177–194. Springer, Heidelberg (2010). https://doi.org/10.1007/978-3-642-17373-8_11
25. Ben-Sasson, E., Chiesa, A., Riabzev, M., Spooner, N., Virza, M., Ward, N.P.: Aurora: transparent succinct arguments for R1CS. In: 38th Annual International Conference on the Theory and Applications of Cryptographic Techniques, EUROCRYPT, pp. 103–128. Springer (2019)
26. Botrel, G., Piellard, T., Housni, Y., Tabaie, A., Gutoski, G., Kubjas, I.: Consensys/gnark-crypto: v0. 9.0 (2023)
27. Bernstein, D.J., Doumen, J., Lange, T., Oosterwijk, J.J.: Faster batch forgery identification. In: International Conference on Cryptology in India, INDOCRYPT, pp. 454–473. Springer (2012)

Fine-Filter: An Effective Defense Against Poisoning Attacks on Frequency Estimation Under LDP

Yuxia Zhou, Qiao Xue[✉], and Youwen Zhu

Nanjing University of Aeronautics and Astronautics, Nanjing, China
qiaoxue@nuaa.edu.cn

Abstract. Local Differential Privacy (LDP) has emerged as a standard framework for privacy-preserving data collection. However, recent work [4] reveals that LDP protocols, e.g., Optimized Unary Encoding (OUE), etc., are vulnerable to data poisoning attacks, where malicious users can send carefully-crafted fake data to alter the estimated frequencies. To defend against such attacks, in this paper, we propose a novel scheme named *Fine-filter*, which serves as a plug-in module deployed on the collector side after data aggregation. In Fine-filter, users are divided into two groups by their reported data patterns. We believe that one group contains all the true users and the other includes both true and malicious users. By comparing the statistic information (e.g., frequency of each item) between two groups, we can locate the corrupt items and identify the malicious users with high confidence. Subsequently, the collector estimates item frequencies from the collected data after removing the data of (identified) malicious users. Experimental results demonstrate that Fine-filter can significantly mitigate the negative impact on estimation caused by poisoning attacks.

Keywords: Local Differential Privacy · Frequency Estimation · Data Poisoning Attack · Defense Mechanism

1 Introduction

Local Differential Privacy (LDP) [7,13] has been widely adopted in privacy-preserving data collection, enabling data aggregators to compute accurate statistics while guaranteeing that no individual's raw data is exposed. Among various LDP mechanisms, the Optimized Unary Encoding (OUE) [21] protocol stands out for its simplicity and utility, and is used in industrial applications such as telemetry and recommendation systems.

Despite these advantages, OUE remains vulnerable to adversarial attacks. Recent studies [4] have shown that malicious users can inject crafted randomized responses to manipulate the aggregate results, compromising the reliability of LDP-based analytics. In particular, Maximal Gain Attack (MGA) [4] demonstrates that a small number of malicious users can significantly inflate the estimated frequency of target items, thereby misleading the data collector. This poses serious risks to downstream applications, especially those relying on accurate frequency information for decision-making.

To address this threat, we propose **Fine-filter**, a lightweight yet effective defense framework tailored to detect and mitigate poisoning attacks under OUE. Fine-filter works by exploiting the inherent statistical inconsistencies introduced by malicious users. Our key insight is that malicious users often generate perturbed reports with atypical patterns—such as an exact number of 1s—that differ from the distribution expected under honest reporting. By partitioning users based on the number of 1s in their reports and analyzing the resulting frequency discrepancies, Fine-filter identifies a candidate set of suspicious target items.

However, attackers can further evade detection by introducing randomness in their attack strategy, giving rise to the adaptive version of MGA (MGA-A) [4]. In this setting, malicious users only perturb a random subset of target items in each report, blurring their statistical signature. To counter this challenge, Fine-filter incorporates a second layer of refinement: it analyzes the frequency of observed bit patterns among suspect users, selects the most common attacker patterns, and identifies consistently inactive bits as noise. This allows us to more accurately estimate the number of real target items and filter out false positives.

In summary, our contributions are as follows:

- We propose **Fine-filter**, a novel mechanism to detect and defend against MGA and MGA-A on OUE, without modifying the original LDP protocol.
- We provide a detailed algorithm and theoretical justification for Fine-filter, and demonstrate through experiment that it significantly reduces the influence of malicious users on frequency estimation.
- Our framework is compatible with existing OUE-based systems, and offers strong empirical resilience against attack strategies.

The rest of this paper is organized as follows. Section 2 introduces the background of Local Differential Privacy and the Optimized Unary Encoding (OUE) protocol, as well as the threat model of data poisoning attacks. Section 3 defines the problem and outlines the goals of defense. Section 4 presents the proposed Fine-filter method, including its algorithm and theoretical analysis. Section 5 provides experimental evaluation on real-world datasets. Section 6 reviews related work, and finally, Sect. 7 concludes the paper.

2 Preliminaries

2.1 Local Differential Privacy

Local differential privacy [7,13], as a variation of differential privacy, offers more rigorous privacy protection compared to traditional differential privacy [9]. Different from differential privacy, under the premise that the data collector is considered untrustworthy, local differential privacy adopts the method where users perturb the data themselves and then send the processed data to the data collector. This enables the data collector to obtain valid analysis results while being oblivious to the true values of the data, thereby guaranteeing that the privacy information of individuals is not leaked while conducting statistical analysis on the data [2,3,6,10].

Definition 1 (ε-Local Differential Privacy). *Given n users, each associated with a record, consider an algorithm M with domain $\mathcal{D}$ and range $\widetilde{\mathcal{D}}$. If for any two records v and v' ($v, v' \in \mathcal{D}$), the mechanism M produces the same output y ($y \subseteq \widetilde{\mathcal{D}}$) satisfying the following inequality:*

$$\Pr[M(v) = y] \leq e^{\varepsilon} \cdot \Pr[M(v') = y],$$

then M ensures ε-local differential privacy.

For localized differential privacy (LDP) protocols, particularly those used for frequency estimation, the process typically involves three main steps: encoding, perturbation, and aggregation.

1. **Encoding:** The user's data is transformed into a representation within the encoding domain $\mathcal{D}$.
2. **Perturbation:** The LDP protocol applies its specific perturbation rules to the encoded data within $\mathcal{D}$, effectively obfuscating the original value. The perturbed data is then transmitted to the central server (the data collector).
3. **Aggregation:** The central server uses the perturbed data from all users to estimate the frequencies of items, enabling statistical analysis while preserving user privacy.

2.2 Optimized Unary Encoding (OUE)

OUE [21] is a widely employed and state-of-the-art LDP protocol for frequency estimation. In OUE, the encoding phase involves converting the user's original data $v_i \in \mathcal{D}$ into a one-hot binary vector of length d, uniquely representing the term v_i such that only the v_i-th position is set to 1 while all other positions are set to 0. This transformation ensures a one-to-one correspondence between the original data and the encoded vector.

During the perturbation phase, the user applies bitwise perturbation to the encoded binary vector. For each bit y_i, the true value is retained with a probability p if it corresponds to the position $i = v$, or it is flipped with a probability q otherwise. This process generates a perturbed binary vector $\mathbf{y} = [y_1, y_2, \ldots, y_d]$, where the probability distribution for each bit is formally expressed as:

$$\Pr(y_i = 1) = \begin{cases} \frac{1}{2} = p, & \text{if } i = v, \\ \frac{1}{e^{\varepsilon}+1} = q, & \text{otherwise.} \end{cases}$$

The perturbation mechanism ensures privacy by adding randomness to the binary vector, while still preserving statistical properties that allow for accurate frequency estimation. Compared to methods that reserve or randomly select values, this bit perturbation approach significantly enhances the accuracy of the protocol, particularly when the data domain $\mathcal{D}$ is large.

In the aggregation phase, the data collector gathers the perturbed binary vectors from all users and counts the frequency of 1 s in each bit position across all vectors. Using these counts, the estimated frequency $\hat{f}_v$ of each item $v \in \mathcal{D}$ is computed according to the standard frequency estimation equation as follows [21]:

$$\hat{f}_v = \frac{\frac{1}{n}\sum_{i=1}^{n} \mathbb{1}_{\mathcal{S}(\mathbf{y}_i)}(v) - q}{p - q}, \tag{1}$$

where $\mathbf{y}_i$ is the perturbed value provided by the i-th user, $\mathcal{S}(\mathbf{y})$ represents the set of items supported by $\mathbf{y}$, and $\mathbb{1}_{\mathcal{S}(\mathbf{y}_i)}(v)$ is the feature function that outputs 1 if and only if $\mathbf{y}_i$ supports item v.

Formally, the feature function $\mathbb{1}_{\mathcal{S}(\mathbf{y})}(v)$ is defined as:

$$\mathbb{1}_{\mathcal{S}(\mathbf{y})}(v) = \begin{cases} 1, & \text{if } v \in \mathcal{S}(\mathbf{y}), \\ 0, & \text{otherwise.} \end{cases} \tag{2}$$

In the OUE protocol, the perturbation value $\mathbf{y}$ supports the term v if and only if the v-th bit of $\mathbf{y}$, denoted as y_v, is equal to 1. Formally, for the OUE protocol, the support set is defined as: $S(\mathbf{y}) = \{v \mid v \in \mathcal{D} \text{ and } y_v = 1\}$.

This approach leverages unary encoding and bitwise perturbation to balance privacy and accuracy effectively, making the OUE protocol particularly well-suited for scenarios with large data domains.

2.3 Poisoning Attacks on OUE: MGA and MGA-A

The Maximal Gain Attack (MGA) [4] targets LDP protocols by injecting forged data to artificially increase the frequency of low-occurrence items, disrupting statistical estimates and potentially misleading data-driven decisions in domains like healthcare or marketing.

We focus on OUE (Optimized Unary Encoding), where the attacker injects m malicious users into the system alongside n genuine users. Each malicious user uploads unperturbed forged data $Y = \{y_{n+1}, \ldots, y_{n+m}\}$, crafted to maximize the frequency gain of a target set $T = \{t_1, \ldots, t_r\}$. The attack gain is defined as:

$$G(Y) = \sum_{t \in T} \mathbb{E}[\hat{f}_{t,a} - \hat{f}_{t,b}], \tag{3}$$

which the attacker maximizes by solving:

$$\max_{Y} G(Y) = \max_{Y} \sum_{t \in T} \mathbb{E}[\Delta \hat{f}_t]. \tag{4}$$

In OUE, the attacker encodes each malicious user's report by setting bits corresponding to all $t \in T$ to 1 and adding $l = \lfloor p + (d-1)q - r \rfloor$ randomly

Table 1. List of key notations

Notation	Description
D	Original data space
d	The number of items, $d = |D|$
v	One item, $v \in D$
n, m	The number of genuine or malicious users
β	the proportion of malicious users, $\beta = \frac{m}{n+m}$
$T = \{t_1, \cdots, t_r\}$	A set of target (corrupt) items
$X = \{x_1, \cdots, x_n\}$	The original values of n genuine users
$\tilde{X} = \{\tilde{x}_1, \cdots, \tilde{x}_n\}$	The reported values from n genuine users
$Y = \{y_1, \cdots, y_m\}$	The reported values from m malicious users
$Z = \{z_1, \cdots, z_{n+m}\}$	The reported values from all users

chosen non-target 1 s to match the expected number of 1 s for genuine users. The total gain becomes:

$$G = \frac{rm}{(n+m)(p-q)} - \frac{\sum_{t \in T} m\left(f_t(p-q) + q\right)}{(n+m)(p-q)}.$$

Adaptive MGA (MGA-A). To evade detection, MGA-A modifies the above strategy by selecting a random subset of $r_s < r$ target items for each malicious user. Each forged vector only sets 1 s at these r_s positions, with l random non-target 1 s added to maintain a natural 1-count pattern. This randomness makes the attack more difficult to distinguish from genuine behavior and enhances stealth.

3 Problem Definition

In this section, the problem definition will be described in terms of both the attack model and the defense objectives. Table 1 lists the key notations used in this paper. This section will introduce the relevant concepts and foundational knowledge that are used throughout this paper.

3.1 Threat Model

We focus on the threat model in prior studies of Maximal Gain Attack against LDP protocols. In what follows, we will discuss the attacker's goals, capabilities, and background knowledge in detail.

Attacker's Goals. The attacker follows the MGA design principle to carry out targeted attacks on the LDP protocol, with the aim of maximizing gain by increasing the frequency of the selected target term.

Attacker's Capabilities and Background Knowledge. We assume that an attacker can control a subset of malicious users within the LDP protocol. These malicious users may either be malicious users injected into the protocol or real users compromised by the attacker. The attacker utilizes these malicious users to submit manipulated data to the server.

Since the LDP protocol is executed on the client side, the attacker has full knowledge of the protocol details employed by genuine users. Specifically, the attacker is aware of the various parameters of the LDP protocol, including the input domain D, the encoding domain $\tilde{D}$, and the privacy budget ε.

3.2 Defense Goal

Our goal is to design an accurate and effective frequency recovery method for OUE subjected to MGA. Since the poisoning frequency is a mixture of n genuine users and m malicious users, intuitively, if malicious users can be identified and removed, the poisoned statistics can be corrected. Based on this idea, we construct the defense model from three roles: genuine users, attacker, and defender.

Genuine Users. Each genuine user i perturbs their private item $x_i \in D$ to $\tilde{x}_i \in \tilde{D}$ following the OUE protocol. Let $X = \{x_1, ..., x_n\}$ and $\tilde{X} = \{\tilde{x}_1, ..., \tilde{x}_n\}$ denote the original and perturbed sets.

Attacker. The attacker controls m users and submits forged data $Y = \{y_1, ..., y_m\}$ crafted using MGA or MGA-A strategy.

Defender. The defender receives all data $Z = \tilde{X} \cup Y = \{z_1, ..., z_{n+m}\}$ and aims to reduce the deviation in frequency estimation due to the attack. The frequency error is:

$$F = \sum_{v \in D} \left| \hat{f}_v^* - \hat{f}_v \right|, \tag{5}$$

where $\hat{f}_v$ is the estimated frequency of item v before the attack and $\hat{f}_v^*$ is the frequency after defense.

4 Fine-Filter

In this section, we develop a theoretical model to defense against MGA, which serves as the foundation for designing an effective defense method. Fine-filter mitigates MGA recovery frequency by formulating rules to identify and filter out malicious users.

4.1 Intuition

Inspired by Eq. (5), we aim to minimize:

$$\min F = \sum_{v \in D} \left| \hat{f}_v^* - \hat{f}_v \right|. \tag{6}$$

We note that when $F(v) = \left| \hat{f}_v^* - \hat{f}_v \right|$ reaches a local minimum for each v, the value in Eq. (6) naturally attains its minimum value. Therefore, we can simplify the problem in Eq. (6) to:

$$\min F(v), \quad \forall v \in D. \tag{7}$$

Based on this, we further simplify the expression for $F(v)$ as follows:

$$F(v) = \left| \hat{f}_v^* - \hat{f}_v \right|$$

$$= \left| \frac{\frac{1}{n'} \sum_{i=1}^{n'} \mathbb{1}_{S(z_i)}(v) - q}{p - q} - \frac{\frac{1}{n} \sum_{i=1}^{n} \mathbb{1}_{S(z_i)}(v) - q}{p - q} \right| \tag{8}$$

where n is the number of users after the defense, $\mathbb{1}_{S(z_i)}(v)$ is the indicator function, and p, q are parameters of the LDP protocol. Since the size of $F(v)$ depends on whether the defender's rejection of malicious users is accurate enough, p and q are not key variables to be considered in this problem.

$$\min_{v \in D} \left| \frac{\sum_{i=1}^{n+m-|R|} \mathbb{1}_{S(z_i)}(v)}{n+m-|R|} - \frac{\sum_{i=1}^{n} \mathbb{1}_{S(z_i)}(v)}{n} \right| \tag{9}$$

where R is the set of users rejected by the defender, accordingly, $|R|$ is the total number of users rejected by the defender. Since the number of malicious users m manipulated by an attack is usually a very small fraction of the total number of users (usually $\frac{m}{n+m} = 0.05$), the number of unqualified' users rejected by a qualified defender should also be small. Therefore, $|m - |R|| \ll n$. That is, when faced with a large number of statistical objects, we have:

$$\frac{1}{n+m-|R|} \approx \frac{1}{n}$$

Based on this, we can approximate the formula above as:

$$\min_{v \in D} \left| \frac{\left(\sum_{z_i \in R_{fp}} \mathbb{1}_{S(z_i)}(v) + \sum_{z_i \in R_{fn}} \mathbb{1}_{S(z_i)}(v) \right)}{n+m-|R|} \right|, \tag{10}$$

where R is the set of rejected users, R_{fp} is the set of genuine users rejected by misjudgment, R_{fn} is the set of malicious users omitted by mistake.

Therefore, in order to solve the optimization problem in Eq. (6), the defender should minimize Eq. (10) by designing an effective screening scheme.

4.2 Defense Against MGA on OUE

In the MGA against the OUE protocol, the attacker designs the number of 1's in the vectors sent by malicious users to evade detection. Specifically, this number is set to match the expected number of 1's in the vectors of genuine users after perturbation: $l = \lfloor p + (d-1)q \rfloor$. In other words, we can assume that all malicious users have an attack weakness with a constant and known number of 1's in their vector.

To address this weakness, we can identify the target item under attack by isolating malicious users from a subset of genuine users through filtering. Specifically, the defender will traverse all users (denote the set of users used as U), count the number of 1's in each user vector, single out those users with the number of 1's equal to $l = \lfloor p + (d-1)q \rfloor$, denote the set of users removed as U_1, and the rest of the set of users as U_2, where we have $U_1 \cap U_2 = \emptyset$ and $U_1 \cup U_2 = U$.

Since the number of 1's in the vector submitted by malicious users is l, we can assume that: all malicious users are contained by the set U_1, i.e., the set U_1 contains all malicious users and some of the genuine users, and all users in U_2 are genuine users, so we approximate the distribution in U_2 as the genuine distribution estimated from genuine users.

Based on the above inference, the server estimates the frequency distributions of users in both U_1 and U_2, and computes the item-wise differences. Since the attacker can target at most l items, the l items with the largest frequency discrepancies are selected as potential attack targets, denoted by $\tilde{T}$. This set may contain both true targets and some false positives. Next, for each user in U_1 and U_2, the 0/1 alignment patterns over the bits in $\tilde{T}$ are extracted. The proportion of each alignment pattern is calculated in both groups, and the difference is computed. The alignment pattern with the largest discrepancy is identified, and the bit positions where it equals 1 are inferred as the true attack targets, denoted as T_I.

Taking Fig. 1 as an example, let U_1 be a mixed set containing both malicious and genuine users, while all users in U_2 are genuine.

Given the selected bit positions $\tilde{T} = [1, 2, 3, 4, 5, 6, 7, 8]$, we extract the 8-bit submission vectors of all users from U_1 and U_2. For each possible bit pattern $\theta \in \{0, 1\}^8$, we compute the frequency of its occurrence in each group as:

$$S_1[\theta] = \frac{|\{u \in U_1 : \text{submission}(u) = \theta\}|}{|U_1|}, \quad S_2[\theta] = \frac{|\{u \in U_2 : \text{submission}(u) = \theta\}|}{|U_2|}$$

The absolute frequency difference is calculated as $\Delta[\theta] = |S_1[\theta] - S_2[\theta]|$, and the bit pattern with the largest discrepancy is identified:

$$\theta^* = \arg\max_\theta \Delta[\theta]$$

As shown in the figure, the pattern $\theta = 10100100$ has the largest difference. The bit positions where θ^* equals 1 (i.e., positions 1, 3, and 6) are inferred to be the target bits under attack, denoted as T_I.

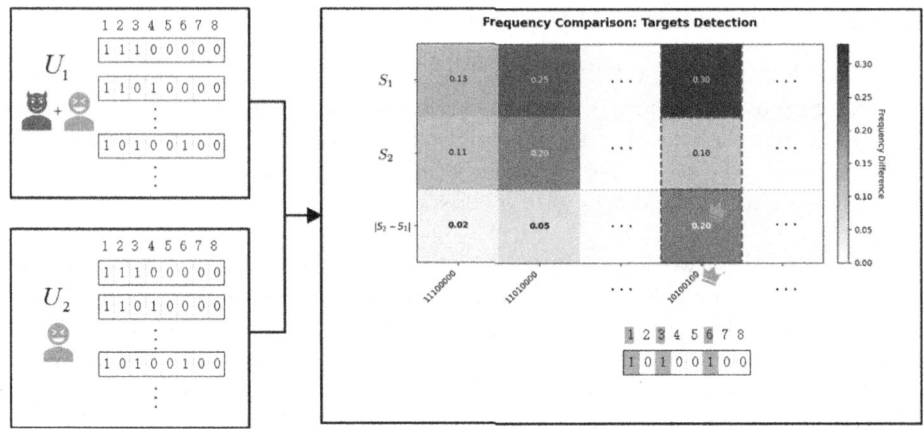

Fig. 1. Fine-filter for MGA-OUE

To isolate the malicious users, we assume that all malicious users report 1 s in these positions. Thus, users in U_1 whose bits in T_I are **not all 1 s** are likely genuine and are moved to U_2, while those whose T_I bits are all 1 s are retained in U_1 and treated as suspected attackers.

Finally, frequency estimation is performed on the updated U_2, and the result is taken as the recovered distribution $\hat{f}_v^*$.

Algorithm 1 describes the server-side post-processing procedure for filtering potential attackers under the MGA-OUE threat model. The process begins by separating users into suspected attackers and assumed genuine users based on the number of 1 s in their reports (Lines 2–3). Next, it identifies candidate target items by comparing frequency estimates between the two user groups (Lines 5–8). In Step 3, the algorithm computes bit alignment patterns over these candidate items and finds the most statistically different pattern between the two groups (Lines 10–13). Based on the identified pattern, it infers the likely target bit positions (Line 14). Finally, it refines the user groups by removing users in U_1 who do not follow the attacker-like pattern and re-estimates frequencies using the updated set of genuine users (Lines 16–17).

4.3 Defense Against MGA-A on OUE

To evade detection, MGA-A introduces an adaptive strategy based on MGA for OUE. Instead of attacking all r target items simultaneously, the attacker selects a random subset of size $r_s < r$ in each user's report. This variation makes the attack less distinguishable from genuine users.

Building on our defense against the standard MGA, we extend the strategy to address this adaptive version. As before, the defender first partitions users into two sets U_1 and U_2 based on the number of 1's in the perturbed vector. Users with exactly $l = \lfloor p + (d-1)q \rfloor$ 1's are placed in U_1, while others go to U_2.

Algorithm 1. Fine-filtering for MGA-OUE

Input: Set of users $U = \{z_1, z_2, \ldots, z_{n+m}\}$ (mixed with genuine and malicious users), upper bound of attack targets l
Output: Recovered frequency distribution $\hat{f}_v^*$
1: **Step 1: Initial separation based on number of 1s**
2: Initialize $U_1 = \{u \in U \mid \sum_{i=1}^{d} z_{u,i} = l\}$ \\Suspected attackers
3: Let $U_2 = U \setminus U_1$ \\Assumed genuine users
4: **Step 2: Identify candidate target items** $\tilde{T}$
5: Estimate frequency $F_1 = \hat{f}_{U_1}(v)$ for all $v \in D$
6: Estimate frequency $F_2 = \hat{f}_{U_2}(v)$ for all $v \in D$
7: Compute $\Delta[v] = |F_1[v] - F_2[v]|$ for all $v \in D$
8: Let $\tilde{T} \leftarrow$ Top-l items with largest $\Delta[v]$
9: **Step 3: Determine target bit positions** T_I
10: For each user $u \in U_1$ and U_2, extract $0/1$ alignment pattern over $\tilde{T}$
11: Compute frequency $S_1[\theta]$ of each pattern θ in U_1
12: Compute frequency $S_2[\theta]$ of each pattern θ in U_2
13: Find $\theta^* = \arg\max_\theta |S_1[\theta] - S_2[\theta]|$
14: Let $T_I \leftarrow \{i \in \tilde{T} \mid \theta_i^* = 1\}$ \\$\theta_i$ is the i-th bit in θ
15: **Step 4: Refine user sets and estimate frequencies**
16: Move users u from U_1 to U_2 if $z_u[T_I] \neq \mathbf{1}$
17: Based on the updated user set U_2, estimate and return the recovered frequencies $\hat{f}_v^*$

Then, similar to OUE-MGA, the defender computes frequency estimates over both U_1 and U_2, and identifies a candidate set of target items $\tilde{T}$ as those with the top-l differences. However, due to random subset selection in MGA-A, $\tilde{T}$ will contain both real targets and noise terms.

To refine this set, we examine the bit patterns (over $\tilde{T}$) of users in U_1, and identify the $\binom{r}{r_s}$ most frequent bit patterns:

$$\mathcal{C}_T^{(r_s)} = \left\{ \mathbf{b} \in \{0,1\}^{|\tilde{T}|} \;\middle|\; \sum_{j=1}^{|\tilde{T}|} b_j = r_s \right\},$$

which correspond to possible malicious user perturbation vectors. Bits that appear as 0 in all such top patterns are treated as error positions and removed. We estimate the true number of target items and the number of active targets per user report as: To estimate the actual number of target items $r = |T|$ and the number of items r_s attacked in each round, the attacker performs an empirical analysis based on the observed perturbed vectors uploaded by malicious users. Let $\tilde{T} \subseteq [d]$ be the set of candidate target items suspected by the attacker. Then we compute:

$$\tilde{r} = |\tilde{T}| - \#\left\{ j \in \tilde{T} : \forall u \in U_1,\; x_u(j) = 0 \right\}$$
$$\tilde{r}_s = \#\left\{ j \in \tilde{T} : \exists u \in U_1,\; x_u(j) = 1 \right\}$$

where $x_u(j)$ denotes the reported value of malicious user $u \in U_1$ at item j, and $\#\{\cdot\}$ represents the cardinality of the set (i.e., the number of its elements). Here, $\tilde{r}$ estimates the number of target items (excluding falsely suspected items with no 1 s), and $\tilde{r}_s$ estimates how many target items were attacked in a given round.

Let the remaining 1-bit positions be T_I, and denote the most frequent patterns as $C_{T_I} \subseteq \{0,1\}^{|T_I|}$. We now refine the user sets: any user in U_1 whose T_I-indexed bits do not match any pattern in C_{T_I} is moved to U_2.

Finally, frequency estimation is performed over the updated U_2, and the result is treated as the recovered frequency $\hat{f}_v^*$. At this point in the set U_2, we have: $\sum_{z_i \in R_{fn}} \mathbb{1}_{S(z_i)}(v) = 0$, $\sum_{z_i \in R_{fp}} \mathbb{1}_{S(z_i)}(v) \approx 0$.

This strategy addresses the key difference in MGA-A—random target subsets per user—while reusing the core principles of the MGA defense (Fig. 2).

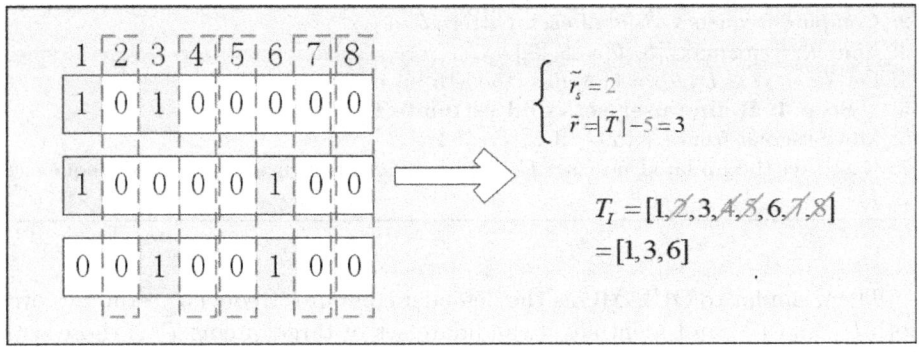

Fig. 2. Fine-filter for MGA-A-OUE

4.4 Discussion

Considering that the defense against MGA-OUE primarily relies on the attacker's characteristic of unifying the number of 1's in a vector to l, if the attacker modifies this characteristic to evade the defense of Fine-filter, the corresponding defense methods should be adjusted accordingly.

Firstly, it is important to note that the attacker sets the number of 1's in the vector to l (the expected number of 1's in a vector submitted by a genuine user) to ensure that the fabricated malicious users closely resemble genuine users, thereby reducing the risk of detection. Based on this premise, if the attacker attempts to evade detection by Fine-filter through altering this fixed value, the deviation from l is expected to be minimal.

Thus, we can expand the window range of U_1 by defining an interval around l, allowing for a controlled variation above and below l. This ensures the maximum inclusion of malicious users while maintaining the value of $\sum_{z_i \in R_{fn}} \mathbb{1}_{S(z_i)}(v)$ within a narrow range.

5 Experimental Evaluation

5.1 Experimental Setup

Datasets. We evaluated our Fine-filter on two real-world datasets, i.e., Fire [18] and IPUMS [17].

- **Fire** [18]: The Fire dataset is a publicly available resource containing fire unit responses to 911 calls from the City's Computer-Aided Dispatch (CAD) system. For our experiments, we used data from 2024, filtered the records by the 'Alarms' type. In this dataset, the unit ID is treated as the item owned by the user, encompassing 315 different item types and 747,554 users.
- **IPUMS** [17]: IPUMS is the world's largest database of individual population data, including microdata samples from U.S. census records (IPUMS-USA), international census records (IPUMS-International), and data from both U.S. and international surveys. We used data from 2024, considering city attributes as items contained within each user's data. The dataset includes 206 items and a total of 1,608,982 users.

LDP Settings. In this study, we consider the OUE protocol for frequency estimation. The details of this protocol are described in Sect. 2.2. In our experiments, we set the default LDP parameter as $\varepsilon = 2$.

Attack Settings. We consider an advanced targeted attack, MGA, and its derivative, MGA-A, with details provided in Sect. 2.3. For the MGA attack, we set the number of target items to $r = 10$. For MGA-A, we set $r_s = 8$, ensuring that each malicious user is independently and randomly sampled from the set of target items T. The ratio of malicious users to the total number of users manipulated by the attacker is $\beta = \frac{m}{n+m} = 0.05$.

5.2 Evaluation Metrics

We use the mean squared error (MSE) as an evaluation metric, which is defined as follows.

Mean Squared Error (MSE). Given original frequencies and aggregated frequencies (poisoned or recovered), we adopt MSE to measure the average error of the frequencies for all items. Specifically,

$$MSE = \frac{1}{d} \sum_{v \in D} \left(\hat{f}_v^* - f_v \right)^2, \tag{11}$$

where d is the size of domain D, f_v is the original frequency of item v in the raw data, $\hat{f}_v^*$ is the recovered aggregated frequency of item v. The smaller the MSE, the more effective the defense scheme will be.

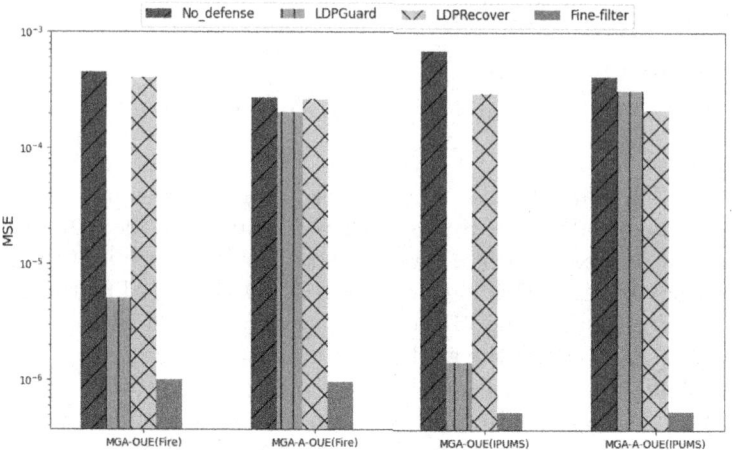

Fig. 3. Performance of different defense schemes on two dataset.

Figure 3 shows the MSE of our Fine-filter and other known defenses (LDP-Guard [11] and LDPRecover [19]) on two real-world datasets. We find that Fine-filter consistently performs best against both MGA and MGA-A, with lower MSE than other defenses.

Specifically, Fine-filter identifies and filters out malicious users by detecting the set of target items under attack and restores the aggregation frequency by rejecting the perturbed data from these malicious users.

Since LDPGuard estimates the percentage of malicious users through two rounds of data collection, the accuracy of its estimation significantly impacts its defense effectiveness, leading to weaker performance compared to Fine-filter. LDPRecover formulates the problem of recovering the aggregation frequency as a constrained inference (CI) problem by deriving the theoretical relationship between the poisoned frequency, the true frequency, and the malicious frequency. However, its theoretical framework relies on the assumption that the dataset is independently and identically distributed (i.i.d.), which limits its effectiveness in certain scenarios. Consequently, its overall performance remains inferior to that of Fine-filter.

5.3 Impact of Parameters

In this subsection, we evaluate the effectiveness of the defense in Fine-filter against MGA and its derivative, MGA-A, by varying the parameters (β, r and ε). In our experiments, we change only one parameter at a time while keeping the others at their default values, i.e., $\beta = 0.05$, $r = 10$, $\varepsilon = 10$, and, specifically for MGA-A, $r_s = 8$. The ranges for these parameters are $\beta \in [0.02, 0.1]$, $r \in [1, 20]$, and $\varepsilon \in [0.5, 2.5]$. It is important to note that r_s always preserves $r_s = r-2$ when the parameter r is varied, so that for MGA-A, $r \in [3, 20]$.

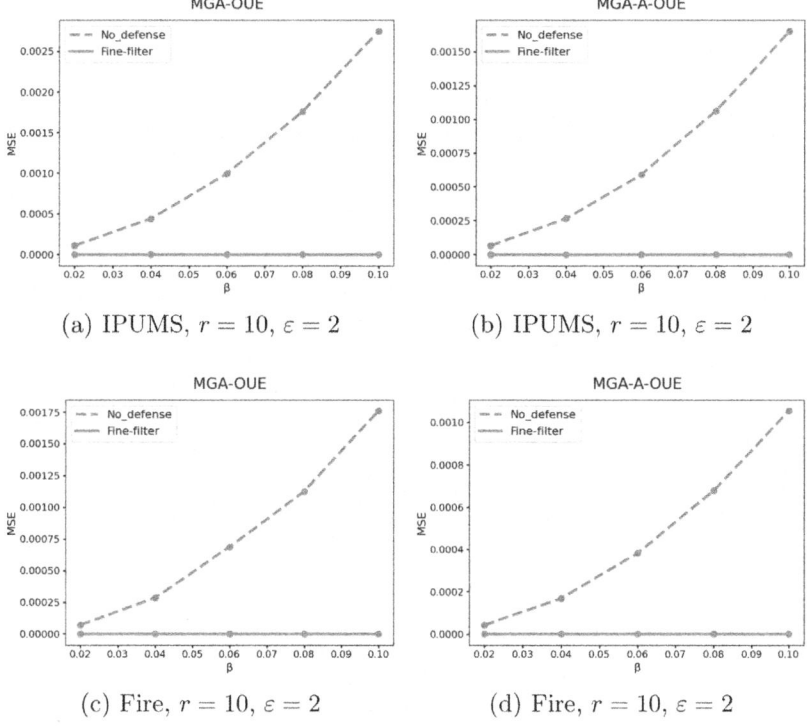

Fig. 4. Impact of β on Fine-filter.

The Impact of the Ratio of Malicious Users β. Figures 4 illustrates the impact of β on the performance of Fine-filter for the OUE protocol under MGA and MGA-A attacks. Experimental results show that the MSE of the attacked OUE increases with rising β, indicating that a higher proportion of malicious users leads to more severe distortion in frequency estimation. However, after applying the Fine-filter defense, the MSE drops significantly. Moreover, Fine-filter demonstrates strong robustness, showing minimal sensitivity to changes in β and consistently maintaining excellent defense performance.

The Impact of the Number of Target Items r. Figures 5 illustrates how the parameter r affects the performance of Fine-filter under MGA and MGA-A attacks on the OUE protocol. Experimental analysis reveals that the MSE of both MGA and MGA-A against OUE increases smoothly as r increases.

This trend arises because the attacker distributes malicious users across a random subset of r targets. As r increases, the attack becomes more dispersed, thereby reducing the concentration of manipulated values and weakening the attack's effectiveness, which leads to an increase in MSE.

After applying Fine-filter as a defense against MGA and MGA-A, the MSE of OUE shows a significant decrease. In particular, when $r > 1$, Fine-filter achieves stable and effective defense performance, with little variation as r increases.

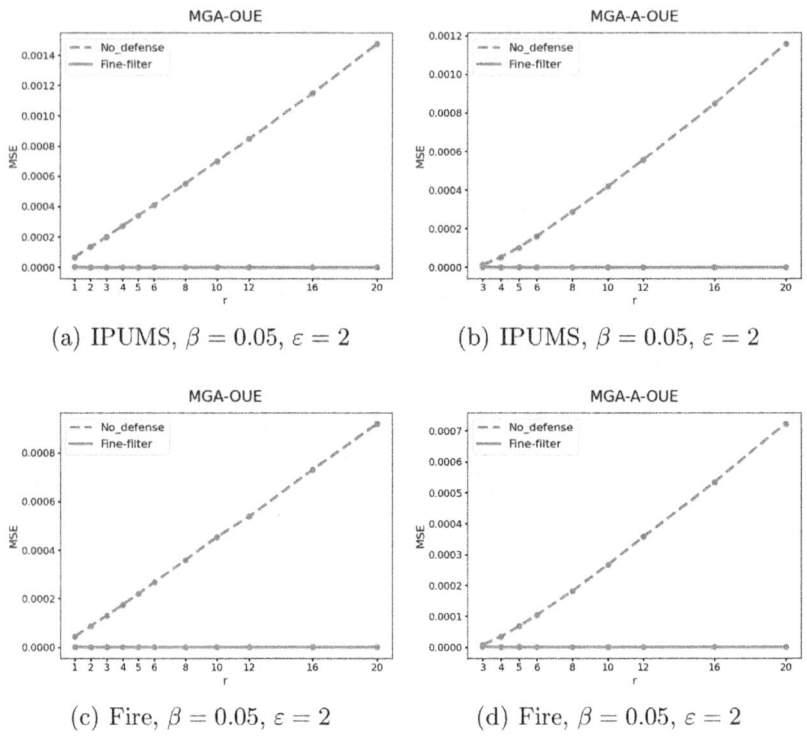

Fig. 5. Impact of r on Fine-filter.

The Impact of the Privacy Budget ε. Figure 6 shows how the privacy budget ε affects the performance of Fine-filter for the OUE protocol under MGA and MGA-A attacks. We observe that as ε increases, the MSE of OUE decreases both before and after applying Fine-filter. This is because OUE is highly sensitive to changes in the privacy budget—larger values of ε increase the number of ones in the post-perturbation vector, making it easier to distinguish between genuine and malicious users. Consequently, a higher privacy budget enhances the effectiveness of the Fine-filter defense.

6 Related Work

6.1 Local Differential Privacy

Local Differential Privacy (LDP) [7,13], a variant of differential privacy, allows untrusted servers to collect and aggregate statistics from distributed users while ensuring provable privacy protection for those users. In LDP, users encode and scramble their private data locally before transmitting them to a central server, which then aggregates statistical items of interest from these scrambled data. Given the growing importance of data statistics and privacy protection, LDP

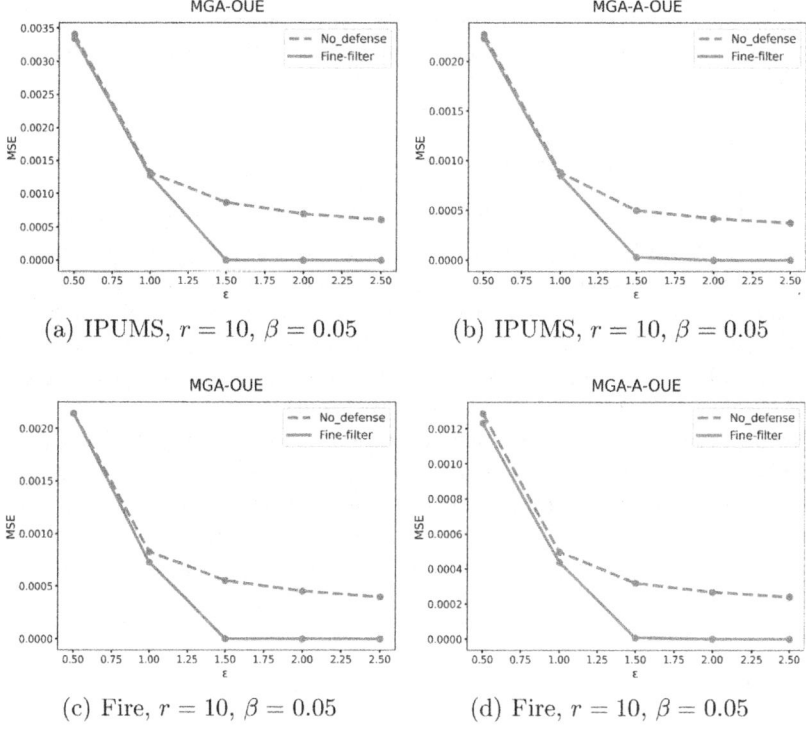

Fig. 6. Impact of ε on Fine-filter.

protocols have been extensively studied by researchers, including applications such as heavy hitters identification [16], range queries [14,15], and more. Notably, research in mean estimation [8,20] and frequency estimation [1,12,21] has been particularly active, with kRR [12], OUE [21], and OLH [21] being some of the more advanced frequency estimation protocols.

6.2 Poisoning Attacks on LDP

From the perspective of attack objectives, data poisoning attacks on LDP protocols can be primarily categorized into two types: target attacks and non-target attacks. In target attacks, the goal of the attacker is to increase the aggregation frequency of a specific target item, effectively promoting it as a popular item [4]. In contrast, in non-target attacks, the attacker aims to reduce the accuracy of the aggregation frequency for all items, thereby degrading the overall statistical results [5].

Regarding attack modes, data poisoning attacks can be categorized into *Input Poisoning Attacks* [4,5] and *Output Poisoning Attacks* [4]. In input poisoning, attackers inject malicious items via compromised users, which are then perturbed by the LDP protocol. In output poisoning, attackers bypass perturbation and directly submit crafted data to the server. Since output poisoning avoids randomization, it typically has a stronger impact on LDP statistics under identical attack conditions.

6.3 Countermeasures Against Poisoning Attacks on LDP

There has been progress in the development of defenses against data poisoning attacks on LDP protocols. Several countermeasures targeting poisoning attacks, specifically against targeted attacks like MGAs, are proposed in [4]. These methods include malicious user detection and conditional probability-based detection. However, these countermeasures are primarily applicable to the OUE and OLH protocols and rely on strong assumptions, such as the server having knowledge of certain details about the attack. In practice, these assumptions are not always valid, making these countermeasures ineffective in most real-world scenarios. Additionally, the frameworks presented in [11,19] offer more comprehensive approaches to defending against data poisoning attacks, but they also contain significant flaws, as they still rely on assumptions about the attacker, such as the defender knowing the proportion of malicious users injected by the attacker into the total user population.

7 Conclusion

In this work, we proposed Fine-filter, an effective defense mechanism against poisoning attacks on frequency estimation under LDP, i.e., OUE. Fine-filter identifies abnormal user behavior by examining the statistical properties of perturbed reports, and dynamically filters users whose data patterns are inconsistent with benign behavior. Our approach successfully detects and mitigates Maximal Gain Attack (MGA) as well as its variant, MGA-A. By bit pattern frequency analysis and subset consistency check, Fine-filter can estimates the number of target items and then identity them with high accuracy. Experimental results demonstrate that Fine-filter recovers highly accurate frequency estimates while maintaining compatibility with OUE protocol.

Acknowledgement. This work is partly supported by the Jiangsu Provincial Key Research and Development Program (No. BE2022068 and BE2022068-1), the Natural Science Foundation of China (No. 62172216 and 62302214), the China Postdoctoral Science Foundation (No. 2023M741684 and GZB20240977), Jiangsu Funding Program for Excellent Postdoctoral Talent, the Fundamental Research Funds for the Central Universities (No. NP2024117), and the Funding of Steady-Supported Guofang Characteristic Subject Fundamental Research Project (No. ILF240061A24).

References

1. Ambainis, A., Jakobsson, M., Lipmaa, H.: Cryptographic randomized response techniques. In: International Workshop on Public Key Cryptography, pp. 425–438. Springer (2004)
2. Bassily, R., Nissim, K., Stemmer, U., Guha Thakurta, A.: Practical locally private heavy hitters. Adv. Neural Inf. Process. Syst. **30** (2017)
3. Bassily, R., Smith, A.: Local, private, efficient protocols for succinct histograms. In: Proceedings of the Forty-Seventh Annual ACM Symposium on Theory of Computing, pp. 127–135 (2015)
4. Cao, X., Jia, J., Gong, N.Z.: Data poisoning attacks to local differential privacy protocols. In: 30th USENIX Security Symposium (USENIX Security 21), pp. 947–964. USENIX Association (2021). https://www.usenix.org/conference/usenixsecurity21/presentation/cao-xiaoyu
5. Cheu, A., Smith, A., Ullman, J.: Manipulation attacks in local differential privacy. In: 2021 IEEE Symposium on Security and Privacy (SP), pp. 883–900. IEEE (2021)
6. Cormode, G., Kulkarni, T., Srivastava, D.: Marginal release under local differential privacy. In: Proceedings of the 2018 International Conference on Management of Data, pp. 131–146 (2018)
7. Duchi, J.C., Jordan, M.I., Wainwright, M.J.: Local privacy and statistical minimax rates (2013). https://doi.org/10.1109/FOCS.2013.53
8. Duchi, J.C., Jordan, M.I., Wainwright, M.J.: Minimax optimal procedures for locally private estimation. J. Am. Stat. Assoc. **113**(521), 182–201 (2018)
9. Dwork, C., McSherry, F., Nissim, K., Smith, A.: Calibrating noise to sensitivity in private data analysis. In: Halevi, S., Rabin, T. (eds.) TCC 2006. LNCS, vol. 3876, pp. 265–284. Springer, Heidelberg (2006). https://doi.org/10.1007/11681878_14
10. Erlingsson, Ú., Pihur, V., Korolova, A.: RAPPOR: randomized aggregatable privacy-preserving ordinal response. In: Proceedings of the 2014 ACM SIGSAC Conference on Computer and Communications Security, pp. 1054–1067 (2014)
11. Huang, K., et al.: LDPGuard: defenses against data poisoning attacks to local differential privacy protocols. IEEE Trans. Knowl. Data Eng. **36**(7), 3195–3209 (2024)
12. Kairouz, P., Oh, S., Viswanath, P.: Extremal mechanisms for local differential privacy. Adv. Neural Inf. Process. Syst. **27** (2014)
13. Kasiviswanathan, S.P., Lee, H.K., Nissim, K., Raskhodnikova, S., Smith, A.: What can we learn privately? SIAM J. Comput. **40**(3), 793–826 (2011)
14. Kulkarni, T.: Answering range queries under local differential privacy. In: Proceedings of the 2019 International Conference on Management of Data, pp. 1832–1834 (2019)
15. Li, C., Hay, M., Miklau, G., Wang, Y.: A data-and workload-aware algorithm for range queries under differential privacy. arXiv preprint arXiv:1410.0265 (2014)
16. Qin, Z., Yang, Y., Yu, T., Khalil, I., Xiao, X., Ren, K.: Heavy hitter estimation over set-valued data with local differential privacy. In: Proceedings of the 2016 ACM SIGSAC Conference on Computer and Communications Security, pp. 192–203 (2016)
17. Ruggles, S., et al.: IPUMS USA: Version 16.0 [dataset] (2024). https://doi.org/10.18128/D010.V16.0
18. San Francisco Open Data Portal: Fire Department and Emergency Medical Services Dispatched Calls for Service (2024). https://data.sfgov.org/Public-Safety/Fire-Department-and-Emergency-Medical-Services-Dis/nuek-vuh3/about_data

19. Sun, X., et al.: LDPRecover: recovering frequencies from poisoning attacks against local differential privacy. In: 2024 IEEE 40th International Conference on Data Engineering (ICDE), pp. 1619–1631. IEEE (2024)
20. Wang, N., et al.: Collecting and analyzing multidimensional data with local differential privacy. In: 2019 IEEE 35th International Conference on Data Engineering (ICDE), pp. 638–649. IEEE (2019)
21. Wang, T., Blocki, J., Li, N., Jha, S.: Locally differentially private protocols for frequency estimation. In: 26th USENIX Security Symposium (USENIX Security 17), pp. 729–745 (2017)

BioVite: Efficient and Compact Privacy-Preserving Biometric Verification via Fully Homomorphic Encryption

Pengfei Zeng[1,2], Han Xia[1,2], and Mingsheng Wang[1,2(✉)]

[1] State Key Laboratory of Cyberspace Security Defense, Institute of Information Engineering, Chinese Academy of Sciences, Beijing, China
{zengpengfei,xiahan}@iie.ac.cn
[2] School of Cyber Security, University of Chinese Academy of Sciences, Beijing, China
wangmingsheng@iie.ac.cn

Abstract. Privacy-preserving biometric verification serves as a secure web application for protecting biometric data, which has garnered public attention in recent years. Fully homomorphic encryption (FHE) enables computation on ciphertexts without accessing the secret key, thereby protecting privacy for clients. However, FHE suffers from significant communication overhead and computational costs, which remain the primary bottlenecks in current FHE-based schemes. In this work, we present BioVite, a novel privacy-preserving biometric verification scheme based on techniques from FHE. By adopting Generalized Learning with Errors (GLWE) encryption with compact parameters and optimizing the membership test, BioVite outperforms state-of-the-art FHE-based verification schemes in both runtime and communication size. It requires only around 0.3 ms and 8.5 KB for 512-dimensional biometric templates during verification. In terms of accuracy and precision, BioVite introduces minimal noise to floating-point similarity computations, and experiments demonstrate that after applying BioVite, the verification accuracy remains comparable to plaintext verification across various face datasets.

Keywords: Privacy-preserving biometric verification · Fully homomorphic encryption · Secure multiparty computation

1 Introduction

Cloud-based biometric verification provides a Web application for human identity verification or multifactor authentication by deploying machine learning as a service (MLaaS) [46]. In an unprotected deployment, the Web service providers store the registered biometric features in the cloud, leading to severe privacy issues [24]. To prevent privacy and sensitive data leakage, approaches such as perturbation [11,33,37,38,52] or cryptographic methods [4,12,16,18,19,26,32,41,

---

This paper is supported by the Strategic Priority Research Program of the Chinese Academy of Sciences under Grant XDB0690200.

© The Author(s), under exclusive license to Springer Nature Singapore Pte Ltd. 2026
J. Han et al. (Eds.): ICICS 2025, LNCS 16217, pp. 503–521, 2026.
https://doi.org/10.1007/978-981-95-3540-8_27

47,53,54] have been explored. Among these approaches, homomorphic encryption reduces communication rounds by allowing arithmetic computations over encrypted data, and homomorphic encryption based biometric authentication has drawn wide interests in recent years [1,3,4,6,14,19,30,31,42,43,50].

Homomorphic Biometric Verification. *Fully homomorphic encryption* (FHE) allows arbitrary operations on encrypted data without decryption. As all operations can be performed on ciphertexts, there is no need to transfer any data except public keys during setup and ciphertexts during evaluation. During biometric verification, biometric features are captured by sensors and uploaded, along with the user's ID. The similarity between the captured features and those stored in the database is calculated and then compared to a predefined threshold. Verification succeeds if the similarity is greater than the threshold. This procedure naturally yields two subtasks when combined with homomorphic encryption: *homomorphic similarity computation* and *homomorphic comparison*. Homomorphic similarity computation is relatively straightforward, and many previous works [42,50] have achieved millisecond-level efficiency, whereas homomorphic comparison remains challenging [30]. Previous works try membership test for solving such problem [5], yet there are still possible optimizations on it.

Client-Server-CSP Architecture. The conventional architecture of FHE is two-party, where the client encrypts data and the server evaluates on the ciphertexts using the public key and evaluation keys from the client. The CPA-security of FHE schemes ensures that the server learns nothing about the input or output data, and this architecture works well in most scenarios. However, in biometric verification, clients can be embedded devices or components of distributed systems. These devices have to face complex network conditions and various ambient statuses, making it highly insecure to store secret keys on the client side.

To mitigate this issue, a *trusted third party* (TTP) or *cryptographic service provider* (CSP) has been introduced [3,16,26]. In this three-party model, the third party manages the secret keys and distributes only the public and evaluation keys to the client and server. After evaluation at the server side, the outputs are transferred to the third party for decryption, whose result is then returned to the client. To further protect this result, ciphertexts are always masked with a random number by the server so that the third party learns nothing from the decrypted (masked) result. The server sends that random number to the client, enabling the recovery of the final evaluation result. However, current schemes utilizing CSP are not enough efficient, as there are heavy communication ($\sim$200 KB) due to ciphertext expansion of FHE, and inefficient parameter selection. We therefore ask a question: *Can we build a FHE-based biometric verification scheme that is more efficient and compact on communication?*

1.1 Our Contributions

Focusing on improving efficiency and reducing communication overhead of privacy-preserving biometric verification, we present BioVite, a purely FHE-based biometric verification scheme. An overview of workflow of BioVite is Fig. 1, and the main features of BioVite are as follows:

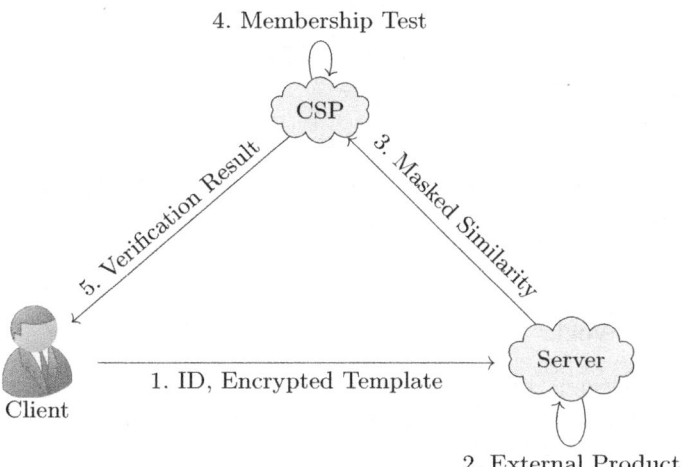

Fig. 1. A system architecture of BioVite. The client sends an encrypted biometric template with ID for verification at the first step. The server performs a fast external product to calculate the similarity between the encrypted templates and queried template from the database. The result is masked and transferred to CSP for membership test. The result of the membership test is the verification result in LWE ciphertext, which is sent back to the client for decryption.

- **GLWE Ciphertexts for Fast Verification.** Previous FHE-based biometric verification schemes adopt RLWE ciphertexts (BGV/BFV/CKKS), which introduce redundancy, leading to inefficiency and increased communication overhead. BioVite employs GLWE ciphertexts, eliminating redundancy and unnecessary computations. Additionally, a pseudo-random number generator and functional key-switching are applied to replace masks with seeds during transferring, reducing communication overhead to approximately 8.5 KB (compared to 70–200 KB in prior work) per verification. The total verification time is only about 0.3 ms, achieving an 80× speedup over recent FHE-based schemes.
- **Optimized Three-party Membership Test.** The homomorphic membership test protocol resolves threshold comparison through private set intersection. Aligning with the message flow in the three-party model, the membership test in BioVite is optimized, achieving a computation cost of approximately 0.06 ms and a ciphertext size of 0.25 KB per verification query.
- **Improving Throughput While Preserving Precision.** We observed that only the most significant bits are useful for threshold comparison, allowing the encoded message to remain close to the error component without compromising accuracy. We provide a highly compact parameter set in Table 2, and the experiments indicate that the verification accuracy under the privacy protection of BioVite remains comparable to the raw FaceNet model (Table 5). Figure 2 illustrates the encoding layout in BioVite.

Fig. 2. BioVite utilizes a compact form of encoding. Since we only need several most significant bits for membership test, the plaintext space can be set close to the error part, while still preserving precisions.

1.2 Related Works

In privacy-preserving biometric verification, obfuscations and cryptographic tools are explored to prevent the leakage of privacy data and calculation results.

Differential privacy based works adding noises over the embedding space to prevent adversaries from recovering information from original templates [11,33]. Some other works construct learning models for verification while still preserving privacy [37,38,52]. These obfuscation-based methods need to balance between accuracy and privacy, thus increasing system complexity.

To achieve a cryptographic level of security, MPC methods are explored. Garbled circuits (GC) [8] can evaluate arbitrary circuits in secure multiparty computation, which, together with oblivious transfer (OT) [44], forms a similarity-and-comparison circuit for solving verification tasks [9,47], or just for comparison [26,47]. Secret sharing-based methods also present very high efficiency and low online communication [12,28].

GC and SS-based methods require multiple rounds during the offline phase due to the need for fresh random budgets (e.g., Beaver's triples [7]). In contrast, FHE-based methods, which only need to send ciphertexts, have no offline communication and only one round of online communication. Paillier cryptosystem is explored first for homomorphic similarity computation [1,4,19,31]. In [43], TFHE scheme to construct a pure FHE-based verification circuit but suffers from poor efficiency. For more efficient homomorphic similarity computation, [6,42] encode template features into the slots of plaintexts, requiring $\lceil \log m \rceil$ homomorphic rotations. Other works encode features on the coefficients of polynomial plaintexts, significantly reducing the computation cost to one single homomorphic multiplication [3,50]. Performing comparison using FHE schemes is challenging, requiring either bootstrapping [43] or many levels for homomorphic evaluation [14,15,30].

In addition to biometric verification, biometric identification is another well-researched task [18,27–29], where no ID is required, and the outputs are yes/no authentication, the best matched template or others. Considering malicious clients or servers, zero-knowledge proofs and commitments are invoked for further security guarantees [4,12,32].

2 Preliminaries

2.1 Notations

In this paper, we denote by $\mathbb{Z}$ the set of integers, $\mathbb{Z}_t$ the zero-centered set $(-t/2, t/2] \cap \mathbb{Z}$ for a positive integer modulus t, $\mathbb{Z}_t^* \subseteq \mathbb{Z}_t$ the set of multiplicative inverses, and $\mathbb{R}$ the set of real numbers. $[t]$ denotes the set $\{0, \cdots, t-1\}$. Column vectors are denoted by bold lower-case letters, e.g., $\boldsymbol{a}$, and its transpose is $\boldsymbol{a}^\top$. For a positive integer q, $[a]_q$ denotes a modulo q resulting in $\mathbb{Z}_q$, $\lfloor x \rceil$ denotes the operation of rounding to the nearest integer for $x \in \mathbb{R}$. $\langle \cdot, \cdot \rangle$ denotes the inner product operator.

2.2 Biometric System

A biometric system for authentication is designed within three steps: feature extraction, similarity computation, and threshold comparison. Before diving into the details of these tasks, we first pin the definition of biometric verification here.

Definition 1 (Biometric Verification). *Biometric verification is a system containing a database of biometric templates, and provides a service which takes as inputs a biometric template with an ID, and outputs a boolean response indicating whether there exists an template labeled with that ID such that these two templates are similar.*

A biometric template is the output of feature extraction via some models [39]. There's another task named biometric identification, which is extract the biometric verification without requiring an input ID. By providing ID, the system needs not find the best template in the database, but only concentrate on that template corresponded to the input ID, therefore providing high efficiency.

Cosine Similarity. We adopt cosine similarity as the metric of the similarity between two m-dimensional vectors $\boldsymbol{a}, \boldsymbol{b}$:

$$\mathrm{CosSim}(\boldsymbol{a}, \boldsymbol{b}) = \left\langle \frac{\boldsymbol{a}}{\|\boldsymbol{a}\|}, \frac{\boldsymbol{b}}{\|\boldsymbol{b}\|} \right\rangle$$

where $\|\boldsymbol{a}\| := \sqrt{\sum_{i=0}^{m-1} a_i^2}$. The distance is then compared to a predefined threshold $\theta \in \mathbb{R}$. A greater similarity indicates that the inputs are closer, suggesting they are from a same person.

Quantization. For a biometric system that stores integer representations of features, quantization shifts the computations to the integer domain. Let $\gamma \in \mathbb{R}$ be the quantization factor; then, the biometric features $\boldsymbol{f} \in \mathbb{R}^m$ are replaced by their quantized counterparts:

$$\lfloor \gamma \cdot \boldsymbol{f} \rceil = (\lfloor \gamma \cdot f_0 \rceil, \cdots, \lfloor \gamma \cdot f_{m-1} \rceil)^\top \in \mathbb{Z}^m$$

The similarity in quadratic form (e.g., squared Euclidean distance and cosine similarity) contains a factor of γ^2. When performing comparison, the threshold also needs quantization as $\lfloor \theta \cdot \gamma^2 \rceil$.

2.3 Fully Homomorphic Encryption

Since Gentry's blueprint [22], lattice-based fully homomorphic encryption schemes have been widely studied [10,13,15,17,21,23]. Current FHE schemes are capable of performing homomorphic addition and multiplication over ciphertexts without decryption. In this work, we adopt the GSW cryptosystem [23] including techniques from TFHE [15] for our protocols. These techniques require a basic understanding of the underlying algebraic structure; thus, we first review the necessary background before delving into the detailed schemes.

Cyclotomic Rings. Let N be a power-of-two, the $2N$-th cyclotomic ring is the extension ring $\mathbb{Z}[\zeta_{2N}]$ where ζ_{2N} is a primitive $2N$-th root of unity. It is isomorphic to the polynomial quotient ring $R := \mathbb{Z}[X]/(X^N + 1)$, and in this paper, we work with the polynomials in R. For a positive integer q, R_q denotes the quotient ring $R/(qR) = \mathbb{Z}_q[X]/(X^N + 1)$. For a vector of ring elements $\boldsymbol{a} \in R^N$, we denote by $a_{i,j}$ the coefficient of X^j for the i-th element in $\boldsymbol{a}$. We denote by $a \leftarrow R_q$ that a is uniformly chosen from R_q, and by $a \leftarrow \chi$ that a is randomly chosen according to a specific distribution χ over R_q.

Generalized Learning with Errors. The generalied learning with errors (GLWE) problem introduced in [10] is a generalization of the learning with errors (LWE) problem [45]. We first review the definition.

Definition 2 (GLWE Problem [10]). *Let $\boldsymbol{s} \in R^n$ be the secret key, and $e \leftarrow \chi$ for some distribution χ over R. A GLWE instance is an $(n+1)$-dimensional vector $(\boldsymbol{a}, b = \langle \boldsymbol{a}, \boldsymbol{s} \rangle + e) \in R_q^{n+1}$ for $\boldsymbol{a} \leftarrow R_q^n$. The decisional GLWE problem asks to distinguish whether a vector over R_q^{n+1} is a GLWE instance or uniformly sampled from R_q^{n+1}.*

Note that n is often referred to as the rank of the GLWE problem. When $n = 1$, such instantiation of GLWE is referred to as the ring learning with errors (RLWE [36]) problem. When the cyclotomic degree $N = 1$, GLWE becomes the (plain) LWE problem [45].

GLWE-based GSW Cryptosystem. We first define the GLWE-based encryption, and then present the construction of GLWE-based GSW (GGSW) ciphertexts upon GLWE instantiations. For $m \in R$, all valid GLWE encryptions of m under the secret key $\boldsymbol{s} \in R^n$ are defined as the set

$$\mathsf{GLWE}_s(\Delta \cdot m) := \{(\boldsymbol{a}, -\langle \boldsymbol{a}, \boldsymbol{s} \rangle + \Delta \cdot m + e) \mid \boldsymbol{a} \leftarrow R_q^n, e \leftarrow \chi\}$$

where χ is an error distribution over R, and Δ is an encoding factor. The decryption for ciphertext $\boldsymbol{c} = (\boldsymbol{a}, b) \in \mathsf{GLWE}_s(m)$ is

$$\mathsf{Dec}(\boldsymbol{s}, \boldsymbol{c}) = \left\lfloor \frac{b + \langle \boldsymbol{a}, \boldsymbol{s} \rangle}{\Delta} \right\rceil_t = m + \left\lfloor \frac{e}{\Delta} \right\rceil_t$$

For a vector of plaintexts $\boldsymbol{m} = (m_0, \cdots, m_{d-1}) \in R^d$, GLWE ciphertexts encrypt each coordinate of $\boldsymbol{m}$ individually:

$$\mathsf{GLWE}_s(\boldsymbol{m}) := (\mathsf{GLWE}_s(m_0), \cdots, \mathsf{GLWE}_s(m_{d-1}))$$

The GGSW encryptions of m under s are a set of GLWE ciphertexts encrypting the message tensored with s and $\mathbf{g} = (1, B_g, \cdots, B_g^{d-1}) \in \mathbb{Z}^d$ where B_g is a positive integer base and $d = \lfloor \log_{B_g} q \rfloor + 1$:

$$\mathsf{GGSW}_s(m) := (\mathsf{GLWE}_s(-m \cdot s \otimes \mathbf{g}), \mathsf{GLWE}_s(m \cdot \mathbf{g}))$$

where $\otimes$ denotes the Kronecker product. As introduced in [15], there exists a homomorphism known as the *external product* between GLWE and GGSW ciphertexts:

$$\begin{aligned} \boxdot : \mathsf{GGSW}_s(m) \times \mathsf{GLWE}_s(m') &\to \mathsf{GLWE}_s(m \cdot m') \\ (\mathbf{C}, \mathbf{c}) &\to \mathbf{g}^{-1}(\mathbf{c}) \cdot \mathbf{C} \end{aligned} \qquad (1)$$

where $\mathbf{g}^{-1}$ is called gadget decomposition that $\langle \mathbf{g}, \mathbf{g}^{-1}(a) \rangle = a$.

Functional Key Switching. The output of the external product between $\mathsf{GGSW}_s(m)$ and $\mathsf{GLWE}_s(m')$ is a GLWE ciphertext in $\mathsf{GLWE}_s(m \cdot m')$. The key insight here is that GGSW ciphertexts encrypt (some encoding of) s, and during the external product, they are used to homomorphically decrypt the GLWE ciphertext. Consequently, the GGSW secret key and GLWE secret key need not be the same, which is developed as a technique called *functional key switching*. We summarize a simplified lemma as follows.

Lemma 1 (Functional Key Switching, Adapted from [15]). *Given a GLWE ciphertext $\mathbf{c} \in \mathsf{GLWE}_{s'}(m')$, and $\mathsf{KS} \in (\mathsf{GLWE}_s(-m \cdot s' \otimes \mathbf{g}), \mathsf{GLWE}_s(m \cdot \mathbf{g}))$, the output of the external product of KS and $\mathbf{c}$ is a GLWE ciphertext $\mathbf{c}' \in \mathsf{GLWE}_s(m \cdot m')$.*

For simplicity, we denote the functional key switching keys by the GGSW ciphertext as well, but they are labeled with the explicit secret keys:

$$\mathsf{GGSW}_s(m; s') := (\mathsf{GLWE}_s(-m \cdot s' \otimes \mathbf{g}), \mathsf{GLWE}_s(m \cdot \mathbf{g}))$$

Note that when $s = s'$, $\mathsf{GGSW}_s(m'; s') = \mathsf{GGSW}_s(m')$. Consequently, the external product can operate on the general forms of GGSW and GLWE ciphertexts:

$$\boxdot : \mathsf{GGSW}_s(m; s') \times \mathsf{GLWE}_{s'}(m') \to \mathsf{GLWE}_s(m \cdot m')$$

Sample Extraction. A GLWE ciphertext contains sufficient data to construct several corresponding LWE ciphertexts. Given a GLWE ciphertext $\mathbf{c} = (\mathbf{a}, b) \in \mathsf{GLWE}_s(m)$, the sample extraction is a function

$$\mathsf{Extract} : \mathsf{GLWE}_s(m) \to \mathsf{LWE}_s(m_0)$$

outputs an LWE ciphertext where m_0 is the constant term of m. We refer to [15] for more details.

2.4 Other Cryptographic Background

Pseudo-random Number Generator. A pseudo-random number generator (PRNG) is a function that takes as input a small length of bit sequence and outputs one with an arbitrary length, which appears random. In this work we select the 128-bit seed as PRNG : $\{0,1\}^{128} \to \{0,1\}^*$.

3 BioVite

In this section, we construct and optimize the building blocks in our scheme: the homomorphic inner product protocol for similarity computation and the homomorphic membership test protocol for threshold comparison.

3.1 Homomorphic Inner Product

BioVite handles cosine similarity, which is computed as the inner product of two normalized vectors. Our homomorphic inner product protocol evaluates this inner product with a single external product and further optimizes it using PRNG, functional key switching, and sample extraction techniques.

Inner Product via External Product. We recall the trick for performing a homomorphic inner product within a single homomorphic multiplication [34,50]. Let $a, b \in R_t$ be two polynomials in the plaintext space; then, the homomorphic multiplication of their ciphertexts yields a new ciphertext encrypting $a \cdot b$ as in Eq. (1). A special polynomial encoding for the data is designed as $a = a_0 + a_1 X + \cdots + a_{N-1} X^{N-1}$, $b = b_0 - b_{N-1} X - \cdots - b_1 X^{N-1}$ and the constant term of $(a \cdot b \mod X^N + 1)$ is $\langle \boldsymbol{a}, \boldsymbol{b} \rangle$. This idea allows us to focus solely on the optimization of homomorphic multiplication, without the need for homomorphic rotations as in [6,42].

Compact Encryption Parameter. Due to security considerations over lattices [2], RLWE-based schemes with a ciphertext modulus of 192 bits require N to be at least 4096 under the Gaussian error distribution with a standard deviation of 3.2, which introduces redundancy since the dimension of the latent space for biometric systems can be empirically smaller (e.g., 512 in FaceNet).

To eliminate this redundancy, GLWE ciphertexts are considered. To encrypt data in a (power-of-2) dimension $m \leq N$, a GLWE ciphertext $(\boldsymbol{a}, b) \in R_q^{n+1}$ requires $m(n+1)$ elements in $\mathbb{Z}_q$, while RLWE requires $2N$. Thus for features in smaller dimension, GLWE ciphertexts are optimal.

The mask part $\boldsymbol{a} \in R_q^n$ of the ciphertext appears to still be burdensome for holding mn coefficients. However, the communication overhead can be compressed using a pseudo-random number generator. Since $\boldsymbol{a}$ is randomly generated during encryption, a random seed can be set for the PRNG to generate it. The seed and the masked polynomial $b \in R_q$ are sufficient to recover $(\boldsymbol{a}, b)$, and only transferring these data is enough.

Furthermore, as we only need several most significant bits for the later threshold comparison, an approximate decryption works well on it. That means the encoding factor Δ for encryption can be set lower while keeping some of the MSBs effective. The concrete parameters and the size of GLWE and PRNG are shown in Table 1.

Table 1. Comparison of storage and communication size for concrete GLWE parameters. Each row achieves a concrete security level of 128-bit by setting $\sigma = 3.2$, the deviation of the Gaussian distribution of noise. The size of the random seed for the PRNG is 128 bits, and #ct indicates the number of ciphertext needed for encrypting a 512-dimensional template.

n	N	$\log q$	Δ	Useful MSB	#ct	Storage	Communication
1	4096	109	2^{90}	2^{19}	1	128	64.03
3	512	64	2^{45}	2^{19}	1	16	4.03
5	256	32	2^{5}	2^{10}	2	12	2.06
3	512	32	2^{5}	2^{10}	1	8	2.03

Functional Key Switching. In the three-party scenario, the client holds the public key pk for encryption, to prevent the potential leakage of the secret key at the client side. Unfortunately, such a configuration fails to exploit the ciphertext compression technique. As the public key is also known by the server, if the random number can be generated at the server side, it can be used to immediately recover the message as $\mathsf{ct} = r \cdot \mathsf{pk} + (0, \Delta \cdot m) + e$. In other words, the compression can only be applied with secret key encryption. To both employ ciphertext compression and protect the secret key for the database, we adopt the functional key switching technique [15].

Say there are two secret keys $s_{\mathsf{csp}}, s_{\mathsf{client}}$, where s_{csp} is held by the CSP for encrypting biometric templates, and s_{client} is held by the client for verification. By functional key switching, the server holds $\mathbf{C} \in \mathsf{GGSW}_{s_{\mathsf{csp}}}(m; s_{\mathsf{client}})$ and when the client sends a ciphertext $\boldsymbol{c} \in \mathsf{GLWE}_{s_{\mathsf{client}}}(m')$, the external product $\mathbf{C} \boxdot \boldsymbol{c}$ is in $\mathsf{GLWE}_{s_{\mathsf{csp}}}(m \cdot m')$ according to Lemma 1. By encoding the data into m and m', the key switching and inner product are performed simultaneously. Since the client is now using secret key encryption, the compression technique is feasible again and communication is reduced as desired.

Sample Extraction. The output of homomorphic multiplication between GGSW and GLWE ciphertexts is still a GLWE ciphertext, whereas we only need the constant term of the polynomial it encrypts. Compared to a heavier decryption of a GLWE ciphertext with complexity of $O(nN \log N)$, decrypting an LWE ciphertext extracted from GLWE only requires $O(nN)$ scalar multiplications in $\mathbb{Z}_q$. Thus, by the sample extraction technique mentioned in Sect. 2.3, the LWE ciphertext extracted from the homomorphic inner product is transferred for decryption.

Table 2. Parameters for evaluation. N is the degree of the underlying cyclotomic ring; n is the GLWE dimension; t is the plaintext modulus; B is the decomposition base; $\log q$ is the number of bits of the ciphertext modulus; σ is the standard deviation of the Gaussian distribution, and λ is the concrete security level. Each parameter set is chosen to ensure at least a 128-bit (classical) security level and a decryption failure probability of less than 2^{-40}.

Inner Product						Membership Test						
N	n	t	$\log q$	σ	B	λ	N	n	t	$\log q$	$\ell - k$	σ	λ
512	3	2^{19}	32	3.2	16	143	128	5	$2^9 + 2^8 + 1$	16	8	3.2	134

3.2 Secure Threshold Comparison

After the homomorphic inner product is performed, the server holds the result that can be decrypted by CSP. To prevent the CSP knowing the similarity, a membership test is launched between CSP and the server for secure homomorphic comparison. Say the partial decryption of the homomorphic inner product is $v \in \mathbb{Z}_q$, a threshold comparison using membership test is to determine whether $v \in \{0, \cdots, q/2\}$. A canonical solution is to encrypt the set in CRT form, which requires NTT encoding for SIMD [5], and then perform the plaintext-ciphertext multiplication. Noticing the asymmetry message flow of the three-party architecture, we therefore propose the membership test without plaintext-ciphertext multiplication.

Compact Membership-Test. Adapt the notation before where $v = \langle \boldsymbol{c}, (\boldsymbol{s}, 1) \rangle$ is the partial decryption, then the membership test problem $v \in \{0, \cdots, q/2\}$ is equivalent to $v + r \in \{0 + r, \cdots, q/2 + r\}$ for some random number $r \in \mathbb{Z}_q$. Note that q may be large, so a rounding operation is performed first to extract the MSBs as $\tilde{v}_r = \lfloor (v+r)/2^k \rfloor$, and the set is $S = \{\lfloor 0 + r \rfloor / 2^k, \cdots, \lfloor q/2 + r \rfloor / 2^k\} \subset \mathbb{Z}_{q/2^k}$ for q is power-of-two. Such set within smaller size can preserve precision: the similarity computed is with factor γ^2, but the real precision is not $1/\gamma^2$ due to rounding error during feature encoding (e.g. when $\gamma = 512$, the precision is empirical precision is about 0.01). Thus we could select several most significant bits to preserve the precision and reduce computation overhead for verification. Here we define the size of reduced set to be $|S|$.

Here's the key point of membership test. Randomly select several numbers $r_0, \cdots, r_{q/2}$ and multiply them into the threshold set to get:

$$\left\{ \left\lfloor \frac{\theta + r}{2^k} \right\rceil \cdot r_0, \cdots, \left\lfloor \frac{q + 2r}{2^{k+1}} \right\rceil \cdot r_{|S|-1} \right\}$$

Then the membership test problem becomes

$$0 \in \left\{ \left\lfloor \frac{\theta + r}{2^k} \right\rceil \cdot r_0, \cdots, \left\lfloor \frac{q + 2r}{2^{k+1}} \right\rceil \cdot r_{|S|-1} \right\} - \left\lfloor \frac{v + r}{2^k} \right\rceil \cdot \{r_0, \cdots, r_{|S|-1}\} \quad (2)$$

Parties:

- *Client:* Semi-honest party who probes new biometric templates for verification.
- *Server:* Semi-honest party who maintains an encrypted database for enrolled biometric templates.
- *CSP:* Semi-honest party who stores the secret key to the encrypted database.

Protocol:

1. **Setup:**
   - CSP generates a GLWE secret key s_{csp} for similarity computation.
   - Client generates a GLWE secret key s_{client}.
   - Client generates a GLWE secret key s_{mt} for membership test, and synchronizes it with the server.

2. **Enrollment:**
   - The client request a fresh GGSW ciphertext encrypting zero from CSP.
   - The client probes new biometric features $\boldsymbol{f} \in \mathbb{R}^N$, and encodes them with a quantization factor γ into a polynomial: $m = \lfloor f_0 \cdot \gamma \rceil - \sum_{i=1}^{N-1} \lfloor f_i \cdot \gamma \rceil X^{N-i}$. An ID is selected for binding this template.
   - The client encodes m into the GGSW plaintext: $\mathbf{C}_{\mathsf{ID}} \in \mathsf{GGSW}_{s_{\mathsf{csp}}}(m; s_{\mathsf{client}})$.
   - The client sends $(\mathsf{ID}, \mathbf{C}_{\mathsf{ID}})$ to the server and the server stores it into the database.

3. **Verification:**
   Client:
   - Probes new biometric features $\boldsymbol{f}' \in \mathbb{R}^N$, and encodes them with γ into a polynomial $m' = \sum_{i=0}^{N} \lfloor f'_i \cdot \gamma \rceil X^i$. Then encrypts it using client's secret key:

   $$\boldsymbol{c}_{\mathsf{ID}} = \mathsf{GLWE.Enc}(s_{\mathsf{client}}, m').$$

   - Sends $(\mathsf{ID}, \boldsymbol{c}_{\mathsf{ID}})$ to the server.

   Server:
   - Extracts $\mathbf{C}_{\mathsf{ID}}$ according to the given ID, and calculates $\boldsymbol{c}_{\mathsf{sim}} = \mathbf{C}_{\mathsf{ID}} \boxdot \boldsymbol{c}_{\mathsf{ID}}$.
   - Uniformly $r \in \mathbb{Z}_q$ and $r_i \in \mathbb{Z}^*_{q/2^k}, i \in [|S|]$ with all non-zero coefficient. Choose a random permutation $\pi : [|S|] \to [|S|]$, then calculates

   $$P_\theta = \sum_{i=0}^{|S|-1} \pi\left(\left\lfloor \frac{\theta + r}{2^k} \right\rceil + i\right) \cdot r_i X^i, \quad P_r = \sum_{i=0}^{|S|-1} r_i X^i$$

   as in Eq. (3), and $\boldsymbol{c}'_{\mathsf{sim}} = \mathsf{Extract}(\boldsymbol{c}_{\mathsf{sim}}) + (0, r)$.
   - Encrypts $\boldsymbol{c}_\theta = \mathsf{GLWE.Enc}(s_{\mathsf{mt}}, P_\theta)$.
   - Sends $(\boldsymbol{c}'_{\mathsf{sim}}, P_r, \boldsymbol{c}_\theta)$ to CSP.

   CSP:
   - Decrypts $\boldsymbol{c}'_{\mathsf{sim}}$ to get v, and calculates $\tilde{v} = \lfloor v/2^k \rceil$.
   - Calculates $\boldsymbol{c}_{\mathsf{mt}} = \boldsymbol{c}_\theta - (0, \tilde{v} \cdot P_r)$.
   - Sends $\boldsymbol{c}_{\mathsf{mt}}$ to the client.

   Client:
   - Decrypts $\boldsymbol{c}_{\mathsf{mt}}$. If there is a zero exists in the former $|S|$ coefficient, then verification passes, else the verification fails.

Fig. 3. The full verification protocol.

Now rewrite them in the ciphertext form, we first let

$$P_\theta = \sum_{i=0}^{|S|-1} \left(\left\lfloor \frac{\theta+r}{2^k} \right\rfloor + i\right) \cdot r_i X^i, \quad P_r = \sum_{i=0}^{|S|-1} r_i X^i \qquad (3)$$

and with the encryptions $c_\theta \in \mathsf{GLWE}(P_\theta)$, all CSP needs to do is to calculate

$$c_\theta - (0, \tilde{v}_r \cdot P_r) \in \mathsf{GLWE}(P_\theta - \tilde{v}_r \cdot P_r) \qquad (4)$$

whose plaintext is the set in Eq. (2). That reduces the computational complexity to scalar-polynomial multiplication, and since the mask part is unchanged, it can be compressed using PRNG, extremely reduces the communication overhead.

We refer to Fig. 3 for the whole scheme, Appendix A for the noise analysis, and Appendix B for security proof for BioVite.

4 Evaluation and Experiments

We evaluate our scheme on the face verification task. The FaceNet [48] is selected for feature extraction, with an embedding dimension of 512. The well-known LFW [25], CFP-FP [49], and AgeDB [40] datasets are selected to test the accuracy. We implement our schemes based on the TFHE-rs [51] library and run experiments on a single core of an Intel Xeon Gold 6230R@2.10GHz processor, hosted by Ubuntu 22.04 LTS. The parameters for evaluation are listed in Table 2. Our code is available on GitHub[1].

Evaluation of BioVite. As shown in Table 3, BioVite is a one-round verification scheme where the message flow is Client → Server → CSP → Client. Due to the sufficient compact parameter and optimization in membership test (as in Eq. (4)), the computation part (os secure inner product and membership test) is extremely efficient (0.02 ms and 0.001 ms, respectively). The other computation parts are encryptions and decryption. By some strategy like pregenerating ciphertexts pool, these time cost can be eliminated, and the online time cost is further reduced. The communication overhead is also very compact, with total 8.5 KB data communicated.

Comparison. We compare our schemes to other FHE-based biometric verification schemes, as shown in Table 4. BioVite outperforms these schemes in both time and communication complexity. These HE-based schemes [1,26] adopt BGV/BFV/CKKS for evaluating inner product, thus producing high communication and computation costs due to redundancy in the plaintext space. Compared to these works, BioVite reduces the total computation overhead to a single external product and a scalar-polynomial multiplication, showing high efficiency and compactness. As for the ciphertext size stored in the database, each of our GGSW requires 128 KB for enrollment, which is same to a single RLWE ciphertext in dimension of 4096.

---

[1] https://github.com/Fainabi/BioVite.

Table 3. Evaluation of the steps in three-party scenario. Time unit is millisecond (ms), and storage unit is kilobyte in binary (KB).

Party	Task	Time	Communication
Client	Encrypts new template	0.237	2.015
Server	Secure inner product	0.020	6.015
	Encrypts membership test messages	0.043	0.265
CSP	Membership test	0.001	0.265
Client	Decrypts verification result	0.020	-
Total		0.324	8.560

Table 4. Benchmark of verification schemes. Template size is normalized to $m = 512$, and the bit count of quantized template elements is 8. Bold digits indicate the best results.

Scheme	Time (ms)	Communication (KB)
[20]	1112	133.16
[1]	1403	73.617
[26]	530	563.2
[5]	27.55	197.6
Ours	**0.324**	**8.560**

Accuracy. We select the LFW [25], CFP-FP [49], and AgeDB [40] datasets for testing the accuracy under different scenario settings. These datasets contain pairs of facial images for verification test. For each pair of the images, they are first passed to FaceNet for extracting embeddings in dimension of 512. These embeddings are then normalized to $[-1, 1]^{512}$ with Euclidean norm of 1. The normalized embeddings are processed and evaluated in two categories:

- *Raw Comparison.* Perform the similarity computation and threshold comparison directly on the embeddings of the test pairs. Then calculate the accuracy according to the ground-truth in the database.
- *Quantization and Rounding on Masked Similarity.* Quantize the normalized embeddings with $\gamma = 512$, round them to integers. The difference of cosine similarity in the last step and threshold, is added with a random number r modulo $t = 2^{19}$, and rounded to reserve the most 8 significant bits. Finally we add the rounded masked similarity with rounded $-r$ to construct the comparison result.

We collect the experiment into Table 5 to show the verification accuracy of BioVite on the LFW, CFP-FP, and AgeDB datasets. BioVite reaches a highly similar accuracy on these datasets.

Table 5. Accuracy on LFW, CFP-FP, and AgeDB datasets. TPR represents true positive rate and TNR represents true negative rate.

Dataset	FaceNet		FaceNet + BioVite	
	TPR	TNR	TPR	TNR
LFW	0.942	0.982	0.942	0.988
CFP-FP	0.945	0.992	0.930	0.995
AgeDB	0.855	0.975	0.839	0.985

A Noise Estimation

In this work, we follow the average-case noise analysis from [15] that models the noise in the ciphertexts during homomorphic operations as independent Gaussian random variables. We begin with the necessary notations.

Notations. For a GLWE ciphertext c, we denote the noise term of c as $\mathsf{Err}(c)$, which is an element in R. Similarly, we denote the noise of a GGSW ciphertext C as $\mathsf{Err}(C)$, which is a vector over R. For $e = e_0 + e_1 X + \cdots + e_{N-1} X^{N-1} \in R$, its ℓ_2-norm and ℓ_∞-norm are defined as $\|e\|_2 = (\sum_{i=0}^{N-1} |e_i|^2)^{1/2}$ and $\|e\|_\infty := \max\{|e_i|\}_{i=0,\ldots,N-1}$. For a vector $\boldsymbol{e} := (e_0, \ldots, e_{d-1}) \in R^d$, its ℓ_p-norm is defined as $\|\boldsymbol{e}\|_p := \max\{\|e_i\|_p\}_{i=0,\ldots,d-1}$ for $p = 2$ or ∞. When modeling all the coefficients of $e \in R$ (or $\boldsymbol{e} \in R^d$) as random variables, we denote the variance of each coefficient as $\mathsf{Var}(e)$ (or $\mathsf{Var}(\boldsymbol{e})$).

The following assumption simplifies bounding the noise variance rather than the ℓ_∞-norm of the error term in the worst-case. Estimating the variance allows for smaller parameters in the homomorphic cryptosystem, as we only need to ensure a sufficiently negligible probability of decryption failure.

Assumption 1 (Independence Heuristic [15]). *All the coefficients of the errors of GLWE or GGSW samples that occur in all the linear combinations we consider are independent and concentrated.*

The following lemmas provide the noise analysis for the homomorphic operations used in our constructions.

Lemma 2 (Functional Key Switching, Adapted From [15]). *Let $c \in \mathsf{GLWE}_{s'}(m')$, $\mathsf{KS} \in (\mathsf{GLWE}_s(-m \cdot s' \otimes \mathbf{g}), \mathsf{GLWE}_s(m \cdot \mathbf{g}))$, and c' be the output of the functional key switching operation. Under Assumption 1, we have*

$$\mathsf{Var}(\mathsf{Err}(c')) \leq (n+1)dN\beta^2 V_{\mathsf{KS}} + (1+nN)\|m\|_2^2 \epsilon^2 + \|m\|_2^2 V_c,$$

where $V_{\mathsf{KS}} := \mathsf{Var}(\mathsf{Err}(\mathsf{KS}))$, $V_c := \mathsf{Var}(\mathsf{Err}(c))$, $\beta = B_g/2$ and ϵ is the approximation error in the approximate gadget decomposition.

Noise Estimation. In the three-party scenario, the noise grows only during the homomorphic inner product, which we summarize in the following theorem.

Theorem 2. *Adopt the notations from Fig. 3, in the three-party scenario, denote $c' = \mathsf{Extract}(\mathbf{C}_{\mathsf{ID}} \boxdot c_{\mathsf{ID}}) + (0, r \cdot \Delta)$ in the homomorphic inner product, then we have $\mathsf{Var}(\mathsf{Err}(c')) \leq (n+1)dN\beta^2 \mathsf{Var}(\mathsf{Err}(\mathbf{C}_{\mathsf{ID}})) + (1+nN)\|m\|_2^2\epsilon^2 + \|m\|_2^2 \mathsf{Var}(\mathsf{Err}(c_{\mathsf{ID}}))$.*

Proof. In the homomorphic inner product, the noise of $\mathbf{C}_{\mathsf{ID}} \boxdot c_{\mathsf{ID}}$ follow directly from Lemma 2. As the sample extract introduces no additional noise, we have the claim.

Decryption Failure Probability. Once we have the estimated noise variance of the output ciphertext after the homomorphic operations, we can calculate the decryption failure probability using the Gaussian tail bound. Specifically, for a ciphertext c with noise variance V, we have

$$\Pr\left[\|\mathsf{Err}(c)\|_\infty > c \cdot \sqrt{V}\right] \leq \mathsf{erfc}(c/\sqrt{2})$$

for any positive real number c. As the decryption procedure fails when $\|\mathsf{Err}(c)\|_\infty > q/(2t)$ for plaintext modulus t, we have

$$\Pr[\|\mathsf{Err}(c)\|_\infty > q/(2t)] \leq \mathsf{erfc}\left(\frac{q}{2t\sqrt{2V}}\right).$$

In this work, we set the decryption failure probability to be less than 2^{-40}, which is sufficiently negligible for real-world applications.

B Security Proof

BioVite is a three-party scheme, containing the client, server, and CSP. We construct the security proof against semi-honest adversaries, which honestly follows the steps of scheme but can calculate anything they want. Following the ideal-real world paradigm, we construct the indistinguishability between simulator and real world views [35]. We first define three simulators:

- $\mathcal{S}_{\mathsf{Client}}$. Inputs: $(s_{\mathsf{client}}, s_{\mathsf{mt}}, m)$; Outputs: $(s_{\mathsf{client}}, s_{\mathsf{mt}}, m; c_{\mathsf{mt}}; c_{\mathsf{ID}})$.
- $\mathcal{S}_{\mathsf{Server}}$. Inputs: $(s_{\mathsf{mt}}, \mathbf{C}_{\mathsf{ID}}; c_{\mathsf{ID}})$; Outputs: $(s_{\mathsf{mt}}, \mathbf{C}_{\mathsf{ID}}; c_{\mathsf{ID}}; c'_{\mathsf{sim}}, P_r, c_\theta, r, r_i;)$.
- $\mathcal{S}_{\mathsf{CSP}}$. Inputs: $(s_{\mathsf{csp}}; c'_{\mathsf{sim}}, P_r, c_\theta)$; Outputs: $(s_{\mathsf{csp}}; c'_{\mathsf{sim}}, P_r, c_\theta; c_{\mathsf{mt}})$.

The views of a simulator are its inputs, outputs and the random number it generates.

Proof for $\mathcal{S}_{\mathsf{Client}}$. Since the client does not contain the enrolled template $\mathbf{C}_{\mathsf{ID}}$ and secret key c_{csp}. It could only generates these data itself, and perform protocol in Fig. 3 in the ideal world. By the hardness of GLWE problem 2, the distribution $\{c_{\mathsf{mt}}\}$ is computational indistinguishable to the uniform distribution. Therefore the real world views $\mathsf{view}_{\mathsf{Client}} = (s_{\mathsf{client}}, s_{\mathsf{mt}}, c_{\mathsf{ID}}, c_{\mathsf{mt}}, m)$ satisfy

$$\{\mathcal{S}_{\mathsf{Client}}(s_{\mathsf{client}}, s_{\mathsf{mt}}, m)\} \stackrel{c}{\equiv} \{\mathsf{view}_{\mathsf{Client}}\}$$

where $\stackrel{c}{\equiv}$ denotes the relation of indistinguishability.

Proof for $\mathcal{S}_\text{Server}$. The only input from the client beside ID is the ciphertext c_ID. In ideal world, the server can only simulate it by generate a new random key s, and use such key for generate a new ciphertext. By applying the hardness of GLWE problem 2, the distribution of such generated ciphertext is indistinguishable to the distribution for input c_ID, therefore the views $\text{view}_\text{Server} = (s_\text{mt}, C_\text{ID}, c_\text{ID}, c_\theta, c'_\text{sim}; P_r, r, \{r_i\})$ satisfy

$$\{\mathcal{S}_\text{Server}(s_\text{mt}, C_\text{ID}; c_\text{ID})\} \stackrel{c}{\equiv} \{\text{view}_\text{Server}\}$$

Proof for $\mathcal{S}_\text{CSP}$. CSP accepts input from the server, in which P_r is uniformly selected and c_θ is encrypted using s_mt. Since CSP does not possess s_mt, the simulator $\mathcal{S}_\text{CSP}$ can only generate it itself and simulate the whole protocol. By the hardness of GLWE problem 2, the ciphertexts generated by s_mt is indistinguishable to uniform distribution. Therefore the views $\text{view}_\text{CSP} = (s_\text{csp}, c'_\text{sim}, P_r, c_\theta, c_\text{mt})$ satisfy

$$\{\mathcal{S}_\text{CSP}(s_\text{csp}; c'_\text{sim}, P_r, c_\theta)\} \stackrel{c}{\equiv} \{\text{view}_\text{Server}\}$$

These three proofs together prove the security of BioVite, in which only the client get the verification result, and no biometric templates leaked.

References

1. Agrawal, S., Badrinarayanan, S., Mukherjee, P., Rindal, P.: Game-set-MATCH: using mobile devices for seamless external-facing biometric matching. In: Ligatti, J., Ou, X., Katz, J., Vigna, G. (eds.) ACM CCS 2020, pp. 1351–1370. ACM Press (2020). https://doi.org/10.1145/3372297.3417287
2. Albrecht, M.R., Player, R., Scott, S.: J. Math. Cryptol. **9**(3), 169–203 (2015)
3. Bai, J., et al.: CryptoMask: privacy-preserving face recognition. In: Wang, D., Yung, M., Liu, Z., Chen, X. (eds.) ICICS 2023. LNCS, vol. 14252, pp. 333–350. Springer, Singapore (2023). https://doi.org/10.1007/978-981-99-7356-9_20
4. Bassit, A., Hahn, F., Peeters, J., Kevenaar, T., Veldhuis, R.N.J., Peter, A.: Fast and accurate likelihood ratio-based biometric verification secure against malicious adversaries. IEEE Trans. Inf. Forensics Secur. **16**, 5045–5060 (2021)
5. Bassit, A., Hahn, F., Veldhuis, R.N.J., Peter, A.: Improved multiplication-free biometric recognition under encryption. IEEE Trans. Biom. Behav. Identity Sci. **6**(3), 314–325 (2024)
6. Bauspieß, P., Olafsson, J., Kolberg, J., Drozdowski, P., Rathgeb, C., Busch, C.: Improved homomorphically encrypted biometric identification using coefficient packing. In: 2022 International Workshop on Biometrics and Forensics (IWBF) (2022)
7. Beaver, D.: Efficient multiparty protocols using circuit randomization. In: Feigenbaum, J. (ed.) CRYPTO 1991. LNCS, vol. 576, pp. 420–432. Springer, Heidelberg (1992). https://doi.org/10.1007/3-540-46766-1_34

8. Beaver, D., Micali, S., Rogaway, P.: The round complexity of secure protocols (extended abstract). In: 22nd ACM STOC, pp. 503–513. ACM Press (1990). https://doi.org/10.1145/100216.100287
9. Blanton, M., Murphy, D.: Privacy preserving biometric authentication for fingerprints and beyond. In: Proceedings of the Fourteenth ACM Conference on Data and Application Security and Privacy, CODASPY 2024 (2024)
10. Brakerski, Z., Gentry, C., Vaikuntanathan, V.: Leveled) fully homomorphic encryption without bootstrapping. In: Goldwasser, S. (ed.) ITCS 2012, pp. 309–325. ACM (2012). https://doi.org/10.1145/2090236.2090262
11. Chamikara, M.A.P., Bertók, P., Khalil, I., Liu, D., Camtepe, S.: Privacy preserving face recognition utilizing differential privacy. Comput. Secur. **97**, 101951 (2020)
12. Cheng, N., Önen, M., Mitrokotsa, A., Chouchane, O., Todisco, M., Ibarrondo, A.: Nomadic: normalising maliciously-secure distance with cosine similarity for two-party biometric authentication. In: Proceedings of the 19th ACM Asia Conference on Computer and Communications Security, ASIA CCS 2024. ACM (2024)
13. Cheon, J.H., Kim, A., Kim, M., Song, Y.: Homomorphic encryption for arithmetic of approximate numbers. In: Takagi, T., Peyrin, T. (eds.) ASIACRYPT 2017. LNCS, vol. 10624, pp. 409–437. Springer, Cham (2017). https://doi.org/10.1007/978-3-319-70694-8_15
14. Cheon, J.H., Kim, D., Kim, D.: Efficient homomorphic comparison methods with optimal complexity. In: Moriai, S., Wang, H. (eds.) ASIACRYPT 2020. LNCS, vol. 12492, pp. 221–256. Springer, Cham (2020). https://doi.org/10.1007/978-3-030-64834-3_8
15. Chillotti, I., Gama, N., Georgieva, M., Izabachène, M.: TFHE: fast fully homomorphic encryption over the torus. J. Cryptol. **33**(1), 34–91 (2019). https://doi.org/10.1007/s00145-019-09319-x
16. Drozdowski, P., Buchmann, N., Rathgeb, C., Margraf, M., Busch, C.: On the application of homomorphic encryption to face identification. In: 2019 International Conference of the Biometrics Special Interest Group, BIOSIG 2019, vol. P-296. LNI, pp. 173–180 (2019)
17. Ducas, L., Micciancio, D.: FHEW: bootstrapping homomorphic encryption in less than a second. In: Oswald, E., Fischlin, M. (eds.) EUROCRYPT 2015. LNCS, vol. 9056, pp. 617–640. Springer, Heidelberg (2015). https://doi.org/10.1007/978-3-662-46800-5_24
18. Engelsma, J.J., Jain, A.K., Boddeti, V.N.: HERS: homomorphically encrypted representation search. IEEE Trans. Biom. Behav. Identity Sci. **4**(3), 349–360 (2022)
19. Erkin, Z., Franz, M., Guajardo, J., Katzenbeisser, S., Lagendijk, I., Toft, T.: Privacy-preserving face recognition. In: Goldberg, I., Atallah, M.J. (eds.) PETS 2009. LNCS, vol. 5672, pp. 235–253. Springer, Heidelberg (2009). https://doi.org/10.1007/978-3-642-03168-7_14
20. Ernst, J., Mitrokotsa, A.: A framework for UC secure privacy preserving biometric authentication using efficient functional encryption. In: Tibouchi, M., Wang, X. (eds.) ACNS 2023, International Conference on Applied Cryptography and Network Security, Part II. LNCS, vol. 13906, pp. 167–196. Springer, Cham (2023). https://doi.org/10.1007/978-3-031-33491-7_7
21. Fan, J., Vercauteren, F.: Somewhat Practical Fully Homomorphic Encryption. Cryptology ePrint Archive, Report 2012/144 (2012). https://eprint.iacr.org/2012/144
22. Gentry, C.: Fully homomorphic encryption using ideal lattices. In: Mitzenmacher, M. (ed.) 41st ACM STOC, pp. 169–178. ACM Press (2009). https://doi.org/10.1145/1536414.1536440

23. Gentry, C., Sahai, A., Waters, B.: Homomorphic encryption from learning with errors: conceptually-simpler, asymptotically-faster, attribute-based. In: Canetti, R., Garay, J.A. (eds.) CRYPTO 2013. LNCS, vol. 8042, pp. 75–92. Springer, Heidelberg (2013). https://doi.org/10.1007/978-3-642-40041-4_5
24. Gomez-Barrero, M., Maiorana, E., Galbally, J., Campisi, P., Fiérrez, J.: Multibiometric template protection based on homomorphic encryption. Pattern Recognit. **67**, 149–163 (2017)
25. Huang, G.B., Ramesh, M., Berg, T., Learned-Miller, E.: Labeled Faces in the Wild: A Database for Studying Face Recognition in Unconstrained Environments. Technical report, University of Massachusetts, Amherst (2007)
26. Huang, H., Wang, L.: Efficient privacy-preserving face identification protocol. IEEE Trans. Serv. Comput. **16**(4), 2632–2641 (2023)
27. Huang, H., Wang, L.: Efficient privacy-preserving face verification scheme. J. Inf. Secur. Appl. **63**, 103055 (2021)
28. Ibarrondo, A., Chabanne, H., Despiegel, V., Önen, M.: Grote: group testing for privacy-preserving face identification. In: Proceedings of the Thirteenth ACM Conference on Data and Application Security and Privacy (2023)
29. Ibarrondo, A., Chabanne, H., Önen, M.: Funshade: function secret sharing for two-party secure thresholded distance evaluation. In: Proceedings of Privacy Enhancing Technologies (2023)
30. Iliashenko, I., Zucca, V.: Faster homomorphic comparison operations for BGV and BFV. PoPETs **2021**(3), 246–264 (2021). https://doi.org/10.2478/popets-2021-0046
31. Im, J., Jeon, S., Lee, M.: Practical privacy-preserving face authentication for smartphones secure against malicious clients. IEEE Trans. Inf. Forensics Secur. **15**, 2386–2401 (2020)
32. Im, J.-H., Jeon, S.-Y., Lee, M.-K.: Practical privacy-preserving face authentication for smartphones secure against malicious clients. IEEE Trans. Inf. Forensics Secur. **15**, 2386–2401 (2020)
33. Ji, J., et al.: Privacy-preserving face recognition with learnable privacy budgets in frequency domain. In: Computer Vision - ECCV 2022 (2022)
34. Juvekar, C., Vaikuntanathan, V., Chandrakasan, A.P.: GAZELLE: a low latency framework for secure neural network inference. In: 27th USENIX Security Symposium, USENIX Security 2018 (2018)
35. Lindell, Y.: How To Simulate It - A Tutorial on the Simulation Proof Technique. Cryptology ePrint Archive, Report 2016/046 (2016). https://eprint.iacr.org/2016/046
36. Lyubashevsky, V., Peikert, C., Regev, O.: On ideal lattices and learning with errors over rings. In: Gilbert, H. (ed.) EUROCRYPT 2010. LNCS, vol. 6110, pp. 1–23. Springer, Heidelberg (2010). https://doi.org/10.1007/978-3-642-13190-5_1
37. Mi, Y., et al.: DuetFace: collaborative privacy-preserving face recognition via channel splitting in the frequency domain. In: MM 2022: The 30th ACM International Conference on Multimedia (2022)
38. Mi, Y., et al.: Privacy-preserving face recognition using random frequency components. In: IEEE/CVF International Conference on Computer Vision, ICCV 2023 (2023)
39. Minaee, S., Abdolrashidi, A., Su, H., Bennamoun, M., Zhang, D.: Biometrics recognition using deep learning: a survey. Artif. Intell. Rev. **56**(8), 8647–8695 (2023)
40. Moschoglou, S., Papaioannou, A., Sagonas, C., Deng, J., Kotsia, I., Zafeiriou, S.: AgeDB: the first manually collected, in-the-wild age database. In: 2017 IEEE Con-

ference on Computer Vision and Pattern Recognition Workshops, CVPR Workshops (2017)
41. Nandakumar, K., Jain, A.K.: Biometric template protection: bridging the performance gap between theory and practice. IEEE Signal Process. Mag. **32**(5), 88–100 (2015)
42. Boddeti, V.N.: Secure face matching using fully homomorphic encryption. In: 2018 IEEE 9th International Conference on Biometrics Theory, Applications and Systems (BTAS) (2018)
43. Pradel, G., Mitchell, C.: Privacy-preserving biometric matching using homomorphic encryption. In: 2021 IEEE 20th International Conference on Trust, Security and Privacy in Computing and Communications (Trust-Com) (2021)
44. Rabin, M.O.: How To Exchange Secrets with Oblivious Transfer. Cryptology ePrint Archive, Report 2005/187 (2005). https://eprint.iacr.org/2005/187
45. Regev, O.: On lattices, learning with errors, random linear codes, and cryptography. In: Gabow, H.N., Fagin, R. (eds.) 37th ACM STOC, pp. 84–93. ACM Press (2005). https://doi.org/10.1145/1060590.1060603
46. Ribeiro, M., Grolinger, K., Capretz, M.A.: MLaaS: machine learning as a service. In: 2015 IEEE 14th International Conference on Machine Learning and Applications (ICMLA), pp. 896–902 (2015)
47. Sadeghi, A., Schneider, T., Wehrenberg, I.: Efficient privacy-preserving face recognition. In: Information, Security and Cryptology - ICISC 2009, Revised Selected Papers, pp. 229–244 (2009)
48. Schroff, F., Kalenichenko, D., Philbin, J.: FaceNet: a unified embedding for face recognition and clustering. In: IEEE Conference on Computer Vision and Pattern Recognition, CVPR 2015 (2015)
49. Sengupta, S., Chen, J., Castillo, C.D., Patel, V.M., Chellappa, R., Jacobs, D.W.: Frontal to profile face verification in the wild. In: 2016 IEEE Winter Conference on Applications of Computer Vision, WACV 2016 (2016)
50. Tamiya, H., Isshiki, T., Mori, K., Obana, S., Ohki, T.: Improved postquantum-secure face template protection system based on packed homomorphic encryption. In: 2021 International Conference of the Biometrics Special Interest Group (BIOSIG), pp. 1–5 (2021)
51. Zama. TFHE-rs: A Pure Rust Implementation of the TFHE Scheme for Boolean and Integer Arithmetics Over Encrypted Data (2022). https://github.com/zama-ai/tfhe-rs
52. Zhang, H., et al.: Validating privacy-preserving face recognition under a minimum assumption. In: Proceedings of the IEEE/CVF Conference on Computer Vision and Pattern Recognition (CVPR), pp. 12205–12214 (2024)
53. Zhang, R., Yan, Z.: A survey on biometric authentication: toward secure and privacy-preserving identification. IEEE Access **7**, 5994–6009 (2019)
54. Zhou, K., Ren, J.: PassBio: privacy-preserving user-centric biometric authentication. IEEE Trans. Inf. Forensics Secur. **13**(12), 3050–3063 (2018)

Authentication and Authorization

Circulation Control Model and Administration for Geospatial Data

Heng Li[1,2,3,4], Fenghua Li[1,2,3], Yunchuan Guo[1,2,3], Lingcui Zhang[1,2,3(✉)], Xiao Wang[1,2,3], and Ziyan Zhou[1,2,3]

[1] Institute of Information Engineering, Chinese Academy of Sciences, Beijing, China
zhanglingcui@iie.ac.cn
[2] School of Cyber Security, University of Chinese Academy of Sciences, Beijing, China
[3] State Key Laboratory of Cyberspace Security Defense, Beijing, China
[4] National Geomatics Center of China, Ministry of Natural Resources of the People's Republic of China, Beijing, China

Abstract. Geospatial data constitutes a critical strategic resource and an emerging factor of production. This data encompasses geographic entities about the entire Earth's surface, capturing detailed features and properties. Through coordinate systems, it achieves precise spatial positioning while revealing dynamic temporal variations. Since it integrates extensive spatial coverage with rich attribute records, this data often reaches large volumes. Existing access control models and mechanisms, however, struggle to regulate how this data is used as it circulates through governance and application scenes. To tackle these challenges in the complex network environment of the real-time intelligent service (PNTRC) network for cross-domain geospatial data, a Geospatial Data Circulation Control (GDCC) model and administrative model were developed, based on the Cyberspace Oriented Access Control (CoAC) model. It provides tailored implementation mechanisms for geospatial data circulation control in various scenes and formally defines and analyzes its functions and methods using Z language. Analysis and evaluation demonstrate that the proposed circulation control mechanism effectively ensures controlled transmission, usage and extended control throughout the entire lifecycle of cross-domain geospatial data circulation.

Keywords: Access Control Model · Data Usage Control · Cross-domain Data Elements Circulation · Geographic Information Security · Public Data Security Governance

1 Introduction

With the rapid advancement of emerging technologies such as 5G/6G, cloud computing, the Internet of Things (IoT), artificial intelligence, big data, and intelligent connected vehicles, systems like the Global Navigation Satellite System (GNSS), Remote Sensing (RS), and Geographic Information System (GIS) are becoming increasingly prevalent. GIS and related fields are now widely applied, driving a sharp rise in the demand for spatiotemporal big data and geographic information services. However, certain casual or unregulated actions can pose risks to a nation's sovereignty, security, and interests. For instance, many common social networking, travel, and other applications require access to or record users' location data, enabling the precise tracking of user activity patterns and the identification of sensitive geographic information. Without adequate security measures, this creates potential vulnerabilities. Similarly, intelligent connected vehicles collect, store, transmit, and process spatial data, including vehicle coordinates and surrounding road infrastructure. If such geographic information is misused, it could pose significant threats to national security.

Researchers [1–4] often employ techniques such as digital watermarking, data encryption, access control, and blockchain to safeguard geospatial data and prevent the leakage, theft, or tampering of geographic information. While these methods effectively protect geographic information within a "bounded" single-domain context, they fall short in addressing the challenge of controllable usage throughout the entire lifecycle of "unbounded" cross-domain geospatial data circulation. In recent years, researchers have proposed various access control models tailored to specific network environments. For example, capability-based access control models for the Internet of Things (IoT) have been developed to address the unique characteristics of IoT networks [5]. Similarly, role-based secure and flexible attribute-enabled access control architectures for space-ground integrated networks [6] have been designed. For big data scenes, models like GuardMR [7], Vigiles [8], HeAC [9], and OT-RBAC [10] have been introduced. Additionally, mechanisms such as CKCM (Cryptographic Key Control Mechanism) for HDFS [11] focus on controlling key resources in big data environments. While these access control models effectively address usage control during cross-domain data circulation, they lack mechanisms specifically designed for the complex network environments and unique challenges of cross-domain geospatial data circulation. Issues such as ensuring that cross-domain geospatial data users comply with contractual restrictions, controlling the scope of data usage, preventing unauthorized secondary dissemination, and enabling remote verification of usage control remain largely unresolved and are still in the exploratory stage. The compliance and efficient use of data circulation require that data elements be controllable throughout their lifecycle. The circulation of data elements is characterized by multi-round dynamic transactions, cross-domain data transmission, and adaptive usage control mechanisms [12, 13]. Addressing controllability issues during data circulation is critical [14–16], encompassing aspects such as access, processing, deletion, desensitization, circulation control, border filtering, traceability, violation detection, auditing, and evidence collection.

Li et al. [17, 18] proposed the Cyberspace-Oriented Access Control (CoAC) model, which consists of three components: access request entities, the network/generalized network, and resources (sets). By linking network access control with authorization/authentication in information systems, the CoAC model enables fine-grained control in ubiquitous access scenes, addressing the limitations of traditional access control mechanisms in dynamic and cross-domain environments. With the advent of the digital economy, the trusted circulation of data has become a primary application of CoAC. By integrating real-time context awareness, fine-grained access control, and lifecycle data usage management, the model effectively mitigates security risks caused by the separation of data ownership and management or the repeated/multiple forwarding of information. The control over repeated/multiple forwarding is further defined as Extended Control (ECON), ensuring the secure and flexible management of sensitive data across organizational boundaries.

Building on the CoAC model, we extend its application to the entire process of cross-domain geospatial data circulation and propose the Geospatial Data Circulation Control (GDCC) model and administrative model. The GDCC model in this paper synthesizes the advantages of existing access control models while enabling customizable, fine-grained security policy management and efficient control throughout the data lifecycle across diverse data circulation scenes (modes). The remainder of this paper is organized as follows: Sect. 2 discusses the requirement of circulation control for geospatial data. In Sect. 3, we present a formal cyberspace-oriented, attribute-based model and administrative model for GDCC, and introduce the implementation mechanism. Section 4 highlights some analysis and evaluation s of the proposed model, followed by the conclusion and future work in Sect. 5.

Contributions In summary, we make the following contributions:

1. We propose the GDCC model and administrative model, which by instantiating the CoAC model and extending its application to the entire process of cross-domain geospatial data circulation.
2. We introduce the implementation mechanism of the proposed model and administrative model, which provides tailored implementation mechanisms for usage control in various scenes and formally defines and analyzes its functions and methods using Z language.
3. We evaluate some characteristics of the proposed model and administrative model, including: 1) Confidentiality. All the elements complete the policy execution based on the security attributes, so as to ensure the confidentiality of the data circulation process; 2) Customizable Policy Management. In addition to the unified authorization of scene (mode), each organization or institution in the intra-domain or inter-domain can customize the data access control policy according to its own situation and implement it efficiently; 3) Models Comparison. A detailed functional comparison between the traditional access control model and the GDCC model proposed from several key perspectives to evaluate the superiority.

2 Requirement of Circulation Control for Geospatial Data

Geospatial data refers to electronically or digitally recorded information pertaining to quantitative measurements, qualitative attributes, spatial distributions, interrelationships, and dynamic patterns of geographic system components. It primarily encompasses vector data, raster data, and three-dimensional data formats. Characterized by extensive coverage, high precision, time-series attributes, and massive data volumes, geospatial data presents unique control challenges in circulation processes that conventional access control models fail to adequately address.

2.1 Promoting the Circulation and Utilization of Geospatial Data Element

The compliant, efficient, and secure circulation of massive geospatial data resources holds transformative significance in unlocking the intrinsic value of data elements, empowering the real economy's advancement, stimulating market entities' vitality, catalyzing new development paradigms, and accelerating high-quality socioeconomic growth.

With extensive experience in data element circulation and security research, we emphasize that data is not equivalent to data elements, and there are fundamental differences between open data sharing and data element circulation. Controlled data sharing and exchanging primarily focuses on confidentiality, integrity, access control, and supports cross-system collaboration and mobile operations. In contrast, data trading emphasizes ownership determination, rights transfer, usage verification, and dispute arbitration. Data circulation control mechanism which refers to the use of cross-domain access control to extended control the whole life cycle of data transmission, storage, usage and destruction, to extended control the time, place, subject, behavior and object of data assets.

2.2 Supporting the Construction of Geospatial Data Infrastructure

Data circulation utilization facility is an important part of data infrastructure. Current data circulation utilization technologies primarily encompass Privacy-Preserving Computation (PPC), Trusted Data Spaces (TDS), and Usage Control (UCON).

PPC. It includes Federated Learning (FL), secure Multi-Party Computation (MPC), Trusted Execution Environments (TEEs), and Homomorphic Encryption (HE), primarily address scenes requiring in-domain computation where raw data or data elements (products) remain localized. While effectively enabling value circulation through computational processes, these methods fall short in addressing controlled secure circulation demands for cross-domain geospatial data element (product) distribution scenes.

TDS. It emerges from the International Data Space (IDS) concept, the TDS technology establishes a consensus-based infrastructure connecting multiple stakeholders to facilitate data resource sharing. This framework supports large-scale geospatial data circulation and collaborative utilization, yet faces implementation challenges including difficulties in constructing trusted ecosystems and high operational costs for fragmented, small-volume geospatial data transactions.

UCON. It refers to the use of technical means to control the transmission, storage, use and destruction of data, so as to control the time, place, subject, behavior and object of data assets. This technology highly relies on blockchain smart contract technology, which converts the data usage control intention of data rights and interests' subjects into smart contract terms that can be machine-readable and processed, and solves the prerequisite problem of geospatial data controllability with a huge cost.

Geospatial data, intrinsically linked to geographic coordinates, is characterized by its multidimensional integrated representation, inherent heterogeneity, massive scale, high precision, cross-domain interoperability, and heightened security and privacy requirements. Therefore, the pervasive cross-domain circulation of massive geospatial data in complex network environments presents a critical research challenge: designing a sophisticated circulation control model that simultaneously satisfies: 1) granular control requirements for data element (product) distribution across varying operational scales; 2) comprehensive lifecycle management capabilities for cross-domain geospatial data circulation within intricate network ecosystems.

In summary, building geospatial data circulation control model to ensure the controllability of geospatial data transmission, usage and extended, technical measures must be implemented, such as linking data usage policies with transaction contracts, binding data usage policies to data security, enforcing trusted execution of data usage policies, and enabling remote verification of policy execution. These measures address issues of unauthorized in-domain access and uncontrolled out-of-domain usage. They enable usage control and remote verification of geospatial data at the user level, ensuring that data usage complies with contractual agreements following the transaction.

3 The Geospatial Data Circulation Control Model

This section introduces the GDCC model and administrative model for managing the real-time intelligent service network of space information, known as the PNTRC network [19].

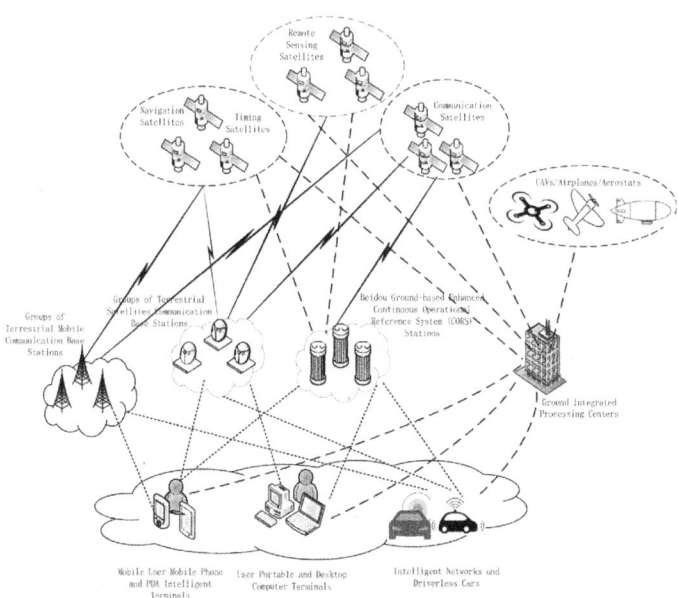

Fig. 1. The PNTRC network architecture.

The PNTRC network underpins the entire lifecycle management of geospatial data—including collection, publication, processing, usage, storage, and destruction—as shown in Fig. 1. In this context, "P" stands for Positioning, "N" for Navigation, "T" for Timing, "R" for Remote Sensing, and "C" for Communication.

This complex network integrates communication satellite systems, navigation satellite systems, remote sensing satellite systems, and the GPS/Beidou ground-based enhanced Continuous Operational Reference System (CORS). It achieves the objectives of "one satellite, multiple uses; multi-satellite networking; multi-network integration; and intelligent services". By deeply coupling the "sky network" with the "ground network", it enables real-time delivery of geospatial and communication data to users' mobile phones, intelligent connected vehicles, mobile terminals, computer network equipment, and base stations based on their needs. This integration meets the intelligent service demands across various scenes of geospatial data circulation and usage.

3.1 Mapping from CoAC to GDCC

The following paragraph introduces a usage control mechanism for geospatial data circulation within the PNTRC complex network environment. This mechanism maps the access control elements of the CoAC model to the entire lifecycle of geospatial data circulation the GDCC model is constructed, as shown in Fig. 2. The model is composed of three key components: the access request entity, the cyberspace environment, and data resource circulation.

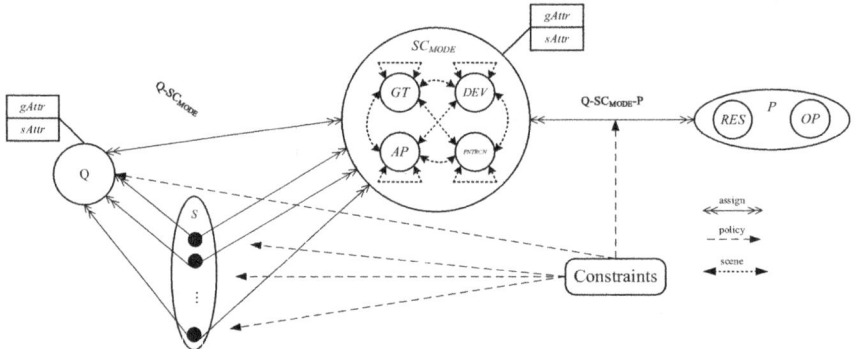

Fig. 2. The Geospatial Data Circulation Control Model.

Accessing the Requested Entity. It encompasses users (regular and administrative) and sessions.

Access Request Entity Quantity(Q). The users accessing geospatial data resources include end-users, providers, and managers. A session refers to the process of accessing geospatial data resources via a client, involving basic services such as navigation, timing, remote sensing, and communication, as well as higher-level application services. The initiator of access to geospatial data resources includes user u and session s, denoted as $q = <u, s>$. The set of all access requesting entities is denoted as Q, and the set of permissions obtained is denoted as P. The general attribute ($gAttr$) of the geospatial data circulation access request entity includes the username (*qName*), the user's operation and maintenance management domain (*qDom*), and other optional general attributes (*qExtGAttr*). The security attribute (*sAttr*) includes authentication method (*qAuth*), login password (*qPwd*), login certificate (*qCert*), public/private key pair (*qKeys*), password expiration time (*qExpTime*), and other optional security attributes (*qExtSAttr*). The operation and the *qDom* of a user are mainly distinguished by the operation and maintenance management party to which the user belongs in the process of geospatial data circulation, and the *sAttr* is also subject to the configuration of the operation and maintenance management party to which the geospatial data belongs. Generic attributes and security attributes support optional attributes, which can be added and set by users, providers or managers of geospatial data access for subsequent extended access control policy configuration. The *gAttr* and *sAttr* of geospatial data stream access request entities can be expressed as follows, respectively.

$Q.gAttr = <qName, qDom, qExtGAttr, ...>$
$Q.sAttr = <qAuth, qPwd, qCert, qKeys, qExpTime, qExtSAttr, ...>$

Cyberspace Environment. Similar to the CoAC model, the GDCC cyberspace environment includes elements such as Generalized Temporal (*GT*), Access Point (*AP*), Access Device (*DEV*), and the PNTRC network (*PNTRCN*).

General Temporal (GT). This refers to the date and time information associated with the environmental elements when a geospatial data circulation access request is initiated. In the geospatial data circulation environment, temporal information consists of two components: the time interval and the duration, collectively represented as *GT*. Where *interval* $\in 2GT^{IN}$ represents the start datetime and end datetime; duration represents the duration in days (*d*), months (*m*), years (*y*), seconds (*s*), minutes (*min*), and hours (*h*).

$GT = \{<interval, duration> \mid interval \in 2GT^{IN}, duration \in R^+\}.$

Access Point (AP). This refers to the Spatial Location (*SL*) or Network Identity (*NI*) of the initial entry point in the routing network when the geospatial data circulation access request entity first connects to the PNTRC network. The *gAttr* of PNTRC network access point includes geographic location (*aGPS*), access rate (*aRate*), access type (*aType*), access policy (*aPoli*), MAC address (*aMAC*), IP address (*aIPv4*), port (*aPort*), etc. The *sAttr* includes the secure transport protocol (*aSecProt*), etc. The *gAttr* and *sAttr* of the access point can be expressed as follows respectively. Among them, the *aSecProt* supports HTTP, HTTP/TLS v1.0, etc.

$AP.gAttr = <aGPS, aRate, aType, aPoli, aMac, aIPv4, aPort, ...>$
$AP.sAttr = <aSecProt,...>$

Access Device (DEV). This refers to the devices used for accessing resources within the PNTRC network during geospatial data circulation. These devices primarily include mobile phones, intelligent connected vehicle sensors, mobile terminals (such as handheld RTK/GPS measurement instruments and Beidou navigation instruments), tablets, laptops, desktop computers, servers, CORS base station receivers, space-based satellite network terminals, ground-based satellite network terminals, Ka-band high-capacity broadband terminals, and Ku-band (FDMA/TDMA) terminals, among others. The *gAttr* of access device *DEV* include operating system (*dOS*), processor (*dCPU*), memory (*dMem*), hard disk (*dDisk*), application (*dApp*), etc. The *sAttr* of a device includes security level (*dSecLev*), security software module (*dSecSMod*), security hardware module (*dSecHMod*), etc. The *gAttr* and *sAttr* of an access device can be expressed as follows, respectively. The access device dev is represented by the triple <*dID, dev.gAttr, dev.sAttr*>. Where *dID* is the access device *ID*.

$DEV.gAttr = <dOS, dCPU, dMem, dDisk, dApp, ...>$
$DEV.sAttr = <dSecLev, dSecSMo, dSecHMod, ...>$

PNTRC Network (PNTRCN). This refers to the carrier of geospatial data dissemination, encompassing all information transmission channels, including the space-based satellite network, ground-based node network, application server network, and application client network. The PNTRC Space Satellite Network (*SSN*) is composed of multiple space satellite nodes (*SSNO*) positioned in geosynchronous orbit. These nodes are interconnected through navigation satellites, timing satellites, remote sensing satellites, and communication satellites. Each satellite node serves as a carrier for positioning and navigation, satellite timing, remote sensing monitoring, data relay, routing exchange, and information storage, processing, and fusion. The PNTRC Ground Node Network (*GNN*) consists of various interconnected ground-based nodes, including Continuous Operational Reference System (*CORS*) reference station nodes (*CORSN*), ground backbone nodes (*GBN*), Ku-band broadband satellite gateway stations (*KUG*), Ka-band high-capacity broadband satellite gateway stations (*KAG*), and S-band satellite gateway stations (*SG*). These nodes are linked through a high-speed terrestrial backbone network. The CORS base station nodes include the Beidou ground base station system and data processing centers. The data processing center generates the network RTK error correction model through network adjustment solutions, enabling centimeter-level positioning accuracy for users accessing differential enhancement services via a wireless login server. The ground backbone nodes serve as gateways and information hubs, performing critical functions such as network control, resource management, protocol conversion, information processing, fusion, and sharing, while ensuring connectivity with other ground systems. As a resource provider, the PNTRC Application Server End (*ASE*) Network is composed of distributed nodes deployed in data processing centers, serving as the core infrastructure for positioning and navigation, satellite timing, remote sensing monitoring data storage, analysis, and computation. This network includes management nodes (*ManNode*), data nodes (*DataNode*), computation nodes (*CompNode*), authentication nodes (*CertNode*), cryptographic service nodes (*CrypNode*), and other server nodes (*SN*). As resource consumers, the PNTRC Application Client End (*ACE*) Network consists of client nodes (*CN*) that connect to and utilize the ASE network. Most client nodes operate independently, without interconnection, and establish connections exclusively with the *ASE* network. At a macro level, the entire PNTRC network can be represented by an indirect graph $G_{PNTRCN} = (V_{PNTRCN}, E_{PNTRCN})$, whose vertices include space-based satellite network V_{SSN}, ground-based node network V_{GNN}, application server network V_{ASE} and application client network V_{ACE}. The edge is composed of $E_{SSN}, E_{GNN}, E_{ASE},$ and E_{ACE}, i.e. The vertex attribute in PNTRC network *PNTRCN* is the union of all vertex attributes in $G_{SSN}, G_{GNN}, G_{ASE}$ and G_{ACE}, and the edge attribute is the union of all edge attributes in $G_{SSN}, G_{GNN}, G_{ASE}$ and G_{ACE}.

$$V_{PNTRCN} = V_{SSN} \cup V_{GNN} \cup V_{ASE} \cup V_{ACE}$$

Circulation Control Model and Administration for Geospatial Data   533

$EPNTRCN = ESSN \cup EGNN \cup EASE \cup EACE$

PNTRC Space-based Satellite Network (SSN). PNTRC space-based satellite network can be expressed as an indirect connected graph $GSSN = (VSSN, ESSN)$, where $VSSN = \{ssno1,... Ssnom\}$ is the vertex set of the graph, which represents the set of space-based satellite nodes. $Ssnoi$ represents the ith space-based satellite node, $1 \leq i < M, M > 3$. $ESSN = \{<ssnoi, ssnoi+1 > |1 \leq i \leq M, ssnoM+1 = ssno1\}$ is the edge set, which represents the transmission link between the space-based satellite nodes. Any vertex in $GSSN$ is only connected to the two preceding and following vertices, which means that the PNTRC space-based satellite node in the synchronous orbit only communicates with the preceding and following satellite nodes. For brevity, *essn* is used to denote the edge of the space-based satellite network. The common attributes of PNTRC space-based satellite nodes include the controller (*ssnController*), whether the gateway station is visible (*ssnVisi*), transmission protocol (*ssnProt*), computing ability (*ssnCompAbility*), storage ability (*ssnStoreCapa*), function (*ssnFunc*), and empty Number of idle channels (*ssnFreeChanNum*), etc. The security attributes of space-based satellite nodes include encryption mode (*ssnEncTpe*), supported secure transport protocol (*ssnSecProt*), etc. Among them, *ssnController* indicates who controls the satellite node, including high orbit communication satellite, low orbit navigation/timing/remote sensing satellite control or ground control, *ssnProt* includes Satellite Transport Protocol (STP), etc. *ssnFunc* includes positioning and navigation, satellite timing, remote sensing monitoring, data relay, routing exchange, information storage, processing and fusion, etc. The *gAttr* and *sAttr* of a vertex are expressed as follows, respectively. The *gAttr* of *essn* include channel type (*eType*), communication bandwidth (*eWidth*), quality of service (*eQos*), physical link layer protocol (*ePhyProt*), routing protocol (*eRoutProt*), network layer protocol (*eNetProt*), transport layer protocol (*eTranProt*), and communication frequency Segments (*eFreg*), etc. The security attributes of satellite network transmission include security level (*eSecLev*), encryption mode (*eEncTpe*), supported secure transmission protocol (*eSecProt*), etc. The *gAttr* and *sAttr* of an edge are expressed as follows, respectively. *eType* consists of two classes: Communication channel (*CommChan*) and control channel (*ControlChan*), *CommChan* includes positioning and navigation communication, satellite timing communication, remote sensing monitoring communication and communication transmission, etc., *ControlChan* includes positioning and navigation control, satellite timing control, remote sensing monitoring control and communication control, etc. *ePhyProt* includes Laor and Dra, etc. *eNetProt* includes IP and DTN, etc. *eRoutProt* includes hierarchical QoS routing protocol (HQRP) and Location Assisted Demand Protocol (LAOD), etc. location-assisted on demand protocol), *eTransprot* includes TCP and UDP, *eFreq* includes L-band and S-band.

$SSN.gAttr = <ssnController, ssnVisi, ssnProt, ssnCompAbility, ssnStoreCapa, ssnFunc, ssnFreeChanNum, ...>$
$SSN.sAttr = <ssnEnc, ssnSecProt, ...>$
$ESSN.gAttr = <eType, eWidth, eQos, ePhyProt, eRoutProt, eNetProt, eTranProt, eFreg, ...>$
$ESSN.sAttr = <eSecLevel, eEncTpe, eSecProt, ...>$

PNTRC Ground Node Network (GNN). The ground-based node network can be represented by an indirect graph $G_{GNN} = (V_{GNN}, E_{GNN})$, where $V_{GNN} = V_{CORSN} \cup V_{GBN} \cup V_{KUG} \cup V_{KAG} \cup V_{SS}$ is the vertex set of the graph. V_{CORSN}, V_{GBN}, V_{KUG}, V_{KAG}, and V_{SG} denote the corresponding nodes of CORS base station node, ground-based backbone node, Ku broadband satellite gateway station, Ka large-capacity broadband satellite gateway station, and S satellite gateway station, respectively. E_{GNN} is the edge set of the complete graph formed by V_{GNN}, that is, $E_{GNN} = \{<V_{CORSN}, V_{GBN}>, <V_{CORSN}, V_{KUG}>, <V_{CORSN}, V_{KAG}>, <V_{CORSN}, V_{SG}>, <V_{GBN}, V_{KUG}>, <V_{GBN}, V_{KAG}>, <V_{GBN}, V_{SG}>, <V_{KUG}, V_{KAG}>, <V_{KUG}, V_{SG}>\}$, where, $<V_{CORSN}, V_{GBN}> \subseteq \{<corsn, gbn> | corsn \in V_{CORSN}, gbn \in V_{GBN}\}$ means that the CORS base station node is connected to the ground-based backbone node, and so on. $<V_{CORSN}, V_{KUG}>, <V_{CORSN}, V_{KAG}>, <V_{CORSN}, V_{SG}>, <V_{GBN}, V_{KUG}>, <V_{GBN}, V_{KAG}>, <V_{GBN}, V_{SG}>, <V_{KUG}, V_{KAG}>, <V_{KUG}, V_{SG}>, <V_{KAG}, V_{SG}>$, etc. The attribute types of vertices and edges in the PNTRC ground-based node network are the same as those in the PNTRC space-based satellite network, but the attribute values are different, so the attribute types will not be described here.

PNTRC Application Server End network (ASE). The PNTRC application server end network can be expressed as an indirect connected graph $G_{ASE} = (V_{ASE}, E_{ASE})$, where $V_{ASE} = \{sn_1, \ldots sn_M\}$ is the vertex set of the graph, which represents the set of server nodes at the PNTRC application server end network, sn_i represents the ith server node, $1 \le i \le M, M > 3$; $E_{ASE} = \{<sn_i, sn_{i+1}> | 1 \le i \le M, sn_{M+1} = sn_1\}$ is the edge set, which represents the transmission link between server nodes. For brevity, the *AES* network is used to denote edges at the server side of PNTRC application. The PNTRC applies the *gAttr* of the server node *sn* on the server side includes heartbeat connection (*snHeart*), time synchronization (*snTime*), CPU remainder (*snCPU*), memory remainder (*snMem*), disk remainder (*snDisk*), service status (*snServStatus*), role instance (*snRoleInst*), and so on. The *sAttr* of the server node *sn* on the application server side of PNTRC includes authentication method (*snAuth*), encryption method (*snEncTpe*), etc. The *gAttr* and *sAttr* of the server node of the PNTRC application server can be expressed as follows, respectively. The *gAttr* of network link *ase* in PNTRC application server includes communication bandwidth (*eWidth*), physical link layer protocol (*ePhyProt*), network layer protocol (*eNetProt*), transport layer protocol (*eTranProt*), etc. The *sAttr* of network link *ase* in PNTRC application server includes secure transport protocol (*eSecProt*), encryption mode (*eEncType*), etc. The *gAttr* and *sAttr* of the server-side network link *ase* of PNTRC application can be expressed as follows, respectively. Among them, *eSecProt* supports HTTP, HTTP/TLS v1.0, etc. The encryption mode *eEncType* supports the national secret SM2/3/4 and TLS_DHE_RSA_WITH_AES_EDE_CBC_SHA256.

SN.gAttr = <*snHeart, snTime, snCPU, snMem, snDisk, snServStatus, snRoleInst, ...* >
SN.sAttr = <*snAuth, snCryp ...* >
ESN.gAttr = <*eWidth, ePhyProt, eNetProt, eTranProt, ...* >
ESN.sAttr = <*eSecProt, eEncType, ...* >

PNTRC Application Client End network (ACE). The PNTRC application client network is represented by an indirect graph $G_{ACE} = (V_{ACE}, E_{ACE})$, where $V_{ACE} = \{cn_1, \ldots$

Cn_k} is the vertex set of the graph, which represents the set of PNTRC application client nodes, cn_i represents the ith client node, $1 \le i \le K$, $K \ge 1$; $E_{ACE} = \{<cn_i, cn_{i+1}> | 1 \le i < M, cn_{M+1} = cn_1\}$ is the edge set, which represents the transmission link between client nodes. The attribute types of the nodes and links in the *ACE* network are the same as those of the nodes and links in the *ASE* network, and the attribute values are different, so the attribute types are not repeated here.

Data Resource Circulation. The GDCC data resource circulation encompasses resources (*RES*) and scene (*SC*). Resources (*RES*) are the objects accessed by the geospatial data circulation access request entity within the PNTRC network, including management and application data. Management data consists of monitoring data, status data, application data, and broadcast data. Application data can be categorized based on its circulation direction into forward data and backward data, and by its content into location coordinates, spatiotemporal data files, and high-precision maps. Scene (*SC*) represents the essential conditions required for a geospatial data circulation access request entity to initiate a session and access data resources within the PNTRC network to obtain permissions.

Resource (RES). It can be represented by a tuple $< rid, rcnt >$, Where *rid* represents the unique identity of *res* and *rcnt* represents the content of *res*. the common attributes of resources used by geospatial data circulation in PNTRC network include resource owner (*rOwner*), resource type (*rType*), resource access policy (*rAccessPoli*), resource precision (*rPrecision*), resource current (*rCurrency*), resource scale (*rScale*), resource access policy (*rAccessPoli*). Resource Storage Location (*rLoc*). The security attributes of geospatial data circulation resources include security level (*rSecLev*), permitted operation (*rAllowedOper*), encryption mode (*rEncTpe*), number of circulation (*rAllowedRound*), number of available remaining (*rSurplusRound*), and right of deletion (*rDelet*), etc. The *gAttr* and *sAttr* of *RES* can be expressed as follows, respectively. among them, *rType* includes management class data (*rManData*) and application class data (*rAppData*).

RES.gAttr $= < $rOwner, rType, rAccessPoli, rPrecision, rCurrency, rScale, rLoc,... $>$

RES.sAttr $= <$ rSecLev, rAllowedOper, rEncIjype, rAllowedRound, rSurplusRound, rDeleteRight... $>$

Definition 1 Operation (OP). In the geospatial data circulation environment, operations refer to the actions performed by an access request entity to interact with data resources within the PNTRC network. Authorization for geospatial data circulation is achieved through a series of continuous actions. Based on the type of application data resources, operations can be categorized into the following types: Positioning Navigation Timing Operation (*PNTO*), Remote Sensing Monitoring Operation (*RSMO*), Autonomous Driving Operation (*ADO*) and Communication Transmission Operation (*CTO*). Each operation type encompasses multiple permitted actions. A single permitted operation (*rAllowedOper*) performed by the access request entity may involve and affect multiple resources (*res*).

Scene (SC). It should include the generalized tense, access point, access device, network and its attributes, as well as the attributes of access request entity q and data resource *res*, denoted as *sc*, represented by a five-tuple $< t, a, d, PNTRCn, attr >$. Here, $t \in GT$, $a \in AP$, $d \in DEV$, $PNTRCn \in PNTRCN$, and $attr \in \{gAttr \cup sAttr\}$.

Definition 2 Mode (MODE). The geospatial data circulation mode refers to the method employed by an access requesting entity to access data resources within the PNTRC network, enabling the achievement of geospatial data circulation and usage objectives. Geospatial Data Opening (*GDO*), Geospatial Data Sharing (*GDS*), Geospatial Data Trading (*GDT*) and Geospatial Data Exchange (*GDE*) different modes correspond to varying usage control scenes for geospatial data circulation. It is important to note that, unlike the scene mapping in the CoAC model, the scenes in the GDCC mechanism encompass the attributes of all elements except the four core elements. Scenes and modes are tightly coupled, with scenes exhibiting a partial order relationship and the ability to contain sub-scenes.

3.2 The GDCC Administrative Model

The core challenge of the GDCC mechanism is to efficiently authorize, authenticate, and control geospatial data, particularly ensuring that geospatial data involved in multi-round transactions adheres to iterative and extended usage control as stipulated by the contractual agreements between data providers and consumers. To address this issue, this paper draws on the privilege management approach from the CoAC model. It establishes usage control policies for different geospatial data circulation environments by leveraging scenes, attributes, and modes. These policies enable the assignment and revocation of user access rights (P, privileges) to geospatial data resources. A scene is defined by four key elements: generalized temporal, access point, access device, and network, as shown in Fig. 3.

The Geospatial Data Circulation Control Administrative Model in the PNTRC complex network environment establishes a propagation chain for "who, when, where, and what processing is performed on geospatial data" within and across domains. This involves configuring, revoking, and updating the attributes, scenes, modes, operations, and permissions associated with access request entities accessing specific geospatial data resources via designated access points, devices, and networks during a given time period. The model consists of three main sub-models: attribute-scene (mode) administrative (*SCAD*), attribute-scene (mode)-permission administrative (*SCPAD*), user-attribute-scene (mode) administrative (*USCAD*). Additionally, it requires supporting management models such as User Identity and Access Management (*UIAM*) and Session Management (*SM*) to ensure comprehensive functionality.

Definition 3 Administrative Scene (ADSC). The administrative scene is represented by the quadruple $<adminGT, admintAP, adminDEV, adminPNTRCN>$, which means that the geospatial data circulation control management entity accesses the device through *adminDEV*, from the *admintAP* specific access point, at a specific time in *adminGT*. Scenes for administration via *adminPNTRCN* specific networks.

Circulation Control Model and Administration for Geospatial Data 537

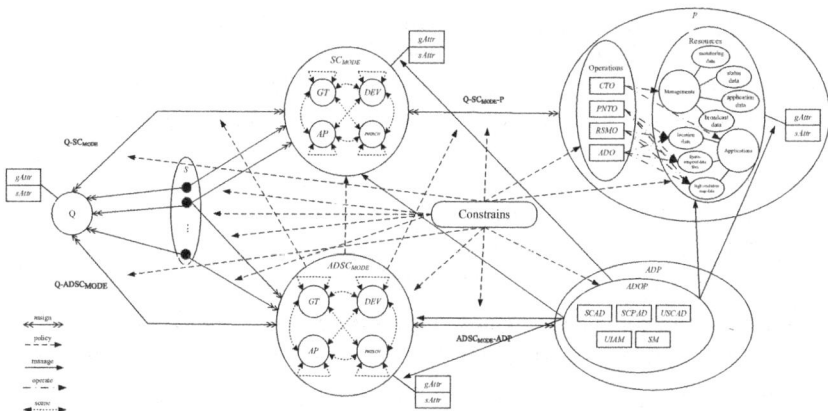

Fig. 3. The Geospatial Data Circulation Control Administrative Model.

SCAD. It involves the management of attribute metadata for all elements and scene components, serving as the foundation for permission management and user scene (mode) management.

SCPAD. This refers to the process of assigning and revoking permissions for a defined scene (mode).

USCAD. This involves managing a user's attribute values and assigning a scene (mode) when the user submits an access request.

UIAM. Its purpose is to manage user identities and access, provide unified identity authentication, and support session management as well as user-attribute scene (mode) management.

SM. It is responsible for creating or closing a session for each access request that successfully passes user identity authentication, following the authentication of the geospatial data circulation access request entity, and prior to obtaining scene permissions.

3.3 The Implementation Mechanism of GDCC Model and Administrative Model

Attributes are the union of the access request entity, resource attributes, and scenes. During permission assignment, the system predefines which operations can be performed on which resources, under which modes, and in which scenes (completing the "scene (mode)-permission" allocation). By default, any scene (mode) without assigned permissions denies access. When an access request entity (Q) initiates a request to access a resource (RES) in the PNTRC complex network for geospatial data circulation, the permission policy decision point first identifies the scene (mode) the user is in (i.e., "user-scene (mode)" determination). It then matches this scene (mode) against the predefined "scene-permission" allocation. Based on the permissions associated with the matched scene (mode), the decision point determines whether Q is authorized to perform a specific operation on RES. To ensure precise implementation of the GDCC mechanism, the core functions required for its operation are defined as follows:

AttrCollect (Q.gAttr, GT.gAttr, AP.gAttr, DEV.gAttr, PNTRCN.gAttr, RES.gAttr, Q.sAttr, GT.sAttr, AP.sAttr, DEV.sAttr, PNTRCN.sAttr, RES.sAttr) → {gAttr_value ∪ sAttr_value}
AttrCheck (Q.gAttr, GT.gAttr, AP.gAttr, DEV.gAttr, PNTRCN.gAttr, RES.gAttr, Q.sAttr, GT.sAttr, AP.sAttr, DEV.sAttr, PNTRCN.sAttr, RES.sAttr, gAttr_value ∪ sAttr_value) → {true, false}
scCreate (gAttr_value ∪ sAttr_value) → {q.sccreate}
scMatch (Q.SC ⊆ SC) → {true, false}
scPermissionAssign (SC, P) → {true, false}
scPermissionRevoke (SC, Q.P) → {true, false}

GDCC Implementation Mechanism in Different Scene (Mode). It includes geospatial data exchange, opening, trading and sharing.

Geospatial Data Exchange (GDE). In this scene (mode), the data involved in circulation primarily consists of classified or sensitive geographic information that remains within the domain. Therefore, the GDCC implementation focuses solely on preventing unauthorized access to geographic information data within the domain. The formal implementation of creating a *GDE* scene (mode) is shown in Fig. 4 by Z language.

createScene (SCENE$_{MODE}$?, q?:Quantity, rsc?:Resource, (t, ap, dev, PNTRCn)?:Name, result!:Boolean) ◁

sc=selectFunc ((t, ap, dev, PNTRCn), getAttr(q, rsc))

if sc$_{mode}$ ∉ SCENE$_{MODE}$

then SCENE$_{GDE}$=SCENE$_{MODE}$∪{sc$_{mode}$}, result=true

else result=false ▷

Fig. 4. The implementation function for creating a *GDE* scene (mode) by Z language.

Geospatial Data Opening (GDO). In this scene (mode), the circulation data primarily consists of public geographic information data. Therefore, in addition to ensuring data security within the domain, the GDCC implementation must address issues related to the trusted circulation and use of public geographic information data across domains. This includes ensuring security binding with cross-domain circulation and usage policies, enabling dynamic authorization, and supporting other authorized operations. The formal implementations of permission function are shown in Fig. 5 and Fig. 6 by Z language.

assignScenePermission (PERMISSION?, sc$_{mode}$?:Scene, p?:Permission, result!: Boolean) ◁

if sc$_{mode}$∈ SCENE$_{MODE}$ and p ∉ PERMISSION

then PERMISSION$_{GDO}$=PERMISSION∪{p}, result=true

else result=false ▷

Fig. 5. The permission enforcement function for assigning the *GDO* scene (mode) by Z language.

revokeScenePermission (PERMISSION?, p?: PERMISSION, result!: Boolean) ◁

if p ∈ (PERMISSION

then PERMISSION$_{GDO}$=PERMISSION\{p}, result=true

else result=false▷

Fig. 6. The permission revocation function for revoking the *GDO* scene (mode) by Z language.

Geospatial Data Trading (GDT). In this scene (mode), the data being circulated and utilized primarily consists of public product data elements that are exchanged across multiple levels, regions, systems, departments, and business domains. Therefore, the implementation of GDCC must address the challenge of usage control for iterative and extended operations during the multi-round, cross-domain transactions of geospatial data, in accordance with the contractual agreements between data element providers and consumers. On one hand, GDCC employs a standardized approach to describe the expected sharing intentions of geospatial data owners. This includes specifying conditions (e.g., *rAllowedRound* and *rSurplusRound* for allowable transaction rounds), permissions (e.g., *rAllowedOper* for authorized operations), and obligations for geospatial data users (e.g., *rLoc*, which mandates data storage at a specified location to continue circulation, or *rDeleteRight*, requiring data deletion after expiration). On the other hand, GDCC constructs intra-domain and inter-domain data transmission chains that incorporate factors such as "who, when, where, and what actions are performed with data elements" based on multi-source geospatial data usage logs. This enables the implementation of a hybrid authorization mechanism that combines sharing intentions and data element transmission chain attributes, ensuring controlled circulation and usage of geospatial data throughout its entire lifecycle.

Geospatial Data Sharing (GDS). In this scene (mode), the data used in circulation primarily consists of customized geographic information applications and service data, without involving the ownership transfer of data elements. Therefore, the implementation of GDCC emphasizes the specialized configuration of policy execution points, context processing, policy decision points, policy trust points, and policy management points. In practical applications, the *GDS* scene (mode) typically employs off-site circulation methods, such as offline signing of confidentiality agreements or issuing authorized user licenses. These methods are standard and will not be elaborated upon here.

Analysis of GDCC Administrative Process. Given the complex semantics of GDCC administrative functions, relying solely on natural language for descriptions can lead to ambiguity. To address this, it is essential to establish a standardized method for representing the data owner's intended data-sharing expectations. This involves designing a set of corresponding management models and functions to ensure the safe and efficient operation of the GDCC mechanism across various scenes (modes). This section analyzes the geospatial data circulation management process and employs the Z formal specification language [20] (referred to as "Z language") to provide a formalized description of the management functions. Similar to CoAC, the detailed description of each function is not given in this paper. For details, please refer to the function description of cross-network access control mechanism in complex network environment.

SCAD. These include functions such as creating attribute metadata *createAttrMeta()*, deleting attribute metadata *deleteAttrMeta()*, selecting a scene (mode) description *selectFunc()*, creating a scene (mode) *createScene()*, modifying or updating a scene *modifyScene()*, and deleting a scene (mode) *deleteScene()*, among others.

SCPAD. It includes the *assignScenePermission()* function and the *revokeScenePermission()* function.

USCAD. It includes functions such as adding or modifying attribute values *modifyAttr()*, selecting or retrieving attributes *getAttr()*, querying attribute values *getAttrValue()*, verifying the time *verifyT()*, verifying the access point *verifyAP()*, verifying the access device *verifyDEV()*, verifying the network *verifyPNTRCN()*, and others.

UIAM. It includes functions such as user registration *qSignUp()*, user logout *qLogOff()*, password/certificate initialization, updating, and revocation *updateCert()*, password/certificate authentication *authCert()*, user list query *getQList()*, and others.

SM. These include functions such as creating a session *createQS()* for authenticated users, assigning scenes *assignSCS()* to sessions, and closing sessions *closeQS()*.

4 Analysis and Evaluation

It is important to clarify that in the model presented in this paper, assigning permissions grants access, revoking permissions denies access, and in the absence of a defined scene (mode) or permission, access is rejected by default. Properties and scenes (modes) do not support inheritance. When scenes overlap within the same mode, authorization is granted, and the combined result of the overlapping scenes is also permitted, eliminating the possibility of permission conflicts.

The GDCC model and administrative model proposed in this paper is specifically tailored to PNTRC networks and their attributes, effectively meeting confidentiality requirements while supporting policy customization. Additionally, we conduct a detailed functional comparison between traditional access control models and the GDCC model from multiple critical perspectives including roles, temporal aspects, authorization mechanisms, permission adjustments, circulation control authentications, and circulation scenes (modes). In order to systematically evaluate the theoretical advancements and practical superiority of the proposed model and administrative model.

4.1 Confidentiality

The security attributes of access points, resources, access devices, and networks define parameters such as security levels, encryption methods, and secure transmission protocols. Security levels encompass the network, resource, and transmission channel security where the access points and devices operate. Through policy enforcement, the GDCC model and administrative model ensures that geospatial data with a lower security level can only be imported into higher-security networks in a one-way manner and accessed solely by users with corresponding roles. Simultaneously, it prevents geospatial data from higher-security networks from circulating into lower-security networks, thereby maintaining confidentiality.

4.2 Customizable Policy Management

The access control policy for geospatial data resources, defined by the data owner, is included in the generic attributes of the resource. This allows for integrated authorization through scenes (modes), while enabling each organization or institution whether intra-domain or inter-domain to customize access control policies for geospatial data according to their specific requirements. The geospatial data circulation scene is tightly coupled with its circulation mode. The GDCC model and administrative model predefines specific circulation modes (e.g., *GDO*, *GDS*, *GDT*, and *GDE* scene (mode)) for respective operational contexts during access control permission allocation, while ensuring data usage control policies consistently adhere to contractual agreements. Thus, the established "scene (circulation mode)-permission" mapping is maintained throughout the entire lifecycle of geospatial data resource utilization.

4.3 Models Comparison

Given the unique characteristics of geospatial data, currently it lacks a suitable access control framework capable of systematically addressing the challenges of controlled transmission, usage, and extended comprehensive lifecycle management during data circulation.

Table 1. The functional comparison of the access control models supporting the corresponding policy management.

Model	Role	Temporal	Authorization	Permissions Adjustment		Circulation Control Authentication			Circulation Control Scene (Mode)
				Dynamic	Adaptive	Transmission Control	Usage Control	Extended Control	
RBAC	√	×	√	√	×	×	×	×	×
TRBAC	√	√	√	√	√	×	×	×	×
ABAC	√	√	√	√	l	×	×	×	×
ABE	×	√	√	×	×	×	×	×	×
TrustBAC	×	×	√	√	×	×	×	×	×
AcBAC	×	√	√	√	√	×	×	×	×
UCON	√	√	√	√	√	×	√	×	×
CoAC	√	√	√	√	√	√	√	√	×
GDCC	√	√	√	√	√	√	√	√	√

As shown in the comparative analysis in Table 1, Existing access control frameworks, including fundamental models such as the Role-Based Access Control (RBAC) [21], the

Attribute-Based Access Control (ABAC) [22], the Attribute-Based Encryption (ABE) [23], and the Usage Control (UCON) [24, 25], along with their derivatives such as the Task Role-Based Access Control (TRBAC) [26], the Ciphertext Policy Attribute-Based Encryption (CP-ABE) [27], the Key Policy Attribute-Based Encryption (KP-ABE) [28], the Trust Based Access Control (TrustBAC) [29], and the Action Based Access Control (AcBAC) [30], typically focus on isolated aspects of access control. Notably, the CoAC model specifically targets cross-domain data access control in complex network environments. The symbol "$\sqrt{}$" denotes that the access control model supports the corresponding policy management functionality, "×" indicates the absence of such support, and "l" signifies optional support for the functionality (i.e., the model provides configurable but non-mandatory implementation of this feature).

5 Conclusion and Future Work

With the widespread application of high-temporal, high-spatial, and high-spectral resolution remote sensing imagery, along with surveying and mapping remote sensing technologies in areas such as Real 3D China, global geographic information resources, and marine mapping, the significance of geospatial data in scenes involving exchange, opening, trading, and sharing has become increasingly evident. This highlights the urgent need to establish data usage control standards and trusted execution mechanisms for the entire lifecycle of geospatial data elements circulation.

In this paper, to address the complexities of the PNTRC network environment, we propose an access control mechanism for geospatial data circulation and presents the system model and administrative model by mapping the CoAC model. The administrative functions and methods within the model are formally described using Z language, and specific implementation mechanisms are outlined for various usage scenes of geospatial data circulation. This ensures the controlled usage of cross-domain geospatial data throughout its entire circulation lifecycle effectively.

We are currently in the preliminary stages of a collaborative case study to apply the GDCC model to ensure the secure circulation control of geospatial data. The primary objective and similar work revolve around establishing regulatory-compliant circulation control mechanisms for geospatial data within PNTRC network, while ensuring synergistic optimization of transmission integrity and utilization efficacy. Subsequent investigations should prioritize the implementation of performance benchmarking experiments in operational environments, particularly focusing on computational overhead and latency within large-scale distributed architectures, to rigorously validate the GDCC model's operational efficiency.

Acknowledgments. This work is supported by the National Key R&D Program of China No. 2023YFB3106505, and the National Natural Science Foundation of China No. U24A20240 & No. 62441226.

References

1. Zhu, C.Q.: Research progress of geographic data digital watermarking and encryption control technology. Acta Geodaet. Cartogr. Sinica **61**(10), 1609–1619 (2017)

2. Mao, J., Zhu, C.Q., Guo, J.F., et al.: Geospatial data access control threat model and countermeasures. Surv. Mapp. Sci. **43**(2), 88–94 (2018)
3. Le Thik, T., Dang, T.K., Kuonen, P., et al.: STRoBAC-spatial temporal role based access control. In: Nguyen, N.T., Hoang, K., Jędrzejowicz, P. (eds.) ICCCI 2012. LNCS, vol. 7654, pp. 201–211. Springer, Heidelberg (2012). https://doi.org/10.1007/978-3-642-34707-8_21
4. Ji, C., Li, X.Q., Liu, Q.: Structured data protection method in geographic information system. Inf. Netw. Secur. **15**(11), 71–76 (2015)
5. Samarati, P., de Vimercati, S.C.: Access control: policies, models, and mechanisms. In: Focardi, R., Gorrieri, R. (eds.) FOSAD 2000. LNCS, vol. 2171, pp. 137–196. Springer, Heidelberg (2000). https://doi.org/10.1007/3-540-45608-2_3
6. Zhang, L.C., Xu, Y.B., Li, F.H., et al.: Dynamic security-empowering architecture for space-ground integration information network. J. Commun. **42**(09), 87–95 (2021)
7. Ulusoy, H., Colombo, P., Ferrari, E., et al.: GuardMR: fine-grained security policy enforcement for MapReduce systems. In: Proceedings of the 10th ACM Symposium on Information, Computer and Communications Security (ASIACCS), pp. 285–296. ACM Press, New York (2015)
8. Ulusoy, H., Kantarcioglu, M., Pattuk, E., et al.: Vigiles: fine-grained access control for mapreduce systems. In: Proceedings of 2014 IEEE International Congress on Big Data (IEEE BigData), pp: 40–47. IEEE Press, Piscataway (2014)
9. Gupta, M., Patwa, F., Sandhu, R.: POSTER: access control model for the hadoop ecosystem. In: Proceedings of the 22nd ACM on Symposium on Access Control Models and Technologies (SACMAT), pp. 125–127. ACM Press, New York (2017)
10. Gupta, M., Patwa, F., Sandhu, R.: Object-tagged RBAC model for the hadoop ecosystem. In: Livraga, G., Zhu, S. (eds.) DBSec 2017. LNCS, vol. 10359, pp. 63–81. Springer, Heidelberg (2017). https://doi.org/10.1007/978-3-319-61176-1_4
11. Jin, W., Li, F.H., Yu, M.J., et al.: HDFS-oriented cryptographic key resource control mechanism. J. Commun. **43**(09), 27–41 (2022)
12. Li, F.H., Li, H., Niu, B., et al.: Research category and future development trend of data elements circulation and security. J. Commun. **45**(05), 1–11 (2024)
13. Baldi G Diaz-Tellez, Y., Dimitrakos, T., et al.: Session-dependent usage control for big data. J. Internet Serv. Inf. Secur. **10**(3), 76–92 (2020)
14. Cheung, H., Yang, C., et al.: New smart-grid operation-based network access control. In: The IEEE International Conference on Energy Conversion Congress and Exposition (ECCE), pp. 1203–1207. IEEE (2015)
15. Bernabe, B., Ramos, J., Gomez, A.F.S., et al.: TACIoT: multidimensional trust-aware access control system for the internet of things. Soft. Comput. **20**(5), 1763–1779 (2016)
16. Qi, H., Ma, H., Li, J., et al.: Access control model based on role and attribute and its applications on space-ground integration networks. In: The IEEE International Conference on Computer Science and Network Technology (ICCSNT), pp. 1118–1122. IEEE (2015)
17. Li, F.H., Wang, Y.C., Yin, L.H., et al.: Novel cyberspace-oriented access control model. J. Commun. **37**(05), 9–20 (2016)
18. Li, F.H., Chen, T.Z., Wang, Z., et al.: Cross-network access control mechanism for complex network environment. J. Commun. **39**(02), 1–10 (2018)
19. Li, D.R., Shen, X.: Research on the development strategy of real time and intelligent space-based information service system in China. Strategic Study CAE **22**(02), 138–143 (2020)
20. OASIS Open: OASIS eXtensible Access Control Markup Language (XACML) TC version 3.0. https://groups.oasis-open.org/communities/docs.oasis-open.org. Accessed December 2022
21. Sandhu, R.S., Coyne, E.J., Feinstein, H.L., et al.: Role-based access control models. Computer **29**(2), 38–47 (1996)

22. Hu, V.C., Ferraiolo, D., Kuhn, R., et al.: Guide to attribute based access control (ABAC) definition and considerations. NIST Spec. Publ. **800**(162), 1–47 (2014)
23. Sahai, A., Waters, B.: Fuzzy identity-based encryption. In: Cramer, R. (ed.) EUROCRYPT 2005. LNCS, vol. 3494, pp. 457–473. Springer, Heidelberg (2005). https://doi.org/10.1007/11426639_27
24. Park, J., Sandhu, R.: Towards usage control models: beyond traditional access control. In: 7th ACM Symposium on Access Control Models and Technologies (SACMAT), pp. 1–8. ACM Press, New York (2002)
25. Park, J., Sandhu, R.: The UCON ABC usage control model. In: ACM Transactions on Information and System Security (TISSEC), pp. 60–96. ACM Press, New York (2004)
26. Knorr, K.: Dynamic access control through Petri net workflows. In: Proceedings of the Proceedings 16th Annual Computer Security Applications Conference (ACSAC), pp. 1–9. IEEE (2001)
27. Bethencourt, J., Sahai, A., Waters, B.: Ciphertext-policy attribute-based encryption. In: IEEE Symposium on Security and Privacy (S&P), pp. 1–14. IEEE (2007)
28. Goyal, V., Pandey, O., Sahai, A., et al.: Attribute-based encryption for fine-grained access control of encrypted data. In: ACM Conference on Computer and Communications Security (CCS), pp. 89–98. ACM Press, Alexandria (2006)
29. Almenárez, F., Marín, A., Campo, C., García R., C.: TrustAC: trust-based access control for pervasive devices. In: Hutter, D., Ullmann, M. (eds.) SPC 2005. LNCS, vol. 3450, pp. 225–238. Springer, Heidelberg (2005). https://doi.org/10.1007/11414360_22
30. Bertino, E., Ghinita, G., Kamra, A.: Access control for databases: concepts and systems. Found. Trends Databases **3**(1–2), 1–148 (2011)

Identifying Unusual Personal Data in Mobile Apps for Better Privacy Compliance Check

Jiatao Cheng[1], Yuhong Nan[1(✉)], Xueqiang Wang[2], Zhefan Chen[1], and Yuliang Zhang[3]

[1] Sun Yat-sen University, Guangzhou, China
{chengjt6,nanyh,chenzhf55}@mail2.sysu.edu.cn
[2] University of Central Florida, Orlando, USA
[3] City University of Hong Kong, Hong Kong, China
xueqiang.wang.01@gmail.com, yzhang5785-c@my.cityu.edu.hk

Abstract. Automatically identifying a user's personal data in mobile apps is crucial for various downstream tasks in mobile privacy, such as detecting privacy compliance issues like unexpected data collection and sharing. The general approaches are pinpointing the in-app user personal data by comparing text similarity with a limited number of keywords (e.g., location, phone number). However, the expression of personal data has constantly been evolving. For example, *"pronouns"* is used to refer to gender identity nowadays. This type of *Unusual Personal Data* (UPD for short) is hardly covered by previous mechanisms. In this paper we propose FICO, a new framework to identify in-app UPD for mobile apps through F̲ine-grained C̲ontext-aware UI analysis. In particular, to uncover UPD, we developed a new methodology that leverages the *contextual affinity* of UI elements, bridging the knowledge gap caused by the evolving nature of natural language expressions. From experimental evaluation and measurement, we present that FICO is capable of handling *Unusual Personal Data* and enhancing privacy compliance.

Keywords: Personal Data · Privacy Compliance · UI Analysis

1 Introduction

As mobile phones continue to evolve, they empower users to perform a wide array of tasks through mobile applications (apps). However, the rich functionalities offered by these apps also enable developers to collect diverse and nuanced personal data, much of which is less frequently addressed in privacy literature. This type of data is referred to as unusual personal data (UPD). For example, fitness apps may request specific body measurements, such as chest, hip, and waist circumferences, while social networking apps might seek information on race, relationship status, and education.

The extensive collection of personal data has prompted growing concerns regarding privacy risks [9,15,32]. Identifying UPD within app user interfaces (UIs) presents significant challenges, particularly due to the difficulty in distinguishing personal data from non-personal data (e.g., general app settings). Moreover, the changing expression of natural literature brings the risk of overlooking UPD presented in previously unrecognized formats.

Previous studies addressing this issue primarily utilize static analysis of app UIs [1,12,18,20]. They typically identify user input elements and determine whether these correspond to personal data by comparing the displayed information against a predefined list of privacy-related keywords. Unfortunately, this method is increasingly ineffective in the rapidly changing landscape of mobile app UIs. As app functionalities diversify, the representation of personal data also evolves, rendering keyword lists quickly outdated. Compiling a comprehensive list of keywords to encompass all forms of personal data is not only labor-intensive but may also be impractical. Consequently, existing approaches primarily identify general personal data (GPD for short), which constitutes only a subset of the data collected by apps. This oversight raises concerns regarding the soundness and reliability of these methods.

In this paper, we propose FiCo, a new end-to-end approach that identifies in-app user UPD with high coverage. Specifically, FiCo considers that existing approaches [1,18] (i.e., keyword-dependent) fail to scale and cover diverse personal data, FiCo first identifies general personal data (GPD) based on UI semantics, such as the attributes of UI views and their description or text labels. Subsequently, FiCo exploits contextual affinity to extend GPD discovery to contextually related UIs, thereby effectively identifying UPD. This context-based extension facilitates comprehensive UPD coverage.

We evaluated FiCo on 200 top-popular apps in the wild to demonstrate its effectiveness. FiCo reported 348 items of UPD, with a precision of 93.5% and a recall of 91.6%. Notably, we found that previous works, UiRef [1] and UIPicker [18], miss over 40% of the data.

To further demonstrate its effectiveness in enhancing privacy compliance, we use FiCo to check whether the mobile apps in our dataset properly disclose the usage of the UPD. Our research showed that 112 out of the 200 apps (56%) have privacy compliance issues because they fail to disclose all user UPD in their privacy policies, which means the compliance issue of UPD is pervasive in the current ecosystem of mobile apps, and they are neglected by most app developers.

The contributions of this paper are as follows:

- We first highlight the universally neglected problem of UPD in personal data identification. To solve this problem, we provide an in-depth understanding of the contextual affinity of UIs, and propose an innovative algorithm for recognizing UPD.
- We built a prototype for personal data detection, and evaluated it on 200 apps from major app stores to demonstrate its effectiveness. The results suggest

that FiCo holds significant promise for enhancing various privacy-critical applications.
- We release the code of FiCo and a list of UPD. All code and data can be found at: https://github.com/SnoopyTlion/ICICS-FiCo.

The rest of this paper is organized as follows: Sect. 2 describes the scope of personal data, and uses a motivating example to illustrate the goals and challenges of our research. Section 3 provides the design of FiCo, and Sect. 4 presents the implementation details. Section 5 provides the experimental setup and evaluation of FiCo. Section 6 measures the privacy compliance in the current Android application ecosystem. Section 7 discusses the application scenarios of FiCo, its limitations, and future work. Section 8 covers related work, and Sect. 9 concludes the paper.

2 Motivation

2.1 UPD as Part of Personal Data

With the irreversible trend toward personalization and intelligence in mobile apps, the personal data processed by these apps has evolved and expanded significantly. Some of this data, although private, has not been commonly known or well-studied in prior research, which we refer to as UPD in our study. Drawing from observations of real-world apps, this study confirms that a piece of data in apps is classified as UPD based on its association with commonly used privacy keywords in previous research (denoted as K_p). For example, consider the term *"pronouns"* in Fig. 1. This term is not recorded in K_p, meaning it has not been used by prior research to detect personal data. However, we observe that *"pronouns"* does refer to the gender identity of a person, as indicated by online documents [13,14,22,31]. Thus, we confirm that the term is UPD since it is not in K_p and appears in online documents that link it to known personal data.

2.2 Motivating Example

We use the following motivating example to illustrate our research. Figure 1 displays UIs collecting UPD and other personal data from the TikTok app. On top of the UI in Fig. 1 (a), the user can modify a series of profile data, such as *"Name"* and *"Bio"*, which are commonly known as GPD. The UI also allows the user to modify her gender identity by launching another UI (bottom part of Fig. 1 (b)), enabling the user to specify free-form identities within an editable text view. Interestingly, the app does not use common words, such as *"gender"*, to describe gender identity but opts for the less commonly seen word *"Pronouns"*.

2.3 Goals and Challenges

The primary objective of FiCo is to facilitate the automated analysis of in-app UPD collections like (*"Pronouns"*). To achieve this goal, it is essential to

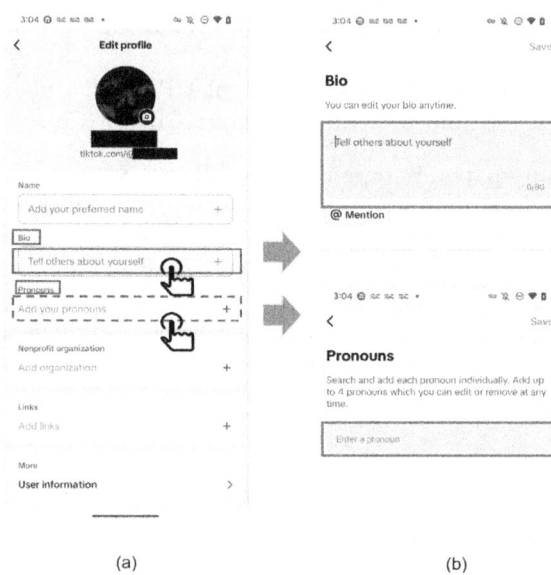

Fig. 1. An example of an item of UPD (i.e., pronouns) collected in TikTok.

adopt an approach that provides *high effectiveness (i.e. coverage) of in-app UPD identification*. Current approaches that rely on predefined system API lists [5] or privacy-related keyword lists [18–20] do not adequately address this requirement, given the diverse data points that are contextual to the functionalities of different applications. As the example shown in Fig. 1, the use of unusual text labels (e.g., *pronouns*) to describe personal data makes it difficult for existing techniques to effectively capture these data items.

Another challenge lies in conducting the analysis within an acceptable cost. Given the widespread adoption of large language models (LLMs) in recent years, many approaches depend on LLM capabilities to address complex problems effectively. However, as we will further show in our evaluation (in Sect. 5), the associated costs of utilizing LLMs cannot be overlooked. The financial cost of using commercial APIs (e.g., OpenAI) often limits their accessibility and generalizability. Additionally, the high volume of API requests leads to considerable time delays, which can further hinder the efficiency of application UI exploration. In general, the above challenges pose a significant barrier to the adoption of LLMs for automated analysis in this task. Therefore, our approach aims to find a lightweight, off-the-shelf solution that can better facilitate privacy compliance analysis.

3 FICO Design

3.1 Overview

In this study, we introduce an approach called FICO to automatically identify user UPD. The overview of our approach is illustrated in Fig. 2. FICO takes the

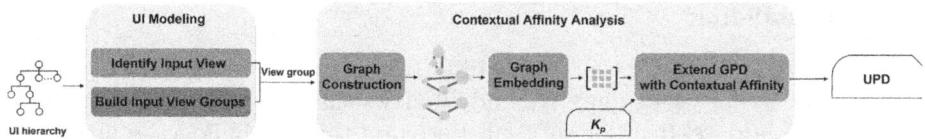

Fig. 2. Overview of FiCo.

UI hierarchy as input, models it as view groups, and then conducts contextual affinity analysis to identify UPD.

To identify the in-app UPD with high coverage. we leverage the observation that UI views with similar functionalities, such as collecting user personal data, often share similar contextual information. This observation is aligned with research in the UI design domain [26,34], which reports that *views with similar functionalities are often organized following the same rules (e.g., view layout) and placed close to each other.* Inspired by this, we take the contextual information and organizational rules of views into consideration and propose *contextual affinity* to assist the UPD identification. Previous keyword-dependent approaches [1,18] perform personal data identification in low coverage and the UPD is missed from their report. We use the views reported by the keyword-dependent approaches as a starting point and utilize contextual affinity to expand the discovery to other unusual views collecting UPD. In this way, the terms for UPD do not have to be included in the pre-defined keyword list. Thus, FiCo identifies UPD in a promising way, overcoming the limitations of keyword-dependent approaches and ensuring high coverage.

Take the analysis for TikTok in Fig. 1 (a) and (b) as an example. Firstly, according to the layout and organizational rules, we identify two input view groups. The first input view group contains the purple-boxed text view *"Bio"*, the red-boxed clickable view *"Tell others about yourself"* in Fig. 1 (a), and the green-boxed input view (i.e., EditText) in the upper half of Fig. 1 (b). The second input view group contains the text view *"Pronouns"*, the clickable view *"Add your pronouns"*, and the input view in the lower half of Fig. 1 (b). Through UI semantics analysis similar to prior keyword-dependent approaches, we can tell that the first input view group collects user general personal data (GPD) since it uses a common privacy-related keyword, i.e., *"Bio"*, as its label. We may not determine whether the second input view group collects personal data or not since *"Pronouns"* does not appear in any privacy-related keyword lists. However, given that the two input view groups are contextually close to each other, e.g., situated nearby and exhibit highly similar UI patterns and contexts, we can determine that the second group also collects personal data (i.e., *"Pronouns"*) from users. Consequently, by considering the contextual affinity of UIs, we developed a novel methodology that not only identifies GPD based on UI semantics but also extends the data to a broader domain and identifies UPD.

3.2 UI Modeling

To analyze the personal data collected by UIs, an essential step is to determine which views accept user input. We find all views that are able to capture user's input from Android UI framework [4]. Based on this, we build view groups for further analysis.

As illustrated in Fig. 1, multiple individual views serving the same purpose can be grouped together. For example, the two views in Fig. 1 (a) labeled "*Pronouns*" and "*Add your pronouns*", and the green-boxed view in the lower half of Fig. 1 (b) collectively serve the purpose of guiding users to input their pronouns. Identifying and analyzing the views per group would provide more semantic information about whether user personal data is collected by individual views in the group, and also enable more robust detection of views that are contextually close to each other, thereby expanding the detection of GPD to UPD.

In this study, we model view groups using three intuitive strategies inspired by prior researches [1,18]. First, when given an input view, we examine its parent view within the same UI that encapsulates the input view, along with any other views within the parent view, i.e., sibling views of the input view. We add these sibling views to the input view group if none of them are input views (a strong indicator that the sibling views are created solely for describing the input view).

Second, we check the preceding view that leads to the rendering of the input view when users navigate through the apps, and add it to the input view group if the preceding view does not lead to any other input views. For example, the red-boxed view "*Add your pronouns*" in Fig. 1 (a) only results in the rendering of one input view, i.e., the *EditText* the bottom part of Fig. 1 (b). Therefore, it is considered in the same group as the input view.

Third, we adapt methods proposed in UiRef [1] to cluster the associated views in the input view group. Particularly, UiRef reports that app developers often organize UI views following specific patterns, e.g., a *TextView* is frequently aligned to the input view it describes at either a horizontal or vertical angle. Therefore, we leverage this observation to associate *TextViews* with an input view, even though they are not covered by the same parent view.

Finally, we treat all text labels in the view groups as the semantics of the view groups because they collectively describe the semantics of the input view.

3.3 Contextual Affinity Analysis

Using input view groups built from the previous step as input, we can perform semantic analysis to detect which of the groups collect personal data, i.e., by checking whether the text labels of the input view groups contain any of the privacy-related keywords in K_p. Here, we denote the output, i.e., the set of view groups known to collect personal data, as $S_k = \{e \mid e = (vg, dt)\}$, where vg and dt are the input view group represented by a series of view ids, and the content of personal data being collected, respectively. The other input view groups that were unknown to collect personal data are denoted as $\mathcal{S}_{un} = \{e \mid e = (vg, dt)\}$, where dt is empty.

FICO achieves better coverage on personal data since it can expand the detection from $\mathcal{S}_k$ to $\mathcal{S}_{un}$, thus include UPD. As shown in Algorithm 1, we can implement a procedure called ContextualExtension, which takes the $\mathcal{S}_k$ and $\mathcal{S}_{un}$ as input, and outputs all input view groups that collect personal data, denoted by $\mathcal{D}$. Specifically, for each element (e_k) in $\mathcal{S}_k$, we iterate over $\mathcal{S}_{un}$ to identify another element (e_{un}) that has an input view group within the same UI as the view group in e_k (Line 4 - 15). For the input view groups in the two elements $(e_k.vg$ and $e_{un}.vg)$, we embed them and generate their vector representations based upon the properties and layouts of the views, denoted by v_k and v_{un} (Line 7 and Line 9). After that, we calculate the contextual affinity of the two input view groups by calculating the similarities of the two vectors (Line 10). High similarity indicates that the input view group in e_{un} also collects user personal data, so we find the e_{un} that has the highest similarity with e_k (Line 11 - 13) and store it into $\mathcal{S}_{ex}$ (Line 14 - 15). In this way, we add UPD into our discovery. Then we use the newly discovered view groups in $\mathcal{S}_{ex}$ to extend more personal data (Line 16 - 17). In that case, we detect the potential content of the data by identifying the object nouns in the text labels of the view groups in $\mathcal{S}_k$ (Line 18). This is achieved using natural language processing techniques, such as POS tagging for text labels in short phrases and dependency parsing for sentences. From this process, FICO can uncover both GPD and UPD in UIs.

Below, we elaborate on two key steps in the algorithm.

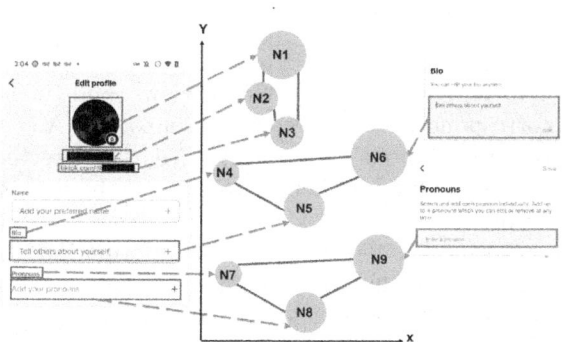

Fig. 3. A simplified graph representation of the motivating example. Each node is an individual view. Each edge is a representation of how two individual views are organized into a view group.

Embeddings of Input View Groups. The individual views and their relationships within an input view group are analogous to a weighted graph, where the views resemble graph nodes and the relative layout of views can be represented by the weights of edges. Figure 3 shows the much-simplified graph representations to help visualize the input view groups in the motivating example.

Algorithm 1. The process of extending user personal data using contextual affinity.

Input: view groups known to collect personal data $\mathcal{S}_k$, view groups unknown to collect personal data $\mathcal{S}_{un}$
Output: all view groups collecting user personal data $\mathcal{D}$
ProcedureContextualExtension($\mathcal{S}_k$, $\mathcal{S}_{un}$)
1: $\mathcal{S}_{ex} \leftarrow \emptyset$
2: $\triangle \leftarrow \mathcal{S}_k$
3: **while** $\triangle$ is not $\emptyset$ **do**
4:     **for** e_k in $\triangle$ **do**
5:        $affinity \leftarrow -1$
6:        $e_{ex} \leftarrow \emptyset$
7:        $v_k \leftarrow$ EmbedViewGroup($e_k.vg$)
8:        **for** e_{un} in $\mathcal{S}_{un}$ **do**
9:           $v_{un} \leftarrow$ EmbedViewGroup($e_{un}.vg$)
10:          $affinity_t \leftarrow$ CalculSmilarity(v_k, v_{un})
11:          **if** $affinity <= affinity_t$ **then**
12:             $affinity \leftarrow affinity_t$
13:             $e_{ex} \leftarrow e_{un}$
14:        **if** v_{ex} is not $\emptyset$ **then**
15:           $\mathcal{S}_{ex} \leftarrow \mathcal{S}_{ex} \cup \{e_{ex}\}$
16:     $\triangle \leftarrow \mathcal{S}_{ex} \setminus \mathcal{S}_k$
17:     $\mathcal{S}_k \leftarrow \mathcal{S}_k \cup \mathcal{S}_{ex}$
18: $\mathcal{D} \leftarrow \mathcal{S}_k$
19: **return** $\mathcal{D}$

The nodes *N1* to *N9* represent nine individual views, and the nodes are located in three graphs corresponding to three view groups: *N1-N3*, *N4-N6*, and *N7-N9*. The view group represented by graph *N4-N6* was detected to collect user personal data, i.e., "*bio*". Our goal here is to expand the detection to the other unknown view groups.

Each input view group has four key pieces of information: the coordinates of individual views (location of the graph nodes), the size of the individual views (size of the nodes), the relative distance between individual views (length of the edges), and the alignment of individual views (slope of the edges). With this information, we can identify that the view group represented by graph *N7-N9* is similar to that represented by *N4-N6*, and therefore may also collect user personal data. However, the graph *N1-N3* is less similar to *N4-N6* and may not collect user personal data. Therefore, we perform embeddings for the input view groups by constructing vectors from the above pieces of information. Specifically, assuming that an input view group consists of n individual views numbered from 1 to n. For the i-th individual view, we can extract a series of numeric features, including the margins from vertical boundaries and horizontal boundaries (x_i and y_i) of the view in the screen, the width (w_i) and the height (h_i) of the view. We further calculate the coordinates of the centroid of the view, which is denoted by xc_i and yc_i. Based on the above numeric features, we can create a 6-dimensional vector that incorporates the average size, location, alignment, and distances of the individual views in an input view group. In Eq. 1, we show the detailed method for constructing the vector. In particular, we use alignment (g) and relative distance (d) to capture the arrangement of individual views in the view group.

$$v = \begin{cases} w = \frac{1}{n}\sum_{i=1}^{n} w_i \\ h = \frac{1}{n}\sum_{i=1}^{n} h_i \\ x = \frac{1}{n}\sum_{i=1}^{n} x_i \\ y = \frac{1}{n}\sum_{i=1}^{n} y_i \\ g = \frac{2}{n*(n-1)}\sum_{i=1}^{n}\sum_{j=i+1}^{n}(xc_i - xc_j)/(yc_i - yc_j) \\ d = \frac{2}{n*(n-1)}\sum_{i=1}^{n}\sum_{j=i+1}^{n}\sqrt{(xc_i - xc_j)^2 + (yc_i - yc_j)^2} \end{cases} \quad (1)$$

Extend Personal Data Identification. Given an input view group known to collect personal data, we extend the identification of personal data to other groups by checking their highest contextual affinity to the input view group. Before the calculation, we implement min-max normalization on each dimension of all vectors to avoid comparison in different magnitudes. Then we calculate the cosine similarity [24] of two vectors corresponding to different view groups. In this study, we used a similarity threshold (θ), where a similarity lower than θ indicates that a new input view group does not have enough contextual affinity to be extended. We leave the process of determining θ in Sect. 4.

4 Implementation

We implemented a prototype of FICO with around 1k lines of code in Python. As mentioned in Sect. 1, we use K_p to assist the UI semantics analysis (in Sect. 3.2). To construct K_p, a manual review of research papers published within the last five years in prominent security and privacy venues was conducted. This review identified publications focusing on personal data within the context of user privacy, resulting in a keyword list (denoted as K_p) comprising 1,970 established keywords. For NLP analysis of UI semantics, we use Spacy [29] and HanLP [11]. The dynamic app exercise process is implemented based on Fastbot2 [16], a state-of-the-art tool for dynamic App exploration. We dynamically dump the UI hierarchy and use it to perform personal data identification and extract the first non-empty field among "*text*", "*hint-text*", "*content-description*", and "*resource-id*" to serve as textual information.

We deploy FICO on a Windows desktop PC with an 8-core, 2.30 GHz CPU and 32GB RAM. The apps are tested on an Android 13.0 Pixel4 device with an 8-core CPU and 4GB RAM. Each app is tested for 1 h using Fastbot2 [16].

Threshold for Personal Data Extension. To search for a θ value that enables us to extend the identification with high accuracy, we conducted an

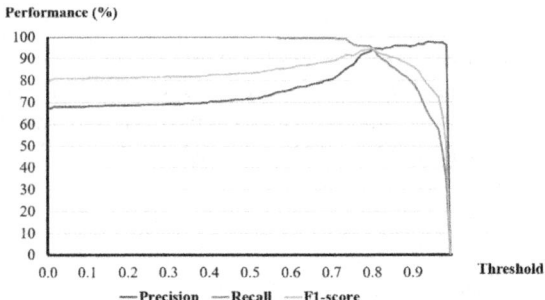

Fig. 4. Performance of contextual analysis under different threshold.

analysis involving two sets of input view groups. These sets consist of pairs that should be classified as either closely related or not closely related. More specifically, we randomly selected 40 apps and collected 300 pairs of input view groups, with each pair within the same UI, and confirmed to collect personal data. These pairs were then added to the set S_{close}. Additionally, we gathered 150 pairs of input view groups, with each pair under the same UI path but only one group collects personal data, forming the set S_{far}. We calculate the vector similarity for all pairs of input view groups in S_{close} and S_{far}. We searched the θ value in the range of 0 and 1, with a step of 0.01. A pair in S_{far} that has a similarity larger than θ leads to a false positive, while a pair in S_{close} that has a similarity lower than θ leads to a false negative. For each θ value, we calculate precision and recall for the extension, and present the results in Fig. 4. Our analysis results show that the value 0.79 can be an optimal θ value since it achieves a best F1-score of 96.9% with a precision of 96.3% and a recall of 97.5%. Thus, we use this threshold value in the implementation of the prototype.

5 Evaluation

We evaluate the effectiveness of FICO by answering the following research questions:
RQ1: What is the performance for FICO in identifying UPD from mobile apps?
RQ2: What are the benefits of the contextual affinity adopted by FICO?
RQ3: How does FICO outperform other tools in comparison?

5.1 Evaluation Setup

Dataset. To evaluate FICO, we manually labeled a dataset (as D_{App}) that consists of 200 mobile apps. Specifically, the dataset can be divided into two parts. 100 apps are randomly selected from Google Play's most popular 1,000 apps under the region of the US. The other 100 apps are selected from Tencent App Store's most popular 1,000 apps under the region of China. Overall, the average download amount of these apps is around 400M. In this way, our dataset includes apps in the two most-spoken languages: English and Chinese.

Ground Truth Labeling. We manually reviewed the aforementioned 200 apps by running each app and exploring their UIs to collect the ground truth of user personal data. Specifically, we invite three Ph.D. students working on mobile privacy to participate in this labeling process. Each student manually explored the app and recorded the personal data items they found. Each app is given a 30-minute time budget. Overall, the Cohens Kappa coefficient [8] is 78.9%. For the discrepancies between the labeling results among the three students, we ask them to discuss until they reach an agreement. A personal data item will only be recorded in ground truth if it is unanimously recognized as personal data by all students. In this process, we use K_p and criteria in Sect. 2.1 as a guideline for labeling UPD and GPD.

Overall, we have identified a total of 380 UPD items as our ground truth (see Table 1 for details). Note that the same personal data appearing in different UIs of the same app will be counted only once. In addition to these items, 1,972 are GPD labeled as well, which are used to demonstrate the soundness of FiCo for all types of personal data.

5.2 RQ1: Performance on UPD identification

Table 1. Performance for identifying UPD.

D_{app}	Ground Truth	TP	FP	FN	Precision	Recall	F1-score
Apps-US	187	172	9	15	95.0%	92.0%	93.5%
Apps-CN	193	176	15	17	92.1%	91.2%	91.6%
Total	380	348	24	32	93.5%	91.6%	92.6%

Table 1 shows the overall effectiveness of FiCo for UPD identification on the D_{App} dataset. Here, a true positive (TP) means FiCo correctly reports the UPD labeled in our ground truth, a false negative (FN) means FiCo missed reporting the UPD in our ground truth, a false positive (FP) means FiCo incorrectly reports a data item which is not in our ground truth as UPD. Note that we did not calculate the true negatives (i.e., data irrelevant to user privacy), as it is rather difficult to comprehensively label each UI element in mobile apps. Overall, FiCo achieves a precision of 93.5% and a recall of 91.6% (with 92.6% F1-score).

The performance demonstrates that FiCo is highly effective in addressing the task of identifying UPD. Furthermore, a comparison of results across apps from different countries reveals that FICO achieves consistently similar performance. This consistency suggests that our approach is robust and generalizable to apps, regardless of variations in the textual semantics describing app data.

We analyze false positives and false negatives in FiCo in the following.

False Positives. Most of the false positives in FICO are caused by the inaccurate POS tagging for short terms shown in the app UI. For example, the POS tagging in Scapy [29] incorrectly interprets a verb as a noun (*"edit"* in the term *"edit page"*), which misleads the further analysis. Another type of false positives is caused by a lack of fine-grained parsing of textual descriptions. For example, the input box *"password"* is described with *"You should enter more than 6 digits"*. Due to the coarse-grained dependency-relation parsing over the sentence, FICO mistakenly includes *"digits"* as a user's personal data. These limitations could be addressed by integrating more advanced NLP models and techniques. While such specific NLP mechanisms could effectively reduce false positives, they are beyond the scope of our current study.

False Negatives. Most of the false negatives are caused by Fastbot2 [16]. Specifically, we found that Fastbot2 may miss some app-customized UIs, leading to certain privacy-related UI elements not being analyzed by FICO. Additionally, some false negatives are due to the limited UI exploration within the given time budget. Although FICO is specifically tailored for privacy-related content, it may still miss some privacy-related UIs if their paths are relatively long and deep.

5.3 RQ2: Contribution of Contextual Affinity

To support our argument that the use of contextual affinity strategy aligns with the real-world UI implementation, we evaluate whether two UI elements having high contextual affinity indeed have similar functions. We first locate view groups that contain user personal data based on the predefined K_p. Next, we examine whether the surrounding view groups that share the highest similarity also contain user personal data. For this analysis, we randomly sampled 696 view groups corresponding to the collection of personal data. Our analysis showed that 96.3% (670/696) of the surrounding view groups sharing the highest similarity (which is also greater than the threshold) contain personal data. Therefore, applying contextual affinity in FICO has the potential to significantly expand the identification of personal data.

5.4 RQ3: Comparison with Existing Tools

To further demonstrate the advantages of FICO, we compare its performance with two existing tools for privacy data identification: UIPicker [18] and UiRef [1]. Both tools rely on NLP techniques to calculate semantic similarity between their predefined keywords and the text in UI elements. As they statically parse UI hierarchies from apps' resource files, we extract their NLP components and provide them with dynamically rendered UI hierarchies to ensure a fair comparison.

Table 2 presents the evaluation results of different tools on D_{App}, demonstrating that FICO achieves a significantly lower number of false negatives, with a recall of 91.6%, compared to UIPicker and UiRef, which achieve recalls of 36.3%

and 59.5%, respectively. These results indicate that semantic similarity-based approaches struggle to effectively handle UPD. This limitation arises because the expression of words evolves rapidly in the real world, making it difficult for semantic approaches to adapt to these changes. In contrast, FiCo leverages contextual affinity to expand its knowledge to a broader domain, effectively bridging the knowledge gap caused by evolving word expressions.

Table 2. Performance of different tools for UPD and GPD identification.

Performance	UPD						GPD					
	TP	FP	FN	Precision	Recall	F1-score	TP	FP	FN	Precision	Recall	F1-score
UIPicker	138	6	242	95.8%	36.3%	52.7%	1,392	103	580	93.1%	70.6%	80.3%
UiRef	226	17	154	93.0%	59.5%	72.6%	1,612	218	360	88.1%	81.4%	84.7%
FiCo	348	24	32	93.5%	91.6%	92.6%	1,771	340	201	83.8%	89.8%	86.7%

Table 3. Identified top-10 non-compliant apps in disclosing user personal data.

Package name	Selected non-compliant user personal data
com.tinder	sleeping habits, relationship status, children plan, ellipses
com.linkedin.android	pronunciation, interests, kid, pronouns
com.taobao.idlefish	fiduciary, liquidated damages, litigation cost
com.instagram.android	description category, Ad preferences
com.airbnb.android	government ID, suite (address), profession
tv.danmaku.bili	mainland travel permit for HK. and MO. residents
net.csdn.csdnplus	admission Date, date of graduation
ctrip.android.view	floor (address), house number (address)
com.hnjc.dl	exercise frequnce, diet habit, height
com.tieyou.train.ark	tax ID number, invoice, admission date

To evaluate whether FiCo can also outperform in GPD identification, we compare its performance with that of UIPicker and UiRef. As shown in Table 2, FiCo achieves the highest recall and F1-score among the three tools. This demonstrates that the contextual affinity utilized by FICO outperforms the semantic similarity-based approaches employed by UIPicker and UiRef.

Lastly, we also compare FiCo with commercial LLMs in terms of financial cost. To evaluate the financial cost of LLM in our task, we randomly selected 20

UI hierarchies from our dataset and used OpenAI's tokenizer [23] to calculate the number of input tokens. We found that LLM requires an average of 11k tokens to process the content of each UI hierarchy. According to OpenAI's pricing, the average price of the model is $1.8 per 1 million input tokens. In our dataset, each app has an average of 3,600 UI hierarchies that need to be processed. This means that if we use LLM to complete our task, each app would require at least $70 in expenditure, which is an unacceptable cost. In contrast, FiCo does not incur any financial cost, making it a more cost-effective solution for this application.

6 Measurement

We leverage FiCo to perform a consistency analysis by comparing the personal data identified on app UIs and in the privacy policies to detect privacy non-compliance, more specifically, missing disclosure of personal data. We use a state-of-the-art tool, i.e., ATPChecker [37], to analyze app privacy policies and collect the set of personal data that apps claim to collect or share (denoted by S_{claim}). For each personal data point identified on app UIs (D_{ui}), we check whether it appears in S_{claim}. A problem here is that personal data on app UIs may not appear exactly as in privacy policies. To address the problem, we compared all synonyms of D_{ui} to S_{claim}, and also checked whether any data point in S_{claim} is a hypernym of D_{ui}, which is defined in the ontology of PolicyLint [2].

6.1 Analysis Results

In D_{App}, we found that 170 apps fail to disclose all personal data in S_{claim}, resulting in a total of 1,162 missed data points (an average of 6.8 per app). Among the 170 apps, 112 apps fail to disclose all UPD, totaling 328 missed data items (an average of 2.9 per app). Table 3 presents 10 example applications along with instances of non-compliant data.

To better understand the influence of UPD on compliance issues, we compared the proportion of missing data between GPD and UPD. For each app, the missing proportion of personal data was calculated as follows:

- For UPD, the proportion was calculated by dividing the amount of missing UPD in D_{ui} by the total amount of UPD in D_{ui}.
- For GPD, the missing proportion was calculated by dividing the amount of missing GPD by the total amount of GPD in D_{ui}.

Our findings reveal that the average missing proportion for GPD is 58.9%, while for UPD, it is significantly higher at 86.3%.

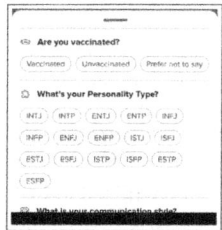  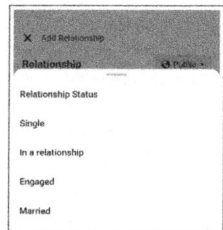

(a) COVID vaccine status collected by Tinder. (b) Identity information collected by Bigo. (c) Relationship status collected by Facebook.

Fig. 5. Representative examples of non-compliant user personal data in different popular apps.

> **Information with special protections**
>
> You might choose to provide information about your religious views, your sexual orientation, political views, health, racial or ethnic origin, philosophical beliefs or trade union membership. These and other types of information could have special protections under the laws of your jurisdiction.

Fig. 6. Statements in Facebook's privacy policy, the personal data *"relationship status"* is missed.

6.2 Case Studies

We examine several popular apps and discuss the privacy compliance issues identified by FICo.

• **Tinder** fails to disclose the largest number of personal data points (i.e., 59 items) in our dataset. We found that it collects personal data such as the ***COVID-19 vaccination status*** (Fig. 5 (a)), body type, and smoking habits, all of which are data that should have been clearly declared in advance. Unfortunately, the most relevant privacy policy statement potentially related to such data is a vague description of personal data as *"such as basic profile details and the types of people you'd like to meet."*

• **Bigo Live** is a live streaming app with 500M+ installs on Google Play. As shown in Fig. 5 (b), the app collects user's ***racial identity*** in its profile page, such as Asian and Black which is indeed sensitive to app users. Unfortunately, its privacy policy does not make any disclosure of this data point.

• **Facebook** has been under a lot of scrutiny in terms of privacy and security. However, while Facebook has invested a significant amount of effort into meeting privacy compliance requirements, FICo identifies that the ***relationship status*** ("single" or "married" shown in Fig. 5 (c)) is missed in Facebook's privacy policy statements, even though a lot of other personal data under the same granularity are actually disclosed (Fig. 6).

7 Discussion and Limitation

7.1 Application Scenarios of FICO

In addition to scrutinizing missing data disclosures in the app's privacy policies (see Sect. 6), FICO proves versatile in various compliance check scenarios. For instance, FICO can help identify misuse of personal data in terms of data processing purposes. Additionally, FICO can be used by privacy researchers and industry stakeholders for detailed analysis of app privacy behavior, such as fingerprinting in-app personal data and using it for further analysis, such as tracking data sharing.

7.2 Limitation and Future Work

FICO has several limitations that require further refinement. FICO assumes that app UI design adheres to common practices in software UX/UI development, as discussed in prior research [26,34]. However, if app UIs are designed without following these practices, FICO may fail to identify the personal data. Furthermore, despite cross-validation by multiple students, labeling the personal data in the ground truth may still be influenced by subjective judgment and the limited time available for manual app analysis. We believe that involving more privacy experts could further improve the accuracy of the ground truth.

8 Related Work

Our work is closely related to the topics of privacy protection for mobile apps. Over the past years, especially after the implementation of GDPR, the analysis of mobile apps has been a hot topic in the domain of privacy protection. A series of researches [3,7,17,30,32,35,36] has studied the inconsistency between the claimed data usage in privacy policies or GUI and the actual behaviors of mobile apps. For example, Xi et al. [33] reveal the UI-to-behavior mismatch using machine learning and static program analysis. Nan et al. [17] aim to achieve better coverage of sensitive data in the IoT apps by analyzing the semantics in code. Besides, some prior studies [6,10,21,25,27,28] detected whether the behaviors of mobile apps indeed violate privacy legislation such as GDPR, CCPA, and COPPA. For example, Reyes et al. [25] analyzed child-directed apps to determine if they violated COPPA by collecting identifiable information from children. Du et al. [10] performed static data flow analysis to determine if the opt-out UIs of mobile apps are consistent with the actual app behavior. Santhanam et al. [28] make an effort to detect the violation of GDPR "right to be forgotten". Different from previous works, FICO enhances the coverage of identified UPD by exploiting the rich contextual information in apps' UIs.

9 Conclusion

In this paper, we propose FiCo, a new framework to identify the in-app UPD for mobile apps through UI analysis. To enhance the coverage, we developed a novel methodology that identifies UPD based on the contextual affinity of UIs.

Acknowledgments. Yuhong Nan was supported in part by the National Natural Science Foundation of China (62202510), Guangdong Basic and Applied Basic Research Foundation (2023A1515012971), Guangdong Zhujiang Talent Program (No. 2023QN10X561) and the Fifth Electronic Research Institute of the Ministry of Industry and Information Technology Key Laboratory Open Project (No. HK202403639).

Disclosure of Interests. The authors have no competing interests to declare that are relevant to the content of this article.

References

1. Andow, B., Acharya, A., Li, D., Enck, W., Singh, K., Xie, T.: UiRef: analysis of sensitive user inputs in android applications. In: Proceedings of the 10th ACM Conference on Security and Privacy in Wireless and Mobile Networks, pp. 23–34 (2017)
2. Andow, B., et .: {PolicyLint}: investigating internal privacy policy contradictions on google play. In: 28th USENIX Security Symposium (USENIX Security 19), pp. 585–602 (2019)
3. Andow, B., et al.: Actions speak louder than words:{Entity-Sensitive} privacy policy and data flow analysis with {PoliCheck}. In: 29th USENIX Security Symposium (USENIX Security 20), pp. 985–1002 (2020)
4. Android: Android mobile app developer tools – android developers. https://developer.android.com/ (2024)
5. Arzt, S., Rasthofer, S., Bodden, E.: Susi: A tool for the fully automated classification and categorization of android sources and sinks. University of Darmstadt, Tech. Rep. TUDCS-2013-0114 (2013)
6. Bui, D., Tang, B., Shin, K.G.: Do opt-outs really opt me out? In: Proceedings of the 2022 ACM SIGSAC Conference on Computer and Communications Security, pp. 425–439 (2022)
7. Bui, D., Yao, Y., Shin, K.G., Choi, J.M., Shin, J.: Consistency analysis of data-usage purposes in mobile apps. In: Proceedings of the 2021 ACM SIGSAC Conference on Computer and Communications Security, pp. 2824–2843 (2021)
8. Cohen, J.: A coefficient of agreement for nominal scales. Educ. Psychol. Measur. **20**(1), 37–46 (1960)
9. Dong, Z., Wang, L., Xie, H., Xu, G., Wang, H.: Privacy analysis of period tracking mobile apps in the post-roe v. wade era. In: Proceedings of the 37th IEEE/ACM International Conference on Automated Software Engineering, pp. 1–6 (2022)
10. Du, X., Yang, Z., Lin, J., Cao, Y., Yang, M.: Withdrawing is believing? Detecting inconsistencies between withdrawal choices and third-party data collections in mobile apps. In: 2024 IEEE Symposium on Security and Privacy (SP), pp. 14–14. IEEE Computer Society (2023)
11. HanLP: hankcs/HanLP. https://github.com/hankcs/HanLP (2024)

12. Huang, J., Li, Z., Xiao, X., Wu, Z., Lu, K., Zhang, X., Jiang, G.: {SUPOR}: precise and scalable sensitive user input detection for android apps. In: 24th USENIX Security Symposium (USENIX Security 15), pp. 977–992 (2015)
13. Instagram: Instagram profile pronouns: Here's how to display your gender identity. https://www.fastcompany.com/90635735/instagram-profile-pronouns-heres-how-to-display-your-gender-identity (2023)
14. LGBTQIA: LGBTQIA resource center - pronouns & inclusive language. https://lgbtqia.ucdavis.edu/educated/pronouns-inclusive-language (2023)
15. Li, S., et al.: Collect responsibly but deliver arbitrarily? A study on cross-user privacy leakage in mobile apps. In: Proceedings of the 2022 ACM SIGSAC Conference on Computer and Communications Security, pp. 1887–1900 (2022)
16. Lv, Z., Peng, C., Zhang, Z., Su, T., Liu, K., Yang, P.: Fastbot2: reusable automated model-based GUI testing for android enhanced by reinforcement learning. In: Proceedings of the 37th IEEE/ACM International Conference on Automated Software Engineering (ASE 2022) (2022)
17. Nan, Y., et al.: Are you spying on me?{Large-Scale} analysis on {IoT} data exposure through companion apps. In: 32nd USENIX Security Symposium (USENIX Security 23), pp. 6665–6682 (2023)
18. Nan, Y., Yang, M., Yang, Z., Zhou, S., Gu, G., Wang, X.: {UIPicker}:{User-Input} privacy identification in mobile applications. In: 24th USENIX Security Symposium (USENIX Security 15), pp. 993–1008 (2015)
19. Nan, Y., Yang, Z., Wang, X., Zhang, Y., Zhu, D., Yang, M.: Finding clues for your secrets: Semantics-driven, learning-based privacy discovery in mobile apps. In: NDSS (2018)
20. Nan, Y., et al.: Identifying user-input privacy in mobile applications at a large scale. IEEE Trans. Inf. Forensics Secur. **12**(3), 647–661 (2016)
21. Nguyen, T.T., Backes, M., Marnau, N., Stock, B.: Share first, ask later (or never?) studying violations of {GDPR's} explicit consent in android apps. In: 30th USENIX Security Symposium (USENIX Security 21), pp. 3667–3684 (2021)
22. NPR: A guide to understanding gender identity and pronouns: NPR. https://www.npr.org/2021/06/02/996319297/gender-identity-pronouns-expression-guide-lgbtq (2021)
23. OpenAI: Tokenizer. https://platform.openai.com/tokenizer (2024)
24. Rahutomo, F., Kitasuka, T., Aritsugi, M., et al.: Semantic cosine similarity. In: The 7th international student conference on advanced science and technology ICAST. vol. 4, p. 1. University of Seoul South Korea (2012)
25. Reyes, I., et al.: "won't somebody think of the children?" examining COPPA compliance at scale. In: The 18th Privacy Enhancing Technologies Symposium (PETS 2018) (2018)
26. S1T2: Apply crap to design — S1T2 blog. https://s1t2.com/blog/step-1-generously-apply-crap-to-design (2023)
27. Samarin, N., et al.: Lessons in VCR repair: Compliance of android app developers with the California consumer privacy act (CCPA). arXiv preprint arXiv:2304.00944 (2023)
28. Santhanam, P., Dang, H., Shan, Z., Neamtiu, I.: Scraping sticky leftovers: app user information left on servers after account deletion. In: 2022 IEEE Symposium on Security and Privacy (SP), pp. 2145–2160. IEEE (2022)
29. spaCy: spacy · industrial-strength natural language processing in python. https://spacy.io/ (2024)

30. Tan, Z., Song, W.: PTPDroid: detecting violated user privacy disclosures to third-parties of android apps. In: 2023 IEEE/ACM 45th International Conference on Software Engineering (ICSE), pp. 473–485. IEEE (2023)
31. TikTok: Pronouns in profile & live events — creator portal — TikTok. https://www.tiktok.com/creators/creator-portal/en-us/product-feature-updates/january-2022/ (2023)
32. Wang, J., et al.: Understanding malicious cross-library data harvesting on android. In: 30th USENIX Security Symposium (USENIX Security 21), pp. 4133–4150 (2021)
33. Xi, S., et al.: Deepintent: deep icon-behavior learning for detecting intention-behavior discrepancy in mobile apps. In: Proceedings of the 2019 ACM SIGSAC Conference on Computer and Communications Security, pp. 2421–2436 (2019)
34. Xie, M., Xing, Z., Feng, S., Xu, X., Zhu, L., Chen, C.: Psychologically-inspired, unsupervised inference of perceptual groups of GUI widgets from GUI images. In: Proceedings of the 30th ACM Joint European Software Engineering Conference and Symposium on the Foundations of Software Engineering, pp. 332–343 (2022)
35. Yu, L., et al.: PPChecker: towards accessing the trustworthiness of android apps' privacy policies. IEEE Trans. Softw. Eng. **47**(2), 221–242 (2018)
36. Zhang, X., Wang, Y., Zhang, X., Huang, Z., Zhang, L., Yang, M.: Understanding privacy over-collection in WeChat sub-app ecosystem. arXiv preprint arXiv:2306.08391 (2023)
37. Zhao, K., et al.: Demystifying privacy policy of third-party libraries in mobile apps. In: 2023 IEEE/ACM 45th International Conference on Software Engineering (ICSE), pp. 1583–1595. IEEE (2023)

Why Biting the Bait? Understanding Bait and Switch UI Dark Patterns in Mobile Apps

Yixi Lin[1], Yue Xu[1], Zitong Yao[1], Yuhong Nan[1(✉)], Queping Kong[1,2,3], and Xueqiang Wang[4]

[1] Sun Yat-sen University, Guangzhou, China
{linyx98,xuyue73,yaozt,kongqp}@mail2.sysu.edu.cn, nanyh@mail.sysu.edu.cn
[2] Guangzhou Forensic Science Institute, Guangzhou, China
[3] Cyber Police Division, Guangzhou Public Security Bureau, Guangzhou, China
[4] University of Central Florida, Orlando, USA
xueqiang.wang@ucf.edu

Abstract. Mobile applications (apps) heavily rely on user interfaces (UIs) to enhance user experiences. However, despite these benefits, many UIs have been reported to incorporate deceptive practices, collectively known as *dark patterns*, to manipulate users' decision-making processes. One highly manipulative pattern undermining user autonomy is the "Bait and Switch" (*BnS*) pattern, which initially presents bait (e.g., low-cost services) to attract users, then deceptively switches to different, often unwanted content (e.g., high-priced products or UIs that gather user personal data). Through the manipulation of user behaviors, *BnS* raises serious concerns related to security, privacy, financial safety, and more. A systematic identification, analysis, and understanding of *BnS* in mobile platforms have not yet been conducted.

To bridge this gap, we propose BAITHUNTER – a general analytical framework for *BnS*. BAITHUNTER is driven by the key observation that *BnS* usually exhibits inconsistent semantics between the bait and the subsequent content. It leverages a novel combination of image processing techniques and large language models (LLMs) to capture and reason about the semantic features of app UIs. Our evaluation results demonstrate that BAITHUNTER is highly effective in identifying *BnS*, with a precision of 0.867, a recall of 0.938, and an F1-score of 0.901. We applied BAITHUNTER to a dataset containing 306 U.S. and 306 Chinese apps and found 774 potential instances of *BnS*, with 270 (44.1%) of the apps containing at least one *BnS* instance. Analyzing these *BnS* instances makes it possible to better characterize, understand how much today's users are manipulated by the apps, and for what (illicit) purposes, e.g., unexpected data collection and use, and illicit profits for the app developers, etc.

Keywords: Dark patterns · Mobile security · Usable security

Y. Lin, Y. Xu and Z. Yao—These authors contributed equally to this work.

1 Introduction

Most modern software applications (apps), including those on mobile platforms, offer user interfaces (UIs) that allow end users to interact with them. However, despite their many benefits, app UIs have also provided opportunities for malicious apps to exploit users. For instance, some app developers design misleading UIs to deceive users into taking unintended actions, thereby gaining illicit benefits such as harvesting more personal data. Such practices are often referred to as *dark patterns* (or *deceptive atterns*), a concept introduced by Harry Brignull [5] in 2010, whose taxonomy and definitions have recently been extensively discussed [21,30,31].

The "*Bait and Switch*" patterns (*BnS*) [16], among the various types of dark patterns, have become a serious concern for the mobile community, as they are specifically designed to compromise user autonomy. *BnS* initially promises one desirable service to attract users and then deceptively replaces it with a different, and often harmful, option. Figure 1 shows an example of the *BnS* found in MaoDou [17], a popular educational app for kids in China (developed by Gaotu, the third-largest after-school education company). The app first presents a UI that notifies child users of the "free poetry lessons" and induces them to click the "Unlock now" button (the UI on the left in Fig. 1), which appears to activate the free lessons. Upon clicking the button, however, the users are shown a purchase or subscription UI (the UI on the right). The shift from free lessons to purchase or subscription results in perceived inconsistency between user's click intention and the semantics of post-click UI in the *BnS* example. Due to the limited cognitive development and comprehension abilities of children users, they are highly susceptible to falling for such deceptive practice of the app, which raises serious ethical and legal concerns.

Fig. 1. An example of Bait and Switch pattern (*BnS*).

Automatic *BnS* Detection. Multiple governmental entities, such as the U.S. Federal Trade Commission (FTC) [16] and the European Union (EU) [11], have enacted regulations pertaining to *BnS*, with a concerted effort to mitigate the proliferation of these unethical practices. To this end, there is a strong need to automatically identify *BnS*, which can effectively faciliate compliance check.

While some researchers have explored relevant techniques to detect dark patterns in mobile apps, these approaches [6,29,35] are not sufficient for detecting *BnS* due to inherent limitations. Particularly, *BnS requires an understanding of the context across multiple UIs, as it features semantic inconsistencies between what the prior UI promises and what the subsequent UIs actually offer*. For example, UIGuard [6] and AidUI [29] analyze individual UIs to identify various dark patterns, while Raju et al. [35] focus on alerting users to dark patterns through source code analysis for cues like flashy visuals or abnormal behaviors. The former cannot inherently identify *BnS* because *BnS* often span multiple interacting UIs, and the latter suffers from high false positive alarms.

Our Work. In this study, we first build BAITHUNTER, a generic framework for detecting *BnS* in mobile apps based on the analysis of semantic inconsistencies across the UIs before and after user's click action. Building the framework presents two challenges. First, how to identify users' actual perception of the pre-click UI, i.e., what they would normally expect in subsequent UIs, considering the wide range of UI information that affects user perception. Second, how to generically reason about the (in)consistency of the UI transition from the pre-click UI to the post-click UI.

In BAITHUNTER, we address the above challenges by strategically designing the extraction and analysis of UI features related to *BnS*, together with new applications of LLMs for generically determining inconsistent UI semantics. BAITHUNTER starts with those salient UI elements that attract users' attention. For example, a click button with high-contrast background colors (e.g., the button "Unlock now" in Fig. 1. BAITHUNTER extracts a number of UI features that affect *BnS*, including textual information, font size, and their color contrast with the background information through CV and OCR techniques. Second, with the above extracted features, BAITHUNTER tries to summarize user perceptions of the UI. More specifically, BAITHUNTER extracts the user-sensable information highly associated with user clicks. For example, texts with small size and low background contrasts are filtered out, as it is rather difficult to raise awareness of app users among the various texts shown on the UI screen. Third, BAITHUNTER takes advantage of large language models (LLMs) to understand UI semantics and detect their inconsistencies. For each suspicious *BnS* UI transition, BAITHUNTER provides the LLM with the user-sensable semantics, the original content shown on UI screen, and asks the LLM whether the transition between the two UIs are rational. In this way, the LLM tells whether the UI pair belongs to *BnS*, as well as the corresponding justifications (e.g., which type of *BnS* it belongs to).

Our evaluation on a manually labeled dataset consists of 200 instances showed that BAITHUNTER achieves a precision of 0.867 and a recall of 0.938. To facilitate future research, we have released the source code and data of BAITHUNTER at https://github.com/Lin1Yx/BaitHunter.

Large-Scale Measurement. To better understand the pervasiveness and the ecosystem of *BnS* in real-world, we analyze a total number of 612 randomly selected apps across 18 app categories (Sect. 6), covering two dominant regions (i.e., U.S. and China). Our findings reveal a broad distribution of *BnS* across different app categories. These identified *BnS* are further categorized into five unique *BnS* implementation scenarios. For example, a large number of *BnS* are designed for attracting app downloads, boosting shopping quantity, and enhancing user engagement. These tactics are primarily designed to benefiting developers and their commercial partners (e.g., advertisers) at the potential cost of user experience. Our analysis provides new insights on how these dark patterns affect users' normal interaction and app usability.

Our research and findings has broad applications in ensuring ethical and legal compliance within the mobile ecosystem. For example, app developers can employ BAITHUNTER for self-assessment, proactively identifying and rectifying any non-compliant or deceptive elements within their apps. Similarly, regulatory authorities can utilize BAITHUNTER for thorough compliance checks, ensuring the app market remains free from *BnS* practices. By vetting apps for compliance, regulatory bodies uphold consumer rights and maintain a fair and transparent app marketplace.

In summary, our contributions can be summarized as follows:

- We conduct the first systematic analysis of *BnS* on mobile apps. Our work highlights the urgent need for automated detection and auditing of *BnS*.
- We design and implement BAITHUNTER, the first generic framework that leverages a combination techniques (i.e., OCR, CV and LLM) to reason about the UI and detect the *BnS* risks in mobile apps.
- We perform an empirical analysis on 612 random-chosen mobile apps to study *BnS* ecosystem in U.S. and China. The results revealed a number of findings regarding the pervasiveness of *BnS*.

2 Background

Bait and Switch (*BnS*). Prior efforts have demonstrated the consequences of *BnS*, particularly through building a taxonomy for defining and categorizing dark patterns [9,21,28] and reporting individual instances [24,39]. For example, Di et al. [9] define *BnS* as a sneaking pattern that *seems to have a specific result but instead causes another, unwanted outcome*. Brignull's dark patterns website [24] presents several *BnS* instances, such as a recruiting platform that promises to offer "free access to jobs paying over £50,000" while requiring users to give away extra sensitive data (e.g., detailed resumes) and pay for accessing detailed job descriptions [24].

Motivating Examples. We present two motivating examples to show the key characteristics of *BnS*, and highlights the challenges for automatic detection.

Example 1: False Promise as Baits. A false promise occurs when the content displayed after clicking completely diverges from the user's expectations. For example, in Fig. 2, the pre-click UI prompts the user to withdraw cash (1.18 RMB) immediately. However, upon clicking the UI, the user is directed to another UI for opening a different app, i.e., *Baidu Cloud*. In this example, the pre-click UI offers a false promise that is semantically inconsistent with the post-click UI.

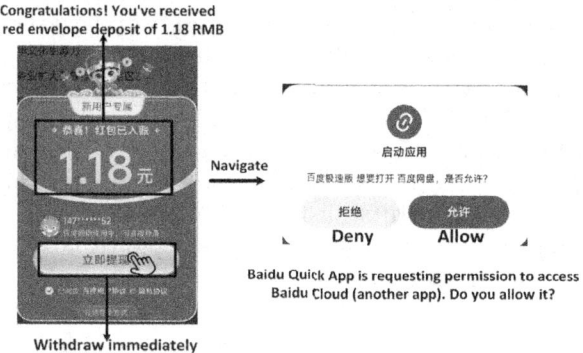

Fig. 2. An example of False Promise. The pre-click UI offers a 1.18 RMB withdrawal but redirects to another app.

Example 2: Misleading Promise as Baits. A variant of the above example appears in two main forms. First, although the post-click UI content is announced on the pre-click UI, it is deliberately obscured by adding misleading details. Second, the pre-click UI content is used to mislead users into engaging with activities in the post-click UI.

Fig. 3. An example of Misleading Promise. The pre-click UI offers VIP membership, while the pre-click UI hides requirements like creating an account and saving money with small, background-colored text.

For example, in Fig. 3, the post-click UI allows the user to open a bank account. An ethical design choice for the pre-click UI would be to prominently highlight the option for "*new users to open an account.*" However, in this example, this option is presented in a small, gray font, making it barely noticeable to users. Instead, the pre-click UI emphasizes an option to "*save money and get member rewards every day.*" Users eager to receive member rewards (from Youku Video [42], the Chinese counterpart of YouTube) are highly likely to be affected by this misleading promise.

Challenges in Automatic Detection. A straight forward way to detect *BnS* could be directly asking LLM and obtain the results. Unfortunately, our preliminary analysis showed such an approach reports both high false positives and false negatives, due to the complexity of UI layout and the various elements it contains. Firstly, an app UI is composed of diverse features, including not only text content but also visual elements such as font size and background color, all of which affect user's understanding and perception to the presented content. These information could be missed by the LLM model due to the lack of background knowledge and reasoning capabilities. In the meantime, fully rely on LLM is not economically practical, as advanced LLMs are significantly more costly. This is particularly concerned in our task, as an app contains quite a lot of UI transitions, but rather a few of them could be instances of *BnS*. Therefore, it is necessary to develop a more light-weight but effective solution to capture the critical features as a premise to understand the UI semantics, as these information are critical for identifying *BnS*.

Problem Statement. BAITHUNTER aims at providing a generic approach to detect *BnS* based on UI screenshots. Therefore, the key capability of reasoning about *BnS* is intentionally designed to be platform-independent (e.g., supporting both Android and iOS apps). In the meantime, BAITHUNTER does not consider adversary that intentionally evade the feature extraction through deep learning models, such as OCR model for text extraction. BAITHUNTER can be easily adapted to alternative, more advanced models to defend against these attacks. As an end-to-end system, we assume the input of BAITHUNTER are the two UI screens before and after UI transition. We adopt the basic app exploration framework (e.g., UiAutomator) to detect *BnS*. More advanced UI exploration techniques are out-of-scope in our research.

Our analysis of *BnS* are based on apps in U.S. and China, the two largest app markets globally. This focus is due to the considerable market size and presence of advanced digital ecosystems, which are prevalent in these regions. However, the methodology of BAITHUNTER are universally applicable, as the features of *BnS*, specifically the inconsistent semantics of the bait and the subsequent content, remains uniform worldwide.

3 Methodology

Figure 4 shows an overview of BAITHUNTER, including its input, output, and intermediate components. The input of BAITHUNTER is two UI screenshots (as

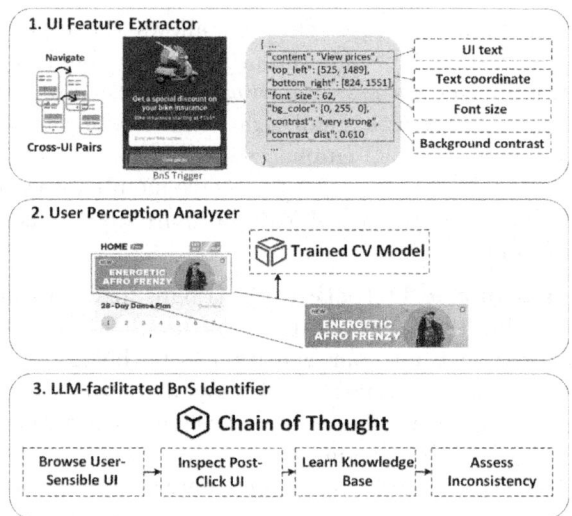

Fig. 4. Overview of BAITHUNTER.

a UI pair) with a clickable UI elements (e.g., a confirmation button) that triggers the UI transition. Then, BAITHUNTER detects *BnS* through the following three components: (1) a *UI Feature Extractor* that extracts the text and visual features related to *BnS*, (2) a *User Perception Analyzer* that highlights the user-sensible information from the pre-click UI, and (3) an *LLM-facilitated BnS Identifier* that justify whether the provided UI pair is indeed with *BnS* based on the provided semantics of the two UI screenshots. The final output of BAITHUNTER is the identification result showing whether the given UI pair is with *BnS*, together with the corresponding reason. For example, the previous UI does not contain any information about the second UI.

3.1 UI Feature Extraction

The UI feature extraction process begins with an arbitrary UI screenshot as input and outputs a set of features critical for detection, with *BnS* trigger identification as a prerequisite. This process consists of two key steps: (1) UI text extraction, and (2) Visual feature extraction. Specifically, BAITHUNTER first identifies clickable elements that may act as BnS triggers. Once a trigger is detected, it extracts UI features such as displayed texts, font size, and layout details (e.g., the coordinates of each UI element), as previously mentioned in Sect. 1.

Unlike previous studies [35,43] that rely on parsing platform-specific layout files (e.g., Android XML), our approach adopts a more general and robust computer vision (CV)-based method. In practice, UI implementations vary significantly across platforms and are often obfuscated, making it difficult to reliably extract structural features such as view hierarchies. In contrast, the visual appearance of key elements–such as button-like regions or prominent texts–remains consistent from a user's perspective. By leveraging visual layout features, BAITHUNTER is able to identify trigger elements and extract meaningful features in a platform-independent manner.

***BnS* Trigger Identification.** BAITHUNTER identifies clickable elements (i.e., buttons) that are highly likely to trigger *BnS*. Given the complex functionality of modern mobile apps, not every UI elements, or UI transitions are necessarily related to *BnS*. In fact, most of the clickable elements on an UI screen are providing legitimate functions. Therefore, it is necessary to selectively analyze those UI transitions that are more likely involved with *BnS*.

In fact, most *BnS* are triggered by clickable elements, i.e., baits, that are visually significant to app users compared to other UI elements. Therefore, BAITHUNTER identified those visually significant, clickable UI elements based on their color contrasts with their surrounding background. For example, the button "ENERGETIC AFRO FRENZY" in Fig. 5 is identified as a potential *BnS* trigger due to its clear boundaries presented on the UI screen. We achieve this by training a customized Yolo [38] CV model on manually verified samples.

UI Text Extraction. For each screenshot, BAITHUNTER first extracts all UI texts based on OCR techniques. We adopts ChineseOCR [4], a lightweight, efficient tool that supports texts in both Chinese and English. The output here is a sequence of sentences (short terms) of all UI texts shown on the screen.

Visual Feature Extraction. BAITHUNTER extracts a series of basic visual features that affect user perceptions, such as Text coordinate, font size, and color contrast between text and its surrounding background. For each UI, we summarize the features in JSON as shown in the first module of Fig. 4.

- **Text coordinate.** We use the coordinates of the upper-left text boundary reported by OCR as the UI text coordinates. For example, the UI element "View prices" in Fig. 4 (green button in the first box), the OCR tool reports its relative coordinates (e.g., top_left (525, 1489), bottom_right [824, 1511]) on the UI screen.
- **Font size.** The font size is defined by the maximum value of the text box height and the average character width, both in pixels. Specifically, the text box height and width are obtained from the bounding box coordinates provided by OCR, while the average character width is derived by dividing the text box width by the number of characters in the text. Additionally, to avoid inaccuracy caused by multi-line texts, we used text box data prior to the box merge process to complete this calculation.

3.2 User Perception Analysis

The second component is designed to analyze and extract user-perceptible information from the UI associated with the clicked button. As discussed in Sect. 1, the nature of *BnS* suggests that the information displayed on the UI screen

Fig. 5. An example of identified grouped UI elements (in the red square). (Color figure online)

may not always correspond to what app users perceive, due to a variety of visual and textual manipulation techniques employed by app developers. For instance, text that is very small in size and has low contrast with the background may go unnoticed by users among more visually prominent text (e.g., large font size). Additionally, in a UI with a wealth of content, text that is more closely related to the clicked button tends to be more noticeable to users, and vice versa.

We consider user obtain perceptions when click from two metrics (1) grouped layout elements and (2) visually apparent elements among the UI screen.

Layout-Based Element Grouping. For pre-click UIs, we chose YOLO [38] to detect groups of related UI elements that are spatially related to the clicked elements. The YOLO model takes the UI screenshot and the user's click coordinates as inputs, and predicts a bounding box that localizes the region visually associated with the click. UI elements falling within this bounding box are considered perceptible to the user during interaction.

Visual-Feature Based Element Filtering. Given a set of grouped UI elements, BAITHUNTER evaluates the user-perceptibility of each element by applying predefined thresholds to collected UI features. Specifically, the font size of an element must be at least 65% of the font size of the clicked element. This threshold was empirically derived based on a manual analysis conducted by 10 human participants, who each reviewed 100 UI screens from popular mobile applications. During this process, elements with font sizes below 65% of the clicked element's font size were consistently perceived as visually secondary or less noticeable. Additionally, the color contrast between the text and its background must exceed a threshold of 0.3, aligning with WCAG's minimum contrast requirements [1]. This filtering process involves a simple value comparison of these visual features. The result is a list of UI Feature that are deemed perceptible and relevant to the user for the given UI screen.

3.3 LLM-Facilitated *BnS* Identification

For each suspicious BnS UI transition triggered by a button click, BAITHUNTER needs to understand the user-perceived information and original features presented on the pre- and post-click UIs to assess the rationality of the transition.

This assessment requires nuanced semantic understanding, which is both labor-intensive and time-consuming when done manually. In contrast, LLMs can perform this task more efficiently. To this end, this component comprises two key parts: (1) a knowledge base tailored for few-shot learning by the LLM, and (2) a well-crafted prompt designed to discern whether a given UI pair shows semantic consistencies (i.e., *BnS*).

BnS Knowledge-Base for Few-Shot Learning. The *BnS* knowledge base is composed of descriptions of the two categories of *BnS* of interest (i.e., False Promise and Misleading Promise, as elaborated in Sect. 2), along with an illustrative example for each category. For each example, we extracted the UI features of the pre-click and post-click UIs of its *BnS*.

Prompt Design and Optimization. As part of few-shot learning, we prompt the LLM models with the user-sensable UI features and rationale for why they are considered *BnS*. In particular, the prompts use the *Chain of Thought (CoT)* [40] approach, where BAITHUNTER simulates human cognitive processes to reason about *BnS*. The CoT-based prompts replicate the user's navigation process within an app, elaborating the sequence of pre- and post-click UIs and assessing inconsistencies between them. BAITHUNTER categorizes each pair of pre- and post-click UIs as either "Normal" or one of the *BnS* categories described in Sect. 2, and provides the reasons for the classification.

This method enhances BAITHUNTER's analytical accuracy by mirroring the mental reasoning behind user interactions and ensuring a thorough understanding of intent, leading to more precise identification of *BnS*. Figure 8 in the Appendix lists the prompts used for detecting *BnS*.

4 Implementation

In this section, we elaborate on the implementation details of BAITHUNTER.

App Exploration and UI Collection. To collect Cross-UI pairs, we employ the UiAutomator2 [19] - the most widely used automated testing framework. Before starting the automated exploration, we ensure that the application is already logged in. This allows us to effectively cover a variety of common scenarios. Throughout the automated traversal process, we systematically iterated through all clickable components within the application. Concurrently, we documented the coordinates of the click component, screenshots before and after navigation, and navigation relationships. To avoid redundant testing as well as ensure the coverage of potential *BnS* cases, we set the maximum testing time to 1 h.

YOLO Model Training. As discussed in Sects. 3.1 and 3.2, we utilize the YOLO model for two key tasks: BnS trigger extraction and layout-based element grouping. The first task is aimed at identifying potential BnS triggers, which are visually significant and clickable UI elements. The second task focuses on identifying groups of UI elements that are closely associated with clickable buttons, thus facilitating the analysis of user click perception.

We trained two YOLO models on a set of 282 UI screenshots, each annotated for a specific task. The screenshots were randomly collected from the top 150 apps on Wandoujia [34], initially totaling 300 images. To reduce bias, we used the Structural Similarity Index (SSIM, on a scale of 0 to 1, where 1 indicates identical images) to remove highly similar screenshots (SSIM > 0.9), resulting in 282 unique images. The dataset was split into training and test sets at a 6:4 ratio to ensure sufficient data for model training while maintaining a robust evaluation set.

For layout-based element grouping, we annotated groupable UI elements (e.g., buttons and their associated components) using the labelImg [25] tool, focusing on visually and functionally related elements within the same group. For BnS trigger extraction, the same screenshots were re-annotated to label visually prominent, clickable UI elements (e.g., high-contrast buttons) likely to trigger BnS transitions.

Adoption of LLMs. As described in Sect. 3.3, we expect LLMs to demonstrate superior analytical capabilities when analyzing given UI data and understanding *BnS* criterion. We initially used GPT-4 [33] by OpenAI, a widely recognized LLM with strong comprehension capabilities. GPT-4 exhibits outstanding performance in detecting *BnS* within U.S.. However, in the context of China, which features unique cultural phenomena such as "red envelopes", Qwen [2] by Alibaba Cloud–primarily trained in Chinese and English–proved more adept at handling OCR challenges like stylized fonts and overlapping text layers. To optimize performance, we used GPT-4 for U.S. apps and Qwen for Chinese apps.

Notably, BAITHUNTERś framework is designed to leverage the semantic understanding of LLMs and our misleading feature extraction methods, rather than being tied to any specific model. This flexibility allows for the integration of more advanced LLMs in the future, further enhancing the detection of BnS patterns.

5 Evaluation

Evaluation Setup and Dataset. To build our Ground Truth dataset for evaluating BAITHUNTER's effectiveness, we collect a dataset of real-world apps and a Cross-UI Pairs dataset from these apps through a large-scale analysis.

- **APP dataset** (D_{app}). We first crawled top 162 popular apps from Wandoujia [34], one of the most popular Android app stores in China, in November 2023, using a Redmi 10 smartphone running Android 11. These apps encompass 13 categories, ranging from banking and instant messaging to social media and utility applications.
- **Cross-UI Pairs dataset** (D_{cui}). Existing dark pattern datasets are limited to single UIs, prompting us to construct a Cross-UI Pairs dataset (D_{cui}) spanning multiple UIs. Using the collection method described in Sect. 4, we gathered UI pairs from various apps in D_{app}. To ensure diversity, we capped the contribution of any single app to a maximum of three randomly selected

pairs. This approach ensures that our dataset is not overly dependent on any single application. Two authors manually annotated the data, with disagreements resolved through discussion with a third author. The data was manually categorized into three types: "Normal" (no BnS strategy), "False Promise," and "Misleading Promise." The final dataset comprises 100 positive samples ("Normal") and 100 negative samples, evenly divided into 50 "False Promise" and 50 "Misleading Promise" instances.

Evaluation Metrics. As described in Sect. 3.3, BAITHUNTER classifies each Cross-UI pair into one of the three categories: "Normal", "False Promise" or "Misleading Promise". As a three-classification task, we evaluate the effectiveness of BAITHUNTER with three metrics, i.e., Precision (TP/(TP+FP)), where TP represents True Positives and FP represents False Positives), Recall (TP/(TP+FN)), where FN represents False Negatives, and F1-score (2*TP/(2*TP+FP+FN)).

5.1 Overall Effectiveness

As shown in Table 1, BAITHUNTER can detect *BnS* cases with a precision of 86.7%, a recall of 93.8%, and an F1-score of 90.1%. Notably, BAITHUNTER performs generally better for False Promise, the most significant inconsistency, with precision (91.0%), recall (94.2%), and F1-score (92.6%) .We also assessed the runtime of BAITHUNTER, which averages 7.2 s for full-process detection, with the LLM contributing the majority of the time consumption (5.2 s).

Table 1. Overall Effectiveness of BAITHUNTER

Cross-UI pair type	Precision	Recall	F1-Score
Normal	0.957	0.890	0.922
False Promise	0.910	0.942	0.926
Misleading Promise	0.824	0.933	0.875

Reasons for Mis-Classifications. By manually inspecting the misclassified data from BAITHUNTER, we found that inaccurate text extraction is the primary reason for misclassifications.

The accuracy of the UI Feature Extraction's output is critical as it directly influences the LLM's determination of *BnS* risks. Errors within this output, particularly those caused by OCR, can significantly degrade performance. OCR's challenge in accurately recognizing text from artistic fonts or overlaying elements is notable.

As is mentioned earlier, Qwen has a certain level of capacity to handle these errors. Still, it cannot rectify all the recognition mistakes in complex scenarios.

5.2 LLM-Facilitated *BnS* Identification

In addition to overall effectiveness, we also evaluate the diverse design decisions of LLM-facilitated *BnS* Identification, which utilizes LLMs with a carefully designed prompt. We explore how adjusting the temperature hyperparameter, which controls the smoothness of word choice probabilities, influences LLM performance in BAITHUNTER. Performance is evaluated using the average values of Precision, Recall, and F1-Score, which serve as the basis for comparing BAITHUNTER's effectiveness across different temperature settings.

Our results indicate that BAITHUNTER achieves optimal performance when the temperature hyperparameter of Qwen is set to 1.5 and that of GPT-4 is set to 0.7. Importantly, the optimal value for the hyperparameter may vary with the choice of different LLMs. Furthermore, since the performance of our system does not depend on the specific LLM used, researchers are advised to choose the LLM more suitable for their linguistic contexts.

6 Understanding *BnS* in the Wild

In this section, we pivot to analyzing the general ecosystem of *BnS* in China and the U.S., two regions that have dominate mobile app markets.

Scope and Measurement Setup. We randomly select apps from all 18 categories in Tencent App Store [37] and Google Play [18]. From each app store, we randomly collect 17 apps from each category to be the dataset D_{rdm}, which means 612 apps in total. We utilize BAITHUNTER to run and test each app for at most 1 h and limit the maximum number of traversal layers to 4. Eventually, from D_{rdm}, we collect 4,984 UI pairs from Chinese apps and 4,135 UI pairs from U.S. apps. These UI pairs then become the basic dataset of the following research. Note that BAITHUNTER cannot go through every single UI in the apps, our results only provide the lower bound of the real *BnS* ecology. The real *BnS* ecology may be more malicious than what we have detected.

Table 2. Prevalence of *BnS* in the Wild

BnS	*BnS* instance (UI pair)		App	
	China	U.S.	China	U.S.
Misleading Promise	387 (73.2%)	153 (64.4%)	152 (92.7%)	87 (82.1%)
False Promise	142 (26.8%)	92 (37.6%)	57 (34.8%)	41 (38.7%)
Total	529	245	164	106

6.1 Types of *BnS*

As shown in Table 2, 10.6% of UI pairs from Chinese apps and 6.0% of UI pairs from U.S. apps are *BnS* cases. This result strongly highlights the widespread prevalence of *BnS* within the app ecosystem. Additionally, in both countries, the patterns using Misleading Promise as Bait are significantly more common than False Promise, suggesting that *BnS* are more employed to obscure important information rather than misaligning content expectations. The lower frequency of False Promise patterns could be attributed to the greater difficulty in achieving them, as they require a complete content mismatch between UIs.

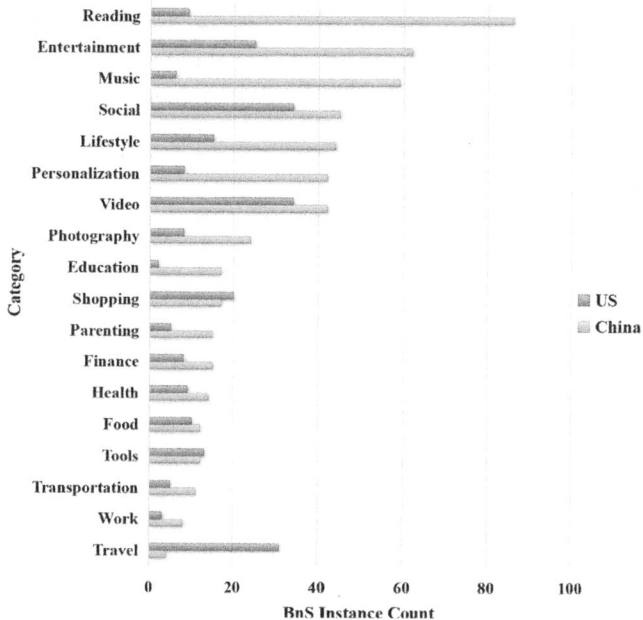

Fig. 6. Distribution of *BnS* Instances among App Categories.

Figure 6 shows the distribution of *BnS* among app categories. Notably, Chinese apps in categories such as Reading, Entertainment, and Music are more likely to contain *BnS*. Meanwhile, U.S. apps in Social, Video, and Travel show higher usage of *BnS*. This may be because these apps primarily make profits by monetizing user traffic or the number of ad clicks. Thus, they are more ambitious in monitoring users by implementing more *BnS* to gain benefits.

We also noticed that, in most app categories, Chinese apps incorporate more *BnS* than U.S. apps. This may be caused by different level of regulation enforcements in the two countries. For example, the Federal Trade Commission (FTC) has been focusing on dark patterns in apps and penalizing a number of apps with violation practices [12,14], especially those in scenarios like program subscription and fraud purchase. Consequently, U.S. apps may implement fewer *BnS*.

6.2 *BnS* Integration Scenarios

In this subsection, we summarize and analyze five distinct *BnS* integration scenarios in mobile apps, based on an official report from the FTC [13]. As illustrated in Table 3, these scenarios primarily aim to boost user engagement and maximize profits, benefiting app developers while adversely affecting app users (Fig. 7).

Table 3. Distribution of *BnS* Integration Scenarios.

Scenarios	Misleading Promise	False Promise	Total
Ad Engagement	206 (26.6%)	164 (21.2%)	370 (47.8%)
Shopping Quantity	123 (15.9%)	23 (3.0%)	146 (18.9%)
Active Users	108 (14.0%)	32 (4.1%)	140 (18.2%)
Program Subscription	85 (11.0%)	14 (1.8%)	99 (12.8%)
Personal Data Collection	13 (1.7%)	2 (0.3%)	15 (2.0%)

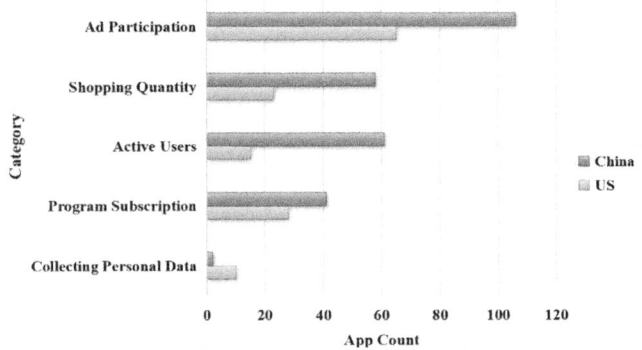

Fig. 7. Number of apps affected by different *BnS* implementation scenarios.

Increasing ad Engagement. We discovered that 170 (27.8%) of the apps use 370 (47.8%) *BnS* to attract users to click (view) ads. In this case, developers use baits like rewards to induce users clicking on specially-designed elements and then users are redirected to a UI of advertisement. In China, such a misleading strategy is clearly prohibited by the government [36]. However, we still find 106 (62.3%) Chinese apps using *BnS* to increase ad participation.

Boosting In-app Shopping Quantity. In 81 (13.2%) tested apps, 146 (18.9%) *BnS* patterns are used to fuel in-app purchases. This scenario may appear because developers can take a portion from orders. For example, developers may claim a $30 coupon on the pre-click UI while on the post-click UI, users can only receive a coupon of $5. Meanwhile, post-click UIs are always filled with personalized products to maximize user engagement and spending.

Enhancing Active Users. 140 (18.2%) *BnS* instances in 76 (12.4%) apps are found to enhance active app users using misleading or false baits. In some cases, *BnS* is used to maintain a continuous engagement with users. For example, some apps use *BnS* to induce users to contact the customer service representative. In other cases, users are misled to participate in continuous activities for rewards. For instance, a pre-click UI may declare an immediate reward, while users have to sign-in everyday for two weeks to actually receive the reward. With these tricks, developers can effectively encourage active app engagement.

Expanding Program Subscription. 99 (12.8%) of the detected *BnS* patterns are adopted by 70 (11.4%) apps to increase program subscription. Some apps claim that users can get membership "for free", while users must subscribe to an auto-renewal program. In another example, when normally exploring apps, users find themselves required to purchase for further usage, as they access members-only functionalities. We consider these cases to be malicious, since they disrespect users' right to know. For U.S. apps, this design may violate the FTC Act by unfairly charging consumers without their consent [15].

Inducing Personal Data Collection. 12 (2.0%) apps even employ 15 (2.0%) *BnS* patterns to gain users' permissions for accessing or even sharing personal data like ID card information. Although the cases are relatively limited, their severity should not be underestimated. They flagrantly contravene the regulations stipulated by MIIT [32] and FTC [13]. This scenario also poses profound threats to user privacy and security, underscoring the urgent need for increased vigilance for extensive personal data protection.

7 Discussion and Limitation

Countermeasures Against *BnS*. Based on our analysis of *BnS*, we propose three strategies to mitigate the emerging *BnS* threats. First, *BnS* may be introduced by third parties like ad producers. Thus, app developers should scrutinize embedded advertisements. Second, app stores should pay more attention to *BnS* in in-store apps, perhaps utilizing effective tools like BAITHUNTER to assist detection. We also encourage relevant administrations to apply more enforcement to reduce the negative impact of *BnS*. Noticing the complexity of real mobile ecosystem, regulating *BnS* requires the cooperation of the whole system.

Ecosystems Outside Analyzed Regions. While our analysis focuses on China and the U.S., we also briefly examined other regions and their regulations [10,20]. We found that *BnS* distribution varies regionally–for example, baits like free palmistry readings are more effective in some Middle Eastern countries. However, *BnS* implementations globally share a key similarity: content mismatch in UIs. This suggests BAITHUNTER can be effective across ecosystems.

Additionally, our summary of *BnS* integration scenarios should apply broadly, as they stem from developers' profit-driven motives. Thus, while *BnS* distributions differ, the underlying ecosystem dynamics remain consistent worldwide.

Limitation. BAITHUNTER is currently limited to Android since we utilize specific automation tools specifically designed for Android. However, BAITHUNTER are designed to be operating system-independent in analyzing *BnS*, leveraging the consistent features of *BnS* and UI uniformity across platforms. Thus, by integrating automation tools like XCTest [3] and Appium [8], future work could extend BAITHUNTER to other platforms with minimal adjustments.

8 Related Work

Dark Patterns in Mobile Apps. Since Harry Brignull proposed the ontology for dark patterns [5], researchers [21,22,31] have noticed the negative impact of *BnS* on user experience. A number of prior research [26,27,30] studied various dark patterns in diverse scenarios. Mathur et al. [30] studied 11K shopping websites, showing how dark patterns deceive users into making additional purchases. Liu et al. [26] designed MadDroid for automated detection of devious ad contents, revealing that roughly 6% of 40,000 tested Android apps delivered devious ad contents. Long et al. [27] measured dark patterns in top 150 apps and mini-apps in China, providing guidelines for assessment. Unlike these studies that are limited to dark patterns within one single UI, our work focuses on dark patterns related to UI transitions. To this end, our work summarized typical types of *BnS*, and proposed a new framework for automatic identification.

Automated UI Analysis. From a broader perspective, our work can be categorized as a type of automated UI analysis in mobile apps. Actually, researchers have paid considerable attention to this field. Hao et al. [23] developed a framework PUMA to automatically analyze performance, security, and correctness properties of the mobile apps. QoE Doctor [7] uses UI automation techniques to replay QoE-related user behavior, and measures the user-perceived latency directly from UI changes. UIED [41] integrates multiple detection methods, including traditional CV methods and deep learning models, to automatically handle various complex GUI images for accurate GUI elements detection. Mansur et al. [29] introduce AidUI, a novel automated approach to detect, classify and localize visual and textual cues in application screenshots, signifying the presence of unique UI dark patterns.

9 Conclusion

This paper presents the first comprehensive analysis of Bait and Switch Patterns (*BnS*) in mobile applications. We propose BAITHUNTER, an effective framework for identifying and combating with *BnS* based on Large Language Model. The evaluation results of the whole framework indicate its overall effectiveness and practicality. We also conduct the first large-scale analysis the ecology of *BnS* in 612 randomly selected Chinese or U.S.-based apps.

Acknowledgments. We thank all anonymous reviewers for their valuable comments and suggestions. Yuhong Nan was supported in part by the National Natural Science Foundation of China (62202510), the Guangdong Basic and Applied Basic Research Foundation (2023A1515012971) and the Guangdong Zhujiang Talent Program (No. 2023QN10X561).

A Prompt Design

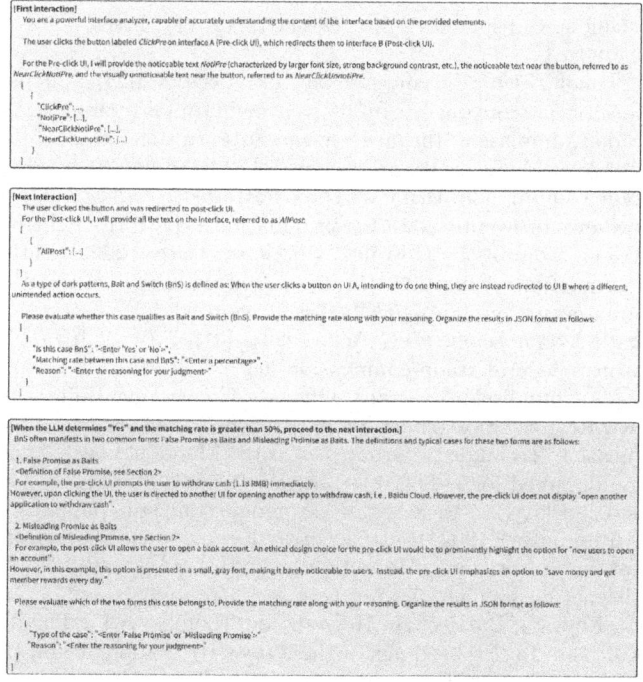

Fig. 8. Prompt for LLM Interaction

References

1. Accessibility Guidelines Working Group: Understanding SC 1.4.3: Contrast (Minimum) (Level AA) (2024). https://www.w3.org/WAI/WCAG22/Understanding/contrast-minimum.html
2. Alibaba Cloud: Qwen (2024). https://tongyi.aliyun.com/qianwen/

3. Apple Inc.: XCTest (2024). https://developer.apple.com/documentation/xctest
4. Bai Kaiyin: ChineseOCR (2024). https://github.com/chineseocr/chineseocr
5. Brignull, H.: Dark patterns (2018). http://darkpatterns.org/
6. Chen, J., et al.: Unveiling the tricks: automated detection of dark patterns in mobile applications. In: Proceedings of the 36th Annual ACM Symposium on User Interface Software and Technology, pp. 1–20 (2023)
7. Chen, Q.A., et al.: QoE doctor: diagnosing mobile app QoE with automated UI control and cross-layer analysis. In: Proceedings of the 2014 Conference on Internet Measurement Conference, IMC '14, pp. 151–164. Association for Computing Machinery, New York (2014). https://doi.org/10.1145/2663716.2663726
8. Dan Cuellar: Appium (2024). https://appium.io/docs/en/latest/
9. Di Geronimo, L., Braz, L., Fregnan, E., Palomba, F., Bacchelli, A.: UI dark patterns and where to find them: a study on mobile applications and user perception. In: Proceedings of the 2020 CHI Conference on Human Factors in Computing Systems, pp. 1–14 (2020)
10. European Union: Consumer Rights Directive (2011). https://eur-lex.europa.eu/eli/dir/2011/83/oj
11. European Union: Unfair commercial practices (2024). https://europa.eu/youreurope/citizens/consumers/unfair-treatment/unfair-commercial-practices/
12. Federal Trade Commission: Illegal Program Subscription (2018). https://www.ftc.gov/legal-library/browse/cases-proceedings/172-3186-age-learning-inc-abcmouse
13. Federal Trade Commission: Bringing Dark Patterns to Light (2022). https://www.ftc.gov/news-events/events/2021/04/bringing-dark-patterns-light-ftc-workshop
14. Federal Trade Commission: Fraud Purchase Case (2022). https://www.ftc.gov/business-guidance/blog/2023/11/ftc-california-allege-cri-genetics-made-deceptive-dna-accuracy-claims-falsified-reviews-used
15. Federal Trade Commission: FTC Act (2024). https://www.ftc.gov/legal-library/browse/statutes/federal-trade-commission-act
16. Federal Trade Commission: Penalty Offenses Concerning Bait & Switch (2024). https://www.ftc.gov/enforcement/penalty-offenses/bait-switch
17. Gaotu Education Technology Group Co., Ltd: Maodou Loves Ancient Poetry, a Chinese app designed for kids (2024). https://feihua100.com/
18. Google: Google Play (2024). https://play.google.com/store/apps
19. Google: UiAutomator2 (2024). https://github.com/openatx/uiautomator2
20. Government of Norway: Marketing Control Act (2018). https://www.forbrukertilsynet.no/english/the-marketing-control-act
21. Gray, C.M., Kou, Y., Battles, B., Hoggatt, J., Toombs, A.L.: The dark (patterns) side of UX design. In: Proceedings of the 2018 CHI Conference on Human Factors in Computing Systems, CHI '18, pp. 1–14. Association for Computing Machinery, New York (2018). https://doi.org/10.1145/3173574.3174108
22. Greenberg, S., Boring, S., Vermeulen, J., Dostal, J.: Dark patterns in proxemic interactions: a critical perspective. In: Proceedings of the 2014 Conference on Designing Interactive Systems, DIS '14, pp. 523–532. Association for Computing Machinery, New York (2014). https://doi.org/10.1145/2598510.2598541
23. Hao, S., Liu, B., Nath, S., Halfond, W.G., Govindan, R.: PUMA: programmable UI-automation for large-scale dynamic analysis of mobile apps. In: Proceedings of the 12th Annual International Conference on Mobile Systems, Applications, and Services, MobiSys '14, pp. 204–217. Association for Computing Machinery, New York (2014). https://doi.org/10.1145/2594368.2594390
24. Harry Brignull: Bait and Switch (2012). https://old.deceptive.design/bait_and_switch/

25. HumanSignal: labelImg (2024). https://github.com/HumanSignal/labelImg
26. Liu, T., et al.: Maddroid: characterizing and detecting devious ad contents for android apps. In: Proceedings of the Web Conference 2020, pp. 1715–1726 (2020)
27. Long, M., Xu, Y., Wu, J., Ou, Q., Nan, Y.: Understanding dark UI patterns in the mobile ecosystem: a case study of apps in China. In: Proceedings of the 2023 ACM Workshop on Secure and Trustworthy Superapps, SaTS '23, pp. 33–40. Association for Computing Machinery, New York (2023). https://doi.org/10.1145/3605762.3624431
28. Luguri, J., Strahilevitz, L.J.: Shining a light on dark patterns. J. Legal Anal. **13**(1), 43–109 (2021)
29. Mansur, S.H., Salma, S., Awofisayo, D., Moran, K.: Aidui: toward automated recognition of dark patterns in user interfaces. In: 2023 IEEE/ACM 45th International Conference on Software Engineering (ICSE), pp. 1958–1970. IEEE (2023)
30. Mathur, A., et al.: Dark patterns at scale: findings from a crawl of 11K shopping websites. Proc. ACM Hum.-Comput. Interact. **3**(CSCW), 1–32 (2019)
31. Mathur, A., Kshirsagar, M., Mayer, J.: What makes a dark pattern... dark? design attributes, normative considerations, and measurement methods. In: Proceedings of the 2021 CHI Conference on Human Factors in Computing Systems, CHI '21. Association for Computing Machinery, New York (2021).https://doi.org/10.1145/3411764.3445610
32. Ministry of Industry and Information Technology: Identification Method of App's Illegal Collection and Use of Personal Information (2023). https://www.miit.gov.cn/xwdt/gxdt/sjdt/art/2020/art_7ec9786084224c0db9628de5becb1085.html
33. OpenAI: GPT-4 (2023). https://openai.com/gpt-4
34. Pea Labs: Wandoujia (2023). https://www.wandoujia.com
35. Raju, S.H., Waris, S.F., Adinarayna, S., Jadala, V.C., Rao, G.S.: Smart dark pattern detection: making aware of misleading patterns through the intended app. In: Shakya, S., Balas, V.E., Kamolphiwong, S., Du, K.-L. (eds.) Sentimental Analysis and Deep Learning. AISC, vol. 1408, pp. 933–947. Springer, Singapore (2022). https://doi.org/10.1007/978-981-16-5157-1_72
36. State Administration for Market Regulation: Measures for the Management of Internet Advertising (2023). https://www.gov.cn/gongbao/2023/issue_10506/202306/content_6885261.html
37. Tencent: Tencent App Store (2024). https://sj.qq.com/
38. Ultralytics: YOLOv5 (2024). https://github.com/ultralytics/yolov5
39. UXP2 Lab: UXP2 Dark Patterns (2024). https://darkpatterns.uxp2.com/
40. Wei, J., et al.: Chain-of-thought prompting elicits reasoning in large language models. Adv. Neural. Inf. Process. Syst. **35**, 24824–24837 (2022)
41. Xie, M., Feng, S., Xing, Z., Chen, J., Chen, C.: UIED: a hybrid tool for GUI element detection. In: Proceedings of the 28th ACM Joint Meeting on European Software Engineering Conference and Symposium on the Foundations of Software Engineering, ESEC/FSE 2020, pp. 1655–1659. Association for Computing Machinery, New York (2020). https://doi.org/10.1145/3368089.3417940
42. Youku Tudou Inc.: Youku Video (2024). https://youku.com
43. Zhou, H., et al.: UI obfuscation and its effects on automated UI analysis for android apps. In: Proceedings of the 35th IEEE/ACM International Conference on Automated Software Engineering, pp. 199–210 (2020)

Author Index

A
An, Chen I-286
Au, Man Ho I-218, I-273

B
Bai, Shuangjie II-442
Bai, Ye I-237
Beozzo, Emanuele II-274

C
Cai, Jintao I-119
Cai, Lijun III-213
Cai, Liujia I-100
Cao, Yun III-533
Chen, Chenghao III-134
Chen, Danwei II-463
Chen, Hua III-57
Chen, Jie I-3
Chen, Kai III-330
Chen, Kun III-295
Chen, Liquan I-203
Chen, Rongmao I-119
Chen, Tieming II-531
Chen, Wenyi I-100
Chen, Xinjian I-306
Chen, Yifei I-387
Chen, Yige III-457
Chen, Zhefan I-545
Chen, Zhili II-39
Cheng, Guang III-438
Cheng, Jiatao I-545
Cheng, Ke III-398
Cheng, Ruoxi III-277
Chu, Qiaohan I-3
Conti, Mauro I-387
Crispo, Bruno II-274
Cui, Xiaohui II-501
Cunha, Luís II-274

D
Dai, Guangxiang III-253
Dai, Jun I-329
Deng, Li III-438
Deng, Shanlin III-379
Deng, Xianwen III-476
Ding, Xiaosong II-367
Ding, Yizhong III-277
Ding, Yong II-482, III-493
Ding, Zhaoyun II-513
Du, Qiuyan I-3
Du, Yang III-359
Duan, Ao II-181

F
Fabien, Eyezo'o Benjamin II-347
Fan, Fengrui II-403
Fan, Jialiang II-3
Fan, Jingjing I-218, I-273
Fan, Limin III-57
Fan, Tianyuan I-160
Fang, Junbin I-40
Fang, Sixin III-398
Fang, Wenbo III-195
Fotos, Nikolaos II-255
Fu, Haocheng III-533
Fu, Qiang I-405

G
Gao, Guoju III-359
Gao, Jianbin II-60, II-347
Gao, Qiyuan I-424
Gao, Ya I-444
Gao, Yiwen I-181
Garrett, Ian Y. III-3
Gerdes, Ryan M. III-3
Giannetsos, Thanassis II-255
Gong, Junqing II-39
Grisafi, Michele II-274
Gu, Dawu III-22, III-134
Gu, Qi I-463

Guan, Yewei   I-366
Gui, Ling   III-213
Guo, Fuchun   I-62
Guo, Hua   I-366
Guo, Linhai   II-79
Guo, Peiyuan   II-313
Guo, Yunchuan   I-525
Guo, Zerui   I-22

H

Hao, Zhize   II-139
He, Debiao   I-237
He, Jingnan   I-286
Hong, Cheng   III-22
Hu, Xiaoming   II-442
Hua, Baojian   III-77, III-379
Huang, Chanying   II-387, III-312
Huang, He   III-359
Huang, Hongxian   III-533
Huang, Huafeng   III-175
Huang, Luqi   I-62
Huang, Mingming   II-100
Huang, Qiong   I-22, I-306
Huang, Tianyi   III-134
Huang, Yawen   II-424
Huang, Yichi   I-119

J

Jia, Huajie   III-457
Jia, Xiangkun   III-175
Jiang, Fenghua   I-100
Jiang, Yinghua   I-203
Jiang, Yongzhen   III-97
Jiang, Yufeng   II-293
Jiang, Zhuochen   III-77
Jiang, Zoe L.   I-40
Jiang, Zoe Lin   I-273
Jin, Hua   II-313
Jonathan, Anto Leoba   II-60

K

Kan, Haocheng   II-331
Karas, Dimitrios S.   II-255
Kong, Queping   I-564
Krontiris, Ioannis   II-255

L

Lai, Junzuo   I-141
Leaticia, Kuiche Sop Brinda   II-60

Lei, Jian   II-293
Li, Chang   II-21
Li, Chaoyue   II-236
Li, Fenghua   I-525
Li, Haoran   III-97
Li, Heng   I-525
Li, Jiangfeng   II-79
Li, Meng   I-387
Li, Minghang   II-3, II-119
Li, Qi   III-115
Li, Shikang   II-403
Li, Xiangman   III-553
Li, Yanting   I-306
Li, Yiwei   II-403
Li, Yu   I-347
Li, Yuantong   II-367
Li, Yumei   I-62
Li, Zhenyu   III-493
Li, Zhuangwei   I-387
Liang, Hai   II-482, III-493
Liang, Minzhi   I-203
Liao, Huimei   II-100
Lin, Hao   II-442
Lin, Wangqun   II-513
Lin, Xiaocong   I-463
Lin, Yixi   I-564
Lin, Zixing   III-573
Liu, Botao   I-82
Liu, Changjun   III-533
Liu, Feng   III-232
Liu, Guanxu   II-79
Liu, Hai   II-216
Liu, Hongjia   III-97
Liu, Jianghua   II-293
Liu, Jianwei   III-41
Liu, Junxiu   I-405
Liu, Mingliang   III-379
Liu, Nianlu   III-195
Liu, Qi   I-424
Liu, Tao   II-139
Liu, Wenmao   I-463
Liu, Xiaoying   III-511
Liu, Xinzheng   II-513
Liu, Yan   II-442
Liu, Yi   I-141
Liu, Yuejun   I-181
Liu, Yueling   III-493
Liu, Ziyao   I-286
Lu, Binqin   II-21
Lu, Siqi   I-100

Lu, Xianhui I-286
Lu, Xingye I-218
Lu, Zhitong III-330
Luo, Decun II-3
Luo, Guibo II-331
Luo, Jiang II-550
Luo, Jun I-203
Luo, Min I-237
Luo, Pingbin I-306
Luo, Xiaomin III-340
Luo, Yuling I-405
Lv, Meiyang III-533
Lv, Mingqi II-531

M
Ma, Duohe III-253
Ma, Sha I-22
Meng, Weizhi II-255
Mi, Wei II-100
Mihaljević, Miodrag J. III-340
Min, Xuyan I-203
Mu, Chao I-444

N
Nan, Yuhong I-545, I-564
Ni, Jianbing III-553
Niu, Weina III-419
Nong, Junxiang I-424

O
Oliveira, Daniel II-274
Ouattara, Koffi Ismael II-255
Ouyang, Xue I-405

P
Pan, Jiageng III-97
Peng, Cong I-237
Peng, Hongye II-216
Peng, Tao III-213
Peng, Yadong II-216
Pinto, Sandro II-274

Q
Qian, Haifeng II-39
Qian, Jin I-203
Qiao, Yan I-387
Qin, Bo II-3, II-119
Qin, Kailun III-134
Qin, ShaoHua II-550

Qin, Sheng I-405
Qin, Zheng III-398
Qiu, Dongyan II-181
Qiu, Xuebo II-531
Qu, Shipei III-22

R
Richard, Befoum Stephane II-347
Rossini, Mulenga Mukupa II-347
Rui, Li III-573

S
Saydiev, Bektemir II-501
Schreiber, Maximilian III-153
Shang, Tao III-41
Shao, Wei III-340
Shen, Gang I-160
Shi, Rong III-232
Shi, Yang II-79
Shi, Yipeng III-134
Song, Qijie II-531
Song, Yaolong III-573
Song, Yipeng I-141
Su, Purui III-175
Sun, Xiaomeng III-115
Sun, Xiaoyan I-329
Sun, Yi II-100
Sun, Yinxia I-347
Sun, Yu-E III-359
Susilo, Willy I-62, I-119

T
Tan, Zejiu I-273
Tang, Junwei III-213
Tang, Ming I-82
Tang, Shijie II-482
Tang, Tianyi III-553
Tang, Zelin I-366
Tang, Zhengzhou III-457
Tao, Jun II-198
Tao, Yu II-159
Tao, Yuhan II-463
Teng, Minyu II-79
Tian, Maoze II-216
Tippe, Pascal III-153
Tu, Zhengzhou I-40

V
Victor, Kombou II-60, II-347

W

Wang, An III-57
Wang, Chaoyun II-387
Wang, Han III-493
Wang, Jiabei I-181
Wang, Lianhai III-340
Wang, Luping I-3
Wang, Mingsheng I-257, I-503
Wang, Peng III-253
Wang, Qin II-3
Wang, Qizheng III-340
Wang, Shuai II-403
Wang, Wen III-232
Wang, Wenhao I-347
Wang, Xiangyu III-419
Wang, Xiao I-525
Wang, Xiaofen II-367
Wang, Xiuhua II-403
Wang, Xueqiang I-545, I-564
Wang, Xuyu I-329
Wang, Yijun III-476
Wang, Yiyang III-295
Wang, Yongjuan I-100, III-419
Wang, Yuanyuan III-57
Wang, Yuntao I-160
Wang, Yuxuan III-22
Wang, Yuzhu I-160
Wang, Zhangrui I-405
Wang, Zhe II-550
Wang, Zhiqiang III-277
Wang, Zibo III-175
Wei, Bohang II-119
Wu, Chenchen I-463
Wu, Qianhong I-424, II-3, II-119
Wu, Shaoqian III-476
Wu, Si III-379
Wu, Xiabai III-295
Wu, Xiaodong III-553
Wu, Xiaoming I-444

X

Xia, Han I-503
Xia, Hu II-60
Xia, Qi II-60, II-347
Xiao, Shan III-312
Xie, Jintao II-79
Xie, Min I-40
Xiong, Shihong II-119
Xu, Chenhao II-293
Xu, Dazhi I-181
Xu, Haowen I-329
Xu, Jian III-97
Xu, Jun III-359
Xu, Lei II-293
Xu, Qing II-367
Xu, Shujiang III-340
Xu, Xiaolong II-236
Xu, Yue I-564
Xu, Zhen III-330
Xue, Qiao I-484
Xue, Zhi III-476

Y

Yan, Chuping III-511
Yan, Jia III-175
Yan, Kedong II-387, III-312
Yan, Lianglin I-257
Yan, Yu III-419
Yang, Anjia I-141
Yang, Changsong II-482, III-493
Yang, Huibo II-139
Yang, Kaijie I-366
Yang, Kaixuan II-21
Yang, Ming I-444
Yang, Qian III-330
Yang, Shaojun I-119
Yang, Wenjie I-119
Yang, Yang II-119
Yang, Yi III-175
Yang, Yiming III-57
Yang, Yuchen I-3
Yang, Zhichao I-237
Yao, Zitong I-564
Ye, Aoshuang III-213
Yi, Xiaowei III-533
Yiu, Siu Ming I-218, I-273
You, Weijing I-463
Yu, Jintong III-22
Yu, Qingyuan III-511
Yu, Yong I-40, III-553
Yuan, Boshi III-134
Yuan, Qingjun III-419
Yuan, Shaowei III-277

Z

Zeng, Pengfei I-257, I-503
Zhan, Mingwei III-476
Zhang, Bo II-367
Zhang, Chi I-273, II-482, III-22, III-134

Zhang, Futai   I-119
Zhang, Hanwen   II-331
Zhang, Huaicong   II-424
Zhang, Huan   III-312
Zhang, Jixin   III-398
Zhang, Ke   II-367
Zhang, Kun   III-41
Zhang, Lingcui   I-525
Zhang, Linlin   III-195
Zhang, Mingwu   I-160, III-398
Zhang, Peng   II-181
Zhang, Shuhui   III-340
Zhang, Wenyang   II-21
Zhang, Wenying   III-115
Zhang, Xiaodan   II-100
Zhang, Xiaolin   III-134
Zhang, Xuman   III-438
Zhang, Xuyang   II-313
Zhang, Yongming   II-236
Zhang, Yuan   I-347
Zhang, Yuanjing   III-41
Zhang, Yuliang   I-545
Zhao, Faqi   III-232
Zhao, Kai   III-195
Zhao, Ruijie   III-476

Zhao, Runze   I-100
Zhao, Tianya   I-329
Zheng, Lei   II-367
Zheng, Yu   II-403
Zhong, Sheng   I-347
Zhou, Feng   III-57
Zhou, Guoqiao   III-232
Zhou, Lu   II-159
Zhou, Shouchen   II-159
Zhou, Yongbin   I-181
Zhou, Yuxia   I-484
Zhou, Ziyan   I-525
Zhu, Fei   III-213
Zhu, Huijuan   II-21
Zhu, Tiantian   II-531
Zhu, Yanbei   III-419
Zhu, Youwen   I-484
Zhu, Yuesheng   II-331
Zhuang, Chaofeng   II-39
Zou, Hao   II-198
Zou, Xianglu   II-181
Zu, Siyuan   I-405
Zukaib, Umer   II-501
Zuo, Cong   II-293

Made in the USA
Monee, IL
03 May 2026